AF323290

DATA MINING METHODS

Second Edition

DATA MINING METHODS
Second Edition

Rajan Chattamvelli

Alpha Science International Ltd.
Oxford, U.K.

Data Mining Methods
Second Edition
582 pgs. | 64 figs. | 122 tbls.

Rajan Chattamvelli
Associate Professor
Department of Information Technology
Periyar Maniammai University
Thanjavur, Tamil Nadu

ALPHA SCIENCE INTERNATIONAL LTD.
7200 The Quorum, Oxford Business Park North
Garsington Road, Oxford OX4 2JZ, U.K.

www.alphasci.com

Printed from the camera-ready copy provided by the Author.

ISBN 978-1-78332-219-0

Preface

The first edition of this book was well-received by the academic community. Suggestions and comments made by various readers prompted me to bring out a second edition. Major changes include the addition of new chapters on regression, and text mining and splitting of the data warehousing and OLAP chapter into two. Some of the chapters have been thoroughly revised – especially the OLAP, neural networks, web mining and support vector machines. Most of the typing errors that crept into the first edition have been removed.

During the course of six years, several URL links that appeared in the first edition have either moved to new locations or ceased to exist. The second edition has removed as many of such links as possible, and added new links. A major change in the second edition is the inclusion of software program links at the end of most chapters in tabular form. This will help the readers to choose the most appropriate software link quickly. The reference list at the end of each chapter is also thoroughly revised. Some new exercises have been added in a few chapters.

Text Mining is a hot area of research as majority of data on the web is of text form. In addition, text data arises from scanned documents, dictations, talks and lectures. The last chapter introduces text mining and its applications. Software programs for text mining are given at the end of that chapter.

Any suggestions or comments for improvement are welcome. All suggestions should be sent to **dmmbook@gmail.com**. An up-to-date errata will be made available on an ongoing basis.

Rajan Chattamvelli

Table of Contents

List of Figures

List of Tables

List of Algorithms

1

Basic Concepts in Data Mining

Chapter objectives

- Understand what is data

- Understand data categories

- Describe the standard data scales (Nominal, Ordinal, Interval, Ratio)

- Introduce extended data types

- Distinguish databases and datawarehouses

- Introduce data mining

- Describe supervised and unsupervised learning

- Understand data mining approaches

- Describe some applications of data mining

1.1 Introduction

Data mining is a multi-billion dollar global market that is gaining in popularity. Effective utilisation of past data has helped many companies in huge money and time savings, and in keeping abreast of competitors. Data mining is an inter-disciplinary field, which originated from statistics, data visualisation, databases, and machine learning. There are many learning algorithms used in data mining – association rules, decision trees, neural networks, genetic algorithms, support vector machines etc. Anyone with a basic understanding of data visualisation techniques, statistics (probability, summary measures, regression and correlation, least squares principle, cluster analysis), matrices and geometry can easily get started with data mining. More important is an understanding of scales of measurement, data preparation and transformation techniques, data storage technologies (data bases and datawarehouses), and online analytical processing (OLAP).

1.2 Data Scales

A measurement is the process of assigning a number, a label or other identifying attribute to a variable. This measurement can be done by humans or by machines. The variable values are often restricted to be in a specific range. As examples, assigning marks (in the range 0 to 100) to each student, finding the height (in inches) of persons, income of customers etc record a value in a specific range depending upon the unit chosen. Labels are usually assigned by visual inspection, or by observing some properties. For instance, assigning hair colour, skin colour, gender (male or female), age groups etc are based upon visual properties. Assigning the marital status, smoking and drinking habits etc of a person are done by other properties or habits. All these measurements and observations generate data of various kinds. These measurements are intended for communication, interpretation or subsequent processing by humans or computers. Any object that could generate data is called an entity. The entities can be humans, animate or inanimate objects, processes, actions etc.

Definition 1 A measurement of variables on entities is called data (singular is *datum*).
Each entity has various kinds of attributes. The attributes of interest will vary for different applications. For example, in a study to find out how marks of a student depend upon family income, the hair colour, religion etc are unimportant attributes. In an application to automatically detect the age group and geographic origin of a person appearing in video images captured by a surveillance camera, the skin and hair colours, stature, type of spectacles worn, moustache etc are important attributes, and dress colour, carry bags etc are unimportant attributes.

Definition 2 An attribute is a measurable or observable property of an entity. An identifying attribute is a single attribute or a combination of attributes to uniquely distinguish different entities.

All attribute values of an entity are stored together with each entity. Entities are identified by unique ID-numbers, labels, or other identifying attributes (eg: social security numbers, PAN, credit card numbers, roll numbers). Small companies and educational institutions use the last name (surname or family name) and first name (given name) of persons as attributes for unique identification. But in larger domains (at state or country level) there could exist multiple persons with the same last and first names. Similarly, few duplicate names exist in telephone directories of large cities. In such situations, either a combination of attributes is used to disambiguate the duplicates, or unique ID numbers are assigned to each entity. As an example, the transaction ID, along with the date-stamp (date and time of transaction)[1] are used in temporal data mining to identify a multitude of transactions at large super markets that typically have many sales counters.

1.2.1 Data vs Information

Data are basic facts on an entity. When too many entities generate data, it may convey some meaningful information. Hence information can be regarded as 'organised data' about one or more subjects of interest that convey some meaning. Information can also be derived from summarising data, displaying data and transforming data. Information is an essential ingredient

---

[1]This is called *date-stamping*, which can be applied to numeric or text data. Numeric date-stamping converts the date and time to unique integers, which are used as the prefix and transaction IDs are then concatenated at the end of it as suffix to generate a unique ID. Text date-stamping concatenates the date and time with other text data to generate a text string.

of discovering knowledge from data. The aim in data mining is to derive hitherto unknown knowledge in the form of patterns, trends, anomalies and outliers, discontinuities and deviations, or similarities from voluminous data of various kinds.

A good understanding of data scales (also called scales of measurement) is important to get a good grasp of data mining models. With the advent of computers, these scales have been refined and standardised. This is due to the fact that computers can record all kinds of data for storage and manipulation.

1.2.2 Data Types

Computers distinguish various types of data using built-in data types. These data types vary a lot among various operating systems (OS), different programming languages, and databases within the same OS. But the fundamental data types are more or less the same in modern programming languages.

Most programming languages have *byte, short, int, long, float,* and *double* data types for numeric data (*byte* type can also store characters). These types are used to optimally allocate just enough memory, and improve the scope of data range. Real numbers can be stored in float or double data types. In addition to the data types in standard programming languages, the databases have their own data types that are optimised for data storage and processing. For example, the clob type is used to store large character objects (like dictations) and blob type is used to store large binary objects. As these are close analogues of programming language data types, we will not discuss it further.

1.3 Data Categories

There are three types of data called operational data, nonoperational data and metadata. Operational data are used by enterprises in routine business applications. These are updated in real-time, and can be operated on by add, update and delete operations. Examples are sales data, accounting data, billing data etc. Non-operational data do not change quite often. Examples are demographic data of students, customer's location or identification information etc. Metadata are data about data. It is a higher-level identification and summary information about stored (computer) data. Metadata are associated with a group of entities, one or more parent data files or databases. For instance, database metadata can contain data about the record length, number of fields in each record, file and record sizes, index names and fields etc. Metadata can be stored along with the parent data (as in database metadata), in separate data files or as embedded data. The metadata can be centralised (stored in a single location) or distributed (stored in a distributed network). They are used in datawarehouses and datamarts to speedup multiple user access and updation to terabytes of data, and to enforce data consistency (see chapter 4).

In addition to providing statistical information about their parent files, they can also provide OS controlled parameters like last access and updating dates, date of creation, ownership etc of their parent files. These are called administrative metadata. More than one metadata file may be associated with a single data file. Triggering mechanisms are used to update metadata files when their parent files are changed. They can also be updated manually by humans or software.

1.4   Scales of Measurement

A scale of measurement is used to record the data using a proper unit of measurement. The most frequently used scales of measurements are Nominal, Ordinal, Interval and Ratio (NOIR)[2] scale. The NOIR scale is the fundamental building block on which the extended data types are built (see §1.5 in page 1-13).

Definition 3 Quantitative and categorical data
Quantitative data are measured on a numeric scale. They can be whole integers, real numbers, fractions or complex numbers. Because a scale of measurement is involved, they are expressed in a unit of measurement.

Quantitative data can be real or complex type. Complex data have real and imaginary parts. We seldom come across complex data in data mining applications. Hence no further mention of it will be made in the rest of the chapter. Real data can be measured on a discrete scale or continuous scale. Discrete data give rise to integers (positive or negative along with zero). Examples of discrete data are the number of items ordered by a customer, total marks obtained by a student, number of votes of a candidate, total members in a family, total visitors to a web site etc. These can either be counts or numbers truncated to nearest integer (eg: marks 75.3 is truncated to nearest integer 75). Continuous data can take any real value. For example, the weight of an article, total service time (expressed in minutes), the size of a file (in kilobytes), student GPA etc can take any decimal value when the number of decimal places are restricted. The average daily temperature of a city can take positive or negative value depending upon the latitude of the city and time of the year (temperatures of some cities can become sub-zero during winter time).

Categorical data can take one of a finite number of labels. Data miners can conveniently choose numbers, letters (in any alphabet), strings etc as labels. A simple example is gender={Male,Female}, with two categories. These categories can also be coded as {'M','F'}, {0,1} etc[3]. Standard labels exist for some categorical variables. Examples are blood groups ={A,B,AB,O}, Rh Factor={+, -} (see below). But the labels given to categorical variables does not matter in most data mining algorithms because they are used to distinguish the categories among themselves. An exception is the class labels used in binary support vector machines (SVM), which are always chosen as {+1, -1}.

1.4.1   Standard Scales of Measurement

The NOIR data scale was introduced by Stevens [SS46] as a convenient method to classify data. We will call it the standard scale to distinguish it from extended scales of measurement used to store text data, audio data, image data and multimedia data.

1.4.2   Nominal Scale

As the name implies, the nominal scale is used to label data categories using a consistent naming convention. The labels can be numbers, letters, strings, enumerated constants or other keyboard symbols. The variables that give rise to nominal data are called nominal variables. Examples

---

[2]Noir is a French word that means 'black', 'dark' etc

[3]The single letters chosen depends upon the natural language. For example, it is coded as {'H','D'} Herren (Male) and Damen (Female) in German; {'M','H'} (Macho, Hembra) in Spanish; {'M','V'} (Man, Vrouw) in Dutch; {'M','F'} in Italian (Maschio, Femmina) & French (Male, Femelle).

are gender, smoking habits, race, eye colour, country code, languages known etc. They can be numerically coded by starting with 0 or 1. The codes used to label the data do not have numerical significance.

Austrian biologist Landsteiner[a] (1868-1943) invented the categorisation of human blood as A,B,AB,O (called *ABO grouping*) in 1900 [LK00]. It has applications in blood transfusion and organ transplantations. The ABO grouping gets its name from the presence or absence of A and B antigens (Antigens are chemical compounds recognised by the blood as 'foreign'. The AB group has both A and B antigens present, whereas the O group has neither) in red blood cells. Landsteiner won the medicine Nobel prize in 1930 for his invention that saved millions of human lives (www.nobelprize.org/educational_games/medicine/landsteiner/readmore.html). Landsteiner and Alex Wiener invented the Rhesus (Rh) antigens (that are coded as '+' and '-') in 1937 (It gets its name because it is present in the blood of Rhesus monkeys). This primary categorisation has been extended to many sub-categories recently. For a long time, it was thought that the blood group is time-invariant. Extensive computerisation of medical data has already proved that the human blood type can change rarely over time (especially in sub-categories). As an example, malignant mutations (in patients with acute myelocytic leukemia) has resulted in sub-group changes, but the change is more prevalent in Rh phenotypes (due to medical procedures and rarely due to transfusion of wrong Rh subgroups. An O- person can donate blood to any other type, and an AB+ (called a universal receiver) can receive blood of any type). The glycosidase enzyme synthesised from some bacterias has the ability to strip off the A & B antigens from blood cells to transform 'A', 'B' and 'AB' blood types to 'O' type [LS07]. See ibgrl.blood.co.uk/ISBT%20Pages/ISBTHome.htm, icr.org/article/3647, en.wikipedia.org/wiki/Human_blood_group_systems, answers.com/topic/blood-type etc.

---

[a](He discovered only 3 blood types [A,B,O]. The type AB was invented by von Decastello & Sturli in 1902.)

Nominal data are also known as categorical data, although 'categorical' is a broader term that includes nominal and ordinal data. They are assigned unambiguous labels to distinguish between various values.

Definition 4 A variable that takes a value among a set of mutually exclusive codes that have no logical order is known as a nominal variable. To be meaningful, the number of categories should be two or more, but countably finite.

Example 1 The human blood groups are categorised as {A,B,O,AB}, and the Rhesus (Rh) factor is categorised as {'+', '−'}. Both these are nominal data. The labels '+' and '−' are chosen for Rh factor historically, but it could be any two distinct symbols. These labels do not have any relative ordering among themselves. Therefore we cannot say that blood group B is better, or worse than group A. They are simple distinguishing labels to identify each individual (for transfusion and organ transplantation purposes) based upon their blood type. Because differences exist between two persons with the same blood group, but with different Rh factors, they are combined as in A+, B−, O−, AB+ etc. We could also recode these combinations using another scale or label.

Example 2 The nationality (country of birth) of a person can be coded using the complete country name, using a two letter code (as used in Internet addressing), using a number etc. There is no meaning in comparing (numerically or using relational operators) the nationalities of two individuals. Hence it is a nominal variable (there are around 232 countries, of which 192 are UN

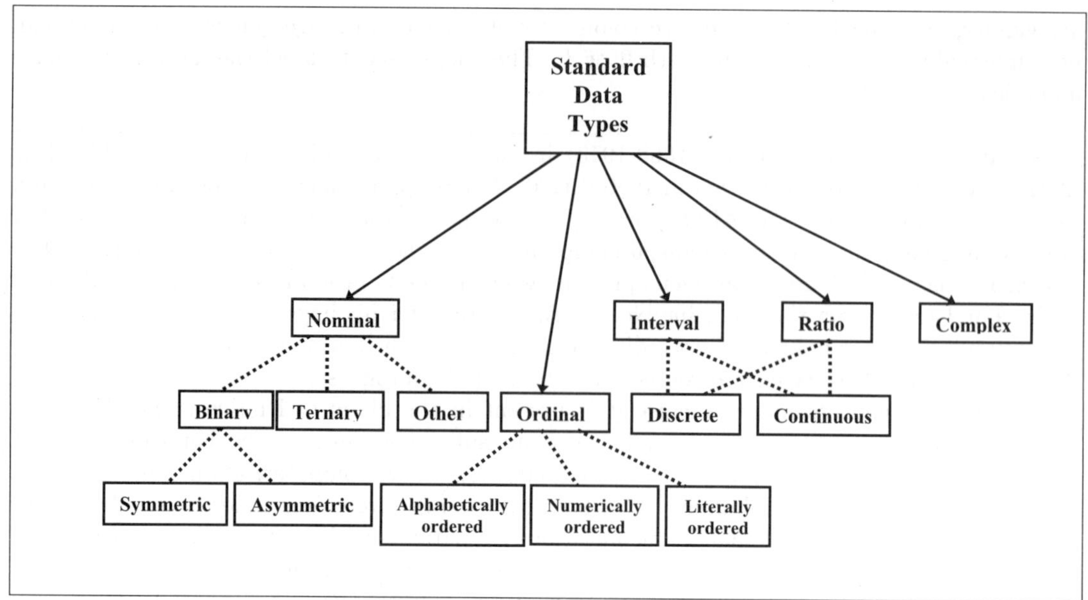

Figure 1.1: The Standard Scales of Measurement

recognized as of this writing. See en.wikipedia.org/wiki/United_Nations_recognized_countries). Other examples are – programming languages known to an applicant, majoring subject of a candidate, vehicle makes, models and exterior colours etc.

Example 3 As data are transmitted over communication channels in the form of bits, it is customary to choose binary strings to code attribute values. For instance, the status of a data packet sent by your computer={status unknown='00', lost='01', in transit='10', delivered='11'}. Here the size of each code is 2 because we have only 4 categories. If there are n categories, we must choose minimum size (length) m such that 2^m equals or exceeds n. This is expressed as m=$\lceil \log_2(n) \rceil$. As an example, if there are 9 categories, we have m=4 because 3 bits can represent up to 8 categories only.

1.4.2.1   Coding of Nominal Variables

Before coding a nominal variable, the total possible number of categories must be known. Thus we cannot use a letter of the alphabet to represent each country in the world. That is why the Internet uses a two letter combination to uniquely identify each country (eg: 'id'=Indonesia, 'il'=Israel, 'in'=India, 'iq'=Iraq, 'ir' = Iran, 'is'=Iceland, 'it'=Italy etc)[4]. Similarly, favorite food={'A'=American, 'C'=Chinese, 'I'=Indian, 'J'=Japanese,'P'=Punjabi, 'O'=Others} will suffice in most situations, if the total number of categories does not exceed 26. When it exceeds 26, either upper and lowercase letters or numbers may be used, or better yet, two letter combinations utilised.

---

[4]See ftp.ics.uci.edu/pub/websoft/wwwstat/country-codes.txt, www.ethnologue.com/codes/CountryCodes.tab, or www.greenwichmeantime.com for a complete list

The labels or codes chosen for various values should be consistent and uniform as far as possible (The blood-group is an exception in which we choose single letters for 'A','B','O', and two letters for 'AB' (because it started much before the advent of computer era)). It is not advisable to mix letters and numbers/digits to code different values of a variable. Consider the problem of coding all the languages spoken in the world. As there are thousands of languages, at least a 3 letter combination (with total 26^3 codes possible) is needed to uniquely code them (see www.ethnologue.com/codes/LanguageCodes.tab for 3 letter codes of all 7325 language identifiers of the world (This list contains lots of non-existent and duplicate codes). All codes need not be of fixed length. As an example, country calling codes (international telephone) are numerically coded as 3 digits in most countries (see www.countrycallingcodes.com/countrylist.php? country=A etc.). When the city codes are added to make direct calls to any city of the world, the 3 digits are insufficient[5], in which case we get variable length codes.

Summary statistics suitable for nominal data are the mode, and contingency correlation. Since the numerical values assigned to nominal data are used only for unique identification, it is meaningless to compute the mean of it (See §3.7 for an exception). As there are no ordering among the elements, the median is also meaningless. Arithmetic $(+,-,*,/)$ and logical operations $(<,>,!=)$ are not permitted on nominal data. The allowed operations are access (you can read and check its value), count() and re-coding (into another non-overlapping symbol set). They can be graphed using pie-charts or bar-charts.

1.4.3 Binary Variable

Definition 5 A nominal variable with exactly two mutually exclusive categories that have no logical order is known as binary variable[6], and the corresponding datum is called binary data.

Example 4 Gender={male,female}, condition of a device={working, nonworking}, login attempt={success, failure}, status of a switch={on, off} etc are all binary variables.

Example 5 In some cases, a clear cut boundary between the two cases may not exist, but are assumed. For instance, eating habit={vegetarian, non-vegetarian} is considered as binary. If a person rarely eats some type of non-veg food, we need to decide where such cases have to be grouped. When data are of higher order scales (interval or ratio type), it is much easier to come up with a cutoff point. As an example, fever={mild, severe} can be decided upon by a cutoff point as fever is measured on a numeric scale. Similarly, body mass index (BMI)={normal, abnormal} is decided upon by a cutoff (usually 24).

1.4.3.1 Coding of Binary Variables

Binary data can be coded using letters, numbers, special symbols, names, strings or even sentences (with two or more words). For example, flight={arrived, not yet arrived} is binary data. We could of course use one word or abbreviations when an option is more than one word. Hence flight={arrived, late} convey the same meaning as above. It is common practice in computer programming to code them as numeric, character or enumerated string types. Some of the data mining algorithms use the {0,1} coding for binary data. Examples are logistic regression, neural

---

[5]Land-line and cell phone codes can be found at www.greenwichmeantime.com/XXX/YYY/telephone.htm where XXX is the continent name and YYY is the country name (eg: www.greenwichmeantime.com/asia/japan/telephone.htm)

[6]Other names are *boolean variable* or a 0/1 *indicator variable* or *dichotomous* variable.

networks. In an online e-commerce transaction, if a customer has ordered and bought an item, we will assign the code 1 to the corresponding item and if the customer ordered, but did not buy it, we will denote it as 0. The binary support vector machines (SVM) [chapter 13] use $\{-1,+1\}$ labeling for the two classes into which an item is classified. The dummy binary variables used in some optimisation problems always assume the values $\{0, 1\}$. Other possible codes are $\{'+', '-'\}$, {true,false}, {yes,no}, {on,off}, {success, failure}, {online,offline}, {insured, uninsured} etc. If coded as 0 and 1, the mean of binary data gives the proportion of items labeled as 1 in our sample.

1.4.3.2 Symmetric vs Asymmetric Binary Variables

Different choices of a binary variable may have unequal importance in some applications. In a study to correlate the risk of a disease among smokers and non-smokers, we could label smoker=1 and non-smoker=0. Similarly, in a study to find out if there is any association between body fatness and high blood pressure, we could code overfat people=1, and others=0.

Definition 6 Symmetric binary variable
If the two choices of a binary variable have *equal importance*, it is called symmetric binary variable. Examples are gender={male, female}, Rh factor=$\{'+', '-'\}$.

Definition 7 Asymmetric binary variable
If the two choices of a binary variable have *unequal importance*, it is called asymmetric binary variable.

Examples are login attempt={success,failure}, transaction type={genuine, fraudulent}, status of a message={delivered, lost}, passenger type ={regular, occasional}, occupancy status of an apartment={free, occupied}, or {rented, owned}.

Whether to use a binary variable as symmetric or asymmetric depends upon what we intend to prove, or on the research hypothesis. If our hypothesis concerns the transaction amount involved in fraudulent attempts, it is beneficial to code fraudulent=1 and genuine=0.

Example 6 Combining nominal variables
Fine-grained categorisation may not be required in some applications. Suppose the gender is coded as gender={'m'=Male, 'f'=Female} and marital status is coded as marital={'S'=Single, 'M'=Married, 'D'=Divorced, 'W'=Widowed}. If a research study distinguishes between marital statuses of males and females, we could combine both categories as status={'1'=Single Male, '2'= Single Female, '3'=Married Male, ..., '7'=Widower, '8'=Widow}
One-to-one data transformation can be used to recode nominal variables (say from alphabetic codes to numeric codes, if the number of categories are limited). Many-to-one transformation (on sub-groups) can be used to reduce the number of codes, but is not recommended.

Example 7 Nominal dichotomisation
Data of other types can be dichotomised to nominal scale. For example, the speed of a vehicle (which is a ratio variable, discussed below) can be categorised as {slow, fast} using a cut-off speed. Similarly, result={fail, pass}, symptom={fever,no fever}, height={short, tall} utilise a single cut-off value. The proper cutoff value is derived from domain knowledge or past data. It may or may not be universal. The cutoff to be used in the result example above is institution dependent, and the cutoff for presence or absence of fever is universal ($98.4°F$). Nominal variables are analysed using contingency tables. More details can be found in chapter 3.

Example 8 Which of the following binary variables are symmetric?
a) gender b) knows driving c) drinking habits d) owns computer
Solution: Gender is a symmetric variable and we can assign male=1 and female=0 or *vice versa*. But this can become an asymmetric variable in those research studies that specifically target one particular sex. Other variables are most often asymmetric as in most research studies we distinguish between the alternatives. For instance, an applicant who knows driving may be preferred for some job openings, a medical study may be to correlate drinking habits among high risk and low risk symptoms, a person without computer may be preferred for a computer sales campaign.

1.4.4   Ternary Variables

Definition 8 If a nominal variable assumes values in exactly three mutually exclusive categories that do not have any logical order, it is called a ternary variable.

Example 9 Primary color={R,G,B}, where R=Red, G=Green, B=Blue. All other colours are formed by combining the primary colours in various proportions.

Example 10 A working traffic light can have three values (or states) as Traffic light={Green, Yellow, Red}. Hence traffic signal states (during a fixed time period to include the transitions) can be considered a ternary variable, although there is an order of appearance among the three values.

As in the case of binary variables, the codes for ternary variables can be numeric, character or string type. For example, furnishing_status={fully furnished, partially furnished, unfurnished} and eating habits={pure veg, occasional nonveg, non-vegetarian}. Some binary variables that have doubtful or ambiguous cases can be converted into ternary variables. For instance, the symptom of a disease may be {present, absent, irrelevant} in a research study.

Nominal scale is the most basic level of measurement as they are identifying labels or tags to distinguish various choices. If they are numerically coded, the numbers have no significance other than to classify the data values. Nominal variables can be analysed in terms of its constituent categories. Nominal data can be visualised using line charts, bar charts or pie charts. The idea of symmetric and asymmetric binary variables can be extended to ternary variables too. If all choices of a ternary variable have equal importance, it is called a symmetric ternary variable. Otherwise it is asymmetric.

Example 11 Explain why you cannot use the median as a summary measure for nominal data.
Solution: The median is defined only for ordinal or higher scales of data as it is the middle value (after data are arranged in increasing order). Nominal data do not have any order between the elements. Thus even if the data are numerically coded, the median is meaningless.

1.4.5   Ordinal Scale

Definition 9 Ordered nominal data are known as ordinal data and the variable that generates it is called ordinal variable.
The values assumed by an ordinal variable can be ordered among themselves as each pair of values can be compared literally or using relational operators ($>, \geq, <, \leq, \neq$). For example, shirt size = {small, medium, large, extra large} is an ordinal variable because an order among various

pairs is evident. The labels or codes given to ordinal variables can either be meaningful words (in any language), letters of an alphabet or numbers in any base. The digits $\{0,1\}$ used in binary variables is a special case of ordinal data when the magnitudes are taken into account.

If an ordinal variable has 4 values, we could code it using two binary digits as follows: severity of accident=$\{$mild='00', severe='01', critical='10', deadly='11'$\}$. Ordinal data can be ordered alphabetically, numerically or literally. Numerical ordering is used when all choices are coded as numbers. Circularly ordered data (called cyclic ordinal data) are a special case in which an ordinal variable repeats its values continuously and cyclically. Hence a natural flow or transition exists between the last value and first value (). The traffic signal values mentioned above is an example. Other uni-dimensional examples are season=$\{$spring, summer, fall, winter$\}$, time=$\{$AM, PM$\}$, time of day=$\{$morning, noon, evening, night$\}$, weekdays, etc. Combining multiple cyclic ordinal variables give rise to multi-dimensional cyclic variables. The seasons and time examples given above depends upon the geographic location of the place of observations. When two geographically well-spearated regions are considered simultaneously, we get multi-dimensional cyclic variable.

Example 12 Ordinal data transformation
Ordinal variables can be transformed to higher scales of measurement. But it is more common to transform higher scales of measurements to ordinal scale. For example, suppose the age of a patient is recorded as a decimal number between 1 and 100. We could define a cutoff interval for various age categories and re-code it as patient $=\{$child, teenager, youth, adult, senile$\}$, which is an ordinal variable because the various pairs can be compared among themselves. When higher order scales are converted into ordinal scale, there is loss of information because each value in a fixed interval is mapped to a single code as in this example.

1.4.5.1 Allowed Operations

Since various elements can be ordered among themselves, the usual 'relational operators' $<$ (less than or precedes) and $>$ (more than or succeeds) can be used on ordinal data. These operators have the usual meaning when the data are coded on a numeric scale. For alphabetic data stored in a computer, these operators revert to precedence order in the language used by the codepage of the OS. If the codes are strings, an alphabetical string comparison is carried out.

Example 13 All types of ratings (movie rating, web site rating, ratings of services, tastes, likeness etc), rating of books and magazines, political parties, tourist spots, cities, countries, etc. are ordinal data. These ratings can be coded alphabetically or numerically, although the latter is preferred in some data mining algorithms. For example, the ratings $=\{$'poor', 'fair', 'good', 'excellent'$\}$ can be numerically coded as 0 through 3 or 1 through 4. In both the codings, the natural order $poor < fair < good < excellent$ is preserved. But, the difference 'excellent minus good' does not equal 'fair minus poor'.

Summary measures *mode* and *median* can be used on ordinal data (see figure 3.4, pp.3-19). Addition and subtraction are meaningless, but data can be ordered or ranked (numerically, alphabetically or literally). Hence we can find any of the percentiles of ordinal data. Calculations based on order are permitted (count, classify, rank, find largest, find smallest, find median). Spearman's ρ can be used as a measure of the strength of association between two sets of ordinal data. This is computed using the ranks of observations in each group [CR12].

Codes for categorical variables should be chosen such that the margin between different values are well defined and disjoint. If no categories have preference over the others, it is called symmetric coding. Nominal and ordinal variables are together called qualitative variable. Ordinal data can also be graphed using line charts, bar charts and pie charts.

1.4.6 Interval Scale

Definition 10 Ordinal data with the additional property that differences between any two values represent equal difference in the amount of the characteristic measured is called interval data. The zero point (origin) for interval data is completely arbitrary.

Hence all ordinal data with well-defined intervals are interval data. Interval data are measured on a numeric scale. The difference between any pair of data values matters, but the ratio of different values is meaningless because there is no 'natural zero'.

Example 14 Earthquake intensities are measured on the Richter scale. An earthquake of magnitude 8.2 is not twice as strong as another of 4.1 magnitude. The reason is that the magnitude is found as the logarithm of amplitude recorded by seismographs.

Example 15 Calendar dates can be subtracted from each other to get the number of days in between. It does not make sense to divide one calendar date by another. The zero point, year 0 AD in Julian and Gregorian calendars are arbitrary. Hence it is an interval data.

Example 16 Fever is an interval variable with its origin as 98.4° Fahrenheit for humans (if the body temperature is 98.4°F, there is no fever, but if it is above the normal (due to medical reasons), we say that there is fever. This origin temperature differs among various creatures). We cannot say that a person with temperature 102.4°F has 'twice' the fever of another person with 100.4°F.

Other examples are atmospheric pressure and humidity, location coordinates using longitude and latitude, etc. The temperature scale in Fahrenheit and Centigrade are interval data because 0 degrees Fahrenheit or Celsius does not represent zero (absence of) heat.

1.4.6.1 Allowed Operations on Interval Data

We can add and subtract constants to or from interval data. Addition and subtraction can also be performed on interval data pairs. Negation (changing the sign) and multiplication by a constant (scaling) are permitted. All operations on ordinal data defined in the previous section are also valid. Permissible transformations are linear or affine transformations (cx+d) where c and d are any real constants, and $c \neq 0$. Since such a transformation applies a change of origin and scale to the data, the origin and unit of measurement of the data are arbitrary. Other one-to-one nonlinear transformations (log, exp, etc) can also be applied.

1.4.6.2 Interval Data Transformations

The interval data can be transformed to nominal or ordinal scale, but with loss of information. For example, the temperature of a city can be classified as hot, mild or cold using appropriate cutoff temperatures. Similarly, a suitable temperature bracketing can be used to categorise the temperature of a liquid as {cold, lukewarm, tepid, warm, hot}, which is an ordinal datum because the values can be ordered. Interval data can also be scaled using change of origin and scale techniques. A common example is the transformation from Fahrenheit to Celsius scale[7], using C=(F-32) * 5/9 = .5555 * F - 17.777 approximately, and F=C2 + 32 −C2/10, where C2=2∗C. Thus if C=180, C2=360 and (32−C2/10)=32−36=−4, so that F=360−4=356° F.

The interval data gets its name from the fact that differences (which are called intervals if the scale of measurement is numeric) between various values are comparable. They can be graphed using histograms and frequency polygons or frequency curves.

1.4.7   Ratio Scale

Definition 11 Interval data with a clear definition of 0 are called ratio data. In this scale, both differences between data values and ratios of (nonzero) data pairs are meaningful. Variables that give rise to ratio data are called 'ratio variables'.

Example 17 If the height of a son is 2 feet 8 inches and his father is twice as tall, the same relationship holds whether the height is measured in metric or decimal scale. Other examples are total family income, age of a person or article, speed of vehicles or CPU speed of a computer, price of an article etc. All counts that include a zero are considered to be ratio variables. As examples, number of files downloaded from the internet, number of likes received for a social media posting, number of scores earned by a person in sports events (like cricket) are all ratio variables.

1.4.7.1   Operations on Ratio Data

All arithmetic operations on interval data are applicable to ratio data. In addition, multiplication, division, squaring, square roots etc are allowed. Permissible transformations are any linear transformation (aX+b) where (a,b)$\in \mathbb{R}$ and $a \neq 0$. Interval and ratio data are together called quantitative data.

A clear distinction between the interval and ratio scales are not made in most data mining applications because they are both stored in the same data type (float, double etc). Several data mining algorithms use the concept of distances (which measure dissimilarities). The popular distance metrics are Euclidean distance, city-block (Manhattan) distance, Mahalanobis distance and Minkowski distance. All these measures can be computed on ratio data because a lack of distance (distance=0) implies that the data are identical. In some stream mining applications, both categorical and quantitative variables are intermixed in a stream. The distance in such cases is computed by separating out them into categorical and quantitative, and computing the distance measure separately and taking a linear combination of these distances. The interval data transformations are discussed in detail in chapter 3. See also [HM01], [PD99].

1.4.8   Choosing the Correct Scale

One of the first challenges for researchers and data miners is the most appropriate choice of the scale of measurement for each variable under consideration. This may be obvious for most variables. We could start by asking the question whether it should be chosen as categorical (as various categories) or quantitative (measured as a number). If it is categorical, we could further subdivide it as nominal if there is no specific order among the values or as ordinal if there they can be ordered. If it is nominal, we could check the number of possible values that the variable

---

[7]See www.convert-me.com/en/, www.onlineconversion.com

could assume to decide whether it is binary, ternary or other. If the data are quantitative, we must decide whether it is interval or ratio type. Both these types of data are measured in multiple units. A simple check can be used to distinguish between the two for numeric data. If the 0 value matches in all popular units of measurement, it is ratio data, and otherwise it is interval data. For instance 0^oF$\neq 0^o$C$\neq 0^o$K. Use the figure 1 in page 1-6 as a general guideline. A rule-of-thumb is to descend as far as possible to decide the data type and number of categories needed.

1.5 Extended Scales of Measurement

The NOIR data taxonomy is the building block on which the extended data types are built. There are many other types of data in common use, especially in digital computing. The prominent of them are Text, Audio, Video, Image, Graphics, Animation, Multimedia (TAVIGAM) data. They can be structured (eg: text data (HTML, XML)) or unstructured (eg: audio). Another categorisation as Spatial, Temporal and Other (STO) is also popular.

1.5.1 Nonstandard Data

Text data are comprised of a stream of characters in any encoding (codepage in a native language), where each character in a codepage is an array of bits. Examples of text data include email messages, web page contents, medical and legal transcription data, chat room data and data generated by various activities like customer reviews of a product, business correspondences, service requests, advertisements, tweets on twitter sites etc. In the categorical variables section, we have used labels that are of text form. But, text data are different from text labels. As they are formed by concatenating characters (and other symbols) of any language encoding, they carry semantic information. Even single English characters like 'a' and 'I' that appear in a text passage have a semantic interpretation.

Other data types on the web (audio, video, graphics, animation, multimedia) are comprised of streams. They may also be compressed or uncompressed and of various formats. Audio data can be stored in structured form using compression schemes. Mining these types of data is greatly facilitated by transformation into another domain. For example, audio data can be transformed into the frequency domain using Discrete Fourier Transform (DFT) to reveal frequency characteristics. Similarly, images can be mined by transforming the data using Discrete Cosine Transform (DCT). They are further discussed in chapter 14.

1.5.2 Numeric Data Discretisation

Definition 12 The process of dividing the range of continuous data into meaningful non-overlapping intervals is called discretisation.

The intervals are usually contiguous (without gaps). It is used to reduce the data size, and to group a large number of data into manageable number of groups. There are many techniques for discretising numeric data. The popular methods are equal interval binning, equal frequency binning and entropy based discretisation. As the name implies, the equal interval binning divides the range (span) of data into equal width intervals. The process of dividing numeric variable into binary is called binarisation.

Example 18 Consider the total time taken by an e-commerce customer, from the time of login to the time of a firm order. This time interval will vary greatly depending on a number of factors

– education level, prior ordering experience, network connectivity and bandwidth, language skills etc. We could discretise this time into equal width intervals (say less than 1 minute, from 61 seconds to 120 seconds, etc). The equal frequency binning can have variable class widths but utilise the frequency of observations falling in each class to be nearly identical.

1.5.2.1   Entropy Based Discretisation

Consider a sample S which is divided into two non-overlapping and non-empty subsets S_1 and S_2. The entropy of S is defined as $I(S) = -\sum_{i=1}^{n} p_i \log(p_i)$ where p_i is the probability of class S_i. The entropy function after partitioning is $I(S,T) = (|S_1|/S)I(S_1) + (|S_2|/S)I(S_2)$. Since different partitions give rise to different entropies, we have to choose that partition which minimises the entropy over all possible partitions. This can be found by trial and error or by recursion. In the recursive case, the process is continued until $I(S) - I(S,T)$ is greater than a prespecified constant.

1.6   Databases and Data Warehouses

Definition 13 A database is a structured repository of meaningful data organised in an optimal way to minimise storage redundancy, to improve data security at multiple levels, and for fast updates and retrievals by multiple users.

Most databases arrange the data in multiple related tables to facilitate multiple users to access and update the data seamlessly. All databases use index files created using the primary keys for optimal access speed. Depending upon the data organisation scheme, the databases may be categorised as follows:

1. Hierarchical databases
   Data are organised in a hierarchy (usually as a tree).

2. Network databases
   Data are organised as a network of records using pointers.

3. Relational databases
   Data are organised as tables that are related using primary/secondary keys.

4. Object oriented (and object-relational) databases
   Data are organised as related entities (called 'objects' in object oriented programming) with private data and public or private methods (member functions).

5. Multidimensional databases
   Data are stored as data cubes and accessed using OLAP tools.

6. Embedded databases
   Embedded databases are built into microchips or operating systems. One example is the SQLite built into the Android operating system that runs some smart phones, smart watches and many other gadgets.

Databases are designed for fast accesses (reads and writes) over a shared network by a large number of transactional users who update the data at atomic levels. Access restrictions can be applied at field level, record level, table level or entire database level. The data in databases are accessed using structured query languages (SQL) or its variants. Most databases have the facility to generate custom-built reports. Data for reports can be selected either by an SQL statement or using the Query by Example (QBE) method. The reporting facilities of databases are far too insufficient for data mining purposes. Nevertheless, virtual datawarehouses (chapter 4) are built using data residing in databases to create an illusion to the data miner as if they work with real datawarehouses. This is accomplished using middleware programs that populate datacubes using back-end data residing in tables, and updating the cube data according to the user actions in OLAP or its variants.

1.6.1 Data Warehouses

Definition 14 A data warehouse (DW) is an integrated, time-variant, and non-volatile collection of summarised historical data repository, specifically structured for fast querying and reporting with an intention to make business decisions.

The process of extracting, transforming, cleansing, summarising and consolidating operational data into informational data at a central repository to facilitate decision support is called data warehousing. Data collected from different sources are made consistent, cleansed and formatted to get rid off null or redundant values and stored in highly summarised form for fast querying. A datamart is a smaller focused subset of a datawarehouse, designed for a specific line of business. The OLAP technology is used to slice and dice volumes of data stored in datawarehouses and data marts. Variants of SQL are used to access the data in DW and datamarts (an SQL statement underlie each of the OLAP operations like slicing, dicing, etc.).

1.7 Data Mining

Definition 15 Data mining is the process of extracting hitherto unknown and potentially useful patterns, trends, anomalies and rules from stored historical data for business promotion, decision making or classification.

Data mining is an inter-disciplinary field with roots in enterprise decision support [JG98].

The results obtained by a data mining process are used in making business decisions and short-term predictions. It has diversified into many other fields that have no business context. For example, the SVM is used to give a categorical label to unseen data instances using a model obtained from a set of labeled training data. This has applications in pattern classification, text categorisation, sign recognition etc more than in business. Similarly, genetic algorithms (chapter 10) and neural networks (chapter 11) are used for optimisation of empirically observed functions under constraints. Hence data mining is gaining a broader meaning and intuition than mere business decision making.

Available data for a mining session can be divided into three groups – training data, test data and result validation data. There are no hard and fast rules regarding such a division. If a data mining model does not need to be 'pruned', a data miner can simply divide the entire data into training and test sets and build a model using the training data, which can be validated using the test data. Training data are random samples of available data, used to develop a data mining model. This model is tested for accuracy and conformity using the test data. During the validation phase, the miner could also adjust the model by minimising an error criterion. Validation data are used to estimate the error to see how well the models perform under general conditions.

1.7.1 Exploratory Data Analysis vs Data Mining

Exploratory Data Analysis (EDA) is a similar technique for summarising and identifying patterns in data. But EDA is often applied on small-volume data generated by statistical sampling, by direct observations or controlled measurements, and analysed using purely statistical techniques. Online interaction with data is much less or even non-existent in EDA[8]. Most often a single statistical model (like a regression model, discriminant model, or clustering model) is built from the data and utilized for predictions, classification, hypothesis verification or process control. Almost all of the EDA models have distributional assumptions on the data (and random error variation). Some of the data mining models (eg: decision trees, SVM, neural networks) do not have any data assumptions. Data collection and analysis are quite often carried out by the same person in EDA, so that they are fully familiar with what to expect as a result of the EDA. In addition, EDA includes tried and tested techniques for single and multiple hypothesis verification (as in t-tests, ANOVA, etc), etimation of unknown population parameters, and building confidence intervals for them, which are not considered in data mining methods. Large volumes of historical data (typically of the order of millions) that are highly summarised for fast querying are often used in data mining. Data miners have a variety of tools at their disposal for digging into the data, including some of the statistical techniques like least-squares, clustering, outlier detection, etc. The results of data mining are put to use for business advantage, cost reduction, etc and are repetitive in nature.

1.8 Supervised and Unsupervised Learning

Data mining models[9] can be classified as supervised, semi-supervised or unsupervised. Analogously, statistical models can also be classified as such. The following section gives the definition of each and then compares some of the well-known statistical models as well.

1.8.1 Supervised Models

Definition 16 A model built from labeled training data that is subsequently used for a data mining task (like classification, prediction or estimation) on new (unseen data) is called a supervised learning model. In other words, labeled data supplied by a user are used to build a model with parameters and error information. This data (called training data) are then discarded, leaving behind a data mining model with parameters. The test data are then used by this model for prediction, classification and result validation etc (table 1.1).

Definition 17 A model with parameters or error information built from a random sample of available labeled data that is subsequently used for prediction, classification or conformity checks is called a supervised learning model.

All supervised models use a subset of the available data called *training data* to build a data mining model, usually in a single pass through the data. It additionally provides information on training errors. The training data are subsequently discarded (ignored), leaving us with a model (with parameters estimated from training data) for subsequent use. They use labeled training data, and may require additional user input during the training phase. They can be used for prediction or classification. Examples are logistic regression (LR) models, decision

---

[8]Most EDA tools have a filtering capability to select subsets of data, which is the equivalent of a user query in data mining to slice and dice data

[9]as well as EDA models

trees (DT), artificial neural networks (ANN), and SVM. The LR and DT are both considered as statistical models as they originated as statistical techniques, much before the emergence of data mining (table 1.2). The NN is considered as supervised model when used for classification (but it has other uses which are better grouped as unsupervised learning). The binary SVM uses two classes that are always labeled as $\{\pm 1\}$, which are symmetrically combined to form the constraints. Multi-class SVMs use any numeric class labels as discussed in chapter 13. All SVM models estimate the unknown parameters (slope and intercept term) of an n-dimensional (n-D) hyperplane using the available training data. The Latent Semantic Analysis (LSA) is a supervised model as we need to form the term by document matrix (TD-matrix) using a unique label for each document (which are used to index the columns of the TD matrix. These labels can be numbers or strings). A truncated singular value decomposition of the TD-matrix results in our learning model. The LSA can be used for information retrieval (IR), classification and clustering. When used for IR, the user queries are at first vectorised (converted into a vector) using the keywords present in the query, and then projected into the reduced rank feature space. A sufficient number of vectors are then filtered from this feature space using a similarity metric. These matching vectors are then coverted to the most similar vectors in the original input space. The document labels are used to identify matching documents in the input space.

Labeled training data may be difficult or time consuming to obtain, as it requires human effort. If unlabeled data are relatively easy to obtain, we could use a mixture of labeled and unlabeled data to train a classification model. This is called semi-supervised learning. Examples are Expectation Maximisation (EM) with generative mixture models, Gaussian random fields model and graph based methods.

1.8.2 Unsupervised Models

Unsupervised models do not consider the data as training and test data, but the entire available data are used to build a global model with parameters, error information or some particular property and the model is valid only for the data from which it was derived. Examples are clustering models.

Definition 18 A model built from unlabeled training data to automatically extract patterns, trends, rules, clusters or outliers, or detect anomalies from voluminous data is called an unsupervised learning model.

Unsupervised models always use unlabeled training data and often does not require additional parameter settings. Examples are association rule mining (ARM)[10] and data or attribute clustering[11]. Unsupervised learning usually involves multiple passes through the data to produce a result. Other examples are link analysis, principal component analysis (PCA), and independent component analysis (ICA). Some of the algorithms mentioned in subsequent chapters can be used for both supervised and unsupervised learning.

Example 19 Distinguish between a supervised and unsupervised classification.
Supervised classification uses labeled training data. The class to which a training datum belongs is known *apriori*. This data are used to build a data mining model, using which unlabeled data can be classified. Examples are SVM, neural networks and decision trees. Unsupervised classification methods use unlabeled training data, and obtains the classifier automatically. Examples

---

[10]ARM algorithms can accept a minimum support and confidence to discard uninteresting rules, and come up with a concise set of meaningful rules.

[11]Some of the clustering algorithms also accepts additional parameters like number of clusters to be found.

Table 1.1: Supervised vs Unsupervised Data Mining Models

Type	Example
Supervised	Decision Trees, SVM, SVR, Neural networks (for classification)
Unsupervised	Association rule mining, cluster analysis
Both	Genetic Algorithms, Web Mining Models

Table 1.2: Supervised vs Unsupervised Statistical Models

Type	Example
Supervised	Logistic regression, decision trees, Discrete Discriminant Analysis
Unsupervised	Multiple Linear Regression, Canonical Correlation Analysis ANOVA, MANOVA, ANCOVA, Factor Analysis

are clustering models, nearest-neighbour models. Both the techniques can accept parameters from the user. For example, the k-means clustering and its variants (chapter 9) accept a best guess on the number of clusters (k) present in the data.

1.9   Steps in Data Mining

- Data gathering and filtering
  Data gathering step identifies the source of data to be used for creating a data mining model. Operational data may contain redundant or duplicate information, attributes not needed for decision support etc. Data filtering simply extracts needed or essential attributes and records (rows) from operational data.

- Data standardisation
  All categorical variables are brought to a standard coding scheme (eg:blood groups A, B, O,AB may be coded as '00', '01', '10', '11')[12]. The date and time, currency formatting characters, etc are also standardised. Categorical variables are reduced to the lowest denomination possible by collapsing unwanted categories.

- Data cleansing
  Cleansing (also called data hygiene) is used to load datawarehouses and datamarts from raw data such that it is consistent, and conforms to the overall business processes of an organisation. Incorrect and inaccurate data can cause decision making errors. All data that violate the rules are either discarded or transformed using a cleansing operation (§4.4.2). All records with missing data values are either discarded or supplied with default values. All inconsistent data values are made consistent. All formatting characters are standardised (eg:use of decimal point and comma). All types of errors in data are corrected.

---

[12]The A,B,O,AB blood types are known in Russian republics by the Roman numerals I, II, III, IV.

- Loading of filtered transaction data into the data warehouse
  Transaction data are most often kept in databases as normalised tables. Since dataware-houses often need all relevant information, a de-normalisation step may be required before data are loaded. For example, a patient's demographic information is usually kept in a separate table, and related with other tables using a primary key like patient_id. Extract-Transform-Load (ETL) utilities are used to load the data into datawarehouses (§4.3).

- Summarising data
  Data warehouses contain highly summarised data to make multiple simultaneous retrievals quick and efficient. Data are pre-summarised and collapsed into analysis categories and stored in OLAP structures called data cubes.

- Security and user management
  Datawarehousing environment typically have a large number of users who access it over wired or wireless connections. Although the data are 'read-only' there are other security threats like jamming, session hijacking, processor overloading etc that must be addressed. These can be done at the hardware and software levels or a combination of them. User management using authentication and authorisation must also be in place to detect in-truders (§4.8).

- Data analysis and visualisation
  Most data mining tools come bundled with data analysis and proprietary visualisation tools. Data visualisation can be used as a prelude to choose the correct model. As a simple example, a plot of the positive and negative classes in binary SVM can tell us if the data are linearly separable; or if not, how much separability violation is present. This knowledge can be used to select the most appropriate SVM model.

1.9.1 Data Mining Approaches

Data mining can take on different approaches and build different models depending upon the type of data involved and the objectives (verification or validation driven vs discovery driven mining). In the validation driven approach, the data miner hypothesises some model or pattern and use the data to either accept the hypothesis or reject it at a suitable level of significance. This approach can use anything from a simple query to complex multi-dimensional analyses and visualisation tools. In the discovery driven approach, the data miner looks for hitherto unknown patterns and trends in the data.

- Association rules
  An association rule is a casual relationship among categorical variables or quantitative variables that have been discretised, and the affinity of which is specified using a user sup-plied confidence and support thresholds. Association rules are built from related activities. Association rule mining (ARM) algorithms extract rules using a user specified minimum confidence and support. They may be uni-dimensional or multi-dimensional. This is fully discussed in chapter 7.

- Clustering
  A technique to break a large heterogeneous population into small number of homogeneous groups is called clustering. Clusters can be found using one of the standard algorithms (eg: hierarchical or partitional algorithm), or obtained using SVM, Kohonen networks or decision trees (numeric data), or by Latent Semantic Clustering (LSC) (for text, audio, image, video data).

- Classification

  A technique to organise unlabeled data into a finite number of distinct classes or categories using a model built from labeled training data is known as classification. Examples include decision trees, neural networks, and SVM.

- Regression

  The dependency of one or more independent predictor variables on a single response variable is modeled using regression. Support vector regression (SVR) is a technique that uses a linear or nonlinear regression band using SVM.

- Optimisation

  Optimisation models use one or more objective functions that are maximised or minimised using one or more constraints, both of which can be either well-defined mathematical functions or empirically defined (as in genetic algorithms(GA)). GAs are optimisation techniques for solving complex problems based upon the principle of natural selection in biological evolution.

- Web and text mining

  Mining useful information from text and other extended data types on the web are called web and text mining. The process of labeling documents using their semantic content into distinct categories is known as text categorisation.

1.9.2   Data Mining Query Language (DMQL)

DMQL is an SQL-like query language to facilitate data mining activities. For example, to mine association rules, we specify the input attributes (columns), conditions on the rows and the confidence and support thresholds as follows:

```
mine association_rule
in relevance to attribute1 [,attribute2]*
from datastore
where   condition1 [relop condition2]*
with  support threshold  =  minsupport
with  confidence  threshold  = minconfidence
```

where attribute* are the input variables to be used, conditions are used to filter out the records, minsupport and minconfidence are fractional values greater than 0.

1.10   Some Applications

Most of the large companies with global presence are migrating to datawarehousing and data mining. Data mining has diversified into many business and non-business fields, including bioinformatics, higher education, national security, and government. In most companies, customers come and go - old customers leave due to various reasons (customer attrition) and new customers are always coming. When a customer leaves, the company may be interested in knowing if the move is temporary or permanent, which other existing customers are likely to move, what are the reasons and patterns that lead to the move, what are the chances of lost customers coming back and so on. All these can be answered using data mining techniques. In the following paragraphs, we briefly mention some of the popular applications. Other applications are scattered throughout the text in subsequent chapters.

1.10.1 Banking

Banks deal with multitutudes of customers like individuals, families, business enterprises, government agencies, other banks and financial institutions in different parts of the world. Customer data are mined in valuating various customers, providing additional credit facilities to credit worthy customers, identifying loan defaulting customers, etc. As information vary widely among these groups, segmented data mining models are built. Banks also use data mining techniques to automatically detect fraudulent transactions, credit card thefts, risk analysis, investment banking, cards services, asset evaluations, cash flow analysis, predicting sensitive market directions, pricing strategies in competitive market, cross-sectional and time series analysis, detecting money laundering, deceptive stock transactions, contingent claim analysis, resource planning, repayment time modeling for loan customers from different socio-economic strata etc. Banks use monthly expenditure data of credit card holders to suggest products and services to its members. Various other applications are given in subsequent chapters.

1.10.2 Communications

Communication can be initiated and controlled by humans or by machines. Telephone call data are mined to decide call routing requirements, and best routing strategies, etc. They can also provide information on various carriers and in upgrading existing facilities. Data from customer complaint calls, emergency calls, and calls during peak hours on holidays or special occasions are also being mined to improve services. Calls routed through special numbers, or calling schemes can identify lifestyle categories of customers. Mining monthly billing data can identify socio-economic patterns or customers who share similar characteristics. Computer communication data can be mined to decide optimal bandwidth expansions, establishing communication proxy servers, optimal routing of packets etc.

1.10.3 Government

Data mining is used by many government agencies to improve services or performance, cut down expenses, detect fraud in real-time, analyse scientific and research information, develop infrastructure, for counter-intelligence, and in national security. Most popular modeling involve prediction, classification, clustering, evolution analysis, trend detection and anomaly detection, in addition to automatic summarisation. Industrial and domestic energy consumption data may be mined to decide about locations and capacities of future energy sources or plants. Identifying factors that affect crop yield can maximise agricultural output. The tax filing and tax pre-payment data can be mined to decide the optimal staffing requirements and identifying surpluses. It is also being used in finance and economy monitoring, e-governance, appraisals, electronic surveillance, healthcare and transport management. Other areas include defense, labor and social welfare, employment services, immigration programs and foreign services, best practices in governance, etc.

1.10.4 Hospitals

Large hospitals generate plenty of data about past patients, treatment records, disease prevalence etc. Patient's prior treatment records are being utilised to decide about patient specific treatment plans. Past inpatient data can be mined to estimate resource requirements for each disease category. Patient symptoms for various seasons or time of the year can be mined to decide about proper course of actions or in issuing advance warnings about impeding illnesses.

Mining the past data can also help in optimal stocking of essential medicines and other resources at correct levels in clinics, pharmacies and emergency rooms. Patient data can be analysed in a multitude of sub-categories using age, economic strata, education level, treatment plans, insurance policies, ethnicity, and various medical symptoms. Other applications include health insurance fraud and abuse detection, alternate treatment plans, risk factor analysis, etc.

1.10.5   Insurance

Insurance is a thriving business in most democracies. The most successful of these businesses are auto insurance, health insurance and fire insurance. Insurance companies utilise past claim data in deciding about premium rates for various categories. Data mining techniques are applied to assess risks and identify customer profiles. Who are the people at high risks of a fatal auto accident? Which patient profiles submit high medical claims? What are the fire disaster rates involving loss of human life and heavy property damage? These are some of the technical challenges for which data mining can provide valuable answers. The claim frequencies, claim amounts for loss of life, loss of property etc are probabilistic in nature and can be modeled by statistical distributions. Claims data contain dozens of attributes of various scales, some of which are correlated. Available past data may have missing values. Identifying groups of individuals in various categories can be facilitated using a well-trained genetic model. These models can then be used to classify new customers, catch fraudulent insurance claims, detect trends, predict insurer attrition, streamline operations and offer better services. Insurance data can be mined to determine premiums for various categories, in speedy processing of claims, managing investment, customer relation management etc.

1.10.6   Sports

Sport coaches use data mining techniques to mine for winning (and losing) strategies and play patterns. Softwares (eg: IBM Advanced Scout) use stored video images of winning moves to analyse the movements of players to predict orchestrated plays and strategies, and come up with statistical summaries and other useful information. Perhaps some day the data mining software will catch match fixings by detecting unusual deviations in player strategies.

1.10.7   Miscellaneous

There are lot many other applications to data mining, some of which are discussed in subsequent chapters. Data mining techniques are being used to advantage to detect fraud in financial transactions, analyse credit card usage patterns, and detect illegal use of cards. Advertising companies use datamining to optimally allocate budget for different media (various newspapers, magazines, TV, web etc) by mining past sales revenues. Diet manufacturers for over-fat people uses data mining of online and tele-survey data to find most popular magazines read, TV programs watched and web sites visited by overweight/overfat people in various categories. E-commerce companies use data mining for customer categorisation.

1.11   Summary

This chapter gave an overview of various data typologies and some extended data types. It also introduced the basic concepts in data mining, that are discussed further in subsequent chapters. From the business perspective, data mining can be used to control costs, to maximise revenues, to minimise delays, to retain customers, to streamline business processes, in fraud detection (credit cards, telephone cards, insurance cards, etc) among many other things. Benefits of data mining are not limited to commercial fields. It spans a broad range of other fields including astronomy, bioinformatics, data communications, medical sciences, natural sciences etc. The aim of DM is to build a simple, small and easily understood model from a large amount of data, irrespective of the data origin. An impromptu trained data miner may simply wade through verbiage of data, while a properly trained person can quickly come up with potentially useful nuggets of information in a record time.

1.12   Exercises

1. Assume that you have a heterogeneous group of people to each of whom the following questions are asked. Write down the type of variable (in NOIR typology) you will use, and possible values (for nominal and ordinal variables) as answers to each of the following:

   a) What is your current religious affiliation?

   b) What is the severity of your fever?

   c) Did the symptom completely vanish with the medication?

   d) What letter-grade did you get in English?

   e) What is your family's income?

   f) What newspapers do you read?

   g) Which TV channels do you watch every week?

   h) Which languages can you read?

   i) What is your grade point average?

   j) What is your mother-tongue?

   k) What is the power of your eye glasses (spectacles)?

   l) How many pages did you read in your book?

   m) What is your typing speed in words per minute?

2. What type of data in NOIR typology are the following:

   a) customer rating of a service with maximum score 10

   b) defaulting loan amount of a customer

   c) Jersey number of a player (to distinguish each player in a team)

   d) duration of a movie (in minutes)

   e) answer choices for a multiple-choice question

   f) week days coded as 0=Sunday, 1=Monday, .. 6=Saturday

   g) systolic blood pressure of a patient

   h) resolution in dots per inch (dpi) of printers

   i) branches of studies in a college

   j) amount of sugar in a cup of orange juice

k) number of orders of an e-com customer

l) wind classifications as (breeze, gale, storm, hurricane).

3. Identify the type of data, and assign single-letter codes for each of the following:

a) hair color ={black, red, gray, white}

b) credit card={VISA, Mastercard, Bankcard}

c) sales region={urban, rural, village}

d) auto insurance={minimum liability, normal, premium, comprehensive}

e) hobbies={chess, music, movies, numismatics, philately, traveling }

f) floppy={3M, Dysan, IBM, Maxell, Sony, Verbatim}

g) drink={milk, fruit juice, tea, coffee, cola}

4. What is the least informative of the scales?

A) Nominal B) Ordinal C) Interval D) Ratio

5. On what type of data (N,O,I,R) can you compute each of the following?

1) harmonic mean   2) median of a sample

3) range   4) inter-quartile range

5) maximum of a sample.

6. In a study to attract *uninsured* persons to a new insurance scheme, how will you code the *insured* variable?.

7. Identify each of the binary variables as symmetric or asymmetric:

a) person=(left handed, right handed) b) gender=(male, female)

c) diabetic=(yes,no) d) frequent flyer=(yes,no) e) fever=(low,high).

8. What central tendency measures are most appropriate for nominal and ordinal data?

9. Can a nominal variable be used to index a database column? When will it be effective?

10. Identify each of the following variables as discrete or continuous:

1) formatted capacity of a computer disk

2) data transmission rate of an internet connection

3) number of words per minute typed by a clerk

4) foreign currency exchange rate on any day

5) total spam mail received per day

6) number of e-com orders from an IP address

7) total score accumulated by an online game player

8) calorific value of a soft drink

9) odometer reading of a car

10) GDP of a country

11. What type of variables are the following?

    (a) year={regular year, leap year}, (b) day={work day, weekend, holiday}, (c) entertainment ={TV, radio, music, movies}, (d) blood pressure ={normal, abnormal}, (e) travel time to school (f) total spending on food per month.

12. Are the following nominal variables? Explain why or why not.

    a) geographic position={longitude, latitude, altitude}

    b) ENT_consultation={Ear, Nose, Throat}

    c) coffee={water, coffee powder, milk, sugar}

    d) tea type={light, medium, strong}

13. Which of the following are data mining models?

    1) association rules 2) clustering 3) data compression

    4) encryption 5) regression 6) decision trees

14. What are some of the advantages of data mining?

15. From which disciplines or fields did data mining originate from?

16. What are the permissible operations on nominal data?

17. Which of the following are sources of analysis for data mining?

    a) credit card transactions b) discount coupons, c) complaint calls, d) emergency calls.

18. Classify each of the following data mining techniques as supervised, semi-supervised or unsupervised.
    a) Association rules b) multiple linear regression c) neural networks d) decision trees e) support vector machines f) text segmentation

19. What are the allowed operations on cyclic ordinal variable?

20. Distinguish between symmetric and asymmetric binary variables. How do you decide the coding scheme to be used (say 0 and 1) for asymmetric binary variable?

21. Which data scales in NOIR typology are together called categorical data?

22. Distinguish between qualitative data and quantitative data. Which of them is prominant in higher-level programming languages? Can you represent qualitative data using variable declarations of quantitative type?

23. What are the valid operations allowed on Interval and Ratio data?

24. Give 4 examples of extended data types.

25. Give 3 formats in which text data occur on the web. What are some activities that result in (or generate) text data?

26. What is data discretisation? What are various approaches available?

27. Generate 50 random numbers (n) in the range $(-10, +10)$ and recode it into a "standard precipitation index" using table 1.3.

Table 1.3: Standard Precipitation Index

SPI value	Description	Code
≤ -2	Extremely dry	XD
-1.5 to -1.99	Severely dry	SD
-1.0 to -1.49	Moderately dry	MD
-0.99 to 0.99	Nearly normal	NN
1.0 to 1.49	Moderately wet	MW
1.5 to 1.99	Very wet	VW
≥ 2	Extremely wet	XW

28. If there are 56 variable values for a nominal variable, what coding scheme will you use?

29. Distinguish supervised and unsupervised data mining models. Give examples of each in education, medical sciences, insurance.

30. What are the major steps in data mining? What are the major issues in DM?

31. What are the popular classification of data mining systems?

32. Compare data mining and knowledge discovery models.

33. What are data mining metrics?

34. Describe data mining security. What are the challenges in implementing it?.

1.12.0.1 References

[CW08] Chakrabarti, S., Witten, I.H.,et.al (2008). Data mining – know it all, Elsevier.

[CR12] Chattamvelli, R. (2012). Statistical Algorithms, Narosa publishing house, New Delhi; Alpha science international, Oxford, UK.

[HK06] Han, J., Kamber, M. (2012) Data Mining: concepts and techniques, 2nd ed., Morgan Kaufmann.

[HM01] Hand, D., Mannila, H., Smith, P.(2001). *Principles of data mining*, MIT Press, MA.

[HP97] Hosking, J.R., Pednault E.P., Sudan, M (1997). A statistical perspective on data mining, Future generation computing systems, special issue on data mining, 1997.

[JG98] Jorgensen, M., Gentleman, R. (1998). Data mining, *Chance*, 11,34-39 & 42.

[MG99] Mackinnon, M. J., Glick, N. (1999). Data mining and knowledge discovery in databases: an overview,. Australian and New Zealand Journal of Statistics 41: 255-275.

[PD99] Pyle, D.(1999). *Data Preparation for data mining*, Morgan Kaufmann.

[SS46] Stevens, S.S. (1946). On the theory of scales of measurement, *Science*, 103, 677-680.

2
Data Visualisation Techniques

Chapter objectives

- Understand different data display diagrams

- Understand charts and graphs

- Explain visualisation categories

- Describe one-variable and overlay charts

- Describe two-variable diagrams

- Explain hierarchical charts

- Describe cause-and-effect diagrams

- Describe multivariate data visualisation

2.1 Graphics and visualisation

Definition 2.1 Data visualisation is a tool to convert "raw data" into meaningful images or graphics for effortless human comprehension or communication. It is an effective technique to graphically display large volumes of data, either using the extraneous features of objects that generated the data like color, size, various categories or using their inherent relationships like dependency of one variable on another (table 2.2). This chapter discusses visualisation from a data mining perspective. Our primary aim is to dig out patterns and trends hidden in cleansed voluminous data. Graphic display devices are the most appropriate for visualising multi-dimensional data. When dimensionality is small (say 3 or 4), other display devices (like printers and plotters) may also be used to visualise data as a series of two dimensional images or graphs by fixing all but two dimensions. For example, consider a three dimensional data set in which one of the dimensions is a categorical variable *race* that assumes 6 values {Buddhist, Christian, Hindu, Jain, Muslim, Sikh}; and the other two variables (family income, total travel spending) are quantitative. By fixing each of the 6 values of race, we can display the slice of data for the other two variables (income, travel spending) as a two dimensional graph. Such plots are called matrix-plots. If there are sufficient number of data for each of the categories, we could infer meaningful relationships among the quantitative variables.

Data Visualisation Techniques (DVT) were prevalent even before the advent of computers[1]. Most of the data display graphs (barchart, pie-chart etc.) were in use a few centuries ago. Scientific visualisation is an interdisciplinary field that utilises data structures, computer graphics, etc. It flourished as a separate discipline when high-speed display devices became commonplace. Initially, computer displays were very limited in size and resolutions. With the advent of Graphic Processing Units (GPU) in 1970's, and large sized high-resolution screens[2] in 1980's, tremendous progress has been made in visualisation technology. These techniques are widely used in image processing, computational science, medicine, engineering, data mining etc.

Data visualisation tools provide a single snapshot of complex relationships that are otherwise incomprehensible or difficult to interpret. The aim is to effectively convey or communicate the information content of (large volumes of) data as accurately as possible using graphical representations with minimal effort for human cognition. A side by side display of two data sets facilitates visual comparison, whereas a scatterplot of two correlated variables depict interdependency. Scatterplots also provide more insight into the type of relationships like linear, curvilinear or piece-wise linear relationships or discontinuity within the span of data range. It can tremendously help in choosing the correct data mining model to be used. For example, the linear regression model is ineffective when just a few of the variables interact nonlinearly. Similarly, clustering model is most effective when there are distinct clusters present. In the absence of clear demarcating boundaries, different clustering algorithms will produce different results. Visualisation softwares display data in various forms, and allow a user to manipulate the data (eg: scale (enlarge or reduce), rotate) interactively. Advanced visualisation softwares in addition have many types of operations like projection, roll-up and drill-down, slicing, dicing, etc that allows a data miner to search for patterns and trends in narrowly focused sub-dimensions. These are compute-intensive tasks for large data.

The input to the visualisation can come from a variety of sources like stored data in spreadsheets, databases, datawarehouses or datamarts, external devices (scanners, communication ports, pipes etc), electronic sensors, smart-dusts (motes)[3], medical devices etc. All visualisation softwares are not appropriate for data mining purposes, but most data mining softwares have in-built visualisation capabilities.

2.1.1 Power of Visualisation

Data may be visualised in spatial or temporal dimensions. DVT can reduce multidimensional relationships into finer levels of details in lower dimensions easily and efficiently. Modern visualisation tools offer extensive options to sift through high volumes of data to find unexpected relationships. As an example, the OLAP visualisation tools have inbuilt operations called slicing and dicing to explore data in sub-dimensions. By mapping or reducing higher dimensional dependencies into lower dimensions can result in valuable insight into the structural aspects of the data, or on the dependency of a combination of variables on others. It is important to use the most appropriate model first; followed by different diagrams or charts for each situation.

---

[1] A London doctor (Dr.John Snow[1813-1858]) used the overlaid spatial visualisation technique by mapping the 500 cholera deaths of August-September 1854, to contaminated water supply and leaking cesspool areas of Soho neighborhood of London suburb. See www.ph.ucla.edu/epi/snow.html for details.

[2] Graphical Processing Units are chips optimised for high-speed graphics tasks, and work in tandem with CPU

[3] Smart-dusts are miniaturised data collection devices. See [CR11] chapter 8 for a detailed discussion.

For example, all hierarchical clustering algorithms produce a dendrogram as output. A density-based clustering algorithm may produce a KD-Tree or an R*-Tree instead, which has different information content. Approximate data distributions can be understood using a histogram. A pie-chart serves the same purpose in emphasising the relative magnitude or proportions of several categories, but the categories in a pie-chart can appear in any order. These underline the importance of the most appropriate visualisation model, charts and graphs for each situation.

2.2   Summarisation vs Visualisation

As mentioned above, the same data can be presented in many forms. Summarisation is a special form of presentation in which summary information is arranged in tabular or graphical form. In several analysis problems, we may be satisfied with the summary information rather than a graphical display. The summary of Yes and No responses among males and females in a survey can be conveniently put in a 2x2 table as follows: Data to be visualised is most often

Table 2.1: Yes and No Responses in a Survey

	Male	Female
Yes	230	180
No	75	105

numeric (in NOIR scale). Other types of data like spatial data, text data in multiple forms, etc can also be visualised in various forms. Some popular data analysis and visualisation software programs keep summary information in tabular form. For example, tables underlie bar-charts, pie-charts, pictograms, Pareto diagrams, and many hierarchical diagrams. Tabular data can be visualised in multiple forms. The data in table 2.1 can be visualised row-wise or column-wise. Thus summarisation (in tabular or other forms) can be a prelude to data visualisation.

2.2.1   Tables

Tables are valuable tools for data presentation, and are often used as a first step in some of the other data display and analysis techniques. The input and expected output values of neural networks, attribute values and class labels of SVM training data, decision tables, term-by-document matrix of LSI are all represented as tables (see [CR11]). Tables can be simple or structured. They can organise data in two or higher-dimensions. Higher dimensional tables are visualised in two dimensions by fixing all dimensions except any two. They can present numeric or non-numeric data (dates, text, images, graphics).

Two-dimensional tables are organised as rows and columns. The intersection of a row and a column is called a *cell* or *element* of the table. The value stored in a cell is known as *cell value* or *table entry*. Each cell is uniquely identified by its row and column indices[4] where data

---

[4]In computer programming, tables are represented as arrays. In programming languages like C, C++, Java etc, the cell indices start at 0, but in data analysis, the cells are numbered starting with 1.

values are kept. Dimension of a table is the *actual number* of data cells in it, and is expressed in the form (r x c) where r is the number of rows and c is the number of columns. Although row and/or column labels and captions are used to describe the table data, they are not counted in the dimension of tables. They are part of typographic information to present data in a convincing form.[5] A table with just one column (but with more than one row) is known as a list (or column vector in Mathematics). An advantage of tables generated on computer screens is that the individual cells can store URLs to other tables, or point to external resources like video clips, songs, and other resources. Another classification of tables as primary and derived tables also exists. The interpretation or comparison of tabular data is more clear and concise, making it an indispensable tool for data display, and summary analysis. They organise, classify or summarise data in a form that is easily comprehensible. A related concept is called matrix. The elements of a table could be heterogeneous whereas all elements of a matrix are chosen from a homogeneous set. In other words, matrix elements are at the same granularity level, whereas granularity is imposed along the rows or columns in tables [CR12].

Each table cell can contain data of NOIR typology, or of extended types as discussed below. In other words, the cell values can be numeric or non-numeric. If the variable is nominal, each category is usually represented by a single row or column. For ordinal variables, the corresponding row/column can represent either a single value or a group of values. In the case of interval or ratio data, each row/column represents a range of values. Tables used in computing (as in web pages) can even have URLs, images, video clips, and graphics as table entries. A table may be kept in sorted order, if the variable is ordinal or higher scale. This improves visual comparison greatly, and is a popular technique in database derived tables[6].

Table 2.2: Scientific vs Information Visualisation

Scientific	Informational
emphasise on entities and attributes	focus on multiple observations or groups
data from samples, databases	data from external entities/groups
inter-relationships	nominal relationships
fine-grained attention	coarse-grained attention

Summary tables often contain numerical data, along with summary measures. These measures are conveniently kept as last row or column, but could also summarise blocks of rows/columns. Each table cell itself could also summarise data in parent table(s). Summary statistics like totals, means, variances, quartiles, ranges etc may also appear along appropriate columns or rows.

---

[5]Other typographic attributes include cell margins, fonts and typefaces, captions, spacings, subheadings, footnotes, boxing, alignment and justifications, display notation (decimal, scientific), precision etc). The labels may also be symbolic constants clearly marked as legends.

[6]Database derived tables obtain their data from databases, datawarehouses or datamarts. Results are displayed in tabular format with additional embellishments like ordering, grouping, summary counts etc

2.3   Graphics

Graphics are pictorial representations of data to *visually highlight* facts and relationships. They are a step ahead of tabular presentations, and is the preferred choice in exploratory data mining. Static displays used to depict positioning and formatting of groups of data are called presentation graphics. An advantage of graphics is that no detailed technical knowledge is needed to comprehend them (exceptions exist in some multivariate diagrams and plots as in Andrew plots (pp.2-21)). Computer generated graphics can also be linked to the parent data source that was used to generate it in such a way that when the parent data changes, the graphics are automatically updated. Thus dynamic data updates are instantaneously reflected in graphics[7]. This is especially useful in collaborative data mining.

Table 2.3: Monthly expenditures in $ (left), and Reason for return of forms (right)

item	category	amount
1	Rent	$600
2	Food	$400
3	Travel/gas	$310
4	Insurance	$250
5	Phone bill	$60
6	Clothing	$60
7	Electricity	$50
8	Utilities	$40
9	Other	$30
	Total	$ 1800

Reason for return of tax forms	Total Count
Missing entries	375
Incorrect values	158
Erasures	44
Unclear writings	10
Displaced entries	9
Missing Attachments	3
Others	1
Total	600

2.3.1   Scientific vs Information Visualisation

Scientific visualisation is used to display physical data features in sufficient detail (table 2.2). A magnified window with a user selected focus may be be coupled for visualising focused data in minute detail. Information visualisation is used to view unstructured or abstract high-dimensional data, or to highlight critical features used in user-guided exploratory investigation. Cartographic visualisation is mainly used to view boundaries of geographic regions, but also finds applications in terrain visualisation, geospatial interpretation, mapping regions with common characteristics, etc. Visualising the micro-bubble flow (magnetic bubble imaging) inside an artery, three dimensional volumetric visualisations like computer aided tomographic image, molecular structure of a drug etc are examples of scientific visualisation. Visualising the LAN interconnectivity, link structures (URLs) of a web pages are examples of abstract information visualisation. Statistical graphs (like pie-charts, bar charts and Pareto diagrams) and diagrams (like dendrogram used in hierarchical clustering, Andrew plots, CART, etc) display structure in the data rather than details. A good graph should have clarity, brevity, lack of distortion and instant comprehension.

---

[7]This is called Object Linking and Embedding (OLE)

2.4 One Variable Diagrams

Single variable diagrams are used to convey univariate information. The variable may be discrete or continuous. Continuous variables are graphed either by grouping them into classes (discretised; see chapter 1), or using a sliding window (as in time series, signal display etc). These are the primitive building blocks on which advanced graphs and charts are built upon. In the following section, we briefly describe the most popular charts and graphs.

2.4.1 Line Charts

These charts are used to display data over an interval. The interval is usually consecutive (fig. 2.1). Successive data values are joined together by a straight line (hence the name 'line chart'). It is useful to visually assess data variation and jumps. More than one series can be combined into a single line chart using either different colours, or visual patterns for drawing the lines (like dotted line, dashed line, bold line etc) (fig. 2.2). They are useful in detecting trends and

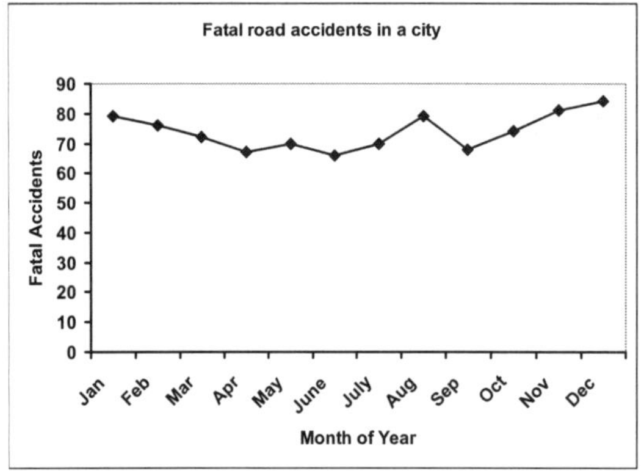

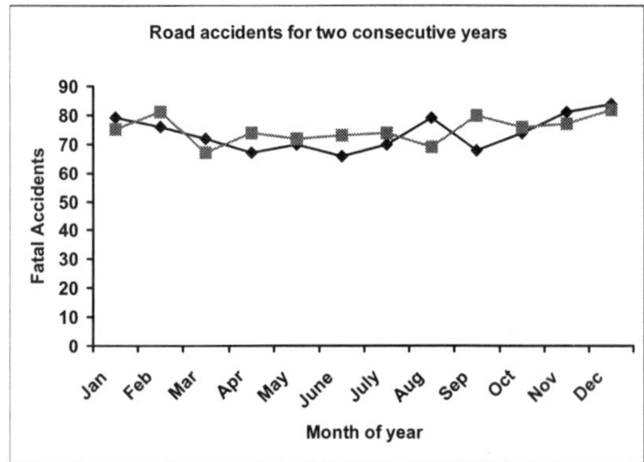

Figure 2.1: Simple line chart

Figure 2.2: Overlaid line chart

variations in multiple series. Figure 2.1 gives the number of fatal accidents during a 12 months period, and figure 2.2 gives a super-imposed line-chart for two consecutive years.

2.4.2 Bar Charts

Bar charts display categorical data distribution. In other words, they are used to display data values spaced out on the base-axis (which is the X-axis if the graph is drawn vertically, and Y-axis if drawn horizontally). Figure 2.3 gives the number of visitors to a web site during a week (as vertical bar chart). Here weekdays are (cyclic) categorical variables. Bars are drawn without touching each other (in contrast to histograms in which the bars touch adjacent ones if data distribution does not have gaps or discontinuities). A segmented bar chart (also called component bar chart, or stacked column charts) is an extension in which each bar is further subdivided as proportionate rectangles using another variable (which is usually categorical). The country-wise sale of items of various types (books in different subjects, services of various

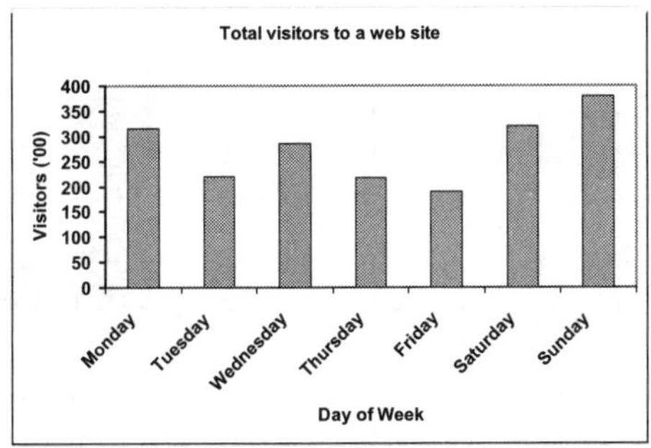

Figure 2.3: Vertical bar chart

types) can be represented as a segmented bar chart in which the axis may be labeled with country (with bars proportionally segmented using sales amounts of subjects) or axis labeled with subjects (with bars proportionally segmented using total sales for countries). Other examples are country-wise medal distribution in games, advertisement expenses for TV, web and printed media for several years. To convey meaningful information, the segmenting variable must be categorical (eg: total medals bagged by *countries*, total advertisement expenditures for various *media* like web, TV, radio, printed media).

2.4.3   Histogram

A histogram pictorially displays the frequency count or proportion (fraction) of a continuous variable falling in various intervals[8]. The intervals are usually continuous (continuous value assumption)[9], and of fixed size (uniform spread assumption). Each spike of a histogram can represent any measurable quantity like average income, weight, expenditure etc. First and last intervals may either be closed or open ended. Histograms used in data mining applications can be 2- or higher dimensional.

The first step in creating a histogram is to decide on the class intervals (which are usually of fixed size) and then accumulate the quantity of interest falling in each class into a table. When sample size is small, slight variations in the class widths can result in fluctuations in the frequencies.

If the entire data set can be partitioned into ≥ 2 groups using a nominal variable, each set can be plotted separately using a histogram. Such plots are known as *conditional histograms* since the value of the nominal variable is fixed for each of them (see trellis-plots below). This is useful when there are a few categories for a nominal variable, and too many data values.

---

[8]Intervals are also known as classes, segments, categories, bins, or groups

[9]which distinguishes it from bar charts, in which the variable is qualitative (nominal or ordinal), and the spikes usually do not touch each other

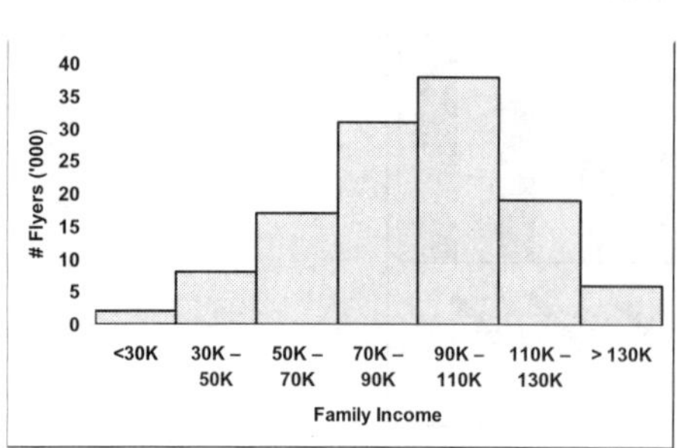
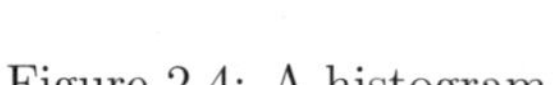

Figure 2.4: A histogram.

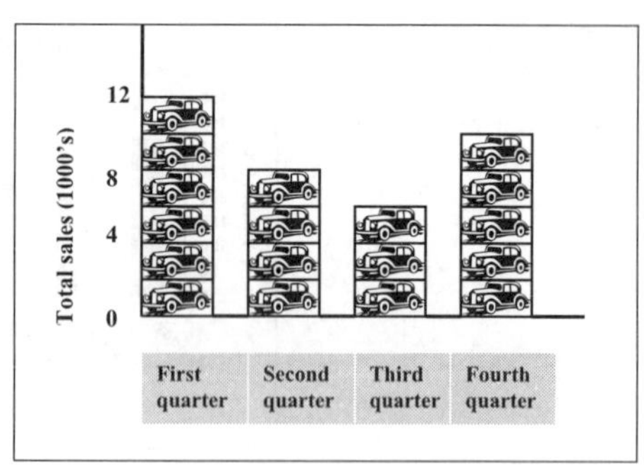

Figure 2.5: A pictogram.

2.4.3.1   Desirable Qualities of a Histogram

Different data miners may create different histograms from the same data, since the number of classes and scaling of variables are arbitrary. A few important points to consider while creating a histogram are given below.

- Reasonable number of bins (classes)
  There are no hard and fast rules regarding the number of classes, although from a comprehension viewpoint, the classes should be between 2 and 15. Too many classes can result in misinterpretation of the data.

- Each class should have the same class width
  The counts or proportions falling in various intervals can be compared among themselves if the class widths are the same.

- Class intervals should preferably be continuous (without gaps)
  This depends upon the data distributions. When discontinuities are widely separated, either different histograms are built for each continuous group, or the empty groups are merged into one group to improve the look.

2.4.4   Marginal Plots

These are combination plots comprising of a regular scatter-plot of two quantitative variables as the base. The name comes from the fact that the marginal distribution of variables are also drawn at the right and top margins (by default). Two barcharts, dot-plots or histograms that capture the frequency distribution of the variables along the X and Y axes are augmented as shown in figure 2.12 in pp.2-14. They can throw insight on the approximate relationship between two variables, and detect outliers if any. The name is derived from the fact that these augmented plots are drawn on the (right and top) margins. Thus the marginal plot provides more information on data distribution over and above the individual plots (they are 3-in-1 diagrams). The scale of variables are quantitative, but the unit need not be equal. Hence the variables could be standardised to have a better look. It can be used to check if the data are genuine, were derived by a random process, or by an artificial algorithm (like a sinusoidal data generator).

2.4.5 Pictogram

If the frequency in a barchart or histogram corresponds to an identifiable object, entity or event that can be represented by an icon or picture, we may fill the size or area by pictures itself. The resulting graph is called a pictogram. Examples include population of cities, or number of visitors to a web site represented by people figures; sales volume of an auto represented by pictures of autos (see figure 2.5) etc. An analogous technique for tables is called iconographic visualisation, in which each row or column of a data table is represented by an icon or glyph. An advantage of a pictogram is that it is easy to interpret even by non-specialists as it has an improved look and comprehension. A disadvantage is that unless the top portion is properly aligned, the pictogram's icons may be cut in part (unless it is too small) and may look uninteresting (a solution is to scale it to fit the fractional slot). Different icons can be used to distinguish different categories of an area chart. As the sectors of a pie chart are adjacent, each sector can be drawn with a picture or icon that is different from its neighboring sectors to produce a picto-pie chart.

2.4.6 Time Charts

Charts in which the variable of interest is time are known as time charts. Examples are – the average price of a stock during various weeks or months, the average temperature of a city during a week, number of visitors to a website during hourly intervals. Similarly, the occurrence of major earthquakes throughout the past century or number of deaths due to natural disasters like tsunamis, landslides, epidemics etc during past decade can be conveniently represented by time charts. Chart annotations (descriptive messages added on top of a graph in the form of 'callouts' to make it easier to comprehend) are often used in such charts. If the data on the time line is absent or very sparse, those portions can be deleted to improve the look of the graph. The frequency axis is scaled properly to avoid data values cluttering close to the time axis or gets spread too much. Run-sequence plots are special types of timecharts in which one variable is time and is usually used for sequenced data like stock market prices.

2.4.6.1 Temporal Histograms

In temporal histograms, one of the axes (usually X-axis) represents time. For example, suppose a company wishes to represent total advertisement expenditures and total sales revenue for each of several consecutive time periods (quarters or years). These can be drawn as a side-by-side temporal histogram. Other examples are total visitors to a web site in hourly intervals, total students enrolled by semester, total vehicles sold per month, etc.

2.4.6.2 Spatial Histograms

These are histograms with the base of the spikes aligned with a spatial 2D-map. Consider an e-commerce site with world-wide customers. The country-wise sale of items can be represented as in figure 2.6, in which a properly aligned geographic diagram is used to represent the regions.

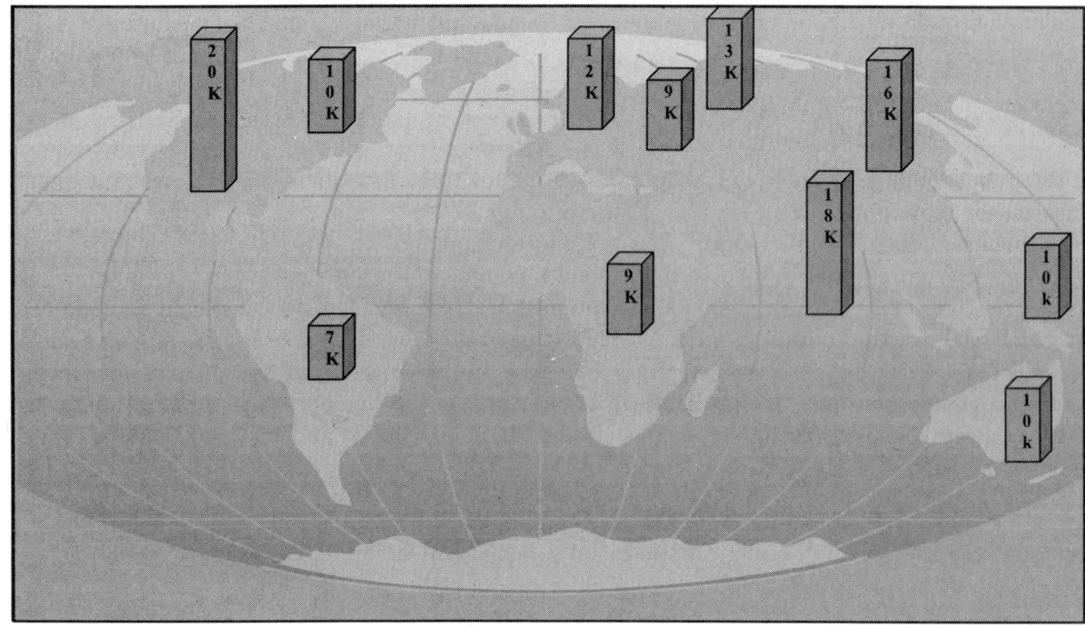

Figure 2.6: A spatial histogram.

These are also called surface plot diagrams. They are used when entities that generate data are spatially distributed as in web commerce, long distance communications (telephone, email), political surveys, opinion polls, epidemics, natural disasters etc. The base diagram need not necessarily be geographical. In medical studies it could be the human body or parts thereof, in astronomical studies it could be various regions of the sky, and in electronics it could be transistor distributions on a device/board. Spatial histograms can also be redrawn as regular histograms with the bases labeled with aligning map locations. The advantage of spatial alignment is faster comprehension. Spatial pictogram is an extension in which the spikes are drawn using pictures as in a pictogram. If the spikes in figure 2.6 represent book sales, each spike is drawn using the image of a stack of books. This allows multiple data series to be plotted onto the same base diagram (eg: sales of books and iPhones in different countries). Graph annotations in the form of 'callouts' can be used for better comprehension of multiple spatial histograms.

2.4.7   Pareto diagrams

The Pareto diagrams (also called Pareto charts)[10] are special types of histograms in which the spikes are ordered in descending order of the data values (longest spike at left and shortest at the right).

---

[10] An Italian economist Vilfredo Pareto (1848-1923) is the originator of the Pareto chart, which is used mainly by economists to prioritise activities on an importance scale. The Pareto principle states that "80% of the wealth is held by only 20% of the people". Although the number split may vary from 80/20 rule, the essence of the rule is to graphically emphasise the relative importance of differences among different categories or to distinguish "vital few activities from the trivial many".

It is used to order the root causes in decreasing order of relative importance (When two frequencies are equal, they may either be ordered on a secondary criterion or chosen arbitrarily). The Pareto diagram is usually drawn such that the horizontal axis represents the problem or cause (in decreasing order of importance) and the left vertical axis represents the frequency of occurrence or cost associated with each problem or cause, and the right vertical axis represents percentages. To construct a Pareto chart, 1) identify the problematic categories which will be

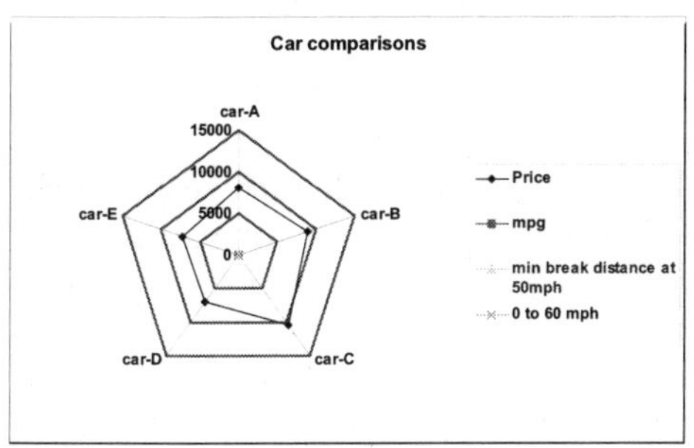

Figure 2.7: A Radar chart.

Figure 2.8: An exploded Pie chart.

displayed on X-axis, 2) decide the measure to be used on Y-axis (count, time, quantity, cost, defect causes, revenue loss). Scale it up or down if values are too small or too large. 3) Sort the categories in decreasing order of Y-axis measure (count in our example) and 4) plot the data using a bar chart (or histogram) 5) Since we are interested in the percentages, mark the right side vertical axis of the graph from 0 to 100% (with 0 at bottom and 100 at the top) and find the cumulative percentage of each successive category. 6) Draw a line connecting the cumulative frequency of various categories. 7) Identify the vital few from trivial many by drawing a horizontal line from the right vertical axis to cut the cumulative frequencies. Typical applications include customer complaints tracking, software bugs removal, tracking incomplete application forms (applications for admission, tax applications etc).

The graphical representations discussed above assume an axis-mapping for each dimension. The bar charts are usually drawn with vertical bars as shown in figure 2.3. But we can also draw them horizontally to suit individual situation. This is especially suited for printing or plotting large graphs in landscape mode. The orientation is unimportant for pie-charts. The bar charts, histograms, scatterplots etc can be drawn in linear or nonlinear scales on any of the axes. For example, the X axis can have logarithmic scale and Y axis can have linear scale. This type of charts is useful in some data mining applications where one or more variables increase in powers of 10. However, they are most often used in engineering applications where rate of changes occur logarithmically, exponentially or some other nonlinear way.

There are two popular generalisations of Pareto charts – multi-level (hierarchical) Pareto charts and weighted Pareto charts. Suppose the categories in the Pareto chart can be further divided into subcategories. In such situations, we can apply the Pareto principle at two hierarchical levels. As an example, suppose all errors and incomplete entries in tax applications are to be remedied. The categories may be 1) missing entries 2) incomprehensible handwritings,

3) erased entries, 4) unclear entries/inappropriate values 5) displaced entries (entering phone number in name row, entering tax due in the row for business income etc or entering every entry off by one step or row) and 6) other faults. This gives us the first level Pareto chart. The missing entries[11] can further be classified into 1) missing tax-ID, 2) missing numbers, 3) missing addresses, 4) missing contact details (phone, email etc), 5) missing totals, 6) missing signatures, 7) missing dates, 8) other missing entries. These form the second level Pareto chart. Such sub-category decomposition may be carried out for other first level data like unclear entries (unclear due to erasures, smudges and dirts, foldings, torn parts, burned portions, etc).

2.4.8   Pie-charts

These are so called because they are drawn as sectors of a circle in which each sector-angle is obtained as a fraction of the total angle of a circle - namely 2π radians or equivalently 360 degrees. A pie chart is used to visually depict the relative size or proportion of various categories that make up the whole 360 degrees of a circle. If many small categories exist in a pie chart, they can be combined and drawn as a single sector of the chart. This single sector can be linked and drawn as a separate pie chart, to view them as large sectors. There are two ways to draw the Pie charts - first one in which all the sectors cling together as a single circle, and the second one in which one or more sectors protrude as in figure 2.8. This technique called *exploding* has a exquisite visual impact. Either selected sectors (to emphasize particular categories) or all sectors can be exploded as in figure 2.8. Pie charts can also be three dimensional.

Different icons can be used to distinguish different categories of a pie-chart. As the sectors of a pie chart are adjacent, each sector can be drawn with a picture or icon that is different from its neighbouring sectors to produce a *picto-pie chart*. Because the relative magnitudes of various categories determine the size of each sector, large numbers (in each category) may be scaled (by dividing by a nonzero constant) to ease the calculation. A generalisation of Pie-chart is called a doughnut chart in which instead of a single series, we represent multiple series drawn in separate rings. The series with less number of classes is preferably chosen for interior sector to have better visual perception.

Because the human eye is better in judging linear displays (histograms, bar charts etc) than judging relative areas represented as angular sectors of a circle, Pie charts are unsuitable for people who are *not* mathematically oriented. Even slight variations in frequencies of adjacent spikes of a histogram or bar-chart can easily be caught, but when converted to sectors of a circle, the slight differences are not easy to spot, unless they are numerically labeled (even if different colours or patterns are used for adjacent sectors, all pie-chart sectors must be numerically labeled). Sectors of a pie-chart can be drawn in any desired order (clockwise or counter-clockwise), and adjacent spikes (classes) of a barchart/histogram need not be drawn as adjacent sectors of a pie-chart. The number of classes in a Pie-chart should preferably be between 2 and 15. A 'sorted pie chart' is one in which adjacent sectors are ordered according to their relative areas (which represent corresponding frequencies or percentages). This is meaningful only when the number of sectors is four or more (If there are 3 sectors, one of the [clock-wise or anti-clock-wise] orientations will automatically make it sorted, using a proper starting point).

---

[11]the no longer alive column indicates that the applicant has passed away after sending in the application.

Example 2.1 Compute the sector angles of a pie-chart for the data in table 2.4 in page 2-5.
Solution: We first sum the counts to obtain 600, using which the proportions can be found.
These proportions are multiplied by 360 degrees to get sector sizes as shown in table 2.4.

Table 2.4: Sizes of sectors to construct a pie-chart for return of tax forms

Reason	Count	fraction	angle
Missing entries	375	0.6250	225
Incorrect values	158	0.2633	94.80
Erasures	44	0.0733	26.40
Unclear writings	10	0.0167	6.00
Displaced entries	9	0.0150	5.40
Missing Attachments	3	0.0050	1.80
Others	1	0.001667	0.60
Total	600	1	360

2.4.9 Radar charts

These are also called polar charts or spider-charts because they are laid out radially using two
or more categories. Each category is assigned its own axis drawn through the center. The
chart is interpreted by focusing on the center point, and traversing outwards. For example, to
visually compare five different cars on 4 qualities (price, miles per gallon, minimum breaking
distance at various speeds, time to reach 60mph from rest) we could draw a radar chart, and
use various colours for each attribute. These attributes may be scaled to a proper range, and
assigned to separate axes. Undirected lines are used to connect various values. Other examples
include comparing sports teams, individual players, stock portfolios, different prescriptions for
same illnesses, different exposures to media by separate groups or persons, advertisement costs
in various media vs sales revenue, fund spending in different projects, budgeted vs spent funds
under various categories or departments, etc. It is also used in many psychological analyses (score
interactions in various types of tests, (ambition,personality,experience) interactions), etc. The
information content in a radar chart can be captured into a table, but the visual representation
is more appealing to compare and contrast.
Stacked area radar chart is a variant in which the areas in between two series are given uniform
colour or pattern.

2.4.10 Frequency polygons and frequency curves

A frequency polygon is constructed by connecting the top points of (vertical) histogram bars
by a piece-wise continuous straight line. The information content of frequency polygons is the
same as that presented by a histogram, but it helps in visualising the frequency variations in a
slightly better form, and can be used as a prelude to drawing frequency curves, which are drawn
by smoothing out the piece-wise linear curve. Frequency curves can throw enough light on the

distribution of sample points (eg: to check whether the parent population is normal or skewed). They can also be higher dimensional.

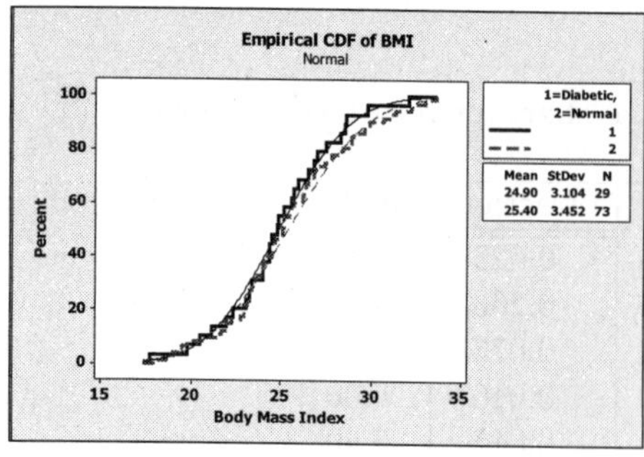

Figure 2.9: An empirical CDF plot.

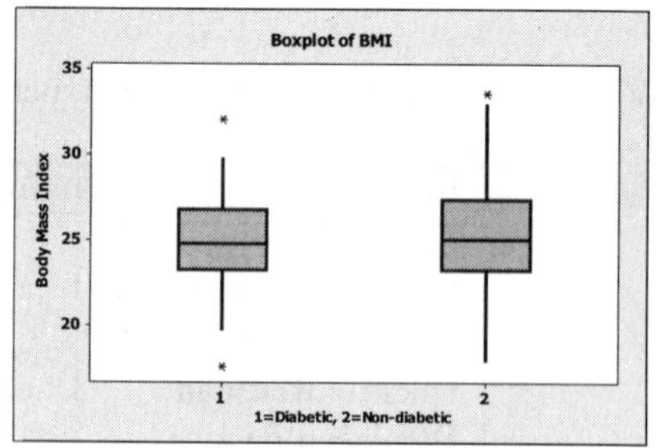

Figure 2.10: BMI of Non-diabetics.

2.4.11   Stem-and-leaf plots

These plots are used to represent a small set of numbers, text or strings that have some characteristic in common. In the case of numbers, the leading digit is used to group related items together. For strings, we use common prefixes for grouping. The rest of the data items are ordered in ascending (or alphabetical) order to ease comprehension. The left column is called the *stem* and the other column is the *leaf*. A double-sided stem and leaf plot uses two stems (one at left and the other at right) and leaves as above.

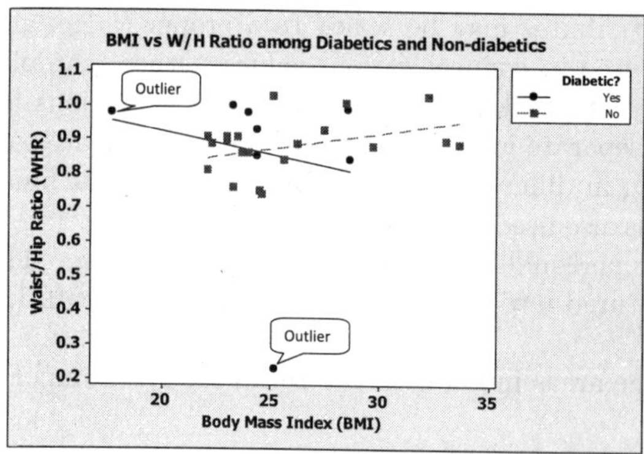

Figure 2.11: Summary for BMI.

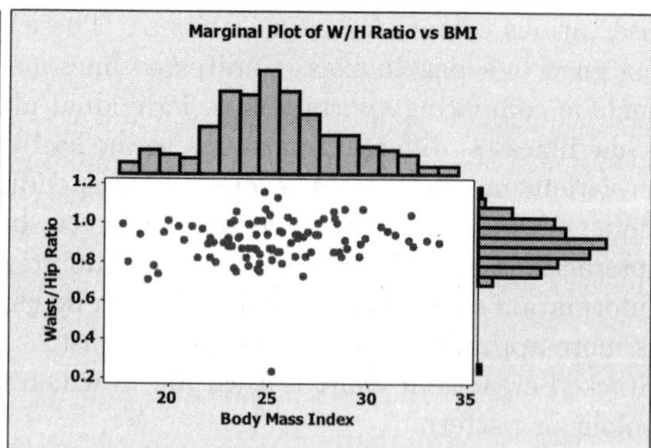

Figure 2.12: W/H of Non-diabetics.

Example 2.2 The following data represent the temperature readings at 20 different cities on a particular summer day. Prepare the stem-and leaf plot.
25, 31, 28, 34, 19, 22, 27, 31, 28, 30, 18, 34, 32, 25, 27, 33, 29, 32, 30, 37.

Because each number is two digits long, we need only a single stem. The stems $\{1,2,3\}$ are kept in sorted order to facilitate the plot. From the plot, we infer that the parent population is probably skewed to the left (because there are 2 data values less than 20, 8 sample values between 20 and 29, 10 data values between 30 and 39).

1	8 9
2	2 4 5 5 7 7 8 9
3	0 0 1 1 2 3 3 4 5 7

Numeric data items used in stem and leaf plots can be in any radix. It is easy to find the mean of the numbers in a stem and leaf plot. Assuming that the numbers are decimal and less than 100, the sum of the second row of table 2.2 is given by $8 * 20 + (2 + 4 + 5 + 5 + 7 + 7 + 8 + 9)$ where 8 is the number of leaves. For single digit numbers, we prepend a '0' which is the stem (eg: 5 becomes 05). Similarly find the sum of other rows, sum them up and divide by the total number of leaves. If the numbers have 3 digits (less than 1000), the stem should be multiplied by 10^2, the first digit of the leaf multiplied by 10. As noted above, all data values are made equal width by prepending required number of 0's (eg: 7 becomes 007, 63 become 063, etc). For other radixes, replace the multiplier 10 by the appropriate base. For example, octal stem and leaf plots should use 8^k instead of 10^k as multiplier. If the stems are kept in sorted order and leaves are normalised to have equal width, the *approximate* frequency distribution can be inferred from the stem and leaf plot (whether the parent population is normal, bell shaped, exponential (cliff type), saw-toothed (multi-modal), bimodal, skewed, uniform or truncated distributions can be inferred).

2.4.12   Overlay charts

These charts are obtained by overlaying one or more charts on top of another. The Pareto chart in figure 2.13 in page 2-16 is a self-overlay chart with a histogram and a frequency curve derived from the same data. Barcharts and line charts, or barcharts and area charts can easily be overlaid. Similarly, a scatterplot and a moving average chart can be overlaid for time series data (a scatterplot and a line chart can be overlaid as in linear regression). When two overlaid data sequences are different, calibration of one variable is done on the left vertical axis, and calibration of the other variable is done on the right axis. These are explained using a legend (as count in thousands on left axis, sales in millions on right axis). Different colours or patterns may be used to render exquisite appearance to various overlaid graphs. For instance, number of two-wheelers sold (barchart in blue) and total amount of sales (line chart in green) for various years can be overlaid. This helps in comparing the two trends in a single snapshot. Other examples are comparing advertisement expenditures and revenues, total visitors and total orders to a web site etc.

2.5   Multi-variable Diagrams

Diagrams produced from two or more variables are called multi-variate diagrams. These variables can be a mix of categorical and continuous types. In the following discussion, it is assumed that at least one variable is continuous.

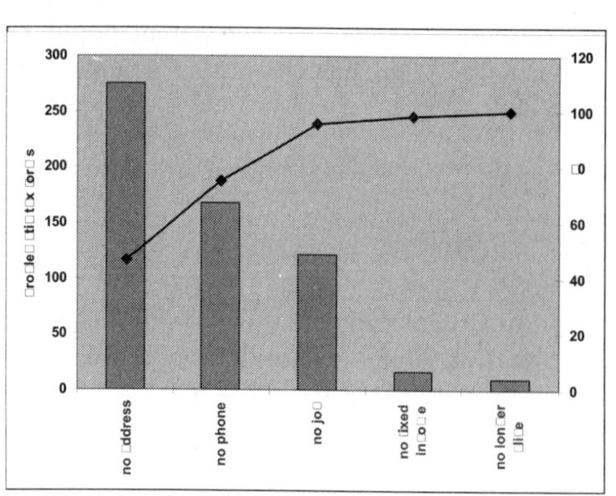

Figure 2.13: A Pareto-chart.

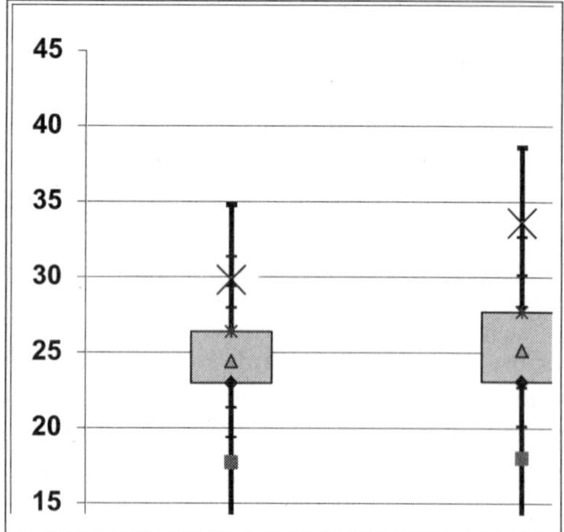

Figure 2.14: A Box plot.

2.5.1   Scatterplot

A scatterplot is a graphical representation of two or more continuous variables. Each variable is assigned to one of the axes, and each data point is represented by a single symbol (a dot, a star or an x mark). Two-dimensional scatterplots are also known as *scattergrams*. These plots can give valuable information about the strength of relationships between the variables of interest, and also the type of relationships. It can be used to detect outliers and clusters present in the data (see below). Too many data points (say > 500) in a scatterplot can clutter up the results, making it difficult to detect any hidden patterns. In some cases, scaling up one of the variables can result in more revealing relationships. Plotting multiple points that share the same coordinates (in 2D) is a challenging problem. This phenomenon is called data point overlapping or data collision. There are many methods to resolve the overlap problem:

1. Linear mapping of brightness
   Since the human eye is more sensitive to luma than chroma, the overlapping points can be assigned different luminocity depending on the overlap count.

2. Use different colours
   The chromatic colours (r,g) are the easiest as they do not use brightness information. The respective values can be derived as a function of R,G,B values using the formulae:$r = R/(R + G + B)$ and $g = G/(R + G + B)$. Because r+g+b $= 1$, the blue component is found as b=1-r-g. Then linear scaling is used to plot colliding points with different colour values.

3. Use a bubble chart
   The bubble chart (§2.5.2) weighs each displayed point differently. Each overlapping point

will weigh it to increase the size of the bubble proportionally so that multiple collisions can easily be detected.

4. Use higher-dimensional scatterplots
   Multi-dimensional scatterplots have the facility to rotate the display area through various angles, so that the overlapping points can easily be viewed.

5. Use online querying
   This is an add-on feature for the user to query a selected point or region of a scatterplot (using a pointing device or by entering the corresponding coordinates) when it is displayed on computer console or other LCD devices. Overlapping points, if any, can be separately analysed in detail (eg:in a popup window).

Too few data points (say < 10) in a scatterplot may lead to wrong conclusions. It is important to balance the number of points appropriately to draw reasonable conclusions. Typically, 30 to 50 random points will suffice to understand the patterns using a single scatterplot. But in data mining applications, this number is much higher. Hence trellis-scatterplots may be employed by conditioning on one or more of the other variables (preferably categorical) [MD98]. Scatterplots can throw insight into special types of relationships like linear, nonlinear, discontinuous, or piece-wise continuous between variables. In other words, they are useful to check if variables are pair-wise linearly or nonlinearly related. They can detect if data are discontinuous, or piece-wise continuous. They can reveal distinct clusters and outliers in data. They can also validate

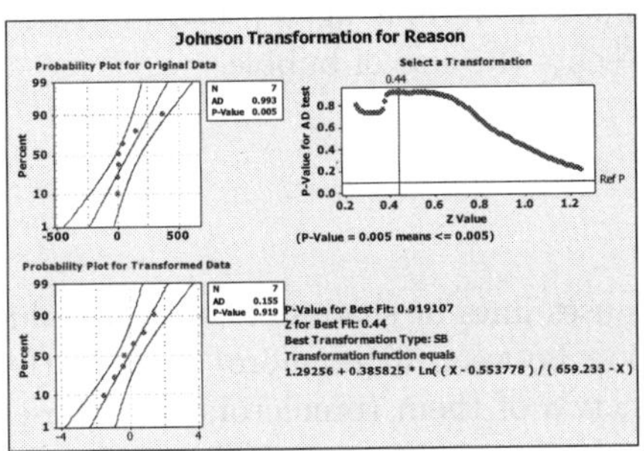

Figure 2.15: Johnson trans. plot (Reason for return of Tax forms).

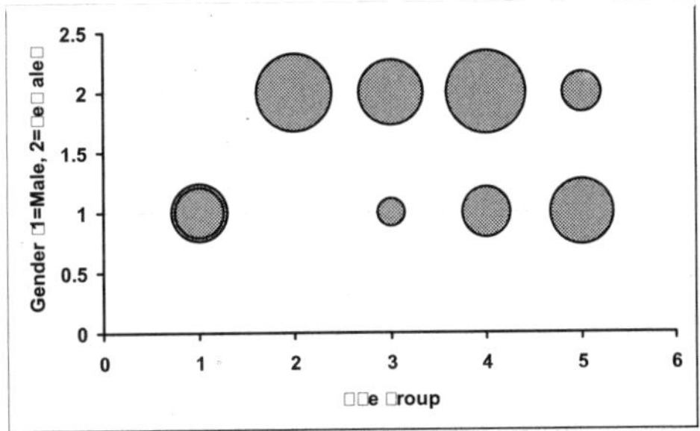

Figure 2.16: A bubble chart for orders

homoscedasticity or heteroscedasticity of variances. When there are n (>2) variables to be plotted, we could produce a matrix plot where each plot fixes 2 out of n variables and ignores all other variables temporarily to produce $\binom{n}{2}$ scatterplots. For example if there are 3 variables X, Y, Z the matrix plot will contain 3 plots corresponding to (X,Y), (X,Z) and (Y,Z). See [CM84], [MD02] for more information.

2.5.2 Bubble chart

A *bubble chart* is an extension of the scatterplot in which a third variable is present, which is assumed to be uncorrelated with the other two variables. Instead of representing the correlated

Table 2.5: Items purchases at a store (for bubble chart)

Age grp	Gender code	Items bought	Age grp	Gender code	Items bought
1	1	4	5	2	2
2	2	2	3	1	1
3	2	5	1	1	3
4	1	3	4	2	8
2	2	7	5	1	5

variable pairs by a dot or an x mark, we draw a bubble (circle in 2-D, hypersphere in n-D) whose relative size indicates the magnitude of the third variable. In other words, bubble charts add an extra quantitative variable to emphasise the relative size to a display. As in the case of marginal plot, a dot-plot or histogram can be augmented for each of the categories at the right margin to reveal the individual distribution in each category.

This idea can be extended to higher dimensions by replacing circles with spheres or cubes. A shading hue may also be used to weigh multiple variables within a bubble (two equal circles with different colours or intensities will indicate two different weighting). Figure 2.5.1 gives the bubble chart for 5 males (=1) and 5 females(=2) customers in various age groups (1=15-20, 2=20-25, etc) where the number of items purchased represents the size of bubble.

2.5.3   Contour plots

Another extension of scatter plot, known as *contour plots* uses lines or shades of constant value instead of discrete points to explore the potential relation between 3 (or more) quantitative variables in NOIR typology. When there are 3 variables, two of them (predictors) are chosen for X and Y axes, and the values of the third variable is plotted as a contour. These are color coded for visual comprehension (and a table of legends are provided to explicate the intervals or grouping of shades), where darker regions or points indicate higher Z-variable values. If there exist clear-cut dense regions in the contour, they may be demarcated by a curve. It is more effective in data visualisation when there are large numbers of heterogeneous data points available. It finds applications in mapping the prevalence of epidemics, household income distributions, in medical imaging, in various geographical explorations like mapping the sea levels, soil compositions, sea temperatures, rainfall, air pollutants and in mappings magnetic fields, electron density maps etc. As the distribution of X and Y cannot be inferred from a contour plot, a marginal plot may be augmented to have better insight. This plot is irreversible — if a contour plot is given, it is not in general possible to reconstruct the numbers from it (whereas histograms, Pareto charts, piecharts etc are reversible).

Two dimensional contour plots comprise of contours (closed curves) that are smooth and non-overlapping and are useful in geographical depictions. Three dimensional extension of contour plots are known as *iso-surfaces*.

2.6 Hierarchical charts

All of the charts discussed above are non-hierarchical (except multi-level Pareto charts). In hierarchical charts, data are organised in a hierarchy based on values of variables or structure of data. Examples include decision trees, organic chemical structures, family trees, employee hierarchy, and disease hierarchies. Trees are nonlinear data structures that are popular in computer science (chapter 5).

2.6.1 Polar trees

Polar trees are drawn by reconfiguring the tree nodes along circles or closed contours such that all branches that are equidistant from the root have their positions aligned on the same contour. Instead of portraying the root at the top (or left), we simply portray it as the root at the center.

2.6.2 Cause and effect diagrams

They are also called fish-bone diagram or Ishikawa diagram. The central arrow (in figure 2.17) is called a *spine*. They use a structured approach to analyse the effects resulting from a single cause. They can depict, identify and sort-out various causes of a problem. The fishbone diagram is the same as binary trees in which some arcs are emphasised.

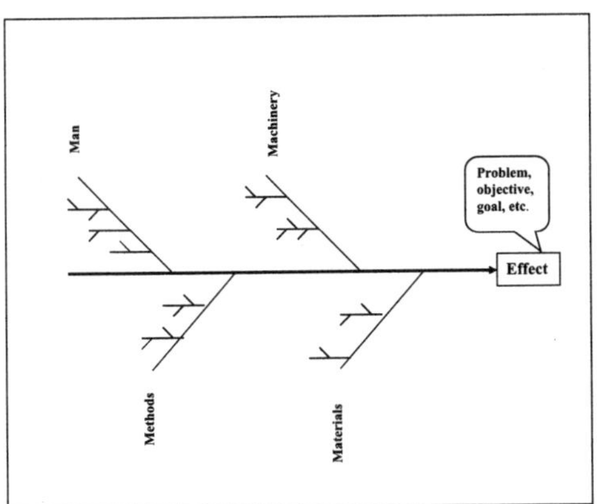

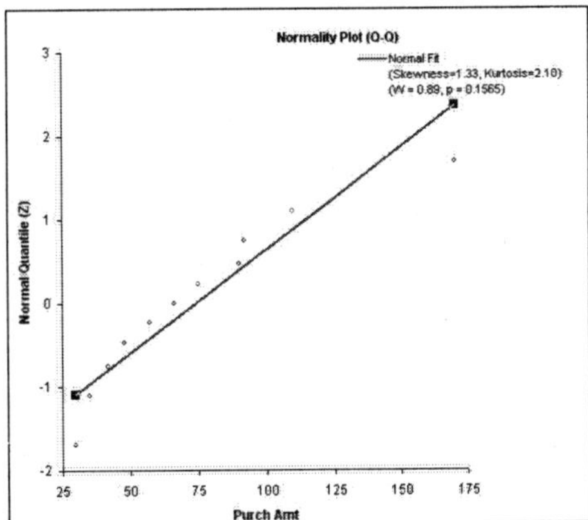

Figure 2.17: A 4M fishbone diagram.　　Figure 2.18: The Q-Q plot (table 2.6)

There are many ways to structure the fishbone diagram. Two popular ways are called 4M approach (methods, manpower, materials, machinery), and 4P approach (policies, procedures, people, plant) which represent the spikes originating from the spine. The spikes are subdivided according to various causes. A lesser-known model is 3M & P in which the major spikes correspond to (methods, materials, machines, people).

2.6.3 Trellis plots

These are higher dimensional plots to visualise multivariate interactions between explanatory variables acting on a response variable ([MT95],[MD98]). The entire display area is divided into panels in which a subset of the data is graphed (scatterplot, boxplot, dot plot etc). All trellis[12] plots have the same scale to ease visual comparison. The trellis-plot framework comprises of display layouts, panels for individual display, orderly enumeration of conditioning variables, panel ordering and axis labeling [MD98]. The conditioning variables should preferably be categorical. They may be ordered using natural values or the plots ordered using statistics of each individual plot (eg: mean of the X variable). Continuous variables are subdivided into intervals and each interval considered as conditioning variable.

2.6.4 Q-Q Plots

A basic assumption in many statistical analyses is that the data are drawn from a normal population. A test for normality can be carried out either analytically or graphically. The Q-Q (Quantile-Quantile)-plots is a graphical tool for checking the normality assumption of the parent population.[13] The quantiles are plotted along the X-axis and order statistic of the sample along Y-axis.

Table 2.6: Customer age vs purchase amounts

Cust age	22	45	31	26	21	19	30	25	42	23	51
purchase amount	170	90	75	92	42	30	66	48	35	57	110

Example 2.3 Table 2.6 gives the data on purchases made by customers at an e-commerce site. The Q-Q plot for it is shown in figure 2.18.

The Q-Q plots can also be used to check if two samples have come from identical populations. Samples are normalised using a common transformation (eg: Z-transform) before the quantiles are found. Quantiles of the first sample are then plotted against the quantiles of the second sample (optionally, a straight reference line with slope 1 (that makes 45-degrees to each axis) is also drawn). If the two samples are from the same parent population, there will be a perfect alignment of the quantiles along the reference line. Departures from the reference line indicate the falsity of our hypothesis (that the samples are *not* from the same parent population).

The main advantage of Q-Q plots is that the sample sizes need not be equal. It can easily be extended to multiple dimensions and many distributional aspects can be tested simultaneously. Normal probability plots (NPP) are very similar, except that NPP uses expected value of the kth order statistic along the X-axis. If the points clutter around a straight line, it is an indication of normality of the parent population.

---

[12]The literal meaning of trellis is a grid-like ladder used to climb trees, buildings and wall-like structures.

[13]The data may also be smoothed out using a power transformation as $y_j = \frac{x_j^p - 1}{p}$ (if $p \neq 0$) or as $y_j = \log(x_j)$ (if $p = 0$) where p is a suitably chosen constant.

2.6.5 Chernoff plots

Chernoff plots (also called Chernoff faces) are iconic plots of multivariate data in which facial characteristics (head eccentricity, eye eccentricity, pupil size, eyebrow shape, angle, height from eye, eye size, shape, separation, nose and ear size and position, mouth shape/curvature, size and orifice, position) and emotional expressions are used to represent variations in the data. It does not specify actual data values, which is a limitation of it in data visualisation. In addition, all possible states are difficult to remember, thereby limiting its use for heterogeneous data display.

2.6.6 Andrew plots

Andrew's plot[AD72] is used to visualise structure in high dimensional quantitative data. It is a two dimensional projection of multivariate data that preserves the mean, variance, and L_2-norm; and groups similar points with distance proximity. It has applications in outlier detection, clustering, correspondence analysis, contingency tables [KN01], quality control, artificial neural network training [GM00], and robotics. It transforms multivariate data using a parameterised function (called Andrew's curve):

$$f(t) = x_1/\sqrt{2} + x_2 \sin(t) + x_3 \cos(t) + x_4 \sin(2t) + x_5 \cos(2t) + .. \tag{2.1}$$

where the data vector $x = (x_1, x_2, ..x_d)$ is multiplied by the weight vector $[\frac{1}{\sqrt{2}}, \sin(t), \cos(t), \sin(2t), \cos(2t), \cdots]$ and $t \in [-\pi, \pi]$ (the missing operator is *). In other words, it weighs a multivariate observation vector X, except the first element, alternatively by sin(kt) and cosine(kt), for k=1,2,$\cdots$,d. As the overall shape of the plotted curve is determined by low frequency terms, they are associated with most important attributes.

Andrews curves are infinitely differentiable and higher order derivatives do not involve the first observation. It can be considered as an L_2 distance preserving static plot that groups similar points with distance proximity. A disadvantage of Andrews plots is that the odd observations altogether vanish near t=0 (because sin(kt)=0 when t=0). Alternatively, we could replace the even weights by $[\sin(kt)+\cos(kt)]/\sqrt{2}$ and odd weights (except first) by $[\sin(kt)-\cos(kt)]/\sqrt{2}$ as done in [KN01]. These weights also have the disadvantage that if x_{2k} and x_{2k+1} are nearly equal, $x_{2k}[\sin(kt) + \cos(kt)]/\sqrt{2} + x_{2k+1}[\sin(kt) - \cos(kt)]/\sqrt{2}] \simeq \sqrt{2}x_{2k}\sin(kt)$ which again vanishes in the neighborhood of t=0. In addition, $[\sin(kt)-\cos(kt)]$ vanishes in the neighborhoods of kt = $(2j+1)\pi/4$, j=0, 1, 2, etc. A slight modification of Andrew's curve could remedy this situation. Since the weight of the first observation = $1/\sqrt{2}$=sin(45), we could change the origin of the arguments and express it in the alternate form: $f_x(t) = x_1 \sin(\pi/4) - x_2 \sin(\pi/4 + t) + x_3 \cos(\pi/4 + t) - x_4 \sin(\pi/4 + 2t) + \cdots$. When t is 0, the sines and cosines in this series each evaluate to $1/\sqrt{2}$, providing more insight into the central region of the curve. Andrews plot has been extended to visualise data in multi-dimensions in [WS93].

2.6.7 Box and Whisker Plots

The box and whisker plots (simply known as boxplot) is a graphical depiction of different batches of data that makes use of order statistics along with the mean, min and max (the five number summary – mean, smallest observation, Q1, median, Q3)(or (min, Q_1, median=Q_2, Q_3, max), where Q_1 is the first quartile, and Q_3 is the third quartile. It comprises of a box, and

perpendicular whiskers at both ends of it. It is used to show how location measures are related to spread measures for a sample. Box glyph is a special case that is used to show the spread of data values. The Boxplot is created as follows:
1) Draw a vertical box with bottom and top borders at $Q1$ and $Q3$. 2) Draw the median and mean inside the box. 3) Calculate IQR $= Q_3 - Q_1$. Draw whiskers from each end of the box to most remote points 1.5*IQR away. 4) Connect whiskers to the box using vertical lines. 5) Give correct labels 6) Show outliers using a '*' or 'o' depending upon where they lie.

2.6.8 Stem Plots

A stem plot is used to visualise outliers, if any, in the data. Outlier detection is an important first step in many statistical procedures. A single outlier can affect the regression line, a clustering procedure or a discriminant function. Stemplots are used when the size of a data set is small.

Example 2.4 Table 2.7 gives the data on diabetes patients (diabetic=1, non-diabetic=2). The box-and-whisker plot is shown in figure 2.14 (The plot shows that the BMI of non-diabetics has greater variance than diabetic patients. A possible explanation is that the diabetics are in

Table 2.7: BMI Data on Diabetes Patients

2	2	2	2	2	1	2	2	2	2	2	1	1	2	2
23.3	26.2	25.6	25.1	23.3	23.3	23	23.7	33	29.7	28.5	25.2	22.1	23	28.4
2	2	2	2	2	1	1	1	1	1	2	2	1	2	2
24.6	32.2	33.6	27.4	24.4	17.7	24	24	24	24	24.5	28.6	23.5	22.3	22.1

general choosier, and take more controlled food than normal ones due to medical advice).

2.6.9 Miscellaneous Plots and Charts

A profile plot is a useful data analysis tool for detecting outliers, correlations and coherent clusters of data items. A coplot (conditioning plot)[CW93] is used to reveal the dependence of two variables on a third (conditional) variable, which is arranged in increasing sequence of a conditioning variable. An area chart shows the relative importance of various categories. Flow diagrams are used to show sequential dependencies. The sequence of web pages visited by e-commerce customers starting from the home page and ending with a specific exit page (eg: a firm order page) can be captured using a flow diagram. Other examples are Log-log graphs, Fisher-Pry curves, Ogive curves, PERT diagrams, UML flow diagrams, etc. See [CW93], [MG06], [FB08], [CH08], [TA08],[UT06] for further details.

2.6.10 Visualisation in Data Mining

Visualisation is an important tool in large data mining applications. Patterns, trends, and anomalies hidden in voluminous data can easily be explored using appropriate visualisation tools. These tools come in a variety of forms and functionalities – from the simplest graphics to advanced OLAP visualisation tools. Visualisation tools work directly on the cleansed data available in datawarehouses or datamarts. They build display models for analysts to dig out hidden relationships by interactive manual tuning of multidimensional parameters. The type of data displays used depends on the ingenuity and insightfulness of the analyst. These displays are used either for exploratory data analysis (EDA), confirmatory analysis, or simply to present the results (reporting). Some data mining techniques use statistical models in unsupervised mode to identify relationships among variables. These two techniques – data visualisation in supervised mode and machine-based unsupervised techniques – are often used in OLAP mining.

Most data visualisation tools allow the user to drill down to finer details if needed, which is the distinguishing feature of DVT from traditional data analysis. Other differences between EDS and DVT include the massive volumes of data involved, accuracy of data, and time criticality. The aim of using these tools may either be discovery of hidden patterns/trends, or prediction of future outcomes using discovered patterns from multiple perspectives. The outcome of visualisation can be explanatory or discovery. The end-result of this step is in the form of associations, decision trees (lists), classifications, clusters, or anomalies from expected or common relationships like outliers, discontinuity of trends etc. Statistical concepts play an important role in data mining [FH98], [FH98].

2.7 Data Visualisation Technology

Extracting and analysing hidden relationships, patterns or trends from stored data and presenting it in pictorial form that can be easily interpreted by experts and non-experts alike is accomplished using data visualisation. Creating the analysis models involve consolidating the data as follows:

- Cleanse and summarise data into a datawarehouse or datamart.

- Extract needed data from the source pool into the data visualisation tool environment.

- Explore the data with the data visualisation tools. Interactive visualisation is user-centered and iterates through pattern discovery loops in which the user interacts with backend data using visualisation tools.

- Save the results in proper format for future use

Several data visualisation tools exist to analyse the data. Some of these are overlapping or complementary in nature. An analyst can choose between simple charting tools or more sophisticated graphs and diagrams, to complex and full-fledged business intelligence tools.

2.7.1 Self-organising maps (SOM)

This promising technique aims for dimensionality reduction and similarity grouping (Kohonen, 1989) through self organising Kohonen neural network model. It applies data compression using vector quantisation, and preserves all major topological relations within the training set. It is an unsupervised technique for dynamic data visualisation and data mining. See chapter 8 for details.

2.8   Software for Data Visualisation

Several data visualisation tools exist to analyse the data. Some of these are overlapping or complementary in nature. An analyst can choose between simple charting tools or more sophisticated graphs and diagrams to complex and full-fledged business intelligence tools. The most popular tools for visualisation are as follows:

Table 2.8: Software for Data Visualisation

URL	Name
www.jmp.com	JMP Software
salstat.com	data analysis software
www.minitab.com/Downloads	Minitab
microcal.com	Origin
www.lcb.uu.se/tools/rosetta/index.php	Rosetta
insightful.com/industry/pharm/clinicalgraphics.asp	stat graphics
en.softonic.com	S-plus
sigmaplot.com	Sigmaplot
winsite.com/utilities/Miscellaneous	SOFA statistics for windows
qweas.com	Sysgraph
stata.com	Stata
graphpad.com	InStat, StatMate
winsite.com	Stat view
systat.com	systat
eric.univ-lyon2.fr/~ricco/tanagra/index.htm	Tanagra
tecplot.com	Tecplot
unistat.com	Unistat
maxstat.de	statistical analysis for windows
statgraphics.com	Visualisation software
texasoft.com	Winks SDA software

2.9   Summary

Data visualisation plays a prominent role in data mining, exploratory data analysis, and can be a useful prelude to many of the modeling techniques. It can not only reveal peculiarities in data, but also help the modeler to decide alternate modeling choices or sub-dimensional modeling. This can tremendously save time as it makes the analysis more focused.

2.10   Exercises

1. Mark as true or false:

a) Scatterplots can be used for uni-variate data

b) Dimension of a table is a measure of its overall size

c) Table cells can contain nominal, ordinal, interval or ratio type of data

d) A pictogram can be used for any type of area chart

e) A pie-chart can distinguish between linear and nonlinear relationships

f) The box in a boxplot represents 50% of the data

g) The width of a boxplot has no significance

h) Bubble charts are used to compare one or more categories/variables

i) Q-Q plots can be used to check if samples come from identical populations

j) Chernoff plots are multivariable diagrams

2. Which of the following diagrams are used for nominal data?

   A) scatterplot B) stem & leaf plot C) Pie-chart D) Line chart E) fishbone diagram

3. What is a segmented bar-chart? What type of variable is used for segmenting?

4. What is a conditional histogram? What are some desirable qualities of a histogram?

5. What are time charts? Give 3 examples.

6. When is a spatial histogram more appropriate? Give two examples.

7. What are Pareto diagrams? Give any 2 examples where it is useful.

8. Which data are most suitable for pie-charts?

   A) nominal data B) ordinal data C) interval data D) ratio data

9. What are radar charts? Give 2 examples where they are used.

10. Describe the stem-and-leaf plot. What type of data do they work with?

11. What are two disadvantages of the stem-and-leaf plot?

12. What are fish-bone diagrams? What is another name for it? What are two popular approaches to structure it?

13. What are the differences between bar charts and histograms?

14. What is a Q-Q plot? What data assumptions can you check with it?

15. What are the advantages of Q-Q plots? How are they different from normal probability plots?

16. Which of the following plots are used to reveal the dependence of two variables on a third (conditional) variable?
    (A) scatterplot (B) stem-plot (C) coplot (D) Pareto diagram

Table 2.9: Visitors to a web site ('000)

Sun	Mon	Tue	Wed	Thu	Fri	Sat	Tot
18	9	7.5	10.5	9	14	17	85

17. Develop the pie chart for the data in table 2.9.

18. Describe any 3 scientific fields that use visualisation.

19. Data in table 2.10 gives the number of frauds of various types at a bank. Develop a Pareto chart.

Table 2.10: Different types of fraud

#	Fraud type	total
1	Stolen card	330
2	Duplicate card	138
3	Falsified card application	90
4	Other	42
	Total	600

20. What is a disadvantage of a pictogram? What are its advantages?

21. What is a stem-and-leaf plot? For what type of data is it used?

22. A toy manufacturer is interested in tracking customer complaints using appropriate charts. Data in table 2.11 gives the number of customer complaints for 3 types of toys for kids in the age-group 2 – 6. Develop appropriate charts.

23. What are the various steps to construct a boxplot?. Which measures are needed?

24. Distinguish between temporal and spatial histograms. Can a spatial histogram have a 'base' other than a geographical map?

25. What is a trellis plot? What type of variable (of the NOIR types) is preferred for conditioning?

26. A computer program can fail due to run-time exceptions (which are recoverable) and errors (unrecoverable). Develop a fishbone diagram if the major exceptions are I/O exception, null pointer exception, hardware exception, data exception, memory exception and other exceptions. List a few subcategories of each of the leafs.

27. Describe 2 methods to resolve the overlap problem (data collision) in scatter-plotting numeric data.

Table 2.11: Customer complaints on 3 types of toys

#	Complaint type	toy-1	toy-2	toy-3
1	Physically injurious	412	325	248
2	Allergic materials/parts	291	303	207
3	Easily spoiled parts	112	80	46
4	Easily broken/inoperative	94	78	161
5	Too big size	87	58	32
6	Other	74	56	26
	Total	1070	900	720

28. What are two disadvantages of Chernoff plot?

29. What is a spatial histogram? Describe how you can redraw a spatial histogram as a regular histogram.

30. What roles does visualisation play in data mining? If you have a large number of categorical and quantitative variables in a data mining problem, describe how you will use the coplots (page 2-22) to discover patterns and trends.

31. What is an Andrew's plot used for? What are its advantages over Chernoff plot?

32. What are the popular ways to structure a fish-bone diagram?

33. Describe the data overlap problem in scatterplots. What are the possible solutions to the data overlap on an online display device?

34. How does a spatial histogram differ from classical histograms? What are the practical problems encountered while constructing a spatial histogram?

35. Describe the sorted pie-chart. How can you construct a bar-chart from a pie-chart and *vice versa*.

36. Which diagrams are most useful for nominal data? for ratio data?

37. Describe how tables can help in data visualisation. How will you store the data for a spatial histogram (as in Fig 2.6) in tables?

38. What calibration is used on the left and right margins of a Pareto diagram? What is the minimum and maximum calibration needed on the right margin?

39. Which variables in NOIR typology are used on the X-axis and Y-axis of a marginal dot plot? What does this indicate?

40. A spatial histogram is used to depict spatial data distributions. If this data vary over time at several locations, describe how you will construct a spatio-temporal histogram. Give an example of spatio-temporal data.

41. What can you infer on the data distribution from a Pareto chart if the line segment representing the cumulative frequencies is almost a straight line sloping up?

42. A disadvantage of Andrews plots is that the odd observations altogether vanish near t=0 (because sin(kt)=0 when t=0). A possible solution is to replace the even weights by $[\sin(kt)+\cos(kt)]/\sqrt{2}$, and odd weights (except the first) by $[\sin(kt)\text{-}\cos(kt)]/\sqrt{2}$. Does this modification change the shape of Andrew's plot? Are there alternate solutions?

43. Consider a modified Andrew's curve. Since the weight of the first observation is $\frac{1}{\sqrt{2}}=\sin(45)$, we could change the origin of the arguments, and express it in the alternate form: $f_x(t) = x_1\sin(\pi/4) - x_2\sin(\pi/4+t) + x_3\cos(\pi/4+t) - x_4\sin(\pi/4+2t) + \cdots$. What does the sines and cosines in this series each evaluate to when t is 0? Does this provide more insight into the central region of the curve?.

44. Which diagram is used to display both the joint distribution and marginal distributions of two variables in the same plot? How can you infer if the data are artificial or natural?

45. If there are n variables, how many sub-plots are there in a matrix plot?. What distinguishes the $(i,j)^{th}$ and $(j,i)^{th}$ diagrams in a matrix plot?.

46. How many variables are needed for a contour plot? What are the types of the variables in NOIR typology?

47. What can you infer from a contour plot in each of the following cases? (i) there are many clear-cut dark regions in the plot (ii) there are no separate demarcations at all (the hue is uniform throughout the plot (iii) there is a gradual increase in the hue

48. What are the values used to calibrate X and Y axes of (i) Q-Q plot (ii) Pareto diagram?

2.10.0.1   References

[AD72] Andrews, D. F.(1972) Plots of high-dimensional data, *Biometrics*, 28, 125-136.

[BG98] Blasius, J., Greenacre, M., (eds) (1998), *Visualisation of Categorical Data*, Academic Press, San Diego, CA.

[CR11] Chattamvelli, Rajan (2011) *Data Mining Algorithms*, Narosa publishing house, Daryaganj, New Delhi-2; Alpha science international, Oxford, UK.

[CR12] Chattamvelli, R. (2012). Statistical Algorithms, Narosa publishing house, Daryaganj, New Delhi-2; Alpha science international, Oxford, UK.

[CH08] Chen, C., Härdle, W., Unwin, A. (eds) (2008) *Handbook of data visualisation*, Springer-verlag, (books.google.com).

[CH73] Chernoff,H. (1973) The use of faces to represent points in k-dimensional space graphically. *Journal of the American Statistical Association*, 68,361-368.

[CR75] Chernoff, H., Rizvi, M. H. (1975) Effect on classification error or random permutations of features in representing multivariate data by faces, *Journal of American Statistical Association*, 70, 548-554.

[CB97] Chi, E.H., Barry, P., Riedl, J., Konstan, J. (1997) A spreadsheet approach to information visualisation, *Proc. of the symposium on Information visualisation*, Phoenix, AZ, 17-24.

[CW85] Cleveland, W. S.(1985) *The elements of graphing data*, Wadsworth, Monterey, USA.

[CW93] Cleveland, W. S. (1993) *Visualizing data*, AT&T, Murray Hill, NJ.

[CM84] Cleveland, W. S., McGill, R. (1984) The many faces of the scatterplot, *Journal of the American Statistical Association*, 79, 807-822.

[DMS98] Dhillon, I.S., Modha, D.S., Spangler, W.S. (1998) Visualizing class structure of multidimensional data, Proceedings of 30th Symposium on the Interface: Computer Science and Statistics, vol 30, Minneapolis, MN, 488-493.

[FR81] Flury, B., Riedwyl. H. (1981) Graphical representation of multivariate data by means of asymmetrical faces, *Journal of American Statistical Association*, 76, 757-765.

[FB97] Flury, B.(1997) *A first course in multivariate statistics*, Springer-Verlag, New York.

[FH98] Friedman, J.H. (1998) Data mining and statistics: What's the connection? *Computing science and statistics*, 29(1), 3-9.

[FB08] Fry,Ben (2008) *Visualizing data*, O'Reilly, Sebastopol, CA, USA (books.google.com).

[GM00] Gallagher, M. (2000) *Multi-layer perceptron error surfaces: visualisation, structure and modeling*, Dept. of CSE, University of Queensland, Australia.

[HH00] Hofmann, H. (2000) Exploring Categorical Data: Interactive Mosaic Plots, *Metrika*, 51(1), 11-26.

[HH03] Hofmann, H. (2003) Constructing and Reading Mosaicplots, *Computational Statistics & Data Analysis*: Special Issue on Data Visualisation, 43(4), 565-580.

[KN01] Khattree, R., Naik, D.N. (2001) Andrews plots for multivariate data: some new suggestions and applications, *Journal of Statistical Planning and inference*,

[KOH89] Kohonen, T. (1989) *Self-Organisation and Associative Memory*, 3rd Ed. Berlin: Springer-Verlag.

[MD98] Martin, D.R. (1998) Trellis Displays For Deep Understanding of Financial Data, February 1998 issue of Financial Engineering News, www.fenews.com/fen3/trellis.html

[MT95] Martin, T. (1995). Trellis displays vs interactive graphics, *Computational statistics*, 10(2), 112-127.

[MD02] Murdoch, D.J. (2002) Drawing a Scatterplot, *Chance*, 13(3),53-55.

[MG06] Myatt, G.J. (2006) *Making sense of data: A practical guide to exploratory data analysis and Data Mining*, Wiley, NY (books.google.com).

[TA08] Telea, A.C.(2008). *Data visualisation: principles and practice*, A.K. Peters, Wellesley.

[UT06] Unwin, A., Theus,M., Hofmann, A. (2006) *Graphics of large data sets*, Springer

[WS93] Wegman, E.J. and Shen, J. (1993) Three-Dimensional Andrews Plots and the Grand Tour, *Computing Science and Statistics*, 25,284-288.

3
Probability and Statistics

Chapter objectives

- Introduce the concept of probability

- Understand various ways to represent probability

- Explore the rules of probability

- Explain Venn diagrams and DeMorgan's laws

- Bayes Theorem for conditional probability

- Introduce mathematical expectation

- Define measures of central tendency: mean, median, mode

- Define measures of spread: variance and standard deviation, inter-quartile range

- Understand regression, assumptions in linear regression

- Scatterplots and its advantages

- Describe the least squares principle

- Explain correlation coefficient and its properties

- Distinguish regression and correlation

- Introduce Monte Carlo methods and Contingency Tables

3.1   Introduction

The theory of probability is built upon a strong foundation. It has applications in a wide range of fields like predicting the outcomes of gambling and games of chance[1], estimating survival times of

---

[1] In sports and games, it is also called 'the odds'.

electronic components, genetic inheritance, actuarial science, marketing, stock-market, inventory control etc. There exist many approaches to the study of probability. Here we describe only four of the popular approaches. In the *classical approach*, the notion of probability is developed using a *sample space* of possible outcomes of a random experiment. An *experiment* is a procedure that can be repeated any number of times under identical conditions. The outcomes of an experiment are called events. An outcome (ω) is one among the set of all well-defined possible outcomes (called sample space Ω, which may be finite or countably infinite). The outcomes are assumed to be mutually exclusive[2] (Two events A and B are mutually exclusive if they have no outcomes in common. Symbolically: A $\cap$ B=ϕ where ϕ is the empty set.)

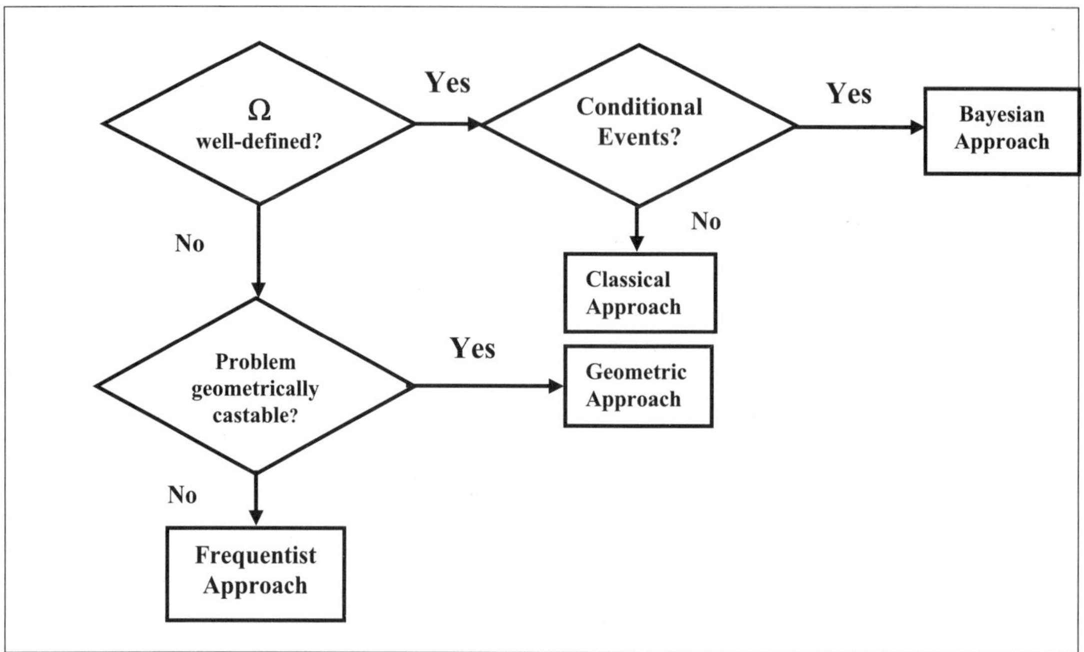

Figure 3.1: Four popular approaches to probability.

For example, if a fair coin is tossed once, the sample space is $\Omega = \{H, T\}$ where H denotes Head and T denotes Tail. For every outcome O, we assign a probability $0 \leq \mathrm{P}(O_i) \leq 1$ such that $\sum_{i \geq 0} \mathrm{P}(O_i) = 1$ (see below for the notation P(O)).

In the *frequentist approach*, the probability P(A) is found as the frequency of occurrence of the event based upon observations or experimental trials. Suppose we observed from past data that 30% of all weekend customers to a store purchase bread. We are interested in the event A='a randomly selected weekend customer will buy bread'. By the frequentist approach P(A)=0.3. In the *Bayesian approach* we reason about beliefs under conditions of uncertainty using a body of knowledge. Subjective or personal probability expresses the faith of a person in an unknown or future event. The *geometrical probability* uses any type of geometrical representations to find the probability of an event. When a trial has an infinite set of equiprobable outcomes that can be parametrised by a subset of $\Re^d$ (d-dimensional space) using a geometric figure like a line

---

[2]In set theoretic notation this implies disjoint sets

segment, a square, an ellipse, a circle, or higher dimensional equivalents, it is possible to express the probability in terms of the figure's attributes. In other words, the geometric approach to probability maps a problem into a well-defined geometric figure (lines, triangles, circles, squares, etc) and uses the properties of these figures (length, circumference, area, volume) to find the probability. In data mining applications, the frequentist approach is the most often used method. Figure 3.1 gives a rough sketch of various approaches to be used. Classical approach can be used if Ω is well-defined and conditional events are not involved. Bayesian approach most often involve conditional events (in some Bayesian models, Ω can still be well-defined)[3].

3.2 Probability

Definition 3.1 Probability is a quantitative measure of uncertainty or chance.

The chance mechanism may either be an event (likelihood of error free transmission of a data packet, chance of winning a game, likelihood that two political contestants will address the same location or share the same podium etc), a random phenomenon (chance that an electronic component will fail in a computer), or an experiment (probability of survival after a surgery, probability that a new drug will be more effective than others).

The probability obtained by classical and geometric approaches are often *time-invariant*, whereas the probability obtained by frequentist and Bayesian approaches may or may not depend upon time. This relies upon whether the events involved are *time-dependent* or not. As an example, the probability that two dice will show up a sum of 5 is $4/36 = 1/9$, irrespective of the time of the experiment. The probability that a customer will buy the solution manual along with the purchase of a book will vary as the book ages in the market. Similarly, the probability that a heart attack will prove fatal for a patient older than 70, will vary as new drugs and treatments are invented. In data mining studies, we tacitly assume that the probability of an event is *constant* over a suitable time period.

Probability is encountered in many data mining models like decision trees, association rules, neural networks, genetic algorithms, web mining etc. For example, the Gini index, and Shannon's entropy measures both use the probability that an item belongs to class k. The *support* and *confidence* measures used in association rule mining are both probabilities. The conditional mutation operation in genetic algorithms works on a randomly generated probability to decide whether a mutation must be carried out or not. The *precision* and *recall* measures used in web mining are actually probabilities of preciseness and completeness of a search.

3.2.1 Different Ways to Express Probability

Any of the following notations can be used for representing the probability:

a) As a proper fraction of the form p/q (eg: 1/3, 2/55)
   This form is used by humans while solving probability problems.
   Example: Probability that a customer to an entertainment site is a student is 3/5.

b) In decimal notation (eg: .66, 0.0018)
   This is the most common form used by software programs for computation and communication, and by humans using calculators.

---

[3]Other approaches include Kolmogorov system of probability, comparative probability, etc

Examples: The probability that a customer visits the same restaurant 2 or more times on a day is .025, the probability is .0001 that a light bulb will fail during the first one hundred hours of operation.

c) In scientific notation (eg: 66×10^{-2}, 1.80E-3)
Scientific notation is used when the resulting probability is very small (with several leading zeroes) because it can save space. For instance, the probability .000001 can be represented as 1.0E-6 where 1.0 is the characteristic, and -6 is the mantissa (here $E-n$ means 10^{-n}). Example: Probability that a customer will return an electronic gadget within 3 days of its sale is $2.5 \times 10^{-3} = .0025$, probability that a passenger will take air insurance before flight is 1.8 E-4=.00018 etc.

d) As a percentage (eg: 66%, .18%)
Percentage form of probability is usually used in conversations or writings (especially in the news media).
Example: Probability that shoes and sox will be sold together at a shoe-store is 40%. Probability that a customer to a supermarket will pay by credit card is 65%.

e) In words (as *"two out of fifteen"* or *"one in ten"*).
This form of probability is extensively used in conversations.
Examples: Chances are *'fifty-fifty'* that a visitor to a book publisher's web site is a student, there is *'one in five'* chance that a medical insurance customer will opt for whole family insurance.

f) In pictorial form
Probabilities can be represented in graphical and pictorial forms. This method is usually used for random variables and geometric probability problems.

g) As area under a curve
As the area under a statistical distribution always sum up to one, we could represent probabilities as left tail area under such curves.
Examples: The probability of rejecting statistical hypotheses are represented as areas under a suitable curve (this depends upon a test-statistic used to build our hypothesis). Examples of extensively tabulated continuous distributions for this purpose are the standard normal distribution and Student's t-distribution.

Almost certain event has probability close to one, while uncertain or unlikely event has probability near 0. A probability of .5 implies a "fifty-fifty" chance for the occurrence or non-occurrence of an event. Assuming that all our decimal numbers are positive, we could store any decimal number in just 2 memory locations (one for storing p, and the other for storing q). Signed decimals need an extra one bit to store the sign as 0 for positive and 1 for negative. As an unsigned int type can store numbers between 0 and 65535 in just 2 bytes of memory, we could represent a great majority of fractions that we encounter in practice using this method, provided that both the numerator and denominator are less than 65536. We could use the unsigned long int data type (4 bytes of memory) when larger numbers are involved, as it can store up to 4294967295.

Example 3.1 A shop short-lists 120 customers for a sales promotional offer from among their large clientele. If there are 60 prizes offered to the short-listed group, what is the probability for each customer to win a prize?
Solution: As there are 60 prizes and 120 customers, there is a $60/120 = 50\%$ chance for each

customer to win a prize (provided each customer can get at most one prize). In other words, chances are "fifty-fifty" that each customer will win a prize.

Example 3.2 In a tournament involving two teams A and B, 4 out of 12 games were won by team-A and the rest won by team-B. What is the probability for team B to win a new game against team A?
Solution: As team-B won 8 games out of 12, chances that team-B will win a new match against team-A is $8/12 = 2/3$.

3.2.2 A Notation for Probability

Probability of an event 'A' is denoted by P(A) (or as $\pi(A)$) where the enclosing parentheses contain an event, a label, a token or a random variable. The argument could also be a subscripted variable, a set, an expression involving sets, (using set theoretic operations) or conditional probability expressions like P(A|B). The same notation is used, irrespective of the approach (classical, frequentist, geometric or Bayesian). Events may either be specified explicitly or can be abbreviated by a label. In those programming languages that have the enumeration data type, we could also represent them using an enumeration. For example, enum season={Spring, Summer, Fall, Winter}; defines an enumeration. If the probability of an epidemic vary over the seasons, we could define them using this definition. As another example, consider the experiment of throwing a fair (unbiased) coin. Assuming that the coin does not end up on its side, the probability of Head showing up is 1/2, and the probability of Tail showing up is also 1/2. These are represented as P(Head) = P(Tail) = 1/2. Alternatively, we can specify a label A='Head showing up' so that P(A)=1/2. Since these are the only two possibilities', we call it mutually exclusive and exhaustive. The complementary event 'Tail showing up' is denoted by $\overline{A}$. So, we may also represent the above probability statements as $P(A) = 1/2 = P(\overline{A})$.

Event labels are usually denoted by capital letters (English, Greek etc). We can also combine events and denote them by other labels or abbreviated letters. For example, let 'A' denote the event that a supermarket customer buys alcoholic drinks, 'B' denote the event that the customer buys bakery item, and 'M' denote the event that customer buys meat. We can combine these events and denote it by another letter; or combine their individual labels so that 'AM' denotes that the customer buys alcoholic drinks and meat etc. We will use the single letter notation (i) to avoid confusion between 'AND'ing ($\cap$) and 'OR'ing ($\cup$) of various events (The missing operator between 'A' and 'M' is assumed to be the $\cap$ operator.), (ii) due to the possibility of different combinations (AMB, ABM, BAM, BMA, MAB, MBA), all denoting that the customer has bought all three items. Let X denote the event that a customer buys alcoholic drinks *and* bakery items. Then $P(X) = P(A \cap B)$. Let Y denote the event that the customer buys alcoholic drinks *or* bakery items. Then $P(Y) = P(A \cup B)$.

The probability of a random variable is denoted by p(x) or as p(X=x) (In Statistics, f(x) denotes a probability density function (pdf), which for a fixed value of x (say c) gives the probability that the random variable takes the value c[4]. The cumulative distribution function (cdf), denoted as F(x), is actually the probability of a random variable X taking values less than or equal to x (ie: $F(x) = P(X \leq x)$). We use "lowercase-p" notation for probability of random variables, and f(x), g(x) etc for pdf.

Probabilities can also be denoted without the parentheses (different types of brackets like simple, square and curly brackets) using the subscripted notation P_k. The subscript notation

---

[4]for continuous random variables, the integral $\int_{c-\delta x}^{c+\delta x} f(x)dx$ gives the corresponding probability

is more popular in some fields like stochastic processes, decision trees, genetic algorithms and neural networks. For instance, probability that an item belongs to a particular class in decision tree induction is denoted by P_k using Gini index or Shannon's entropy. Similarly P_m and P_x denote the mutation and crossover probabilities in GA.

When probability is expressed as a fraction, it is the usual practice to reduce it into the proper form (by making the numerator and denominator without common factors). For instance, P(A)=6/15 can be reduced to P(A)=2/5, since 3 is a common factor of the numerator and denominator. It goes without saying that the reduction be applied even in conversational form of probability. Hence *'chances are three-out-of-fifteen'* is better worded as *'chances are one-in-five'*. Similarly, when the result is an expression involving fractions, it is a good idea to reduce it to a single fraction. As an example, P(A) = (3/5 - 2/7)/(3/7) can be reduced to 11/35 x 7/3 = 11/15. Algorithms to convert decimal or scientific form to fractional form (p/q) can be found in [SC15].

People in different fields tend to have their own likes and dislikes to express any decimal number. Perhaps an electronic engineer may represent it in terms of inductors and capacitors, an electrical engineer in terms of currents and voltages, whereas a nuclear scientist may represent it in terms of properties of protons, neutrons and electrons. Among the various representations, the fractional form p/q is the most useful form preferred by statisticians as it is easier to remember. We know that a number in single precision (with 8 decimal places) needs 4 bytes of computer memory (depending upon the programming language and OS used). Single precision is far too insufficient for decimals that repeat beyond the 8th digit. With double precision, we can store up to 16 decimal places. As shown above, even this is far too insufficient to represent some of the small fractions like 7/29 without truncation. If we truncate it at the wrong decimal place (say 8th or 16th place), the resulting fraction won't come even close to 7 in the numerator and 29 in the denominator. Assuming that all our decimal numbers are positive, we could store any decimal number in just 2 unsigned memory locations (one for storing p, and the other for storing q). Signed decimals need an extra one bit to store the sign as 0 for positive and 1 for negative. As an unsigned int type can store numbers between 0 and 65535, we could represent a great majority of fractions encountered in practice using this method, provided that both the numerator and denominator are less than 65536. Because computations require further program instructions to operate upon these unsigned integers, it is more of use for persistent storage of decimals. But in those situations where we have to work with millions of decimal numbers, the fractional form as p and q will pay-off provided that the size of the total program code does not exceed any extra memory that were needed, had we used arrays of float or double type to store them.

The P() notation is also used in many other contexts. For instance, the probability generating function (pgf) of a random variable is denoted as P(t), where t is a scalar for univariate distributions, and a vector for multivariate distributions. Optionally, the t can be followed by the parameters of a distribution (eg: $P(t,p)=(q+pt)^n$, where q=1-p). Similarly, point operations on image pixels in computer graphics are denoted by P(opr)).

For the coin tossing experiment considered above, assume that the probability of Head showing up is p, and probability of Tail showing up is q = 1-p. Then the probability of Head or Tail may be represented in compact form in an unknown variable x as $p(x) = p^x q^{1-x}$ where x takes the values 0 or 1. To make it a probability function (pdf = f(x)), we have to define it for every possible value of x. Thus we modify the definition as follows:

$$f(x) = \begin{cases} p^x q^{1-x} & \text{for} \quad x = 0, 1; \\ 0 & \text{otherwise.} \end{cases}$$

Hence all of the symbols P(x), p(x), f(x), g(x), F(x), etc denote probabilities in various contexts. When a distribution has parameters, they may also be incorporated into these symbols. For instance, $f(x;\lambda) = e^{-\lambda}\lambda^x/x!$ denote a Poisson distribution with parameter $\lambda > 0$.

> The events associated with a probability statement can be given a short language description ('a Head shows up', 'a customer is a student'), a label or token. A label is a short placeholder or abbreviation for an event. By convention, it is chosen as capital letters of an alphabet. The label method is the preferred choice due to its ability to generalise to event interactions like unions (A∪B), intersections (A∩B), complements ($\overline{A}$), difference (A-B), etc (table 3.1). Labels or names may also be given to *special* probabilities. For instance, in hypothesis testing, *power* denotes the probability of rejecting a null hypothesis when alternative hypothesis (H_a) is true, 'Type I error' (α) denotes the probability of rejecting a true null hypothesis, and 'Type II error' (β) denotes the probability of accepting a false null hypothesis.

3.2.3   Methods of Counting

Lemma 3.1 If one thing (or activity) can be done in 'm' ways and another in 'n' ways, the two together can be done in m*n different ways.

Example 3.3 An online birthday card service offers 10 varieties of styles for printing text, and 5 varieties of background images. Total how many varieties of cards can be printed?
By the lemma, there are 10*5 = 50 varieties of cards that can be printed.

3.2.3.1   Independence of Events

The independence of events is a useful concept to data miners looking for patterns and trends in voluminous data. Although the event is assumed as abstract in the following discussion, it is applicable to variates that are governed by a probability law. If two variates are uncorrelated, it implies that they are independent. But the reverse is not true (two or more variates can be independent, but still can have a small correlation among them). For instance, the height of a student and marks in an exam are two independent variates. But in any sample of size ≥ 3, we will find a small correlation between them (for a sample of size n=2, there is perfect correlation ($\pm$ 1) between them).

Definition 3.2 Two events 'A' and 'B' are independent if the occurrence of either of them is not influenced by prior knowledge about the occurrence of the other event. Symbolically, we express this as P(A and B) = P(A ⋒ B) =P(A)P(B) (the missing operator between P(A)P(B) is the multiplier ($*$) operator).
This can also be expressed as P(A)= P(A|B) (see below).

Example 3.4 At an online store, let 'A' denote the event that the customer orders pop-music, and 'B' denote the event that the customer has red hair. Let P(A) =.28 and P(B)=.30. The probability that a red-hair customer will order pop-music is P(A)*P(B) = .28 * .30 = .084, since the events 'A' and 'B' are independent.

3.3   Rules of Probability

Rule 1 Probability is always between 0 and 1 ($0 \leq P(A) \leq 1$).
In section 3.2, we have seen various ways to express probability. All of the methods described there (except the percentage method) maps the probability into the interval [0,1].

Rule 2 Probability of the entire sample space is 1. That is $P(\Omega)=1$.
The proof follows trivially because the probability of all the events occurring is certainty.

Rule 3 Probability of occurrence of either of two *disjoint* events is the sum of their individual probabilities (ie: $P(A \cup B)=P(A)+P(B)$).

Example 3.5 Let X denote the event that an e-commerce customer connects to a web site using Internet Explorer with probability .4, and Y denote the event that the browser used is Navigator with probability .36. Then the probability that a randomly chosen customer uses either of these browsers is $P(X \cup Y) = .4+.36 = .76$.
This rule can be extended to any number of disjoint events. Let $A_1, A_2, \cdots, A_n$ be disjoint events. Then $P(A_1 \cup A_2 \cdots \cup A_n)=\sum_{i=1}^{n} P(A_i)$.

Rule 4 Product Rule
If A and B are two independent events, the probability of occurrence of both of these events is the product of their individual probabilities:- $P(A \cap B) = P(A)P(B)$.
Proof: Since the events are independent, the occurrence of A has nothing to do with the occurrence of B. The probability of occurrences of A and B is the product of their individual probabilities. This rule can be generalised to any number of independent events as $P(A_1 A_2 \cdots A_n)= P(A_1)P(A_2)..P(A_n)$

Example 3.6 In the above example, U denote the event that a client machine runs Windows OS with probability .6, and V denote the event that a client machine runs Linux with probability .3. The probability that a client from Windows OS connects to the web site using Explorer browser is $.4 * .6 = 0.24$

Rule 5 Sum rule
The probability of occurrence of either of two events (not necessarily independent) is $P(A \cup B) = P(A) + P(B) - P(A \cap B)$
Proof follows trivially using the principle of inclusion and exclusion. Let 'X' denote the common intersection of events A and B ($X=A \cap B$). Then $P(A)+P(B)$ will contain the 'X' portion twice. So we need to subtract it once to get $P(A \cup B)$.

Example 3.7 In the above example, if 18% of the clients run Windows OS and use Internet Explorer, what is the probability that a randomly chosen client uses Windows OS or Internet Explorer?
Solution: Let A denote the event 'client uses Windows OS', and B denote the event 'client connects using Internet Explorer'. Then $P(A) = .6$, $P(B) = .4$ (from above examples). $P(A \cap B)$ is given as $18\% = .18$. Substituting these values in Rule 5, we get $P(A \cup B) = P(A) + P(B) - P(A \cap B) = .6 + .4 - .18 = .82$

Rule 6 Complement rule
The probability of non-occurrence of an event is the complement of the probability of occurrence.
Symbolically, $P(\overline{A})=1-P(A)$.
The complement of an event comprises of all events in the sample space Ω except the event.
Since the probability of the sample space is 1, it follows that the probability of the event union
the probability of its complement is 1. Symbolically $P(A) + P(\overline{A}) = 1$, from which the result
follows.

If the probability of occurrence of an event is estimated before the trial, it is known as
'apriori' probability. For instance, the probability of occurrence of a 'Head' in an unbiased coin
tossing experiment is $1/2$. In this example, we assume that the outcomes are equally likely. The
predicted probability of an outcome using a mathematical model is called theoretical probability.

An 'empirical' probability is estimated after an experimental trial using known or observed
frequencies of outcomes. Here the assumption is that the trials are independent. It may also be
estimated using a computer simulation.

Table 3.1: Some common symbols in probability

Symbol	Description	Probabilistic interpretation
Ω	set of all outcomes	Sample Space
ω	a member of set	An outcome
A	subset of Ω	An event (an outcome in A occurs)
$\overline{A}$, A^c, A'	complement of A	No outcomes in A occurs
$A \cup B$	union of sets	An outcome in either A or B occurs
$A \cap B$	intersection of sets	Both A and B occurs
$A - B$	difference of sets	Event in A but not in B occurs

Experimental probability is derived numerically through the use of existing or simulated
data. In the example above, if we toss the coin 100 times and observes the number of Head's
that turn up, we could find the experimental probability of observing a Head.

Example 3.8 The number of male and female customers visiting a supermarket between 1PM
and 2PM on each working day of a week are given in table 3.2. Find   (i) Probability of a
randomly chosen customer on any working day between 1PM and 2PM to be a male.   (ii)
Probability that a randomly chosen customer on Thursday between 1PM and 2PM is a female.
These answers follow directly from the table as (i) $35/65 = 7/13$ and (ii) $6/11$.

Table 3.2: Customers to a store

Day	Males	Females	Total	Day	Males	Females	Total
Monday	11	3	14	Thursday	5	6	11
Tuesday	6	8	14	Friday	6	7	13
Wednesday	7	6	13	Total	35	30	65

3.3.1 Probability Model

A probability model is a triplet $\mathbb{P} = (\Omega, S, p(x))$ that mathematically represents a random phenomenon, where Ω is the sample space, S is a set of events associated with an experiment, and p(x) is the probability associated with each event in S such that $\sum_i p(x_i) = 1$.

3.3.2 Entropy vs Probability

Data communication works by exchanging data packets between two sources (like two computers, two wireless devices etc). A packet may either be delivered to the destination with or without errors, or lost in transit. The receiver may need to know how certain it is to receive a packet from various sources. A measure of this uncertainty is known as entropy. The information content h is measured in bits and is defined as $h(x) = \log_2(1/w)$ where w is the weight of x. Small entropy values indicate presence of structure and large entropy values indicate randomness. For example, if a big file needs to be downloaded from the Internet, and if the same file is available at multiple locations, the download can take place in parallel with one piece from one location, another piece from another location etc as done by some download programs. Here p_i's are the probabilities of receiving the data without errors from i^{th} download site, and they are independent. Probability and entropy are inversely related in the sense that the probability of certainty is 1 while entropy of certainty is 0. Probability is a measure of the *degree of belief*, and entropy is a measure of the *lack of pattern or organisation*. A related measure is the Kullback-Leibler divergence (also called relative entropy) based upon two pdfs $KL(p||q) = -\sum_x p(x) \log \frac{p(x)}{q(x)}$, which is non-negative (it is zero when p=q). This measure is not symmetric. It is used in text mining and document retrieval. It calculates the relevance of each document with respect to (wrt) a user query, which is necessary in the case of image and multimedia retrievals.

The concept of entropy originated in thermodynamics. Shannon (1916-2001)[a] introduced it in 1948 to information theory [SC48]. It is a measure of the degree of disorder or uncertainty in a system. It has many applications in data mining. For example, the greedy decision tree growing uses entropy minimisation criteria (chapter 5). The agglomerative categorical clustering with entropy criterion (ACE) is a popular entropy-based technique for categorical data clustering (chapter 7). Information gain and mutual information are entropy-based measures that utilise the presence or absence of terms in various documents (chapter 10). Hobbit distance provides bit-wise similarity used in adaptive nearest neighbour class entropy of histograms. An application of entropy to time-stamped document classification can be found in [CZ06].

[a]www2.research.att.com/~njas/doc/shannon.html, cm.bell-labs.com/cm/ms/what/shannonday/paper.html

---

3.4 Venn Diagrams

Event relationships in probability statements are best described pictorially using a Venn diagram[5]. It is a universal notation for representing sets, their possible intersections (common elements) and dependencies (subset/superset). As shown in figure 3.2, the universal set is indicated by a rectangle, and events are represented by circles or ellipses within the rectangle. This is a convenient notation to visualise intersecting events. For example, to visualise customers

---

[5]Venn diagrams were introduced in 1881 by the British mathematician John Venn (1834-1923) for representing set operations. It was later extended by many researchers mentioned above.

who buy bread and milk among those who have purchased bread or milk, we can use a Venn diagram as in figure 3.2.

Figure 3.2: Venn Diagram showing customers who bought milk and bread.

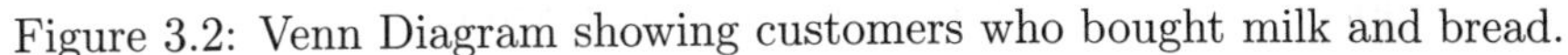

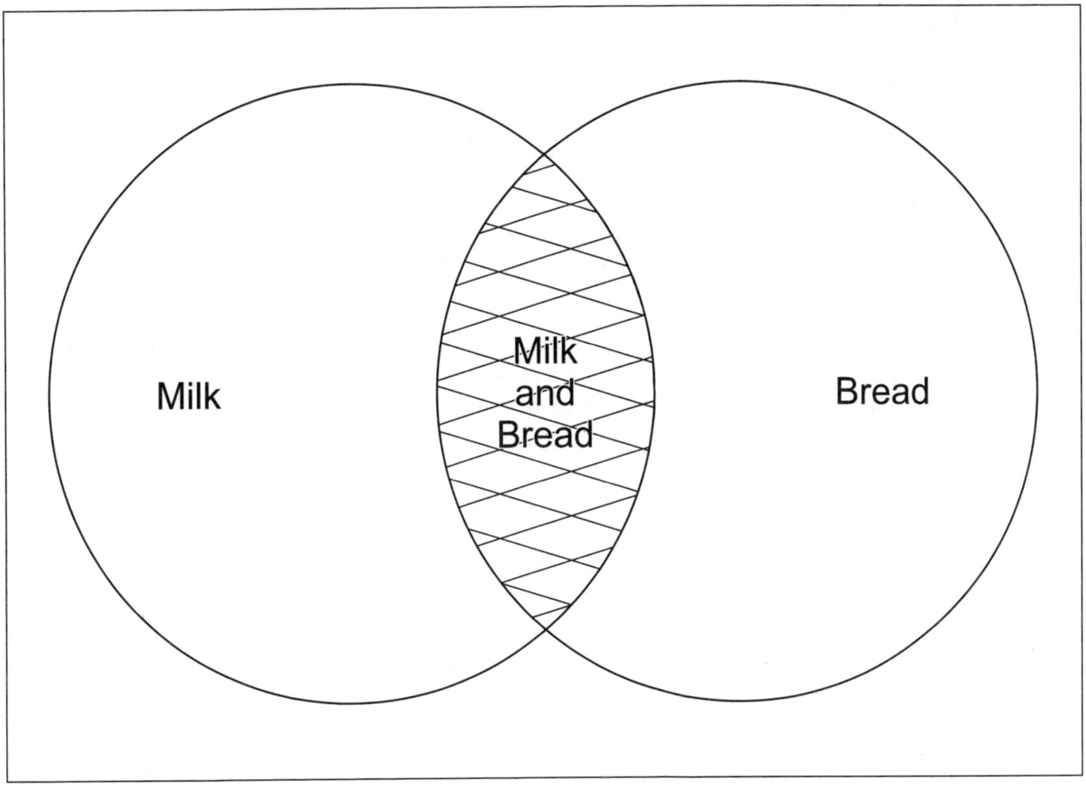

Euler diagrams are an extension of Venn diagrams to represent more than one sample space. Other similar diagrams include the Karnaugh maps, Johnston diagrams, Edwards' Venn Diagrams and Peirce diagrams. Since these are seldom used in data mining applications, we do not discuss them further.

3.4.1 De'Morgan's Laws

These laws relate the complement of compound events in terms of individual complements. This helps to break down complex problems into simpler ones that are easier to solve.

Rule 7 Complement of an intersection is the union of their complements.
Let A and B be two arbitrary events. Then $\overline{A \cap B} = \overline{A} \cup \overline{B}$. This can be easily proved using Venn diagram. Let x be an arbitrary element belonging to the LHS. Then x does not belong to $A \cap B$. Let y be an arbitrary element belonging to RHS. Then it either does not belong to A or it does not belong to B. As $\overline{A}$ includes elements belonging to B and $\overline{B}$ includes elements belonging to A, $\overline{A} \cup \overline{B}$ could include elements in A and B but not in both. In the above figure this is the area outside the crossed portion.

Rule 8 Complement of a Union is the intersection of their complements. Symbolically, $\overline{A \cup B} = \overline{A} \cap \overline{B}$. These rules can be extended to any number of events as follows: $\left(\overline{\cup_{i=1}^{n} A_i}\right) = \cap_{i=1}^{n} \overline{A}_i$, and $\left(\overline{\cap_{i=1}^{n} A_i}\right) = \cup_{i=1}^{n} \overline{A}_i$ We omit the proof as it can be found in many statistics textbooks.

3.5 Bayes Theorem

Conditional probability has its origin in gambling and games of chance. It is the probability of occurrence of an event with prior knowledge or assumption about another event. It finds applications in a variety of fields. For example, a 'virtual seismologist' to warn earthquakes, well before it actually happens, uses conditional probability [CH04]. It is also used in linear discriminant analysis [MG92], to estimate posterior probabilities in research studies, supervised classification techniques using prior information, reconstruction of medical images [GS95], email spam filtering, prediction of sports and games outcomes, etc. The basic ingredient of Bayes theorem is conditional probability, which is a concept based upon two or more events or random variables. Conditional probability is used in many data mining models. The confidence measure used in association rule mining (§6.1.2) can be expressed as a conditional probability. It is also used in genetic algorithms (chapter 8), neural networks (chapter 9), and web mining (chapter 10). The symbol P(A|B) denotes the conditional probability of event A given that event B has occurred, which is evaluated as P(A|B) = P(A $\cap$ B)/P(B). In the case of random variables, f(x|y) = f(x,y) / f(y). This can be represented in terms of hypothesis and evidence as follows:

Lemma 3.2 Probability of Hypothesis given Evidence is the ratio of joint occurrence of Hypothesis and Evidence over probability of Evidence. P(H | E) = P(H $\cap$ E) / P(E).

Corollary 1 The unconditional probability of Hypothesis is the sum of the products of the probability of Hypothesis given evidence, and probability of Evidence; and probability of Hypothesis given no evidence, and probability of no-evidence. P(H) = P(H | E) . P(E) + P(H |$\overline{E}$) . P($\overline{E}$) Here P(H | E) is the posterior probability. These can be obtained from each other with the help of prior probabilities and likelihood as given by Bayes theorem.

3.5.1 Bayes Theorem for Conditional Probability

This theorem is also known as the law of inverse probability. Bayes theorem is used to calculate posterior probability in terms of priors. Conceptually, posterior= Likelihood x prior/evidence where likelihood is estimated from sample data or found by other means. It expresses *aposteriori* probability in terms of *apriori* probabilities using newly acquired information. Symbolically, it can be written as

$$P(A_i|B) = [P(A_i).P(B|A_i)]/[P(A_1).P(B|A_1) + P(A_2).P(B|A_2) + \cdots + P(A_n).P(B|A_n)]$$
$$=[P(A_i).P(B|A_i)]/ \sum[P(A_i).P(B|A_i)]$$ where the summation is taken over all possible events.

    Proof: Let A_i denote possible explanations for a given set of data B. As the data size increases, the probability P(B|A_i) P(A_i) increases.

Consider P(A|B) =P(A$\cap$ B)/P(B) = P(A). P(B|A)/P(B).

Using total probability law, B = BA + B$\overline{A}$ and hence P(B) = P(BA) + P(B$\overline{A}$) =P(A).P(B|A) + P($\overline{A}$).P(B|$\overline{A}$)

Figure 3.3: Bayes Theorem as causes and effects.

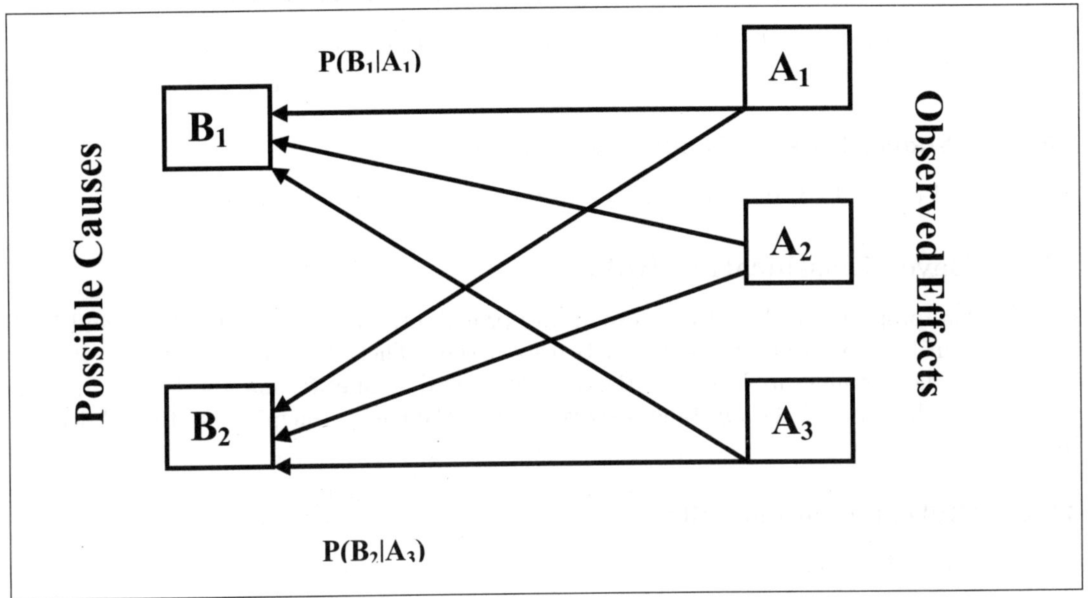

The proof follows for the two event decomposition of A by substitution. If A is decomposed as $A = A_1 \cup A_2 \cup \cdots A_n$, then B can be represented as $B = BA_1 \cup BA_2 \cup \cdots \cup BA_n$. Thus $P(B) = \sum_i P(BA_i) = \sum_i P(A_i).P(B|A_i)$. This proves the theorem for the general case. Bayes theorem can be used to obtain probability of probable causes from observed effects as shown in figure 3.3.

In multiple hypotheses situations, Bayes theorem provides a 'best' estimate for the probability of evidence under the assumption that each hypothesis is true.

Corollary 2 $P(A).P(B|A) = P(B).P(A|B)$

Example 3.9 Consider cash withdrawals at an ATM booth. From analysis of prior fraudulent transaction, a Bank has found that the probability of any transaction to be fraudulent is one in thousand (P(Fraud)=.001), 90% of fraudulent transactions are for amounts above 2000 (ie: P(Amount > 2000|Fraud) = .90), and 99% of cash withdrawals for amounts > 2000 are genuine. Using this information, what is the probability that a transaction is fraudulent, given that the withdrawal amount is 4000?

   Solution: We have P(Fraud)=.001, P(Amount > 2000|Fraud) = .90, P(Amount > 2000|Not Fraud) = .99

By Bayes theorem P(Fraud| Amount>2000) = P(Fraud) * P(Amount > 2000|Fraud)/
[P(Fraud) * P(Amount > 2000|Fraud) + P(Not Fraud) * P(Amount > 2000| Not Fraud)] =
.001 * .90 / [.001 * .90 + .999 * .99] = .0009/(.0009+.98901) = .90917 E-3 = .000909.

3.5.1.1 Odds-Likelihood Ratio Form of Bayes Theorem.

In some applications we are interested in finding the ratio of the likelihoods

$$\frac{P(Hypothesis_1|Evidence)}{P(Hypothesis_2|Evidence)} = \frac{P(Hypothesis_1)P(Evidence|Hypothesis_1)}{P(Hypothesis_2)P(Evidence|Hypothesis_2)} \tag{3.1}$$

3.5.1.2 Product Rule for Conditional Probability

P(AB|C) = P(A|C). P(B|AC)= P(B|C). P(A|BC) where AB denotes A $\cap$ B, etc.

3.5.2 Bayes Classification Rule

Consider two propositions A and B whose apriori probabilities are known. Let U(A) and U(B) denote the utilities of propositions A and B respectively. Then A is preferred over B if U(A) > U(B). In data mining applications, Bayes' rule is used to specify how the learning system updates its beliefs as new data instances arrive. This is the basic principle of statistical decision theory.

3.5.2.1 Rule of Expected Utility

Assuming A as the action and B as the consequence, this rule gives the utility of A as

$$U(A) = P(B|A)U(A \cap B) + P(\overline{B}|A)U(A \cap \overline{B}) \tag{3.2}$$

A disadvantage of this approach is that they depend upon prior probabilities of propositions explicitly. If these are unknown, they need to be estimated (using point estimation, EM algorithm[6], stochastic sampling or parametric approximations) before starting the decision process.

Now consider the problem of classifying a dichotomous attribute using data instances. If the two attribute values are 'Yes' and 'No', we could obtain a measure of entropy using the probabilities of Yes and No responses as

$$E(S) = -p_{yes} \log_2(p_{yes}) - p_{no} \log_2(p_{no}).$$

3.6 Mathematical Expectation

Definition 3.3 Mathematical expectation of a random variable or a function of it is the expected value of the argument in the population. The mathematical expectation of X is denoted by E(X) and is a single number (in univariate case) used to quantify the random variable as $\sum_{i=1}^{n} p_i x_i$ (for discrete case). It is also called the 'Expected Value' of X or the population mean (μ). The random variable X can be discrete or continuous, univariate or multivariate. It appears in decision trees, regression, neural networks, genetic algorithms and SVM.

Consider a discrete univariate random variable X that takes different values x_i with corresponding probabilities p_i for i=1,2,$\cdots$ n. The expected value is found by multiplying each outcome by the corresponding probability, and summing the results over the entire set of outcomes. In other words, the expected value of a random variable X is the weighted average value of X over all possible values of X, where the weights are the corresponding probabilities. If X is

---

[6]Expectation Maximisation(EM) algorithm is used for maximum likelihood estimation of parameters from missing or incomplete data.

Table 3.3: A discrete random variable

x	1	2	3	4	5	6
p(x)	1/12	1/6	1/4	1/4	1/6	1/12

continuous, the expected value is defined as follows:
$E(X) = \int_{\Omega} x f(x) dx$ where Ω denotes the entire range of X.

Example 3.10 A discrete random variable X is given in table 3.3. Find E(X).
 Solution: The expected value can be obtained directly from tabular representations as shown above. We get E(X) = 1*(1/12)+2*(1/6)+3*(1/4)+4*(1/4)+ 5*(1/6) + 6*(1/12) = 3.5

The idea of mathematical expectation is more general than to random variables. It can be applied to any situation where there are sets of alternatives for which the probability of occurrences are known.

3.7   Statistics

Statistics in a broad sense is the branch of science that deals with the collection, classification, tabulation, summarising, analysing and modeling of data, extracting information from data and drawing conclusions from the data. Statistical data mining uses various techniques of statistics to extract features or trends from data. For grasping data mining applications and interpreting the results presented by software programs[7], we must be familiar with some basic concepts of statistics discussed in the following sections.

3.7.1   Population vs Sample

Definition 3.4 The totality of all elements of interest in a study is called the *population*. A statistical population may comprise of animate or inanimate objects, symbols, entities, etc.

Example 3.11 Consider a medical study to find out insulin administration methods by diabetic patients in various geographic and socio-economic groups. Insulin can be taken in oral form, as a nasal spray or as an injection of some sort into the blood stream. In this study, the population is the set of all diabetic patients in the world who take insulin.

Definition 3.5 A random sample is a true representative subset of a population (in the statistical sense).
 The population may be known or unknown, enumerable or un-enumerable. The sample is always known and enumerably finite. The first step in obtaining a sample is identifying the correct population. Each and every element of the population should have an equal chance of being included in a random sample. Every element of the population should be uniquely distinguished using a label or ID number for the sampling purpose. In those cases where this is impossible, a sample can be taken using geometric or other methods. As an example, the microscopic blood examination uses area sampling to estimate various constituents of the blood.

---

[7]links to statistical software programs are given at the end of chapter 2

Example 3.12 In a study to find out customer buying patterns of a consumer product, the population is the set of all customers who have bought that item. This population may or may not be known. For most expensive electronic gadgets and equipments, the manufacturers keep track of customers. But for inexpensive products (pen, calculator, CDs), this information is not recorded. Thus the population can be unknown.

Example 3.13 To find out the number of defects or damages in any manufactured item, the population is the set of all items manufactured. This population is countably finite. The population for a coin tossing experiment in which k consecutive heads appear together for a fixed value of k is countably infinite. Sometimes the population may be impossible to track or enumerate. Some examples are – the population of all stars with a minimum luminosity, population of all computer data files, population of all HTTP requests on the web.

Supervised data mining models use a sample of available data as a training set. Samples can be drawn from a population in many ways. Most popular techniques are simple random sampling, stratified sampling, systematic sampling, and cluster sampling. Computer generated samples are used in Monte-Carlo methods (see §3.12). Samples may be drawn with replacement or without replacement of items already drawn. If the population size is large, this will not make a difference, as the probability of each item to be included in a sample is almost equal.

3.7.2   Parameter vs Statistic

A parameter describes the population of interest. A population can have zero or more parameters. Consider the Cauchy's distribution $f(x) = \frac{1}{\pi} \frac{1}{(1+x^2)}$ for $-\infty < x < \infty$. It has no parameters, although it describes a population[8].

Definition 3.6 A well-defined[9] function of the sample values is called a *statistic*.

The unknown parameters of a population are estimated using a statistic or a function of it. In the most trivial case, a statistic is a function of a single sample observation. Accuracy of this estimate increases with increasing sample size. The population in data mining applications often has multiple parameters. But populations are difficult to work with (due to large size, geographical spread, uncertainties involved, lack of time to enumerate, etc). In addition, datawarehouses may contain only a snapshot of the data collected over a reasonable time period. We then compute a statistic from a representative sample of the population, called a *random sample*.

3.8   Measures of Location

A function of the sample values (a statistic) that summarises the *locational* information into a single number is known as a measure of location. Examples are the mean, median, and mode[10]. These are single numbers for univariate data, and vectors for multivariate data.

---

[8]The more general form of Cauchy's distribution is $f(x; a, b) = \frac{K}{\pi} \frac{1}{a^2+(x-b)^2}$, with 2 parameters a and b.

[9]Here *well-defined* means that all functions of the sample values are not considered as statistics. Consider the function $t = 1/[x_1 + x_3 + \cdots + x_n - (x_2 + x_4 + \cdots + x_{n-1})]$, where $x_1, x_2, \cdots, x_n$ is a random sample of size n. As the sample values are unordered, it could happen that the first sum=second sum in the denominator. This results in the denominator becoming zero and our statistic tending to $1/0 = \infty$. Hence a well-defined statistic should not tend to infinity even for small sample sizes. Other examples can be contrived in terms of exponential and logarithmic functions.

[10]In statistical parlance, 'mean' or 'average' always denotes the 'arithmetic mean'

Even before Statistics emerged as a branch of scientific discipline, the arithmetic mean (average), median, minimum and maximum of a set were widely used in gambling and games of chance. The word *statistic* was coined after the emergence of Statistics. The plural of *statistic* as well as the subject are both called statistics. The mean, variance and so on being arithmetic functions of sample values are examples of statistics (which could also denote numerical facts, as in statistics on trade). The median, minimum and maximum observations are called 'order statistics' because they depend upon the relative order of elements (sample values). A function computed from the ranking of elements is called 'rank order statistic'. For example, Kendall's τ is a rank order statistic obtained from the concordance and discordance pairs.

In some data mining applications, the population changes continuously. So each sample can be considered as a true representative for a finite time period only. The pace of change can be deterministic or random, slow or fast, uniform or nonuniform. Each sample has an inherent time stamp associated with it, that identify when the sample was taken or when it was 'current'. By analysing the sample, important conclusions about the population characteristics can be drawn in a timely and efficient manner.

3.8.1 Mean, Median and Mode

Definition 3.7 The mean of a sample is denoted as $\overline{x}$ and is found by summing all the observations and dividing by the total number of observations. If x_1, $x_2,..x_n$ are the sample values, the sample mean is defined as

$$\overline{x} = (x_1 + x_2 + \cdots + x_n)/n = \frac{1}{n}\sum_{i=1}^{n} x_i.$$

The analogous population mean is denoted by μ. Sample mean provides an unbiased estimate of the population mean as $\hat{\mu} = \overline{x}$. When repeated samples are taken from a population, the means of these samples will clutter around the true population mean μ.

The population mean, median and mode coincides for continuous symmetric (unimodal) distributions. However, this relationship need not hold for samples drawn from above populations (these values will become more or less equal when the sample size is increased beyond a limit). A classical *rule-of-thumb* given in many textbooks is that for right skewed distributions, mean $\geq$ median $\geq$ mode (the reverse relationship holds if it is skewed to the left). This rule was refuted recently for some continuous and discrete distributions in [HP05].

The mean depends upon all sample values, while only relative sizes of sample values are taken into consideration in finding the median and mode. The mean can be used for interval or ratio type data, and the mode can be used for any type of data in the NOIR typology. Median of ungrouped data considers only the ranks of observations, and not the actual values themselves. It can be used for ordinal, interval or ratio data types. A *pre-condition* to compute the mean and median is to know the scale of measurement in NOIR typology. The mean of nominal data is meaningful in one particular case – when a binary variable is coded as 0 and 1, and we wish to find the proportion of sample values that is coded as 1. In this case, the mean will give the correct proportion.

In the following paragraph, we give some theorems and corollaries to find the mean of combined samples and reduced samples (after deleting some observations). These find applications in pruning of data mining models (by adding/deleting observations or replacing one observation with another in a training set), in outlier detection etc.

Theorem 3.1 If $\overline{x}_1$ and $\overline{x}_2$ are the means of two samples of sizes n_1 and n_2 respectively, the mean of the combined sample is given by $\overline{x} = (n_1\overline{x}_1 + n_2\overline{x}_2)/(n_1 + n_2)$.
Proof. Since the mean of the first sample is $\overline{x}_1$, the sum of the observations in it is $n_1 \overline{x}_1$. Similarly, the sum of the observations in the second sample is $n_2 \overline{x}_2$. Summing these expressions gives the sum of all observations from both samples combined as $n_1 \overline{x}_1 + n_2 \overline{x}_2$, from which the grand mean is obtained by dividing it by $(n_1 + n_2)$.

Example 3.14 The mean revenue at an e-commerce store on a Saturday from 40 customers is \$1800 and on the next day from 60 customers is \$2400. What is the mean revenue for that weekend?
Solution: Here we have $\overline{x}_1$=1800, n_1=40, and $\overline{x}_2$=2400, n_2=60. Hence using the above theorem, the combined mean is $(40*1800+60*2400)/100 = 216000/100 = 2160$.
The above theorem can be extended to an arbitrary number of disjoint samples as follows.

Theorem 3.2 If $\overline{x}_i$, i=1,2,..m are the means of m samples of sizes $n_1, n_2, \cdots, n_m$ respectively, the mean of the combined sample is given by $\overline{x} = (n_1\overline{x}_1 + n_2\overline{x}_2 + \cdots + n_m\overline{x}_m)/(n_1 + n_2 + \cdots + n_m)$.
Proof. Let the ith sample be denoted by S_i, and the number of elements in it be $n_i = |S_i|$. As S_i has n_i elements, the sum of the sample values in S_i is $n_i * \overline{x}_i$ (this follows by cross-multiplication from the definition of $\overline{x}_i = \sum_{x_i \in S_i} x_i/n_i$). Summing over every sample gives the sum of all observations as $n_1\overline{x}_1 + n_2\overline{x}_2 + \cdots + n_m\overline{x}_m$. Dividing by the total number of sample values $(\sum_{i=1}^{m} |S_i| = \sum_{i=1}^{m} n_i)$ gives the result.

Corollary 3 If m observations with mean $\overline{x}_m$ are added to a sample of size n with mean $\overline{x}$, the new mean is given by $\overline{x}_{new} = (n\overline{x} + m\overline{x}_m)/(n + m)$.

Corollary 4 If a new observation x_k is added to a sample of size n with mean $\overline{x}$, the new mean is given by $(n\overline{x} + x_k)/(n + 1)$.
Proof: This follows easily by considering x_k as a sample of size 1 with mean itself.
The following two corollaries are useful in updating the mean when a sample is trimmed by deleting one or more observations. This has applications in outlier detection (see §3.10 in page 3-26) and some of the pruning algorithms.

Corollary 5 If an existing observation x_k is removed from a sample of size n with mean $\overline{x}$, the new mean is given by $(n\overline{x} - x_k)/(n - 1)$.

Corollary 6 If m observations with mean $\overline{x}_m$ are removed from a sample of size n with mean $\overline{x}$, the new mean is given by $\overline{x}_{new} = (n\overline{x} - m\overline{x}_m)/(n - m)$.

Corollary 7 If m observations with mean $\overline{x}_m$ are added to a sample of size n with mean $\overline{x}$, the new mean is given by $\overline{x}_{new} = (n\overline{x} + m\overline{x}_m)/(n + m)$.

These corollaries can be used to compute the corrected mean of a sample in which an observation was erroneously recorded. Similarly, situations arise where the mean of a sliding window of observations is continuously required (as in time series, spatial data files, multimedia data files etc). The window is slid at each step by a fixed amount so that the observations in the second window have some overlap with observations in the first window. The above formulae can easily be modified when the linear displacement is in steps other than one.

Figure 3.4: Mean, median or mode?

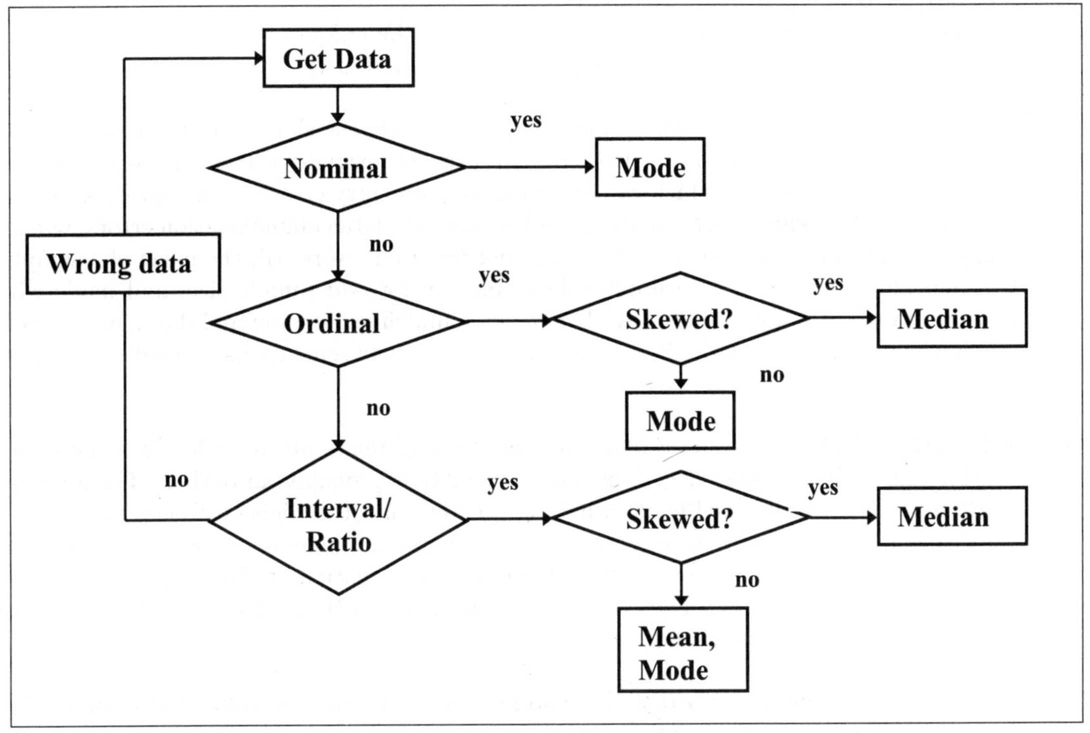

3.8.1.1 Weighted Mean

The weighted mean is obtained by multiplying (or dividing) each observation by an appropriate nonzero weight. If $w_1, w_2, \cdots, w_n$ are the weights associated with $x_1, x_2, \cdots, x_n$ respectively, the weighted mean is given by $\overline{x}_w = (w_1 x_1 + w_2 x_2 + \cdots + w_n x_n)/(w_1 + w_2 + \cdots + w_n) = \sum_{i=1}^{n} w_i x_i / \sum_{i=1}^{n} w_i$. Weighted mean assigns different importance to different sample observations. The mean of grouped data is obtained from this by replacing w_i's with corresponding frequencies f_i's as $\overline{x} = \sum_{i=1}^{n} f_i x_i / F$ where $F = \sum_{i=1}^{n} f_i$. See [SC15] for weighted mean recurrences.

Corollary 8 When all weights are equal, the weighted mean becomes the arithmetic mean.

3.8.1.2 Advantages of Mean

The mean is the most frequently used measure for numeric data due to its desirable properties over other measures. If the population from which the sample is drawn is normal, the distribution of the mean is also normal (even if the population distribution is non-normal, the sampling distribution of the mean will tend to the normal law as the sample size increases. This is called the central limit theorem). Since the sum of the deviation of observations from their mean is zero, it acts as a balancing point for the sample. Symbolically, $\sum_{i=1}^{n}(x_i - \overline{x}) = \sum_{i=1}^{n} x_i - n\overline{x} = n\overline{x} - n\overline{x} = 0$. Some of the advantages of mean are given below:

1) It is easy to compute.
2) It lends itself to further arithmetic treatment (mean of a combined sample can be found from individual sample means)
3) It is always unique (whereas mode of a sample need not be unique)
4) It does not require data sorting (whereas medians are found after data sorting)
The main disadvantages of the mean are as follows:
(i) Since each data item is given the same weight, it is influenced by extreme observations (called outliers, see §3.10). The mean is easily distorted by extreme values (too small or too large than other sample values), which can result due to data entry errors, data capturing device errors (scanner for HB pencil marks, magnetic ink scanners), data communication errors, wrong formatting errors, format conversion errors, and other program errors. (ii) the mean of a sample need not coincide with one of the sample values (median for odd sample size, and mode will always coincide with a sample value), (iii) The mean is undefined for nominal data. It may not be meaningful for ordinal numeric data, since the interval between successive values need not be uniform.

Example 3.15 Suppose the systolic blood pressure of 6 patients are recorded in a database as BP=[114, 122, 110, 201, 132, 129]. The data value 201 is probably an outlier. The average systolic BP for all 6 samples is 134.66, which is greater than the highest observation (132), if outlier 201 is discarded. The mean is pushed up beyond the highest observation excluding the outlier. It does not properly convey the summarised information of the sample, if just one extreme outlier is present in it. The average without 201 is 121.4, which is in between the minimum and maximum of the values.

Lemma 3.3 If a constant c is subtracted (or added) from each sample value, the mean of the transformed variables is linearly displaced by c.
Proof: Let $\bar{x}$ be the mean of the sample X=$\{x_1, x_2, \cdots, x_n\}$. Let $y_i = x_i - c$. Then $\bar{y} = \sum_{i=1}^{n}(x_i - c)/n = 1/n(\sum_{i=1}^{n} x_i - nc) = \bar{x} - c$.

Corollary 9 If each observation is scaled by multiplying (dividing) by a nonzero constant, the mean is given by $\bar{y} = c\bar{x}$ (or $\bar{y} = \bar{x}/c$, if scaled by division).

3.8.2   Median

Definition 3.8 Median of a sample is the middle value when the data are arranged in increasing or decreasing order. If the sample size (number of data items) is odd, there is a unique middle element at $[(n+1)/2]^{th}$ position. If the number of data items is even, the arithmetic mean of the two middle observations at $(n/2)^{th}$ and $(n/2+1)^{th}$ positions is the median. Symbolically:

$$\text{Median} = \begin{cases} x_{(n+1)/2} & \text{if} \quad n \text{ is odd}; \\ .5(x_{(n/2)} + x_{(n/2)+1}) & \text{if} \quad n \text{ is even} \end{cases}$$

In other words, the median equally divides the sample such that 50% of the items are below it. Trivially, the mean and median coincides if the sample size is 1 or 2. The median is used in image processing (smoothing filters of images called (weighted) median-filters are useful in 'noise-suppression'), cluster analysis (k-median clustering), median boxplots, median-based partitioning algorithms (bucket-sort, median-partitioned quick sort), median ranked set sampling and in dispersion measures (mean deviation from the median). The median of grouped data is found in two steps:

1) Find the class to which the median belongs
2) Compute the median as Median = L+ c * (n/2 - M)/f where L is the lower limit of the median class, c is the fixed class interval, n is the sample size, M is the cumulative frequency up to median class and f is the frequency in the median class.

Finding the median of a sample is a well-studied problem in computer science. The median of very large samples could be found efficiently, without completely sorting the data items [BC00].

3.8.2.1 Advantages of Median

1) Median is least influenced by extreme observations (for n>2) because the numerical value of observations are utilised only in arranging the values among themselves.
2) Median can be approximated graphically (from ogive curves)
3) Median is better than the mean for skewed data (see figure 3.4 in page 3-19)
4) If the frequency distribution has open ends (the distribution extends to $\pm\infty$), the median could still be easily computed from the above formula.

The disadvantages of median are as follows:– (i) computational complexity in finding the median is much more than finding the mean, (ii) multiple passes through the data may be required to find the median of unsorted samples (whereas mean and mode can be found in a single pass through the data), (iii) median of a combined sample cannot in general be found from individual medians, (iv) To find the median, the nature (odd or even) of the sample size should be known, whereas this is immaterial for finding the mean and mode (see below).

3.8.3 Mode

Definition 3.9 Mode is defined as that observation which occurs most frequently.
The corresponding value is called *modal value*. If each data item is unique, any of the observations can be taken as the mode. A population with 2 or more modes is called multi-modal (but a multi-modal sample may arise from a uni-modal population, and *vice versa*). The mode of grouped data can be found using a two step procedure:
1) Find the class to which the mode belongs
2) Compute the mode using the formula $Mode = L + c * \delta_u/(\delta_l + \delta_u)$ where L is the lower limit of modal class, δ_l is the difference in frequency between modal class and the class below it, δ_u is the difference in frequency between modal class and the next class above it.

3.8.3.1 Advantages of Mode

i) Mode can be approximated graphically, which is useful for skewed distributions and in multivariate case.
ii) Mode is not influenced by outliers
iii) The modal value coincides with a sample observation (whereas the median for even sample size and mean need not coincide with sample values)
iv) Mode is the most appropriate measure for categorical data.

The biggest disadvantage of mode is that it need not be unique. If there are two maximum frequencies, it is called bimodal. Mode may coincide with the minimum or maximum of the sample (which is not possible for mean for n$\geq$ 2, although it could happen for median. eg: median{2,2,2,3,5}=2, and median{2,4,5,5,5}=5). When large number of data items are missing or has default values, the mode can wrongly get located at the missing value as in the following example. If data are not cleansed, several variables with default values may exist in the source file. For example, some of the database programs assign 0 as a default to numeric variables. Similarly, data files created using programming languages that have default values for program variables may contain many of these values, unless the program specifically replace these with legitimate values. Mode utilises only the value of most frequently occurring observation (max frequency counts) in contrast to the mean that utilises actual values of each and every item in the sample.

Example 3.16 A class has 12 students. The following data gives the number of minutes a student watched a TV program during the past week. [0, 0, 0, 20, 35, 35, 40, 60, 75, 80, 90, 90]. Find the mean, median and mode. Which of these measures is most appropriate in this situation?

Solution: The mean is $525/12 = 43.75$. As n=12, the median is the average of 35 and 40 (6th and 7th observations), and is 37.5. Since three students did not watch the TV program last week, the mode is 0. In this problem, the best measure is the median.

The mean-median-mode inequality (mean $\leq$ median $\leq$ mode) provides an empirical relationship between these three measures[11]. Another approximate relationship exist for bell-shaped distributions as (mean-mode) $\simeq$ 3(mean-median) (mean=median=mode relationship holds in the population for symmetric uni-modal distributions, but these need not be equal for a sample drawn from that population). Median is computationally the most difficult statistic for large data mining applications (see §3.8). However, using an estimate of the mean and mode, the median (for bell shaped curves) can be approximated as median = (2*mean+mode)/3. All of the above measures are expressed in the same unit as the observations.

Arithmetic mean is the frequently used statistic in data mining. For example, the K-means algorithm uses k numbers to group the training set into clusters (the algorithm starts with k arbitrary numbers which can be the averages of an arbitrary k-partition of the training set). In some data mining applications, it is desirable to have *the* observation nearest to the mean (if the mean does not coincide with a sample item) to ease the computation. Such an item is called the *medoid*[12]. The K-medoid clustering algorithm uses 'k' medoids of a sample (the training set is partitioned into k sub-samples, and the mean of each is found. The observations nearest to each mean is chosen from subsamples to get the K-medoids). As median is least influenced by extreme observations, it is used instead of the mean in some algorithms, like the K-median clustering, quadrant correlation etc [CN93]. Median can also be used to normalise the variables. Figure 3.4 in page 3-19 gives a simple rule to decide between the mean, median or mode as an appropriate measure of central tendency.

3.8.4 Geometric Mean

Definition 3.10 Geometric mean (GM) of n observations (none of which are zeros) is defined as

$$\mathrm{GM} = (x_1 x_2 .. x_n)^{1/n} = (\prod_{i=1}^{n} x_i)^{1/n},$$

where $\prod$ denotes the product of the observations. Taking logarithm (to any base) of both sides give $\log(GM) = (1/n) \sum_{i=1}^{n} \log(x_i)$. This shows that log(GM) is the arithmetic mean in the 'log-space'. Because the logarithm is defined only for positive argument, this summary measure

---

[11]This relationship need not hold for highly skewed distributions. eg:f(x) $= e^{-cx}$

[12]The mediod is easy to locate when data are uncorrelated. In the presence of correlation, the *nearness* of a data point can be biased using distance measures other than Mahalanobis distance.

is meaningful only when all observations are positive (If at least one observation is 0, the product will itself be zero). The usual practice in such situations is to omit all zeroes, and find the GM of the remaining values. The GM coincides with the sample observations when all observations are equal. Hence it is most appropriate when the *product* of several numbers combine together to produce a resulting quantity as in rates of changes, exchange rates, compound interests, etc.

As the GM inherently involves the product of observations, the change of origin technique (§3.11.1) is not useful. But the change of scale transformation Y=cX (§3.11.2) provides the relationship given in §3.5 in page 3-29. The GM for grouped data is given by $(\prod_{i=1}^{n}(f_i x_i))^{1/F}$ where $F = \sum_{j=1}^{n} f_j$ is the total frequency.

The GM has many practical utilities. The Fowlkes-Mallow index for measuring the degree of agreement between two groups (regions of cluster space) and clusters utilise the GM of two conditional probabilities:– (i) The conditional probability P(two items are in the same cluster|they belong to same region) and (ii) P(two items are in the same region | they are in the same cluster). It is also used in image enhancements to smooth low contrast images by taking the GM of the surrounding pixels (provided none of the neighbouring pixel values are zeroes). See [GT08] for an application of GM to finance.

Example 3.17 If a stock price has gone up by 10% in the first month, 12% in the second month and 15% in the third month of a year, what is the net increase in price of this stock?
In this type of problems involving rate of increase or decrease, GM is the most suitable measure to be used. It is found as GM $= (1.10 * 1.12 * 1.15)^{1/3} = (1.4168)^{1/3} = 1.12314$. Thus the typical monthly increase is 12.31%.

3.8.5   Harmonic Mean

Definition 3.11 If all observations are nonzero, the reciprocal of the arithmetic mean of the reciprocals of observations is known as harmonic mean (HM). For ungrouped data, it is defined as $HM = n / \sum_{i=1}^{n}(1/x_i)$. The HM of a group of non-zero numbers coincides with them when all the numbers are equal. Hence the HM is used when several (non-zero) numbers combine via reciprocals to produce a resulting quantity, as in the case of finding the mean speed of vehicles that go the same distance (not for the same duration). The HM for grouped data is given by $F / \sum_{i=1}^{n}(f_i/x_i)$ where $F = \sum_{j=1}^{n} f_j$ is the total frequency. The HM finds applications in data clustering (k-harmonic means algorithm). The F-score used in text mining is the harmonic mean of precision and recall (chapter 10). As in the case of GM, the change of origin transformation is meaningless. The change of scale transformation for HM is given in §3.6 in page 3-29.

A simple inequality exists between the three popular means as: (AM$\geq$GM$\geq$HM). If there are just two observations in a sample, these measures coincides only when both of them are equal (see exercise 21). When each of the sample values are weighted using the same set of weights, the above inequality is preserved [MP03]. See [SC15] for details.

3.8.6   Which Mean is Most Appropriate?

Many other means also exist in addition to the popular means discussed above. One example is the quadratic mean defined as $\sqrt{\sum_i x_i^2 / n}$. As data mining involves a variety of numeric data, we need to select the most appropriate mean for each numeric variable. A simple rule is to ask "How do the numbers naturally combine to produce a resultant value?".

1. If the numbers combine *additively* to produce a resultant value, use the arithmetic mean. Examples are currency amounts like profits, sales prices, insurance premiums, overdue amounts, incomes, etc; quantities measured on a scale like heights, weights, thickness, rainfall, snowfall etc.

2. If non-zero numbers combine *multiplicatively* to produce a resultant value, use the geometric mean. In other words, if the logarithm (to any base) of several numbers combine additively, use the GM. Examples are rates of changes like successive discounts; price or stock market increases and decreases; successive size changes (enlargements or contractions) of images, graphics; successive volume changes etc.

3. If the *reciprocals* of non-zero numbers combine *additively* to produce a resultant value, use the harmonic mean. Examples are electrical resistance in parallel circuits, average speed of vehicles for the same distance, etc.

4. If the *squares* of several numbers combine *additively* to produce a resultant value, use the quadratic mean. Examples are straight-line distances in n-D (for n$\geq$2), etc.

3.9 Measures of Dispersion

Location measures are far too insufficient to understand the data. Two or more identical samples can come from populations with the same location measure, but with different dispersions.

Definition 3.12 A dispersion measure concisely summarises the extent of spread of observations in a sample. Some of them measures it around a specified location measure (see below).

Most frequently used univariate dispersion measures are the variance and standard deviation, inter-quartile range, mean absolute deviation around the mean and the mean deviation from the median. Among these, the variance, standard deviation (the positive square root of variance), and the mean absolute deviation around the mean use the sample mean ($\bar{x}$) as the pivoting location measure, while mean absolute deviation from the median uses the median as the pivoting location measure. The range, inter-quartile range and quartile deviation do not use a location measure as a pivot. In the following sections, we consider the dispersion measures for univariate data.

3.9.1 Range

Let x_1, $x_2,..x_n$ be 'n' sample values that are arranged in increasing order.

Definition 3.13 The range of a sample is the difference between the largest and smallest observation of the sample. Symbolically, R=$(x_n - x_1)$.
We need only the smallest and largest observations of a sample to compute the range. This can be obtained in a single pass through the data (unless the data are sorted, in which case we can easily pick out the smallest and largest observations in two fetches). Range has applications in statistical quality control (SQC), data plotting, data discretisation, data transformations etc. For instance, the min-max transformation in (§3.11.4) uses the data range in the denominator.

The range is easy to compute and easy to interpret. The distribution of sample range can be derived if the parent population to which the sample belongs is known. The biggest disadvantage of range is that it is extremely sensitive to outliers (on both extremes). As it does not utilise every observation of a sample, it cannot distinguish between skewed distributions that have the same range. It is applicable to ordinal and higher scales of measurement. Range is unaffected by a change of origin data transformation (see theorem 3.4 in page 3-28).

3.9.2 Inter-Quartile Range

Definition 3.14 The inter-quartile range (IQR) is defined as $(Q_3 - Q_1)$ where Q_3 and Q_1 are the upper and lower quartiles (Q_1 is that value below which one-forth of the observations will fall, and Q_3 is that value below which three-forth of the observations will fall, after the sample is arranged in ascending order).
One half of IQR is called the quartile deviation (QD). It is unaffected by outliers, and provides supplementary information on the spread of observations around the center of the sample. It is used in boxplots to visually detect outliers.

3.9.3 Mean Absolute Deviation

Definition 3.15 The mean absolute deviation is defined as $\sum_{i=1}^{n} |x_i - \bar{x}|/n$, where $\bar{x}$ is the sample mean. It is also called mean (absolute) deviation from the mean.
As in the case of the arithmetic mean, this measure uses each and every observation of the sample. It is affected by outliers, but not as much as the range. A related statistic is median absolute deviation (mean absolute deviation around the median) defined as $\sum_{i=1}^{n} |x_i - \text{Median}|/n$, which uses the median as the pivot. Median absolute deviation around the median is the middle value of (sorted) absolute deviations of observations from the median. See [SC15] for novel methods to find the MD of discrete and continuous distributions and for mean deviation generating functions.

Example 3.18 If there are two items (x_1, x_2) in a sample, prove that the mean absolute deviation around the mean and median are both equal to half the sample range.
Solution: The mean absolute deviation around the mean is given by $\frac{1}{2}|x_1 - (x_1 + x_2)/2| + \frac{1}{2}|x_2 - (x_1 + x_2)/2| = \frac{1}{2}(|(x_1 - x_2)/2| + |(x_2 - x_1)/2|)$. Because $|x_1 - x_2| = |x_2 - x_1|$, the above reduces to $\frac{1}{2}|x_2 - x_1|$ which is half the absolute value of range. As the range is the difference between the largest and smallest of the sample values, it is always positive, irrespective of the sign of the data values. Thus $\frac{1}{2}|x_2 - x_1| = \frac{1}{2}(x_2 - x_1)$, which is half the range. Since the median of a sample of size 2 is the mean of the sample values, the second part follows easily from the above.

3.9.4 Variance

Definition 3.16 The variance of a sample is defined as $s^2 = \sum_{i=1}^{n}(x_i - \bar{x})^2/n$ where $\bar{x}$ is the sample mean, and the variance of the population is defined as $\sigma^2 = \sum_{i=1}^{n}(x_i - \mu)^2/N$ where μ is the population mean, n is the sample size, and N is the population size.
Some authors use (n-1) in the denominator of sample variance to indicate the loss of one *degrees of freedom* due to the estimation of the mean from the sample data. We will keep it n in analogy with the sample covariance (see below) that uses n in the denominator although $\bar{x}$ and $\bar{y}$ are estimated from the data. In addition, the sample covariance should reduce to the variance when $y_i's$ are replaced by $x_i's$ (and $\bar{y}$ is replaced by $\bar{x}$).

By expanding the square and summing the resulting terms individually, this could also be computed as

$$s^2 = \left(\sum_{i=1}^{n} x_i^2 - n\overline{x}^2 \right) / n = \sum_{i=1}^{n} x_i^2 / n - \overline{x}^2.$$

The positive square root of variance is called the *standard deviation*. Many other formulas are also available for both the sample and population variances (see [JA01], [JM04]), which are more of theoretical interest than from a computational viewpoint. The main advantages of variance are that (i) it uses all of the sample observations, (ii) it lends itself to further arithmetic operations, (iii) distribution of sample variance is known when the population distribution is known, and (iv) it can be found without data sorting.

A related measure is the coefficient of variation (CV), which is a unit-less ratio-measure defined as $(s/\overline{x})$ x 100. The distribution of *unscaled* CV (ie. $s/\overline{x}$) can be approximated using a central χ^2 distribution, when samples come from a normal population. This is called McKay's approximation, which has a *scaled* Type-II noncentral beta distribution [CR95] with scaling factor $n(1+1/\gamma^2)$, where n is the sample size and γ is the population CV [FV08].

Theorem 3.3 If data are transformed as $y = (x - a)/c$, the variances are related as $s_y^2 = (1/c^2)s_x^2$. **Proof.** The means are clearly related as $\overline{y} = (\overline{x} - a)/c$. Consider $\sum_{i=1}^{n}(y_i - \overline{y})^2 = \sum_{i=1}^{n}((x_i - a)/c - (\overline{x} - a)/c)^2 = 1/c^2 \sum_{i=1}^{n}((x_i - a) - (\overline{x} - a))^2 = 1/c^2 \sum_{i=1}^{n}(x_i - \overline{x})^2$. Dividing both sides by n gives $s_y^2 = (1/c^2)s_x^2$.

3.10 Outliers in Data

Outliers are encountered in several data mining models, including decision trees, cluster analysis, neural networks and SVM.

Definition 3.17 Outliers are extreme observations that are inconsistent wrt other observations.

Data mining models constructed from training data containing outliers may not be well-generalisable. Hence identifying and removing outliers can improve the validity of our model. An observation that is considered as an outlier in one study may not be considered as an outlier in another study. This is because of the fact that an outlier observation may have unusual values for a few of the attributes only. Too many outliers in a sample can throw some insight into the data distribution.

3.10.1 Spatial vs Temporal Outliers

Outliers may be spatial, temporal or neither. In most data mining models, we come across spatial outliers. These are observations with one or more unusual attribute values. Outliers may be detected graphically or analytically. Graphical tests use data visualising in appropriate dimensions or sub-dimensions. It is unreliable in the presence of correlation, which can distort the *distance separation* of outliers from other data. A graph-based algorithm can be used when there are few outliers. Bivariate outliers are easy to detect from the scatter-plots. Special scatterplots like Moran plots ([AL88]), variogram clouds, etc are also available to detect them. Statistical measures are used to detect outliers analytically, using one of the distance measures discussed in chapter 7. Spatial outliers are detected using the 3σ rule. The mean of the sample (excluding the extreme observation) is compared to each observation using the chosen distance measure. If this distance is more than 3 times the standard deviation of the entire sample, the observation is considered as an outlier.[13]

Temporal outliers are observations with unusual data value for temporal attributes. For example, a customer visit to an ATM machine after midnight, or a customer visit to a super market at 2AM are examples of temporal outliers. They are most often used in time series data.

3.10.2 Graphical Detection of Outliers

The boxplot (see chapter 2) is a graphical display for exploratory analysis introduced by Tukey in 1977. It uses box edges and whiskers (hinges) to demarcate potential outliers. It does not use potential outliers in computing the whiskers, leading to the 'masking' effect. Boxplot tags two types of outliers – mild and extremes. A *mild outlier* is an observation outside the range $(Q_1 - 1.5\ IQR, Q_3 + 1.5\ IQR)$. An *extreme outlier* is an observation (if any) that falls beyond the range $(Q_1 - 3\ IQR, Q_3 + 3\ IQR)$ where IQR $= Q_3 - Q_1$.

Multivariate outlier detection is more difficult due to the possibility of correlation among the variables. Multidimensional outliers in the presence of correlation are detected using squared Mahalanobis distance metric $D^2(x, \overline{x}) = (x - \overline{x})'S^{-1}(x - \overline{x}) > k$, where $S_{p \times p}$ is the pooled covariance matrix, and p is the number of attributes in each vector[BI05]. In data mining applications with hundreds of attributes, mixed outliers may be present in spatial and temporal variables. An outlier may be defined with respect to a whole sample or with respect to sub-samples. For example, in a clustering application with a large number of sparse points and few distinct clusters, the sparse points may not fall outside the 3σ rule of the entire sample.

Outliers find applications in detecting fraud, dishonesty, theft, network intrusions, locating process deviations, etc. A non-parametric method using distance of an instance to its nearest neighbour was suggested by Knorr and Ng ([KN98]), which is especially useful for large samples of non-normal data. As the median is more robust than mean for nonnormal (and skewed) data, boxplots are prepared around the median ($|\overline{x} - median|/s > $ k). If the parent distribution is known to be skewed, the boxplot whisker(s) can be adjusted using a quantified measure of skewness.[14] Outliers can provide valuable information in some data mining applications. All extreme outliers can be further analysed using root cause analysis if an assignable cause can be found (measurement error, spurious recording, incorrect specifications, wrong distributional assumption, transformation error, change in a process, random outlier). See [BL94], [AY01] for a list of outlier detection methods.

3.11 Data Transformations

Data transformation is an important step in data cleansing. Datawarehouses and data marts are built from heterogeneous enterprise data spread across databases, spreadsheets, application files (word processors, custom applications etc), pure text files (comma separated, tab separated files), markup files (HTML, WML, XML etc) and external files. Different people often maintain these files. Data transformations are needed to bring uniformity to the data. In addition, it can be used to scale the data to a preferred range. These are discussed in detail in later chapters.

---

[13]The 3σ rule is quite arbitrary. In some applications (like clustering), even a $2\ \sigma$ observation may be considered an outlier if all other points fall in tighter variance contours.

[14]Karl Pearson provided a quantitative measure of skewness as $\eta = (\overline{x} - mode)/s$, which has an expected value close to zero for large samples from symmetric distributions, where s is the sample standard deviation.

3.11.1   Change of Origin

This is a frequently used technique for large data. It is so called because it shifts data points by subtracting (or adding) a constant from each sample observation. If sample values are integers, an integer constant is preferred for a change of origin. This technique is very useful when the data values (that combine additively) are large, and the variability is not so large. Examples are various blood counts, loan amount for new vehicles, number of pages in a textbook, project costs, daily or weekly collection amounts at large businesses etc.

Example 3.19 Apply change of origin method to the data X=$\{115, 128, 110, 104, 133\}$ and find the mean and range.

We subtract 120 (chosen arbitrarily) from each observation to get $x_i' = x_i - 120$ as $X' = \{-5, 8, -10, -16, 13\}$, from which $\sum_i x_i' =$ -31+21 = -10. The mean of X' is $\overline{x}'$=-10/5 = -2. The mean of the original data is 120-2 = 118=$\overline{x}$. As the minimum is -16 and maximum is 13, the data range is 13-(-16)=29. This is also the range of original data (133-104).

Theorem 3.4 The range and quartile deviation (QD) of original data are preserved by a change of origin transformation.

Proof. Let x_1 and x_n be the minimum and maximum of the original data. The change of origin transforms these values to $y_1 = x_1 - k$ and $y_n = x_n - k$, where k can be any real number (positive or negative). The new range is $y_n - y_1 = (x_n - k) - (x_1 - k) = x_n - x_1$. As the change of origin shifts all of the data points uniformly, the quartiles are shifted by the same value k. Hence the new QD is $([Q_3 - k] - [Q_1 - k])/2 = (Q_3 - Q_1)/2$.

Lemma 3.4 Prove that the variance of a sample is unchanged by the change of origin transformation.

Proof: Let $y_i = x_i - c$ be a change of origin transformation where c$\neq$0. Summing the values and dividing by the total number of observations gives $\overline{y} = \overline{x} - c$. We will first prove the result for unscaled variance. Consider $s_y^2 = \sum_{i=1}^{n}(y_i - \overline{y})^2$. Substitute $y_i = x_i - c$ and $\overline{y} = \overline{x} - c$ to get $s_y^2 = \sum_{i=1}^{n}(y_i - \overline{y})^2 = \sum_{i=1}^{n}((x_i - c) - (\overline{x} - c))^2$. The +c and −c cancels out giving $s_y^2 = \sum_{i=1}^{n}(x_i - \overline{x})^2 = s_x^2$. By dividing both sides by proper degrees of freedom gives the result.

3.11.2   Change of Scale

This technique is used to shorten the range of large numbers or lengthen the range of very small numbers. We divide each large observation by the same constant (>1) to get a much smaller range. This is useful when numbers are large and has high variability. Examples are family income, total insured amounts, annual insurance premiums, defaulting loan amounts, advertising expenses in various media and regions etc. Let $x_1, x_2, ..x_n$ be 'n' sample values, and c be a *non-zero* constant. Define $y_i = x_i/c$. If c is less than the minimum of the observations, each of the y_i's are greater than 1, if x_i's are positive. Similarly, if c is greater than the maximum of the observations, each y_i's are less than 1. For values of c between minimum and maximum of the sample, we get values on the real line (positive real line if all x_i's are positive). If all values are small fractions, we may multiply by a constant to scale them up.

Example 3.20 Let X=$\{.001, .006, .0095, .015, .03\}$. By choosing c=1000 and scaling up using $y_i = cx_i$ we get Y=$\{1, 6, 9.5, 15, 30\}$. This is an example of decimal scaling in which the decimal point is moved by multiplying/dividing by a power of 10. The mean of Y is $\overline{y}$=61.5/5 = 12.3, and thus the mean of X is $\overline{x} = \overline{y}/c$=.0123.

Lemma 3.5 Prove that the GM of change of scale transformed data (y-variable) is given by $GM(Y) = c\,GM(X)$, where $y_i = cx_i$, c is a constant (positive or negative).
Proof: $GM(y) = (cx_1.cx_2..cx_n)^{1/n} = c(\prod_{i=1}^{n} x_i)^{1/n} = c * GM(x)$.

Lemma 3.6 If $y_i = cx_i$ where c is a constant (positive or negative), prove that the HM of transformed data (y-variable) is given by $HM(Y) = c * HM(X)$.
Proof: $HM(y) = n/\sum_{i=1}^{n} 1/y_i = n/\sum_{i=1}^{n} 1/(cx_i)$. Taking the constant c to the numerator, this becomes $c*n/\sum_{i=1}^{n} 1/x_i = c * HM(x)$.

3.11.3   Change of Origin and Scale

This is the most frequently used technique to transform data values. Depending upon the constants used to change the origin and scale, a variety of transformed intervals can be obtained.

Theorem 3.5 A sample of values (x) in the range (a,b) can be transformed to a new interval (c,d) by the transformation

$$y_i = c + [(d - c)/(b - a)] * (x_i - a). \tag{3.3}$$

Proof. By putting x=a in the expression gives y = c. Putting x=b gives y=c+[(d-c)/(b-a)]*(b-a) = c+(d-c) = d. Since (x-a)/(b-a) and (y-c)/(d-c) both map points in the respective intervals to the [0,1] range, all intermediate values in (a,b) will get mapped to a value in (c,d) range. This proves the result. This holds even if the open interval (a,b) is replaced by the closed interval [a,b], in which case the transformed values lie in the closed interval [c,d].

An application of this transformation in image processing is called *contrast stretching*. Consider a pixel (a digital picture element[15]) in the range (x_L, x_H). To *contrast stretch* the picture to a higher value (say) (y_L, y_H), we could use the above transformation with x_L=a, x_H=b, y_L=c, y_H=d to get $y_i = y_L + [(y_H - y_L)/(x_H - x_L)] * (x_i - x_L)$.

Corollary 10 Prove that the sample x in the range (a,b) can be transformed to a new interval (c,d) by the transformation

$$y_i = \frac{1}{(b - a)} [(d - c) * x_i + (bc - ad)]. \tag{3.4}$$

Proof: Write $c + [(d - c)/(b - a)] * (x_i - a)$ as $[c-a*(d-c)/(b-a)]+x_i*(d-c)/(b-a)$. Taking (b-a) as the common denominator, the first expression simplifies to (bc-ac-ad+ac)/(b-a). Cancel out 'ac' to get (bc-ad)/(b-a). Combining with the second expression and taking the common denominator outside gives the required result.

Example 3.21 Online time in seconds spent by six e-commerce customers to a floral site for a firm order are [60, 90, 118, 150, 165, 170]. Transform the data to the range [10,60].
Solution: Here a=60, b=170, c=10, d=60. Thus (d-c)/(b-a) = (60-10)/(170-60) = 5/11. Hence the required transformation is y = 10+(5/11)(x-60) = [10,23.6364,36.3636,50.9091,57.7273,60]. See chapter 7, §7.3 for more applications of these transformations.

---

[15]a related term is *voxel* (volume element) which is a unit element of a graphics image in 3 or more dimensions.

Corollary 11 A sample X in $[-k, +k]$ can be transformed to the range $[a,b]$ by the transformation $Y = a + \frac{R}{2k}(x+k)$, where $R = (b-a)$.

Proof: Putting $x = -k$ gives $Y = a$. Putting $x = +k$ gives $Y = a + R = a + (b-a) = b$. All intermediate values are transformed into the interval (a,b). For instance, $x = 0$ gets mapped to $a + R/2 = a + (b-a)/2 = (a+b)/2$. Hence the X values get transformed to the range $[a,b]$.

Example 3.22 The error measurements recorded by a high-precision machine on a product are as follows: $X = \{-.2, .12, .03, -.05, .15, -.04, .20, -.17\}$. Transform the data to the $[0,1]$ range.

Solution: Here $k = .2$ (as the data vary between $-.2$ and $+.2$), $a = 0$, $b = 1$, $R = 1 - 0 = 1$. Hence the required transformation is $Y = 0 + \frac{1}{2*.2}(x + .2) = \frac{1}{.4}(x+.2) = \frac{10}{4}(x+.2) = 2.5*(x+.2)$. The resulting Y vector is $Y = \{0.0, 0.8, 0.575, 0.375, 0.875, 0.4, 1.0, 0.075\}$.

3.11.4   Min-max Transformation

This transformation is used to map any sample values to the interval $[0,1]$. It gets its name from the fact that the sample minimum is used to change the origin, and the sample range is used to change the scale of data.

Theorem 3.6 Any numeric variable x in the interval (x_{min}, x_{max}) can be transformed to a new interval $[0, 1]$ by the transformation $y = (x - x_{min}) / R$, where $R = (x_{max} - x_{min})$ is the range, and x_{min}, x_{max} are the minimum and maximum of the sample values.

Proof. Substituting $x = x_{min}$ gives $y = 0$, and $x = x_{max}$ gives $y = 1$. All intermediate values are transformed to the interval $(0,1)$. Hence the transformed values are mapped to $[0,1]$.

Example 3.23 Transform the above data into the $[0,1]$ range.

Solution: Here $R = (170 - 60) = 110$. We make the transformation $y_k = (x_k - 60)/110$ to get $[0, 0.2727, 0.5273, 0.8182, 0.9545, 1]$.

3.11.5   Transformation to Symmetric Range

Lemma 3.7 A sample X in any finite range can be mapped to the interval $[-1, +1]$ by a simple change of origin and scale transformation.

Proof: Consider the transformation $y = 2*[(x - x_{min})/R] - 1$, where $R = (x_{max} - x_{min})$. When $x = x_{min}$, y becomes -1, and when $x = x_{max}$, $y = +1$. All intermediate values are mapped to points within the interval $[-1, +1]$. Thus the result. This holds even if x_{min} is negative.

Example 3.24 Transform the data $[34, 43, 55, 62, 68, 74]$ to the interval $[-1, +1]$.

Solution: Here the minimum is 34 and maximum is 74. Thus the range is 40. Using lemma 3.7, we get the transformed data as $y = [2*(x-34)/40] - 1 = [-1, -.55, .05, .40, .70, +1]$.

Theorem 3.7 A sample X in any finite range with at least 2 elements can be mapped to the range $Y = [-k, +k]$ by the transformation

$$y_i = \frac{k}{R}[2 * x_i - (x_{min} + x_{max})], \quad \text{where } R = (x_{max} - x_{min}). \tag{3.5}$$

Proof. Putting $x = x_{min}$ gives $y = \frac{k}{R}[2 * x_{min} - (x_{min} + x_{max})] = \frac{k}{R}[x_{min} - x_{max}] = -k$ because $R = (x_{max} - x_{min})$. Putting $x = x_{max}$ gives $y = \frac{k}{R}[2 * x_{max} - (x_{min} + x_{max})] = +k$. All intermediate values are mapped to the interval $(-k, +k)$. For instance $x_c = (x_{min} + x_{max})/2$ gets mapped to 0. This proves the result.

Remark 1 In the particular case when the sample has just two elements, they are mapped exactly to $-k$ and $+k$ respectively.
Proof: Consider a sample of two values (x,y). Rearrange them such that x_{min}=x, x_{max} = y or vice versa. Substituting in equation 3.5, x_{min} gets mapped to $-k$ and x_{max} gets mapped to $+k$.

Example 3.25 Transform the above data into the intervals [-.5, +.5], and [-3, +3].
Solution: Here k=.5, so that k/R = .5/(74-34) = .0125, x_{min}+x_{max}=74+34 = 108, giving the transformation y = .0125*(2*x -108). Resulting y vector is [-0.5,-0.275,0.025,0.20,0.35,0.5]. For the [-3,+3] range, k=3 and k/R = 3/40 = 0.075, giving the transformation y = 0.075*(2*x -108). Resulting values are [-3, -1.65, 0.15, 1.2, 2.1, 3].

Corollary 12 A sample X in any finite range $[x_{min}, x_{max}]$ can be mapped to an arbitrary range [a,b] using the transformation

$$y_i = \frac{1}{R}\left[b*(x - x_{min}) - a*(x - x_{max})\right], \text{where } R = (x_{max} - x_{min}). \qquad (3.6)$$

Proof: Putting x=x_{min} gives y=$\frac{1}{R}[0 - a*(x_{min} - x_{max})]$=$\frac{1}{R}[a*(x_{max} - x_{min})]$=a. Similarly, putting x=x_{max} gives y=$\frac{1}{R}[b*(x_{max} - x_{min}) - 0]$=b. All intermediate values are mapped to the interval (a,b). For example, putting x=$(x_{min} + x_{max})/2$ should give a value within (a,b) range because the arithmetic mean of the minimum and maximum of the sample must always be a number within the range.

$$y = \frac{1}{R}[b*((x_{min} + x_{max})/2 - x_{min}) - a*((x_{min} + x_{max})/2 - x_{max})] \qquad (3.7)$$

$$= \frac{1}{R}\left[\frac{b}{2}*(x_{max} - x_{min}) - \frac{a}{2}*(x_{min} - x_{max})\right] = \frac{1}{R}\left[\frac{b}{2}*R - \frac{a}{2}*(-R)\right] = \frac{(b+a)}{2}.$$

This proves the result, as (a+b)/2 is a number within the range (a,b).

Remark 2 This transformation can be obtained directly from theorem 3.5.

Remark 3 If x=0 is an input data point, it gets mapped to $\frac{1}{R}[b*(0 - x_{min}) - a*(0 - x_{max})]$ $=\frac{1}{R}[a*x_{max} - b*x_{min}]$. Add and subtract $a*x_{min}$ and simplify to get the transformed point as $\frac{1}{R}[a*(x_{max} - x_{min}) - x_{min}*(b - a)]$=a-$\frac{b-a}{R}*x_{min}$. As x=0 is in the input data, x_{min} must be ≤ 0. If x_{min}=0, it is mapped to a. Otherwise ($x_{min} < 0$) the datum x=0 gets mapped to a number greater than 'a' because (b-a) being the range is always greater than 0. For instance, if $x_{min} = -.2$, then x=0 is mapped to a-$\frac{b-a}{R}*(-.2)$=a+.2*(b-a)/R.

Lemma 3.8 A change of scale transformation can be applied prior to the above transformations to get identical results.
Proof: We will prove the lemma for corollary 3.7. Proof follows similarly for other transformations. Consider the change of scale transformation $x' = x/c$ where c is a nonzero constant. Substituting x= cx' in the above, we get $y = \frac{k}{R}[2cx' - (cx'_{min} + cx'_{max})]$. Because R = ($x_{max}$ - x_{min}) = c(x'_{max} - x'_{min}) = cR', the c in the numerator and denominator cancels out, giving $y = \frac{k}{R'}[2x' - (x'_{min} + x'_{max})] = y'$. This could ease the computations when numbers are very large (choose c>>0), or very small (choose c<<0, or equivalently $\frac{1}{c}$ >>0) [The symbol >> means much greater than].

Example 3.26 Transform the data [34000, 43000, 55000, 62000, 68000, 74000] to the interval [-1,+1], and [-5,+5].
Solution: We will choose c=10 000, and divide each value by c to get X'=[3.4, 4.3, 5.5, 6.2, 6.8, 7.4]. Here range R=(7.4-3.4) = 4, k=1 and $(x_{min}+x_{max})$=3.4 + 7.4 = 10.8, giving the transformation y = .25*(2*x -10.8). The resulting y vector is [-1, -.55, .05, .40, .70, +1]. For the second case, we have k=5, so that k/R=5/4=1.25, giving the transformation y = 1.25*(2*x -10.8). The resulting y vector is [-5,-2.75, 0.25, 2.0, 3.5, +5].

3.11.6   Standard Normalisation

This transformation is so called because it is extensively used in statistics in standardising normal scores. Here the origin is changed using the mean of the sample, and the scale is changed using the standard deviation of the sample. Symbolically $y_i = (x_i - \overline{x})/s$ where s is the standard deviation. The resulting values of y are called z-scores and will almost always lie in the interval [-3,+3].

Theorem 3.8 The z-scores remain the same if data are transformed as $y = (x - a)/c$.
Proof. We know from theorem 3.3 that $s_y = (1/c)s_x$. Consider $z_y = (y - \overline{y})/s_y = ((x - a)/c - (\overline{x} - a)/c)/(s_x/c)=((x - a) - (\overline{x} - a))/s_x = (x - \overline{x})/s_x = z_x$. This is also evident from the fact that E(Z)=0 and V(Z)=1 for standard normalised data.

Example 3.27 The sales amounts of 8 customers to a store are X=(32, 80, 56, 75, 69, 26, 44, 50). Apply the standard normalisation:
Solution: The mean is 54 and variance is 390. Thus the transformation y=(x-54)/19.74842, gives Y=(-1.114, 1.31656, .101274, 1.063376, .759554, -1.41783, -.50637, -.20255).

Example 3.28 Annual incomes of five families are X= $\{28600, 40000, 51800, 34100, 75000\}$. Find the z-scores.
Solution: Since the numbers are large, we will apply simple scaling to find the mean and variance. We divide each data by 10 000 to get $S' = \{2.86, 4.0, 5.18, 3.41, 7.5\}$. The mean of scaled data is 22.95/5 = 4.59, and variance is $s^2 = 1/5[139.9582 - 5*4.59^2]$=13.5496/5=2.70992, from which the standard deviation is obtained as 1.64618. The corresponding z-scores are easily found as [-1.0509, -0.3584, 0.3584, -0.7168, 1.7677]. Using theorem 3.8, the corresponding z-scores for S are also the same, which can be verified using the transformation z = (x-45900)/16461.83465.

3.11.7   Nonlinear Transformations

Linear data transformations may be insufficient in some data mining applications. The popular non-linear transformations are square-root transformation, trigonometric and hyperbolic transformations, logarithmic and exponential transformations, power transformations and polynomial transformations. These transformations are used either to stabilise the variance of the data, or to bring the data into one of the well-known distributional form.

3.12   Monte Carlo Methods

There are two popular simulation techniques called *discrete-event* and *Monte Carlo* simulation. Time-dependent events (eg: queuing and inventory systems, stock markets, etc) are simulated using discrete-event simulation. Monte Carlo (MC) methods are used to simulate random or stochastic processes that are not time dependent. An approximate distribution of the population must be known for Monte Carlo methods to work. The basic ingredient of Monte Carlo simulation is a random sample. If the cdf of a random variable is available in analytical form, random numbers can be generated from it using the inverse transformation method. Because the cdf (being a probability, see §3.2.2) assumes values between 0 and 1, and is uniformly increasing for increasing x values, we can invert it to get $X := F^{-1}(U)$ where U is uniformly distributed over (0,1). Using a random number in a fixed interval like [0,1] or [-1,1], we can generate random numbers from most unimodal distributions. The Gaussian (normal) random numbers can be generated using this approach as follows: Consider two independently distributed random numbers x and y in the interval [-1,1] (if a random number generator returns values in any other range, they may be transformed to [-1,+1] range using the transformations given in §3.10.4). Compute $d = x^2 + y^2$. Obviously d lies in the range [0,2]. Continue generating random numbers until a pair with $d \in (0,1]$ is obtained. Using such a pair, find $z_0 = xS$ and $z_1 = yS$ where $S = \sqrt{-2\ln d}$. This method has a polar version, known as Box-Muller method [BM58]: If u_1 and u_2 are independent uniform random variables in the interval [0,1], then $z_0 = V\cos(2\pi u_2)$ and $z_1 = V\sin(2\pi u_2)$ are two random numbers from standard normal where $V=\sqrt{(-2\ln u_1)}$.

Most data mining applications involve a multitude of parameters that follow different statistical laws. Monte-Carlo analysis is based upon distributional assumptions of these unknown parameters. Random samples are drawn from the parameter space and an approximate model is constructed. The output of the model is then compared to the expected values after many simulation runs. If the distribution of parameters are unknown, an appropriate distribution may be assumed from domain knowledge or using other statistical techniques. Normality tests can be used to check if a sample comes from a normal population or not. If the sample size is less than 50, the Shapiro-Wilks test can be used and for samples size more than 50, the D'Agostino test is used. Other popular tests are Kolmogorov-Smirnov test, and Lilliefors test (based on normalised data). The error in simulation can be minimised by utilising a large sample.

3.12.0.1 Components of Monte Carlo Simulation

1. Probability density function

2. Random number generator for desired distribution (Uniformly distributed random numbers can be converted to random numbers from most distributions by a change of variable technique)

3. Sampling rule from the distribution

4. Error estimation

Monte-Carlo methods have many data mining applications. Sequential Monte Carlo methods are used in neural network training. It is used to simulate the approximate sampling distribution of jointly distributed and multivariate test statistics, in Bayesian logistic regression, and genetic algorithms. Computation speed is the greatest barrier in implementing MC methods. Parallel MC libraries are available to speed up the computation.

3.13 Contingency Tables

Definition 3.18 Contingency table is a statistical tool to analyze two or more categorical data, each with at least 2 groups or classes.

Table 3.4: Auto accident break-up among left and right handed drivers

	left handed	right handed	total
Non-fatal	8	141	149
Fatal	3	23	26
Total	11	164	175

It conveniently summarises information (counts, frequencies, etc) on nominal or ordinal data of two (or more) distinct populations or groups into rows and columns of a two-way frequency table with an intent to test the independence of the discrete random variables represented by the rows and columns[16]. The cell entries are most often the frequencies for specific combinations of row and column values. If there are two binary variables, we get a 2x2 contingency table. For instance, the auto accident frequencies of Left & Right-handed Male & Female drivers can be arranged into a 2x2 contingency table. Pearson's chi-square test for homogeneity is used to decide if the frequency counts are identically distributed across the populations. Observed (O_i) and expected (E_i) frequencies are used to compute the χ^2 statistic, which is a ratio measure in NOIR typology. The expected frequency E_i of a cell is found by multiplying the total row and column count for that position and dividing it by the grand total. Mathematically, it is defined as $\chi^2 = \sum_{i=1}^{n}(O_i - E_i)^2/E_i$.

The chi-square statistic for a 2x2 contingency table, if the observed frequencies are a,b,c,d can be found as q = $N(ad - bc)^2/[(a+b)(a+c)(b+d)(c+d)]$. As this expression is symmetric in (a,b) & (c,d) (the expression remains the same when a & c are swapped and simultaneously b & d are swapped) it is immaterial which attribute is used as row or column. The numerator of q being a square is always positive. Hence we will use the right tail of the χ^2 distribution to test the hypothesis. When one or more cell frequencies are less than 5, a correction is applied to the numerator as $N(|ad - bc| - N/2)^2$. A table of the chi-square distribution with degrees of freedom (r-1)*(s-1), where r is the number of rows and s is the number of columns, is consulted to determine the critical values.

Consider the 2x2 data in table 3.4 on auto accidents among left-handed and right-handed people. Our hypothesis is that the fatality of accidents is independent of left- or right-handedness. The degrees of freedom is (2-1)x(2-1) = 1. Expected frequencies are 149x11/175=9.3657, 149x164/175=139.634, 11x26/175=1.634 and 164x26/175=24.3657. We calculate the χ^2 using the above formula as 0.199149 +0.013357+1.141279+0.076549=1.430335. This is compared with the tabulated value of χ^2 (see www.math.upenn.edu/~mgersten/chi_square.html, or www.itl.nist.gov/div898/handbook/eda/section3/eda3674.htm) with 1 degrees of freedom. As 1.43 lies between 1.3233 and 2.7055, with significance levels .25 and .10, we reject the hypothesis at level .25.

3.14   Exercises

1. Mark as true or false:

    a) Range can be computed for any type of data

    b) Arithmetic mean of integer data is always an integer

---

[16]Other uses include interaction and goodness of fit testing, comparing various models that best explain the data distribution

   c) Probability of an impossible event is zero

   d) Mode is the most appropriate measure for nominal data

   e) Median can be approximated graphically

   f) Change of scale is not applicable to find geometric mean

   g) Inter-quartile range is a better spread measure than standard deviation

   h) Range of data is preserved by a change of origin transformation

2. What are the different approaches to probability theory?

3. Explain with examples four different ways to express probability.

4. Identify the population in each of the following:

   1) A study to find out the most commonly misspelled words in a language

   2) A study to correlate typing mistakes among left-handed and right-handed people

   3) A study to find out fraudulent money transactions at a bank.

5. If all values in a sample are the same constant (say c), what is the standard deviation? What is the mean? Does the mode exist?

6. The marks of 8 students in a test are 65, 82, 71, 69, 93, 77, 86 and 89. Transform the data using theorem 3.5 (page 3-29) to the range 0 to 10, and using corollary 3.7 (page 3-30) to the range [-3,+3].

7. When can the mean absolute deviation from the mean become zero?

8. Let A,B,C denote the events that Bread, Egg and Meat has been purchased. Represent each of the following using Venn diagram. a) All customers who have all of them purchased. b) Customers who have purchased bread and meat but not egg. c) customers who have purchased meat and egg but not bread.

9. If the probabilities that a customer purchases a computer book is 1/5, an adventure book is 3/10 and probability that both are purchased in one visit is 1/19, what is the probability that a customer purchases either of them?

10. Prove that the mean of $x_1 = (p+q)/2$ and $x_2 = (p-q)/2$ depends only on p, and the variance depends only on q.

11. If the license plates of a car in a state has 3 letters and 4 numbers, how many total license plates are possible (i) if repetition is allowed, (ii) repetition not allowed in digits?, (iii) repetition not allowed in both digits and letters?

12. The arithmetic mean of 15 customer orders is 54. Find the new (combined) arithmetic mean in each of the following situations: (i) a new order for amount 70 is received, (ii) an order for amount 38 is canceled, (iii) 3 new orders with mean = 56 is received.

13. Find the median and mode of the following data: (i) X=[5,2,3,2,5,5], (ii) Y=[70, 65, 90, 70, 80, 75, 95]

14. Using the scaling transformation (Lemma 4), find the GM of the following data (i) X=[50, 65, 72, 84, 90, 98], (ii) Y=[22000, 25000, 34000, 40000, 63000]

15. Find the mean absolute deviation from the mean and median for the above data.

16. Find the variance of the above data using theorem 3.

17. Find the inter-quartile range IQR $= (Q_3 - Q_1)$ for the data in example 3.16 (page 3-22). Prove that the change of origin transformation preserves the IQR.

18. Mark as True or False:
    a) Probabilities of conditional events can exceed one
    b) The probabilities of outcomes of a random experiment must all add up to one

19. The change of origin transformation is meaningful for which of the means?
    A) arithmetic mean B) geometric mean C) harmonic mean D) all of them

20. Consider a communication source S that needs to send n messages $(m_1, m_2, \cdots, m_n)$ over a noisy data channel. Let the probability that message m_i is delivered to destination without error be p_i. If the messages are sent one after another (sequentially), and are independent (each message is sent independent of previous message), what is the entropy of the receiver? What base will you use for the logarithm?

21. If the Arithmetic and Geometric means of 2 numbers are equal, prove that the numbers themselves are equal

22. If the Arithmetic and Harmonic means of 2 numbers are equal, prove that the numbers themselves are equal

23. If the Geometric and Harmonic means of 2 numbers are equal, prove that the numbers themselves are equal

24. Prove that the change of origin transformation preserves the sample range, and the change of scale transformation Y=cX gives Range(Y) $=$ c* Range(X). Does this relationship hold for quartile deviation, and mean absolute deviation from the mean?

25. A car manufacturer offers 4 types of cars (petrol, diesel, hybrid, and purely electric), each in 5 different exterior colours. Total how many choices does a customer have in selecting a car?

26. Prove that a sample in any finite range can be transformed to the range [-k,+k] using $y_i = \frac{2k}{R}(x - \overline{x}_{1n})$ where $\overline{x}_{1n} = (x_{min} + x_{max})/2$ is the mean of the minimum and maximum of the sample, and R=$(x_{max} - x_{min})$ is the range.

27. What are the two popular types of simulation. In what situations are they used?

28. What are some data mining applications of contingency tables? What statistic is used for contingency table analysis?

3.14.0.2 References

[AY01] Aggarwal, C. C., Yu, P. S. (2001). Outlier detection for high dimensional data, in Proceedings of the 27th ACM SIGMOD international conference on management of data (SIGMOD01), 37-46.

[AL88] Anselin, L. (1988). *Spatial econometrics: methods and models*, Kluwer, Netherlands.

[BL94] Barnett, V., Lewis, T. (1994). *Outliers in statistical data*, Wiley, NY.

[BJ77] Bhattacharya,G., Johnson R.A. (1977). *Statistics - Concepts and Methods*, Wiley, NY.

[BC00] Battiato, S., Cantone, D., Catalano, D., Cincotti, G., Hofri, M. (2000). An efficient algorithm for the approximate median selection problem, 4th Italian conference, CIAC, Rome: Lecture Notes in CS, vol 1767, 226-240, Springer.

[BI05] Ben-Gal, I. (2005). Outlier detection (in *Data mining and knowledge discovery handbook*, Maimon O., Rokach, L. eds.), 131-146, Springer, New York.

[BG58] Box,G. E. P., Muller,M.E.(1958). A note on the generation of random normal deviates, *Annals of mathematical statistics*, 29(2), 610-611.

[CR95] Chattamvelli, R. (1995). A note on the noncentral beta distribution function, *The American statistician*, 49, 231-234.

[CZ06] Chundi, P., Zhang, R., Castellanos, M. (2006). Entropy based measure functions for analyzing time stamped documents, Proc of SIAM international conference, www.siam.org/ meetings/sdm06/

[CN93] Cressie, N. (1993). *Statistics for spatial data*, Wiley, NY.

[CH04] Cua, G. B., Heaton, T. H.(2004). The Virtual Seismologist (VS) method: a Bayesian approach to seismic early warning, American Geophysical Union, Fall meeting 2004, abstract #S21A-0253, adsabs.harvard.edu/abs/2004AGUFM.S21A0253C

[FV08] Forkman,J., Verrill, S. (2008). The distribution of McKay's approximation for the coefficient of variation, *Statistics and probability letters*, 78(1), 10-14 (pub-epsilon.slu.se/317 /01/Forkman_Verrill_080610.pdf)

[GS95] Glad, I., Sebastiani, G. (1995). A Bayesian approach to synthetic magnetic resonance imaging, *Biometrika*, 82, 237-250.

[GT08] Galadima, D.J., Tsaku, N. (2008). The geometric mean model in Finance, *Science world journal*, 3(4), 5-7 (www.scienceworldjournal.org/article/viewFile/4824/3305).

[KN98] Knorr, E., Ng, R. (1998). Algorithms for mining distance-based outliers in large datasets, Proc. of 24th VLDB conference, 392-403, Morgan Kaufman.

[MG92] McLachlan, G.J. (1992). *Discriminant analysis and statistical pattern recognition*, Wiley, NY.

[MP03] Mercer, P.R. (2003). Refined arithmetic, geometric and harmonic mean inequalities, *Rocky mountain journal of mathematics*,33(4),1459-1464 (rmmc.asu.edu/abstracts/rmj/ vol33-4/mercpag1.pdf).

[SC15] Shanmugam, R., Chattamvelli, R. (2015). *Statistics for Scientists and Engineers*, John Wiley, New York.

[SC48] Shannon, C. E. (1948). A mathematical theory of communication (parts I and II), *Bell System Technical Journal*, XXVII:379-423.

[HP05] von Hippel, P.T. (2005). Mean, median, and skew: Correcting a textbook rule, *Journal of statistics education*, 13(2), online: www.amstat.org/publications/jse/v13n2/vonhippel. html

4
Datawarehousing

Chapter objectives

- Understand datawarehouses and datamarts

- Discuss the goals of datawarehousing

- Distinguish datawarehouses and large databases

- Compare datawarehouses and datamarts

- Understand Metadata catalogs

- Review datawarehouse architectures

- Discuss ETL tools

- Describe data cleansing

- Understand Distributed datawarehouses

- Virtual and web-based datawarehouses

- Describe datawarehouse indexing

- Discuss security in datawarehousing

4.1 The Datawarehouse

Definition 4.1 The datawarehouse (DW) is a 'read-only' repository of cleansed, consolidated, non-volatile and integrated historical data used for cost-effective business intelligence and managerial decision making.

It is used by knowledge workers (executives, managers, business and financial analysts), support users, and clients for managerial decision making, and for ad-hoc reporting (see figure 1). All DW users need not be decision makers, though. They may include analysts, users generating reports, post-decision monitoring people, operational personnel, suppliers and clients. Hence some enterprises maintain data unrelated to decision making in the DW to facilitate reporting and specific analysis.

Data warehousing is the process of accumulating operational and legacy data from heterogeneous data sources, transforming, cleansing, normalising and filtering it for creation and incremental uploads into the DW. Querying, reporting, analysis and visualisation tools access this data (see figure 4.1). A DW administrator, who updates the data on a periodic basis, maintains it. DW data are not updated directly by users, but by Extract - Transform - Load (ETL) software on an *ad-hoc* basis. Since data updates happen at different points in time, the data available for querying may not be *current* (in database terminology, *current* means *up-to-the-second*). OnLine Analytical Processing (OLAP) software is used to mine the data in datawarehouses and datamarts (DM) (see §4.2). It is usually built at the company or organisation level, and provides a stable view over a reasonable period of time. Some large companies maintain multiple DW or datamarts for functionally independent activities (purchasing and contracting; marketing, sales, and customer support etc).

The aim in building a DW is to have a global (company-wide) view of historical data maintained with a business context[1]. A DW is not just a dump of the entire data maintained by an organisation, but it is an asset to facilitate enterprise-wide decisions in *the most cost-effective and speedy way*. Datawarehouses are also used to mine hitherto unknown patterns and trends from large historical data. It contains aggregates, and metadata (§4.1.5) in addition to time stamped historical data that optimise read operations (retrievals) rather than updates (Data updates can happen at any time, and at any level (including atomic levels) in OLTP systems, whereas it is far too infrequent in a DW environment).

Only a subset of the data may be essential for strategic decision making, which can be decided upon by consulting with knowledge workers and decision makers. Once created, the data in the DW remains static until incremental changes occurring in source data are propagated through the DW/DM on a periodic basis.

4.1.1 Goals of Data Warehousing

Increasing business intelligence (BI) and improving the strategic business decision making (SBDM) capabilities are the primary goals of enterprise datawarehousing. It is also used to make ad-hoc reports[2]. It finds applications in many other fields like biometrics, genetics, pharmacology, medicine, finance, Customer Resource Management (CRM) and Geographic Information Systems (GIS). Newer applications areas are also reported in the literature [BS97], [WJ05]. The main goals of datawarehousing are:

- Create a single repository of historical data
  Most datawarehouses contain massive amounts of data extracted from heterogeneous sources. These time-stamped historical data have various levels of utility in decision making. Data are maintained at a single location in centralised datawarehouses[3]. In distributed datawarehouses it is maintained (usually at different geographical locations) as independent repositories that are linked together using a network topology (usually the star topology). From the user's perspective, such datawarehouses can be regarded as a single logical repository, which has many advantages from various business angles.

- Share data for strategic decision making
  Datawarehouses provide transparent access to shared data through middleware programs.

---

[1] Although DWs can be built for many other non-business contexts like medical DW, chemical DW, geographical DW, genetics DW etc, we will assume a business DW unless otherwise stated

[2] ad-hoc queries obtain information from a DW without a predefined report format or desired output form.

[3] Unless otherwise stated, a datawarehouse will mean a centralised datawarehouse.

Figure 4.1: Data mining tasks: Analysis, visualisation, and reports generation.

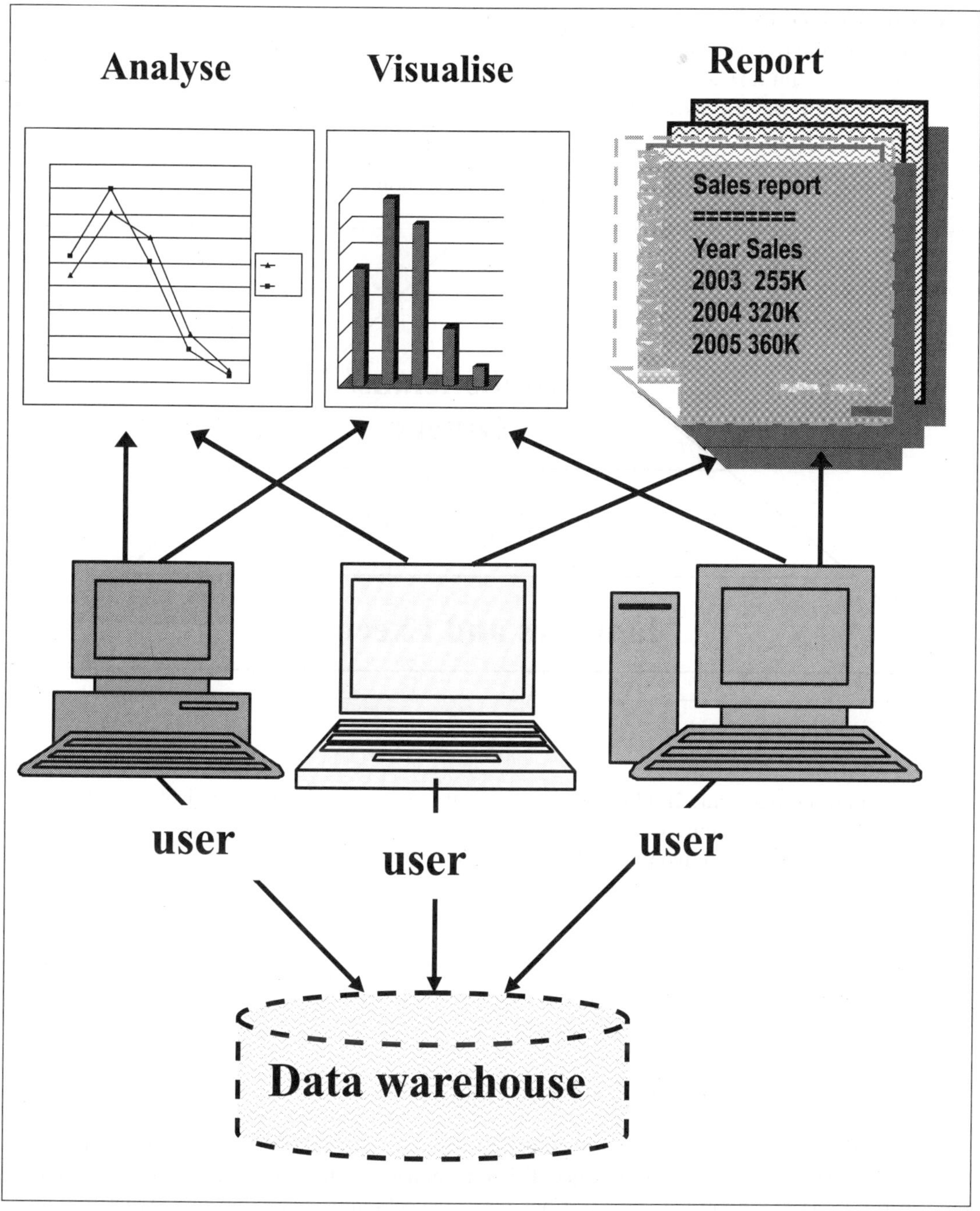

Knowledge workers, analysts and executives simultaneously access data in a DW (see figure

4.2). Thus every user gets a consistent *single window* view of the shared data, which could lead to uniform decisions across an organisation.

- Separate security policies
  Because different database systems, spreadsheets and applications have separate security features, maintaining a consistent and uniform security policy in an OnLine Transaction Processing (OLTP) environment is often cumbersome. Although the users of OLTP and

Figure 4.2: Major datawarehouse users.

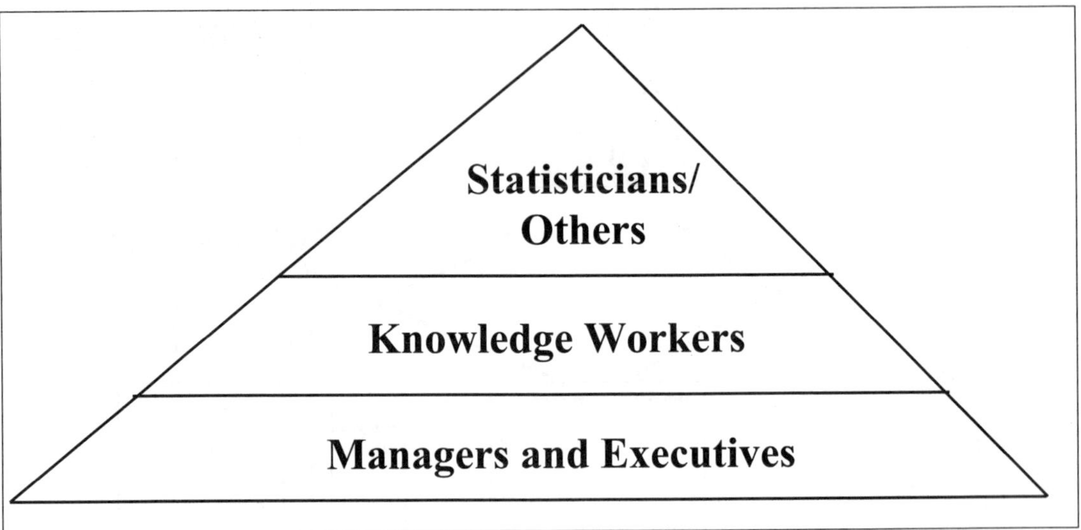

DW may be different, implementing a security policy in the DW environment is simple and flexible. The security can be implemented at various levels of detail in DW/DM. For instance, sensitive financial data that must be protected from malicious users may be kept in datamarts (§4.2), and access restricted through a combination of security policies.

- Provide easy scalability
  There are four types of datawarehouses – virtual DW, centralised DW, distributed DW and hybrid DW. The datawarehouse architecture permits easy scalability in terms of user groups and data (user scalability is easily possible in OLTP systems as well).

4.1.2   Advantages of Data Warehousing

Information is one of the key assets for cutting-edge success in technology-driven businesses. For most enterprises with a long presence, this information is distributed in a multitude of data repositories in different software applications (spreadsheets, databases, flat and structured files). Some of them with decades of existence[4] have migrated through various hardware and software

---

[4]According to estimates, more than 70% of the data of large corporates with decades of presence are still on mainframe computers.

systems, a variety of generational databases, in addition to data repositories in custom-built in-house applications. Integrating all these scattered enterprise data that are potentially useful in business decision making will improve the morale, and could increase the profit in a record time. There are other motives for datawarehousing as well. For instance, a medical datawarehouse could be used for classifications, cluster detection, causations and hypothesis verification. The major advantages of DW/DM are discussed below.

- Data quality and consistency
  Quality of data in a DW is of utmost importance, because sensitive business decisions that could involve on the order of millions of dollars could be based upon the results mined using OLAP. Since data in a DW are read-only, and are captured through ETL programs, high levels of data quality can be maintained. The data staging phase (see fig 4.4.1) ensures highest possible quality at atomic level. This is not the case in OLTP systems that often update the data continuously. If ETL process is standardised for the DW and each data marts, long-term consistency is also possible with minimal effort. The integrity of data is maintained at the highest level. A data dictionary that captures the source of the data, cleansing rules and conditions, data ranges etc can be kept for later reference.

- Maintain cutting-edge business advantage
  Most companies migrate to datawarehousing to leverage accumulated knowledge for cutting-edge business competition. Important factors that lead to slumps in key business measures can be identified on an ongoing basis with the DW. It can easily be tuned for fast decision making, although the boundary between them is vague.

- Restructure the architecture
  Because the data are centrally located, it can easily be restructured as and when required. For example, a star-schema based architecture (§5.4) may need large storage space due to de-normalised tables.[5] This may be converted into snowflake or constellation schemas, which normalises the dimension tables thereby reducing storage space.

> Database normalisation simplifies the design of databases and minimises redundant data storage. It improves the efficiency of database operations, reduces storage space requirements for data, database views and table indexes, and improves throughput by reducing unwanted operations. There are four normal forms for relational database tables. Among these, three are popular (1st, 2nd and 3rd level normal forms). These are built hierarchically from bottom-up. Hence a table in k^{th} normal form is automatically in $(k-1)^{th}$ normal form, but not *vice-versa*.

- Scale easily
  As mentioned above, the data are easily scalable in terms of users and locations. Most of the OLAP mining processes occur at the client side. Hence only system resources limit user scalability. Location scalability is pertinent to distributed datawarehouses. See §4.5 for details.

- Ad-hoc reporting
  Data warehouses are being used on a periodic basis by corporates to monitor key business performance indicators, and unexpected deviations in business. The period of this analysis

---

[5]normalised databases may require more indirections due to extra links introduced. This may in turn increase the execution time of some operations.

will vary greatly among the companies and are dependent upon many internal and external factors (market fluctuations, business process changes, customer attrition rates, foreign exchange rates, etc). Reports are of different types – operational, tactical, and strategic reports. Operational reports support and enhance ongoing operations. Tactical reports monitor and report near-future business processes. Strategic reports monitor long-term business interests and uses aggregation measures. A DW can be used to generate ad-hoc reports using summary/aggregation measures with links to detailed reports at various hierarchical levels of detail (eg: navigable or clickable reports on display devices). Using the DW for tactical reports is not recommended because the data may not be as current as the data in OLTP systems or Operational Data Store (ODS) (§4.1.4). The reports can be used as an information communication tool between the OLAP miner (who may not be the actual decision maker) and senior management who takes important business decisions.

- Other advantages
  Datawarehousing process may also identify redundant data storage and inefficiencies in OLTP systems, which can lead to new business rules or data consistency and cleansing procedures. The volatility of data and criticality of it are important in some decision making situations. Maintaining the entire DW as *current* may be costly. Hence highly volatile data are kept in smaller datamarts which are easy to update.

4.1.3   Datawarehouses vs Databases

A datawarehouse contains cleansed, time stamped, non-volatile historical data devoid of missing and out-of- range values that are organised around *subjects* rather than *transactions*. It typically contains millions or even billions of opportunistic data records. A very large database (VLDB), on the contrary, is a repository of structured data that are often kept in one or more database files/tables. It is characterised by a large size (of the order of terabytes) and workload (large number of concurrent users). The data may either be time-varying (purchasing behavior, expenditures etc) or static (demographic information like sex, religion, mother-tongue; head of household etc). Time-varying data collected over a sufficiently long period of time is the main focus in datawarehousing applications. The data in a VLDB are frequently updated to keep it up-to-date. This is usually done using query languages (eg: SQL) or custom-built applications. Data updates in datawarehouses are often limited to addition of new data records (obtained from ETL after the last time it was updated) and is called *incremental updates*. DW is mainly used by analysts and decision makers, whereas the VLDB users can be anyone involved in querying, updating or maintaining it. Databases can also be updated by automatic data collection devices (eg: magnetic ink scanners, optical character recognition devices) or sensors (optical, pressure, and temperature sensors). Data fields in a VLDB are amenable to data transformations. For example, if the last name, first name and middle initial of a customer are stored in 3 separate fields, a query can concatenate (join together) these fields when a report is produced. Data transformations are used at the time of DW creation, rather than at data mining or reporting stage. Datawarehouses often contain pre-aggregated data, which are loaded into OLAP cubes at the time of cube creation. Pre-aggregation considerably speeds-up the analysis, which is important in performance-critical online mining. Pre-aggregation may also be implemented at datacube level.

4.1.4 Operational Data Stores (ODS)

An ODS is a data repository that is deployed for tactical reasons and contains current, subject-oriented and denormalised data. The ODS consolidates operational data from multiple sources, and has limited update capabilities (add, change, delete). It has a dual purpose in enterprises – 1) to integrate data spread across a multitude of base files into a single and current repository, (2) to serve as a single platform for feeding DW/DM. There are two ways to populate or update data in a DW/DM from ODS. In the *pull approach*, applications running on DW server pulls the data from ODS, using an interface on the ODS schemas. In the *push approach*, applications on the ODS server pushes the data to a DW/DM using an interface on the DW structures.

In the distant past, the ODS and DW were distinct entities. Differences between the two are thinning, and they are converging to be the same thing. Advances in hardware, data communication and software technologies have made it possible to keep DW/DM almost current through data synchronisation with OLTP systems. In addition, limited data updates are possible with user interfaces, making the DW repositories as 'read-write'. All these have thinned the boundary between the ODS and the DW.

4.1.5 Metadata Catalogs

Metadata catalogs have their origin in the 1970's as *data dictionaries* and *schemas*. Data dictionary is similar to metadata and defines the organisation of the databases, descriptions of the structure of tables, views, and the data it contains. They are used by data administration tools to provide insight into the nature of the data, and for support operations. They are also used by incremental-update programs to merge newly acquired data into existing DW/DMs. Relationships among the tables in a database from an application point of view are called the database schema. Large corporates have dozens of databases and tables. Business applications typically access a subset of the tables. Hence there can exist different schema designs for different applications. When these OLTP data are loaded into a DW, there could exist multiple rules and dependencies that are denormalised. Keeping all these information in separate metadata repositories may create housekeeping and performance problems.

Metadata can be stored internally in the respective databases/DWs/DMs, stored in separate files, or in *metadatamarts* (Metadatamarts are micro datamarts that contain structured metadata in a shareable form for seamless operation). Metadata catalogs contain data about metadata files, information about their content, structure, dependencies, business rules, and administrative information. It is a single data repository that captures a multitude of metadata and provides a uniform, consistent and timely access mechanism by integrating multiple repositories [AM97], [SA04], [SE06], [SM99]. They are extremely useful in integrating distributed data warehouses (§4.5). Virtual datawarehouses (§4.5.1) also rely heavily upon metadata catalogs to provide a uniform view of data kept in various data repositories (databases, spreadsheets, custom files, etc), as if they were all kept in a real DW.

The information in metadata files is updated after successful data uploads on a periodic basis. Temporal data like last updating date and time, spatial data like file location, file size etc can change over time. Corporate data warehouses are hosted on high-end parallel computers (symmetric multiprocessors) with *simultaneous update* capabilities. Hence keeping metadata separately may be advantageous in certain situations. Further discussions on metadata can be found in [SA04], [SE06].

4.1.6   The Datawarehousing Team

Datawarehousing is a tenuous and often lengthy process that takes many man months for planning, architecture, design, development, testing, deployment, and user training. There are many people involved in the lifecycle of datawarehouses. Below we give the chief players and their major roles and responsibilities.[6]

- **DW Architect**
  The datawarehouse architect/designer (Data Architect/OLAP Architect/ Business Objects Architect, Domain Expert) decides the overall structure of the DW/DM, including the schema to be used (star, snowflake, constellation), dimensionality modeling to be used, metadata structures and contents, data dictionaries, ETL and security architectures to be used, network topology (of DDW) to be used, etc.

- **DW Administrator**
  The DW administrator is responsible for software selection, general administration, and creation of indexes, views and metadata. In addition, they identify user groups, assign user ids and passwords, manage access permissions, decide security settings, perform load balancing, and ensure quality and uninterrupted connectivity, troubleshoot user and system problems, resolve conflicts arising out of partial data updates.

- **DW Manager**
  The DW manager manages DW implementation projects and translates business requirements into DW designs, manages business critical BI platform, does requirements capture and evaluation, defines implementation roadmaps, estimates project costs, builds and manages a DW team, maintains relationships with prospective users (executives, knowledge workers), manages other DW personnel, and oversees user training.

- **DW Programmer**
  The DW programmer/developer is responsible for general maintenance and support of DW initiative, and writes programs to cleanse, transform and load data (ETL)[7], writes data update programs, handles exceptions and errors, writes support programs, generates and updates reporting systems. In addition, they may transfer/download/merge DW and DM data, and externalise data captured by user operations.

- **DW Users**
  These are the people who query the DW/DM using OLAP tools. The intricacies of a DW (architectures, data hierarchies etc) are hidden from the users. Most of the DW users are managers and executives, followed by knowledge workers, statisticians, information seekers and researchers.

Some large enterprises further split the roles of the team members for more efficient management or implementation of the DW/DM. For instance, a data stager (ETL designer, ETL specialist) is responsible for the staging activities, and fact table / dimensional table staging operations. The data modeler (DW modeling specialist) works in close association with DW architect in designing the dimensional models, a DW quality assurance specialist ensures that strictest quality norms are met during ETL, updation and data transfer operations. A DW analyst is a 'generic' OLAP user who analyses the data. A DW testing team conducts software

---

[6]Some of the roles mentioned in this section can overlap or are exchangeable.

[7]Most of the DW softwares have in-built capabilities for ETL and periodic updates, thereby relieving the DW programmers of this task.

tests. A DW trainer trains the users and brings them up-to-speed in data analysis. A DW technical support specialist provides overall consultancy in technical aspects.

4.1.7 Datawarehouse Architecture

An enterprise architecture describes how data are stored, managed and accessed in complex enterprise-wide information systems. It gives better insight into:
(1) persistent data storage structures, (2) data access mechanism by components and applications, (3) data interfaces by external systems, (4) information needs and application layers.

Figure 4.3: General architectures.

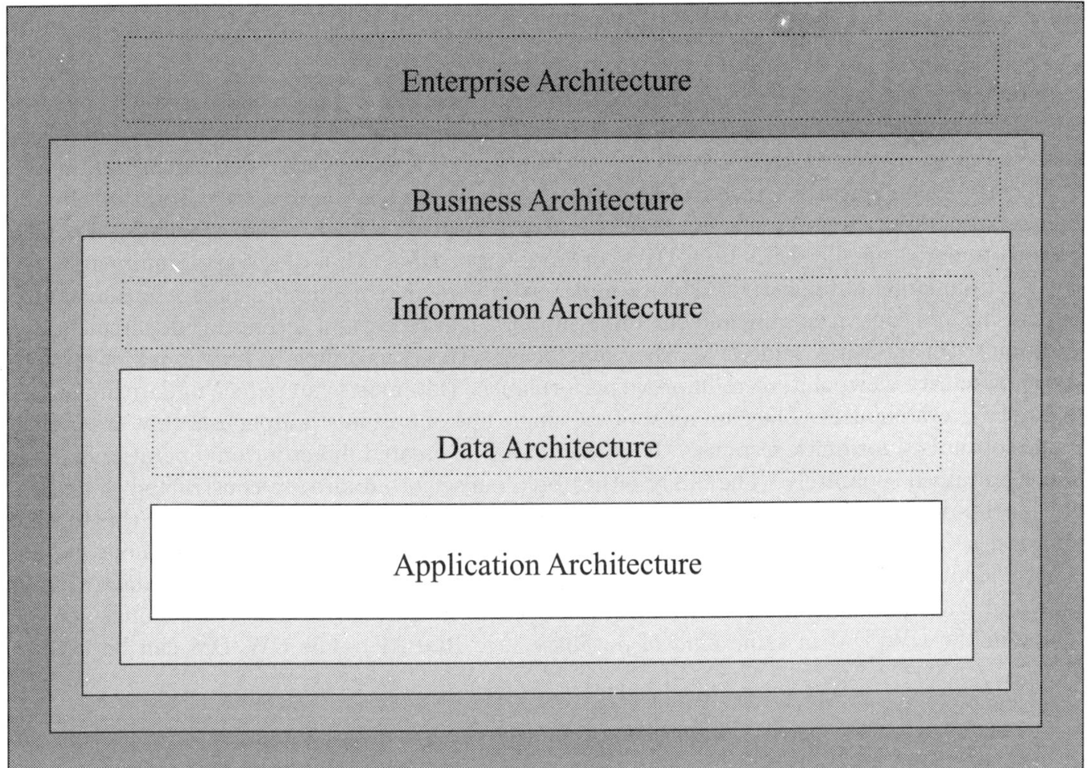

Definition 4.2 A DW architecture gives a high-level view of data organisation, and data access mechanisms in a datawarehouse.

It documents the overall framework for detailed design, development and inter-operability. Its common elements or building blocks are ETL tools (§4.3.1) and data sources, the DW /DMs, OLAP servers, metadata, and optional administration tools. Depending upon the inclusion of an ODS and presence of datamart(s), many architectures exist (see figure 4.3). An architecture with a multi-dimensional datawarehouse and independent datamarts is shown in figure 4.5.

The figure 4.4 gives an architecture with an additional ODS layer behind the datawarehouse. There are two types of architectures - physical and logical architecture. Physical architecture describes the structure and interconnection of tangible objects that are part of datawarehousing, and encompasses the database schema. A logical architecture is a blueprint of the entire data in datawarehouse(s) and datamart(s), irrespective of environments, storage technologies and softwares. These may sit on top of a technical architecture that describe the hardware platforms, network structures, application servers etc.

The logical architecture has 3 sub-layers – datawarehouse layer, business layer and application layer. The datawarehouse layer consists of operational data sources (databases, spreadsheets, text files, application data, external data), data staging, metadata repository and the datawarehouse and optional datamarts.

4.2 Data Marts

A datamart is a subset (of column attributes and/or rows) of a datawarehouse designed to meet specific requirements of user groups. They are more secure, and less costly to build and operate than the massively catalogued DW. They may be tied to a parent DW (dependent datamart), or operate independently (independent datamarts). Datamarts are created from an already functional datawarehouse in the top-down approach. Feature reduction and case reduction may be used to derive smaller DMs from DWs, or from larger DMs (Mini-datamarts (minimarts) are created from larger datamart(s)). For example, a company may maintain separate datamarts for purchasing and contracts, sales and customer support, marketing and accounting departments, in addition to an enterprise wide DW. They may be structured according to functionality, proximity, and security viewpoint, or to improve performance. Datamarts can reduce update anomalies in large datawarehouses. They are most often 'single line of business' warehouses that are performance optimised for quick response. They could also be created directly from operational data, and maintained separately. They may either be a subset of an already constructed datawarehouse (top-down approach), or a precursor to a full-fledged future datawarehouse (bottom-up approach). The data formats, architecture and organisation of data, retrieval methods etc are exactly identical for datamarts and datawarehouses. They could contain summary data at finer levels, in addition to detailed data. If it is derived from a datawarehouse, it is convenient to maintain the DW/DM in same kind of database (eg: ROLAP). The DW/DM can be used in

Table 4.1: Comparison of datawarehouses and datamarts

datawarehouse	datamart
enterprise-wide data	department or group-based, focused data
very large in size	relatively small
open, less secure data	can be highly secured
denormalised, not performance optimised	optimised for performance
virtual DWs mapped to databases	virtual datamart can map to parent DW

conjunction for querying (provided they are not proper subsets) as shown in figure 4.5. If the data rapidly changes over time, a dependent datamart may maintain incremental updates that are periodically merged with the parent datawarehouse. As an example, data pertaining to new customers acquired after last update can be used as incremental updates. Data updating algorithms can work faster with these smaller DMs, thereby reducing processing time. The structure of data in a DW may also change over time. For instance, addition of new products and services, new features to existing products, new functionality requirements, compliance to new standards etc could all necessitate structural changes in DW/DM. In such situations, it is much faster to change the structure of a DM rather than the monstrous DW. Integrated datamarts contain separate physical datamarts that share information. In addition, the DMs can form a hierarchy so that OLAP tools can access data in multiple DMs.

Datamarts can be categorised using the data they contain (atomic datamarts and aggregated datamarts), processes they perform (ad-hoc querying, reporting, analysis, visualisation (§4.1)) or the purpose for which they are maintained (prediction, post-decision monitoring).

Atomic datamarts contain multidimensional data at the lowest level of detail, usually in a star schema. Aggregated datamarts contain aggregated data derived from atomic datamarts, parent datawarehouse or created directly during data-staging. Both of them need not purely contain data of specific types – atomic datamarts could contain aggregate data for enhancing query and *vice versa*. Another categorisation is as analytical, reporting and consolidation datamarts.

4.2.0.1 Advantages of Datamarts

- Smaller size
  A datawarehouse can be bloated with large amounts of unused or seldom used data. Incremental updates seldom check the data usage by OLAP tools, which lead to gradual size increases (this is true for data marts as well with frequent updates, that have been in use for a long time). Separating essential data for frequent querying into smaller datamarts can speed-up the queries, accelerate periodic updates, improve operational efficiency, and ease the maintenance of indexes, metadata etc. In addition, load balancing, quality assurance, and time criticality issues are easy to implement on datamarts.

- Focused functionality
  Datamarts in general have a much more narrow focus than datawarehouses. This can help in logically separating large OLAP user groups into more homogeneous subgroups. This allows administrators to implement stringent security policies for selected data marts, and a more liberal security policy for others. Expensive data mining software packages can be exclusively installed and operated on smaller user groups on focussed datamarts.

- Stringer security restrictions
  Datawarehouses may contain secure and open data. Security can be implemented at datawarehouse level, or various levels of datacubes (from coarser cube level to narrower cell levels). When the number of data miners is large, this type of security architecture could become cumbersome to administer and error prone. By separating sensitive data into datamarts, and associating access control at finer levels for user groups ensures strictest levels of security implementation at datamart level.

- Exploratory data
  A datamart can be constructed exclusively for exploratory data analysis (EDA). Supervised data mining models discussed in subsequent chapters assume a 'training set' and a 'test set' for building and validating the models. This is easy to implement using datamarts and minimarts.

- Data updates
  The DW being a complete full-fledged repository will naturally have more users than a datamart that is more focused. Hence data updates in datamarts may be more practical than updating directly in the datawarehouse. This is especially true for DWs being accessed by users in different hemi-spheres. Moreover, if a subset of data rapidly changes over time, it may be more convenient to capture it into a separate data mart rather than keep it along with non-temporal data in a DW.

- Noisy data
  If noise (impurity) is present in the data, it may be beneficial to separate it into a datamart. Various noise removal techniques can be tried and tested on this datamart, which could eventually be merged to the DW.

- Scalability
  New datamarts can easily be created and incorporated into existing datamarts. They are scalable in terms of users and data. Distributed data marts are used by companies with global business presence, and by organisations with collaborative research with shared data.

4.2.1   Building a Datawarehouse

Setting up a datawarehouse from scratch is a major investment in terms of hardware, software, middlewares and man-hours of work involved. Datawarehouse modeling is the process of gathering requirements, analysis and validation of the modeling process [IG03]. As in the case of VLDB, a roadmap is used to implement any datawarehouse project. This roadmap will give the sequence of activities, and time-line for completion. It also gives the details on the induction and effort utilisation of various DW personnel mentioned in §4.1.6, as well as the hardware and software acquisitions. Maintaining, and updating a DW on an ongoing basis is relatively inexpensive in terms of time and effort. Datawarehouses are often built per enterprise. Data of different qualities, inconsistent formats and various encoding are extracted from heterogeneous sources (databases, spreadsheets, flat files, application files, etc), and brought together into a single repository in a common format (see figure 4.4). Each DW is created using a layered approach in which each layer of data are derived from the previous layer. Data consistency, integrity, and conformance checks are performed in each layer. Building a DW could become quite complex when there are many dimension tables and fact tables involved (see below). The fact tables are usually kept in third or higher normal form. Star-schema dimension tables are kept in second normal form for fast query response. In a snowflake schema, the dimension tables are also normalised into third or higher normal form so as to have a typical relational database design. When new records (with unique keys) are added to the dimension tables, a corresponding record must be entered into the fact table to avoid the 'dangling pointer' problem (referential integrity in database terminology).

4.2.2   Building Datamarts

Datamarts are built using bottom-up or top-down approach. In the bottom-up approach, smaller prototypical data marts of historical data are built first, which are put to use for OLAP miners. These DMs are later combined to create the full-fledged DW. The top-down approach is analogous to the design of RDBMS using the water-fall approach (planning, requirements gathering and analysis, warehouse schema design, metadata design, data cleansing, loading and

Figure 4.4: A datawarehouse fed from ODS.

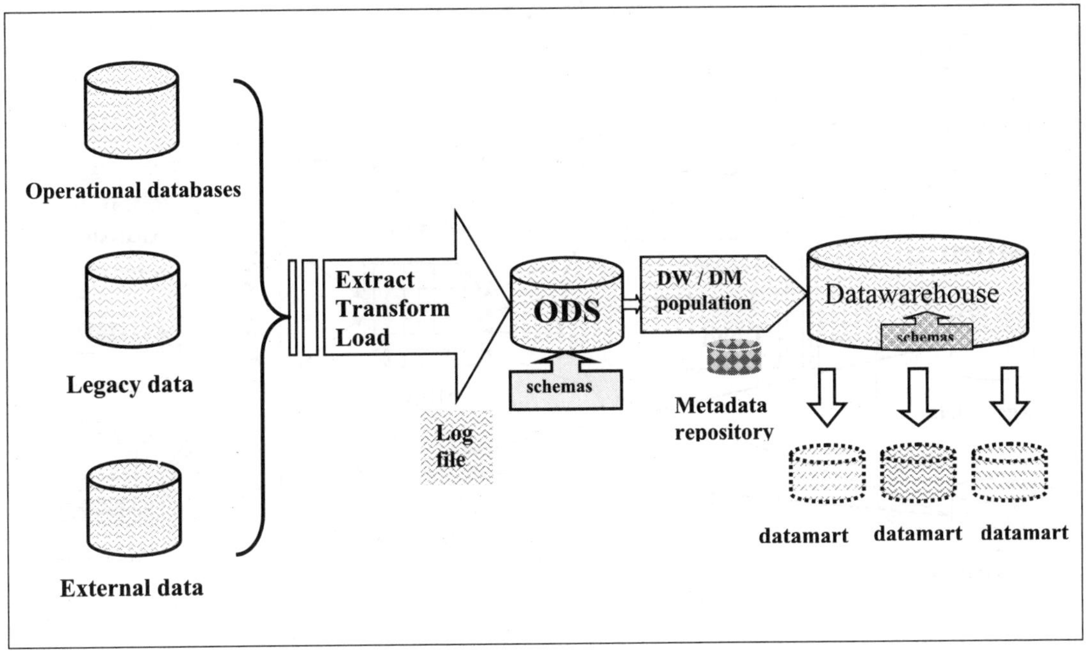

integrating, schema testing, integrity testing, load testing, deployment). In this approach, all requirements are captured and the schema and tables are formalised. Other structured approaches like the 'spiral approach', which decomposes a system in terms of smaller functional subsystems, may also be employed. The hybrid approach combines top-down and bottom-up approaches. Datawarehouses can exclude data that are irrelevant to decision making (ODS can contain irrelevant data). Relevancy of data is determined using the time stamp (of the data) or by the attribute values.

The temporal dimension allows analysts to model business trends of the past. Time series data are aggregated over appropriate intervals (weeks, months, quarters, etc), and could lose their significance over the passage of time. Similarly, payroll data of past employees, entertainment data etc may be irrelevant to decision making, if they are more than a few years old. Hence they may be time-filtered using a cutoff date or appropriate interval(s). Likewise, demographic attributes, contact information of customers, bill payment dates, business correspondence dates etc (column attributes) may not be relevant to decision making. Datamarts provide further filtering by separating out data (rows) and attributes (columns) for focused functionality or certain periods of time.

4.3 ETL (Extract, Transform, Load)

Definition 4.3 The process of extracting data from OLTP systems and external sources, cleansing, customising, checking for compatibility issues and loading it into a DW/DM is called ETL. Most often it migrates data from multiple data sources into one or more destinations (dataware-

Figure 4.5: Independent datamarts and a multi-dimensional datawarehouse.

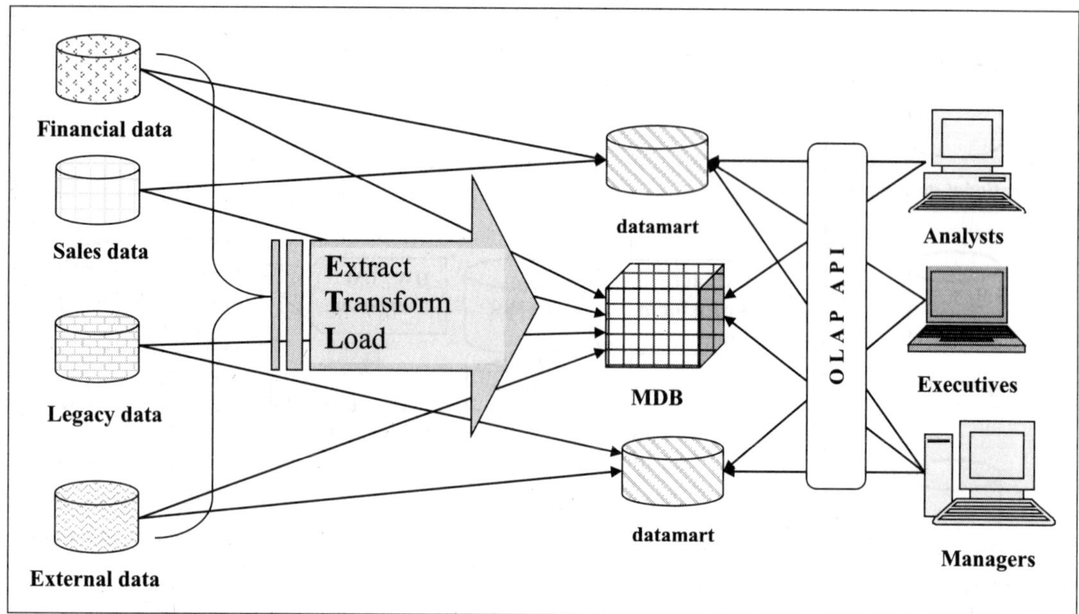

house or datamarts) after proper transformations (see §4.4.4).

It can work in various modes - bulk update mode, incremental update mode or trickle mode, which can be set by the administrator in a 'load window' (of some ETL software). The DW backend tools or special ETL tools are used for data loading. Incremental mode uses configuration files, status indicators, metadata catalogs etc to update a DW/datamart using newly extracted data (since last update) to make it *current*.

4.3.1   ETL Tools

Software programs for the ETL process are called ETL tools. Developer-friendly ETL tools simplify the whole process by eliminating complex coding, and by facilitating the loading from multiple sources.

They are also used in the incremental loading of delta of changes (since last upload) on a periodic basis. ETL tools are also available for semi-structured data types like text files in various formats [VR05]. A short list of ETL tools appears in table 4.2. Kettle ETL is a java API available from www.kettle.be/en/api.htm, See www.stylusstudio.com/etl/ for Stylus Studio tool, www.fileguru.com/apps/etl_tools_download for a list of ETL, testing and data uploading tools; www.etltool.com for a comparison of ETL tools, www.etltool.com/etlbooksandstuff.htm for books and articles related to ETL, and datawarehouse.ittoolbox.com for several discussion groups related to ETL.

4.4   Data Staging

This is a major pre-processing step in building a DW. In some organisations, it is also the most time consuming because much manual intervention is needed at various stages. Possible inconsistencies could exist in the formats, semantics, and content of various data kept in different systems. Data transformations will depend upon the type of data being extracted, and geographical locations from where data were collected. For instance, date and time formats and the use of comma instead of a period for separating whole and part of currency amounts differ among various countries[8]. Staging involves several steps described below.

Table 4.2: A short list of popular ETL Tools

Tool Name	Company Name
Informatica	Informatica Corporation
DT/Studio	Embarcadero Technologies
DataStage	IBM
Ab Initio	Ab Initio Software Corporation
Data Junction	Pervasive Software
Pentaho Kettle	Pentaho.com
Talend	talend.com
Oracle Warehouse Builder	Oracle Corporation
Microsoft SQL Server	Microsoft
Redbrick loader	IBM

4.4.1   Data Extraction

This is the first step in data staging. It identifies the data source (using the URN (Uniform Resource Name) if data are web-based, and using the fully qualified name for other data). It reads the raw data in various (low level) formats, and converts them into a uniform (high level) format for further transformation. Because the data resides in different files and forms, metadata proves to be useful in this step. Information on proprietary formats used by software vendors for their various products are also useful for seamless data extraction. Equipment malfunctions, program errors, out-of-range checking errors, data transmission errors, codepage differences etc can result in erroneous or missing data. For instance, if an optical scanner from a printed sheet with HB pencil marks captures the patient data in a hospital, scanner errors, incomplete and out-of-place pencil marks, etc could result in missing values.

4.4.2   Data Cleansing

The cleansing makes data more consistent using checks and validations. Pitfalls of defective data include wrong conclusions on mined data, profit loss, extra cleansing expenditures, loss of customer credibility, delays in deployment, and compliance problems. Cleansing stage converts the data into standard formats (by fixing data ranges, field sizes, encodings, precisions) so that they can be ordered, merged and aggregated. This step is crucial because the operational and

---

[8]Currency amounts use a comma in place of a (period) . in Germany and Switzerland, so that an amount 64,50 is actually 64.50

Figure 4.6: ETL architecture.

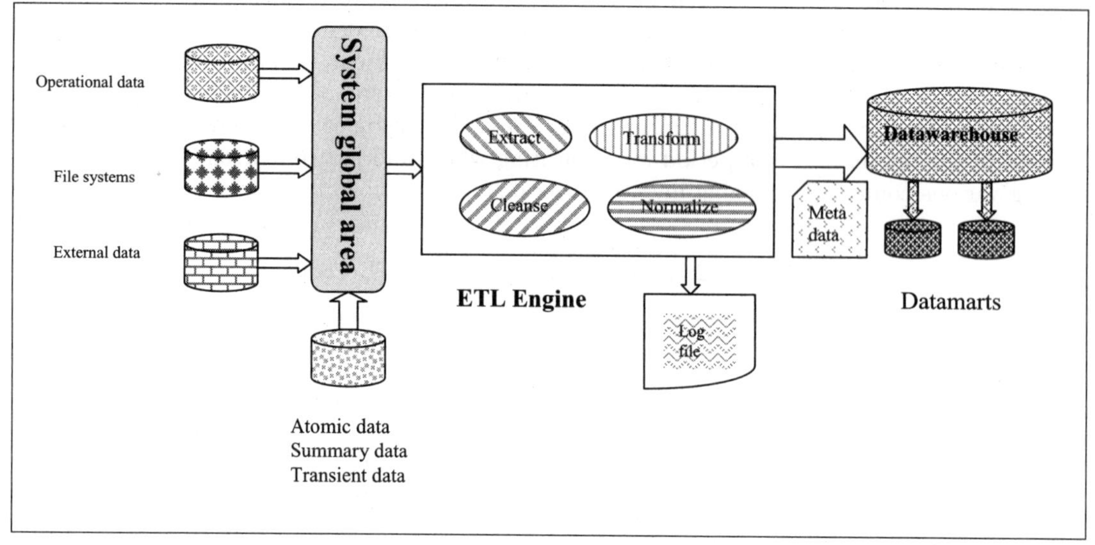

legacy systems, flat files etc are built by different people over long periods of time. The formats and storage of currency amounts in databases could differ from that kept in a spreadsheet, MS-Word table, or in an XML file. Another problem is data inconsistency and impurity. As an added advantage, this stage may reveal sources of erroneous and outlier data, and new business rules to be incorporated into OLTP systems.

Example 4.1 A business system codes the gender as ('M'=Male, 'F'=Female). The data captured by a CGI program over the web use ('m'=Male, 'f'=Female), a spreadsheet codes it as ('1'=Male, '0'=Female) and an XML file codes it as 'male' and 'female'. These codes also vary among different languages and countries[9]. A data cleansing rule can convert all of these data into a standard code (which is enterprise dependent, and fixed by the DW manager).

Example 4.2 The date and time are kept differently in various OLTP systems. Most databases have a 'date' data type, whereas spreadsheets represent date category under various formats of different lengths. Hence cleansing the data require thorough knowledge about the data source, its geographic location, formatting conventions, validity domains and its linkages to other data. The most frequent date transformation is to one of the standard formats "dd/mm/yy" or "mm/dd/yy". This is unsuitable in shared global DW/DM accessed from various countries with conflicting date formats. An alternative in such situations is to store the date as an integer number of days elapsed after an epoch date (say January 1, AD 1900), and use country specific wrappers for local format conversion at the application layer.

---

[9]'H'=Herren (Male), 'D'=Damen (Female) in German. But in several languages, both of them starts with the same letter as in (Monsieur, Madame) in French, (Signore, Signori) Italian, (Señor,Señorita) in Spanish.

4.4.3 Replacing Missing Values

Data values can remain missing due to many reasons. If data are collected over the web or through dialog forms, some of the values that are not entered by a user can remain missing. Most data capture methods check for essential information only, and assume default values for missing fields. Assigning default values to missing or noisy data may not always work correctly.

4.4.3.1 Direct Estimation Techniques

These techniques use statistical measures or heuristic to assign reasonable estimates to missing data. The arithmetic mean or mode can be used as a default value for numeric missing data. Consider an Alumni web site of a college that captures the data using a web form. Assume that some of the alumni skip the *professional experience (in years)* field, but the majority of them supplies it correctly. We can replace the missing data by the mode or average (arithmetic mean) professional experience of all the alumni in the database. The mode is the preferred choice for categorical data.

4.4.3.2 Class-based Estimation Techniques

Accuracy of estimation can be improved by identifying a class to which the entity whose missing value is to be estimated is identified. In the alumni example, we can replace missing professional experience by the average experience of his/her classmates. This is a more meaningful estimate as most alumni starts working immediately after graduation. Similarly, if the age of a diabetic patient is missing, we identify the individual as male or female, and estimate the age of that sub-category as the replacement value. If the data are categorical, the mode is used as the default value (mode of the most similar record with common traits). For example, if the marital status of a college student is missing, it can be replaced by 'S' for single, if most college students are unmarried. If the employment status of an undergraduate bank loan applicant is missing, it is replaced by the mode of (non-missing) the employment status of all undergraduate bank loan applicants. As mentioned above, the class-specific mode is more appropriate than the global mode for missing values.

4.4.3.3 Other Estimation Techniques

Sometimes we may be able to estimate a meaningful value using a regression model obtained from a correlated variable. Consider an online insurance system that calculates the probable premium of potential customers. If a customer does not supply the years of driving experience, we could estimate it from his/her age and family income because these are highly correlated with how many years the customer has been driving (as the minimum age for getting a driving license vary in different countries, a better estimate can be obtained using the predicates $(age - C_0)$ and $(income - C_1)$ where C_0 is the country-specific minimum age limit for driving, and C_1 is the minimum family income of people who owns a vehicle. Another example is the price of used vehicles advertised for sale. If this information is missing, we could estimate it using the age of the vehicle, and mileage (odometer reading) as they are highly correlated. As different vehicles depreciate differently, a class-specific feature mean (using manufacturer, make and model) may be used as the missing value. These are highly accurate when the correlation is high. Latent semantic indexing (chapter 14) based estimation techniques are available for non-numeric missing data. See [TJ08] for an estimation technique for missing data in decision trees.

Some software packages represent missing values by blanks, while some others have special constants. For instance, a spreadsheet stores blanks in empty cells, while databases store 'null' for text fields. Different domains, countries and industries have their own "best choices" for missing values. For example, a missing race is replaced by an 'H' in a Hindu majority country. Our assumption is that the number of missing values is relatively small. When there are large numbers of missing values, a statistical law based upon observed frequency distributions in various classes may have to be used.

4.4.4   Data Transformation

Chapter 3 discussed several popular mathematical data transformation techniques. Populating a datawarehouse with data captured from a multitude of sources may require other types of important transformations discussed below:

1. Simple transformations

   These transformations are applied at the attribute level. They may either be re-coding of values, range transformations, standardisations, changing the data types, sizes or formats. Re-coding '0' to Female and '1' to Male is a simple transformation. Changing the type for the number of days since an account is overdue from an unsigned integer to an integer (so that a -3 indicates that the account will be overdue in 3 more days time, and a $+$ve number like 10 indicates that it is already overdue for 10 days) is a data type transformation. Converting the standard time (in AM/PM format) to the range 0 to 24 is a range transformation.

2. Decimal Scaling

   The scaling transformation is applied to numeric data to reduce the range of values to a predefined interval. The popular interval choices are [0,1] and [-1,1] (see §3.10). If each data item is positive, dividing by the maximum possible value scales all transformed values to the [0,1] range. But if none of the data values are close to zero, the range may be narrower than [0,1] (or [-1,+1]). As an example, if the marks obtained by students are in the range 50 to 95, dividing by 95 scales the data to the range [.529,1.0].

   Example Given the data X=[2, 8, 10, 13, 17, 20], apply the scaling transformation to the range [0,1].
   Here the maximum is 20. Dividing each number by 20 gives X'=[.1, .4, .5, .65, .85, 1.0].

3. Min-max Normalisation

   This is a change of origin and scale transformation that converts the data to [0,1] range. It is mathematically represented as $z[i] = (x[i] - x_{min})/(x_{max} - x_{min})$ where x_{min} and x_{max} are the minimum and maximum of the values. See chapter 3 (§3.10) for details.

4. Smoothing

   This transformation is applicable to decimal data (with fractions). Smoothing adjusts the value to the nearest integer or boundary. In other words, $x_{new} = \lfloor x \rfloor$ if fractional part is less than .5 and $x_{new} = \lceil x \rceil = \lfloor x \rfloor + 1$ otherwise[10], where x is a real number.

   Example Consider the X=$\{80.5, 24.95, 56.30, 40.80\}$. The smoothing process rounds each element according to above formula to give X'=$\{81, 25, 56, 41\}$.

---

[10] $\lfloor x \rfloor$ denotes the biggest integer less than x, $\lceil x \rceil$ denotes the smallest integer greater than x

5. Scrubbing
   This is a cleansing technique to improve 'data hygiene' by fixing incorrectly formatted data and incomplete data, resolving synonym and alias, out-of-date, redundant, and incomplete data. These types of errors may result from (1) errors in data entry (typos, incompatible keyboards[11], spelling variations, equipment errors, data communication errors, (2) incompatible data coding standards (3) range checking errors, etc.

   As an example, suppose a database records a customer name as "William", and a spreadsheet captures it as "Bill". Such synonyms can be caught using a table lookup procedure. They are most often used for text data.

6. Normalisation
   There are many data normalisation techniques used during a DW creation. The min-max normalisation described above, is usually applied to non-negative data, but could easily be modified for any range. Other normalisation techniques are given in chapter 3 (§3.10).

7. Other Transformations
   Special transformations exist for images, graphics, videos etc. For instance, rotation, translation, elastic distortion, scaling (zooming or panning) are often employed for image alignment, image scaling and other normalisations. Rigid transformations that preserve the shapes, affine and nonlinear transformations which do not preserve the shape and angles are utilised for image and spatial data.

The scaling transformation can be expressed in matrix form as

$$\begin{bmatrix} x' \\ y' \end{bmatrix} = \begin{bmatrix} s_x & 0 \\ 0 & s_y \end{bmatrix} \begin{bmatrix} x \\ y \end{bmatrix}$$

where s_x and s_y are the scaling factors in x and y directions and the rotation transformation as:

$$\begin{bmatrix} x' \\ y' \end{bmatrix} = \begin{bmatrix} \cos(\theta) & -\sin(\theta) \\ \sin(\theta) & \cos(\theta) \end{bmatrix} \begin{bmatrix} x \\ y \end{bmatrix}$$

where θ is the counter-clockwise angle of rotation. An advantage of the matrix representation is that the reverse transformations can easily be obtained using matrix inverses.

The data transformations and cleansing rules vary for different data types. Some of these rules are available in software maintenance and user manuals, program documentations, etc. Data kept in database files (eg: MS Access file) may have input masks, validation rules, formats etc from which some of the cleansing rules can be extracted. Other rules can be obtained from domain experts, data standards, 'best practices' in industries, etc.

4.5   Distributed Datawarehouses

Most organisations maintain a monolithic two-tier client/server datawarehouse with all users on a local network, or company-wide intranets. This may not be suitable for some enterprises with business components in various geographical locations. Some enterprises have multiple procurement departments in various countries, production department in another country, sales and customer support departments in various other countries. Maintaining all these data at a central location may not be practical for seamless operation when each of these business components generate mammoth data.

---

[11]Keyboards made in different countries arrange the keys differently. Belgian keyboards swap the (X, Z) keys.

Definition 4.4 Distributed datawarehouses (DDW) are multiple, geographically separated repositories that are joined logically over a high speed communication channel, and transparently accessed from anywhere.

Accessing distributed data transparently as a single centralised system is called warehouse mediation. The architecture of mediation system can have a variety of forms. Each node in a DDW could either be a single DW or multiple datawarehouses and datamarts. The interconnections (fig 4.7) can be static (dedicated) or dynamic (on-demand). In a static interconnected DDW, the node connections are predetermined. For instance, it could be a star topology in which a central DW serves as the master node, and all other nodes communicate among themselves through the master node. User queries that cannot be satisfied by the local DW (as determined by metadata), will get routed through the master node to other nodes. The number of DDWs/datamarts involved, type of datawarehouses (OLAP, MOLAP, ROLAP etc) or communication topology are hidden from the user. Hence this method permits data scaling easily by adding new *location-independent* DW at any time and publishing the information (in the name service) to appropriate nodes (datamarts can also be removed dynamically from the network due to too much load, security violation or other reasons). The "secure core" service of DDW provides a reliable way of authentication of client requests in the distributed network, thereby eliminating third party malicious requests.

Another architecture uses a centralised DW and multiple distributed datamarts that communicate with the central datawarehouse. One of the primary motives in designing a DDW is the separation of data in such a way that the inter-location requests are minimal. In the vertical approach, the data are separated using functional decomposition (accounting, human resources, marketing, sales, etc may have their own datamarts), data proximity, or performance speed. The data proximity architecture follows a 'data nearness' approach in which the warehouse is created using the distributed nature of the organisation. An advantage of this approach is the ease of updating. Business co-operations, company mergers etc may also create a need for integrating geographically distributed datawarehouses. Instead of a flat structure (in which each of the warehouses have equal importance), one could also build up a distributed hyper-tier hierarchy of relationships among the data warehouses. The combined metadata of each of them may be replicated and kept on each location to facilitate query redirections. Partitioning algorithms can be developed using the access patterns to optimally restructure existing data or to introduce new virtual datamarts or materialised views. A virtual warehouse or datamart of limited scope may be created using the data in a DDW to facilitate inter-DW queries. The OLAP council has created benchmarks (ABP-1) to validate query results in a DDW.

4.5.0.1   Advantages of DDW

- High scalability
  Once operational, the DDW are highly scalable. New clusters can easily be created and integrated into existing architecture seamlessly. In addition, nodes may be removed dynamically without affecting the rest of the components.

- Replication and mirroring
  Datawarehouse replication is used to reduce communication bottlenecks, improve reliability and 24-hour availability. Replication can be local or remote, but mirroring is almost always local. The difference between the two is that mirrored data are always current (they are updated and synchronised after every operation that changes one of the peers).

- Integrated dataflows

Figure 4.7: Distributed datawarehouse architecture.

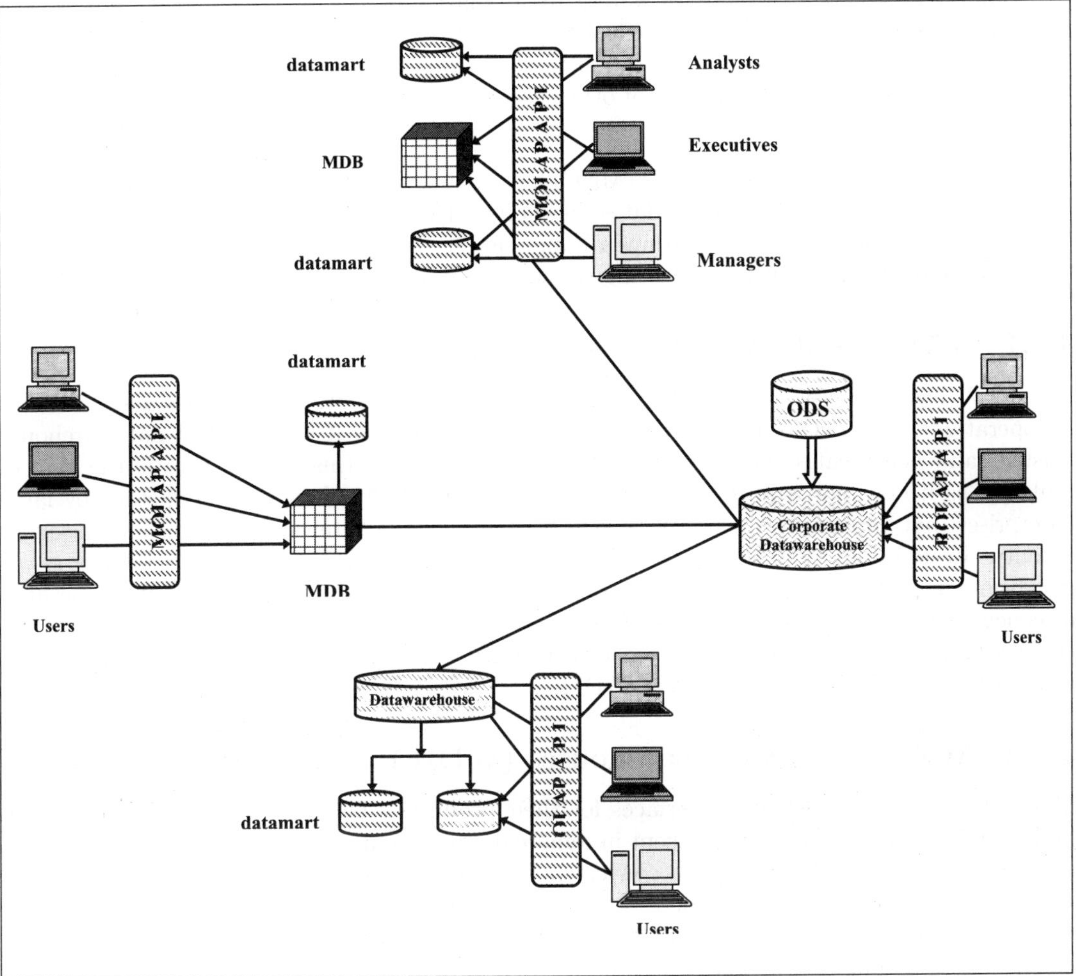

Distributed queries are validated using the metadata, before they are redirected to appropriate nodes. Hence data flows between DDW occur in an integrated and efficient manner.

- Workload balancing
  Although the DDWs are hosted on high-end servers, bandwidth problems can cause an imbalance on the workload. In addition, different peak loads at various nodes may result due to many reasons (geographical location, festive activities etc). The workload balancing technique is used among cooperating DDWs to off-load or redirect a part of the work to another DW. This results in rapid response times during peak mining activities. Workload balancing algorithms can improve query response times and reduce network congestion problems.

- Multiple mining software
  OLAP mining softwares have different qualities and interfacing capabilities with other data mining tools. Some of these softwares can also have different foreign language interfaces.[12] Hence users in different geographical locations may have personal preferences regarding the softwares to be used. In addition, text data may contain different codepages which may not be needed by majority of other users.

- All-time availability
  Several multi-national companies (MNC) have global presence with offices in different parts of the world. Since the data updates takes place during night times in one location, access by other locations may create update anomalies. To reduce such errors, the warehouse can be distributed without affecting the workflow at any of the locations.

4.5.1   Virtual Data Warehouses (VDW)

The VDW has no physical datawarehouse or datamart(s) present, but provides a set of views on operational data. OLAP tools can still mine the data, as if the physical datawarehouse existed (and hence called virtual). The biggest advantage of this architecture is the ease of building, as it does not require large investments. It may be used as a training stage by large enterprises that implement a real DW, and in situations where datawarehousing is an infrequent activity. In addition, multiple VDW can be deployed with different security architectures. A disadvantage of VDW is that it could degrade the performance of many OLTP systems because of concurrent access. Moreover, most VDW softwares have limited data cleansing capability, thereby limiting the conclusions that can be drawn from an analysis. Administrators use virtual warehouse catalogs to manage the VDW/DM.

4.5.2   Web-based Data Warehouses (WDW)

Most OLAP vendors offer web interfaces for their products. This has in turn resulted in web based retrieval and analysis of data kept in remote datawarehouses. An ETL tool to extract web data employs wrappers, parsers, and converters to extract markup data (HTML, XML, WML), email messages, data from discussion groups, news site data, archive data, spatial data and other types of embedded data (figure 4.8). The WDW provides global access to the enterprise data on a 24x7 basis from anywhere on the Internet. In addition, various natural language interfaces, keyboards in regional languages and touch-screens can be used to seamlessly access the data from different countries. It extends the DW capability cost-effectively at the expense of compromising data security. Security concerns can be overcome to an extent using secure communication protocols, encryption and compression techniques. Popular DW softwares on the web include Essbase (Arbor Software), Web Agent (Oracle Corporation), Web Enabled Tools (SAS), WebIntelligence (Business Objects), DecisionWeb (Comshare Commander), WebPlan Enterprise Planning Systems, Brio web warehouse (Brio Technology), FishFrame (www.FishFrame.org), etc.

4.6   Spatial Datawarehouses (SDW)

The SDW contain spatial data (census data, GIS data, remote sensing data, location data) along with non-spatial data and are queried by SOLAP. They are used to mine for spatial trends and

---

[12]If the language interface is the only requirement, it can be solved at the application layer, without a DDW.

Figure 4.8: Creating a datawarehouse using web data.

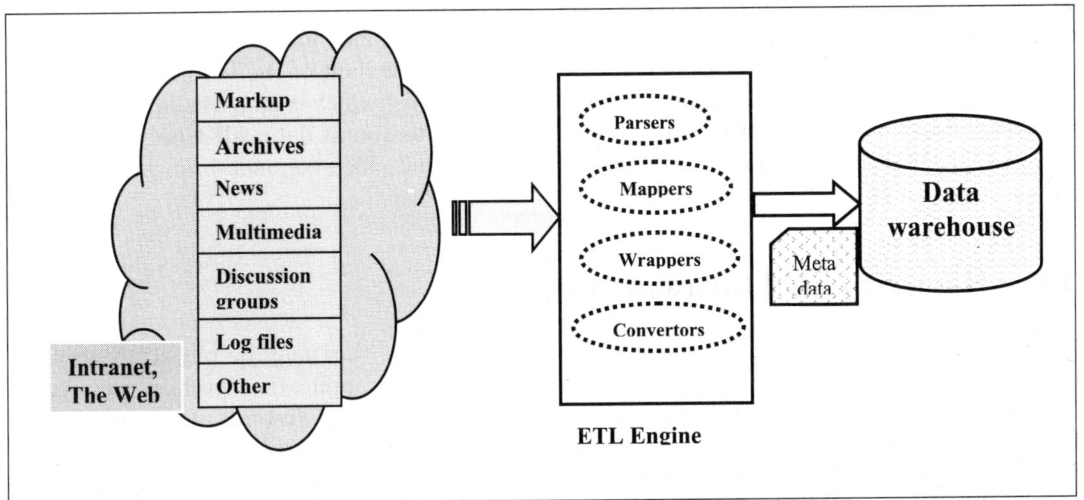

patterns (with respect to a fixed frame of reference). They provide insight into spatial relationships (meets, contains, adjacent to, similar to, close-by, crosses, intersects etc). If spatial boundaries are not well defined, inter-operability problems may arise (due to cross-border, cross-sector, cross-type or overlap of data). Generalised spatial operations have also been proposed in the literature (simplification, spatial aggregation, elimination, refinement, displacement etc). Spatial hierarchies are used for roll-up and drill-down operations (see 5.5 in page 5-19). A spatial query may include spatial measures (and spatial ranges) that utilise spatial algorithms for fast response [SC03]. Separate spatial indices are created to speed-up the analysis. Common spatial operators are distances (Euclidean or citiblock), curvatures, areas, etc and spatial measures are centroids, geometric unions. Spatial attributes can form a hierarchy (part-of). A SDW can answer queries in location-based services like:

1) Who are the primary care providers closest to a patient's home or office?
2) Which regions have out-of-stock products most frequently?
3) Which locations have maximum service back-logs?
4) What are the closest stores or service points within 4 miles to a customer's current location?

They are increasingly being used in market research, telemarketing, medical insurance, customer service, and strategic planning. Examples of SDW are CubeSTOR (cubeWerx), JMap, etc. More details can be found in [BY99], [MZ07].

4.7   DW Indexing

Datawarehouses use bitmap indexes or its variants. Some of the commercial RDBMS also use bitmap indexes (eg: Oracle, Informix, Sybase). A simple bitmap index uses a vector of values for each attribute (the info being kept in metadata). For instance, a region attribute that can assume the values South, West, North, East will have 4 bitmap vectors for it, one for each region. Similarly, a bitmap for a working day (Monday to Friday) will have 5-bit vectors. One of the advantages of bitmap indexing is its ability to find quick intersections in a database [PT02]. Spatio-temporal DW requires special indexing techniques using spatial access methods because they contain time-variant multi-dimensional locational data that have spatial hierarchies. Fast accessing of DWs is made possible using a variant of the B-tree called R-trees [GA84] that utilise minimum bounding boxes (MBB), R*-Tree which is an extension of B+-trees to multidimensional space [BK90], TB-Trees that utilise temporal data, aR-trees [PT02] that augment tree nodes with summarised data, or HR-Trees that share common branches to reduce redundancy [NS98].

4.8   Security in Datawarehousing

OLTP systems can be secured at various levels. For instance, they may be physically secured, access restricted at computer (or OS) level, network level and application level on the servers. Thus plenty of options exist for stringent security control of OLTP systems. DWs are created from data in operational and legacy systems. Security policies are implemented starting from the early stages of DW design [BB00]. Privacy laws, secrecy of (financial, personal) information, and confidentiality of other types of information are the primary concerns for implementing a security policy. In addition, a security policy can thwart attacks by third-party malicious users (intruders, impersonators, trojan horses etc). If data from the DW are made available to clients, a client-specific security infrastructure must be in place. Security loopholes may exist in the OS, middleware programs, application programs or utilities. New loopholes can emerge over time, that require additional security layers. These features must be seamlessly implemented without creating performance bottlenecks, down times, etc.

A DW needs a robust and layered security policy due to the heterogeneity of data collected from multiple sources with different security stamps. It can be implemented at different levels – at the DW or datamart level, middleware and application layer. Popular security features include access control lists (ACL), encryption, audit trails, intrusion and misuse detection using firewalls. Audit trails track the data usage in a DW and is used for tuning and usage querying. They are turned off during data updates. Implementing an enterprise-wide layered security architecture on a centralised DW is simpler in terms of cost, effort, and ease of managing. The security features are implemented in back-end, middleware and front-end software. Since most of the DWs are accessed from company-wide internal networks (intranets), the security attacks originate from internal employees. Security violations can be monitored continuously by automated tools that log such events and alert administrators promptly.

Opinions vary among DW specialists regarding the security architecture of datamarts. Some argue that different security policies on multiple datamarts could become unwieldy to manage. Others argue that compartmentalised datamarts can be hosted on separate servers and separate security policies can be implemented at varying strengths depending upon the sensitivity of data.

4.9   Exercises

1. Mark as true or false:
   (a) Datawarehouses can be created from data marts using a bottom-up approach.
   (b) ETL tools are primarily meant for metadata creation

    (c) Datamarts are built solely due to security reasons
    (d) Virtual DW operations heavily depend upon metadata
    (e) Proprietary vendor information is useful during data integration

2. What are the important decisions in the design of data warehouse?

3. Describe a data warehouse architecture and its relation to operational data store with a neat diagram.

4. What is the dimension of the results returned by the SELECT COUNT(*) query of SQL (on a database table)? What is the dimension of results obtained by the GROUP BY operator?

5. Describe metadata and its uses.

6. If an operator ALL is defined on a dimension of a data warehouse that selects all attributes of that dimension, what does the following commands return?
a) SELECT (ALL, ALL, $\cdots$, ALL, COUNT(*))
b) SELECT (ALL, ALL, $\cdots$, ALL, COUNT DISTINCT(*))

7. Describe scrubbing and how it is related to data hygiene.

8. How is the missing value problem dealt with during data staging? What are the missing value estimation techniques?

9. Describe the ETL architecture using a diagram. What are the major components of an ETL engine?

10. Give any 3 examples of ETL tools.

11. What is meant by noisy data? How are data denoised?

12. What is metadata? How is it managed?

13. What are the different reports categories?

14. What is location independence in distributed datawarehouses? How is it implemented?

15. What are the ways to populate a DW/DM from ODS?

16. What are the categorisations of datamarts?

17. What are the advantages of datamarts over a DW?

18. What are the different data updating modes available?

19. Describe data smoothing with an example.

20. What are the advantages of distributed datawarehouses?

21. Describe virtual datamarts and its uses.

22. Describe different types of datawarehouses.

23. Distinguish between operational data store and a DW.

24. What is workload balancing in DDW? What are the advantages?

25. What is a spatial datawarehouse? What are some of the operations in it?

4.9.0.1 References

[AM97] Anahory, S., Murray, D.(1997). *Data warehousing in the real world*, Addison-wesley.

[BK90] Beckmann, N., Kriegel, H., Schneider, R, Seeger, B. (1990). The R*-Tree: An efficient and robust access method for points and rectangles, SIGMOD.

[BY99] Bedard, Y. (1999). Principles of spatial database analysis and design, Ch:29 in *Geographical Information systems: Principles, techniques, applications and management* (Longley, P.A., Goodchild, M.F., Maguire, D.J., Rhind, D.W. eds), 413-424, John Wiley, NY.

[BS97] Berson, A., Smith, S.J. (1997). *Data warehousing, data mining and OLAP*, McGraw-Hill, NY.

[BL04] Berry, M.J.A., Linoff,G.S.(2004). *Data mining techniques*, 2nd edition, Wiley, NY.

[BB00] Bhargava, B. (2000). Security in data warehousing, *Proc of the 3rd data warehousing and knowledge discovery*, DAWAK,

[CD97] Chaudhuri, S., Dayal, U. (1997). An overview of data warehousing and OLAP technology, *ACM SIGMOD record*, 26(1), 65-74.

[IG03] Imhoff, G., Geiger, (2003). *Mastering data warehouse design: Relational and dimensional techniques*, Wiley.

[MZ07] Malinowski, E., Zimanyi, E. (2007). Spatial data warehouses: some solutions and unresolved problems, *IEEE international workshop on databases for next generation researchers*, SWOD 2007, 1-6.

[NS98] Nascimento, M., Silva, J. (1998). Towards historical R-trees, ACM SAC.

[PT02] Papadias, D., Tao, Y., Kalnis, P., Zhang, J. (2002). Indexing spatio-temporal data warehouses, *Proceedings of 18th International Conference on Data Engineering*. IEEE Computer Society, 166-175.

[SA04] Sen, A. (2004). Metadata management: past, present and future, *Decision support systems*, 37(1), 151-173.

[SE06] Shankaranarayanan,G., Even,A (2006). The metadata enigma, *Communications of the ACM*, 49(2), 88-94, Feb '06.

[SM99] Stohr, T., Muller, R., Rahm, E. (1999). An integrative and uniform model for metadata management in data warehousing environments, *Proc of the international workshop on design and management of data warehouses*, Heidelberg, Germany.

[TJ08] Twala,B., Jones, M.C., Hand, D.J. (2008). Good methods for coping with missing data in decision trees, *Pattern Recognition Letters*, 29, 950-956.

[VR05] Viana,N., Raminhos,R., Moura-Pires, J. (2005). A Real Time Data Extraction, Transformation and Loading Solution for Semi-structured Text Files. *Progress in Artificial Intelligence*, Lecture Notes in Computer Science 3808, (Carlos Bento, Amlcar Cardoso, Gal Dias (Eds.)), Springer, 383-394.

[WJ05] Wang, J. (2005). *Encyclopedia of Data Warehousing and Mining*, 2nd Edition, Idea group, PA.

5
Online Analytical Processing

CHAPTER OBJECTIVES

- Understand OLAP and related technologies

- Distinguish OLAP and OLTP systems

- Advantages of OLAP/OLAM

- Understand Data cubes and cuboids

- Describe star, snowflake and constellation schemas

- Explore OLAP cube operations

- Explain mobile OLAP

- Compare OLAP and Statistical Databases

- Multimedia OLAP

- Applications of OLAP

5.1   What is OLAP?

OnLine Analytical Processing (OLAP) is a decision support tool used by managers, executives and knowledge workers for fast and complex analysis requirements from different business perspectives on data stored in OLAP compliant software systems.

Definition 5.1 OLAP is the process and methodology for interactive processing of historical multi-dimensional data stored in a computer,[1] with an intent to gain insight into the data to reveal trends and anomalies through a variety of views for possible decision making.

OLAP is also used by government agencies (eg: homeland security, public utilities, tax departments) and researchers in various fields (eg: genomics, medical sciences). It uses data

---

[1]Data for OLAP usually resides in datawarehouses or datamarts. In virtual datawarehousing, it may be kept in ODS or databases.

granularity to dig into the multidimensional data. The granularity is a measure of the level of detail of a fact table – more the level of detail $\Rightarrow$ more the granularity. Low granularity refers to aggregated or summarised data, and high granularity levels refer to near transactional data. It may be considered as a supervised data analysis and mining tool that provides dynamic views of historical summary data at finer levels of granularity.

The raw data are transformed into strategic 'in-time' information with a variety of views by the data miner. It gives more insight into complex data dependencies through fast and interactive access, than available with conventional software programs (OLTP systems, spreadsheets[2], etc). It is an advancement of, and a significant shift from the traditional paradigm of 'query processing and results display' in flat tabular format. OLAP uses many layers (data access layer, presentation layer etc) and simultaneously caters to multiple user requests. It offers fast response to queries, irrespective of the size and complexity of backend data. The process of mining OLAP data is called OnLine Analytical Mining (OLAM). The OLAP/OLAM evolved from several concepts in databases, decision support systems, machine learning, OLTP, EIS, data visualisation and data structures. The software that enables higher level querying idiom is also called OLAP.

5.1.1 OLAP Data Cubes

The data used by OLAP are usually voluminous and are stored in datawarehouses, datamarts or databases (centralised or distributed). Multi-dimensional OLAP arranges data as hypercubes[3] and provides visual insight into finer dimensions of the data. This helps in exploring possible structural relationships in the data, and avoids the painstakingly hard task of multiple analyses or modeling in sub-dimensions as done in OLTP. OLAP facilitates data analysis by aggregating[4] or grouping data in different dimensions (either using categorical variables or quantitative variables arranged in appropriate intervals.

Example 5.1 Find some of the possible dimensions in a large supermarket that uses OLAP. List some typical queries.

Solution 5.1 Customer gender, marital status, occupation category, age (grouped into classes like 0-15, 15-25, etc), diet preferences can be the dimensions of customers in a super market). Some typical queries are: (1) Find total sales revenue generated by all customers in the age

---

[2]A spreadsheet is conceptually a two-dimensional array with the data cells arranged in rows and columns. Each cell is uniquely identified by its row and column numbers/labels. Some companies store multiple data clusters in separate rectangular regions inside a single spreadsheet due to the sheer size on the dimensions of the spreadsheet available.

[3]Cubes in 3 dimensions are easy to visualise. Although cubes in higher dimensions are called hypercubes, it is a general definition that includes lower dimensions as well. Hence a one dimensional hypercube is a single point, two dimensional hypercube is a square. To avoid confusion between hypercubes used in parallel architectures, we will use the term *datacube* in the rest of the chapter. It is not a 'cube' in the strict geometric sense because the dimensions need not be equal.

[4]The aggregation operation (pp.5-11) computes a single value (or summary vector of values) from a larger set (multiset). Full pre-aggregation require large storage space, and no pre-aggregation results in slow response times.

group 15-25, who have no income. (2) Find top 10 best selling products to female customers with at least 2 kids in the family during a particular quarter. (3) Select products whose sales to customers of various diet preferences dropped from last quarter.

Dimensions define the edges of a datacube, and grouping algebra (analogous to relational algebra of RDBMS) is used to query the cube data. Each query in the grouping algebra defined over an appropriate schema (eg: star or snowflake) is a tuple $< P, o >$ where P is a pattern and 'o' is an operation (typically an aggregation operation).

OLAP uses slicing and dicing of data, drilling up and down through hierarchies of dimensions (eg: in time, location, product, and salesperson dimensions) for pattern discovery. Hence it is useful for analysing inter-data dependencies like sales by regions, fiscal periods, salesmen, products, promotional offers, etc. The OLAP datacube can become extremely complex when there are many dimensions involved, since every permutation of dimensions is taken into consideration. By fixing a value for a dimension, the n-dimensional data cube reduces to an (n-1) dimensional cuboid.

E.F. Codd (1923-2003), the inventor of the well-known relational database technology has formulated 12 rules for a true OLAP database. These are called Codd's rule for OLAP [CC93]. It is more of interest to OLAP software developers than to data miners and end-users.

5.1.1.1 Codd's twelve rules for OLAP

- ♣ **Multidimensional view**: Discovering patterns and trends hidden in multivariate data involve inter and intra-dimensional search and cross-dimensional data analysis. Hence the data miner should be able to view retrieved data in a multidimensional conceptual space and use special operators (see §5.5) to manipulate the data.

- ♣ **Transparency to user**: Data miners should be able to view the backend data transparently as an open system architecture, irrespective of the storage scheme used[5] (see §5.4), and whether it comes from homogeneous or heterogeneous data stores.

- ♣ **Accessibility**: The OLAP tool should act as a proxy for the OLAP miner and hide all accessibility issues using certificates, access permissions, access control lists etc. In addition, the tool should bring the needed data using dynamic queries that map to selections and operations of the miner.

- ♣ **Consistent reporting**: The OLAP tool should allow multiple users to produce consistent reports using user-specified conditions in any sub-dimensions, irrespective of the data volume, number of data sources or dimension counts involved.

- ♣ **Client/Server architecture**: The C/S architecture is a popular paradigm in which multiple clients connect simultaneously to a back-end server and make information requests. OLAP mining sessions should be able to support many data miners, each of whose data requests are mapped to separate datacubes. This makes it easy to scale OLAP to hundreds of users.

- ♣ **Generic Dimensionality**: From the OLAP miners point of view, all dimensions are equally accessible in its structure and operational capabilities.

---

[5]ROLAP, HOLAP, MOLAP, SOLAP can be regarded as storage options that are hidden from OLAP miners

♣ **Dynamic sparse matrix handling capability**: Because some of the cross-dimensional data could be sparse, the OLAP tool should be able to optimally manage any complex subset of the available dimensional combinations.

♣ **Multi-user support**: Concurrency control mechanisms should allow multiple users to create models from the same or overlapping back-end data with integrity and security constraints.

♣ **Cross-dimensional operations**: The cross-dimensional operations are carried out using cube metadata. The OLAP tool should allow the user to perform any type of complex operations smoothly, and not just cross-tabulations along any of the available dimensions within the consolidation paths. The variety of OLAP operations without restriction on their order of application should allow the complete exploration of any subset of the dimensions.

♣ **Intuitive data manipulation**: The OLAP miner should be able to *point-and-click*, or *drag-and-drop* the data directly for manipulating along any type of consolidation paths (including reversals), without the help of menus, popups and icons.

♣ **Flexible reporting**: In addition to the visualisation capabilities, the OLAP tool provide flexible reports. Reports can be produced from the datacube content in any order to any level of detail allowed within the dimension.

♣ **Unlimited levels of dimension for aggregation**: The OLAP tool should allow a reasonable number of dimensions within the analytical model to suit the mining requirements in various domains (here the word 'unlimited' does not mean it literally, but indicates only sufficient number of dimensions).

5.1.2   OLAP vs OLTP

OnLine Transaction Processing (OLTP) systems are intended to process database transactions using a data manipulation language (like SQL insert, update, delete, purge etc). Examples of OLTP systems are:– Airline reservation system, Stock trading system, Banking system, E-commerce system, Insurance claims processing system.

OLTP systems are used in day-to-day operations for transaction processing and querying using small repetitive transactions. Most of the OLTP systems follow the three-tiered MVC (Model-View-Controller) architecture. Users interact with the system through the presentation layer (View). Application layer(Controller) covers the program logic. The data layer (Model) manipulates (store,update,retrieve) the data in back-end databases. They typically involve many databases (or tables) that are created using the ER model of database design. They contain *most current* data, as updates can happen at any time, and at any level including atomic levels (one bit in a database is updated). They use indexing[6] and hashing for fast data access. The indexes use primary or secondary keys present in database tables (columns) or functions thereof. The indexes are usually built by a DBA (or users with update privileges), and updated when data in corresponding columns are changed, or records in tables are inserted/deleted or other structural changes occur in a table.

---

[6]Popular relational databases use the B-tree and its variants for indexing, which is unsuitable in OLAP due to multidimensionality of views.

Table 5.1: A comparison of OLAP and OLTP platforms

	OLTP	OLAP
Users	clerks, operators	Knowledge workers
User load	thousands	hundreds
Usage	repetitive	ad-hoc
Data	up-to-date, detailed	historical, summarised
Data access	few records	thousands to millions of records
Data operations	read-write	scans
Data uploads	frequent	occasional
Data level detail	atomic	aggregated
DB size	Mega to Gigabytes	Giga, Tera to Peta bytes
Purpose	daily operations	decision support
Architecture	application oriented	subject oriented
Nature of query	low computation	computationally intensive
Query constraints	minimal	multiple constraints
Query sessions	short running	long running, repetitive

OLTP operations typically access a few tuples (although OLTP also has operations like SELECT COUNT(*) or UPDATE tableName SET colval=newval, etc that accesses an entire table). OLTP is customer oriented and has narrowly focused data. A major step in OLTP systems design is requirements gathering from prospective users (to prepare the SRS). OLTP transactions (such as INSERT and UPDATE) are designed to be highly efficient due to their frequent use. Tables are normalised to reduce database redundancies, and improve performance and throughput. The very design that makes OLTP systems fast and accurate, with facilities for rollback in case a transaction cannot be successfully committed, makes them unsuitable for large interactive data analysis. Progress in this direction has been made with object oriented databases (OODB) and object relational databases (ORDB). For example, [GP00] discusses the OLAP++ federated tool that integrates OLAP cube data with OODB.

OLAP databases are kept separate from OLTP databases. The OLAP *thick client* application is organised into 3 tiers – data server tier, application tier and client tier. The application tier includes OLAP API's and tools built using these APIs. The OLAP *thin client* is grouped into 2 tiers – data server tier and application tier. The OLAP uses star, snowflake schema or multi-dimensional data model as described in §5.4(p.5-15) and has facilities for aggregation and summarisation for fast query response (eg: total sales per store, total sales by salesperson, average sales amount per day). Aggregation over computed categories of a dimension is called an *OLAP histogram* [HR96].[7]

OLAP works with huge amounts of data (of the order of terabytes) and are optimised for multidimensional operations [TE02]. It could involve hundreds to thousands of attributes of different data types in the dimensional tables. The OLAP operations manipulate data organised

---

[7]In data visualisation and statistics, the histogram is a type of graph that displays the frequencies in distinct non-overlapping classes (usually of equal width).

in multiple levels of abstraction, and hence are computationally very intensive. OLAP rapidly analyses voluminous data using multiple constraints which may take much longer on an SQL based OLTP platform. The challenge in OLAP query processing is to process the complex queries with a partial scan of the data. It is a semi-supervised learning model in which query formulation is driven by the interactive selections of the OLAP miner.[8] In contrast to OLTP systems with hundreds of users, OLAP typically have few users and long running query sessions.

5.1.3   OLAP Data and Indexes

There are two types of OLAP data called *fact data* and *dimension data*. Fact data represent facts (sales amount, total transactions, tax paid). Dimension data encapsulate finer data about entities. Dimensions are the important attributes used as a basis for business decisions. They may be prioritised in different domains and arranged in high to low utility using a numeric measure. For example, a sales manager may prioritise the dimensions as sales regions, seasons, sales persons, products, etc; whereas an advertising manager may prioritise it as sales regions, time (months, years etc), advertisement media, and the number of repeat exposures of an advertisement.

As the name implies, a bitmap is an array of bits that is a map of a collection of values of a parent attribute (say 'A'). A bitmap in databases has the same size N as the number of records in a database. The attribute A is assumed to be categorical which takes k distinct values. Accordingly there exist k bitmaps (each of size N) for the attribute. Table 5.2 gives the data on 7 students. The gender is binary (M=Male, F=Female); the religion in our case is ternary (C=Christian, H=Hindu, M=Muslim), and the Major subject in this set is ternary. Hence the gender bitmap consists of two vectors of size 7 each. The religion and major subject bitmaps have 3 vectors each. If missing values occur for any of the attributes, it will count as a new value so that there could be a Null-value bitmap. This is shown in the last column where a null value is indicated as N/A (Not Available). When the atribute is binary, we can generate the bitmaps by bit complementing. For instance, the female bitmap 0011010 is obtained from male bitmap 1100101 by reversing each 0 to a 1; and a 1 to a zero. Thus selection of records (SELECT command of SQL) matching multiple conditions becomes easy to implement using logical operators. If the attribute is ternary or higher, the bitmap for last attribute can be automatically generated if the values for all others are known. As an example, add (logical OR) CSE and EE bitmaps in the last column to get 1110101, from which the bitmap for IT follows easily as 0001010. Counting of records (SELECT COUNT(*) of SQL) is also easy as it involves logical operations on bitmaps. Count of the number of Males is the same as the number of 1's in the corresponding bitmap. A disadvantage of bitmaps is that it is very much dependent on the order of records in a database. All bitmap indexes must be updated when old records are deleted or new records are added to a database, or records are swapped.

Most OLAP operations are read-only, without record locking or logging operations as performed in OLTP systems. Once the cuboid is populated with the data, most subsequent operations access the data contained in the cuboid (user datacubes are separately indexed for fast retrieval). Because the B-Tree indexing (used in some OLTP databases) is very inefficient in providing multiple views and is memory hungry, OLAP datawarehouses utilise bitmap indexing

---

[8]It can also be used as a supervised learning model.

Table 5.2: Bitmap Index Illustration

#	Name	Gender	Religion	Major	Hosteller
0	Arun	Male	Hindu	CSE	No
1	Brill	Male	Christian	EE	Yes
2	Christi	Female	Christian	CSE	N/A
3	Diana	Female	Hindu	IT	Yes
4	Nissar	Male	Muslim	EE	N/A
5	Subaida	Female	Muslim	IT	No
6	Tom	Male	Christian	CSE	Yes
	Bitmaps	1100101 (M) 0011010 (F)	0110001(C) 100100(H) 0000110(M)	1010001 (CSE) 0100100 (EE) 0001010 (IT)	0101001 (Yes) 1000010 (No) 0010100 (N/A)

technique [WM99]. They are easier to implement and their operations involve reading large blocks of data bits that are operated upon by bit-wise operators. Variants of the bit-map indexing techniques include interval encoded bitmaps [CI99] that utilise the sparcity in bitmaps, hierarchical bit-map indexes that speed-up retrieval. The disadvantages of bit-map indexing are the large space requirement and time complexity in insert/update operations. However, these operations are much less frequent in OLAP than in OLTP. Moreover, the indexing and other system functions are seldom performed by an OLAP user. A comparison of tree-based and bitmap indices can be found in [JL99].

5.1.4   Advantages of OLAP

Many fortune 500 companies are utilising the OLAP technology over and above other data mining tools for competitive advantage in the dynamic market. It reveals relations and trends across key business dimensions, and helps in monitoring unexpected ups and downs in business. Those factors that influence the instant decision-making process are automatically available online with OLAP. The major advantages of OLAP are summarised below:

- Focused data

  Many large companies maintain data in heterogeneous databases, spreadsheets, flat files (text files, hypertext files, data streams), structured files (audio, video files of varying formats) or other application files in diverse environments (Windows, UNIX, Linux etc). Integrating these heterogeneous data into OLAP systems is a challenging task that may require wrappers, middleware, metadata dictionaries, database-to-database conversion software[9] or custom built application programs. These data are consolidated into datawarehouses or datamarts that are queried by multiple users simultaneously. The OLAP servers are separate from transaction processing servers because they have different functionalities. Thus

---

[9]See www.dbconvert.com that supports MS-Access, Oracle, Foxpro, Postgres, SQlite, Firebird, MS SQL, etc; DBMS/Copy (www.databaseconversionsoftware.com) that supports almost all databases; Gnu PSPP (www.gnu.org/software/PSPP).

a focused datawarehouse kept exclusively for querying by knowledge workers will not affect the performance of OLTP systems that are often used by clerks and other end-users.

- Flexibility
  OLTP systems may use either a single-vendor backend database (Oracle, DB2, Informix), or keep data in separate databases. These are structured into tables, views and indices that are native to the vendor product. The OLAP systems are built on datawarehouses or datamarts without the complexities involved in tables and relations between them. Some OLAP tools permit the analysed data to be saved (export functionality to external tools and applications, or clipboard copying) in various formats for further processing or integrating with other data mining tools. A cube data markup language (CDML) may be used to save one user's data, which may be used in another session of the same user or by another user. Since OLAP tools vary greatly in their features, functionalities, and analysis capabilities, multiple OLAP tools can be deployed on a datawarehouse. Integrating OLAP data seamlessly across these tools is done using metadata warehouses, XML and CDML.

- New modeling dimensions
  Most OLAP vendors use the cube model with analytical capabilities for data representation. In closed loop business modeling (CLBM), the output of an analysis phase is fed back to improve business processes. Newer techniques like CPM (see §5.7.1, p.5-26) have been introduced to enhance the capabilities of the cube model. See references [CJ06], [MA03], [MV05] for further details.

- Efficient analysis
  OLAP operations allow an experienced data miner to find out correlations and causations from voluminous historical data. Most online analyses are performed by repetitive user interactions. Hence at coarser levels, it may be advantageous to provide an *approximate answer* to user queries in some situations [JM00]. Because approximate queries take lesser time to run, it may improve throughput without affecting accuracy at finer levels. An experienced OLAP miner can reveal trends and patterns much more quickly and efficiently than is possible through conventional techniques discussed in previous chapters. Moreover, since OLAP analysis does not partition the data as 'training' and 'test' sets, insights gained can immediately be put to use in decision making. It has fast learning times, and is easier to integrate with specialised data mining tools than possible in OLTP systems. Cube data may also be shared among multiple users, thereby setting the stage for collaborative mining.

- Scalability
  OLAP based systems are much more scalable than OLTP based systems. They are scalable in terms of users and data. Large concurrent user groups can easily be created and amended to OLAP. Multiple datamarts or distributed datawarehouses can be amended at any time, without affecting current OLAP miners.

5.2 Data Cubes and Cuboids

OLAP is used for online analysis of large volumes of data with an intention to make an immediate business decision based upon new findings of the analysis. The traditional query and report models used in most databases (including relational databases) display information as two-dimensional table(s) (if there are more than 1 rows returned). The OLAP goes one step further in capturing the model as a cuboid or cross tables (or crosstabs for short)[10] and allowing the user to mine the resulting data interactively (see below). Since OLAP considers many dimensions simultaneously, data are usually viewed in selected sub-dimensions. To analyse the sales revenue of various products in different geographical regions over a period of time, a few of the dimensions can be held constant. The total sales may depend upon the number of salespersons or stores in each region, total advertisement spending over a time period, marketing strategies, incentives and discounts offered and time of the year.

Datacubes are built from user selected data in dimension tables. The datacube captures data of the power set (set of all subsets) of the aggregates of various dimensions. If each of the 'n' dimensions contains n_i categories, the total number of super aggregate values are $\prod(n_i + 1)$ where $\prod$ denotes the product operator and the '+1' denotes the extra category of 'ALL' values. Power cubes can incorporate calculations, conditions, filtering, and business rules. Materialised views are used to store pre-computed query results with an aim to improve complex querying. Important numerical measures are pre-aggregated during datacube construction. Data may be viewed in selected dimensions to extract any hidden relationships and other dimensional combinations explored, if needed.

Each user query maps the data to a cuboid and these may differ from user to user, as also in different queries/sessions of the same user. Dimensions of a cuboid are represented by distinct categorical variables present in the dimension tables. Quantitative variables like time or income range are categorised using an appropriate interval representing one category. The base cuboid for a query holds the lowest level of summarisation. Topmost 0-D cuboid that holds the highest-level of summarisation is called *apex cuboid*. Individual cells in the data cube store values at the intersection of coordinates defined by the edges of the cube at specified coordinates.[11]

When put to business use, OLAP cubes are used to reveal trends, new patterns and structures in data sets.[12] They can also be used to verify hypotheses. For example, if a discovered pattern has resulted in radical changes in the business processes (and changed some data distributions) of an organisation, does the newly updated datawarehouse still support the previous pattern? Modeling is applied online across dimensions, through hierarchies of underlying level of detail of data aggregation to reveal hidden patterns and trends. Hence the ability and training of the

---

[10]Crosstabs are two-dimensional grids similar to spreadsheets. The word cuboid is a misnomer as it creates the impression that it is a 3 dimensional representation. A better name might have been 'hyper-cuboid' as it could involve any finite number of dimensions.

[11]This is analogous to the way multi-dimensional arrays are implemented in programming languages, except that array indexing in some languages start at 0 and most languages use a number to identify the index values.

[12]The data cube is a conceptual model, which can be used to produce reports of various complexities [RG99]. Trivially, it can also be used during software setup time, to test various functionalities like externalisation and cube data transmission or storage.

OLAP miner plays an important role in successful interpretation of the results from complex data. See [RK97] for discussion on hierarchical data cubes and [KS04] for clustered data cubes.

5.2.1   Dimensional Modeling

Dimensional modeling is the design concept used in OLAP technology ([VS99], [VS00]). As explained above, the data are contained in two types of tables called 'Fact Table' and 'Dimension Table'. The meaning conveyed by a single instance of a fact table is called its *grain*. The attributes in the fact table are divided into two groups called *dimension attributes* $(d_1, d_2, \cdots, d_m)$ and *measure attributes* $(m_1, m_2, \cdots, m_k)$. Examples of dimensional attributes are customer id, location key, branch id and time (of transaction). Arithmetic operations are not allowed on dimensional attributes (but we could use count() and exist() operations). Measure attributes can be subjected to arithmetic and other operations. Examples of measure attributes are profit, total price, average sales amount. Dimensional attributes can have a natural domain hierarchy (parent/child, is-a, part-of, etc). These hierarchies are used in drill-down operations.

Example 5.2 Identify the following attributes as dimension attributes and measure attributes:–
(i) item number, (ii) tax paid, (iii) time to deliver item, (iv) product id

Solution 5.2 Item number and product id are both used to identify an item, and are thus dimensional attributes.

5.2.2   Concept Hierarchy (CH)

Definition 5.2 A hierarchical mapping from distinct low level hierarchy to higher level is called a concept hierarchy.
A name hierarchy can include {lastname, middle initial, first name} and an address hierarchy can have the attributes {name, street, city, state, country zip}. Hierarchies can exist as 'is-a' type of aggregation, 'has-a' type or 'part-of' type. An example of an is-a hierarchy is {car, land vehicle, vehicle}. Street, city, state, country forms a 'part-of' geographical hierarchy.

5.2.2.1   Fact Table

Fact tables contain numeric measurements, metrics or facts of business processes along with foreign keys used to relate dimension tables (foreign keys in the fact table(s) are nothing but the primary keys of dimensions tables. It is through these links that navigation to various dimensions are made possible). As examples, average monthly sales, total VAT[13] paid are fact table data. The level of detail of the fact table is known as its granularity. Examples of foreign keys of dimension tables are product_id, customer_id, time key (see figure 5.1).

---

[13]VAT is an abbreviation for Value Added Tax.

5.2.2.2 Dimension Table

A dimension in OLAP represents a categorical attribute with 2 or more distinct levels. Various columns in a dimension table are known as dimensional attributes. Dimension tables capture the context of the measurements for 2 or more dimensional attributes. Each dimension table contains data for one dimension. They may be related in different ways as parent/child, as hierarchies or as dependencies. For example, the attributes of location dimension in customer table are street address, city, state, country, zip code etc. The salesperson salary may be dependent upon employment type (hourly employee, salaried employee, contract employee), years in service and hours worked (see fig 5.2). A calculated attribute is obtained from other members' values. Queries constrain these attributes to fixed values or ranges of values. Typical enterprise dimensions are time, products, geographical regions, advertisement media, sales channels, etc.

5.2.2.3 Additive Facts

Most measure attributes are amenable to arithmetic operations. They may be additive, semi-additive (sub-additive) or non-additive.

Definition 5.3 Additive facts are columns in fact tables that can be added across various dimensions.

Example 5.3 Sales amount is an additive fact, since summing the amount across different dimensions yield the total sales amount.

Definition 5.4 Semi-additive facts can be aggregated across some, but not all dimensions. They are usually nonadditive across time dimension.

Example 5.4 Monthly balance and VAT are semi-additive.

Definition 5.5 Nonadditive facts cannot be added across dimensions, because it does not make any sense in aggregating them.

Example 5.5 The customer's age, height, weight, blood pressure, blood sugar level, etc are nonadditive. Similarly temperature of a city, atmospheric pressure and humidity, physical properties of materials like density, illumination, etc are also nonadditive. In figure 5.1, payment type is a non-additive measure attribute. See exercises for further examples.[14]

5.3 Aggregation measures

The fast speed with which OLAP processes "group-by queries" is due to the statistical information stored in cube dimensions using aggregation measures. A good understanding of these

---

[14]Some authors define a fact as sub-additive, if it is additive along some of the dimensions. Customer count may be additive along a few, but not all of the dimensions.

measures is useful in obtaining a better insight into the complex data dependencies along cube dimensions. Aggregate Join Indexes (AJI) are used to speedup the querying using pre-calculated aggregates and joins. The aggregation measures used in the cube can be categorized as follows:[15]

5.3.1   Distributive

Definition 5.6 An aggregation measure is *distributive* if it can be computed[16] from distinct lower dimensional cubes.

The domain of values are partitioned into disjoint subsets, and each subset is individually aggregated. If the final result obtained by aggregating the partial results is the same as the result obtained without partitioning, the measure is called distributive. In other words, the aggregate of an n-dimensional cube can be expressed in terms of two aggregates of (n-m) dimensional and m-dimensional cubes (and the process repeated) where m is an integer between 1 and n-1 (between 1 and $\lfloor n/2 \rfloor$ to be exact, due to symmetry). As one-dimensional examples,

$$\text{sum}(x_1, x_2, \cdots, x_n) = x_n + \text{sum}(x_1, x_2, \cdots, x_{n-1}) = x_1 + \text{sum}(x_2, x_3, \cdots, x_n) \qquad (5.1)$$
$$= \text{sum}(x_1, x_2, \cdots, x_m) + \text{sum}(x_{m+1}, x_{m+2}, \cdots, x_n)$$
$$= \text{sum}(\text{sum}(x_1, x_2, \cdots, x_m), \text{sum}(x_{m+1}, x_{m+2}, \cdots, x_n)).$$

and left-recursive computation of count measure:

$$\text{count}(x_1, x_2, \cdots, x_n) = 1 + \text{count}(x_1, x_2, \cdots, x_{n-1}). \qquad (5.2)$$

If zeros and duplicate values are distinguished, this could also be written as

$$\text{count}(x_1, x_2, \cdots, x_n) = m + \text{count}(x_1, x_2, \cdots, x_{n-m}), \qquad (5.3)$$

or equivalently

$$\text{count}(x_1, x_2, \cdots, x_n) = \text{count}(x_1, x_2, \cdots, x_m) + \text{count}(x_{m+1}, x_{m+2}, \cdots, x_n) \qquad (5.4)$$
$$= \text{sum}(\text{count}(x_1, x_2, \cdots, x_m), \text{count}(x_{m+1}, x_{m+2}, \cdots, x_n)).$$

As another example,

$$\min(x_1, x_2, \cdots, x_n) = \min\left(\min(x_1, x_2, \cdots, x_{n-1}), x_n\right) \qquad (5.5)$$
$$= \min\left(\min(x_1, x_2, \cdots, x_m), \min(x_{m+1}, x_{m+2}, \cdots, x_n)\right).$$

Replace 'min' with 'max' in this expression to get the corresponding expression for the maximum of a set. Mathematically, a distributive aggregation function can be written as $F(A_{i,j}) = G(F(R_i)) = G(F(C_j))$ where $A_{i,j}$ are elements of a 2-dimensional data cube, $R_i = \{A_{i,j}, j = 1, 2, \cdots\}$ and $C_j = \{A_{i,j}, i = 1, 2, \cdots\}$ represent all values along individual dimensions. For

---

[15]we will synonymously use aggregation measures and aggregation functions to mean the same thing, although a measure can be obtained using more than one function as shown here.

[16]Distributive and algebraic measures can be computed recursively. The recursion may be direct or indirect as discussed below.

finding the minimum, F = G = min and for finding the sum, F=G=sum. But to find the count, F = count, G = sum, as shown in equation (5.5), in which case we get indirect recursion. This can be generalised to n-dimensional data cubes. Other examples of distributive measures are the geometric operations (union, intersection etc of bounded geometric regions/boxes) used in spatial OLAP (SOLAP).

5.3.2 Algebraic

Definition 5.7 An aggregation measure is *algebraic* if a constant number of aggregates of sub-dimensional data cubes can give the measure value of an n-dimensional data cube.

An example is the average (arithmetic mean), recursively defined as:

$$\text{avg}(x_1, x_2, \cdots, x_n) = ((n-1)/n)\,\text{avg}(x_1, x_2, \cdots, x_{n-1}) + x_n/n \qquad (5.6)$$
$$= (m/n)\,\text{avg}(x_1, x_2, \cdots, x_m) + ((n-m)/n)\,\text{avg}(x_{m+1}, x_{m+2}, \cdots, x_n).$$

In other words, an algebraic aggregate measure can be obtained from M distributive aggregate measures:
$\text{avg}(x_1, x_2, \cdots, x_n) = \text{sum}(x_1, x_2, \cdots, x_n)/\text{count}(x_1, x_2, \cdots, x_n)$ where both sum() and count() are distributive. Mathematically, we call an aggregation measure F() to be algebraic if $F((X_{i,j}) = H((G(X_{i,j})|i = i, 2, \cdots))|j = 1, 2, \cdots))$. Other examples are variance, covariance, maxN[17], MinN, and centroid operator used in spatial OLAP.

Example 5.6 The geometric mean (GM) of $(x_1, x_2, \cdots, x_n)$ can be found as
$\log\,(\text{G}(x_1, x_2, \cdots, x_n)){=}(1/n)\,\log\,(x_1.x_2.\cdots.x_n)$. Using log(ab)=log(a)+log(b), the RHS becomes $\log(x_n)/\text{n} + (\text{n-1})/\text{n*}\log\,(\text{G}(x_1, x_2, \cdots, x_{n-1}))$. Hence the GM is algebraic in the log-space.

5.3.3 Holistic

Definition 5.8 An aggregation measure is *holistic*[18] if its value for an n-dimensional cube cannot be computed using lower dimensional subcubes.

In other words, there does not exist an algebraic function to combine the measures for sub-dimensions to produce the value for sought dimension [GB96]. Examples are the mode, median, quartile deviation, coefficient of variation and correlation coefficient. The holistic group can be further categorised into *unique holistic, non-unique holistic* and *ambiguous holistic*.

---

[17]This measure returns the nth maximum of a set of numeric data.
[18]The literal meaning of holistic is "something that deals with the whole, and not just the parts of it"

Median, quartiles and percentiles are rank-based measures. If the medians (or quartiles) of the disjoint subsets (sub-samples) of a sample are known, it is *not* possible to combine them to get the median of the whole sample, except in trivial situations. Mode is a calculus-based (maximum frequency) measure. The mode of a set *cannot* be found, in general, from the modes of subsets, because it could be non-unique (Even if it is unique, when the modal values get split among multiple subsets, it may not even figure out in the subset modes. For example, suppose the modal mark in a class is 72 with frequency 3, and assume that there are 2 students with 70 marks. If the set of students are split arbitrarily into 3 sub-samples, it could happen that all 72's end-up in separate groups while both 70's end up in one group). The range of S can be computed from the range of subsets with some extra information. Because the range is the difference between the maximum and minimum of observations, either of these extremes (for each subset) could be found if the other and the range are known (max = min+range). The range of a set can be found if the range and one of the extremes of each subset are known. Let the triplet (r_i, e_i, t_i) respectively denote the range, extreme and type of extreme of each subset. Here e_i denotes either the minimum or the maximum of ith subset, and $t_i=0$ if e_i is minimum and $t_i=1$ if e_i is maximum. Then the range of S is $\max(e_i + r_i|_{t_i=0},\ e_i|_{t_i=1})$ - $\min(e_i|_{t_i=0},\ e_i - r_i|_{t_i=1})$. Hence the *disambiguating factor (or disambiguating statistic in some cases)* needed for computing the range of S is the tuple (e_i, t_i) for each subset. Alternatively, if the minimum and maximum of each subset are known, the range of S is R $=(\max(max_i)$ - $\min(min_i))$. The sample variance, which is an algebraic measure, also needs each individual sub-sample size and arithmetic means as disambiguating factors for combining (pooling) them.

5.3.3.1 Unique holistic

Definition 5.9 A measure is *unique holistic* if it is holistic, and has a unique value (without using disambiguation) for any given sample.

For a numeric sample S=$(x_1, x_2, \cdots, x_n)$, the range (R) is a unique number, which is the difference between the maximum and minimum of observations (symbolically R=$(x_{(n)} - x_{(1)})$ where $x_{(1)}$ is the minimum (first order-statistic) and $x_{(n)}$ is the maximum). Range of a set (R) cannot be found using the ranges (R_i's) of partitioned subsets (unless the minimum and maximum of each subset are known). Hence R is a unique holistic measure (even if the minimum and maximum observations occur multiple times, there is only one range). Other examples are the mean absolute deviation, and the bivariate measures Pearson's correlation coefficient (r), Spearman's ρ, and Kendall's τ.

5.3.3.2 Non-unique holistic

Definition 5.10 A measure is *non-unique holistic* if it is holistic, and a unique value does not always exist without using disambiguation.

The median of observations $(x_1, x_2, \cdots, x_n)$ is the middle value after arranging the data values in ascending (or descending/alphabetic) order. If 'n' is odd, there is a unique median. Otherwise there are two medians and the arithmetic mean (if data are numeric) of the middle values (which may not be a data value in the sample) is taken as the median. *The median of a*

Table 5.3: Examples of various types of measures

Measure type	examples
Distributive	sum, count, max, min
Algebraic	mean, variance, covariance, maxN, minN, spatial centroid
Unique holistic	range,CV,Pearson's r,Spearman's ρ,Kendall's τ
Non-unique holistic	median, IQR, quartile deviation, percentiles, mode

sample cannot be found from the medians of partitioned sub-samples. Median is defined for ordinal data too. Hence the median is a non-unique holistic measure. For instance, if the usability of a web site, or quality of a service is coded on an ordinal scale as {S=Superb, E=Excellent, G=Good, F=Fair, N=Not good, P=Poor}, the median opinion score will be that code below which 50% of the respondents cast their votes. If this frequency does not tally exactly on a code, the median is not unique (due to the symmetry of the definition of median, it will be the code above which 50% cast their votes). However, by re-coding the above values on a numeric scale, median can be approximated as a real number. In such situations, the mode seems to be a better measure than the median. By the same argument, the inter-quartile range IQR = $(Q_3 - Q_1)$ and quartile deviation QD = IQR/2 are also non-unique holistic.

The mode of a set of observations is that value which occurs most frequently. It tells us nothing about the frequency count of 'modal value'. If the maximum frequency occurs for two or more data values, any of them can be taken as the mode. If each observation is different (there are no duplicates), any of the observations can be regarded as the mode. *The mode (M) of a sample S cannot in general be found from modes (M_i) of sub-samples.* In addition, if a sample has more than one 'distinct peaks' (as revealed by a histogram or frequency polygon), it is called a *multi-modal* (with multiple modes) sample. Hence the mode is a non-unique holistic measure. However, it can be disambiguated in some situations if the maximum among the modal frequencies of multiple modes is regarded as *the* mode. This is applicable to multivariate case too. If the modal frequencies are approximately equal, or if the parent distribution is cusp-shaped, the disambiguated mode may not be appropriate.

Measures of various categories may be combined using arithmetic operations. The coefficient of variation CV=100*$\bar{x}$/s where s denotes the sample standard deviation is a unique holistic measure because it returns a unique real number for real data (it is holistic because CV of a set cannot in general be obtained from partitioned subsets). Similarly (mean - median) is a nonunique holistic measure due to the presence of median. See exercises for further details.

5.4   OLAP schemas

The most popular OLAP schemas are the star-schema, snowflake schema, and fact constellation schema.

5.4.1   Star Schema

The OLAP datawarehouses use a star schema, snowflake schema or object based schema and represent the static relationships between entities of interest. The star schema has one or more central tables[19] called fact tables that contain numeric attributes and summary measures. Facts are related to dimensions through table joins, with which facts are browsed across a combination of dimensions. The cube metadata are used to check consistency of data and safety of operations, which results in quick query response. Metadata wrappers are employed to integrate different metadata representations.

Figure 5.1: The star-schema for sales fact.

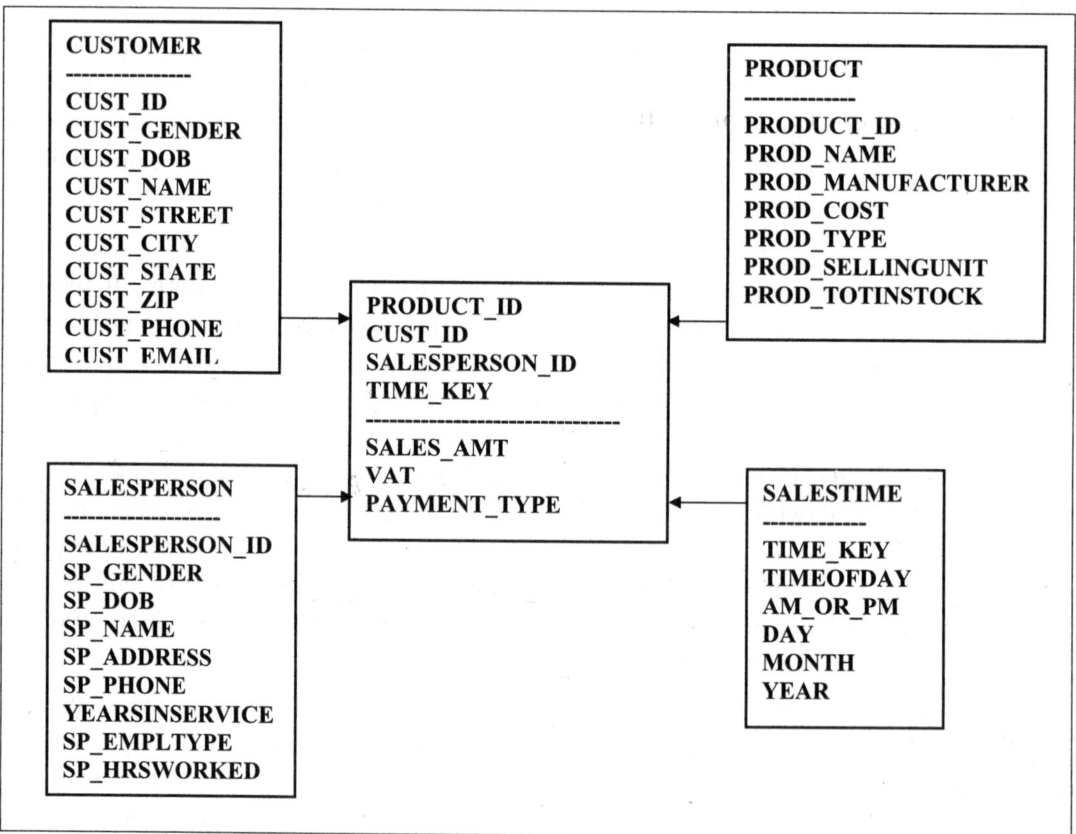

The main advantages of star schema are that (i) they are easy to understand, (ii) they have better performance and (iii) they result in smaller query execution times.

---

[19]These are more meaningful in ROLAP and HOLAP than in MOLAP. Most star schemas have a single fact table and multiple dimension tables.

5.4.2   Snowflake Schema

Dimension keys are used to relate fact table to dimension tables. The snowflake schema is a refinement of star schema in which the dimension tables are split using database normalisation, resulting in a snowflake pattern. This reduces redundancies and eliminates duplicate entries in the constructed cuboid. Naturally, the snowflake schema needs more joins than the star schema (if the resulting dimensions are more) and hence can perform poorly than the star schema. A compromise is to use a partially normalised snowflake schema (mixed schema) in which only large dimension tables with high redundancy are normalised. The dimension tables of snowflake schema could contain primary keys (to relate them with fact table) and foreign keys (to relate them to other dimension tables). The type of analysis performed quite often is the deciding factor whether to opt for star or snowflake schemas, or a hybrid of these. This may or may not be fully known at design time. Hence a compromise should be reached during design time between space efficiency[20] and performance issue.

5.4.3   Fact constellation Schema

Fact constellation schema is an extension of star schema in which multiple fact tables share (one or more) dimension tables.

In figure 5.3, there are two fact tables – sales fact and supplier fact. Sales fact table captures the attributes of a sale, and supplier fact table captures all details of suppliers. Dimension tables 'address' and 'product' are shared among the fact tables. This schema can be further extended, and is left as an exercise. An advantage of fact-constellation schema is the storage efficiency, which is a critical issue in very large data warehousing applications.

5.4.4   OLAP Queries

OLAP queries (also called decision support queries) are complex information requests to discover patterns and trends in vast amounts of data. Summary measures about events and attributes of objects are used to explore dimensional axes using OLAP operations mentioned in §5.5. Extensions to the SQL standard (SQL99 onwards) contain several commands to facilitate OLAP queries. The query engine dynamically converts each user action in OLAP into appropriate SQL statements. The OLAP cubes are created using the CREATE_CUBE statement, the general syntax of which is as follows:

CREATE_CUBE (cube_name, display_name, cube_owner, description)

The cube_name and owner are specified in the above command to dynamically drop unwanted cubes afterwards using the following command:

DROP_CUBE (cube_owner, cube_name);

The DMQL uses a slightly different syntax as follows:

DEFINE CUBE <cube_name>[<dimension_list>]:<measure_list>

---

[20]The star schema takes more space than snowflake schema.

Figure 5.2: The snowflake-schema used in ROLAP.

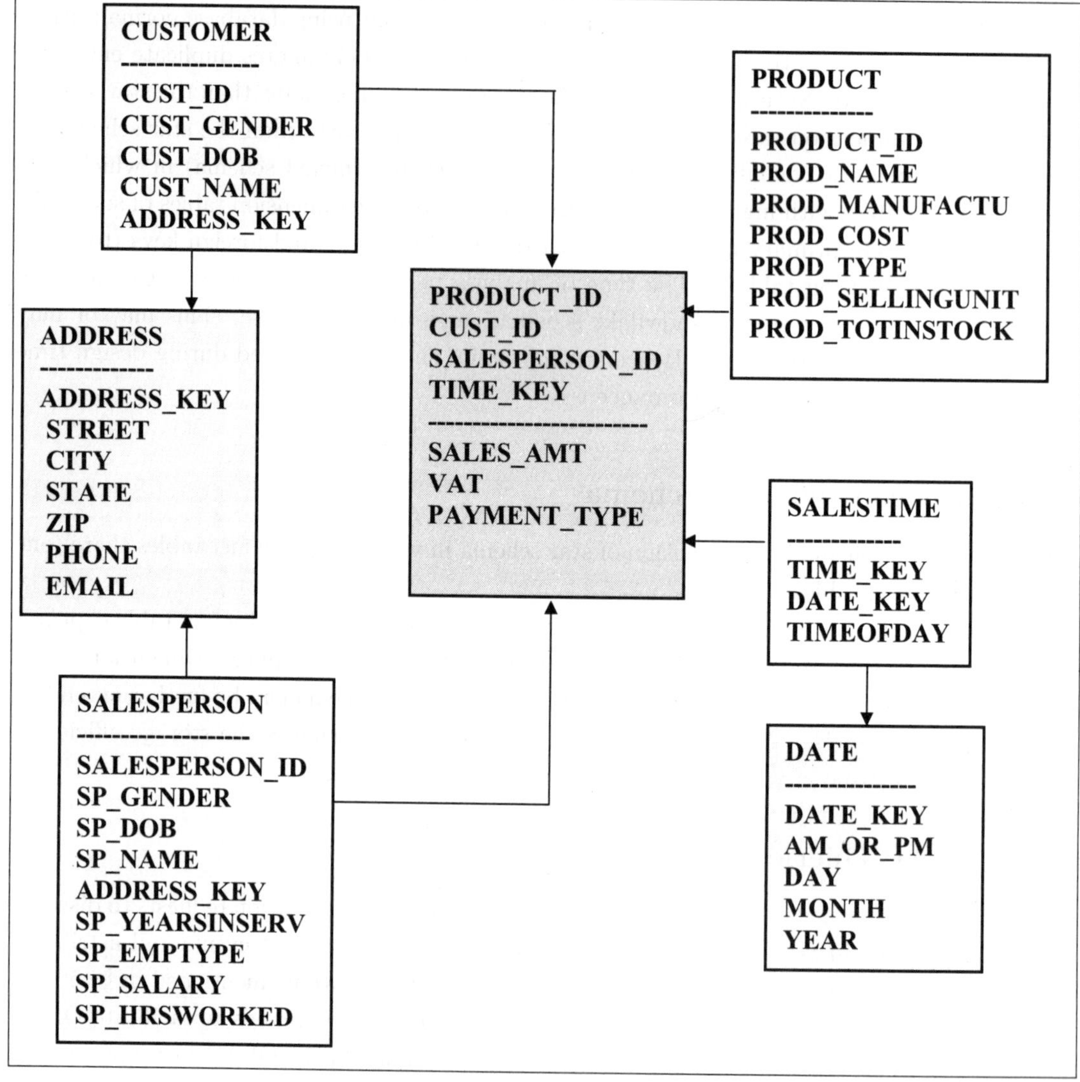

Once a data cube has been created, the create dimension statement can be used as follows:
CREATE_DIMENSION (dimension_name, display_name, dimension_owner, dimension_type, description).
The DMQL uses the following syntax:
DEFINE DIMENSION<dimension_name> AS (<attribute or sub-dimension list>)

Example 5.7 To define the time dimension in DMQL, we can use the following statements:

Figure 5.3: The fact constellation schema.

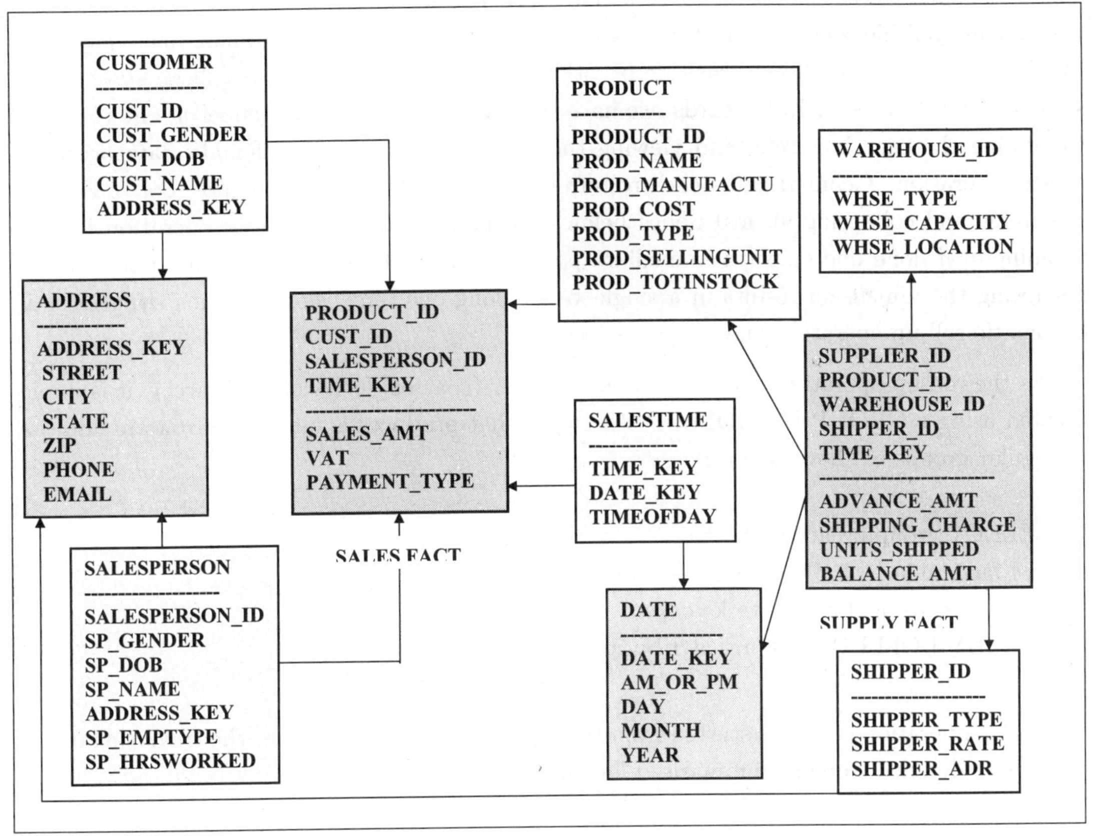

DEFINE DIMENSION time AS(time_key,day,week,month,quarter,year)

5.5   OLAP Operations

A datacube is populated with the data matching a user query. An OLAP user need to dig into the nooks and crannies of the datacube to locate any patterns, trends, anomalies or outliers in the data. The most common user-initiated operations are *roll-up, drill-down, slicing, dicing, pivoting* and *drill-through*. Other operations include *ranking* (sorting in ascending or descending order), *drill-across*, and inter-cube operations like *stacking* (one beneath the other), *racking* (side by side juxtaposition), *merging*, etc. Common operations for alphanumeric dimensional attributes are *equal, precedes, succeeds, in between* and *ranking*. We briefly review the most common operations below:

5.5.1   Roll-up

This is a macroscopic operation to view data at a *coarser* levels of granularity by collapsing the dimensional hierarchy. Consider a payment hierarchy, in which customers make payments by VISA, Mastercard, checks, cash or by other means. Since ladies and college students (in some countries) have exclusive cards, we have split the hierarchy using gender at the top level. Collapsing the card hierarchy and viewing the data under 'male' and 'female' categories is a roll-up operation. Geometrically, each roll-up operation traverses towards the root of the tree hierarchy, and collapsing all leaf nodes below that node. Thus the roll-up operation is more meaningful if done one step (to immediate parent node of the tree) at a time. On occasion, collapsing the dimension results in a single value along one dimension. In such situations, an automatic roll-up operation may be carried out.

As the roll-up operation traverses towards higher (coarser) levels of a hierarchy, it is implemented using a GROUP BY ROLLUP statement that produces a subtotal of rows in addition to regular groups as shown below:

```
SELECT grouping_attributes
FROM fact_table (<JOIN dimension_tables>*)
WHERE dimensional_attribute_k=const
GROUP BY ROLLUP(grouping_attribute)
```

Rollup operation is always started from the bottom up. In other words, the traversing occurs from more detailed to more summarised levels.

5.5.2   Drill-down

The drill-down operation traverses from a coarser to finer levels of granularity. Drilling down breaks an aggregate into its constituents to deeper levels of consolidation. This microscopic operation allows a magnified view of the data along a fixed dimension. For instance, a sales manager may view the *quarterly* sales revenues rather than *yearly* sales revenues. An advertisement manager may wish to correlate ad-costs under printed media (newspapers, magazines, etc) to sales volume variations in different geographic regions over a period of time, rather than under all media. A medical specialist may be more interested in viewing patient data under various diabetes categories or other disease sub-categories rather than major disease categories.

This operation is geometrically interpreted as a traversal from a parent node (most summarised) to the child nodes (most detailed) of the tree hierarchy (which reveals data at *finer* levels of granularity). Hence it also is more meaningful if carried out step by step (analyse in immediate sub-dimensions). Viewing the payment details under credit card is a drill-down operation. An analyst may do a drill-down followed by a roll-up to navigate to sub-dimensions to explore any dependencies in those levels.

This could be expressed as an SQL statement below. The WHERE clause constrains one or

Figure 5.4: An example datacube representing sales figures.

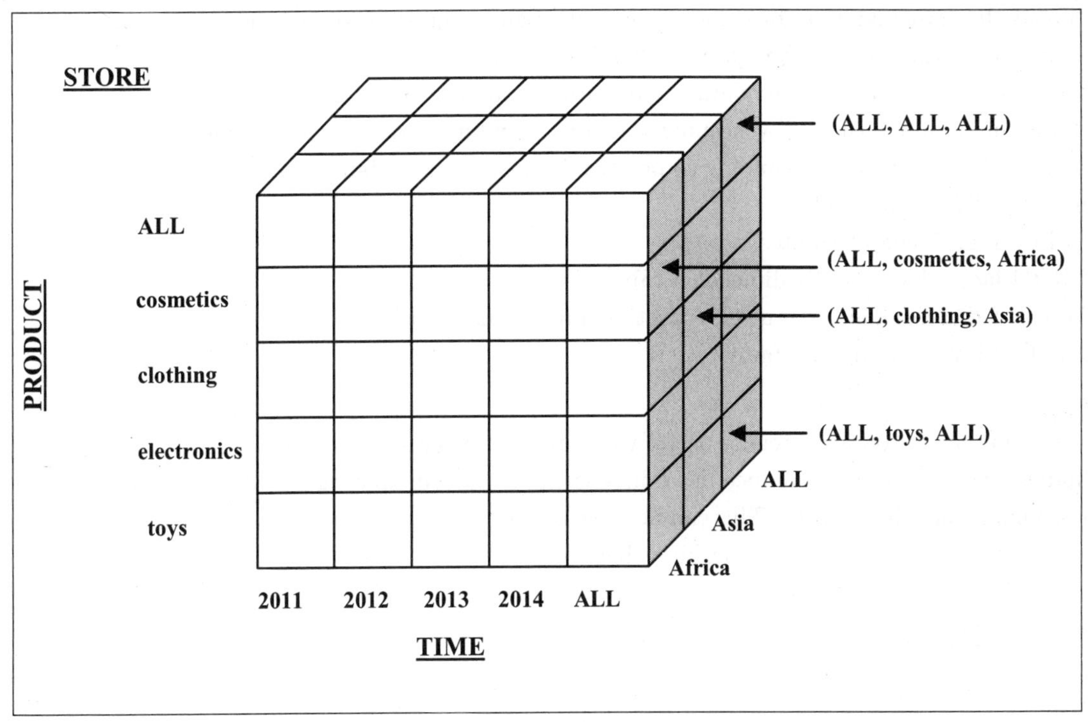

more of the dimensional attributes.

SELECT grouping_attributes

FROM fact_table (<JOIN dimension_tables>*)

WHERE dimensional_attribute_k=constant

5.5.3 Slicing

Slicing operation fixes one value of a dimension, and selects all corresponding values of other dimensions. Geometric interpretation of slicing is a cut through the cuboid. A slice is a 'regular' subset of a cuboid from an end-user's perspective. By *regular* it is meant that slices are always taken such that the subset is perpendicular to coordinate planes forming the faces of the cuboid in which the variable(s) are held constant and parallel to all other planes[21]. From the programming point of view, the slicing operation fixes one or more dimensional attribute(s). Slicing is performed by fixing the value of a variable. Data omitted from a slice are the set of all values (in figure 5.4) for which $x \neq 2006$. Hence slicing is analogous to specifying a WHERE condition

---

[21]Axes of the cuboid are always assumed to be orthogonal. If the dimensionality is greater than 3, the resulting slice is also a cuboid

on an SQL statement (SELECT * from Y, Z WHERE X=2006). Slicing operation filters out the geometric intersection of a plane (defined by X=2006 in above example) with the cells of the cuboid. In most investigations, the planes are axially aligned. A data cube contains an "ALL" attribute that holds the aggregate result for each dimension. This slice is important in mining for correlations present in the data (all other slices are useful in mining for partial correlations, if any, in sub-dimensions (by keeping the corresponding variable fixed)). Slicing operation is more difficult and time consuming to perform in unstructured databases. It is equivalent to the following SQL statement:

SELECT grouping_attributes

FROM fact_table (<JOIN dimension_tables>*)

WHERE dimensional_attribute_k=const[| <dimensional_attribute_j=const>]

GROUP BY grouping_attribute

Slicing operation is especially useful when data distributions are sparse in some dimensions, and an analyst wishes to reduce data dimensionality by eliminating sub dimensions that have less importance than others. This can be done by locking important dimensions and temporarily ignoring sparse dimensions by masking them out.

5.5.4 Dicing

Dicing is the projection of data onto a nonempty subset of dimensions. It restricts the range of values along selected dimensions. Slice and dice operations are the generalisations of select and project operators of SQL.

5.5.5 Pivoting

The geometric interpretation of *pivoting* is a re-orientation of the data cube through rotations. Usually, this operation is applied after a slicing operation to view particular faces of the cuboid. If the slicing plane is not axially aligned, a pivot operation may be performed prior to the slicing operation for proper alignment. It is also used to compare two or more sub-cubes.

Operation	Alias	SQL Equiv	Description
Rollup	Aggregation		Decrease level of detail
Drill down	Aggregation	Nested SQL	Increase level of detail
Slicing	select	WHERE	Fix a dimension, view data $\forall$ other dimensions
Dicing	project	GROUP BY	restriction of dimension

OLAP has the capability to execute operations across full dimensions. As an example, "Profit = Sales amount - marketing expenditure - advertisement expenditure - tax paid" when executed within a datacube will calculate the profit for all combinations of region dimension, seasons dimension and time dimension. A user can then slice the OLAP cube to reveal individual combinations (say profit in South region during summer of 2010), or aggregate it over various dimensional combinations. This avoids the painstakingly hard task of multiple analyses in individual dimensional combinations, and is a time-saver.

5.6 OLAP variants

Several variants of OLAP exist, the popular ones being ROLAP, MOLAP and HOLAP. In the following sections, we briefly review them and provide a few of the OLAP related acronyms.

5.6.1 ROLAP

The Relational OLAP (ROLAP) provides a multidimensional view of data stored in relational databases and can use the star or snowflake schema (or a variant of them mentioned in §5.4). The relational database that stores OLAP data are also called ROLAP. These databases can contain conventional data types (numeric, text, binary objects) as well as audio, graphics, video, multimedia data and other forms of structured data. ROLAP can utilise metadata information in fast retrieval of required data subsets from relational databases.

If there are 'd' dimensions, the ROLAP may be implemented as 2^d relational tables, and queried using ordinary SQL. If the datacube is sparse, the corresponding tables may be stored in compressed form (or discarded, if all values are zeros). This may create natural indexing problems, and could create an overhead on the analytics when the number of dimensions are large.

ROLAP has a highly scalable architecture that utilises OLAP middleware and services. Advances in parallel relational technology are being used for high performance, layered ROLAP.

The automatic updates of cube data to reflect the dynamic changes in the ROLAP database is called *real-time* OLAP. Real-time cubes are useful for making critical business decisions less error prone. Different analysts take different amounts of mining time depending upon their abilities, (wired and wireless) network congestion problems and device capabilities (desktop, palmtop, smart phone, PDA). Even for experienced users, cube mining is seldom instantaneous. Hence real-time updates of cube data can save much time in critical decision making situations. Datawarehouse/datamart designs could include a notification strategy with the cube server that alert data updates. Since the cube server keeps track of all 'live' online cubes and the data they contain, triggering mechanisms can promptly update cube data in online client applications. When client applications are finished, or a new OLAP mining session is started, this information is updated in the cube server to avoid redundancies. However, datawarehouses and datamarts in most applications (especially in business enterprises) are static over a period of time and administrators often run data update programs during off-peak hours or overnight. This eliminates the need for costly online updates of cube data, some of which may be useless for the OLAP miner.

5.6.2 MOLAP

The Multidimensional OLAP (MOLAP) is used to analyse data stored in multi-dimensional databases (MDBMS) and achieves its performance through summary measures[22], caching mechanisms, bitmap indexing and analytical processing [RM03]. A cube server manages the data

---

[22]Aggregates are organised as a hierarchy and precalculated at load time

cubes created by various users simultaneously (each user may work with different data cubes). Data loading into MOLAP is much more involved than in ROLAP. MOLAP servers have impressive online querying performance, that is well suited for large data mining applications. Oracle Express server, Essbase (Arbors software), Visual SQL-Designer (DeskArsenal software) are examples of MOLAP databases. Cube data may be sparse when dimensionality is large. Efficient cube storage schemes are discussed in [KL01], [KS02].

5.6.3 HOLAP

HOLAP (Hybrid OLAP) combines the best features of both ROLAP and MOLAP. Since MOLAP stores the data as data cubes, it is not well-suited for some data types (text, images, audio). These types of data can be stored conveniently in RDBMS (using clob, blob types). It can provide simultaneous analysis of data stored in a multidimensional database and in a relational database(RDBMS). It utilises cube model for aggregate querying and ROLAP tables for relational querying. It can drill through the cube dimensions into the relational data.

5.6.4 DOLAP

DOLAP(Desktop OLAP or Database OLAP) provide multidimensional analysis locally in the client machine on the datacubes collected from relational or multidimensional database servers. The distinguishing feature of DOLAP is its ability to manipulate or post-process cube data on client machines (desktops, laptops, etc). It retrieves the data for data cubes and the analysis can be carried out in client's memory. This is especially useful for offline processing and analysis by mobile users in inexpensive environments. Standard tools like Microsoft Excel can be used to analyse the data. Local metadata can be used to define custom interfaces and it can use open metadata exchange architecture. The overhead involved in compression and decompression of sparse cubes may not be acceptable in real-time mining, although it could be used for externalising cube data, as done in DOLAP. The *Radarcube* is a .NET Winforms and ASP.NET compliant DOLAP product from Radar soft (radar-soft.com/)

5.6.5 SOLAP

The spatial OLAP is used with spatial data warehouses. Spatial data may be categorised as geometric and semantic. Examples of geometric spatial data are the coordinates (cartesian, polar, Global Positioning System (GPS), other navigational position or coordinates [with respect to a reference system] of objects in space) of a point. Semantic spatial data can be aliases, names, or other forms of ids given to distinguish an object. Accordingly there are geometric, semantic and hybrid dimensions in the datacube. Examples of spatial data include satellite telemetry, traffic data (air, road, rail, sea, underwater), etc. SOLAP can be used to compare different classes of data to reveal spatial classifications, spatial clusters, and associations between spatial and other spatial/nonspatial attributes. SOLAP mines data in spatial dimensions and displays the results as maps, albums, or other spatial diagrams. For example, more than 75% of health care data are spatial in nature (locational data of patients, facilities/clinics, previous treatment centers, etc). The same is true in various other disciplines (insurance domains, e-commerce, delivery systems etc). Spatial aggregation operations are used to form the data cube and spatial indexing techniques are used to speed up the mining of pre-aggregated results. An application of SOLAP to emission analysis of industrial pollutants can be found in [MM05], and more details in [SL00], [BT07].

There are other OLAP related acronyms available in the literature, some of which are not well-recognised. Examples are EOLAP (Extended OLAP), POLAP (progressive OLAP; a query decomposition technique to retrieve the results using non-overlapping sub-queries), WOLAP (Web based OLAP, with browser support), JOLAP (an implementation of OLAP by IBM and Oracle). A description of OLAP for temporal multidimensional databases and a Temporal OLAP (TOLAP) as an extension of TSQL2 or SQL/TP can be found in [MV00], [VM02].

5.6.6 OLAP metadata

The data most often used in OLAP are categorical or quantitative type. Hence the OLAP metadata can tremendously help in fast retrieval and finer analysis. Metadata can exist in binary format (embedded), in separate text files or as embedded text in custom tags. The OLAP operations discussed in §5.5 are used for finer levels of querying in focused dimensions. OLAP uses two types of metadata called *technical metadata* and *business metadata*. Technical metadata provide summary information about the data, operations and business rules that are intended for administrators and software tools. Business metadata describes business processes and models (attributes of a graph, chart, table, or cube), calculations (an equation, a data conversion formula) and are intended for knowledge workers. Both types of metadata may be described in static form (as text files) or in functional form (as a macro, API, stored procedures, functions).

5.7 Mobile OLAP

OLAP introduced in 1990's is an analysis and information visualisation technology built upon datawarehouses for decision support. It is mainly used by executives, managers, and business analysts who are almost always on the move. Because these type of users often have high-end smart phones and PDAs, offering wireless OLAP on these devices will make them *mobile decision makers*. OLAP enabled personal data assistants (PDAs) are already available in the market. With the availability of high quality graphics and multimedia compression techniques, OLAP enabled wireless devices (mobile phones, smart phones, pocket PCs, etc) are becoming a reality [BC99], [PR02]. However, Graphical Processing Units (GPU) for hand held devices are also appearing in the market and is set to revolutionise the performance of mobile devices. These chips not only accelerate graphics operations, but also can be used as parallel processors due to the presence of *vertex* and *fragment* processors within them. The received data are first decompressed and rendered with the help of GPU. The Hand-OLAP is a tool to 'OLAP-enable' mobile devices [CF03].

5.7.1 Cube Presentation Model

Various vendors of OLAP have adopted different models based upon the cuboid. Most OLAP systems use a four-layer architecture:– (i) physical layer, (ii) logical layer, (iii) presentation layer and (iv) visualisation layer. The physical layer comprises of the datawarehouse/data marts. The logical layer re-organises user specific data as data cubes. These two layers are implemented on the server side. The presentation layer is a middleware on the application server that manipulates the data (cuboids of any dimension) according to OLAP operations. The visualisation layer sits on the client side and presents the results on 2D screen. A cube presentation model (CPM) based on extended UML/XML is a standard for presentation of OLAP screens that separates the presentation and retrieval layers [MA03], [MV05]. It uses an advanced visualisation technique known as *Table Lens* in human computer interaction [RT97]. It is tailored to cross-tab reports that have facilities for focused highlights. The CPM logical layer comprises of (i) dimensions, (ii) monotone ancestor functions defined on the level hierarchy, (iii) detailed data set DS^k (granulated fact table data), where DS^0 is the set of data at bottom hierarchy (iv) cuboids (comprising of a primary cube and secondary cubes). The XML can be used for information exchange between presentation and visualisation layers as well as between external applications, browsers and add-on tools. Further details on visualisation can be found in [MVS03],[VM06] and the extended UML stereotypes in [MV03], [MA04].

5.7.2 OLAP Server

Because of its simultaneous use by multiple users who expect fast response times, OLAP is hosted on high-end (hardware) servers. OLAP servers are middleware programs that manage user mining sessions and interacts with backend datastores (the front-end is called cube server). They use the client/server architecture and operate on multi-dimensional data structures designed to deliver rapid response to end-users. Consistent availability and rapid real-time response are the crucial factors to be considered in selecting OLAP servers. IBM DB2 OLAP server, Oracle OLAP, SQL Server Analysis services from Microsoft[23], Cognos powerplay are examples.

When GPUs are used for tasks other than graphics related, it is called general purpose GPU (GPGPU)[24]. Most of the OLAP operations (slicing, roll-up, drill-down etc) are performed within the client. The CPU of these device can utilise the GPU for offloading the work with the help of middleware programs (like BROOK for GPU). Cube data may be processed using the stream programming model to speedup the OLAP operations. A multicast layer is used in mobile OLAP application layer to broadcast the user data to selected devices. Since each user works with different cuboids, a subscribe/notify protocol may be used by the server to implement client specific transmission. The broadcast manager uses different data caching techniques for communicating with the variety of mobile devices. One of the challenges faced in wireless transmission is the security (see §5.7.3). Further discussions on wireless transmissions and power consumption issues can be found in [SC02].

---

[23]Microsoft SQL Server 2000 supports MOLAP, ROLAP and HOLAP.
[24]see www.gpgpu.org for details

5.7.3 OLAP security

Being a multi-user, client-server based system that contain strategic information, many security features have been incorporated into the OLAP technology. This includes authorisation and authentication features available in OLTP systems. OLAP is accessed over wired and wireless environments. Separate security and access control mechanisms are implemented in these environments. These involve general datawarehouse security, broadcast manager security and user or group specific security and access control. Another categorisation is as back-end, middle-ware and front-end security.

Cube roles and dimension security are utilised in restricted access to shared cube data. Administrators can assign user specific security for viewing cube data or its partitions (dimensional data, slice data, cell data etc). Security may be restricted from the most granular (cell level) to the most coarser (cube level). Login names and passwords may be used to implement the middleware security. Each user/group is assigned a system generated or user selected id (login name and password). Most OLAP systems are accessed over company-wide intranets. However distributed datawarehouses/datamarts accessed over wide area networks or the web pose severe security threats. Implementing back-end security is more cumbersome than in RDBMS due to the volumetric data. Large datawarehouses can be broken down into smaller datamarts and each of these can implement separate security features. The front-end security is implemented by the cube manager. OLAP access control can be implemented at whole cube level, at different measures, slices of cube or levels of detail. Cube data may also be shared using metadata by exporting it using Cube Data Markup Language (CDML). Other OLAP security issues are discussed in [PP00], [KK97], [SK97].

5.8 OLAP vs Statistical Databases

Statistical data obtained using sampling methods often contain multiple measurements on the subjects. Statistical databases (SDB)[25] originated during the 1970's with the work of theoretical and applied statisticians. Tremendous success of the relational model of E.F. Codd, coupled with the availability of advanced software packages have contributed to the popularity of statistical databases. SDB can use data and metadata (stored in RDBMS or external files) for faster analysis [JB04]. OLAP technology originated from the work of computer scientists in enterprise data mining and decision support. Advances in GPUs, computer storage technologies, human-computer interactions, data compression and multi-tiered architectures have all contributed to the maturity and success of OLAP.

Both SDB and OLAP cubes store multi-dimensional data [RZ96]. However, SDBs are more often used for conceptual modeling and analysis using multivariate models (Multiple Regression, Cluster analysis, Factor analysis, MANOVA, etc) with an implicit assumption about the distribution of parent population (eg: multivariate normal). Statistical analysis utilises visuali-

---

[25]The storage technology used to store and manipulate data as well as the software used for analysing the data will be called SDB.

sation either as an end-result (to see data clusters, if any; fitted least squares regression model, etc) or to check data assumptions (linearity, correlation structure of observations). Although OLAP and statistical databases both support data in multi-dimensional arrays, OLAP is more often used in analysis of enterprise data from a business perspective, whereas SDBs are used for analysis in the life sciences (socio-economic data, medical data, etc). Statistical techniques utilise error minimisation criteria (least mean squares, minimum distance, etc) for building the models whereas OLAP uses summary measures for analysis. Statistical databases use tests of significance to measure the appropriateness of the built model and check how well it performs for different data. These tests are based upon a *statistic* (a function of the sample values; see chapter 3) whose sampling distribution is used to draw conclusions and inferences. OLAP summary measures do not have distributional assumptions (at least from the user's point of view). Some of the statistical data mining (SDM) models are built using a subset of the available data (called training set) and the rest of the data (test set) are used to validate the model. OLAP modeling seldom partition the data into training and test sets. There are separate statistical models for analysis of numeric and categorical data, spatial and temporal data while OLAP can deal with a mix of these data types (categorical data define the dimensions and the summary attributes (computed from quantitative data) are the measures (in fact tables). The dependent and independent variables are identified prior to the analysis in SDM, whereas no such distinction is made in OLAP. Modeling using SDB make assumptions (linearity, nonlinearity, (un)correlated, etc) on the data. Most SDM analyses require user input (number of clusters in some clustering algorithms, number of factors to be extracted, maximum number of iterations to be performed, outlier deviations, distributional parameters etc). Data transformations (log-transform, change of origin and scale, etc) are often used prior to analysis in SDM, whereas they are used at datawarehouse creation time, rather than at the time of analysis in OLAP. Data hierarchies (parent/child, is-a or part-of containment hierarchies) are often ignored in SDM[26] whereas they are fully utilised in OLAP. Most data in SDB are micro-data derived directly from subjects of interest, while OLAP data can be a mix of micro and macro-data. There is a clear distinction between the concepts of 'sample' and 'population' in SDB modeling, whereas it is often irrelevant in OLAP due to the voluminous historical data being analysed (however, OLAP may be carried out on smaller datamarts that are subsets of DW). SDB often organise the data in multidimensional arrays and trees [RZ96] while OLAP uses a combination of arrays, trees, cuboids, indices and hash tables. A comparison of these two approaches can be found in table 5.4. See [SA97], [CL94], [MZ92] for further discussion of SDB.

The simplicity of data representation as cuboids and efficiency of operation are the key benefits of OLAP analysis. A discussion of the hierarchical visualisation techniques for data cubes is given in [MVS03], [MS07].

---

[26]Some of the hierarchical models and algorithms do utilise data hierarchies.

Criteria	SDB	OLAP
Usage	most often single user	almost always multi-user
Users	analysts, researchers	executives, managers
Analysis	online or offline	always online
Data size	small to medium (few Kilobytes)	voluminous (MB to Terabytes)
Data transformations	frequently used	seldom used during analysis
Data type	Numeric (int, float etc)	any (numeric, categorical, other)
Data origin	Micro-data	Macro-data
Data hierarchies	Often ignored	fully utilised
Metadata	seldom used	heavily used
Error minimisation	Least squares, (chosen by user)	software dependent
Security	Data dependent	Always important
User input	often required	seldom required
Precomputation	Minimal	summary measures, precomputed
Data structures	arrays (tables), trees	cuboids, trees, graphs
Data organisation	flat table model	star schema, snowflake schema
Data cleansing	sometimes performed	always performed (at creation time)
Dimensionality	small to medium & dense	large and sparse
Data compression	seldom used[28]	sometimes used
Purpose	inferences,	decision making

Table 5.4: Comparison of Statistical Databases and OLAP

5.9 Data storage, datawarehousing and data mining

Datawarehousing technology is used for fast simultaneous access of data stored in heterogeneous data sources at one (centralised DW) or more (distributed DW) locations. OLAP is an approach to mining detailed, and cleansed data stored in data warehouses or datamarts. It uses summary measures (computed from numeric attributes of stored data) for fast scanning of large number of data records. It transforms stored data into useful information at the fingertips of managers and decision makers using a conceptually simple approach. Each user's OLAP cube contain their view of summary measures extracted from data residing in datamarts or datawarehouses and mapped using one of the schemas. Depending upon the datawarehouse, data may be stored in relational (ROLAP) or multidimensional (MOLAP) databases[27]. Most OLAP tools have advanced exploration and analysis capabilities that provide deeper insight into the data, in addition to the visualisation capabilities, that could reveal correlations and causations. Thus the data mining can proceed in narrower levels using specialised tools and techniques discussed in previous chapters. More details on the connections between the interplay of data storage and data mining can be found in chapter 11.

---

[27]Virtual OLAP can access data in enterprise databases.

5.10   Multimedia OLAP

Mining data that contain a mixture of media and non-media types is called multimedia OLAP. Multimedia datatypes are mined with the help of their metadata that contain useful information. A music store may be interested in finding associations between most often checked out items in terms of producer, theme, lyric etc which are available in metadata files. Mining using internal characteristics may involve a combination of transformations, filtering, feature extractions, selection and pattern matching. The aggregate summary used in numeric data may not be applicable to all multimedia attributes. However a pre-processed and transformed (to other domains like frequency domain) multimedia data may reveal hidden patterns that can be used in multimedia OLAP. Transformations of large volumes of data to frequency domain may be time consuming or impractical at run time. Hence these steps may be carried out prior to the OLAP analysis. But if a few OLAP operations are carried out to reduce the dimensionality of search space, the number of such transformations needed in the 'multimedia dimension' may be substantially reduced. Other dimensionality reduction techniques may also be employed using projection or fusion techniques. As an example, if the data cube for music stores are sliced using singer(s) name and gender(s), the year in which song was published, etc the resulting dimension can be reduced by fusing loudness, notes, beats, lyrics, rhythms to get an aggregated melody. However, the rollup operation from reduced dimensionality to original dimensionality is more difficult because it increases the dimensional attributes in a single step by more than one. Similarly, multimedia images may be progressively mined by first identifying a frequent association or cluster and then progressively mining in sub-dimensions.

5.10.1   Applications

The list of OLAP applications is very extensive. It is used in marketing, advertising, retail sales and profitability analysis, financial and budgeting applications, production planning, resource analysis. Most financial institutions use OLAP for financial modeling, performance analysis, and various decision making purposes. It is also used in airlines, media and entertainment, medicine, pharmaceuticals, insurance, and auto industries, to name a few. Insurance companies use OLAP mining to find out factors that result in cost overruns in healthcare patient groups, customer attritions, identifying new insurance schemes and services etc. E-commerce applications include mining for customer attritions, buying patterns, product performances, market fluctuations.

Most OLAP applications run on enterprise datawarehouses. Applications of OLAP are rapidly growing in a wide range of fields beyond enterprise datawarehouses, including finance and marketing, medicine [KW06], pharmaceutics, engineering, nutrition, fraud detection, toxic hazard analysis, forecasting, geographic data mining, genetics and bio-informatics. In large-scale medical studies, it is typical to have hundreds of variables measured on hundreds to thousands of patients over a period of time which generate copious amounts of data. As data analysis and visualisation tools and techniques can be invoked on selected subdimensions of cube-data, OLAP has already proven to be the most preferred technology by data miners in such situations.

5.10.1.1 An academic application

Almost all academic institutions maintain computer data about enrolled students. Some institutions keep the data in multiple databases under admissions and records, academics, finance and accounts divisions. Some of the data may also be scattered among various departments. In this application, we discuss an OLAP implementation that permit fast and efficient online analysis of student data. Educational administrators and admission committees may need to slice and dice the data by country, geographic region, test scores, financial need, etc. In addition, compatibility issues may have to be considered for assigning hostel room-mates (for unmarried students) and preferences for married students. By consolidating the scattered data into a single datawarehouse proves to be immensely useful in identifying performance indicators across various dimensions, flaws in resource utilisations, optimal budget allocations, predicting student path, performance and student transfer, additional funding requirements, under-utilisation of funds etc.

Table 5.5: List of OLAP Software

URL	Name	C/F/S
community.pentaho.com	BA platform	S
olap4j.org	open java API	F
olap.com	PowerOLAP	F
icCube.com	ic Cube server	S
jedox.com	Jedox OLAP server	F
delphistep.cis.si	pretty OLAP	F
codeload.github.com	kylinOLAP	F
sourceforge.net/projects/mondrian	Mondrian	F
sourceforge.net/projects/phpmyolap	PHP/My SQL Olap	F
instantolap.com/news/free-olap	Instant OLAP	F
radar-soft.com	radarsoft DOLAP	C
reportserver.datenwerke.net	ReportServer	F
www.cognos.com	Cognos	C
jdbc4olap.org	jdbc4olap	F

Notes: PowerOLAP is Excel friendly, ReportServer is a MultiDim OLAP

Student demographics can be indicated as a single dimension, although it includes sex, race, nationality, citizenship(s)[29], physical or mental disabilities (that may require special arrangements), age-groups[30], blood-type, residential status etc. Data for currently enrolled students alone is insufficient in finding a satisfactory answer for above queries, because the trends and patterns that the educators are looking for may be hidden in the past student data. Hence one

---

[29]Nationality indicates the official right (including voting rights, and rights to be elected to govt posts) to be a member of a country, which is established by birth. Citizenship is the state of being a member of a particular country and could be different from nationality, as citizenship can be acquired. Citizenship can also be supra-national (as in EU citizenship) (see en.wikipedia.org/wiki/Nationality, www.eupedia.com/forum/showthread.php?t=21134 for details). These could be the same for most students.

[30]people of various age groups register for web-based diplomas and degrees.

of the architectures discussed in chapter 11 may be utilised in optimally storing current and past student data.

The composite dimensions used in OLAP are arranged hierarchically, so that a roll-up operation takes us from months[31] to semesters and from semesters to years etc. Time is kept as a separate dimension to facilitate analysis and reporting on a yearly basis[32].

This system can easily be extended to other mining tasks related to students. As an example, large universities receive applications from prospective students from home and abroad for admission to various regular and online programs. These data can be mined using OLAP as well (see application section of chapter 8 for more details).

5.11   OLAP Software

Several OLAP related software can be found at the sites given in table 5.5. Commercial choices include Microsoft analysis services (microsoft.com/sql/technologies/analysis/default.mspx), IBM Cognos TM1 (www-01.ibm.com/software/data/cognos/), SAS OLAP server (sas.com), Oracle Essbase (oracle.com), Business Objects, Micro Strategy etc.

5.12   Exercises

1. Mark as true or false:
   (a) OLAP is a supervised learning model
   (b) OLAP can work with both datawarehouses and datamarts
   (c) Aggregation operation can be applied to numerically coded categorical variables
   (d) OLAP miners always start the mining process at the slice with 'ALL' as attributes
   (e) OLAP can be used to find intra-dimensional patterns and trends

2. Which of the following are microscopic operations?
   roll-up, drill-down, pivoting, slicing.

3. Which of the following OLAP operations *does not* use aggregation?
   A) roll-up   B) drill-down   C) slice   D) dice

4. Identify aggregation operators in the following list
   (a) sum (b) minimum of values (c) median (d) mode (e) average (mean) (f) count.

5. What are the values stored in a data cube? Can they be accessed cell-wise?

6. Can each cell of a datacube hold more than one value? What does an empty cell signify?

7. For a sales application, what are the dimensions of the data cube?

---

[31]Per month financial aid roll-up to per semester, and per semester GPA roll-up to annual GPA etc.

[32]As academic and calendar years are different, we have an option to choose time dimension (exercise 22)

8. Identify subdimensions in each of the following dimensions
   a) Time b) Product (product lines, product categories) c) location d) Ad media

9. What indexing technique does OLAP use? What are the advantages of maintaining separate indixes for each OLAP cube?

10. What are the key architectural differences between ROLAP and MOLAP?

11. If the product dimension contain m categories and location dimension has n categories,
    a) what is the dimensionality of the following query?
    SELECT * FROM database
    GROUP BY PRODUCT, LOCATION;
    b) What does the following nested SQL statement return?
    SELECT * FROM
    (SELECT * FROM location
    WHERE location_id = 'ASIA')

12. Explain what is a sparse data cube. How do you interpret the results, if the following query on a data cube returns a NULL value?
    SELECT * FROM products, locations
    WHERE location_id = 'AFRICA'

13. What are the various forms of the data preprocessing? Distinguish differences between preprocessing NOIR data, text data and video/image data.

14. Describe various data types available in data mining.

15. Compare and contrast OLAP and OLTP. How is OLAP different from traditional database querying? What conceptual data model does OLAP use?

16. Explain what is meant by the 'curse of dimensionality' in OLAP. Discuss the importance of establishing a standardized data mining query language. What are the potential benefits and challenges involved?

17. Give an application of SOLAP. How are the results of SOLAP displayed?

18. Identify each of the following facts as additive, semi-additive or non-additive (a) account balance (b) sales amount (c) unit price (d) VAT

19. What aggregation measure (algebraic, distributive, holistic) does each of the following belong to?
    a) The mean deviation from the median of a set of attributes $(x_1, x_2, \cdots, x_n)$ with median M for a dimension $MD = \sum_{i=1}^{n} |x_i - M|$.
    b) Minimum of a set of observations recursively defined as $\min(x_1, x_2, \cdots, x_n) = \min(min(x_1, x_2, \cdots, x_{n-1}), x_n)$.
    c) Maximum of a set of observations recursively defined as $\max(x_1, x_2, \cdots, x_n) =$

$\max(max(x_1, x_2, \cdots, x_{n-1}), x_n)$.

d) Range of a set of observations defined as $\text{range}(x_1, x_2, \cdots, x_n) = \max(x_1, x_2, \cdots, x_n) - \min(x_1, x_2, \cdots, x_n)$.

e) An alternating-sign sum: $(x_1 - x_2 + x_3 - \cdots + (-1)^{(n-1)} x_n)$

f) Sum of squares (SS) of a set of observations recursively defined as $SS(x_1, x_2, \cdots, x_n) = \text{sum}(SS(x_1, x_2, \cdots, x_k), SS(x_{k+1}, x_{k+2}, \cdots, x_m), SS(x_m, x_{m+1}, \cdots, x_n))$.

20. Classify the following measures as unique holistic, nonunique holistic or ambiguous holistic:

    a) Variance (s^2) b) Median c) (Mean - Median) d) Correlation coefficient (r) e) square of r (r^2) f) Coefficient of variation CV=$\overline{x}/s$. g) mean deviation from median $\sum_{i=1}^{n} |x_i - median|$. If the GM of distinct subsets ($S_1, S_2, \cdots, S_k$) of respective cardinalities ($m_1, m_2, \cdots, m_k$) of a set S=($S_1 \cup S_2 \cup \cdots S_k$) are ($GM_1, GM_2, \cdots, GM_k$), the GM of S can be found as $GM = (GM_1^{m_1} * GM_2^{m_2} \cdots * GM_k^{m_k})^{1/(m_1 + m_2 + \cdots + m_k)}$. By taking logarithm of both sides, we get $\log(\text{GM}) = 1/(m_1 + m_2 + \cdots + m_k)[\sum_{i=1}^{k} log(GM_i)]$.

21. Consider a medical diagnostic hierarchy in which the patients are categorised under major illnesses (Heart, Lung diseases, Cancers, Diabetics, etc; group everything else under 'Other' category). Assume a diabetic sub-hierarchy (type-1 to type-4 diabetic categories). Draw the tree diagram and add gender, ethnicity, as additional dimensions. Explain with examples, how the roll-up, drill-down and slicing operations can be implemented.

22. The time dimension in the academic application facilitates yearly analysis and reporting. Discuss how you will distinguish between calendar year and logical academic year (that often starts in late August or early September and ends in July at most US universities) which is more or less fixed for a university from year to year.

23. Write short notes on:

    (i) Data mining in retail industry (ii) Data mining in insurance (iii) Data mining in higher education

24. Extend the academic system to manage prospective student data. Replace the 'student standing' dimension with the program of study (undergraduate, graduate, special) and discuss how you will distinguish between regular (on-campus), correspondence and web-based online programs of study. Add marital-status, smoking and drinking habits as prospective student attributes to assign compatible students as room-mates, for those requiring hostel accommodation. What does each cell of the datacube store for prospective students?

25. Add extra-curricular activities as a separate dimension. Populate the above data cube with sample data, and try to answer the following using OLAP mining:

    (1) Does there exist any correlation between (i) the student age and course loads? (ii) marital-status and time to graduation? (iii) hours employed and time to graduation? (iv) Extra-curricular activities and course load?

26. A banking application

    Banking is another area where the benefits of OLAP mining can save lots of time and resources. Even small banks and financial institutions use a variety of networks for efficient operations and to keep abreast of competitors. Most banks offer different kinds of plastic cards – cash and credit cards of various types and preferences. They have various types of accounts (savings, checking, money-market, fixed deposits etc) and different types of loans (vehicle, house, property, etc). These attributes along with the sex, race, marital status, age-group, employment-group, education levels and family size(s) will form the dimensions. In this exercise, you will create a data cube for banking and populate the datacube using SQL statements. Mine the data to find out hidden patterns.

27. Compare and contrast OLAP and OLTP. How is OLAP different from traditional database querying? What conceptual data model does OLAP use? Explain what is meant by the 'curse of dimensionality' in OLAP.

28. What is a bitmap index? What are its advantages? How do you use the bitmap index to execute the SELECT COUNT(*) command?

29. What are the variants of the bitmap indexing technique? Discuss the advantages and disadvantages of bitmap indexing and its variants.

30. What are the security concerns in OLAP. At what levels can security be implemented?

5.12.0.2   References

[AM97] Anahory, S., Murray, D.(1997). *Data warehousing in the real world*, Addison-wesley.

[BT07] Bimonte,S., Tchounikine,A., Miquel,M. (2007) Spatial OLAP: Open issues and a Web based prototype, 10th AGILE International Conference on Geographic Information Science, Aalborg, Denmark, 8-11 May 2007.

[BC99] Brewster, S.A., Cryer, P.G. (1999) Maximizing screen-space on mobile computing devices, Proceedings of ACM SIGCHI conference, 224-225.

[CJ06] Celko, J. (2006) Analytics and OLAP in SQL, Morgan Kaufmann.

[CI99] Chan, C.Y., Ioannidis, Y.E. (1999) An efficient bitmap encoding scheme for selection queries, SIGMOD conference, 215-226.

[CL94] Cicchetti, R., Lakhal, L. (1994) Matrix relation for statistical database management, Proc of 4th international conf. on extending database technology, 31-44, Springer-NY.

[CC93] Codd, E.F., Codd, S.B., Salley, C.T. (1993) Providing OLAP to user-analysts: an IT mandate, Codd and Date Inc.

[CF03] Cuzzocrea, A., Furfaro, F., Sacca, D. (2003) Hand-OLAP: a system for delivering OLAP services on handheld devices, ISADS, Pisa, Italy, 213-224.

[EO81] Eggers, S.J., Olken, F., Shoshani, A. (1981) A compression technique for large statistical databases, VLDB, 424-434.

[GC99] Goil, S.,Choudhary,A.(1999) A Parallel Scalable Infrastructure for OLAP and Data Mining, Proc. International Data Engineering and Applications Symposium (IDEAS'99), Montreal, August 1999.

[GB96] Gray, J., Bosworth, A., Layman, A., Pirahesh, H. (1996) Data cube: a relational aggregation operator generalizing group-by, cross-tab, and sub-totals, Proc of 12th international conference on data engineering (ICDE), 152-159.

[GP00] Gu, J., Pedersen, T.B.,Shoshani, A. (2000) OLAP++: Powerful and easy-to-use federations of OLAP and Object Databases, Proceedings of the 26th International Conference on Very Large Data Bases, 599-602.

[HJ98] Han,J. (1998) Towards on-line analytical mining in large databases, ACM SIGMOD Record, 27(1), 97-107, March 1998.

[HK06] Han, J., Kamber, M. (2006) Data Mining: concepts and techniques, 2nd ed., Morgan Kaufmann.

[HR96] Harinarayan, V., Rajaraman, A., Ullman, J.D. (1996) Implementing data cubes efficiently, Proc of SIGMOD 96, 205-216.

[JM00] Jermaine, C., Miller, R.J. (2000) Approximate query answering in high-dimensional data cubes, ACM SIGMOD workshop on research issues in data mining and knowledge discovery, 31-36.

[JB04] Johanis, P., Bellerose, P. (2004) Use of standardized metadata to find, select and access statistical data, Joint UNECE/Eurostat work session on statistical metadata, Geneva, Feb 9-11, 2004.

[JL99] Jurgens, M., Lenz, H.J. (1999) Tree based indexes vs bitmap indexes: a performance study, Intl workshop on design and management of data warehouses, Heidelberg,78-87.

[KS04] Karayannidis,N., Sellis,T., Kouvaras, Y.(2004) CUBE File: A File Structure for hierarchically clustered OLAP Cubes. In Proc. of the EDBT'04 International Conference, Heraklion, Greece.

[KW06] Kaur, H., Wasan, S.K. (2006) Empirical study on applications of data mining techniques in healthcare, Journal of Computer Science, 2(2), 194-200.

[KL01] Kim, M., Lim, Y. (2001) A Z-index based MOLAP cube storage scheme, Ewha Institute of Science and Technology research report series EIST-CSE-01002, 2001.

[KS02] Kim, M., Song, J. (2002) An efficient ROLAP cube generation scheme, Journal of Korea Information Science Society, 29(2), 99-109.

[KK97] Kirkgoze, R., Katic, N., Stolba, M., Tjoa, A.M. (1997) A security concept for OLAP. Proc. of 8th international workshop on database and expert systems (DEXA'97), 619-626.

[LT01] Lee,Y., Tak,S., Park,E.K., Stach,J. (2001). Secure and sharable data warehouse/OLAP model for E-commerce, In *Proc. of high performance computing symposium*, 170-177.

[MA03] Maniatis, A.S. (2003) OLAP presentation modeling with UML and XML, Proceedings of BCI 2003, 232-241, Thessaloniki, Greece.

[MV03] Maniatis, A.S.,Vassiliadis, P., Skiadopoulos, S, Vassiliou, Y. (2003) CPM: A Cube Presentation Model for OLAP, Lecture notes in computer science, Vol 2737, 4-13, Springer-Berlin.

[MA04] Maniatis, A. (2004) The Case for Mobile OLAP, Proc. of the international conf. in pervasive information management (PIM '04), Heraklion-Crete, Greece, March 18.

[MV05] Maniatis,A.S., Vassiliadis,P., Skiadopoulos,S.,Vassiliou,Y., Mavrogonatos,G., Michalarias, I.(2005) A presentation model and non-traditional visualisation for OLAP, International journal of Data Warehousing and Data Mining, 1(1), Idea Group Publishing, 136.

[MVS03] Maniatis,A.S, Vassiliadis,P., Skiadopoulos,S.,Vassiliou,Y.(2003) Advanced visualisation for OLAP, Proceedings of the 6th ACM international workshop on Data warehousing and OLAP (DOLAP'03), New Orleans, 9-16.

[MS07] Mansmann, S., Scholl, M.H. (2007) Exploring OLAP aggregates with hierarchical visualisation techniques, SAC'07, March 11-15, Seoul, 1067-1073.

[MM05] Matias, R., Moura-Pires, J. (2005) Spatial online analytical processing (SOLAP): A tool to analyze the emission of pollutants in industrial installations. EPIA 2005 – 12th Portuguese Conference on Artificial Intelligence, 5th International Workshop on Extraction of Knowledge from Databases & Warehouse (Bento,C.,Cardoso,A., Dias,G. (Eds.)).

[MV00] Mendelzon, A.O., Vaisman, A.A. (2000) Temporal queries in OLAP, VLDB, 242-253.

[MZ92] Michalewics, Z. (1992) Statistical and scientific databases, Ellis Horwood.

[PR02] Paelke, V., Reimann, C., Rosenbach, W. (2002) A visualisation design repository for mobile devices, Proc of 2nd ACM international conference on computer graphics, virtual reality, visualisation and interaction in Africa, Cape town, 57-62.

[PP00] Priebe, T., Pernul, G. (2000) Towards OLAP security design - survey and research issues, Proc of 3rd ACM international workshop on data warehousing and OLAP, Washington DC, Nov 10, 2000.

[RM03] Rafanelli, M. (2003) Multidimensional databases - problems and solutions, Idea group, London.

[RT97] Rao, R., Tenev, T. (1997) Extending Table Lens to multidimensional data and OLAP operations, CODATA Euro-American workshop on data & info. visualisation, Paris.

[RZ96] Roten, D., Zhao, J.L. (1996) Extendible arrays for statistical databases and OLAP applications, 8th international conference on statistical and scientific database management (SSDBM), 108-117.

[RK97] Roussopoulos, N., Kotidis, Y., Roussopoulos, M. (1997) Cubetree: organisation of and bulk incremental updates on the data cube, Proceedings of SIGMOD 97.

[RG99] Ruf,T., Grlich,J., Reinfels,I.(1999) Dealing with complex reports in OLAP applications, Proc. of the First Intl Conf. on Data Warehousing & Knowledge Discovery, 41-54.

[SC02] Sharaf, M.A., Chrysanthis, P.K. (2002) Semantic-based delivery of OLAP summary tables in wireless environments, Proceedings of the 11th CIKM, McLean, Virginia.

[SL00] Shekhar, S., Lu, C.T., Tan, X., Chawla, S., Vatsavai, R.R. (2000) Map Cube: A visualisation tool for spatial datawarehouses, Dept of CS, University of Minnesota.

[SA97] Shoshani, A. (1997) OLAP and statistical databases: similarities and differences, ACM symposium on principles of database systems, Tucson, AZ, 185-196.

[SK97] Stolba, M., Kirkgoze, R., Katic, N., Tjoa, A.M. (1997) A security concept for OLAP, Proc. of 8th intl. workshop on database and expert sys application, IEEE comp. press.

[TS06] Tan, V.K.P, Steinbach, M., Kumar, V. (2006) Introduction to data mining: concepts and techniques, Addison Wesley.

[TE02] Thomsen, E. (2002) OLAP solutions: building multi-dimensional info. systems, Wiley.

[VM02] Vaisman, A.A., Mendelzon, A.O. (2002) A temporal query language for OLAP: implementation and a case study, LNCS 2397,(Grahne, G., Ghelli, G., eds.),78-96, Springer.

[VS99] Vassiliadis,P., Sellis,T.(1999) A survey of logical models for OLAP databases, ACM SIGMOD Record, 28(4), 64-69, Dec. 1999.

[VS00] Vassiliadis,P., Skiadopoulos,S. (2000) Modelling and optimisation issues for Multidimensional Databases, Proceedings of the 12th International Conference on Advanced Information Systems Engineering, 482-497, June 05-09, 2000.

[VM06] Vinnik, S., Mansmann, F. (2006) From analysis to interactive exploration: Building visual hierarchies from OLAP cubes, EDBT 2006: Proceedings of the 10th international conference on extending database technology, 496-514.

[WM99] Wu, M.C.(1999) Query optimisation for selections using bitmaps, Proceedings of the international ACM SIGMOD conference, 227-238, June 1999.

6
Decision Trees

Chapter objectives

- Discuss the advantages and disadvantages of decision trees

- Describe popular algorithms for decision trees

- Explain production rule generation from decision trees

- Understand pruning of trees

- Introduce fuzzy decision trees

- Illustrate applications of decision trees

6.1   Graph Theory

Decision trees (DT) are nonlinear supervised learning models that use the concept of trees. A tree in computer science is a data representation model that originated from graph theory. The first part of this chapter introduces the unfamiliar reader to the bare essentials of graph theory needed for a good understanding of subsequent chapters. Those familiar with graph theory may skip to §6.3 in page 6-7. Ample data mining applications of graph theory are given throughout this section.

Graph theory is a branch of mathematics that deals with relationships (eg: adjacency, incidence, hierarchical) between elements of a finite set. The Swiss mathematician Leonhard Euler (1707-1783) introduced it in 1736 in connection with the famous Königsberg bridge problem,[1] and is considered to be the father of Graph Theory. Many data mining models use graphs and trees. For instance, the dependency relationships in spatial data mining may be represented as a graph. The influence diagram that captures event relationships of a DT in a compact form is a directed graph. Dependency relationships among spatial objects can be represented by a graph. Cross-purchase behavior of common items bought by customers together can be represented as a weighted undirected graph. The Uniform Resource Locator (URL) cross-link connectivity among web pages can be represented as a directed or undirected graph, and it could result in a disconnected graph (multi-graph or forest) for short durations. Unidirectional 'voting' in link

---

[1]He proved that a closed tour is impossible if the graph contains nodes of odd degree. The number of nodes of odd degree in a nontrivial graph is even. This is called Euler's lemma. See www.biographybase.com/biography/Euler_Leonhard.html, scienceworld.wolfram.com/biography/Euler.html.

mining can be represented by a directed graph. The multi-category support vector machines (SVM) can be solved using the directed acyclic graph method. Inter-cluster similarities when data contain two or more distinct clusters can be represented by an undirected graph. The term-by-document matrix (TD-matrix) in Latent Semantic Indexing (LSI) can be represented as a bipartite graph from the set of terms to the set of documents.

Graphs are defined on a nonempty set of elements, which are most often finite. The elements of the set are called nodes or vertices. It can be anything distinguishable – objects of various kinds, attributes of objects, various internal states of a machine or process, articles purchased, political parties, persons etc. The nodes can either be a homogeneous set (eg: airports, web users, traffic junctions, wireless access points, etc) or heterogeneous (operators and machines, jobs and applicants, salesmen and geographic regions, users and web browser software etc). Thus we may categorise graphs as homogeneous if all nodes belong to the same type of elements and as heterogeneous otherwise. The TD-matrix in LSI is a heterogeneous bipartite graph as one set of elements represents the terms and the other set represents the documents. A pictorial depiction of the totality of all such relationships for a particular problem is useful for visual comprehension. Most of the graphs that we encounter are 2-dimensional. Such graphs that can be drawn on a plane (paper, screen, plotting devices) are called planar graphs. Nonplanar graphs are those that cannot be drawn in a plane without edge crossing.

Definition 6.1 A *graph* is a pictorial representation of the relationships between the nodes, including self-loops, if any.

The simplest possible graph is a singleton node (without loops). In dynamic-link mining applications, we start with a singleton node (which is the parent page) and successively develop a graph as links are established to other web pages. Most common relationships are adjacency, incidence, hierarchical or inclusion type of dependencies. Each relationship pair is represented by an arc or an edge (and self relationships are represented by a loop.[2]) An edge can be drawn either as a straight line, an arc of a circle or using any other geometric shape. In our study, the shape of an edge is immaterial in characterising the graph.[3] An edge is directed if the direction of connectivity is important. For example, if a link exists between from web page A to another page B, it is represented as a directed arc. A directed graph has at least one directed edge. When edge directions are unimportant, it is called an undirected graph. For instance, the items purchased together in a supermarket can be represented as an undirected graph. The following is an abstract definition of graphs using node and arc sets.

Definition 6.2 A graph is a triplet G=(V,E,o) where V and E are finite sets and o is a binary relation defined on VxE satisfying $E \subseteq [V]^2$ such that $V \cap E = \phi$.

The elements of the set E are called edges (or arcs) and they are 'two element' (need not be unique) subset of nodes in V. For instance, if a term (in LSI) is contained in m documents, we draw m distinct edges from the node that represents the term to each of the document nodes. The binary relation o is also called incidence function. If the edges of a graph are assigned weights, it is called a weighted graph. For example, an airline connectivity graph can be weighted with the distance between airports, time of flight, ticket fare etc.

Definition 6.3 A graph without loops is called a *simple graph* and one with loops is called a *multi-graph*. Self-loops are always undirected.

---

[2] A self loop is a special form of adjacency relationship to indicate that a node is related to itself.

[3] In some engineering applications, each edge has a definite shape.

Example 6.1 At online floral sites, customers order flowers to be sent to friends or family members. This information among a group of frequent customers can be represented as a graph. If a customer orders a flower to be sent to self, it is a loop in the graph.

Definition 6.4 If each edge of a graph is directed, it is known as a totally *directed graph* (or digraph). If all edges are undirected, it is called an *undirected graph*. If it has directed and undirected edges, it is called a *mixed graph.*

The edges are undirected when the relationship is symmetric. Examples are inter-cluster distances, reachability of two users on a network, etc. Graph traversal algorithms use the number of incident arcs at each of the nodes for moving from one node to another connected node.

Definition 6.5 The degree of a node is the number of arcs incident on it. If the graph is directed, we can distinguish between in-degree (number of incoming arcs) and out-degree (number of outgoing arcs).

Example 6.2 The hyperlinks graph (in web mining) between documents is an example of a directed graph. A hyperlink from document-i to document-j does not necessarily imply a link in the opposite direction. A link from one place to another of the same document is represented as a self-loop. The in-degree is the total number of documents that points to the current document and the out-degree is the total number of external hyperlinks in the current document.

If two nodes are directly related, it is called an adjacency relationship. For example, if there is a direct flight between two airports, we draw an arc between those to denote the connectivity relationship. Relationships can be spatial or temporal. Other types of dependencies like precedence relationships, priorities, likeness etc could also be represented as graphs. The nodes at the ends of an edge are called incident nodes. If a node is directly related to many other nodes, it can be depicted as an incidence relationship.

Definition 6.6 A *path* is a succession of edges between two fixed nodes. A closed path is called a *cycle*. A (directed) graph without cycles is called *acyclic.*

The following are some of the paths that we encounter in the rest of the book. The reachability of one web page from another can be represented as a path. At least one path should exist between the input nodes and output nodes of a neural network. The paths from the root to each of the leaves of a decision tree result in classification rules.

Definition 6.7 A graph is *fully connected* if each node in the graph is connected to every other node. Hence every fully connected graph with n nodes, without loops has $n(n-1)/2$ edges.

A disconnected graph has one or more subgraphs (components) that are not connected. The components can be singleton nodes as well. Examples are 1) a document without hyperlinks to other documents, 2) a visitor who visits just one web page, 3) a super market item that is not purchased in combination with other items etc.

6.1.1 Drawing Graphs

To draw a planar graph, we place each node arbitrarily in the XY plane, and make connections between related nodes. Two nodes are connected by an edge if there is a direct relationship between those nodes. The initial placement of the nodes determines the shape (and size) of the graph. Re-positioning the nodes appropriately can minimise arc crossings. Graph size (size of edges and nodes) is unimportant in our study, so long as the edges are properly weighted. For example, in an airline network, the edge lengths need not correspond to the distance between the airports[4]. A tree is a particular type of graph, which can be drawn top-down or bottom-up. To draw it conveniently, start with the root node and work towards the leaves (as in the case of building a DT from production rules). Sometimes we may have to start from the bottom (leaves) and work upwards. This is called bottom-up approach.

Example 6.3 Consider an e-commerce site that sells books and videos. A data mining specialist is interested in analysing the cross-purchase behavior over a fixed period of time, say on an evening. A customer who buys a Fiction item is also likely to buy a Novel. The edge weights in figure 6.1 represent the number of persons who buys books in different fields. On a particular Friday, it could happen that all customers who bought astronomy books did not have any items purchased from other fields.

Figure 6.1: Disconnected graph showing book purchase relationship.

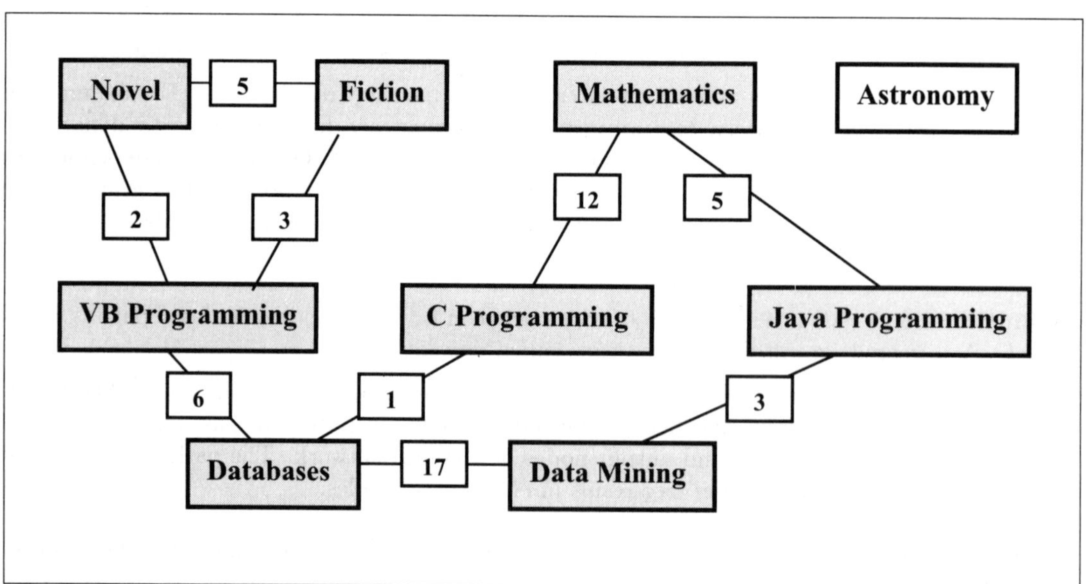

The above could be a disconnected graph for a particular time instance (which may become connected over an extended period of time). From the data mining point of view, we are interested in the entire past data rather than a particular snapshot of time interval. Such a graph can be used to purchase and stock new items, to identify fast-moving items, to cross-link items based upon customer preferences and to identify items that need not be stocked.

---

[4]For visual emphasis, the nodes with maximum degrees can be drawn on a larger scale than those with lower degrees. Similarly in a distribution network, arcs with maximum flow could be darkened. These are not standard notations.

6.1.2 Bipartite Graphs

A bipartite graph is a useful tool in data mining. It can be used to relate two nominal or ordinal sets or variables, or to relate one nominal set with an ordinal set[5]. For example, the preference of 3 categories of items (books, readymade clothes, food, jewellery, shoes) among various races (Hindus, Christians, Muslims, Sikhs, etc.) can be modelled by a bipartite graph. Other examples are gender vs product brands (nominal x nominal), Internet client browser vs total files downloaded (nominal x ratio), number of times computer user logged on vs number of mail messages read (as in public email services); or number of online access vs number of news items read (as in online news services) (both being counts are ratio x ratio type). Bipartite graphs can relate elements in two independent (distinct) heterogeneous disjoint sets as in matching applicants and job vacancies, or assigning workers to machines. For example, the purchasing behavior of customers can be represented as a bipartite graph from the set of all customers to the set of all items sold. As another example, consider the employees of a company who are divided into managers and subordinates. The relationship 'manages' can be modelled as a bipartite graph between these two sets if managers do not manage other managers. Bipartite graphs are also used in many other fields including image clustering, content based image retrieval, marketing research, matching problems, neural networks (especially in Kohonen networks), link mining, etc.

Definition 6.8 A graph G=(V,E,o) is bipartite if the vertex set can be partitioned into two disjoint sets $V = X \cup Y$ (and $X \cap Y = \phi$) such that each edge in E has one endpoint in X and the other end point in Y.

Example 6.4 An online site wishes to analyse the popularity of various browser among users in various continents. This information can be captured as a bipartite graph in which X={firefox, explorer, navigator, chrome, etc.} and Y={Africa, Americas, Asia, Australia, Europe, etc}. As there are thousands of users from each country, we will denote all connections from x_i to y_j by a single weighted line, where the weight indicates how many users from country y_j use browser x_i on a regular basis.

If every cycle of a graph (with two or more nodes) is of even length, it is bipartite (but the converse is not true) A bipartite graph is *complete* if each node in X is connected to every other node in Y. The adjacency matrix representation of an undirected complete bipartite graph will have ones as all entries (as there are m*n edges). To distinguish edge directions (of directed bipartite graphs), we could represent edges from X to Y by +1 and edges from Y to X by -1. By convention, undirected complete bipartite graphs are denoted by $K_{m,n}$. When m=n, we simply drop one of the subscripts (eg: K_3). The information content of a bipartite graph is the same as in a mxn matrix where m and n are the number of categories (cardinalities of the sets X and Y). The above definition helps us to visually identify bipartite graphs, which are usually drawn with the two sets of elements arranged into two columns (left-right) or two rows (up-down).

Example 6.5 Consider a car company that offers three models in three different exterior colours. Each edge in this graph can be labelled with the sales volume of the (model, colour) pair. This graph is not a complete bipartite graph, since some of the edges are missing.

This type of graph gives us valuable information on customer colour preferences and can be applied in many other fields like colouring of toys, bags, caps, sports items, kids-wear, and other consumer goods in different varieties. To be of any practical use, both m and n should be ≥ 2.

---

[5]They can also relate quantitative data using intervals as categories or counts which are ratio-type variables.

Either of them can also be countably large. As an example, if a restaurant (or super market) has 5 people in service and thousands of customers (some of whom are repeat visits), the servicing information can be represented as a bipartite graph. In a web mining application, the content types can be pure text documents, web pages, log files, audio and video files etc and the keys used to retrieve the information can be text strings, pictures, audio and video clips, etc. This can be represented as a bipartite graph.

6.1.2.1 Constructing Bipartite Graphs

To construct a bipartite graph from data records, (a) identify the node sets X and Y and order them if needed. (b) Draw an arc between $x_i \in X$ and $y_i \in Y$ if they are related. (c) Indicate arc directions (directed or undirected) and weighting (if any) (d) If there are unconnected nodes in either X or Y, drop them. If there are disconnected subgraphs, separate them out as independent bipartite graphs. As mentioned before, the term by document matrix used in LSI can be considered as a bipartite graph in which X=set of terms and Y=set of documents (an arc from X to Y indicates that the term is present in the document). In large LSI applications, this reduces to bipartite forests (disconnected bipartite graphs) and this information can be used to cluster similar documents.

6.2   Trees

In computer science parlance, trees are nonlinear (or hierarchical) data representation and modeling tools. It has applications in sorting and searching, data structures, algorithms design, compilers, DBMS, AI, natural language processing, etc. Trees are best understood using a graphical depiction of it with nodes and leaves, where nodes are the connecting points and leaves are the branches. In most tree-based algorithms, the processing (condition checking) takes place at the internal nodes.

Definition 6.9 A *tree* is a graph in which there exists a unique path between any pair of nodes[6].

This implies that the tree is a single connected graph without loops. The *root* of a tree is a special node which is the first vertex created in the top-down tree creation process. It is also the *starting point* for most of the tree traversal, searching and tree based classification algorithms (see decision trees below). A disconnected graph in which each component is a tree is called a *forest*. The *depth* of a node is the number of edges from the root to that node.

The depth of a tree is the maximum depth of leaf nodes (terminal nodes without branches are called leaf nodes). It is also called height of the tree. By convention, the height of a trivial tree with a single node is zero. The difference between the heights of left and right subtrees of a non-leaf node is called its *balancing factor*. We will try to make the balancing factor small in most data mining applications. This is especially important while reconstructing a DT from the set of all production rules. A tree is skewed if the depth is more on one side than the other.

Trees can also be directed or undirected. Undirected trees are encountered in most practical applications. The number of edges incident from top (or left) is called the incoming edge, and the edges towards the bottom (or right) are called outgoing edges. The number of incoming edges for every node of a tree (except the root) is always one. The outgoing edges for leaf nodes

---

[6]From the data structure point of view, a tree is a non-linear data structure in which a relationship (like precedence) exists between values stored at a node and its child node

are zeros. The outgoing edges of a node can represent different categories, decision alternatives or partitioning.

6.2.1 Binary Trees

Binary trees are encountered in many computer fields – sorting and searching, AI, data structures, algorithms, etc. They are used in several divide-and-conquer and branch-and-bound algorithms.

Definition 6.10 If each node in a non-trivial tree has either zero, one or two outgoing edges, it is called a binary tree.

Binary trees can be mapped to several parallel architectures like hypercube, meshes, etc. The dendrogram used in cluster analysis is a *rooted binary tree*, in which (i) the root represents the whole data to be clustered (S) (ii) each leaf node represents single data items (iii) each internal node represents a nonempty subset of S. Some of the DT induction algorithms also create binary trees (see table 10.10). If *each node* has either zero (leaf nodes) or two child nodes, it is called a complete binary tree.

6.2.1.1 Drawing Trees

There is no universally accepted representation for drawing trees. The nodes may be represented by dots, circles, ellipses or rectangles. A circle or ellipse is better in giving a name or label to the nodes. Alternatively, we could draw all internal nodes using one shape (say ellipse) and all leaf nodes using another shape (say rectangles). This will convey more meaning to decision trees. Unlike in a graph, the edges of a tree are almost always drawn as a line segment[7]. Each edge may be given one or more labels. The labels at each rooted level must be unique, but may be the same at different levels (see figure 6.11). For example, if two or more dichotomous variables have 'Yes' or 'No' values, a DT constructed out of them will have the same labels at different levels. Trees are usually drawn from top to bottom (with the root at the top) or from left to right (with root at the left). To save space, some software packages draw tree branches as horizontal lines.

6.3 Decision Trees

Definition 6.11 A decision tree (DT) is a nonparametric classification and prediction model organised in the form of a rooted tree with at least 2 levels, and at least 2 branches at one or more levels that have two types of nodes called decision nodes and class nodes.

The DT is a *supervised* data mining model, which originated in managerial decision theory, gambling and theory of games. The input to a DT algorithm is the labelled training data and the output is the hierarchical structure hidden in the training data. A characteristic of DT is that it decomposes a complex decision making problem into smaller manageable sub-problems.

---

[7]some software packages draw tree edges as two touching line segments

Complex decisions are based upon a large number of factors. These factors are represented by simple binary digits (0 and 1), categorical variables, integers, reals, complex numbers or structured data types. Variables are categorical or quantitative in most data mining applications. In web and text mining, we also come across structured data. Building a DT from data starts by identifying the correct attributes and data records. These are filtered (separated out) into a flat file (without any hierarchical structure on it), with each row representing data about one subject. Attribute (columns) values can be comma separated or tab separated. The chosen format depends upon the software being used for DT induction. The class labels (categories) into which samples get assigned should be known for the training data. As a convenience, the class labels are chosen as the last column (the last data item on each record). The order in which data records are arranged is immaterial, but a proper rearrangement can speed-up the tree induction algorithms. Records with missing values are preferably moved to the end (towards the bottom) of the flat file. All the classes must be mutually exclusive and collectively exhaustive. In other words, each data item should belong unambiguously to a single class. The number of cases should be more than the total number of classes. Data must be sufficient for a reasonable number of splits. In other words, each class should have a reasonable number of representative records. This could be a problem in some domains (eg: in medical studies that have a large number of negatives and a few positives). Hence a stratified sampling technique may be better suited than simple random sampling to select the training data records.

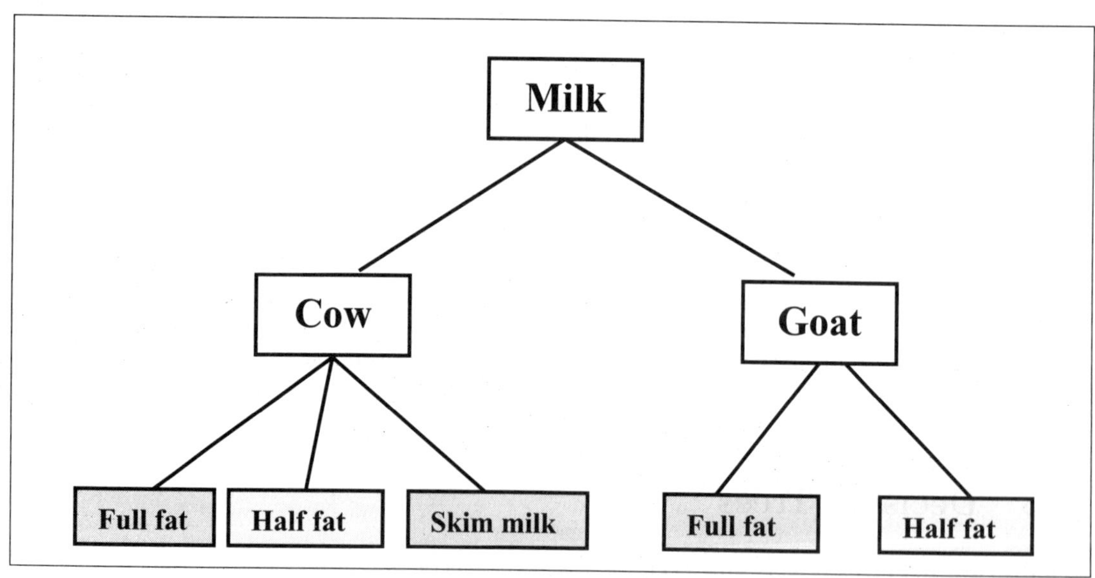

Figure 6.2: A tree hierarchy for milk

The DT partitions a (flat file) data set sequentially to maximise dependent variable differences, thereby reducing the dimensionality of search space at each successive step. This corresponds to partitioning the data into sets of rows using column attribute values. It splits a big row set into smaller ones iteratively. Conceptually, a DT can be viewed as comprising of internal nodes that represent data split points using a condition on a selected attribute, branches (or arcs) that are labelled with attribute values or uncertainties that represent distinct decision rule components used for classification (pp.6-13) or prediction, and leaf nodes that represent the classes or outcomes. The output attributes are usually categorical because our aim is either to classify each new instance or to predict the outcome.

Each node of a DT has an associated purity index[8], and size which is the number of data items under consideration for the split at that node. If the nodes are numbered from top to bottom sequentially, we will denote the size of node i by s_i.

6.3.1   Chance and Terminal Nodes

A node of the DT with at least one child node is called an internal node or non-leaf node. There is at least one sub-tree (which in the trivial case is a pair of nodes connected by an edge) rooted at an internal node. Thus each internal node is a decision point where some conditions are tested using available attribute values. Hence they are also called *decision nodes, chance nodes* or *nonterminals.* All nodes except the root node have a single incoming edge (root node has no incoming edges as it is the starting point). All nodes without children are called leaf nodes (they have no outgoing edges). The leaf nodes (terminal nodes) are the classes into which an item will be classified. To be meaningful, each DT should have at least 2 distinct terminal nodes. If the probability of a predicted class (at leaf node) is 1, it is called totally pure. The classification rule for a class is the path traversed from the root node of a DT to the leaf. There are as many classification rules as there are leaf nodes. The set of all such classification rules is called the production rules for the DT (see §6.3.5). These production rules can be translated into computer program statements (after optimising them if the depth is more than 1).

6.3.2   Advantages of Decision Trees

There are many advantages to DT model over other data mining models. At the conceptual level, the DTs are nonlinear data mining models that perform better than linear models. There are no multi-collinearity problems for correlated variables in DT. In addition, the production rules obtained from a DT can be combined to easily approximate global decision regions from simpler local decision regions. Below we summarise some of the important advantages of decision trees.

- They are easy to understand and interpret
  A DT clearly labels each of the alternatives using the split criteria. Hence it is easy for experts and non-experts alike to understand and interpret them in a practical situation.

- Easy to build and evaluate and hence relatively inexpensive to compute.
  Decision trees are built using training data sets that are small in size. Thus the computational complexity to build a DT is O(mn) where m is the number of attributes and n is the size of the training set. Once built, the tree can be used to classify any number of data sets. The worst case time complexity of a decision is $O(d)$ where d is the maximum depth of the DT.

- Works well without domain experts
  Making decisions is a routine activity in some businesses. Some of the decisions are daring due to the risk factors involved. The DT can be used to make difficult decisions in the absence of a domain expert. A DT may also be used for important rule generations

---

[8]A node is totally pure if all instances belong to the same class.

(see below). DTs eliminate human biases (eg: in credit approvals) in risky classification problems.

- Works for categorical and quantitative data
  As discussed in chapter 1, the interval and ratio types of data can be categorised using various discretisation techniques. They can also be binarised using statistical measures to give binary DTs. Hence the DT model works for both types of data. They can represent any boolean function.

- Useful for classification and regression
  The primary purpose of a DT is to classify an item into one of the known classes. They can also be used for prediction. For example, by mining the past data on loan defaulting customers, a bank can predict the chances that a new loan applicant would likely default on a repayment. DTs can also be used to classify a completed loan applications as likely to be approved or denied (with a certain probability of approval). This is useful to give an advance instant message to online loan applicants. Similarly using a DT, a doctor can predict the likelihood (probability) that a high-risk patient will have a heart attack.

- DT can grow to any depth
  The depth of a DT depends upon the number of attributes (while the breadth depends upon the number of categories or intervals). Since there are a finite number of attributes in any decision making situation, the tree can attain a maximum depth of the number of test attributes.

- Production rules can be obtained directly from a DT
  Production rules are obtained by traversing from the root to the leaves (or in reverse) of the tree. The total number of leaves is an upper bound on the number of production rules.

- Attributes can be chosen in any desired order in a DT
  Although the DT is constructed using an information maximising heuristic (see §6.3.6), the attributes that constitute a production rule can be chosen in any desired order. This may be useful in parallel classification of multiple test-sets. Branches that represent the outcomes of the conditions tested at a chance node can be arranged in any desired order.

- They are relatively faster than other classification models
  Classifying a new item using a DT takes at most d comparisons where d is the maximal depth of the tree. Other classification models (sequential comparisons, neural nets, linear discriminants) either require specialised knowledge or are more time consuming than the DT approach (The support vector machines (SVM) is an exception).

- It is easy to do pruning
  If the predicted class label differs from the actual one, it is called a misclassification. A DT will require pruning to reduce misclassification errors. Pruning will make minor adjustments to the tree to incorporate the misclassified items. This may be needed when the training set is not a true random sample or when the test set is time varying.

- Works when some data values are null
  Missing data are quite common when data are collected using web forms or by data collection devices like optical scanners, sensors or devices operating in noisy channels. Null values can be accommodated into a DT model by selecting the split criteria as 'null' and 'non-null' or using lazy decision trees (see also [TJ08]).

- Can work with different subsets of features at each node
  The split criteria used at each node can be based either upon a single attribute or a combination of attributes (see below).

- They can identify useless attributes
  All attributes may not contribute (positively or negatively) in making a decision. A DT can help to identify such redundant attributes (see discussion in the box).

- They can identify outliers in the data
  The DT can be used to identify outliers. If we store a count of the number of times a node is traversed when the DT is used to classify an item in the test set, zero counts on a path of a subtree will indicate that the corresponding branch was created by an outlier in the training set. Similarly, too low counts along any path (from root to the leaf) will honestly reflect on probable outlier data. They can also help to identify duplicate data records (if each attribute is ordinal, duplicate data items will traverse same path from root to the leaves).

- They can process huge data sets
  A DT is constructed from a training set of small size. However in dynamic tree pruning algorithms, the tree is created and pruned simultaneously using training and test sets. As the DT is a set of production rules at the conceptual level, we can parallelise the task easily using shared memory or distributed memory architectures.

- Miscellaneous
  As a DT learned from data shows clear pathways for each classification, it is easy to generate random samples for pre-specified classes (as in Monte-Carlo simulation studies), and select attributes for research studies (eg: medical diagnosis). For time-dependent attributes, a DT can reveal the impact of time on decisions. They can work with unimodal as well as multi-modal data. Because the scaling of attributes is permitted, they can work across geographical boundaries (irrespective of the measures used for currencies, distances, volumes, weights etc in various countries). Finally, there are no distributional assumptions (as in logistic regression or discriminant analysis) on either the data or on the misclassification errors in a DT.

Example 6.6 Describe how will you use the DT model when some of the data values are null? **Solution:** Suppose attribute X has null and non-null data values. The corresponding BDT node can use one branch for null values, and the other branch for non-null values. If the DT is non-binary, the constituent components of non-null values can be assigned to each branch. If X is a quantitative variable, it can be categorized using an appropriate data range. In some problems, the null branch can also be combined with one of the other categories. When the training size is large, we could also estimate the null values from other data samples with non-missing values.

6.3.3 Disadvantages of Decision Trees

In spite of their impressive advantages, the DTs have a few disadvantages too.

- Common data scale
  All decision factors should be scaled to common units. This is more of a problem when continuous data are discretised using intervals or other methods, than with categorical variables.

- Deduced rules can be very complex
  Many production rules are possible when the number of attributes and classes are both large. In such situations, it is not easy for humans to interpret the generated rules.

- Decision tree could be sub-optimal
  When data are sparse, the DT can under-fit. Another problem is over fitting, which results in spurious decisions.

- Sequential decisions
  The decisions steps obtained from a DT are inherently sequential. When there are large number of production rules, backtracking is difficult to implement, although not impossible.

- Class overlap problem
  When classes partially overlap, the DT model may need large training data to disambiguate it. When the splitting variable is continuous, the linear decision boundary is a hyper-rectangle (in 3D or hyper-parallelepiped in nD) that are perpendicular to the splitting axes. Due to this property, they are called axis-parallel tree induction algorithms. An extension of this in which a linear combination of node attributes is used as test condition was suggested by Murthy [MS95]. The resulting algorithms are called oblique-DT algorithms. For some data distributions, the DT algorithm can get stuck in a local optima due to this restriction.

- Correlated data
  The DT model is not a good choice when input attributes are highly correlated. Correlation and class overlap are related but distinct concepts. Data can be highly correlated even if they are linearly separable. The DT can still be used in such situations where data are either perfectly linearly separable or have small overlap. Correlation is also a desirable property when the input attributes are independent, but are highly correlated with the output variable (classes). For example, spam-mail filtering systems look for the presence of a set of high frequency keywords in spam mails. There are many types of spam mails – financial fraud (that asks you to enter your bank account number or password, asks you to transfer money, etc), medicine fraud (that offers medicines, hormones etc online at cheap price), prize fraud (that tells you that you have won a prize), etc. Some of these spams can easily be caught by looking for the presence of specific keywords (which are the input attributes) in the mail message. These input attributes can be highly correlated with the output attribute (spam=Y or N). Strongly correlated input attributes can result in short tree branches (or equivalently short production rules). As an example, if a mail message contains "instant money" and "you are the winner", it will most probably be categorized as a financial spam[9].

- Validity period
  The validity period of a DT is not explicitly specified, due to the implicit assumption that it is valid for a reasonable period of time. Time dependent data may sometimes invalidate some of the production rules. Such invalidations can either be purely temporary (as in seasonal variations) or permanent for a reasonable period of time. A simple solution is to separate out the production rules as time dependent and independent ones and to prune the

---

[9]Present day spam-filtering systems do not further categorize the spam mail as financial-spam, insurance-spam, etc. This could tremendously help some users like doctors who order genuine medications online, finance departments who solicit online payment from defaulting customers, etc.

DT periodically using time dependent attributes. This is easy to do when time-dependent attributes appear away from the root (towards the leaves).

- Weak fault tolerance
  Some classification models like neural networks have high fault tolerance. Even if some nodes totally fail or links between nodes are broken, a neural network can still be used for classification. Although DTs are seldom implemented using multiple nodes, parallel implementations of very large DTs may require multiple processes or threads that execute the production rules simultaneously.

6.3.4 Classification

Dividing distinguishable things into discrete categories or distinct groups using one or more measurable or observable properties is called classification. As examples, bank clerks visually classify cheques using routing information, doctors classify terminal patients using chances of survival, and palaeontologists classify skeletons using carbon decay method. The distinct classes are known in advance in all these cases.

Definition 6.12 The task of assigning unlabeled data to one among a finite set of pre-determined classes using an empirical rule learned from labeled training data is called *classification*.

The objective in classification is to build a mapping function that assigns class labels to each new instance (predictive modeling) or to verify the appropriateness of class labels already assigned. The categorical class labels y (of training data) are assumed to be known for each of the training data (X,y). Each data tuple is assumed to belong unambiguously to one of several disjoint classes $(C_1, C_2, .., C_n)$. The DT learns the classification rules using the training data so as to minimise the classification error. They are multi-stage hierarchical classifiers in which the rejection of unfavorable cases occur rapidly at each of the intermediate nodes.

Decision Tree Classifiers (DTC) find applications in automatic character and speech recognition, signature and handwriting recognition, computer aided medical diagnosis, signal classifications (eg: signals with or without noise, echo, etc.), expert systems, network intrusion detection (eg: remote user is an intruder or not), marketing (eg: if a planned marketing campaign will be a success or not), astronomy (eg: automatic classification of galaxies as elliptical, spiral, lenticular or irregular[10]), spam-mail filtering etc. In direct marketing applications, one is usually interested in identifying likely customers for a product from a very large population of potential customers using a predictive model. In medical diagnosis they are used to identify patients with symptoms of a particular illness and to avoid or minimise errors in diagnosis.

If each node of a DTC has at most two splits, it is called a Binary Decision Tree Classifier (BDTC). A binary split leads to exactly two courses of actions, and are almost always based upon a single attribute. A multi-way split leads to multiple decisions when the production rules are mutually exclusive (figure 6.4) [KL01]. In other words, the rules corresponding to identical class labels can be combined. This is especially useful when an attribute has too many categories.

Algorithm 1 Hunt's Heuristic for tree construction
Hunt's CLS method recursively splits the available data to build a decision tree by starting with

---

[10]See en.wikipedia.org/wiki/Galaxy_morpholigal_classification, deep-space-astronomy.suite101.com, galaxyzoo.org, etc

the entire data at the root. Let $C = (C_1, C_2, \cdots, C_n)$ be the set of known class labels. At each step k, let R_k denote the set of training records at node k.

substep k_1: If all of the records in R_k belong to the same class y, then label node k as leaf and return (terminate further recursive splitting). If all records in R_k have the same attribute values (X), then terminate further splits and label node k as leaf with the maximum frequency class (in case of a tie, iteratively search up for maximum frequency in parent nodes).

substep k_2: Find the optimal split that will partition the training instances into homogeneous subsets using an attribute test. Recursively apply the split to each of the subsets.

When a decision is based on probabilities, the mathematical expectation can be used as a decision criteria if the stakes at risk are small, and the apriori probabilities are available for each alternative. In *more is better* situations we choose the maximum expected value, and in *less is better* situations we choose the minimum expected value. The certain amount for each alternative is called its *certainty equivalent*. If certainty equivalent and expected value are equal, the alternative is risk neutral. The process of computing expected value of a decision from the leaves to the root is called decision tree rollback.

6.3.5   Production Rules

During the construction of a DT, we accumulate the split criteria at each of the internal nodes. If the splitting attribute is categorical, the criteria use the discrete values along each branch. If it is continuous, the criteria use a continuum range along each edge. For example, if the splitting attribute is gender={Male,Female}, one branch will check for Male and the other branch will check for Female. If the variable is family income, different branches will check the range of income in various brackets. Production rules are obtained by combining the split criteria in an orderly manner from the root to the leaves (or in reverse). The conditions at all internal nodes are simply combined using 'AND' operator. The root-to-leaf traversal is also called top-to-bottom order because trees are almost always drawn with the root at the top. It is the recommended method for DT traversal due to the *divide-and-conquer* nature of the subtrees generated in this process.

Definition 6.13 Production rules (PR) are literal descriptions of rules that can be extracted from a DT by 'AND'ing together the conditions at each of the internal nodes along a path from the root to the leaf or in reverse.

More concretely, production rules convert the hierarchical decisions implicit in a DT into an explicit set of as many rules as there are paths from the root to the leaves of a DT. Conversely, a DT is a compaction of production rules optimally arranged to have a common prefix (path) from the root to the leaves.

They are of the form:– If <antecedent condition> Then <consequent statement> where each <antecedent condition> is either a simple expression or a complex expression 'ANDed' together (using & operator in programming languages). These do not distinguish between test conditions near the root of the DT and those near the leaves. The resulting expressions are called decision lists (DL), which are usually kept ordered (according to size or importance). The consequent determines the class to which a data item will get classified. The DL can be saved (externalised) to permanent storage and reused later, without explicitly reconstructing the DT in memory (this can be done in any programming language or using some DT software). In other words, the DT need be built once, and the PRs extracted from the built tree. They can be rearranged to make the classification task faster for high depth decision trees.

A DT may not classify all of the data accurately. We expect it to predict the class labels of unseen data *as correctly as possible*. A learning algorithm is well-generalisable if it could correctly predict the class labels. If we get stuck during the classification stage (tree traversal), or reach a wrong terminal node, we have a case of misclassification. The decision tree misclassification errors (DTME) may be used to decide if and when the DT need to be pruned. Theoretically, the DTME lies between 0 and 1. If all of the test data are correctly classified, the DTME is zero (it will never be 1 because the training data are drawn randomly from the same distribution that generates the data to be classified). Typically they lie in the range [0,.1] or narrower range. It can also be expressed as a percentage of the total number of test cases. For instance, if 5 out of 100 test cases are misclassified, the error is 5%. The quantity acc = (1-DTME) is called the accuracy of classification.

If multiple data are to be classified, we can sort the data on the primary attribute (at the root), within which sorting is done on the next level attributes and so on. Data samples with common attribute values will then require less number of comparisons. In those situations where the DT is continuously used on an ongoing basis, the DTME may be constantly kept track of to determine if there are any time-dependent attributes, and to decide if a pruning is appropriate. An analysis of the model is helpful when the DTME either fluctuates or tend to rise.

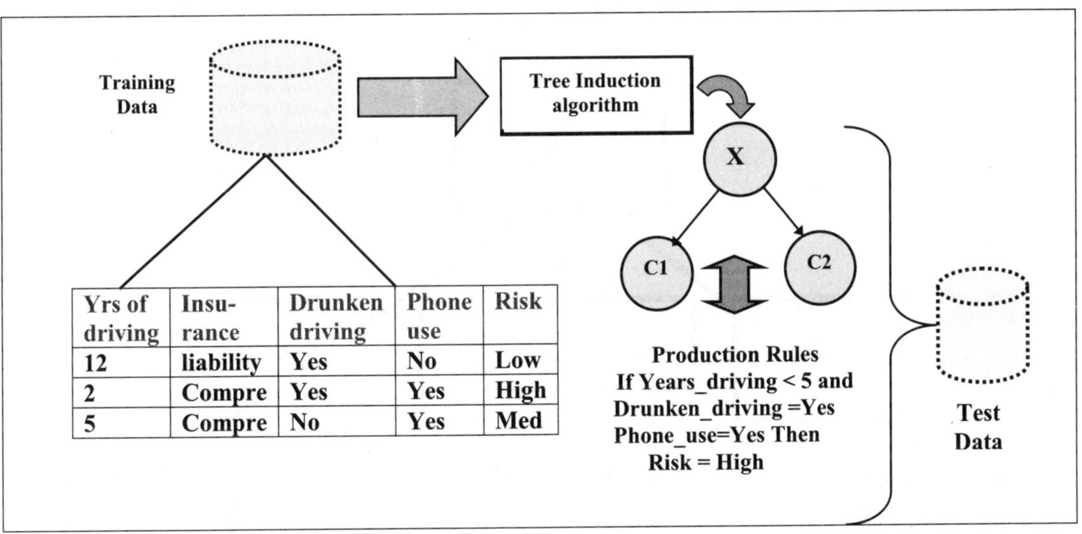

Yrs of driving	Insurance	Drunken driving	Phone use	Risk
12	liability	Yes	No	Low
2	Compre	Yes	Yes	High
5	Compre	No	Yes	Med

Figure 6.3: An overall view of the decision tree model

Those rules that have a large likelihood of being satisfied are evaluated first. They may also be optimised to run on a parallel computer that executes multiple instructions simultaneously.

There is a one-to-one correspondence between a DT and the set of all production rules extracted from it (if we are given the set of production rules, we can create the DT and *vice versa*). The extraction of production rules from a DT is called tree-to-text (or tree to program code) rule translation. To create a DT from the production rules, we must identify the root node. The DT is then constructed in a top-down manner (see below). The rules generated at each of the internal nodes during the tree building process are stored at their respective nodes. When the tree is completely built, it can be put to practical use in making a decision. This is accomplished by starting at the root node with new data. The root condition is tested at first.

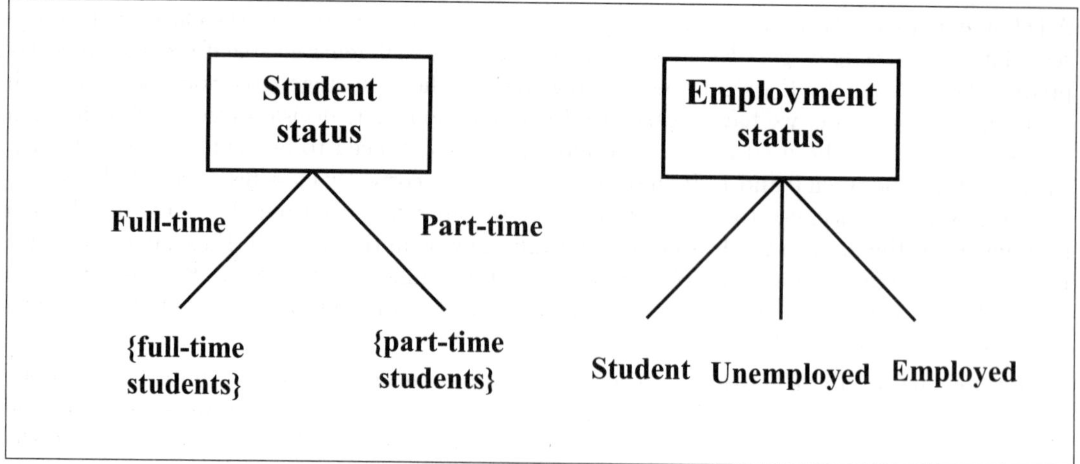

Figure 6.4: Binary and multi-way splits of a tree

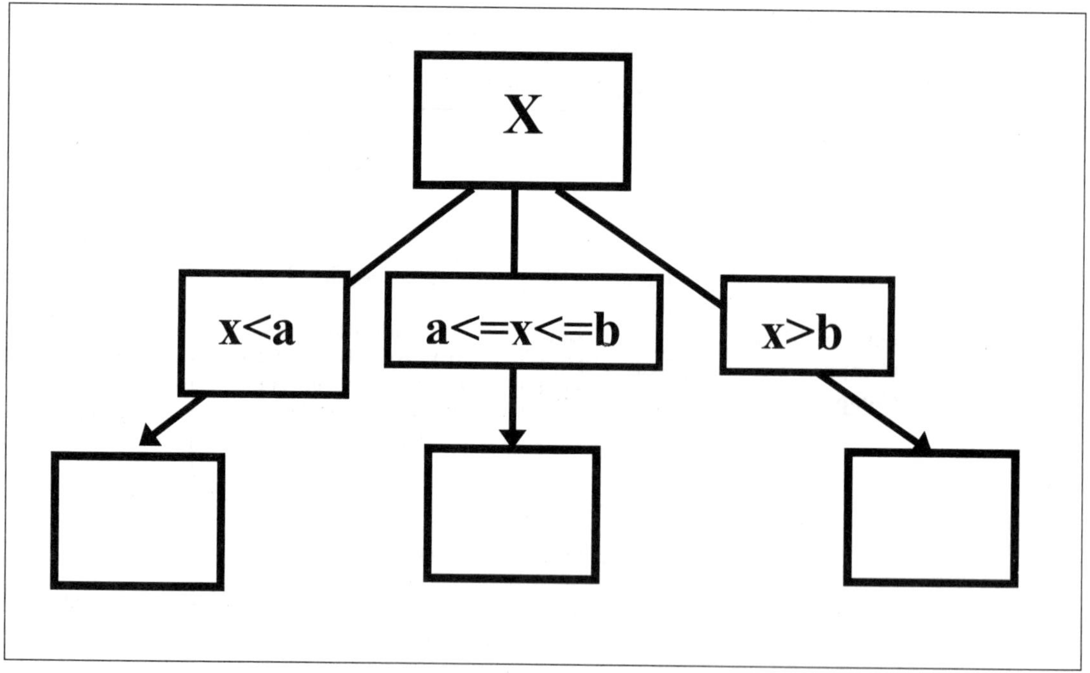

Figure 6.5: Ternary decision tree

Based upon the outcome at the root, we traverse one of the arcs or edges (and ignore all other arcs at that node) using the decision rules that satisfy the data, until we reach one of the leaf nodes. New datum is then assigned the class label of the reached leaf node. Such decision trees are called DT classifiers (DTC). If every internal node has at most two children, it is called a binary decision tree classifier (BDTC).

Theorem 6.1 Any multi-way DTC can be converted into a BDTC
Proof. If each node of a DT has **at most** two children, it is known as a BDT. Hence we will prove the theorem for ternary tree (see figure 6.5). Proof follows automatically by induction for higher order trees. Case 1: Split is based upon a quantitative variable
Without loss of generality, we assume that the first and last splits are open-ended as (x<a and x>b) (if they are close-ended, we can easily make them open ended. For eg: if 0<x<a, we make it as x<a). If there are 3 resulting splits, we could either collapse the second and third or collapse the first and second splits. In the first case we have (x<a, x≥a) as the split at root level. The right branch is further split as (a≤x≤b,x>b). The second case is similarly dealt with. When there are a large number of splits, we could use a divide and conquer strategy to split it more or less evenly. This will result in DT with lesser depths. In figure 6.6, we have splits

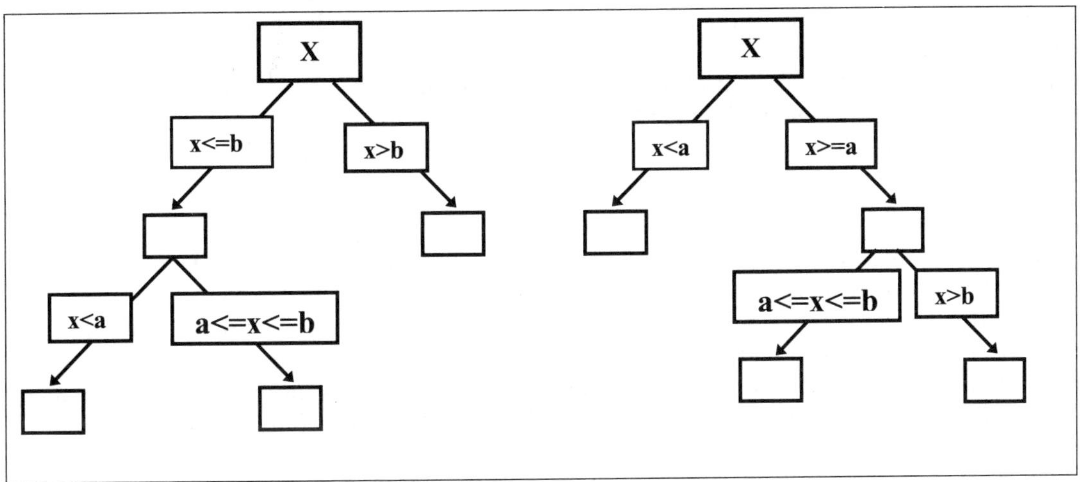

Figure 6.6: Split using 'b' first (left) and 'a' first (right)

corresponding to $x < a$ and $x \geq a$, and the second branch is split into $a \leq x \leq b$ and $x > b$.

Case 2: Split is based upon a categorical variable
The procedure in this case is exactly the same as in the continuous case. Any one of the categories is taken as left branch and the union of the other categories as right branch, which is then split further into its sub-categories. See exercise 11.

6.3.6   Building a DT

Decision trees can be built top-down, bottom-up or using special approaches like growing-pruning or a hybrid approach. The top-down approach is the most widely accepted method. Each training datum is comprised of a set of non-overlapping data attributes and a class label. The class labels in DT are assumed to be nominal, and can be coded alphabetically or numerically[11].

---

[11]The class labels are always chosen as numeric in SVM, as the SVM classifier is a linear combination of support vector points and class labels. Note also that the class labels must be numeric (real-valued) for regression trees.

We start with the entire data set at the root, and recursively split the data instances as we grow the tree, so that a single class predominate as we progress towards the tree construction. At each step, an attribute is identified and a value for that attribute that splits the data is chosen so as to minimise the diversity of class labels in the resulting subsets. The splits induced by an internal node using numeric attributes geometrically represent hyper parallelepipeds parallel to the coordinate axes. As an example, the splits using *traffic penalty points* in $\{[0 \text{ to } 5], [5 \text{ to } 10] \text{ and } [>10]\}$ ranges in the figure 6.7 in page 6-27 divides the input space into three regions, the first two of which are closed hyper-parallelepipeds, and the third is a hyper parallelepiped extending to the highest penalty point in the training data. The third one is further subdivided into two using the *years of driving* attribute values $\{[0 \text{ to } 2] \text{ and } [>2]\}$.

If at any stage, two (or more) attributes have the same highest information gain, we can break the tie by choosing the attribute with more categories as the winner [FU92]. This will result in smaller sub-samples. As noted before, a node is called *pure* if either it has just one training instance, or if all training instances belong to the same class. If there are at least n ≥ 2 training instances that belong to two or more classes, that node is called *impure*. The recursive splitting is continued until all the nodes become pure. We expect this to happen if the training data size is larger than the data dimensionality (number of dependent attributes). This is not a necessary condition though, as it also depends upon the number of classes. When there are very few classes, all the terminal nodes may become pure even if the training data size is smaller than the data dimensionality. Accuracy of classification or prediction won't be 100% if at least one lowest level internal node is impure. Such is an indication of insufficient or improper training data, or high levels of class overlaps.

Faster decisions are possible using a balanced tree with minimum depth. The depth of a DT is directly correlated with the extent of class overlap and the number of classes, and inversely correlated with the number of dependent attributes. In general, smaller trees result when the number of classes is small, and classes do not overlap (in which case some of the attributes may be redundant). We should have an orderly way of splitting the relevant data, as there are many trees that can be built from it. This is accomplished using distance (dissimilarity) measures, information theory based measures, and dependence measures.

A rough-set based algorithm that uses a pre-processing step and its comparison with ID3 algorithm is given in [YR07]. A look-ahead type LSID3 algorithm that uses gain-k principle (ID3 is gain-1 algorithm) performs significantly better when data are heterogeneous.

6.4   Measures for Node Splitting

During the tree growing stage, we have a set of (one or more) records (data instances) at the internal nodes. If the number of records is one or all records belong to the same class, we stop further splitting as it becomes a decision node. Otherwise we have to choose an attribute to optimally split the data so as to have maximum homogeneity in the resulting subsets. There are many measures (called *impurity measures*) that are used to split the data. It returns a real number using the data distribution properties. The most popular among these are the (i) Gini's index (also known as Gini coefficient), (ii) entropy measure, (iii) chi-square measure, and (iv) minimum classification error measure. Other lesser known measures include Bhattacharya distance, Kolmogorov-Smirnov distance etc. In the following section, we briefly describe the popular measures and their properties.

6.4.1 Gini's Index Measure (GIM)

Gini's index is a ratio measure with values in the interval $[0,1]$ used to measure the discriminatory power of rating systems. For a data set S with m distinct classes, the simple Gini index is

$$\text{Gini(S, m)} = 1 - \sum_{k=1}^{m} P_k^2, \tag{6.1}$$

where P_k is the probability that an item belongs to class k (P_k is the relative frequency or proportion of class k), and m is the total number of classes. When the number of classes is large, some of the P_k's can be small. The maximum of Gini(S,m) occurs when each of the probabilities are equal (because $\sum_{k=1}^{m} P_k^2$ is then minimum), with maximum value 1-(1/m) (in this case the number of data instances should be an exact multiple of the number of classes). The minimum occurs when all instances belong to the same class, with minimum value 0 (because one of the p_k's will be 1 and all others zeros).

The attribute that provides the minimum Gini index is chosen for splits. An advantage of Gini measure over entropy measure is that it does not require logarithm to the base 2 (which is unavailable in most programming languages, although it could be computed indirectly as $\log_2(x) = \log_e(x)/\log_e(2.0) = \log_{10}(x)/\log_{10}(2.0)$ where $\log_e(2.0)$=.69314718 (or equivalently $1/\log_e(2.0)$=1.442695) and $\log_{10}(2.0)$=.301029996 (or $1/\log_{10}(2.0)$=3.32192809)). Disadvantages of Gini index are that it emphasises nearly equal-sized splits and could perform poorly when there are a large number of classes. Other versions of Gini index include the normalised symmetric and asymmetric Gini index [BK02] and modified Gini index [KI95].

Example 6.7 Prove that the Gini index is $2p_1q_1$ for the binary case, and it attains its maximum when p_1=.5 with maximum value .5.

Solution 1 Let p_1 and p_2 be the proportions in the binary case (just 2 classes) so that $p_2 = 1 - p_1 = q_1$. Substitute for p_2 in the Gini index to get Gini(S,2)=$1-(p_1^2+p_2^2)$=$1-p_1^2-(1-p_1)^2$. Expand the second term and cancel out the constants to get $-2p_1^2 + 2p_1$. Taking $2p_1$ as a common factor, this simplifies to $2p_1(-p_1+1) = 2p_1p_2 = 2p_1q_1$. As this is symmetric in p_1 and q_1, the Gini index is identical when p_1 and q_1 are swapped. As an example, suppose we have 9 data records at a node with 4 positives (P) and 5 negatives (N). Suppose a split results in 6 records in one subset S_1, and 3 records in the other subset S_2, such that there are 2 positives and 4 negatives (2P,4N) in S_1, and (2P,1N) in S_2. The Gini index for both subsets is $2*(1/3)*(2/3)$ = 4/9. In the general case (with m>2 classes), when the probabilities of two or more subsets are just the permutations of the probabilities in another subset, we could skip the expensive computation by finding its Gini index for one of these subsets (this argument applies to the entropy measure too). Write the above as f(p_1)=$2p_1(1 - p_1) = 2p_1 - 2p_1^2$, and differentiate wrt p_1 to get $\frac{\partial f(p_1)}{\partial p_1} = 2 - 4p_1$. Equating to zero, and solving for p_1 gives $p_1 = 1/2$. The second derivative is -4, showing that this corresponds to the maximum. Substituting in f(p_1) gives the maximum value as $2*.5*(1-.5) = 2/4 = 1/2$.

6.4.2 Minimum Classification Error Measure

This measure defined as MCE = 1-$\max_k p_k$, where p_k is the relative frequency of kth class. It ignores all but the maximum of the proportions. When the maximum is not unique (two or more classes have the same maximum) the split is ambiguous. The minimum MCE occurs when all instances belong to the same class (because the maximum frequency in that case is 1). The maximum occurs when all P'_ks are equal with maximum value $(1-\frac{1}{m})$. This proves that the Gini index and MCE attain the same minimum and maximum values. The biggest advantage of this metric is that it is easy to compute. The disadvantage is that the maximum need not be unique, and is sensitive to outliers (chapter 2).

6.4.3   Shannon's Entropy Measure (SEM)

Entropy measures the 'lack of order' in a system. It was introduced in communication theory by C.E. Shannon (1916-2001). The entropy of a data set S with m classes is defined as

$$\text{Entropy(S)} = \sum_{k=1}^{m} P_k \, \log_2(1/P_k) = -\sum_{k=1}^{m} P_k \, \log_2(P_k), \tag{6.2}$$

where P_k is the relative frequency (proportion) of class k (Last expression above is obtained by using $\log(a/b) = \log(a) - \log(b)$ and $\log_2(1) = 0$). It is measured in units of *bits*[12]. The base of the logarithm is 2 in communication theory. Irrespective of the base of the logarithm, the DT obtained will be the same. Each of the $\log_2(P_k)$ terms are ≤ 0 (and the negative sign outside the summation makes the entropy positive). The entropy is zero when the sample is pure (there is no uncertainty in a system, and each of the elements belong to the same class so that k=1 and $P_k \log_2(P_k)$ is zero (because $\log(1)=0$), and $0 \log(0)$ is assumed as 0). If p'_ks are of the form p/q, we could also write it as

$$-p_k * \log_2 p_k = -(p/q) * \log_2(p/q) = +(p/q) * \log_2(q/p), \tag{6.3}$$

where we have used the fact that $1/(p/q)=q/p$ and $\log(1)=0$. Gini Index and Entropy measure are both symmetric in the proportions p_k. This implies that each of the classes are given an equal chance to decide on the test attribute for the current split. Weighted entropy measures can be constructed when some classes have high priority than others (as in some medical test results with a large number of negatives). A disadvantage of the entropy measure is that it can create memory overflows when probabilities are close to zero or equivalently there are no items belonging to a particular class (we can simply ignore that class from the computation). As an example, the *Cell-phone-use*='No' value in table 6.11 at page 6-28 has the counts (0, 0, 1) under the Low, Medium and High classes. The entropy is calculated only for the High class as $1*\log(1)$ = 0. (when probabilities are exactly zero, we can simply ignore it from the computation).

Example 6.8 Prove that the entropy is (i) one when there are just 2 classes with equal probabilities, (ii) 1.5 when there are 3 classes with probabilities 1/4, 1/4, 1/2 in any combination, (iii) 1.5849625 when there are 3 classes with equal probabilities.
Solution. When there are just 2 classes with equal probabilities, the entropy is $-\frac{1}{2} *\log_2(\frac{1}{2})-\frac{1}{2} * \log_2(\frac{1}{2}) = -\log_2(\frac{1}{2}) = +\log_2(2) = 1$ using equation 6.3. (ii) If there are 3 classes with probabilities 1/4, 1/4, 1/2, the entropy is $-\frac{1}{4} *\log_2(\frac{1}{4})-\frac{1}{4} *\log_2(\frac{1}{4})-\frac{1}{2} *\log_2(\frac{1}{2}) = +\frac{1}{2}\log_2(4)+\frac{1}{2}\log_2(2) = +\frac{1}{2} * 2 + \frac{1}{2} * 1 = 1.50$ using equation 6.3, (iii) the entropy is $-3 * \frac{1}{3} * \log_2(\frac{1}{3}) = +\log_2(3) = 1.5849625$.

Replacing $\log_2(P_k)$ in 6.2 by $\log_2(e) * \log_e(P_k)$ (or by $\log_2(10) * \log_{10}(P_k)$), we see that $\log_2(e) = 1.442695$ (or $\log_2(10)=3.321928$) occurs as a *constant* multiplier in each of the entropies. Thus the entropy calculations can be considerably simplified by replacing the base of

---

[12]The unit is in fact immaterial because other impurity measures like Gini index and MCE do not use a unit.

the logarithm to e or 10, as convenient. As this corresponds to a scaling of the entropy, the unit remains *bits*. As we seek that attribute with minimum entropy or maximum gain, the parent node must also use the same base for the logarithm in Entropy(S). Obviously, a constant multiplier will not alter the minimum or maximum among all the attributes considered for a split. This implies that the multiplier can totally be ignored, and any desired logarithm used for the computation[13]. In other words, the decision tree obtained will remain the same irrespective of the base of the logarithm used in entropy calculations.

Theorem 6.2 Prove that the MCE $\leq$ Gini index for the same set of data instances with at least 2 classes.

Proof. We need to prove that $1\text{-max}_k\ p_k \leq 1 - \sum_{k=1}^{m} p_k^2$. Denote $\max_k\ p_k$ by p_M. Canceling out the constants and changing the sign on both sides, this becomes $\sum_{k=1}^{m} p_k^2 \leq p_M$. Move p_M from the RHS to the LHS, and combine with its matching pair p_M in the summation to get $\sum_{k=1,k\neq M}^{m} p_k^2 + (p_M^2 - p_M) \leq 0$ (if the maximum is not unique, combine p_M with any one of the matching pairs on the LHS). Complete the square in the second term, and simplify to get $\sum_{k=1,k\neq M}^{m} p_k^2 + (p_M - \frac{1}{2})^2 - \frac{1}{4} \leq 0$. Taking $\frac{1}{4}$ to the RHS results in $\sum_{k=1,k\neq M}^{m} p_k^2 + (p_M - \frac{1}{2})^2 \leq \frac{1}{4}$. The LHS being a sum of squares of numbers, each of which is <1, is obviously ≥ 0 (it could be proven that the minimum of the LHS is $\frac{1}{4m}$. We know that a sum of squares assumes the minimum when each of them are equal. This implies that $p_1 = p_2 = \cdots = p_M - \frac{1}{2} = c$. But as $\sum_k p_k = 1$, we get $c = \frac{1}{2m}$. Substituting in $\sum_{k=1,k\neq M}^{m} p_k^2 + (p_M^2 - p_M)$ we get the minimum value as $(m-1) * \frac{1}{4m^2} + \frac{1}{4m^2} = \frac{1}{4m}$). It attains the maximum when one of the p_k's is 1 (note that we have assumed p_M to be the maximum among the p_k's, so that the LHS is maximum when $p_M = 1$), and all other p_k's are zeros. The maximum value is obviously $0 + (\frac{1}{2})^2 = 1/4$. In all the other cases it is between $(\frac{1}{4m}, \frac{1}{4})$. For example, if $p_M = .99$ we get $(p_M - \frac{1}{2})^2 = .2401$, and $\sum_{k=1,k\neq M}^{m} p_k^2$ is upper bounded by $(0.01)^2 = .0001$. Hence the sum is less than $.25$ (if $p_M = .9999$ we get the sum as 0.24990). This proves the result.

6.4.4   Gain and Impurity

The entropy measure simply gives us a number ≥ 0. Our aim at each of the internal nodes is to decide the best attribute to be used to split the data records. The Gain of an attribute v is defined as Gain(S,V)=Entropy(S)$-\sum_{v\in V}(|S_v|/|S|)$ Entropy(S_v) where S_v is a subset of S for which attribute V assumes the value v and $|S|$ denotes the cardinality. If one of the other impurity measures (Gini index or MCE) is used instead of entropy measure, we get Gain(S,V)=I(S)$-\sum_{v\in V}(|S_v|/|S|)$I(v), where I(v) denotes the impurity measure of each subset resulting from the split. Another popular split criterion is called "twoing" rule in which an exhaustive search is carried out to find two classes that makeup more than 50% of the data. The change of impurity measure is then computed as the maximisation problem $\underset{x_j \leq x_j^R, j=1..M}{argmax} \left(\frac{P_l P_r}{4} \sum_{k=1}^{c} |p(k|t_l) - p(k|t_r)|^2 \right)$, where $p(k|t)$ is the conditional probability of class k in node t and P_l, P_r are the probabilities of left and right sub-trees. A disadvantage of this rule is the computational complexity involved in large data sets.

---

[13]The search box of google.com can be used as a calculator. Simply type expressions like -(1/3)*log(1/3)-(2/3)*log(2/3) in the search box (and press Enter key). Google uses ln() for log to the base e, and log() for log to the base 10. See www.google.com/help/calculator.html for other operators.

> Variable selection is an important activity in some domains like medical sciences, economics, business and finance. This is often done after a DT model is built from training data, and validated using test data. As the variable chosen for the split at the root of a DT optimally distinguishes between the subsets of records represented by nodes at the next level, it is the most important attribute in terms of discriminatory power. We then move to the next level and rearrange the decision nodes (terminal nodes at subsequent levels are ignored) in descending order using one of the node-splitting measures discussed in §6.4 in page 6-18). This is continued to subsequent levels where there are internal nodes. But the discriminatory power of variables decreases as we move towards the bottom of a DT. This could tail-off rapidly when a large number of variables are present. Hence a cut-off threshold based upon the conditional impurity measure at depth k may be used to stop accumulating further variables. This information is useful in pruning a DT. If a few of the variables have a large discriminatory power, it is an indication that the tree can be pruned without serious misclassification errors.

6.5   Induction Algorithms

Decision trees are constructed using cleansed and labeled training data with known class labels. This process of building a DT from flat files[14] is called decision tree learning or induction ([QJ86]).

Definition 6.14 The process of iteratively building a hierarchical tree structure hidden in labeled training data (X,y) using an information gain heuristic that splits the available data (X) at each step optimally into more homogeneous subsets until further splits are either impossible or undesirable is called the DT induction.

Many algorithms have been proposed for DT construction (see table 10.10 in page 10-36). Most of these algorithms are based upon Hunt's tree construction principle (pg. 6-13). The tree building algorithms find the optimal tree using a 'one-step at a time' lookahead without backtracking in the solution space that contains all possible DTs. A rough-set based algorithm that uses a pre-processing step and its comparison with ID3 algorithm is given in [YR07]. A look-ahead type LSID3 algorithm that uses gain-k principle (ID3 is gain-1 algorithm) performs significantly better when the data are heterogeneous.

6.5.1   ID3 Algorithm

One of the earliest algorithms for building Decision Trees is the ID3 (Interactive Dichotomizer-3), which uses the concept of information gain. It uses a greedy tree growing approach using an information entropy minimisation criteria.[15] The entropy measure is utilised at each node to build a top-down tree. In its pure form, it is a non-backtracking algorithm. It starts by assigning all training data to the root node of the tree to be built. Using a chosen attribute (called classifying or test attribute), tree branches are created and the sample at the node is partitioned into sub-samples. This step is repeated top-down recursively at each of the child nodes until any of the following conditions is true:

1. there is only one data record

   In this case there are no further options and we assign the corresponding class label.

---

[14]Here flat-file does not mean that data are always taken from a computer file. It could also come from DW/DM, online data collection devices, or could be spread across a network like the web.

[15]The splitting criteria can be based upon entropy, expected probability, Gini index, minimum chi-square, etc.

Algorithm 6.1 ID3 Algorithm

1: Input a training set (S) of labelled data instances in flat file format (initially all at the root, which is the first node created) where X is a categorical data vector and y are the corresponding class labels.
2: Special checks:
begin
If S is empty, return error code
If S has a single element or
If all items in S belong to the same class C,
return root as the single leaf node of the tree.
end
3: Construction:
Identify the set of input variables (partitioning attributes)
If input variable is quantitative,
first categorise it using discretisation technique.
4: **while** (there are more variables x in the input set) **do**
5:      Select the attribute with highest information gain as test attribute
6:      Grow the branches using the available splits
7:      Update the input set by marking variable x
8:      Select the next variable as x
9: **end while**
10: **return** decision tree

---

2. all data records have the same class
In this case we mark the corresponding node with the class label and terminate.

3. all attribute values are the same
Although rare, in this situation we will assign the most frequent class to the corresponding node. If it is not unique, ties are broken by other heuristics like recursively backtracking to the parent node until the tie is broken.

See [BB97] for an SVM formulation that builds a binary-DT, [BB96] for an Extreme Point Tabu Search method that constructs globally optimal DT, and [BK92] for a linear programming approach to build DTs.

6.5.2   C4.5 Algorithm

The C4.5 algorithm proposed by Quinlan [QJ93] is an extension of the ID3 algorithm in which classification can be based on numeric attributes, and missing or noisy data can be incorporated via pruning [TJ08]. Whereas ID3 uses binary splits, the C4.5 algorithm uses multi-way splits (see table 10.10). It uses post-pruning to reduce the size of the DT. An optimiser combines the generated rules to eliminate redundancies. C5.0 is an improvement of C4.5 that has cross-validation and boosting[16] capabilities.

---

[16]The boosting technique resamples the training data

6.5.3 Extended Tabular Method to Build a Decision Tree

Building a DT from large data can become quite complicated. We give a streamlined tabular method to ease the hand computation. We call it the *Extended Tabular Method* (ETM). For each of the variables considered for split, we form a matrix with rows labeled with the possible values of that variable, and the columns labeled with the total number of classes. This matrix is filled with counts that represent the number of records (data instances) that have variable restriction specified by the rows, and belonging to respective classes (columns). Once this matrix is filled, we extend it by adding a *total* column at the right, followed by one column each for the impurity measure, weight and weighted impurity for that particular row. The weight is directly derived from the observed totals. This is shown in table 6.1.

Table 6.1: Chattamvelli's Extended Tabular Method for Constructing a DT

	Classes					Calculations			
Var. values	C_1	C_2	C_3	$\cdots$	C_m	Total	Imp.	Weight	Wtd. Imp.
$\text{val}_1 | \text{range}_1$	f_{11}	f_{12}	f_{13}	$\cdots$	f_{1m}	$S_1 = \sum_k f_{1k}$	$I(1)$	$w_1 = S_1/S$	$w_1 * I(1)$
$\text{val}_2 | \text{range}_2$	f_{21}	f_{22}	f_{23}	$\cdots$	f_{2m}	$S_2 = \sum_k f_{2k}$	$I(2)$	$w_2 = S_2/S$	$w_2 * I(2)$
$\text{val}_j | \text{range}_j$	f_{j1}	f_{j2}	f_{j3}	$\cdots$	f_{jm}	$S_j = \sum_k f_{jk}$	$I(j)$	$w_j = S_j/S$	$w_j * I(j)$
$\cdots$	$\cdots$	$\cdots$	$\cdots$	$\cdots$	$\cdots$	$\cdots$	$\cdots$	$\cdots$	$\cdots$
$\text{val}_n | \text{range}_n$	f_{n1}	f_{n2}	f_{n3}	$\cdots$	f_{nm}	$S_n = \sum_k f_{nk}$	$I(n)$	$w_n = S_n/S$	$w_n * I(n)$
Sum						$S = \sum_{l=1}^{n} S_l$		$\sum_{l=1}^{n} w_l = 1$	$\sum_l w_l * I(l)$

Computing the impurity measure is quite trivial from this table, as it involves simple multiplications and additions. We could skip all impurity computations if the impurity is zero for an attribute (see below). We exemplify our tabular method below.

Table 6.2: Identifying accident risks among automobile drivers (training data)

#	Attributes						Class
Ser. num.	years of driving	insurance coverage	drunken driving	mobile use	prior accidents	penalty points	accident risk
1	0-2	liability	yes	no	0	0-5	medium
2	2-4	full	no	yes	2	6-10	medium
3	0-2	compre	no	yes	0	0-5	low
4	4-6	full	yes	yes	2	6-10	high
5	0-2	liability	yes	yes	0	>10	high
6	>6	compre	no	yes	0	6-10	medium
7	4-6	full	yes	yes	1	>10	medium
8	2-4	compre	yes	no	0	6-10	high
9	>6	full	no	yes	1	0-5	low

legend: (1) liability: minimum liability insurance, (2) full: full-insurance (for own vehicle), (3) compre: comprehensive insurance (damage for own and other vehicles; victims)

Example 6.9 An insurance company is interested in identifying high risk categories among the drivers. Table 6.2 gives the training data. Attributes considered are: (1) Total number of years

of accumulated driving experience, (2) Type of insurance coverage, (3) Whether the driver has the habit of drunken-driving, (4) Cell-phone usage during driving, (5) Number of prior fatal accidents, (6) Total prior traffic penalty points. Of these, the variables (1) and (5) and (6) are numerical, and all others are categorical. The risk categories into which a test data item is to be classified are *low, medium,* and *high* risks. Fit a DT model using the ETM method.
Solution There are 2 'low', 4 'medium' and 3 'high' risk categories in the training data. All logarithms are evaluated to the base 2. The entropy before splitting is -(2/9) log(2/9) - (3/9) log(3/9) - (4/9) log(4/9) = 1.53049 bits.

Table 6.3: Extended Tabular Method for Years of Driving

Yrs\Class	Low	Medium	High	Total	Impurity	Weight	Wtd Impurity
"0-2"	1	1	1	3	1.5849625	3/9	0.528321
"2-4"	0	1	1	2	1.0000	2/9	0.222222
"4-6"	0	1	1	2	1.0000	2/9	0.222222
">6"	1	1	0	2	1.0000	2/9	0.222222
Sum				9		1.0	1.1949875

Splitting on the *years of driving* attribute gives the data presented in table 6.3. The entropy is found as (3/9)[3 * (-1/3)log(1/3)] + (2/9)[2*((-1/2)log(1/2))]+ (2/9)[2*((-1/2)log(1/2))] + (2/9)[2*((-1/2)log(1/2))] = 0.528321 + 0.222222+ 0.222222 + 0.222222 = 1.1949875 bits. As rows 2, 3 and 4 involve (-1/2)log(1/2), we need compute it just once.

Table 6.4: Extended Tabular Method for Insurance Type

Insurance\Class	Low	Med.	High	Total	Impurity	Weight	Wtd Impurity
Liability	0	1	1	2	1.0000	2/9	0.222222
Full	1	2	1	4	1.5000	4/9	0.666667
Comprehensive	1	1	1	3	1.584963	3/9	0.528321
Sum				9		1.0	1.4172097

Splitting on the *insurance type* gives the entropy as shown in table 6.4. Notice that the first row (0,1,1) and the third row (1,1,1) are already encountered in table 6.3. So, we need not recompute it. The entropy for this split is (2/9)[2 * (-1/2)log(1/2)] +(4/9)[-(1/4)log(1/4)-

Table 6.5: Extended Tabular Method for Drunken Driving

Drunken\Class	Low	Med.	High	Total	Impurity	Weight	Wtd Impurity
No	2	2	0	4	1.0000	4/9	0.444444
Yes	0	2	3	5	0.97095	5/9	0.539417
Sum				9		1.0	0.983861

(2/4)log(2/4)-(1/4)log(1/4)] +(3/9)[3*((-1/3)log(1/3))] = 1.4172097 bits. Splitting on *drunken driving* gives the data in table 6.5. The entropy for this split is 0.444444 + 0.539417 = 0.983861 bits. Splitting on *cell-phone use* attribute gives the data presented in table 6.6. The en-

Table 6.6: Extended Tabular Method for Mobile Use While Driving

Cell-use\Class	Low	Med.	High	Total	Impurity	Weight	Wtd Impurity
No	0	1	1	2	1.0000	2/9	0.222222
Yes	2	3	2	7	1.556657	7/9	1.210732995
Sum				9		1.0	1.432955

tropy is found as $((2/9)[2*(-1/2)\log(1/2)] + (7/9) [-(2/7)\log(2/7)-(3/7)\log(3/7)-(2/7)\log(2/7)]$ $= 0.222222 + 1.210733 = 1.432955$ bits. Splitting on *prior accidents* gives the data in table 6.7. The entropy is $0.8455156+0.222222 + 0.222222 = 1.289960$ bits. Finally, splitting on *traffic*

Table 6.7: Extended Tabular Method for Prior Accidents

Accident\Class	Low	Med.	High	Total	Impurity	Weight	Wtd Impurity
0	1	2	2	5	1.521928	5/9	0.845516
1	1	1	0	2	1.0000	2/9	0.222222
2	0	1	1	2	1.0000	2/9	0.222222
Sum				9		1.0	1.289960

penalty points gives the data in table 6.8. The entropy for this split is $0.3060986 + 0.444444 + 0.222222 = 0.972765$ bits. The prior penalty points has the highest gain (or equivalently the

Table 6.8: Extended Tabular Method for Traffic Penalty Points

Penalty \Class	Low	Med	High	Total	Impurity	Weight	Wtd Impurity
"0-5"	2	1	0	3	0.918295834	3/9	0.3060986
"6-10"	0	2	2	4	1.0000	4/9	0.444444
"> 10"	0	1	1	2	1.0000	2/9	0.222222
Sum				9		1.0	0.972765

lowest impurity) followed by drunken driving (this could be due to sampling error or noise in the data, or due to the fact that the algorithm prefers attributes with multiple categories than those with few categories. In addition, drunken driving and penalty points are positively correlated. Note also that the addition of just a few new records could change the attribute for top split from penalty points to drunken driving as the difference in the gains for them are small). This remains the same, had we used the Gini index (the corresponding Gini indexes for respective variables are 0.555556,0.611111,0.488889,0.6190476,0.577778,0.481481). Choosing the highest

Table 6.9: Entropy and gain for top level split

measure (bits)	years of driving	insurance coverage	drunken driving	mobile use	prior accidents	penalty points
entropy	1.194988	1.41721	0.983861	1.432955	1.28996	0.972701
gain	0.335502	0.113280	0.546629	0.097535	0.2405299	**0.557725**

gain tells us to use *prior penalty points* as the splitting attribute at the top level (table 6.9. This gives rise to a ternary tree with the branches containing data points given in table 6.10. The tree growing is continued to subsequent levels. For example, the root branch for penalty points has 4 data points. The entropy for drunken-driving is zero. We could indeed stop further computations at this point. The results are given in table 6.11 in page 6-28. As this choice results in pure samples, we choose drunken-driving for split and stop growing the tree further at this level. The final tree obtained is given in figure 6.7.

Table 6.10: Data for the second level of growing the DT

#	Attributes						Class
Ser. num.	years of driving	insurance coverage	drunken driving	mobile use	prior ac- cidents	penalty points	accident risk
1	0-2	liability	yes	no	0	0-5	medium
2	0-2	compre	no	yes	0	0-5	low
3	>6	full	no	yes	1	0-5	low
4	2-4	full	no	yes	2	6-10	medium
5	4-6	full	yes	yes	2	6-10	high
6	>6	compre	no	yes	0	6-10	medium
7	2-4	compre	yes	no	0	6-10	high
8	0-2	liability	yes	yes	0	>10	high
9	4-6	full	yes	yes	1	>10	medium

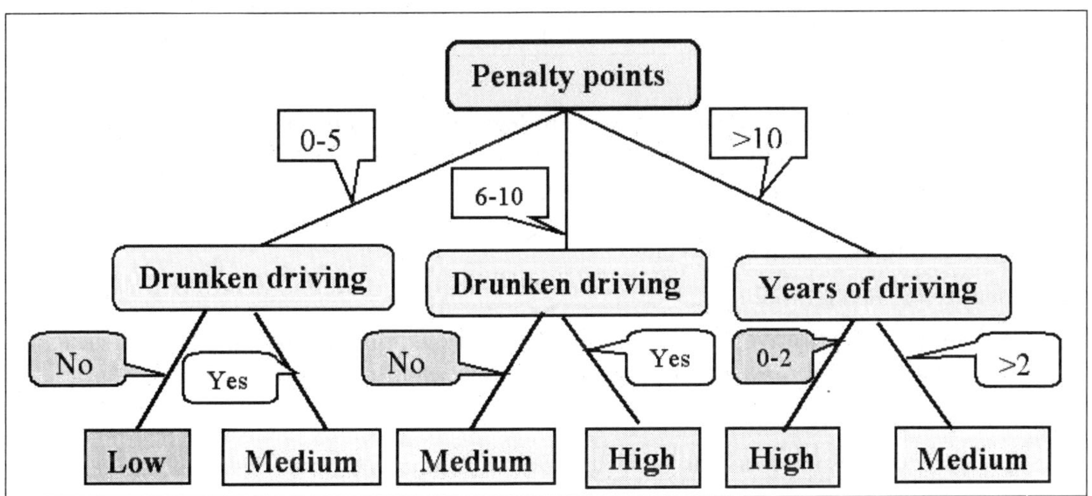

Figure 6.7: DT for predicting accident risk

6.5.4   CHi-squared Automatic Interaction Detector (CHAID)

As the name implies, this is a statistical technique for tree induction ([HJ75], [KG80]) that uses Karl Pearson's χ^2 test for contingency tables. It works for categorical variables (with 2 or more

Table 6.11: Extended Tabular Method at Second Level for DT induction

Attribute	Class			Calculations			
Yrs of driving	Low	Medium	High	Total	Wt	Entropy	Wt*Entropy
"2-4"	0	1	1	2	0.5	0.301029996	0.150514998
"4-6"	0	0	1	1	0.25	0	0
">6"	0	1	0	1	0.25	0	0
Sum				4	1.0		0.150514998
Insurance	Low	Medium	High	Total	Wt	Entropy	Wt*Entropy
full	0	1	1	2	0.5	0.301029996	0.150514998
compre	0	1	1	2	0.5	0.301029996	0.150514998
Sum				4	1.0		0.301029996
Drunk-driving	Low	Medium	High	Total	Wt	Entropy	Wt*Entropy
no	0	2	0	2	0.5	0	0
yes	0	0	2	2	0.5	0	0
Sum				4	1.0		0
Mobile-Use	Low	Medium	High	Total	Wt	Entropy	Wt*Entropy
No	0	0	1	1	0.25	0	0
Yes	0	2	1	3	0.75	0.276434591	0.207325943
Sum				4	1.0		0.207325943
Accidents	Low	Medium	High	Total	Wt	Entropy	Wt*Entropy
0	0	1	1	2	0.5	0.301029996	0.150514998
2	0	1	1	2	0.5	0.301029996	0.150514998
Sum				4	1.0		0.301029996

categories), and can be used as an alternative to logistic regression. Quantitative variables are categorised so as to have more or less an equal number of observations in each category. A rule of thumb in Statistics states that none of the expected cell counts should be less than 1 for the χ^2 test to work well. A merging step is carried out to avoid such pitfalls. The major steps in CHAID algorithm are summarised below:

- Identify predictor variables.

  If there are quantitative variables, use discretisation (binning) techniques to form categories with preferably an equal number of frequencies in each category.

- Conditionally merge the categories.

  If the frequencies are less than 5, then merge with adjacent categories until true.

- Select the best splits at each node.

  Use the smallest p-values computed using the χ^2 statistic to exhaustively search for the best splits.

- Repeat step 3 for each of the generated internal nodes using new attributes until stopping condition is satisfied.

There is no pruning step as it stops growing the DT when a certain condition is met. One disadvantage of CHAID is that it could sometimes result in unrealistically short (small depth) trees. The computational requirements are also higher than for other algorithms. Many other chi-square based algorithms have appeared in the literature including AID, THAID, MAID, XAID etc.

6.5.5 Classification and Regression Tree (CART)

CART is a nonparametric model to choose the most important subset of variables that determine an outcome of interest. Introduced by Breiman *et. al.*(1984)[BF84], it uses the binary recursive partitioning to split data instances into successively homogeneous subsets using Gini diversity measure. It produces regression tree when outcome is continuous, and classification tree when the outcome is categorical. It works better than discriminant analysis when the variables are uncorrelated. Surrogate variables can be used at a node for missing data cases. It can deal with large data sets of high dimensionality. The CART tree (the DT generated by CART) is insensitive to explanatory variable transformations [BF84]. Outliers are easily handled by CART. Another advantage is that a cost matrix can be incorporated for variable selection and cost based pruning. A disadvantage of the standard CART is that the predictions are not probabilistic.

Table 6.12: Comparison of common DT induction algorithms

Algorithm	Attrib.	Split	Measure	Pruning	Year
CHAID	Categorical	multi-way	χ^2	χ^2	1975
CART	General	2-way	Gini	cost complexity	1984
ID3	Binary	2-way	Entropy, χ^2	pre-pruning	1986
FACT	General	multi-way	F-test, LDA	direct stopping	1988
C4.5	General	multi-way	Entropy	error reduction	1993
SPRINT	General	multi-way	Gini	MDL	1996
SLIQ	General	multi-way	Gini	MDL	1996
QUEST	General	2-way	F-test, QDA	cost complexity	1997
PUBLIC	Binary	2-way	Entropy, Gini	MDL	1998
BOAT	General	binary	impurity meas.	Bootstrapping	1999
Rainforest	General	multi-way	generic		2000
CRUISE	Categorical	multi-way	χ^2	cost complexity	2001
GUIDE	General	multi-way	χ^2	MDL	2002
YaDT‡	General	multi-way	Entropy	error reduction	2004

LDA=Linear Discriminant Analysis, QDA=Quadratic DA
Notes: The FACT method uses largest F-statistic for variable selection and LDA for split selection. QUEST uses Anova F-test or Levene F-test for variable selection and quadratic DA, Bonferroni correction for split selection. It uses split selection approach, but has an option for cost complexity pruning. SLIQ uses pre-sorting of training data, breadth-first growth and MDLP-based pruning [MA96]. SPRINT (Schafer 96) has no memory restrictions because it uses attribute list data structures to efficiently manage the memory. BOAT can use a variety of impurity measures and constructs several levels of DT in a single scan of data. PUBLIC (PrUning and BuiLding Integrated in Classification) can use Gini or entropy measure [RS98]. It keeps a node as such without splitting if it is determined using cost criteria that it will be pruned later. Rainforest is a generic algorithm that works with specific algorithms [GR00]. The ID3 algorithm dates back to 1979 that used the χ^2 statistic for split selection and pre-pruning strategy. Later versions of it use various post-pruning methods. ‡YaDT is in fact an efficient implementation of C4.5 like algorithm in C++.

Whereas CHAID, C4.5[QJ93], BOAT [GG99] and C5.0 produce general trees, CART produces binary DTs (see table 10.10). The FACT and its descendant QUEST (Quick, Unbiased, Efficient Statistical Tree) [LS97], [LW09] algorithms[17] have better performance in terms of variable selection bias than CART. The FACT algorithm has a disadvantage that it prefers categorical variables over quantitative variables due to large values of F-statistics. As QUEST algorithm has a large memory (RAM) requirement, it may not work when the training data size is too large. The CART and C4.5 are used for building classification or regression trees, while the QUEST algorithm is often used for regression trees. GUIDE algorithm can construct piece-wise linear regression models. Linear combinations of (quantitative) attribute values can be used in CART, QUEST and GUIDE algorithms. There are many other algorithms like J48 (Weka), FIRM, CAL5, CLOUDS [AR98], LMDT, GOTA, IDX, SUPPORT [CH94], TARGET [GF08] etc that can be used for DT induction.

6.5.6 Misclassification Errors

As mentioned above, the DT is built using the training data. The DT model can then be validated using the test data. If the correct class labels used in the test data are known, we could capture the misclassification information in a 2D table called a *confusion matrix* (table 6.13) ([TS06]). If the actual class label of current data instance is 'i', and the DT predicts that

Table 6.13: Confusion Matrix for m Classes

Actual Class	Predicted Class			
	C_1	C_2	$\cdots$	C_m
C'_1	f_{11}	f_{12}	$\cdots$	f_{1m}
C'_2	f_{21}	f_{22}	$\cdots$	f_{2m}
$\cdots$	$\cdots$	$\cdots$	$\cdots$	$\cdots$
C'_m	f_{m1}	f_{m2}	$\cdots$	f_{mm}

it belongs to class 'j', we add a 1 to the $(i,j)^{th}$ entry. Thus the confusion matrix entries are the number of data instances belonging to class-i being predicted as in class-j. We naturally expect the numbers along the main diagonals to dominate each row (we would expect a diagonally-dominant confusion matrix.). If all diagonal elements are non-zero and all off-diagonal entries are zeros, we have a well-generalisable classifier. This is seldom achievable in the presence of class-overlap. If class-i overlaps with class-j, a small number of items with true class labels C_i may be classified into C_j, and *vice versa*. The extend of this overlap can be measured using the entries C_{ij} and C_{ji}. The sum of the diagonal elements is the total number of correct predictions. Hence the accuracy can be estimated from this matrix as $E = \sum_{i=1}^{m} f_{ii} / (\sum_{i=1}^{m} \sum_{j=1}^{m} f_{ij})$. The statistical significance of the misclassification error for a training data set could be ascertained using Pearson's χ^2 statistic (chapter 3). Several pruning algorithms (described below) uses the confusion matrix.

6.6 Pruning Decision Trees

Once the decision tree has been built, a sensitivity analysis should be performed to test the stability and suitability of the model to variations in the unseen data instances. Expected values

---

[17]QUEST can use Anova F-test, Levene F-test (continuous and categorical), or χ^2 test (nominal data)

of each alternative are evaluated to determine the optimal model. But the decision maker's attitude towards high-risk alternatives can negatively influence the outcome of a sensitivity analysis. Even if a DT perfectly classifies training data (w/o errors), the true error is likely to be higher on new data. Most of the decision tree software packages allow the user to carry out a sensitivity analysis. The technical name for it is *pruning*[18], although we have an implicit assumption that the DT will be modified in some way or other during the pruning phase.

Definition 6.15 The process of adjusting a DT, either during the induction phase or immediately after it is totally built from training data, with an intention to simultaneously reduce the tree size (or depth) and maintain minimal misclassification errors is called decision tree pruning.

This step can reduce tree size by removing unwanted branches or those of little use, thereby increasing the run-time efficiency of classifying new data. A pruning step may not be required in some particular situations – (i) the DT is of shallow depth with just one or two levels (this type of DT is common in text classifications, classification of emails as spam or non-spam, classification of newswire articles into various categories like politics, sports, science and tech, deaths, natural calamities, entertainment, etc), (ii) when the DT is built from a large training data which is a true representative sample from the population, (iii) when the class overlap is minimal or non-existent, (iv) when there are no time-dependent attributes to invalidate the production rules over the passage of time, (v) when test data are minimal and classification is infrequent (as in classification of rare fossils), or in a combination of these cases.

A DT may overfit the data due to many reasons (insufficient training data, outliers in training data, skewed parent population, nonlinear class overlaps etc). If the tree is overfitted, prediction errors can't be accurately estimated. Pruning is the answer to reduce the classification errors in such situations. The type of pruning can differ when the DT is used for classification, and when it is used for prediction or clustering. The popular pruning techniques include cost-complexity pruning (CCP) (CART, QUEST), reduced error pruning (REP), pessimistic error pruning (PEP), minimum error pruning (MEP), and MDL (Minimum Description Length)[19](SLIQ, SPRINT), bootstrapping, etc. The REP and CCP works bottom-up whereas the PEP works top-down.

6.6.1 Pre and Post Pruning

There are two types of pruning techniques known as pre-pruning and post-pruning [RM07]. Each of them have different variants. Pre-pruning occurs when the DT is built step-by-step top-down using a *divide-and-conquer* strategy, and the tree growing is halted when a stopping condition is met. Majority of tree pruning algorithms belong to the second category [KM98].

The post-pruning works in a bottom-up manner on a fully-grown DT with at least 2 levels (an exception is to prune a single level DT with many branches, to one with lesser number of branches, by collapsing or merging 2 or more branches into one). A test set, which is distinct from the training set, is applied to the DT to be pruned and all misclassification errors are noted down. At each iteration, a subtree is identified towards the bottom of the current DT, which when replaced by a single node results in a small estimated error. This process is continued until further pruning is statistically insignificant. Another type of post-pruning works on a finite candidate set of decision trees that are already pruned by different amounts. Each one of them is then evaluated independently using the same test set for significant classification errors. That

---

[18]The literal meaning of *pruning* is 'to cut off branches from a tree so that post-pruned tree will grow better'.
[19]The MDL principle was introduced by Rissanen [RJ78]. See also [QJ89]

tree with minimal classification error or minimal size is chosen as the winner from among them. A variant of the above algorithm uses all of the training samples for tree generation and pruning using cross-validation. The CCP takes into consideration both the number of misclassification errors and the complexity (size) of the resulting tree.

Tree pruning is usually a post-processing step with an intention to minimise overfitting, and to remove redundancies. Decision trees can also be pruned using specialised techniques like integer programming [ZH06], dynamic programming, etc. Dual pruning algorithms have also been proposed in the literature when the attributes have constraints. For further discussions, see [MS95], [BL04]. See §6.9.2 in page 6-36 for a pruning example.

Algorithm 2 MDL Pruning algorithm for a node N

```
If N is a leaf node of the DT, return cost(data(N)+1)
//Let N1 and N2 be the children of node N
cost1 = Prune subtree(N1)
cost2 = Prune subtree(N2)
mincost = min{cost(data(N))+1, cost1 + cost2 + 1 }
```

6.7   Fuzzy Decision Trees

In the standard DT, each instance belongs unambiguously to a class. The Fuzzy decision tree (FDT) is an extension in which a membership function determines the degree of affinity of each instance to various classes. This has the implication that an instance may be assigned to multiple branches of a decision node with a fuzziness measure. Hence during the test phase, the conditions tested at each edge have fuzzy constraints. Weighted FDTs can be constructed using fuzzy weights. In the simplest case, we use different weights for different attributes (columns). Continuous attributes are split using a fixed cutoff value (one cutoff threshold for binary splits, and more than one for higher splits). A data record's continuous attribute taking a value that is in the immediate neighborhood of the threshold may be due to measurement errors or noise. So a sharp split may not be prudent in every situation. The FDT overcomes ambiguity in such situations. When two or more attributes at a node are numeric, a weighted linear combination of the attribute values can be used to decide upon the split criteria. Constraints can be placed on numeric attributes, resulting in a constrained FDT. Fuzzy queries in which one is interested in the set of classes to which a new instance can belong to could be answered by an FDT. The FDTs perform better than standard DTs when the data contain noise (impurity). See [AZ98],[BW99],[CW09],[OW03], and [YS95] for further info.

6.8   Decision Tables

A decision table is a hierarchical structure akin to decision trees, except that data are enumerated into a table using a pair of attributes, rather than a single attribute. They work for both categorical and quantitative variables. Quantitative variables should be categorised using the discretisation technique discussed in chapter 1. Too fine-grained or too coarse discretisation can result in loss of information and sub-optimal performance.

The following steps offer some guidelines to developing decision tables:

1. Determine the maximum size of the table

2. Determine the number of conditions that may affect the decision. Combine rows that overlap, for example, conditions that are mutually exclusive. The number of conditions becomes the number of rows in the top half of the decision table.

3. Determine the number of possible actions that can be taken. This becomes the number of rows in the lower half of the decision table.

4. Determine the number of condition alternatives for each condition. In the simplest form of decision table, there would be two alternatives (Y or N) for each condition. In an extended-entry table, there may be many alternatives for each condition.

5. Calculate the maximum number of columns in the decision table by multiplying the number of alternatives for each condition. Eliminate any impossible situations, inconsistencies, or redundancies.

6.9 Applications

There are a large number of applications to DT in various fields. In this section we briefly describe a few applications and give references for the interested readers to explore it further.

6.9.1 Fraud Detection

Fraud detection is increasingly becoming a necessity due to the large number of uncaught frauds. Fraudulent financial transaction amounts to billions of dollars every year throughout the world. Fraud prevention is different from fraud detection, as the former is pre-transaction safety, and the latter is used during or immediately after a transaction. Many softwares use advanced authorisation technology to prevent fraud, which is slow and computationally intensive.

Table 6.14: ATM cash withdrawal data

#		Attributes				Class
Ser. num.	card id	amount involved	PIN attempts	total retries	time taken	fraudulent transaction
1	id1	1500	2	1	60	No
2	id2	800	3	0	90	Yes
3	id3	250	1	0	70	No
4	id1	1200	2	0	50	No
5	id4	3000	2	2	110	Yes
6	id5	1000	1	1	40	No
7	id6	2700	1	1	76	Yes
8	id3	2800	1	2	55	Yes
9	id7	3400	2	1	75	Yes
10	id8	900	2	1	80	No
11	id9	500	2	2	68	Yes
12	id10	2300	1	0	63	No

An interesting pattern in fraud is that if a (cash) card identity is stolen, the misnomer tries to make as much money as possible without bothering about the *normal* purchasing behavior of the true owner of the card (which may be unknown to the misnomer, until caught). The statistical distribution of the amount of fraudulent transaction in any fixed time intervals is highly skewed, whereas the distribution of the number of fraudulent transactions in a fixed time interval is approximately Poisson distributed. A hybrid of DTs and ANNs can be used to spot unusual financial transactions. Using the true card holder's past activities, a DT can be built. But this is impractical because of the large number of cards in circulation. A solution is to build a weighted DT to predict with confidence (by assigning probabilities) that a transaction might possibly be fraudulent. ANNs can be built easily using the production rules obtained from the constructed DT. Compared to the loss resulting from undetected frauds in the past, a threshold level (on the amount or number of transactions per day) can be set to stop suspicious fraudulent transactions. The suspicious transactions can also be partitioned based upon card type, bank code, time of transaction etc to gain more insight using multiple base models built for various categories. Transactional data[20] (card ID, amount involved, date and time of transaction, number of password (PIN) attempts, total time taken to key in all information) are used to train the DT. In the table §6.18, we have indicated only the 'Yes' and 'No' classes, but in practice a third option 'Cannot judge' can also be added (for example when the card is newly issued and used for the first-time or the card was unused for a long time etc). The 'pin attempts' and 'total retries' are correlated (a retry may occur even after the PIN is entered correctly when the user presses the abort). The time taken is given in seconds. A Java-based system [SP97]

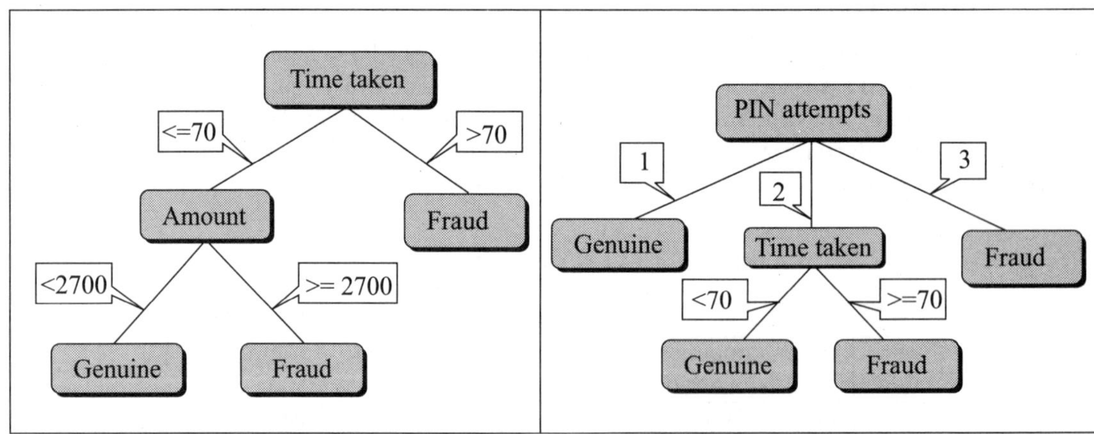

Figure 6.8: DT to catch fraudulent transactions

called JAM uses stratification at multiple levels to detect and report fraud. Other applications include money laundering detection, payment card fraud, mobile telecomm, ratings (sales, profit margin, return on investments, composite ratings, share market earnings, competitive pricing etc), financial forecasting, and business.

Example 6.10 Table 6.18 gives a random selection of cash withdrawal data from an ATM for 12 customers. Using the ETM introduced in page 6-24, fit a decision tree to predict if a future transaction is likely to be a fraud.

Solution. There are 6 frauds and 6 genuine transactions (some of the transactions are repeats of the same customer). Table 6.15 gives the calculations using the ETM method, where the

---

[20]Some of these may not be applicable for online and merchant based transactions.

Table 6.15: Extended Tabular Method for ATM Cash Withdrawal Data

Attribute	Class		Calculations						
Amount	No	Yes	Total	Gini	Entropy	Weight	Wtd Gini	Wtd Entropy
"$\leq$2300"	6	2	8	0.375	0.81128	8/12	0.25	0.540852
">2300"	0	4	4	0	0	4/12	0	0
Sum	6	6	12			1.0	**0.25**	**0.540852**
Pin attempts	No	Yes	Total	Gini	Entropy	Weight	Wtd Gini	Wtd Entropy
1	3	2	5	0.48	0.970951	5/12	0.20	0.404563
2	3	3	6	0.5	1	6/12	0.25	0.50
3	0	1	1	0	0	1/12	0	0
Sum	6	6	12			1.00	**0.45**	**0.904563**
Retries	No	Yes	Total	Gini	Entropy	Weight	Wtd Gini	Wtd Entropy
0	3	1	4	0.375	0.811278	4/12	0.125	0.270426
1	3	2	5	0.480	0.970951	5/12	0.20	0.404563
2	0	3	3	0	0	3/12	0	0
Sum	6	6	12			1.00	**0.325**	**0.674989**
Time taken	No	Yes	Total	Gini	Entropy	Weight	Wtd Gini	Wtd Entropy
"$\leq$60"	3	1	4	0.375	0.811278	4/12	0.125	0.270426
"60-70"	2	1	3	0.444	0.389975	3/12	0.111111	0.09749375
">70"	1	4	5	0.320	0.721928	5/12	0.133333	0.300803
Sum	6	6	12			1.0	**0.369444**	**0.668723**
Time taken	No	Yes	Total	Gini	Entropy	Weight	Wtd Gini	Wtd Entropy
"$\leq$70"	5	2	7	0.408	0.863121	7/12	0.238095	0.503487
">70"	1	4	5	0.320	0.721928	5/12	0.133333	0.300803
Sum	6	6	12			1.00	**0.371429**	**0.804290**

last two columns give the weighted Gini index and weighted entropy. The attributes *amount withdrawn* and *time taken* for completing the user input (in seconds, before the transaction is approved) are continuous. Both the Gini index and entropy for such attributes will depend very much upon the number of levels of discretisation. We have exemplified it for the time taken attribute by splitting it into 3 classes {$\leq$60, 60-70, >70} and 2 classes {$\leq$70, >70}. The minimum entropy is 0.540852 for the amount of transaction attribute with 2 binning levels. Thus it is taken as the split attribute at the root of the tree. Had we chosen the Gini index as the split measure, the attribute chosen for the top level split would have remained the same (with a minimum Gini index value of 0.25). Table 6.17 gives the data for the second level. There are two minimum measures corresponding to number of PIN attempts and number of retries. But as the non-zero measure occurs for PIN attempts=2, we take that variable for the next split. Continuing the process with the next level, we get the DT in figure 6.9.1. With this data, the misclassification error is zero.[21] The split values may differ slightly, when the same problem is solved by one of the commercial software packages. XLminer gave the split value for time taken as 65, and the number of attempts as 1.50, instead. The partitioned data after the first split is given in table 6.16.

---

[21]If a large training set is used, there will be significant misclassification error.

Table 6.16: ATM cash withdrawal data for the second level (amount$\leq$2300 vs amount$>$2300)

card id	amount involved	PIN attempts	total retries	time taken	fraudulent transaction
id1	1500	2	1	60	No
id2	800	3	0	90	Yes
id3	250	1	0	70	No
id1	1200	2	0	50	No
id5	1000	1	1	40	No
id8	900	2	1	80	No
id9	500	2	2	68	Yes
id10	2300	1	0	63	No
id4	3000	2	2	110	Yes
id6	2700	1	1	76	Yes
id3	2800	1	2	55	Yes
id7	3400	2	1	75	Yes

Table 6.17: Extended Tabular Method for ATM Cash Withdrawal Data - Level 2

Attribute	Class		Calculations					
Pin attempts	No	Yes	Total	Gini	Entropy	Weight	Wtd Gini	Wtd Entropy
1	3	0	3	0	0	3/8	0	0
2	3	1	4	0.375	0.811278	4/8	0.1875	0.405639
3	0	1	1	0	0	1/8	0	0
Sum	6	2	8			1	0.1875	0.405639
Retries	No	Yes	Total	Gini	Entropy	Weight	Wtd Gini	Wtd Entropy
0	3	1	4	0.375	0.811278	4/8	0.1875	0.405639
1	3	0	3	0	0	3/8	0	0
2	0	1	1	0	0	1/8	0	0
Sum	6	2	8			1	0.1875	0.405639
Time taken	No	Yes	Total	Gini	Entropy	Weight	Wtd Gini	Wtd Entropy
"$\leq$60"	3	0	3	0	0	3/8	0	0
"$>$60"	3	2	5	0.48	0.970951	5/8	0.3	0.606844
Sum	6	2	8			1	0.3	0.606844
Time taken	No	Yes	Total	Gini	Entropy	Weight	Wtd Gini	Wtd Entropy
"$\leq$70"	5	1	6	0.2778	0.650022	6/8	0.208333	0.487517
"$>$70"	1	1	2	0.5	1	2/8	0.125	0.25
Sum	6	2	8			1	0.333333	0.737517

6.9.2 Pruning ATM Decision Tree

Test (or validation) data can be used to identify and remove the least reliable branches from the built DT. Assume that we have another 10 validation data records, for which the classes are known. We put these data through the built DT, and note down the predicted and actual

Table 6.18: Validation data for ATM cash withdrawal

#		Attributes				Class
Ser. num.	card id	amount involved	PIN attempts	total retries	time taken	fraudulent transaction
1	id1	1500	2	1	60	No
2	id2	800	3	0	90	Yes
3	id3	250	1	0	70	No
4	id1	1200	2	0	50	No
5	id4	3000	2	2	110	Yes
6	id5	1000	1	1	40	No
7	id6	2700	1	1	76	Yes
8	id3	2800	1	2	55	Yes
9	id7	3400	2	1	75	Yes
10	id8	900	2	1	80	No
11	id9	500	2	2	68	Yes
12	id10	2300	1	0	63	No

classes. Some of the leaf nodes may perfectly classify their corresponding data records. But there could exist a few leaf nodes which contain training examples that do not belong to that class. The error rate of such leaf nodes (if any) is the proportion of misclassified examples. A subtree may contain zero or more such leaf nodes. Even if a subtree has all perfect leaf nodes (none of them have misclassified examples), we could prune it if one or more of the classes have very low frequencies compared to the others. For instance suppose a subtree has 3 leaf nodes L_1, L_2, and L_3 that predict the classes C_1, C_2, and C_3 with (correct, incorrect) classification counts (18,0), (40,2), (2,1). The misclassification errors for L_2, and L_3 are 2/40 and 1/2 respectively. This shows that the leaf L_3 has low frequency and the production rule for that leaf is not very reliable. If the leaf L_3 misclassify an instance into C_2, we could merge L_2, and L_3 into a single leaf L_{23}, resulting in a binary sub-tree. The misclassification error for the merged node is then $\leq 3/42$. We can compute the error rate for each internal node that has at least one leaf that misclassify data. If there is just one such leaf node, the misclassification error for the node is easy to find. If there are ≥ 2 leaf nodes that misclassify data, the expected error rate for the node is the weighed average of the error rates at the leaves weighted by the number of xxx.

6.10   Software for Decision Trees

There are many software packages for decision tree analysis.

Weka (Waikato Environment for Knowledge Analysis) is a Java-based free software suite from the University of Waikato, that has the induction algorithm embedded in it[22]. YaDT(Yet another Decision Tree) is an in-memory implementation of entropy-based DT algorithm in C++, with import facilities into various external formats including text, XML, binary, and predictive modeling markup language (PMML) formats ((www.di.unipi.it/~ruggieri/YaDT/YaDT1.2.5.pdf).). The SPSS AnswerTree incorporates CHAID, Exhaustive CHAID, CART and QUEST), and the data mining software SAS Decision Tree (SAS Enterprise Miner), IBM Intelligent Miner

---

[22]Weka implements an extended version of C4.5 called C4.8 (J4.8 or J48) that reduces overfitting.

Table 6.19: Software for Decision Trees

URL	Name	C/F/S
borgelt.net/ida.html	DT	F
www.dtreg.com	DTREG	C
eric.univ-lyon2.fr/~ricco/spina.html	Spina classifier	F
www.gatree.com	GATree	F
www.di.unipi.it/~ruggieri/YaDT/	YaDT	F
www.resample.com	XLMiner	S
www.rulequest.com	See 5	F
www.sgi.com/tech/mlc/docs.html	MLC++	F
www.treeage.com	TreeAge Pro	C
treeplan.com	TreePlan	S
www.cs.waikato.ac.nz/ml/weka	Weka	F
www.spss.com	AnswerTree	C
sas.com	SAS Enterprise Miner	C

Notes: C=Commercial, F=Free, S=Shareware, GATree is based on Genetic Algorithms (GAs)

etc. ODBCMine is a software to build decision trees from data residing in ODBC databases (www.intsysr.com/odbcmine.zip, www.reg.net).

6.11   Exercises

1. Mark as true or false:
   a) The DT model works only for categorical variables
   b) There are no distributional assumptions on training data
   c) Class-overlap is taken care of in a DT model
   d) Faster decisions are possible using a balanced tree with minimum depth.
   e) An unbalanced DT can always be converted into a balanced one
   f) Decision trees are linear classification models
   g) Decision trees can grow to any depth
   h) Time dependent data cannot be used in a DT model
   i) Outliers are easily handled by DT
   j) A leaf-node of a DT has 1 incoming and zero outgoing edges
   k) Internal (decision) node of a DT can have a single outgoing edge.

2. If there are n input attributes, (i) what is the maximum depth of the decision tree? (ii) How many leaves are there for a decision tree with n input attributes?

3. What does the root node of a decision tree represent?

4. If there are m classes, what is the maximum possible value of $Gini(S) = 1 - \sum_{k=1}^{m} P_k^2$? When does it attain the maximum?

5. Describe any 5 advantages of using a decision tree.

6. What are the disadvantages of using a decision tree?

7. Describe how you will identify attributes that contribute too little in DT classification.

8. What is the computational complexity of building a DT model if there are m attributes, and n training data records?

9. Suppose you want to generate random samples from one (or more) pre-specified classes. Is it possible to utilise the information abstracted in a DT model to efficiently generate random data for a class?

10. Describe any three measures that can be used as the splitting criteria at a tree node.

11. Show how the ternary DTC can be split into a BDTC for categorical data.

12. If a DT has a single 3-way split at the root, how many binary DT can you build from it?

13. When does the Gini diversity index attain its maximum and minimum values?

14. Is it possible to construct a decision tree if you are given all corresponding production rules? Is the generated DT unique?

15. Which of the following criteria are valid to stop growing a DT?
A) When the instances at a node are not linearly separate
B) If the (tree) depth of current node is above a threshold
C) If the number of cases in current node is below a threshold
D) If the information gain at a node is insignificant
E) If the current node is pure

16. Which of the following cannot be handled by a DT?
A) Missing (null) data values B) outlier observations C) complex attributes D) linear combinations of values.

17. The DT in figure 6.7 uses two identical split criteria at level 1 from the root. Is it possible to prune the tree? By considering the binary decision on penalty ≤ 10 and penalty > 10, construct an equivalent binary DT.

18. Modify the data in table 6.2 by adding a Gender, Age and driver-side airbag availability as new attributes. Develop a decision tree to predict accident risk for the new data.

19. What is the numeric range of each of the following coefficients? a) Gini index b) entropy c) χ^2 statistic.

20. What are the different types of pruning algorithms available? What will happen to the misclassification rate in a pruned tree?

21. Consider the data in table 6.20 that pertains to magazine subscription. Last column contains various classes (categorical attributes) into which the different cases fall. Develop a decision tree using the Extended Tabular Method (p.6-24).

22. Prove that the Gini index is $2[p_1(1-p_1)+p_2(1-p_2)-p_1p_2]$ for the ternary case (3 classes with respective proportions p_1, p_2, and p_3 such that $p_1+p_2+p_3=1$).

23. What is the *generalisation capability* of a classifier? Which of the classifiers (Naive Bayes classifier, neural networks, SVM, DT, LDA) is best in terms of generalisation?

Table 6.20: Magazine subscription data

Gender	Age (in yrs)	Income ('000)	Educ level	Profession/ employment	subscription status
Male	17	0	HS	Nil	Y
Male	26	32	Grad	Programmer	Y
Female	29	37	Grad	Nurse	N
Female	22	12	und-Grad	Student	N
Male	26	32	Diploma	Farmer	Y
Female	35	29	Diploma	Nurse	Y
Female	41	47	Grad	Farmer	N
Male	20	0	HS	Student	N
Male	39	42	Grad	Programmer	Y
Female	50	38	und-Grad	Clerk	Y

24. What are the pre-conditions on the data instances for the Hunt's heuristic for tree construction to work?

25. Identify the following graphs as homogeneous or heterogeneous:
(a) Königsberg bridge problem graph (b) job-scheduling graph with m jobs and n applicants (c) items bought together graph in a super market (d) term-by-document matrix graph in LSI.

26. An online floral shop accepts orders from customers for flowers to be sent to their friends and family members. What type of graph will result if the senders and receivers are represented as nodes?

27. Consider the Shannon's entropy measure $E = -\sum_{k=1}^{m} p_k \ln(p_k)$. Prove that $-\sum_{k=1}^{m} q_k \ln(q_k) = \ln(c) - (E/c)$ where $q_k = c * p_k$ for any constant $c \neq 0$. Prove that the corresponding relationship for Gini index is $G = \frac{1}{c^2} \sum_{k=1}^{m} q_k(c - q_k) = 1 - \frac{1}{c^2} \sum_{k=1}^{m} q_k^2$.

28. What is a confusion matrix? How is it useful in analysing misclassification errors? Does the dimension of it depends on the training data size?

6.11.0.1 References

[AR98] Alsabti, K., Ranka, S., Singh, V. (1998) CLOUDS: A decision tree classifier for large datasets, *Proc. of the 4th international conference on knowledge discovery and data mining*,2-8, New York.

[AZ98] Apolloni,B., Zamponi,G., Zanaboni,A.M.(1998) Learning fuzzy decision trees, *Neural Networks*, 11, 885-895.

[BK92] Bennett, K.P. (1992) Decision Tree construction via Linear Programming, *Proceedings of the 4th midwest artificial intelligence and cognitive science society conference*, Utica, Illinois, 97-101 (www.rpi.edu/~bennek/papers/DecisionTree.pdf)

[BB96] Bennett, K.P., Blue, J.A. (1996) *Optimal Decision Trees*, R.P.I Math report no. 214, Rensselaer Polytechnic Institute, Troy, NY.

[BB97] Bennett, K.P., Blue, J.A. (1997) *A support vector machine approach to decision trees*, R.P.I Math report No. 97-100, Rensselaer Polytechnic Institute, Troy, NY.

[BL04] Berry, M.J.A., Linoff, G.S. (2004) *Data mining techniques for marketing, sales and customer relationship management*, Wiley, NY.

[BW99] Boyen,X., Wehenkel,L.(1999) Automatic induction of fuzzy decision trees and its applications to power system security assessment, *Fuzzy Sets and Systems*, 102, 3-19.

[BF84] Breiman, L, Friedman, J. H., Olshen,R. A. & Stone,C. J. (1984) *Classification And Regression Trees*, Wadsworth International Group, CA, USA.

[BK02] Borgelt, C., Kruse, R. (2002) *Graphical models – methods for data analysis and mining*, Wiley, England.

[CH94] Chaudhuri, P., Huang, M.C., Loh, W.-Y., Yao, R. (1994) Piecewise-polynomial regression trees, *Statistica Sinica*, 4, 143-167.

[CW09] Chen,Y-L., Wang,T., Wang,B.S., Li,Z-J.(2009) A survey of fuzzy decision tree classifier, *Fuzzy Information Engineering*, 2, 149-159.

[FU92] Fayyad, U. et al.(1992) The attribute selection problem in Decision Tree generation, In 105th national conference on AI (AAAI-92)

[GG99] Gehrke, J.,Ganti, V., Ramakrishnan, R., Loh,W.Y.(1999) BOAT - Optimistic decision tree construction, In *Proc. of the ACM SIGMOD conference on management of data*, June 1999 (http://citeseer.ist.psu.edu/gehrke99boat.html).

[GR00] Gehrke, J., Ramakrishnan, R., Ganti, V. (2000) Rainforest - a framework for fast decision tree construction of large datasets, Data mining and knowledge discovery, 4(2/3), 127-162 (cs.cornell.edu/johannes/papers/1998/vldb1998-rainforest.pdf).

[GF08] Gray, J.B., Fan, G. (2008) Classification tree analysis using TARGET, *Computational Statistics and Data Analysis*, 52(3), 1362-1372 (www.sciencedirect.com).

[HJ75] Hartigan, J.A. (1975) *Cluster analysis*, John Wiley, NY.

[KG80] Kass, G. V.(1980) An exploratory technique for investigating large quantities of categorical data, *Journal of Applied Statistics*, 29 (2), 119-127.

[KI95] Kononenko, I. (1995) On biases in estimating multi-valued attributes, *Proc. of first international conference on knowledge discovery and data mining*, 1034-1040, Menlo Park.

[KL01] Kim, H., Loh, W.Y. (2001) Classification trees with unbiased multiway splits, *Journal of the American Statistical Association*, 96, 589-604 (www.stat.wisc.edu/~loh/cruise.html).

[KM98] Kearns, M., Mansour, Y (1998) A fast, bottom-up decision tree pruning algorithm with near-optimal generalization, *Proc. of the int. conf. on machine learning*,269-277.

[LS97] Loh, W.-Y., Shih, Y.-S. (1997) Split selection methods for classification trees, *Statistica Sinica*, 7(4), 815-840.

[LW09] Loh, W.Y. (2009) Improving the precision of classification trees, *The Annals of Applied Statistics*, 3(4), 1710-1737 (www.stat.wisc.edu/~loh/treeprogs/guide/aoas260.pdf).

[MA96] Mehta,M., Agrawal,R., Rissanen,J.(1996) SLIQ: A fast scalable classifier for data mining, *Proc. of 5th int. conf. on extending database technology*,Avignon,France.

[MJ89] Mingers, J. (1989) An empirical comparison of pruning methods for Decision Tree induction, *Machine Learning*, 4, 227-243.

[MS95] Murthy, S.K. (1995) *On growing better decision trees from data*, Ph.D. dissertation, Dept of Computer Science, JHU, Baltimore, MD, USA (ftp.cs.jhu.edu/pub/ocl/ocl.tar.Z, www.tigr.org/~salzberg/murthy-thesis/thesis.html).

[OW03] Olaru,C., Wehenkel,L. (2003) A complete fuzzy decision tree technique, *Fuzzy Sets and Systems*, 138, 221-254.

[QJ86] Quinlan, J.R. (1986) Induction of decision trees, *Machine Learning*, 1(1), 81-106.

[QJ89] Quinlan,J. R. et al. (1989) Inferring Decision Trees using minimum description length principle, *Information and Computation* 80, 227-248, (Academic Press)

[QJ93] Quinlan, J.R. (1993) *C4.5: programs for machine learning*, Morgan Kaufmann, San Francisco.

[RS98] Rastogi,R., Shim,K. (1998) PUBLIC: A Decision Tree classifier that integrates building and pruning, *Proc. 24th VLDB'98*, 404-415., also in Data mining and knowledge discovery, 4(4), 315-344, 2000.

[RJ78] Rissanen,J. (1978) Modeling by shortest data description, *Automatica*, 14, 465-471.

[RM07] Rokach,L., Maimon,O. (2007) *Data mining with decision trees - Theory and applications*, World scientific publishing co., Singapore.

[RS04] Ruggieri, S. (2004) YaDT: Yet another Decision Tree builder, *Proc. of the 16th international conf. on tools with artificial intelligence* (ICTAI 2004): 260-265. IEEE Press.

[SA96] Schafer, J.C., Agrawal, R., Mehta, M. (1996) SPRINT: A scalable parallel classifier for data mining, *Proc. of 22nd VLDB*, 544-555, Mumbai, India.

[SP97] Stolfo, S.J., Prodromidis, A., Tselepis, S, Lee W, Fan W, Chap P (1997) JAM: Java Agents for Meta Learning over distributed databases, *Proceedings of knowledge discovery and data management*, 74-81.

[TS06] Tan, P.N., Steinbach, M., Kumar, V. (2006) *Introduction to data mining*, Addison Wesley.

[TJ08] Twala,B., Jones, M.C., Hand, D.J. (2008) Good methods for coping with missing data in decision trees, *Pattern Recognition Letters*, 29, 950-956.

[YR07] Yellasiri, R., Raghavendrarao, C., Reddy, V.(2007) Decision tree induction using rough set theory - comparative study, *Journal of theoretical and applied information technology*, 3(4), 110-114, (www.jatit.org/volumes/research-papers/Vol3No4/14vol3no4.pdf).

[YS95] Yuan,Y., Shaw,M.J.(1995) Induction of fuzzy decision trees, *Fuzzy Sets and Systems*, 69, 125-139.

[ZH06] Zhang, Y., Huei-chuen, H. (2006) Decision tree pruning via Integer Programming, Elsevier (http://dollar.biz.uiowa.edu/~zhang12/optpruning.pdf).

7
Association Rules

Chapter objectives

- Understand association rules

- Introduce antecedent and consequent

- Distinguish cross-purchase and sequence-purchase analysis

- Introduce activity indicators

- Review extended association rules

- Explore sparse and rare association rules

- Describe negative association rules

- Define temporal association rules

- Understand multi-level and multi-dimensional association rules

- Describe plan mining

- Introduce fuzzy association rules

- Review algorithms for association rules

- Applications of association rules

7.1 Meaning of Association Rules

The literal meaning of 'association' is a casual dependency, or a spatial or temporal relation, between things, attributes, outcomes, symptoms, occurrences etc. Such associations are very common in data mining. More formally, associations are dependency relationships among variables or items that are hidden in large voluminous data. There is no universally accepted notation for expressing association rules. By convention, we will use a (single or double) arrow to separate the unidirectional or bi-directional association rule. An association rule is an expression of the form LHS$\Rightarrow$RHS (read as LHS implies RHS), where LHS and RHS contain one or more itemsets (categorical variables) without common elements (LHS $\cap$ RHS $= \phi$). The

LHS is called the body or *antecedent*, and RHS is called the head or *consequent* of the rule. Association Rule Mining (ARM) algorithms are used to automatically extract such dependencies from stored data. Attributes that often go together are found by ARM algorithms using a user specified minimum support and confidence (see page 7-6). It is also known as affinity analysis, or *market-basket* analysis. It is used to identify the degree of dependence as well as the strength and confidence among a few attributes or items in a large collection. It is an *unsupervised learning model* that is discovery driven.

Definition 7.1 An association in data mining indicates a *logical dependency* between various attributes, or various properties of an event (like a transaction).

As mentioned above, association rules are implications of the form LHS$\Rightarrow$RHS (s%, c%). Single Consequent Association Rules (SCAR) are those in which the RHS contains just one attribute or item. Single Antecedent Association Rules (SAAR) are those in which the LHS has just one attribute or item. Single Antecedent Single Consequent (SASC) rules have one attribute each on the LHS and RHS. We are often interested in Multi-Antecedent Single Consequent (MASC) or Multi-Antecedent Multi-Consequent (MAMC) rules. Some MAMC rules can be split as several MASC rules.

Association rules obtained may be global, or specific to some geographic region, cultural groups, or could depend upon one or more attributes like age, race, employment specialty, seasons etc. of subjects who generated them. Thus multiple association rules can be mined on the same attributes using class variables, or extrinsic attributes like location of a facility or business, customer qualities like education, employment or income level. For example, an association rule for buying a newspaper and a sports magazine or cine magazine together is restricted to special interest groups (sports persons, sports enthusiasts, etc).

Association rules find applications in supermarkets, e-commerce sites, hospitals, insurance companies, genetics, medicine etc. In purchase domains, they describe how often items were purchased together in the past transactions, so that this affinity can be utilised among those clusters of people (subjects) exhibiting such behaviors. Notice that nothing is assumed about the subjects that generate the data. If this information also is available, we could further refine the association rules and come up with valuable rules in sub-dimensions (like rules among people with various family income brackets, religions, education levels, etc). If such subject attributes are known (eg: subjects or patrons use a smart card issued by the merchant that stores such information) they may be inducted as additional attributes into the data. This is more useful for airlines and travel industry to mine segmented association rules among various strata of customers. We call such induced attributes as foreign or external attributes. These can appear on the LHS or RHS of a rule. By restricting it to be strictly on the LHS, we get sub-dimensional rules. By restricting it to be strictly on the RHS, we get categorical clusters.

7.1.1   Motivation

Transactions at retail shops and outlets generate millions of accumulated retail data records over time. Similarly, patient visits to clinics and hospitals, travel data of passengers and tourists, click-stream data of web users, game players, social networking sites, etc also generate copious amounts of data continuously. These data contain treasures of information, which when put to use can increase the profit, improve the productivity, reduce the time and effort etc in a record time. Shopping behavior of customers are burried inside the voluminous transactions that are captured at the OTC terminals. An algorithm to mine this type of data was proposed by Agrawal et.al. in 1993. This is called the AIS algorithm[1]. This algorithm is limited to only one item in the consequent (RHS), but could have multiple items in the antecedent (MASC type). This was later improved by Agrawal and Srikant in 1994. Their algorithm is called Apriori algorithm for association rule mining.

Supermarkets typically carry tens of thousands of items for sale. If there are 2000 items that could be bought individually or in combination by a customer, there exist $2^{2000} - 1$ possible market baskets (see below). To understand this more clearly, consider 3 items X, Y, and Z. The seven $(2^3 - 1)$ possible purchases are {X, Y, Z, XY, XZ, YZ, and XYZ}, where XY denotes that a customer buys X and Y together in a single trip. A customer seldom purchases majority of items (In some domains like an optician store, a customer could buy majority of items. Similarly in medical domain, a patient could exhibit multiple symptoms like fever, coughing, sneezing, etc). Depending upon the variety of stock (say > 100) and the total number of items in stock, majority of customers purchase between 1 and 30 items in a single trip. This could also differ in music stores, electronic stores etc where the number of items purchased by a customer in a single trip is much less. For example, in a restaurant that offer 20 dishes, a great majority of customers will order 2 or 5 dishes (including drinks) in a single visit (and rarely more than 10 dishes if they come as a family or group of friends).

7.1.2   Uni- and bi- Directional Association

In some fields like medical sciences, insurance etc the distinction between attributes is rather vague. For example, two (or more) symptoms or conditions could co-occur for some of the diseases. These are like symmetric binary variables discussed in chapter 1. But in purchase domains, we could clearly distinguish between uni-directional and bi-directional association rules. If the items purchased have no priority among themselves (in terms of price or durability) it is called bi-directional association rule mining (BARM). It is applicable to two or more items. In BARM, we are interested in *indurable* or *consumable* items that are purchased together. It is not specific to the purchase domain. It occurs in medical sciences, insurance, education and psychology, etc. For instance, a few symptoms may co-occur among students with learning disability. A few common reasons may prompt people to purchase high insurance policies.

The major tasks involved in data mining are categorisation, data and attribute clustering, affinity and link analysis, prediction of individual observations/values or future trends, analysis of trends and deviations, correlational analysis, outlier detection and analysis etc. Some of the above can also be applied to spatial, temporal or spatio-temporal data. Among these, the ARM algorithms are applicable to mine association rules, for classification, attribute clustering, and to a lesser extend for prediction and correlation analysis. For example, if a few items or attributes are found to be highly associated (strong support $>90\%$), we could predict the rest of them that go together or that seldom go together (negative association). This is useful in medical diagnosis using AI techniques, medical or laboratory test requirements etc. The attributes involved can be further analysed using other data mining or statistical models.

Example 7.1 Many of the customers to a supermarket often buy bread and milk together. We can represent it symbolically as Bread => Milk[2]. This is called a binary association rule. In this example, the association is bi-directional or symmetric (because we can express it equivalently as Milk=>Bread). Some customers may buy large quantities of milk or multiple packets of

---

[1]this is formed from the first letter in the names of the three authors (Agrawal-Imielinski-Swami).

[2]This is read as Bread 'implies' Milk

bread. These are tackled by weighted association rule mining (§7.8, page 7-24), in which the items purchased are appropriately weighted by the respective quantities. Data transformation and weighting of itemsets could be used for quantitative attributes.

In uni-directional association rule mining in the purchase domain, we could find the conditional association of an item given that another item has been bought. Items purchased need not always have the same durability, useful life-time or shelf-life. In such cases, the indurable items are more often purchased. Combination-purchases are transactions in which some items are almost always purchased together (hence while mining for combination-purchases, we will eliminate all 1-item transactions as useless (but 1-item transactions are useful in negative association rule mining §7.4.1). As an example, the items X, Y, Z given in the above paragraph are useless in combination purchase analysis). When the number of items purchased is 3 or more, there could be multiple combination purchases in it. For simplicity, we consider combination purchases involving just two items. Here are a few examples:

Example 7.2 Purchasing of computer and printer, classroom whiteboard and ink markers, cell-phone and carrying case, paint and paintbrush, etc are all combination purchases involving two items. These are uni-directional (asymmetric) associations. In these examples, more expensive item is given first. It is natural that customers subsequently buy the less expensive items. This is due to the fact that ink markers run out of ink, or carrying cases wear-out over time. Hence the reverse relationship need not always exist because more expensive items last longer. In some exceptional cases, the durabilities can be in the reverse. For instance, TV and antennas; or computer and UPS are often purchased together. But most antennas and UPS last longer than their companion item (which is more expensive).

7.1.3   Antecedent and Consequent

Every association rule has two parts called the antecedent and consequent that has a sequential dependency among them.

Definition 7.2 The left hand side (LHS) of an association rule is called the *antecedent*, and the right hand side (RHS) is the *consequent*[3]. In the Bread => Milk example, Bread is the antecedent and Milk is the consequent.

Definition 7.3 A rule is a functional dependency between the occurrence of antecedent(s) and consequent(s) or between subsets of data items in a dataset. One or more antecedents can appear on the LHS. In the simplest possible association, we have one attribute each in the LHS and RHS (SASC type). An attribute in the antecedent cannot be a proper subset of another attribute.

Example 7.3 Suppose the market value of a stock 'X' falls by a certain percentage quite often when the stock prices of 'Y' and 'Z' also fall. We could express it using an association rule as (Y, Z) => X. Here the order of 'Y' and 'Z' is unimportant, as they are considered as a group.

As mentioned above, MASC association rules meeting a minimum threshold are useful in understanding the occurrence frequency of a single consequent on multiple antecedents. If there are more than one consequents, it can be split into further association rules. An inter-transaction association is a dependence between itemsets in transactions (repeat visits) of the same customer.

---

[3]The direction of the arrow will tell which is the antecedent.

For example, same customers to a supermarket buy different item combinations in each visit. Similarly customers to e-commerce sites, bookstores, and other businesses generate different inter-transaction association rules. In forward association rules (given above), the antecedent is the frequent itemset. It is also possible to mine backward association rules where the consequent is frequent, is of special importance or of high utility (eg:profit margin of the consequent is higher than the combined profit margins of all antecedents taken together or in combination).

Association rules (for a statistical population) could depend upon many business characteristics like affiliations and brands in stock. As an example, the association rule for cross-purchase of consumer items in a cosmopolitan city's superstore could be different from that in a small town or village store due to the differences in living habits, and customer spending behaviors. An association between soft drinks and popcorn is restricted to some geographical regions (popcorn is not popular in some countries). Likewise, people living in temperate zone tend to buy cigarettes and popcorn more often in winter time than in summer. Long distance travelers (trains, planes) spend more on thick magazines and books than short distance travelers. Families on vacation trips with kids (in some age ranges) spend more on comic and children books. These are just a few instances where an association rule is applicable to particular target groups, and may not be generalisable. Similarly, the confidence and support (explained below) for transactions on same items at different locations may vary. Moreover, presence of duplicate records (like repeat visits of a customer to a supermarket, revisits of patients to a clinic, revisits of users to a web site) could have an impact on the association rules obtained. Hence we will not make any specific assumptions about the data, or entities that generate the data, in the following discussion.

Popular ways to express the rule symbolically are as follows:

```
Rule:   Body → Head[support, confidence]
Rule:   Body → Head[support, confidence]{startTime, endTime}
Rule:   Body → Head[support, confidence]{validityPeriod}
```

Definition 7.4 Association rules are dependency relationships used to identify output attributes (on the RHS) that are strongly related to one or more input attributes (on the LHS) for a particular population.

Attributes on the LHS and RHS are both categorical (most often nominal). Quantitative variables are categorised before the algorithm is run.

Definition 7.5 An itemset is a set of attributes that together encapsulate a data record or transaction.

Let S=$\{i_1, i_2, .., i_m\}$ be the set of all items under consideration. A subset S_k of S with $k > 0$ elements is called a k-itemset. A transaction is a set of one or more itemsets, and is denoted as $T = \{t_1, t_2, .., t_n\}$ where n$\leq$ m. In some domains like supermarkets, insurance, etc n is much less than m. Let X and Y be disjoint subsets of S. An association rule[4] denotes the logical dependence of Y on X as $X => Y$.

7.1.4 Categorical Variables

It is assumed in the above examples that the variables on the LHS are categorical. As noted in chapter 1, these variables assume a finite number of labels that are used only to identify

---

[4]A meaningful association rule requires at least 2 items.

them. For instance, the day of week, gender, education level, race, etc could appear on the LHS. There are many popular approaches to deal with categorical variables. If the number of categories of a variable is large, we decompose the possible values into a hierarchy by grouping alike categories into a separate group. Separate association rules can be mined within each category by partitioning the data into their respective blocks. If two (or more) categories have identical rules, they are merged together. Categorical variables can be of nominal or ordinal type. Ordinal data contain more information than nominal data as the values can be ordered among themselves. This information is useful in arranging the mined association rules more meaningfully. As an example, in cold countries, the sales of cigarette and liquor peak during winter time than in summer as people spend more time in-house. Similarly, some diseases occur more frequently during particular seasons resulting in seasonal variations in the sales of related medicines, or lab tests conducted. After obtaining association rules by one of the algorithms mentioned below, a further analysis (or rearrangement) using ordinal variables is helpful in such situations. In practice, the variables in association rule mining could be quantitative as well. If the minimum and maximum of such variables are exactly known, we could categorise it using one of the techniques discussed in chapter 1.

7.2   Association Rule Measures

Two commonly used measures of association strengths are *confidence* and *support*. Both of them are ratio-type variables in NOIR typology with a range in [0,1] (support can also be specified as a raw number $\leq$ N; see below). Although the zero is well-defined, zero values for them are not much useful in association rule mining. The goal in ARM is to find all rules having support $\geq$ minsupp and confidence $\geq$ minconf where both minsupp and minconf are positive fractions (see below).

7.2.1   Support

To mine any meaningful associations from voluminous data, a user need to specify two parameters called support and confidence. All association rule mining algorithms utilise the support measure first to extract frequent items, followed by the confidence measure.

Definition 7.6 Support is the ratio (or percentage) of the number of itemsets satisfying both antecedent and consequent to the total number of transactions (N).

In other words, it is the fraction (or percentage) of total transactions (N) in which the rule holds [Support = $n(X \cup Y)/N$] (some authors define support as a count (between 1 and N), and the frequency of an itemset as the support divided by the total number of transactions N. ie. Frequency(X)=Support(X)/N). Note that the $\cup$ notation here has a different interpretation from that used in chapter 3. The events associated with probability problems may be conjoint or disjoint. That is why we used the notation $\cup$ to denote the union, and $\cap$ to denote the intersection of events or sets in chapter 3. But we have an implicit assumption in association rule mining that LHS $\cap$ RHS = ϕ. In other words, the LHS and RHS of association rules are disjoint sets. Therefore, the $\cup$ operator is used to mean that X **and** Y occur together in a transaction. We do not use the $\cap$ operator in association rules.

The support value is a measure of interestingness. Unless otherwise stated, an association will mean a "positive affinity" between the items involved. As an example, if 650 out of 1000 customers to a fast food restaurant buy *fries* (=X) and *soft drinks* (=Y) together, the support

(X, Y) = n(X∪Y)/N = 650/1000 = .65. Support is a measure of global significance (prevalence) of the rule that expresses the relative importance (occurrence) of two activities for the entire population. The support count is the occurrence frequency of an itemset in the totality of transactions. As it is expressed as a fraction, the product of the support count and the total number of transactions (N) gives the number of transactions in which the rule holds. An itemset satisfies the minimum support (minsupp) if this count is $\geq$ minsupp*N. We may have to specify an upper and lower bound on the support in rare association mining algorithms (§7.4.3 in page 7-17). All ARM algorithms use a cut-off threshold to filter out potentially useful itemsets.

Definition 7.7 Frequent itemsets are those that dominate the transactions in D, or those that occur most frequently. Frequent itemsets can be found by using a threshold on the frequency (or support). A closed frequent itemset is one for which there is no superset that has the same support (that is present in exactly the same set of transactions). For example, if X occurs in transactions 1,5,10,15 and XY also occurs in 1,5,10,15, then X is not closed. If no other superset of XY have the same set of transactions, then XY is closed. A frequent itemset is maximal iff there are no supersets that is frequent.

7.2.2 Confidence

Definition 7.8 Confidence (strength of evidence) is derived from a subset of the transactions in which two entities (or activities) are related.

It is the proportion of the instances covered by both LHS and RHS of the rule, to that covered by the LHS.

In statistical terms, confidence is the conditional probability (expressed as percentage) that a randomly selected transaction will include all items in the consequent, given that the transaction includes all items in the antecedent. It is the probability that a randomly selected transaction from the entire population containing the antecedent will contain the consequent item together. We can express it either as a fraction or as a percentage. In terms of counts, it is defined as confidence = n(X∪Y)/n(X), where n(X) is the total number of transactions containing item X. The minimum occurs for those items that does not have any affinity with X with minimum value 0. The maximum occurs when each of X and Y occur as a bonded pair, with maximum value 1. As the denominator of confidence is usually less than N, the confidence measure for a set of items is $\geq$ support (c$\geq$s), where the equality holds when the item on the LHS is present in each and every transaction. Because this is rarely the case in practice, c > s generally holds.

Example 7.4 Suppose in a shoe shop, 51 out of 150 male customers who buys shoe also buys sox. Here the subset of interest is 'male customers', who 'buys shoes (X) and sox (Y) together'. This can be expressed as Conf(X,Y) = Supportcount (X∪Y)/Supportcount(X) = n(X∪Y)/n(X) where Supportcount is the number of transactions containing the argument. We are given that n(X∪Y) = 51 and n(X) = 150. Hence confidence = 51/150 = .34 or 34%.

If X is the antecedent and Y is the consequent, the confidence is $P(Y|X) = P(Y \cup X)/P(X)$ (note again that this has exactly the same meaning as in chapter 3 (where we wrote $P(Y|X) = P(Y \cap X)/P(X)$), except that we have used $\cup$ instead of $\cap$ here because X and Y are itemsets (and not events or random variables) with X∩Y=ϕ). It is the ratio (percent) of the number of records that contain X ∪ Y to the number of records that contain only X, so that it is *asymmetric* in X and Y (which means that conf(X,Y) $\neq$ conf(Y,X) unless P(X)=P(Y) or equivalently Supportcount(X) = Supportcount(Y)). A user can specify a minimum support (*minsupp*) and minimum confidence (*minconf*) for association rule mining (see below). An inappropriate choice

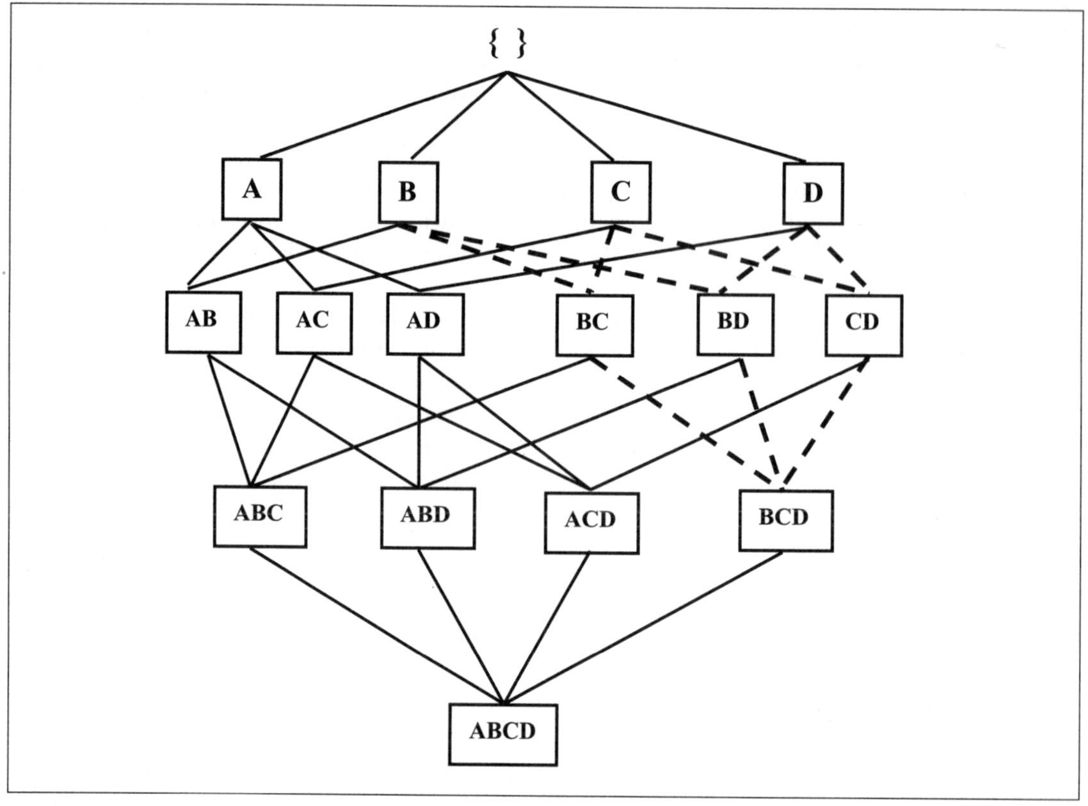

Figure 7.1: The Lattice Graph for Four Items. There exist 2^n-1 possible item combinations for n items, out of which we use 2^n-n-1 to find associations between 2 or more items. Broken lines indicate a subset of item combinations called frequent items, which is what we often look for.

yields either too many, or too few association rules. These thresholds are obtained from domain knowledge, or prior mining sessions. If the data size is too large, we could run the ARM algorithm on a small random sample, and use a *trial-and-error* approach to find approximate values for minsupp and minconf. The average of such values can then be used to find meaningful associations from the entire data. A low support rule can either be due to chance or may indicate a peculiarity of the data. A rule with large support and confidence is called a *strong rule*. Quite often, we look for all such strong rules hidden in the data[5]. The confidence measure could be more important than the support in some applications. Consider the presence (attendance) or absence patterns among a group of students in a class, or members in a committee. Majority of absences occur in pairs – whenever X is absent, Y is also absent. This may either be due to friendship among the absentees, dependency among them or can be a random occurrence. This can be expressed as absent(X→Y)(s%,c%), where X and Y contain one or more persons. In this case, the instructor or convenor may not be interested in the support measure (in how many classes or meetings out of N the persons were absent), but may be more interested in either positive or negative associations among all possible students or members.

---

[5]See en.wikipedia.org/wiki/Association_rule_learning for other related measures.

The aim in market-basket analysis is to find those sets of items most frequently bought together by customers in their past visits from a given set of transactions. Each transaction captures logical information about items bought together. All transactions are equally weighted in the standard ARM algorithm. In some domains like medical sciences, more recent transactions (patient visits) may have a larger influence than distant ones. These can be modeled using an exponentially decreasing weights scheme. Similarly, temporal variations can have an influence on purchasing behavior resulting in time-dependent association rules.

Association rules are a generalisation of production rules generated by decision trees. Each production rule results in just one output attribute (ie. the class to which an item belongs to), whereas an association rule can have one or more output attributes. They relate together intrinsic attributes in homogeneous sub-populations of a hypothetical parent population. They provide highly informative rules by reducing the dimensionality of attributes, and could also provide insight into clusters of entities that exhibit nearly identical behaviors. These rules are then put to use in modeling the parameters of future data, in optimising business processes and in other forms of decision making. The number of combinations of association rules grows rapidly with the number of input attributes. For n distinct attributes there are $\binom{n}{2} + \binom{n}{3} + .. \binom{n}{n} = 2^n - n - 1$ nontrivial rules (we have excluded $\binom{n}{1} = n$ 'one-item' transactions as trivial, and $\binom{n}{0} = 1$ (customers who walks-out without purchasing anything at all, or 'patients' who do not have illnesses are said to belong to the $\binom{n}{0}$ group)). The well-known *apriori* algorithm generates frequent itemset combinations bottom-up.[a] It starts with the minimum number of rules and grows them incrementally at each iteration ([AIS93]).

---

[a] The Apriori principle for association rule mining states that 'Any proper subset of a frequent itemset is frequent'

7.2.3 Lift

The *Lift* measure is defined as Lift(X, Y) = Confidence(X=>Y) / support(Y) = support(X∪Y) /[support(X) *support(Y)].

Some authors define the Lift(X, Y) as the correlation coefficient, which is incorrect. The minimum for Lift occurs when X and Y never co-occur in any of the transactions, with minimum value 0. The maximum for Lift occurs when X and Y always co-occur, with maximum value 1/max(support(X),support(Y)). Therefore, the range for Lift(X,Y) is from [0,1/max(support(X), support(Y))] (whereas the range for correlation is between [-1,+1]). The lift measure finds applications in negative association rule mining.

Example 7.5 Prove that the confidence of a rule and the support are related as Conf(X,Y) =(N/n(X))*support(X,Y), and that Conf(X,Y) $\geq$ support(X,Y). When are they equal?
Solution. We know that Support(X, Y) = number of tuples containing both X and Y/N, where N=total number of transactions. Confidence also has the same numerator, but the denominator of confidence is less than that of support. Symbolically, Conf(X,Y) =n(X∪Y)/n(X). Write this in the form [n(X∪Y)/N]*(N/n(X)). Substitute support(X,Y)=n(X∪Y)/N to get the result. As N/n(X) $\geq$ 1, Conf(X,Y) $\geq$ support(X,Y), or equivalently Conf(X,Y) $\in$ [support(X,Y), 1]. They will be equal only when the item X is present in all of the N transactions.

7.2.4 Cover

Each transaction is assumed to have a unique transaction ID called tid. These are automatically generated at supermarkets, airlines, hospitals etc. It is a concatenation of patient ID and date of visit in hospitals and clinics to make each patient visit unique.

Definition 7.9 The set of identifiers (*tid's*) of transactions in D that support itemsets X is called the *cover* of X. Symbolically, Cover(X) := {tid| (tid,x)$\in$ D, & support(x,D) $\geq$ s}.

A customer who buys a new home computer for the first time will almost always buy a printer with it. One who buys a cell- phone is likely to buy a carrying case with it. In these examples, confidence and support depend upon from where the items are bought. If it is bought from a supermarket that sells thousands of other items, the confidence will be high and support will be low. If the items are purchased from their respective dealer shops, the support also will be high. For instance, a paint-shop found that:– paint=> paint_brush[67%, 83%], with a high confidence. In large domains (large supermarkets, large medical studies, etc) this can be disambiguated by mining in sub-dimensions. As an example, if a large store sells electronic items of various sorts, separate association rules can be mined for computer section, TV section, office equipments section etc. Similar argument applies to various sections of supermarkets that sell a variety of items (electronics, clothing, food, drinks, books etc).

7.3 Association Rule Mining

Definition 7.10 Association rule discovery filters out the rules from a set of transactions. Users specify a minimum confidence and support, and the ARM algorithm outputs those subsets of transactions that contain frequent itemsets.

Adaptive AR Mining Algorithms (AARMA) may adjust the support and/or confidence thresholds on the fly, that results in reduced or increased number of candidate itemsets and rules. Similarly, probabilistic algorithms (like Toivonen's sampling algorithm [TH96], Parthasarathy's progressive sampling algorithm [PS02], intelligent sampling algorithms etc) starts with a reduced support threshold. Reducing the support threshold in general resuls in an increase in the number of frequent itemsets. Sometimes very small changes in support may not result in big jumps in the number of frequent itemsets. These are constant intervals of steadiness for frequent itemsets. This may indicate some peculiarity of the entities that generate the transactions (like family income bracket, age group, etc). In large domains like supermarkets, the graph is continuous for majority of small itemsets, but is sensitive to the *step-function* effect for large itemsets (say k-itemsets for k$\geq$20) and those itemsets containing expensive items. This can be understood by plotting the support threshold (say between s_{min} and 1.0 in steps of .1 or .05) in increasing order on the X-axis, and the number of frequent itemsets found for that threshold along the Y-axis (as a 2D graph), or by adding another dimension (say Z) for the length (size) of itemsets (number of elements k in k-itemset). The length of frequent itemsets too could increase dramatically as the support threshold is reduced. This information can be utilised to speed-up some of the algorithms. A confidence graph can be plotted using confidence (say between c_{min} and 1.0 with an appropriate step) along the X-axis, and the number of rules obtained along the Y-axis. The number of meaningful rules will in general increase when confidence is reduced.

Association rule mining is also called *market-basket analysis*[6], because it originated from finding

Figure 7.2: Cross-purchase graph at a supermarket

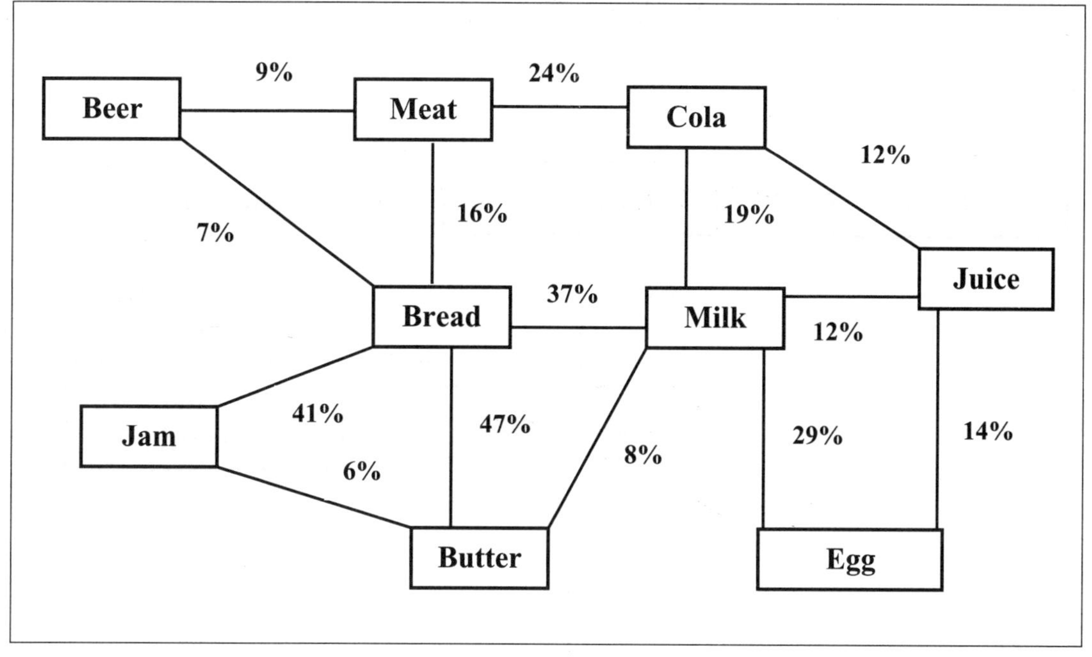

affinity of two (or more) frequent items in business transactions that are captured by point-of-sale (PoS) systems.

After extracting meaningful association rules from past data, we must put it to use immediately. In purchase domains, the items that are found to be 'associated' are arranged or bundled together, displayed in cross-proximity, or enough pointers are provided to minimise the shopping time of customers. This information can also be used to prepare catalogs, arrange pictures in product-offerings web pages, and so on. Cross-purchase behavior can be depicted by a graph in which the nodes represent items, and the edges denote if or how often those items are bought together. Edges of the cross-purchase graph (CPG) can be weighted either wrt the totality of purchases or wrt the subset of transactions that have at least two of these items (single purchases can be represented by self-loops, but these are not considered in ARM algorithms). Figure 7.2 gives a cross-purchase graph that is weighted using the percentage of customers who bought connecting items together. This graph could be disconnected for a small time interval.

Finding strong association rules is important in deriving most benefit from an ARM run. In sales domain, it can be used to set up bundles of packages that customers tend to buy together, cross-sell related items (usually on the web), where a customer who buys one item sees an advertisement for another related item. In medical sciences it is used to find medicine combinations that are most effective, most frequent causative factors and measurements (like lab results, metabolites, etc) that are most related to a symptom.

---

[6]A *market-basket* is a set of items selected by a customer in a single trip. The theory of association rules became popular during 1990's, although the market-basket concept existed long before it.

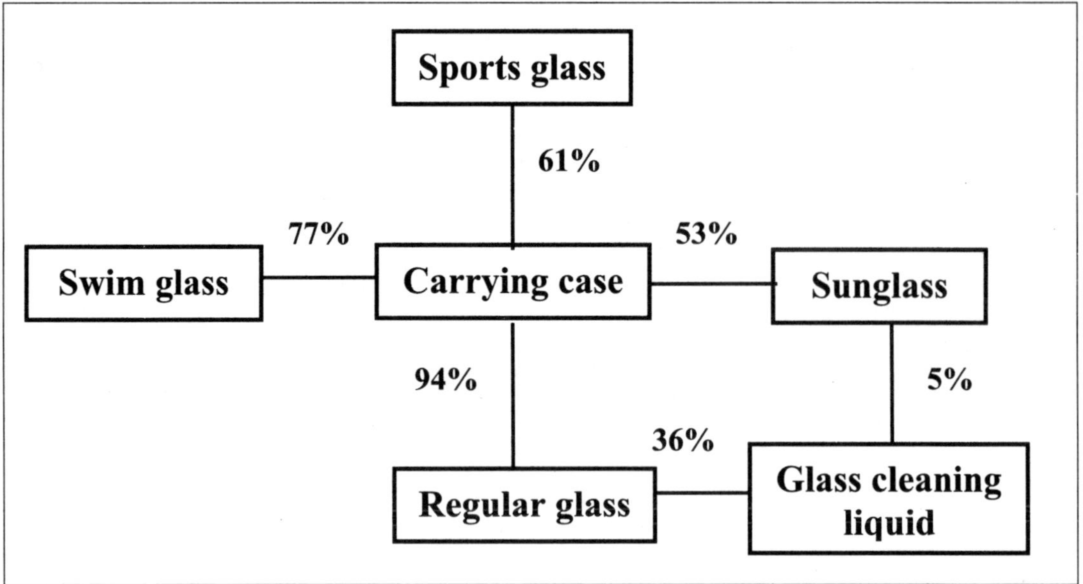

Figure 7.3: Cross purchase at an optical store

7.3.1   Cross-purchase Analysis

This is an unsupervised learning method to find positive affinity among nominal data using minimum support and confidence thresholds. Consider a finite set of items $S_m = \{i_1, i_2, \cdots, i_m\}$ and a set of transactions $T_n = \{t_1, t_2, \cdots, t_n\}$ where (n$\leq$m), and each transaction contain a nontrivial ($m > 0$) subset of items in I. A k-itemset contains exactly k distinct elements of S_m. Market-basket data can be represented in tabular form using asymmetric binary variables (chapter 2) because the weighting for 'item chosen' is different from that for 'item not chosen'. We will always code the item is chosen as 1. Cross-purchase analysis algorithms take such data as input and come up with a small number of meaningful rules that are easy to comprehend or interpret. The totality of possible purchases can be represented as a lattice structure. All ARM algorithms try to locate a boundary between frequent and infrequent items present in the lattice as showin in figure 7.1.

The challenge in Sequence Association Rule Mining (SARM) is to identify the antecedent. This is easy for humans, but not so for computers without additional information. In those situations where both items are 'durable', the item that is more costly is chosen as an antecedent. For example, the computer and printer are both durable and hence computer=> printer. Similarly TV=> Antenna since both are durable. If one of the items is less durable than the other, the more durable item is chosen as the antecedent. Examples are cell-phone=> carrying case, laser printer => toner cartridge. These criteria need not work in other domains. In insurance, physics and chemistry etc they may be related by a *cause and effect*. In medical sciences there are other criteria like 'symptom and prescription', 'condition and procedure' (medical or lab), root-cause and related symptoms.

7.3.2 Sequence-purchase Analysis

Cross purchase analysis mentioned above is a probabilistic occurrence. There are many factors that determine why two or more attributes occur together. In sequence-purchase analysis the focus is on the relative order of transactions. Association rules found in sequence mining are asymmetric (the consequent does not always imply the antecedent). In commercial sequence purchase analysis, we normally arrange the more 'expensive item' on the LHS by convention. In other domains, a utility measure, or score (on ordinal or higher scale) may be used. This is called *asymmetric* sequence analysis. Examples are cell-phone=> carrying case, computer=> printer, (computer, printer)=> UPS, TV=>antenna, laser printer=>toner cartridge, refrigerator=> voltage stabiliser, cigarette-lighter =>cigarette pack.

Functionally dependent events can also be found in other domains. For example {eye test=>lens purchase} or {diabetes=>insulin purchase} are sequence dependencies. Another variant of sequence purchase is the buying patterns in follow-up visits of customers or symptom patterns in revisits of patients to a clinic. This is also used in analysing problems in machine parts replacements, organ transplants, intrusion detection etc ([SS06]).

7.3.3 Activity Indicators

Qualifying association rules with an 'activity indicator' improves readability. These qualifiers can be words or strings in any language that are prepended on the LHS and RHS as follows:
Buys(x,'Bread') => Buys(x,'Jam') [3%,40%]
Buys(x,'Bread') => Buys(x,'Butter') [1%,30%].

This notation not only improves clarity and comprehension, but is also useful if multiple activities are involved in associations.

Example 7.6 An auto-shop offers exchange facility for customers who brings in old vehicles in exchange for new ones. Association rules for 'Buys' and 'Sells' using the attributes *vehicle exchanged,vehicle bought, income, age* and *employment status of customers* are expressed as:
Income(x, [40K, 55K]) ∩ Age (x,[25,30]) ∩ Sells(x,'Sedan') =>Buys(x,'Sports car')[4%,30%].
Income(x, [>50K]) ∩ Age (x,[40,60])∩ employment(x,'white collar') ∩ Sells(x,'Sedan') =>
Buys(x,'with airbag')[4%,30%]

The body of a rule can specify the interval for an attribute using set theoretic notation, as shown above. The above is called multi-dimensional associations (§7.6.2 in page 7-23) because more than one predicates (buys and sells) are involved in it. There are 3 predicates in the antecedent. These can be kept in any order. But when there are multiple association rules, it is beneficial to sort the predicates such that any common predicates come first. The set of rules with common predicates can then be arranged as a tree.

ARM algorithm finds the affinity of attributes by frequent itemsets. Utility-driven association rule mining looks for high-utility (eg: high profit, low risk, importance, health conditions, etc) associations. For instance, if two (or more) items or services are much less frequently sold, but yields high profits compared to other frequently sold items or services, the company may aim to promote these items by mining for high utility rules. Other examples are:– mining for two (or more) symptom combinations that are very rare, but could be fatal; unusual working of components of a machinery, etc. These are called utility driven association rule mining.

Association rules are also used for relating dependent events. In this case, the antecedent and consequent can have a combination of 'predicates' that are related through AND, OR,

NOT operators. The following simple rule (with one predicate on LHS and RHS) captures the information that 53% of the customers who buy bakery items have retired couples or kids in their family.

Buys(x,'Bakery_item') => Family_has(x,'retired_couples or kids')[1%,53%].

7.4   Special Association Rules

The classical algorithms for association rules find the positive affinity of related items. There also exist several special types of associations, the most prominent of which are negative associations, rare associations and sparse associations.

	Support $\geq s_{max}$	Support $\leq s_{min}$
Conf$\geq c_{max}$ (high c_{max})	Frequent Positive rules	Infrequent Negative rules
Conf$\leq c_{min}$ (low c_{min})	Rare Association rules	Infrequent Uninteresting rules

Items	support	cover
Bread	40%	3011, 3012
Bread	40%	3011, 3013
Soda	40%	3012, 3013

7.4.1   Negative Associations

A negative association is the opposite of a positive association. As a purchase domain example, customers who buy coke does not generally buy bottled water. Those who buy coffee do not in general buy tea. These are negative associations. Negative associations can sometimes become obvious using a categorical splitting attribute (eg: gender, seasons, country of customer). No assumptions are made on the attributes that appear in the antecedent, except the scale of measurement. They may either be correlated (for ≥ 2 attributes) or not. The correlation coefficient for such items will always be negative. If this correlation is close to -1, it is called strongly negative associated items. Asociation rules containing strong negative items or attributes are called negative association rules. Strongly negative association rules are almost always symmetric, but the converse need not be true. A rule is strongly closed if X$\Rightarrow$Y always imply that $\neg$X$\Rightarrow\neg$Y (pronounced as 'not X'). Such negative associations are useful in medical sciences in ordering medical or lab tests. If an observed symptoms combination strongly suggests the absence of another illness or condition Y, extra tests for Y can be avoided. This can not only speed-up diagnosis, but also can save time and money.

Binary associations are those that contain just 2 items (one antecedent, and one consequent). We could conveniently arrange the information in a binary association into a 2x2 contigency table. But association rules can have multiple elements in the antecedent and consequent. This gives rise to Multiple Antecedent Single Consequent (MASC), Single Antecedent Multiple Consequent (SAMC), Multiple Antecedent Multiple Consequent (MAMC) etc rules. Each of the antecedent and consequent could contain X or $\neg$X. Accordingly we could obtain Positive Antecedent Negative Consequent (PANC), Negative Antecedent Positive Consequent (NAPC), Negative Antecedent Negative Consequent (NANC) rules, each of which could contain a conjunction of presence or absence of terms resulting in SASC, SAMC, MASC and MAMC types. Some of the negative associations can be symmetric in the variables involved (if X$\rightarrow \neg$Y imply that Y$\rightarrow \neg$X they are negative symmetric associations).

Consider all two-item transactions in a supermarket for simplicity. We could sort all such transactions in descending order of the support value. Then all those pairs of items that have a support count less than a small threshold are rarely associated, and all those that have just one item in the transactions with high support and the other item totally absent are probably strongly negatively associated. If {X} or {X, Y} has support 70%, but {X, Z} does not appear in any transactions at all, we could say that X→ ¬Z. A negative association rule with 3 or more itemsets is *interesting* if one of the proper subsets of it is strongly positively associated. Suppose {Bread, Butter} has support 70%, but {Bread, Butter, PeanutButter} does not appear at all in any transactions. This implies that {Bread, Butter} → ¬PeanutButter, which is an interesting rule. This does not tell us anything about {Bread, PeanutButter} and Butter; or {PeanutButter,Butter} and Bread, although we could surmise that if a customer buys {PeanutButter,Butter}, they are likely to buy Bread, which is a positive rule (because we can semantically interpret the real-world names given to the itemsets. But what if they were implicitly coded as X and Y? Neither a computer nor a human can interpret such encodings to come up with valid rules). Thus we could partition the negative association rule landscape into the following sub-categories:– (i) negative rules that have a proper subset that are strongly positively associated (interesting negative rules), (ii) negative rules that have no proper subsets that are strongly positively associated (weakly interesting rules), (iii) negative rules that contain items which never appear in positive rules with minimum support threshold (uninteresting negative rules). Among these, the uninteresting rules are the most difficult to find as the data to generate them are not present in market-baskets. Thus the challenge is to find the dependency at the negative side of the correlation. Rule effectiveness measures can be defined separately for each of these categories.

Remember that we clearly distinguish between dependent and independent (uncorrelated) variables in regression analysis (chapter 3). To make any meaningful inferences from the mined association rules will require a dependency analysis of the antecedent attributes. Negative associations represent strong non-affinity of events or attributes. As mentioned above, the data needed to mine some negative associations may not be available in market baskets. But some of the negative associations can be derived from the corresponding positive rules. For instance, a person who takes a comprehensive health insurance is unlikely to subscribe to a health magazine. Thus there is a negative association between health insurance amount and health magazine spending. Similarly, loan defaulting customers are unlikely to buy high insurance schemes. An interesting medical application is to ascertain "what diseases are *unlikely* to occur in a patient exhibiting positive symptoms for some other diagnosed diseases". Another stock-market example is given below (example 7.10). A join operation is needed to generate indirect association candidates using a high-support cutoff (for frequent items) and a low-support cutoff (for infrequent items). This join step is more expensive in general than the corresponding join for Apriori candidate generation.

Many Apriori-like algorithms for mining negative associations have appeared in the literature. A brute-force method is to enumerate all negative itemsets for each of the k-itemsets present in the market baskets, and induce them as valid transactions (using a different label). Consider a committee comprising of 6 members (A, B, C, D, E, F) that meets regularly. A convenor wishes to ascertain any negative association among the members. The absence of members A through F in a meeting is represented by (U,V,W,X,Y,Z), so that ¬A=U, ···, ¬F=Z. This simply doubles the attribute space. Assume that A and B are absent in a meeting. This gives us the itemset m={U,V,C,D,E,F}. An algorithm for positive ARM can then be applied on the augmented itemsets. This is clearly impractical in large domains. See INDIRECT [TK02], for between pairs of items was given in Tan and Kumar (2002), Tan et al. (2000).

Example 7.7 A company markets fruit juice in aluminum cans, glass bottles, tetra pack (paper cartels), and plastic cans of various sizes. A store is interested in finding items that are *unlikely* to be bought by customers who buys juice in glass bottles and one litre plastic cans. This is an example of negative association, which may be useful in arranging items in catalogs or in offering cross-purchase discounts.

Simple negative association rules are of the form $\neg X \Rightarrow Y$, $X \Rightarrow \neg Y$, etc[7], etc where $\neg X$ denotes the absence of the itemset X wrt another transaction Y such that $Y \cap X = \Phi$. Multi-negative associations are of the form $\neg X \Rightarrow Y$, $X \Rightarrow \neg Y$, etc where $\neg X$ denotes the absence of the itemset X wrt another transaction Y such that $Y \cap X = \Phi$. Among the above, the NANC is the most difficult to mine, as the data for pure NANC rules like $\neg X \Rightarrow \neg Y$ are never available directly. Note also that $\neg X \Rightarrow \neg Y$ does not always mean that $X \Rightarrow Y$. Depending upon whether the antecedent and consequent contain purely negative attributes or mixed attributes (both present and absent attributes) we could come up with a variety of rules in the negative landscape. Generating such negative rules is computationally more complex than positive rule generation as all possible items absent from the transaction may have to be considered during the candidate generation phase. This is especially true when market baskets are sparce (eg: If a supermarket stores 2000 items and the maximum number of items that any customer purchases in a single trip is 30). Previously found positive association rules can be effectively utilised in narrowing down the search space for interesting negative association rules.

The rule $X \Rightarrow \neg Y$ has support s% if s% of the transactions in T contain itemset X, but do not contain itemset Y. As the confidence depends upon the support of the LHS, we could directly calculate the confidence of the rule $X \Rightarrow \neg Y$ as support(X,$\neg$Y)/support(X). But the confidence of the rule $\neg X \Rightarrow Y$ is calculated as support($\neg$X,Y)/support($\neg$X). A short-cut method exist to compute the support and confidence of simple negative association rules as follows:

Theorem 7.1 Prove that support($X \Rightarrow \neg Y$) = support(X) - support($X \Rightarrow Y$), and confidence($X \Rightarrow \neg Y$) = 1 - confidence($X \Rightarrow Y$) if X and Y are disjoint itemsets. **Proof.** The proof follows easily using the Venn diagram representations. Consider the equality $X = XY + X\overline{Y}$, where X and Y are any arbitrary sets, and the missing operator is $\cap$. As X and Y are assumed to be disjoint, we use the $\cup$ operator for itemsets, and write the above as $X = X \cup Y + X \cup \overline{Y}$. From this we get support(X) = support($X \cup Y$)+support($X \cup \overline{Y}$). Substitute support($X \cup \overline{Y}$)=support($X \Rightarrow \neg Y$), and rearrange to get the above result. More interesting is the corresponding results when the LHS contain negations (non-occurrences). In this case we get support($\neg X \Rightarrow Y$) = support(Y) - support($Y \Rightarrow X$). By writing $Y = Y \cup X + Y \cup \overline{X} = X \cup Y + \overline{X} \cup Y$, we get the result as support($\neg X \cup Y$) = support(Y)- support($X \cup Y$). Divide throughout by support(X) to get the corresponding relationship for confidence. An extension of this rule can be stated as follows:

Theorem 7.2 Prove that support($X \cup \neg Y \cup Z$) = support(XZ)- support($X \cup Y \cup Z$) if X and Y are disjoint itemsets. **Proof.** Combine all positive itemsets into one group, and all negative itemsets inot another group. Let A= XZ and B=Y. Then apply above theorem on A and B to get support($A \cup \neg B$) = support(A)- support($A \cup B$). Now substitute for A and B to get the result.

7.4.2   Sparse Association Rules

In most of the examples given before, there is a steady and continuous or recurrent flow of data (as in day to day customer transactions, continuous visits of patients to a clinic or users worldwide to a web site, click-stream data) that generate copious amount of transactions. Data sparsity is quite common in many domains. Sparse Association rules (SAR) are realised in the following situations:

---

[7]These can also be written as $\not{U} \Rightarrow V$, $X \Rightarrow \not{Y}$, or -U$\Rightarrow$ V, X$\Rightarrow$ -Y, etc.

1. There are too many attributes and too few subjects/transactions

2. Attribute interactions fall in well defined clusters

3. There are too many outliers

An easy way to check for sparsity is to form an nxn matrix in which each row and column corresponds to the attributes. In a medical study, for instance, the rows and columns may be labeled with each of the symptoms (eg: pain at various locations, fever, digestive ailments etc) for *all the patients* under study. Each cell (intersection of rows and columns) denotes the number of patients with both symptoms present. The result is a symmetric matrix with zeroes along the main diagonal. To optimise storage, we may store the total number of patients with that particular symptom along the main diagonal. Pair-wise joint frequencies $[n(P \cap Q)]$ are stored in the upper triangular portion. The lower triangular portion is used to store $[n(Q|P)]$, where $n(X)$ denotes the number of occurrences of the event specified, P and Q are the symptoms, and vertical bar denotes OR operator.

```
Sample data:
Amy   -- {Fever, Cold}
Frank -- {Fever, Headache}
John  -- {Cold, Dizziness, Fever}
Mary  -- {Pneumonia, Cold, Dizziness}
Singh -- {Diarrhea, Pneumonia}
```

For convenience, we arrange the matrix in sorted order of the illness name as in table 7.1.

Table 7.1: Co-occurrence matrix in a medical study

	Cold	Diarrhea	Dizziness	Fever	Headache	Pneumonia
Cold	3	0	2	2	0	1
Diarrhea	4	1	0	0	0	1
Dizziness	3	3	2	1	0	1
Fever	4	4	4	3	1	0
Headache	4	2	3	3	1	0
Pneumonia	4	2	3	5	3	2

We have chosen only the most common symptoms. If we take the totality of symptoms of all patients to a clinic, the matrix will be sparse since some of the symptoms are rare (specific to a small group of patients). To find the association rules, we extract high frequency symptoms first, which correspond to 2-itemsets. After eliminating cases using the minsupp and minconf, we proceed to higher combinations.

Association rules tend to cluster into distinct groups in those situations where the population contains homogeneous subgroups, more so with many attributes under consideration. But these situations may also result in spurious association rules, that happen due to chance.

7.4.3   Rare Associations

In most of the examples cited above, we seek those rules with high confidence or support. Sometimes, we may be interested in rare associations (those with low support and high confidence).

Meaningful inferences can sometimes be drawn from rare associations that could be due to eccentricities among the entities that generate the data. These are like data outliers. Rarity may be either *absolute* or *relative*. Absolute rarity is determined by a cutoff threshold limit. Relative rarity is literally the opposite of frequent associations, and is 'observed' by low relative frequency of occurrence. For example, how often does a person with an annual income below 20K invest huge amounts in real estate or on shares? What factors play a role in the failure of a telecommunications or supply-chain network? There are many such rare associations that occur in practice.

Discovering rare associations using apriori algorithm requires a low support to be set. This will in turn result in a large number of association rules. If the number of attributes is not too large, the tabular approach discussed above will be of use in extracting rare associations by first finding rare 2-itemsets and proceeding as before. Another approach is to separate out all possibly rare cases into a 'rare-group' in the first phase, and then to mine for rare association in this restricted subset. Algorithms to extract rare associations can be found in [WG04], [SM06].

7.4.4   Temporal Association Rule Mining (TARM)

The validity period of association rules obtained by ARM algorithms is not in general constant or even known. Most softwares do not specify a *validity period* for association rules, due to the implicit assumption that it is valid over a reasonable time period. Temporal associations are logical dependencies between attributes or events over a period of time, and are also called time series associations. Examples include association rules obtained from stock market data, web server logs, online auctions, and other time dependent data.

Rules extracted from time series databases in which each transaction is time stamped could be used to predict future associations over similar time intervals. Hence in temporal rule mining, the ordering of the records (itemsets) is important, as in the case of stock market. Temporal patterns signify events ordered by time while temporal rules signify the 'cause and effect' relationships. Trend analysis is another name for temporal rule mining in which long-term, periodic, seasonal and cyclical variations over time are extracted. The time interval may either be fixed (forenoons, nights, wintertime, semester-ending time, festive season, X-mas season etc), over a known period of time, or unspecified. A special case is an association rule extracted over a 'sliding window' of time which is distinguished by a time start (t_s) and a time-end (t_e). The time bracketing works best in those situations where the duration of an activity is short (selling a movie or concert ticket, an e-commerce transaction, buying a share online etc), but may create 'overlap problems' in some applications. Separate rules could also be found for short-term and long-term time intervals. For example, temporal rule mining can find out the profile of customers who make short duration (< 60 seconds) and long duration (> 20 minutes) telephone calls (which may depend upon the time of the year as well). Customer spending on toys and gifts during thanksgivings and X-mas; spending on clothes during festive seasons, start of school year etc are valid for particular time periods, and are known as *periodic associations* as it repeats over fixed time periods. An association between an asthma attack and reflux induced coughing or wheezing of a patient could change over time depending upon the number of asthmatic attacks [AS01]. Some medical symptoms are more prevalent during particular seasons than others, and are often assumed to be valid indefinitely.

To capture the time period in such associations, we introduce the following notation:
Rule:*Body* => *Head*[support, confidence]{validity_period}
where validity period is either explicitly specified, enumerated or extra information like 'periodic

pattern' added to make it complete and unambiguous, or specified using a time interval $[t_s, t_e]$.

Example 7.8 A shop sells gifts and mailing cards. Obviously, the sales will peak during the year-end and on special occasions. A typical temporal association might look like:
Age(X,'18-25') $\cap$ month $\in$ 'December-January') $=>$ spending(X, 'high'){seasonal}.

Such temporal associations are quite common in many domains (medical sciences, health insurance, inventory systems, long-distance telephony, etc). Similar temporal rules exist between accidents and driving experience during winter time due to rain or snow on the road, and between travel and summer time due to summer holidays. Association rules can also be mined with temporal constraints (that bracket the time interval of interest) to eliminate spurious cases (see §7.7 in page 7-23). The weighted temporal association rules can also be used to discard irrelevant cases (whose weights are set to zero).

A problem encountered in temporal association rule mining is the *rule expiry*, which is the invalidation of a previously found association rule over the passage of time. This may either be a temporary phenomena (over seasons, or during particular time intervals) or over an unknown period. As an example, a rule may expire in medical sciences due to the effective control of an epidemic. Similarly, correcting the fault due to defective parts or wrong placement of parts (as in automobile recalls) in a manufacturing environment (machinery, medical equipments, automobiles etc) can result in some rule expiry. Analysing obsolete rules may also benefit businesses in understanding customer behaviors over time, and better control the inventory of such items. Obsolete rules (as in medical studies) may indicate new data to be collected, new treatment plans or prescriptions, or new research studies to be carried out.

7.4.5 Pareto Analysis

Pareto analysis is a formal technique based upon 'Pareto principle'[8] for finding the changes that will give the biggest benefits. It is useful where many possible courses of action are competing with each other. Instead of spreading out the available resources or efforts on tackling multiple problems together, Pareto analysis provides a simple solution:–*Prioritise and conquer* (P&C). Whereas the *divide-and-conquer* (D&C) principle solves a big problem by combining smaller size subproblem solutions, the P&C principle is used to order tasks or subproblems based upon their importance and solving them optimally. A smaller number of rules can explain the inherent associations between a large number of attributes in most situations.

7.4.6 The Inverse Pareto Principle

Does the Pareto principle or the 80/20 rule hold everywhere? Not exactly. There are many situations and phenomena where it holds for a selected population. There also exist situations in which it is just the opposite — 80% of something is related to 80% of a related thing. We call this the inverse Pareto principle. As an example, the 80/20 rule holds for atmospheric pollution by heavy industries in those cities with many heavy industries — "80 percent of the atmospheric pollution is caused by 20 percent of the industries". What about cities with very few manufacturing industries, or about auto-exhaust pollution?. The 80/20 rule of course does not hold in most cities, because the amount of pollution is more or less the same for vehicles of the same type (buses, trucks, 4-wheelers, autos, 2-wheelers, etc). In most cities the law is in the

---

[8]by doing 20% of the work lets you generate 80% of the benefit of a finished job, where 80 and 20 are arbitrary divisions (see §2.2.6).

reverse – "80 percent of the air pollution is caused by 80 percent of the vehicles". The number split will vary depending upon the number of vehicles of various types that ply through a city, which includes transit vehicles. It is found in some cities that "90 percent of the electricity produced are consumed by 90 percent of the patrons", a direct consequence of which is that "90 percent of the revenues come from 90 percent of the patrons".

Consider an ATM machine in a big city that is replenished daily morning with cash. The amount of replenishment is decided for each ATM from past cash withdrawal data. For example, the ATMs near railway stations, movie theatres, banks, bus stations etc are visited by lot many more patrons than those at colleges, and remote locations. Some ATMs have fixed withdrawal options in multiples of a fixed amount. One of the banks found that "In 80 percent of the ATMs in a city, 80 percent of the replenished cash are withdrawn by 80 percent of the patrons". These points should be considered while mining for association rules.

7.4.7 Paired Comparisons Analysis

In Pareto analysis, the alternative options are chosen one at a time using a priority score. The *paired comparison analysis* is an extension of Pareto analysis, in which the priority of many options relative to each other are analysed simultaneously. Since multiple options are evaluated, priorities are assigned with due respect to each of the conflicting options.

When many similar association rules are obtained, prioritising them using a new measure may be beneficial. For instance, a superstore may be more interested in maximising their profit (say 80% of the profit is derived from selling 20% of the items) rather than selling more items with less profit margin, using Pareto principle[9]. If the profit margin differs among the items, the store may be interested in prioritising the association rules for those items having higher margins, fast depreciations or short shelf-lives. Likewise, insurance companies identify and retain customers in low risk categories, levying higher premiums on high-risk categories.

7.4.8 Fuzzy Association Rules

Association rules for continuous variables that have been discretised can be sensitive to the boundary of the split. For example, if the age of a customer is split unevenly, resulting in various age groups, and an association is found between the age group and smoking, a small change in the age of a person can result in a boundary phenomena. Hence an association rule may become invalid over time (as a person ages). If the data used for association rule mining is spread over a long time interval, it will give rise to a 'sharp boundary problem'. Fuzzy Association Rule Mining (FARM) is one of the remedies to this problem in which a data value can contribute to more than one fuzzy set. Fuzzy ARM algorithms are best suited for quantitative variables, as categorising such variables results in "sharp boundary" phenomenon. Relaxing such items to belong to multiple categories with different memberships results in the fuzzy ARM.

The guiding principle of fuzzy classification is the graded classifier memberships. The success depends upon an effective partitioning of input variables as overlapping fuzzy sets, and the application of fuzzy logic. Each data point has a membership function that indicates its degree of membership in each of the overlapping categories. There are many ways to achieve this partition. The simplest is the Ruspini-type partitioning that use triangular membership functions. More complex schemes use smooth tailing functions (similar to tail areas of normal,

---

[9]From the profit maximisation point of view, not only the associations between items, but 1-item (customers buy just one item like liquor) frequency counts are also equally important.

exponential or t-distributions). Fuzziness can be associated with quantitative input variables, or output attributes, or both. Advantages of fuzzy ARM algorithms include an increase in the flexibility, better generalisation capabilities, and robustness. They can also accommodate missing or incomplete data (in both the training set and test sets) using fuzzy logic. In addition, FARM algorithms can sometimes result in substantially less number of association rules than their crisp counterparts.

Consider a medical example, in which a patient exhibits multiple symptoms. Some of the symptoms are observed directly (eg: sneezing, coughing, wheezing) while some others are inferred from measurements or tests (eg: fever, weakness, pain). We consider observed symptoms as crisp and measured or inferred symptoms as fuzzy.

Further discussions can be found in [LX98], [FB02], [GW02].

7.4.9 Plan Mining

A general form of association rule where attributes are non-probabilistically related is known as a plan. The aim of plan mining is to discover all such plans with a specified confidence and support. Data for plan mining comes from repair logs, production runs, working conditions of machinery (eg: flight data of aircraft), etc.

Example 7.9 An aircraft maintenance and repair unit maintains a repair log containing information about all parts that have been replaced or overhauled during each repair run (a repair run can be anywhere from a few hours to a few days). A change of one part may affect the smooth working of many other parts. These parts are not related probabilistically, but have a form of functional dependency. All replaced parts need not be related, but nearer parts may be more related than farther parts. Each part can have multiple attributes that generate data. In the simplest case of one attribute per part, the dependency can be captured into a plan matrix (with diagonal elements zero). A plan mining algorithm can then be used to get important plans with specified confidence and support.

7.5 Generalisations of Association Rules

Association rules have been generalised by various researchers using different criteria. The most popular extensions of association rules are: hierarchical (multilevel) and multi-dimensional association rules, similarity search, constrained association rules and weighted association rules. The multidimensional and multi-level association rule mining can either be interactive or unsupervised. In interactive mining, in addition to the *minsupp* and *minconf*, the user inputs more specific information like dimensionalities, constraints, weights, level-wise support break-ups, etc. In multi-dimensional mining, constraints can be added into 5 categories as described in [HL99] - knowledge type constraints, data constraints, dimension/level constraints, rule constraints and interestingness constraints. There can also be many multilevel hierarchies rooted at different attributes - like different newsgroup hierarchies that programmers visit, or different disease hierarchies in medical studies (diabetes hierarchy (type-1,2,3), pulmonary and cardiovascular hierarchies etc). The aim of similarity search is to extract similar (using a closeness measure) patterns. There are two popular categories called whole sequence matching and subsequence matching. Constrained association rule mining restricts a variable (usually quantitative) into predefined intervals. Constraints may also be placed on confidence and support (§7.7 in page 7-23). Weighted association rules are discussed in §7.8 (in page 7-24).

7.6   Extended Association Rules

The association rules found for one group (geographical region, education level, gender or other categories) may differ from that for another group. For a user specified minimum confidence and support, the ARM algorithm may not always be successful in producing meaningful associations. If there are either too few or too many associations, the min-support and min-confidence need to be revised to get reasonable number of associations. Thus group-wise extraction of association rules may give better insight into logical dependency between attributes of interest in sub-dimensions. Sometimes, splitting the dimensions may reveal hidden associations, as in the following purchase domain examples:

1. Association between the purchase of cigarettes and tartar control toothpaste depends upon the income and education levels.

2. Association between the purchase of cosmetics, shampoo and hair-die depends upon income level, dwelling region or country.

3. Association between purchase of sun-glasses and skin tan lotion depends upon race and skin colour.

In such situations, we add the dimensionality splitting attributes to the consequent to make the association rule narrower, meaningful and generalisable:

cigarette => tartar_toothpaste(income > 40K ∩ education >= 'graduate').

These are called refined association rules, and are helpful in focused marketing, narrowly focused research studies, etc. The end-result of association rule mining is not just the rules found. If a group of persons exhibit identical characteristics as revealed by nearly identical association rules, further research into those groups may reveal hidden patterns and anomalies among the members. A large number of influential association rules may sometimes be due to a small group of entities. For instance, a family with 2 or more kids living with their parents often is a characteristic of some of the influential association rules found in supermarkets. This is more important in medical studies, bioinformatics, insurance and many other fields.

7.6.1   Multi-Level Association Rules (MLAR)

In many practical situations, the items can form a hierarchy like the *whole-part* hierarchies (sub-components of a machine or device, subcategories of an entity etc), *is-a* hierarchy (different brands of a product). Some items can simultaneously belong to multiple hierarchies - product hierarchy, promotional hierarchy, etc. In whole-part hierarchies, one may be interested in finding associations at higher conceptual levels, multiple conceptual levels or at primitive levels where they carry more concrete information. Rules at lower levels are expected to have lower support. Sometimes primitive level associations are hidden using a minimum confidence and support threshold while strong higher-level associations are revealed. MAMC rules are of the form $X \Rightarrow Y$ where both X and Y have 2 or more items.

Examples of *is-a* hierarchy are – the milk categories as milk=(full fat, partially skim (2%), skim-milk), yogurt =(plain, partially cultured, cultured), battery_cell=(eveready, energiser). These can be analysed either using two (or more) association rules or by uniquely identifying each sub category. Hierarchical categories can similarly be analysed using multiple associations for each of the branches of the tree or representing each node of the tree as an input attribute.

A popular algorithm for multi-level rule mining is the top-down progressive deepening in which frequent itemsets are found at each level until no more of them can be found. Then one of the algorithms mentioned in §7.9 is applied at each level. The *minsupp* may either be uniformly specified at each level or can be reduced at lower levels. Rules found on lower levels of the hierarchy may sometimes be redundant due to the already found stronger rules at higher levels. Hence a redundancy-filtering step may be carried out to eliminate such rules. Further discussions can be found in the references [FL96] (gives an object oriented approach to multi-level rule mining), and [SC99].

7.6.2 Multi-Dimensional Association Rules (MDAR)

If an association rule has two or more predicates, it is called multidimensional. In other words, MDAR is an extension of the association rules in single dimensions to multiple predicates (see § 7.3.3 in page 7-13). This approach can also be extended to multi-levels within each dimension. Inter-dimensional rule mining algorithm works within a set of distinct dimensions. Intra-dimensional rule mining works on a set of reference dimensions. Denoting 'and' operator by $\wedge$, we may express this as follows:

age(X,'20-32')$\wedge$ income(X, '30K-60K') => buys(X, 'sedan')[20%, 50%]

Here age and income are continuous variables, which may be discretised (transformed into non-overlapping categories) using a suitable interval.

Example 7.10 Suppose the share price of stock Z goes up whenever the prices of X and Y comes down. We can express this using the above notation as:

down(X) $\wedge$ down(Y) => up(Z)[s%, c%]

7.7 Constrained Association Rules (CAR)

A constraint is a rule that limits or restricts the numerical value, properties or membership of attributes. They may be domain constraint, class constraint or aggregate constraint. While mining for associations, a supermarket with a very large number of daily transactions may put a constraint on the minimum transaction amount to limit the size of itemsets being mined. This is an example of one-sided (lower limit on amount) constraint. Upper limits can also be specified to reduce the outlier observations.

7.7.1 Rule Constraints in Association Rule Mining

There are many kinds of constraints that can be applied:
1) content constraint (Interestingness constraints)
These are constraints in which the content is restricted to some range or group. For instance restricting income levels of employees to be above 20K, amount of gasoline filled up by customers at a petrol station to be above 5 gallons etc are content constraints. These are akin to the conditions specified on the WHERE clause of SQL statements.
2) Knowledge type constraints
This is a high level constraint which specifies the type of knowledge to be mined (clustering, classification, association of various types).
3) Dimension/level constraints
In multi-dimensional association rule mining, the user may restrict the dimensions to have more insight into subspace specific associations.

4) Meta-rule constraints
The rules to be applied are constrained.
5) Data constraints
This restricts the data to be used (attribute restriction)

7.8   Weighted Association Rule Mining (WARM)

In the standard ARM, all items appearing in a transaction are equally weighted (all weights are 1's). If one customer buys 2 dozen cokes and 3 packets of bread in one visit and another customer buys just one coke and one packet of bread, should we weigh both transactions equally? If the first customer visits the store once per week on the average, while the second customer visits 5 times every week, doesn't the 5 separate transactions generated by the second customer get more weight in the mined association rules? This is exactly the reason for weighting the transactions. Similar analogies can be found in medical studies, insurance, financial and other domains. WARM is an extension in which different items are weighted according to an optimality criteria. The weight assigned to an item may indicate either the quantity of item purchased (6 cans of coke, two dozen eggs, 1/2 kg bananas, two packets of chocolate etc) or may be derived from a property of the item. These are examples of *item specific* attribute weighting. Entity (transaction) weighting assign different weights to each entity or transaction. To find association among preferred customers (in supermarkets, airlines, insurance etc), we could apply different weightings to regular and preferred customers. Other variants like vertical weighting using a time span, exponentially decaying weights to the past etc also exist. Low support items may have more profit margins than high support items. In these situations, 'weighted support' may be more important than 'large support' as found by the standard ARM. See [LH01], [TM05], [WY00].

7.9   Algorithms for Association Rules

Data for ARM are often mammoth in size. They are arranged in id:A1,A2,... format where the optional id is transaction id, patient id, customer id, article id, etc and A_i are the attributes, items and so on.

Definition 7.11 The process of mining past data for association rules is known as Association Rule Mining. It is an unsupervised learning algorithm that depends upon a minimum confidence and support to extract meaningful associations from unlabeled data.

The first published algorithm for finding association rules was introduced in [AIS93] and will be referred to as the AIS algorithm. It is limited to only one item in the consequent, and is a multi-pass algorithm over the transactions. It applies a pruning technique during each pass to eliminate unnecessary itemsets using the min-support threshold. Many variants of the AIS algorithm have appeared in the literature, that improve the efficiency using hash-based itemset counting, transaction reduction, partitioning, sampling and other techniques [AM96]. Most of these algorithms are distinguished by the *counting strategy, search strategy* (DFS or BFS) or *direction of search* (top-down or bottom-up). The support measure is successively determined for all itemsets at a specific level of depth in the BFS algorithms. The DFS algorithms on the contrary recursively descends through several depth levels. Tree-based *divide-and-conquer* algorithms have also appeared recently [GG03]. The Apriori algorithm [AS94] is a streamlined version of AIS in which frequent k-itemsets in a pass are used to generate candidate (k+1)-itemsets for the next pass. In purchase domain applications, the database may contain millions of transactions that are unlikely to fit in the memory of some computers to facilitate multiple scans. The partitioned-Apriori is a two-pass algorithm that works on disjoint partitions of the database, each small enough to fit in memory. Locally frequent itemsets are found for each partition (one after another) in the first pass. These are combined in the second pass by finding the support over the entire database.

Another two-pass algorithm is based on the random sampling principle. A large enough sample (simple or stratified) is taken from the database of all transactions. The frequent itemset for the sample is then computed, and infrequent itemset information is kept separately. The database is then scanned to find the support of 'potentially' frequent itemsets found above. If no other frequent itemsets are found, the algorithm stops. Otherwise the potential set is updated and the process is repeated in another pass. The SETM algorithm discussed in [HS95] is a multi-pass algorithm that utilises a unique transaction identifier for itemsets in large databases. The FP(Frequent Pattern)-growth algorithm [HP00] proceeds as the Apriori algorithm by finding frequent 1-itemsets. It then constructs a d-dimensional tree (called FP tree) that uses a compact data structure, and works without the candidate generation process of Apriori algorithm. The Apriori-TID FP-Growth [HP00] algorithm [AS94] is another variant that does not use the database for counting support after the first pass. The Apriori-Hybrid algorithm uses the Apriori algorithm during the initial passes and switches to Apriori-TID in later passes. Other interesting algorithms are CBW [SL04] for low support mining, scalable algorithm Eclat [ZM00], DIC [BM97], algorithm with item-constraints [TS99], [MT94]. The T-Apriori algorithm is used for mining temporal associations. Due to the high computational complexity of associa-

Table 7.2: Comparison of common ARM algorithms

Algorithm name	Search type	no. of passes	Data-structure	Sort Order	Count/ Intersect	Cand gen	Year
AIS	BFS	multiple	List	None	count	Yes	1993
Apriori	BFS	multiple	Tree + HT	Support↓	count	Yes	1994
DHP	BFS	multiple				Yes	1995
SETM	DFS	multiple	None	Bottom-up	intersect	Yes	1995
Apriori Hybrid	BFS	multiple	Tree	Bottom-up	count	Yes	1996
DIC	BFS/DFS	multiple	Trie	Bottom-up	count	Yes	1997
Clique	DFS	Two	BP Graph	None	VTL	Yes	1997
TID-Apriori	BFS	multiple	Tree	Bottom-up	count	Yes	2000
P- Apriori	BFS	multiple	Tree	Bottom-up	count	Yes	2000
FP-Growth	DFS	Two	Fp-Tree	Support↓	count	No	2000
Eclat	DFS	multiple	Lexi Tree	Bottom-up	intersect	Yes	2000
Partitioning	BFS	Two	Hash Table	None	Intersect	Yes	2001
TreeProjection	BFS	Two	Lexi Tree	None		Yes	2001
Dyn. FP-Tree	D&C	Two	Fp-Tree	Support↓	intersect	No	2003
CBW	DFS	multiple	Tree		intersect	Yes	2004

Legend: Cand gen: candidate generation, Lex Tree: Lexicographic tree, HT: Hash Table, BP Graph: Bipartite Graph

tion rule mining from very large transactional databases, many authors have proposed parallel and distributed algorithms for this purpose. Distributed mining of association rules from large databases have been investigated by [AT04], [CH96], [SO05]. Other popular algorithms are: PDM [PC95], CD [AS96], DMA [CH96], CCPD [JP97], CHX [CH98]. An Optimal Distributed Association Mining (ODAM) approach is described in [AT04], and an approximate distributed algorithm can be found in [SO05]. Association rules can also be found from datawarehouses or relational databases using SQL-like Association Rule Mining Query Languages (ARM-QL).

Using a minimum confidence and support thresholds, the rule mining algorithms (see section 7.9 in page 7-24) identify all associations satisfying the specified parameters. It can also be applied to find dependencies between different attributes of the same entity. For instance, a bank may be interested in finding the association between various lender attributes that lead to a default in loan repayment. In identifying frequent flyers, an airline agent could find association between income, credit-card usage, long-distance phone calls, and employment status. Similarly, root causes of learning disabilities among kids, over-spending behaviors among teenagers, fatality of medical symptoms or some drug combinations, etc can be found by ARM algorithms.

7.10 The Apriori Principle

The very first algorithm for association rule mining is called the AIS algorithm [AI93]. The first pass of AIS reads each transaction record, and increments the count of every single item in that transaction. The second pass of AIS uses the results extracted during the first pass: It reads each and every transaction, and checks if one of the frequent singleton itemsets are present in it. If none of the frequent items are present, it is discarded as it cannot be frequent by the apriori principle. Otherwise, candidate itemsets are generated by 'extending' frequent itemsets by concatenating it with every other item present in the transaction to it. As an example, if $\{x\}$ is found to be frequent in the first pass, but $\{y\}$ is not, and the next transaction is $\{x,y,z\}$, then only two candidate 2-itemsets are generated as $\{x,y\}$ and $\{x,z\}$.

Theorem 7.3 Apriori principle: If a k-itemset is frequent, then all of its subsets of ((k-j) itemsets for (j=1,2, $\cdots$, k-1) are also frequent.

One of the earliest available algorithm for association rule mining uses the apriori principle [AI93]. It uses the breadth first search (BFS) with occurrence counts. The notable difference between AIS and Apriori algorithms is that the AIS algorithm extends the candidate itemsets dynamically for each of the transactions, whereas the apriori algorithm generates the candidate set before it looks at each of the transactions. The algorithm is called 'apriori' due to the fact that the apriori properties of itemsets are utilised in finding association rules.

Discovery of association rules is accomplished in two steps - first find items of high support (frequent items) and then find those with high confidence in frequent itemsets.

7.10.1 Apriori algorithm

This is one of the earliest algorithms for ARM. It uses prior knowledge of frequent itemsets acquired in previous pass(es) to narrow down the search space in each pass. This is called Reduce & Conquer (R&C) principle. The classical Apriori algorithm uses a level-by-level search to locate significant association rules using a user supplied minimum support ($minSupp$) **s**, and minimum confidence ($minConf$) **c**. The literal meaning of *high support* is that the itemsets in that transaction occur quite often in D. As the number of transaction records (N) in a database D from which the rules need to be extracted is often unknown prior to the running of the program (an algorithm for ARM), the minSupp and minConf are expressed as raw counts. But, if N is known[10], we could express them either as a percentage or as a fraction in (0,1). This is useful in FP-Growth type algorithms that does a single scan of the entire database D for frequent 1-item transactions, and does a single second scan to construct the FP-Tree. It uses the breadth first search (BFS) with global occurrence counts of itemsets. If I_T is the set of all items, a subset $X=\{i_1, i_2, \cdots, i_k\}$ is a k-itemset, and t=(tid,X) is a *transaction* where tid is a unique transaction identifier[11]. Note that tid is not part of our algorithm, but is used only for identification purposes. Most of the ARM algorithms mentioned below can work without the tid values, but the SETM algorithm [HS95] is an exception (as it uses the tid for large itemsets and candidate itemsets). The order of the items appearing in a transaction are also unimportant. But some of the algorithms discussed below can be speeded up using a proper rearrangement of the items using their occurrence counts or using a lexicographic ordering (useful for hashing). An advantage in ordering itemsets lexicographically is that it is easy to detect and remove duplicate transactions (which is quite common in retail industry). As an example, if 250 customers bought just {Bread, Butter, Milk} in their visits, we could eliminate all duplicates and keep just one of them. Note also that the names given to the items are unimportant. If the total number of items or attributes are ≤ 26, we could very well use the letters of the English alphabet to identify them. After finding the association rules, these symbols can be resubstituted for their real-world names. As an example, if there are 26 different symptoms or conditions that a patient to a particular clinic can exhibit, we encode the symbols by letters A to Z, find the association rules among them and decode them back to the symptoms or conditions. If the total symptoms exceed 26, we could concatenate numbers at the end to form the encodings. The association rules found have nothing to do with our encodings. In other words, if one user encoded them as A1 to A100, and another user used their real names, both of them will obtain the same association rules for the same parent data so long as the cutoff values (c and s) used are the same.

In the pre-processing step, we could eliminate all single itemset transactions (a customer buys a single item, a medical patient exhibits a single symptom, a single faulty component in an equipment, etc). It is beneficial to sort the transactions in increasing order of the number of items (≥ 2) (as each transaction is examined one-at-a-time, an insertion or heap-sort algorithm is okay). Then each transaction t $\in$ D can be rearranged in alphabetical order of itemsets to minimise the search.

The Apriori algorithm proceeds as follows. First find all frequent items (itemsets of size 1) from among those itemsets that contain two or more items (note that as mentioned before, itemsets containing just 1 item are not useful in association rule mining, except in profit maximisation etc). In other words, find all of those items that appear in at least fraction s of the market-baskets. Then find the frequent itemset pairs (tuples), then frequent itemset triples, and so on in an interative level-by-level manner. The candidate set C3 consists of those sets (triplets) {A;B;C} such that all of {A;B}, {A;C}, and {B;C} are in L2. As we proceed to higher levels, the number of itemsets will in general dramatically decrease. For example, suppose there are 1000 possible symptoms or conditions (like pain in various body parts, inflammation of various muscles, abnormal values for various tests etc) that a medical patient can have, but the maximum number of symptoms or conditions any patient can exhibit is less than 10. Here itemsets (symptom combinations) imply that a patient exhibits multiple symptoms/conditions in a single visit. These will vary from visit to visit of the same patient, but there could be a

---

[10]As the support is computed after the first pass through the transactions in D is completed, N can be obtained by that time. This allows users to specify the support as a percentage or fraction.

[11]The TID values can be either unique integers, a string or a time-stamp.

large overlap in the symptoms for the same patient. For instance, a patient with back-pain, high blood pressure, and diabetes is likely to retain most of the conditions in subsequent visits. This information can be utilised in speeding up the algorithm. In the first pass, we could eliminate all duplicate records (into say S_d) and find frequent itemsets with given support and confidence. Then, in an updating pass, we merge itemsets in S_d to come up with the global association rules. In purchase domain, this translates into a customer always buying a few items in each and every visit. As an example, people with kids and pets at home tend to buy baby food, milk and milk-products, candies, and pet food in each visit, in addition to adult needs.

Algorithm 7.1 Apriori Algorithm

1: Input minsupp, minconf thresholds, N=$|D|$ size of DB
2: Step-1: Scan all transactions D, accumulate support count as $L_1 = \{$1-itemsets$\}$
3: Special checks:
   If D is empty, return error code
4: **if** $|L_1| = N$ OR $|L_1| < 2$ **then**
5:       **return** 'Error Code"
6: **end if**
   Eliminate transactions that have support count less than the minsupp.
   Optional: Sort the remaining transactions in increasing order of itemsets,
   Sort itemsets in each transaction in alphabetical (ascending) order of item names
7: Construction:
   Step-0: Eliminate all single item transactions as useless
8: **for** (k = 2; $L_{k-1} \neq \phi$; k++) **do**
      C_k = apriori-gen(L_{k-1});**for** (transactions t $\in$ D) **do**
10:       C_t = subset(C_k, t);
      //Candidates contained in t
11:       **for** (candidates c $\in$ C_t) **do**
12:          c.count++;
13:       **end for**
14:     **end for**
15:    $L_k = \{$ c $\in$ $C_k|$ c.count $\geq$ minsupp $\}$
16: **end for**
17: rulesFound = $\cup_k L_k$
18: **return** rulesFound

If a subset X of L do not satisfy the minimum confidence, we need not check rules with subsets of X on the LHS.

Example 7.11 Find association rules with minconf=.50 and minsupp=.40 for the sales data given below.

Solution: It is convenient to capture the above data into a count (frequency or occurrence) table as shown. We next form the 'support count table' from which confidence can easily be calculated. The first column is the item purchased and second column gives the percentage of customers who bought the item (irrespective of other items purchased).

Tr Id	Itemset
3010	Milk, Bread, Jam
3011	Bread, Butter, Juice
3012	Soda, Bread, Butter
3013	Bread, Juice, Soda
3014	Milk, Juice

Tr Id	Bread	Butter	Jam	Juice	Milk	Soda
3010	1	0	1	0	1	0
3011	1	1	0	1	0	0
3012	1	1	0	0	0	1
3013	1	0	0	1	0	1
3014	0	0	0	1	1	0
Total	4	2	1	3	2	2

This is easily obtained from above table by summing along the columns.

Item	support	cover
Milk	40%	3010, 3014
Bread	80%	3010 - 3013
Soda	40%	3012, 3013
Juice	60%	3011, 3013, 3014
Butter	40%	3011, 3012
Jam	20%	3010

Items	support	cover
Bread, Butter	40%	3011, 3012
Bread, Juice	40%	3011, 3013
Soda, Bread	40%	3012, 3013

Since the support count of Jam is less than 40%, we drop item combinations containing Jam from further consideration. We next form 10 two-item combinations {(Bread,Butter), (Bread,Juice),(Bread,Milk),(Bread,Soda),$\cdots$,(Milk,Soda)}. Support counts of (Butter,Milk) and (Milk, Soda) are zeros as can be seen by taking intersection of corresponding columns. For many of the others, it is one (eg:(Juice,Milk),(Butter,Soda)). All other two-item combinations have support more than 40%, and are shown to the right. To get three-item combinations satisfying our conditions, we could again drop all those itemsets with support less than 40%. There are four 3-item combinations {(Bread,Butter,Juice),(Bread,Butter,Soda), (Bread,Juice,Soda),(Butter, Juice, Soda)}. The corresponding support counts are {1,1,1,0} respectively. Hence the rules satisfying the constraints are as follows:

Rule: Bread => Butter[40%,50%]

since confidence(Bread, Butter) = n(Bread, Butter)/n(Bread) = 2/4 = 50%.

Similarly, Bread => Juice[40%,50%],

Bread => Soda[40%,50%], etc

Association rules in commerce domain are usually generated from volumes of transactions (regular or e-business) using a specified confidence and support thresholds. In practice, the support and confidence rarely round-up to whole integer percentages as in our examples above. Matching the LHS of an association rule gives us a predictor for the association, while matching the RHS of the same rule gives us a recogniser for the association.

7.10.2 Modified Apriori Algorithm

Given a user specified minimum support (minsup) and confidence (minconf), the task is to mine association rules where support and confidence are at least minsup and minconf. It can be broken into two steps: 1. Find all itemsets with support above minsup (find all frequent item sets). 2. For each frequent item set, derive all rules that are supported by that set which have at least minconf. That is: given a frequent item set, X, find all antecedent subsets, A, of X, such that rule, R: A (X-A) has at least minconf. Overall performance of mining association rules is largely determined by step 1. Various algorithms are proposed to discover frequent item sets; Apriori[3] and DHP[6] are the basic two algorithms. The key idea of Apriori algorithm is any subset of a frequent item set must also be frequent. Starting by finding all frequent 1-itemsets

(itemsets with 1 item), we then consider 2-itemsets, and so forth. So during each iteration of the algorithm only candidates found to be frequent in the previous iteration are used to generate a new candidate set during the current iteration. The algorithm terminates when there are no frequent kitemsets. In Apriori algorithm, after the candidate k-itemsets are generated, the transaction database needs to be scanned to count the support for each k-itemset. This is very timeconsuming. In remotely sensed imagery, since transactions are very likely to share patterns, we construct a pattern occurrence counting table, each entry in this table is a pair of a transaction pattern and the number of times it occurs in the original transaction database. Based on this table, later, we do not need to scan the database to count the support for each candidate k-itemset, we only need to sum up the occurring frequencies of all the transaction patterns that contains this set , thus saving a lot of time.

7.11   The FP-Growth Algorithm (FPGA)

The FP(Frequent Pattern)-growth algorithm by Han,et.al.[HP00] is a two-step approach to generate frequent itemsets without candidate-generation. It proceeds as the Apriori algorithm by finding frequent 1-itemsets. It then constructs a tree (called FP tree) that stores the most-frequent items towards the root, and least-frequent items towards the leaves. The biggest advantage of FPGA is that it requires just two scans of the entire database[12]. This is especially useful when the entire set of transactions (D) do not fit in the memory of the computer (see below). A modification of FPGA that builds the FP-tree dynamically using just one scan of the database is available in Gyórödi, *et.al.*[GG03].

7.11.1   FP-Tree

For a better grasp of the FPGA, the reader must be familiar with a few concepts like IHT, FP-trees, item linked lists, etc described below. Let t=(tid,X) be a transaction where X=$\{i_1, i_2, \cdots, i_k\}$ $\subseteq I_T$ is a k-itemset, and D be the set of all transactions.

- FP-Tree (FPT)
  The FP-Tree is a general tree (each node except the leaves can have 1 or more child nodes). This tree is constructed from D using 1-itemsets sorted in descending order of support count (this information is obtained in the first scan of D as described below). Each node of an FP-Tree except the root node is a frequent item. To be more specific, nodes of the FP-Tree are chosen from the set of frequent 1-itemsets (say c) so that *infrequent* 1-itemsets are never considered in building the FP-Tree (they are simply filtered out from each transaction). In addition to the item name (TID), it can also store a prefix count, and an array of pointers (see below). The maximum depth of the FP-Tree is $\leq$ the maximum number of items in any transaction. The exact depth depends upon the chosen support threshold. If this threshold is high, the depth is low and vice versa.

- Item Header Table (IHT)
  This is a table with 3 columns in the standard FP-Tree algorithm to ease the FP-tree traversal and mining. The first column is of type String, second column is of type unsigned int, and the third column is of type pointer to FPNode (ie. FPNode *), where FPNode is a structure that declares the type of each node of the FP-tree. It has as many rows as there are frequent 1-itemsets (c). In other words, there is a one-to-one correspondence

---

[12]Thus the complexity of FP-tree construction is O(2N) where N=|D| is the number of transactions

between the rows of the IHT, and the set of nodes in the FP-tree with the same item name. The first column stores the item names (or item IDs). Because the FPT stores only frequent 1-itemsets, this column of IHT need store only frequent 1-itemset names in *support descending* order. The second column contains the support counts ($\geq$ minsupp), and the last column is a pointer to the nodes in the FP-Tree. As there could exist multiple nodes in the FP-Tree with identical node names, the pointer in the IHT will serve as the head of a linked list through all such nodes. Thus the third column of IHT serves as entry points to c different Singly Linked Lists (SLL). We denote these SLLs by $L_j(fp)$ (or simply as L_j) for each frequent item fp, and for $j=1,\cdots,c$. These linked lists are either built during the FP-Tree building process, or as a separate step. As linked lists are built using the memory address of each node structure, it is computationally efficient to build the IHT during the FP-Tree building process (as each node structure is memory-allocated, we get the address of the memory location to be stored in the SLL). An advantage of building the IHT after the complete construction of the FP-Tree is that the depth information of a node from the root of the FPT can be utilised to decide where that node will appear in the linked list. Smaller conditional FP-Trees are generated if this order is utilised in forming the linked list. This in turn accelerates the algorithm.

- Conditional Pattern Base (CPB)
  A pattern base is a set of item names formed from the paths of the FPT starting with the root R. It can be considered as "subdatabases" that consist of a set of prefix paths in the FP-tree co-occurring with a suffix pattern. The idea is to filter out those transactions that contain items that co-occur with a fixed frequent 1-itemset. As mentioned above, the IHT rows contain frequent 1-itemsets in support descending order. Let f1 denote an arbitrary frequent 1-itemset. As the linked lists $L_j(f1)$ connects all f1 occurrences in the FPT, we can consider item f1 as a fixed suffix pattern, and obtain all transactions in D that contain f1 by collecting all prefix paths of the FPT starting at the root R, and ending in nodes through $L_j(f1)$. To be more specific, if Milk is a frequent itemset, then all transactions that contain Milk will be represented as branches of the FP-Tree starting with root R, and ending in the nodes labeled as Milk (which are part of the SLL $L_j(Milk)$).

- Conditional Pattern Tree (CPT)
  The CPT is a tree formed from the CPB by applying the support threshold. Each CPT will give us a set of 1 or more rules, provided the total support count of frequent items exceed the minsupp threshold. Rules are generated by concatenating the suffix pattern with the frequent patterns generated from a CPT. In general we form (c-1) number of CPT's, some of which do not generate useful rules. If the IHT is processed in the bottom to top order, the very first row (corresponding to the most frequent item in D) is seldom processed[13].

7.11.2   The FP-Tree Construction

Once the L_1 set (frequent 1-itemsets in D in support descending order) has been found, we can build the FP-Tree as follows. First we create an empty root node R labeled "null". It can store a "null" pointer to indicate that it is the root, an array of pointers to child nodes; and optionally store the number of immediate child nodes, maximum depth of the final tree, etc. Subsequent nodes of the tree contain 4 types of data – namely an unsigned integer field (number of itemsets

---

[13]So we need to create the linked lists from 2nd row to the last row of the IHT.

common in the transactions for this node starting with the root (called frequency count)), item name or item ID, one array of pointers to child nodes, and another single pointer (in case the linked list is SLL) that stores the memory address of the next node with the same item label, or a Null if there is no more node in the FPT with the same item-label (Hence we could store this pointer as the first one in the array of pointers of each node, and store the child node pointers subsequently, as the number of child node pointers could vary.). The array of pointers to child nodes allows us to traverse the tree from top to bottom. The number of child nodes could vary from 1 onwards for interior nodes, and is 0 for leaf nodes.

The database of transactions is scanned a second time. As the order of the items appearing in a transaction t is unknown, we need to rearrange the items in t, after it is read, in the sorted order of L_1. This will result in the frequent items of t to appear in decreasing order of support (from left to right). Those items in a transaction that are *not* in the frequent 1-itemset L_1 can safely be dropped (but we cannot drop the entire transaction itself, if it contains at least 2 other frequent itemsets). Let t contain k ≥ 2 items (after dropping infrequent items, if any). Denote the rearranged transaction t as $[t_1, t_{k-1}]$, where t_1 is that item in t with maximum support count. This maximum need not be unique. If there are multiple items with the same maximum, we pick one of them arbitrarily if the FPT is not yet built; and choose that item with the maximum frequency count in the FPT otherwise (which also may rarely clash; in which case one of them is randomly chosen). A branch needs to be created for each of the transactions (with ≥ 2 items), starting from the root node R of the FPT. To decide whether we should create a new branch or maximally utilise one of the already existing branches, we compare the item name t_1 with each of the item names emanating from the root R. If a count of the current children of R is kept at the root, this comparison is easy to do. This count is always $\leq$ m, the maximum number of items or variables. Alternately, we could store the item names immediately below R in sorted order at R, and do a binary search to see if a new item (t_1) is present at R. As this technique can speedup the algorithm, we could use it for the first few levels starting with the root R (say for the first 4 to 5 depths for 1000 items).

If none of them matches (a branch for item name t_1 has not yet been created from R), we add all the items in t (in L_1-order) as a new branch of R, set the frequency count to one each, and set or update the pointers from the third column of IHT to each of the nodes (see below). If one of the branch labels at R matches t_1, we increment the frequency count of that node by one, and then recursively descend down from that node by left shifting t as $[t_2, t_{k-2}]$ until t_{k-2} is empty (if t_{k-2} does not become empty, we add all remaining items in t_{k-2} as a new branch below the last matching node, set corresponding frequency counts of nodes on the new branch to 1 each, and update appropriate SLLs). Thus the paths in the FP-Tree could partially overlap when transactions share common prefix items in L_1 order (if two paths totally overlap, it is a duplicate transaction). There is no path overlap when two transactions have common items, but they are *not* in L_1 order in the beginning.

7.11.3   Creating an Item-Header Table (IHT)

The IHT described above facilitates tree traversal. This table has 3 columns, and as many rows as there are frequent items. For example, in a supermarket that stores 3000 items, the IHT will have a maximum of 3000 rows (assuming that all variables have been categorised). If there are many items with low support count ($<s$), we can discard them and reduce the number of rows of IHT accordingly (note that only frequent items are stored in the FP-tree, and the third column of IHT points to FPT). This happens in medical domains where some of the conditions, symptoms and diseases are rare. It also happens in pharmacies, where some medicines are either

sold in isolation (eg: insulin) or are rarely sold (eg: some skin ointments and eye drops). As mentioned above, the third column of IHT stores a pointer to a node of the FP-tree. This pointer is the head of a linked list that connects item occurrences in the FP-tree via a chain of node-links. The last node in each linked list will of course store a null pointer to indicate that it is the last element in the list. These SLLs are almost always short in some domains (but could be of the order of hundreds in retailing if the number of items for sale is very large). In the best case, the number of links in it could be 1 (denoting a rare purchase or an expensive item). In the worst-case, the number of links in it is the maximum number of levels multiplied by the number of branches at the root R.

7.11.4　Conditional Pattern Bases

Once the FP-Tree and IHT are built, the problem of mining for frequent patterns in the database is greatly simplified. We could simply utilise these two structures to mine the association rules. If the FPT contains a single path P, we could generate the rules very efficiently. If the FPT has a single level, we get only binary rules (SASC type). Otherwise we need to traverse the IHT, and form the CPB. We have built the IHT from frequent 1-itemsets, which we assume as an initial suffix pattern. Let fp be an arbitrary 1-itemset in IHT. We mark all matching nodes in the FP-Tree by traversing the SLL $L_j(fp)$. The set of paths from the root R of FPT to the marked nodes form the conditional patterns. These are used to build the CPT. But we need to do some aggregation to get the CPT from CPB as shown below.

7.11.5　Computational Complexity of ARM

The computational complexity of ARM algorithm depends on both the number of attributes (m) and the number of transactions (n). In most applications $n >> m$ (read as n is much greater than m). Number of attributes may vary anywhere from half a dozen (or less) to a few thousands. In small businesses (like optical stores, shoe and leather shops, music and video shops etc) the number of attributes are small. Thousands of attributes are possible in supermarkets that carry thousands of items. It is also common in some large insurance companies (that keeps track of detailed information on insured, payments, defaults, fraud, attrition etc). A large number of attributes are also common in earth observing satellites that continuously sense atmospheric dynamics, medical and pharmaceutical research using human genome project data, etc. For example, finding association between human genes and diseases by functional single nucleotide polymorphisms (SNPs) with data from various individuals involve several hundreds of variables.

As mentioned before, we will eliminate all 1-itemset transactions since meaningful associations need at least two items (one item transactions are common in purchase domains when a customer buys a single item per trip. It is possible in bookshops, phone shops, computer shops, optical shops etc. In medical domain, it may indicate a single fatal root cause, single symptom, or a condition. In financial domain, it may indicate a single reason for customer attrition, or fluctuations in stock price). All 1-item transactions can be separated out and mined using a utility measure. Counting 1-item transactions is useful in profit maximisation, in inventory management etc. They may also be helpful in speeding up the algorithms as in the following example.

The time complexity of ARM is $O(nm)$ where n is the number of transactions (records), and m is the maximum number of attributes. Computation requirement is maximum during the Frequent Itemset Mining (FIM) phase (discussed below). Many recent algorithms have

Table 7.3: Transactions at an optical store

Trans Id	items sold
100	Regular spex, carrying case
110	Contact lenses
120	Sunglasses, carrying case
130	Regular spex, cleaning liquid
140	Sports glasses
150	Regular spex,carrying case
160	Sports glasses
170	Regular spex
180	Contact lenses, cleaning liquid
190	Swim glasses
200	Regular spex, carrying case cleaning liquid
210	Sports glasses, carrying case

Table 7.4: Count table

Item sold	count	support
Carrying case	5	38.46
Regular spex	4	30.77
cleaning liquid	3	23.08
Contact lenses	1	7.69
Sunglasses	1	7.69
Sports glasses	1	7.69
Total	15	

suggested improvements to the FIM (see 7.9 in page 7-24). If the number of input attributes is small, the time complexity is linear in the number of records.

Example 7.12 Consider an optician who sells regular spectacles (lenses + frames), sunglasses, swim glasses, contact lenses, sports glasses, carrying cases, spex cleaning liquids etc. Some customers buy items of just one type. We have not distinguished between the brand names, costs, sizes, colours or other specifications in the data in table 7.3. There are 5 one-item transactions out of a total of 12 transactions. The support count table is given in table 7.4. After eliminating low support itemsets, there are two 2-item transactions (regular spex, carrying case) with support count 3, and (regular spex, cleaning liquid) with support count 1.

For a *minsupp* of 20%, we get the following rules:
Regular spex => Carrying case [38.46%,75%]
Regular spex => Cleaning liquid [23%,25%]
The last rule is an example of low confidence and low support. In large supermarkets, we get many such rules.

7.12 Applications

Association rules started in the purchase domain. Although rudimentary forms of association rules were in use during the past few decades, it became popular during the 1990's as evidenced by the research work in this field. In this section, we first give examples from this field.

7.12.1 Medical Diagnosis

a) Prescription Mining
Consider a patient who is diagnosed with multiple medical problems. Medicines for one illness may induce other symptoms, or aggravate other illnesses. In addition, drug interactions may occur if multiple drugs are simultaneously administered. In a single prescription, a doctor may be interested in knowing 1) harmful drug interactions 2) induced illnesses and its consequences

3) optimal dosage levels and treatment periods 4) illnesses that are related and unknown but not induced by treatment. By mining past treatment data, extended (§7.6) and generalised association rules (§7.5) can be extracted in sub-dimensions. These are stored in a rules database, which are searched using qualifying attributes. As an example, {abdominal pain, fever, fatigue, lack of appetite, nausia, vomiting, yellow urine}$\rightarrow$ Hepatitis is a valid rule with high confidence.
b) Simptoms Mining
Consider a patient who is diagnosed with multiple medical problems. Medicines for one illness may induce other symptoms, or aggravate other illnesses. In addition, drug interactions may occur if multiple drugs are simultaneously administered. In a single prescription, a doctor may be interested in knowing 1) harmful drug interactions 2) induced illnesses and its consequences 3) optimal dosage levels and treatment periods 4) illnesses that are related and unknown but not induced by treatment. By mining past treatment data, extended (§7.6) and generalised association rules (§7.5) can be extracted in sub-dimensions. These are stored in a rules database, which are searched using qualifying attributes. As an example, {abdominal pain, fever, fatigue, lack of appetite, nausea, vomiting, yellow urine}$\rightarrow$ Hepatitis is a valid rule with high confidence.

Symptoms and conditions of patients at a clinic

PID	Symptoms
P_1	Cold, Sinusitis, Fever, Sore Throat, Swelling
P_2	Pneumonia, Chest Pain, Cold
P_3	Cold, Fever, Bronchitis, Coughing
P_4	Cold,Wheezing,Sore Throat,Nasal Congestion
P_5	Fever, Nasal Congestion, Sore Throat
P_6	Fever, Tonsillitis, Sore Throat
P_7	Cold, Sinusitis, Fever, Sore Throat, Wheezing
P_8	Fever, Sinusitis, Chest Pain
P_9	Swelling,Sore Throat,Cold,Fever,Tonsillitis
P_{10}	Bronchitis, Cold, Coughing
P_{11}	Malaria, Cold, Chest Pain
P_{12}	Pneumonia,Cold,Nasal Congestion,Wheezing

Table 4.7 Frequency counts

Symptom	Abbr	Count
Cold	C	8
Fever	F	7
Sore Throat	ST	6
Chest Pain	PC	3
Nasal Congestion	NC	3
Sinusitis	Si	3
Wheezing	Wz	3
Bronchitis	Bi	2
Coughing	Co	2
Pneumonia	P	2
Swelling	Sw	2
Tonsillitis	T	2
Malaria	M	1

Example 7.13 Data for 12 patients to a clinic are given below. Find the frequent rules for minimum support count minSupp=3.
Solution: We first form the frequency count table. As the cutoff for support is 3, we discard other symptoms (starting with Bronchitis). The frequent 1-itemsets L_1 is {Cold,Fever,Sore Throat,Chest Pain, Nasal Congestion, Sinusitis, Wheezing} with support counts (in descending order) {9,7,6,3,3,3,3}. We have abbreviated Chest Pain as PC because pain at other parts could be abbreviated starting with the letter P. As the first patient's symptoms are not in L_1-order, we rearrange it as $P_1 = $ {Cold, Fever, Sore Throat, Sinusitis}, and drop Swelling, which is infrequent (below the cutoff specified 3). A branch of an FP-tree is immediately created with corresponding counts 1 each. For patient P2, the symptom {Pneumonia} is dropped as it is infrequent. The remaining ones are rearranged as (Cold,Chest Pain). For P3, {Bronchitis,Coughing} are dropped, and {Cold, Fever} in L_1-order is added. As a branch is already there with these labels, we simply increment the count of the respective branches by 1 each. For P4, the frequent symptoms are {Cold,Sore Throat,Nasal Congestion,Wheezing}. As Sore Throat is not there under Cold, we create a new branch as {ST,NC,Wz}. The infrequent symptoms {Bronchitis,

Coughing} are dropped from P10. As 'Cold' is the only remaining symptom, that too is dropped (this results in the count of Cold to be decremented by 1, so that it becomes 8). Continue adding the rest of the patient symptoms to get the FP-Tree shown in figure 7.4. The FPT is mined next. First consider Wheezing as a suffix. Its prefix paths from the root R are $<$C,F,ST,Si:1$>$, $<$C,ST,NC:1$>$, and $<$C,NC:1$>$. This is the CPB of Wheezing. Its conditional FPT consists of just one path $< C : 8 >$. The frequent rules found are {C,NC}$\rightarrow$ Wz:2, {C,ST}$\rightarrow$ Wz:2, and {C}$\rightarrow$ Wz:3. Literally, this means that out of 8 persons who have Cold and Nasal Congestion; 2 of them have Wheezing too. Other rules are found as described before. An interesting rule found

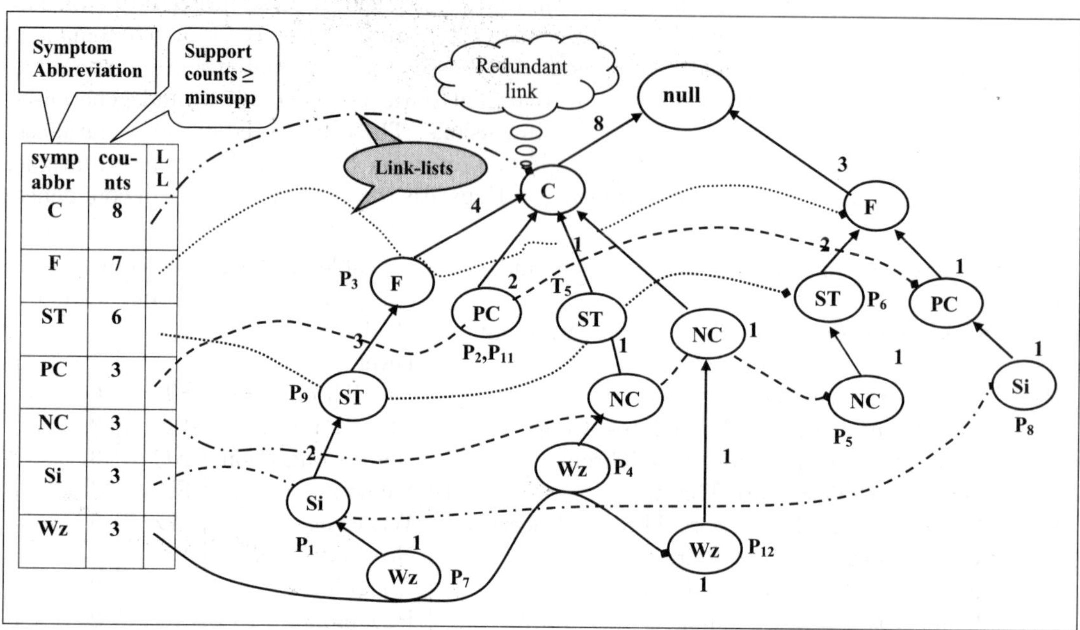

Figure 7.4: The FP-Tree for the medical diagnostic application.

is that out of 8 persons who have Cold, 4 of them (50%) have Fever as well (Confidence({Cold}$\rightarrow$ {Fever})= 4/8=0.50). More interesting is the association between Fever and Sore Throat. Out of 7 Fever patients (from both of the branches), 5 of them have Sore Throat. This has confidence 5/7$\simeq$0.713.

7.12.2 Purchase Domain Application

Association rules have tremendous utility in the purchase domain, where it is used for profit maximisation, time minimisation, optimal (store, catalog) layouts, convenience shopping, combination purchase bundling etc. It can identify cross-purchase pattern of customers or clients by analysing product combinations sold. The rules found are then fruitfully utilised by placing related items in close proximity (or by giving enough indicators to the location of related items). Single packet promotional bundles (eg: notebooks and pen, paint and paintbrush,sunglasses and tan lotions) comprising of frequently purchased items (obtaining by ARM algorithm) may also be offered at discounted prices. Another technique is to give location pointers for related items. For instance, persons who buy baby food are likely to buy toys, milk etc. Obviously, these items

Table 7.5: Some Association rules from FP-Tree

Rule Type	Antecedent	Consequent	Confidence
SASC	Cold	Fever	0.50
	Cold	Sore Throat	0.50
	Fever	Sore Throat	0.714
	Sore Throat	Nasal Congestion	0.666
	Sore Throat	Sinusitis	0.666
SAMC	Cold	Fever, Sore Throat	0.375
	Fever	Sore Throat, Sinusitis	0.50
MASC	Cold,Sore Throat	Wheezing	0.50
	Cold, Fever	Sore Throat	0.75

may not be kept together in most stores. Hence a clear location indicator at both the shelves where these items are stored will help customers who buy either of these items to locate the other item very easily. Association rules can be used to give special offers or discount coupons for related items. If both items are priced high and thus precludes the issuance of discount coupons, a third most frequently bought item that is cheaper than the others as suggested by ARM may be considered, or a combination discount offered on total price. Some stores have gone a step further by providing customers with a 'discount card' or 'bonus card' that is swiped for each purchase made. These cards contain the identity of the clients, accumulated bonus points, etc. Time-sequenced information on product transactions per customer are stored under a unique key (the card id number) and used for identifying high value customers.

Most stores arrange items on shelves with multiple levels and stacks. Using ARM, we can come up with optimal layouts that facilitate convenience shopping. There are several stores where alleys are crowded with customers, especially during peak hours or holidays (eg: chocolates and candy alley). Association rules can help in minimising the shopping time to locate and pickup needed items, in disbursing the customer crowds using alternative layouts, and in determining optimal alley widths, shelf types, levels and stack sizes, display order of items, stock replenishment intervals etc.

7.12.3   Machine Diagnosis

Fault diagnosis is a routine activity in aircrafts, power stations, and most large plants. Faults in one system or unit can trigger faults in many related systems. Diagnosing all possible faults may be a time consuming task that may require multiple technicians, and hours of labor. By mining past data, association rules can be obtained for common and rare faults. They are used as a predictor for fault detection and reveal part dependencies in case of part failures. Note that part-dependencies may not be general but specific to a particular model or type of equipment, so that we may have to use multidimensional constrained ARM. The ARM can save plenty of time and resources in such cases.

7.12.4   Inventory Arrangement

Inventory items are identified using unique id numbers or hierarchical id numbers. These are either bar-coded (as in supermarkets, pharmacies) or marked manually. Most inventories arrange items using the order specified by id number or part of the id number. An alternative arrangement scheme is based upon frequency of usage (fast moving items) that can help in reducing unwanted inventory levels. Association rules are used to figure out item dependencies and items arranged in proximity[14]. This technique will also reduce shelving time substantially.

Table 7.6: Software for Association Rules

URL	Name	C/F
borgelt.net/ida.html	AR software	F
www.cs.umb.edu/~laur/ARMiner/	ARMiner	F
azmy.com	Azmy SuperQuery	C
wizsoft.com/Wizsoft	WizRule	C
adrem.ua.ac.be/~goethals	Apriori algorithm	F
cs.umb.edu/~laur/ARtool/	ARtool	F
fuzzy.cs.uni-magdeburg.de/~borgelt/	Christian Borgelt's Implementations	F
fimi.cs.helsinki.fi	Frequent Itemset Mining Implementation	F
purpleinsight.com	Purple insight mineset	F
www.megaputer.com	Megaputer polyAnalyst	F
www.cs.waikato.ac.nz/ml/weka/	WEKA	F
xore.com	Xaffinity	F
spss.com/clementine/	Clementine Suite	C
rulequest.com	Magnum Opus	C
software.ibm.com/data/intellimine	IBM Intelligent Miner	C

Legend: C=Commercial, F=Free, S=Shareware

7.12.5   Fraud Detection

Association rules, in combination with other data mining models (like decision trees, SVM, neural networks) are finding increasing applications in fraud detection and prevention. Subscription fraud is a serious problem faced by financial and telecommunications industry. Falsified or impersonated documents are usually used to open a new account or start a fraudulent activity. This type of fraud is easy to detect using association rules by applying deviation principles from a 'legitimate customer profile'.

(1) Fake credit card usage and stolen cards
Customer-specific association rules can be used to catch stolen cards. If a credit card is stolen, and is illegally used for a purchase, the association will probably differ from what has been learned before. This may not work for single purchases (in which a customer buys just one item). Hence a single purchase analysis of past data should also be carried out in such situations to identify distinguishing attributes.

(2) Money laundering and other fraudulent activities
Money launderers transfer huge amounts through banks and financial institutions unsuspiciously.

---

[14]In most computerised inventories, humans walk with a printout showing the items and their exact locations, whereas in robot controlled inventories, this method can substantially reduce pickup time.

By mining past data on sending and receiving parties, time and amount of transfer, intervals between two transfers etc, they can be flagged as fraudulent or genuine. As most of the fraudulent transactions often look genuine, other sophisticated data mining models like neural networks are better suited than association rules for this purpose.

(3) Anomaly detection
Unusual happenings and behaviors are called anomalies. A database containing the normal behavior (frequent associations) of each customer should be searched for detecting anomalies. This database must maintain records for all subjects of interest. If there are multiple associations for a customer, it will be prioritised on the support/confidence and stored one after another (as strings in Relational databases)[15].

(4) Impersonations
Impersonations is an increasing problem, not only in financial institutions, but in industry, academia and government. Association rules can detect some impersonations of individuals to obtain credit cards, bank cards, identity cards, driving licenses or passports. A combination of attributes (like education and income levels, present and past employment history, spouse info, past convictions etc) can be used to catch impostors. See [MAA05] for an application to fraud in advertising.

7.13   Software for Association Rules

Many commercial and free software are available to generate association rules. Table 7.6 gives only a few of the dozens of available software.

7.14   Exercises

1. Mark as true or false:
    a) Association rule mining cannot tackle continuous attributes
    b) ARM dependencies are reported as a mathematical relationship
    c) Confidence is always in the interval [0,1]
    d) Affinity analysis is another name for association rule mining
    e) Purchase domain ARM algorithms consider single-item transactions
    f) The *apriori algorithm* generates association rules top down
    g) Extended association rules are obtained by splitting dimensionality
    h) Decreasing values of minSupp will in general result in lesser number of association rules
    i) A rule with high support but a low confidence is not meaningful
    j) If bread and milk have the same support, the confidence(bread $\rightarrow$ milk) and confidence(milk $\rightarrow$ bread) are the same
    k) conf(X,Y) = conf(Y,X) where X and Y are disjoint itemsets
    l) Unsupervised learning is used to discover hidden structure in voluminous data.

2. Find the association rules for the following sales data using minsup=20% and minconf=30%:

---

[15]This is an example of the multi-disciplinary nature of data mining involving database technology, information retrieval, data structures, pattern matching aspects as discussed in Chapter 1

TransactionId	items bought
100	X, Y, Z
200	X, Z, W
400	X, Y, W
700	U, V, W
800	V, Y, Z
900	U, X, Z

3. Prove that support is always $\leq$ confidence.

4. Using data in table 7.1, find the rule for Bread, Butter => Juice

5. A computer in a lab has a 3.5 inch floppy disk, a CD drive, a USB port, and an interface for an external disk. Students who use it has the usage patterns in table 7.7 for external devices. Find all associations with minsupp=.20 and minconf=.25.

Table 7.7: Usage pattern of computer accessories

Student Id	Devices used
100	CD, USB, ext disk
110	CD, USB
120	CD
130	CD, ext disk
140	USB, ext disk
150	Floppy
160	USB
170	CD, USB
180	ext disk
190	CD
200	Floppy, ext disk
210	CD, USB, ext disk

6. Describe weighted association rule mining. In what ways can it improve over the standard ARM algorithm?

7. Give examples of association rules for each of the 4 combinations {support, confidence} x {high,low} (high support and high confidence, etc).

8. If a measure called interest is defined as Interest$(X \rightarrow Y) = |X \cap Y|/(|X| * |Y|)$, what is its range? How does it compare with confidence measure?

9. Give an example of a multi-level association rule in each of the following domains:
   (a) Market-basket analysis (b) air insurance (c) medical diagnosis

10. Give examples for high support at higher levels.

11. If high denotes $> 80\%$ and low denotes $< 15\%$, how will you interpret the following:
   a) high support and low confidence
   b) low support and high confidence
   c) low support and low confidence

12. How is *Lift* measure expressed in terms of confidence and support?. If the confidence is .8 and support is .5, what is the *Lift?*.

13. Compare and contrast the production rules obtained from decision trees with the association rules.

14. How are the antecedent and consequent identified in sequence association rule mining?

15. What are the advantages of adding 'activity indicators' in association rules?

16. What does higher value of 'lift' indicate? What are its bounds (lower and upper)?.

17. Describe the Apriori algorithm. How many major steps does it use?

18. How can the Apriori algorithm be modified to find rare associations? Distinguish between absolute rarity and relative rarity.

19. Why are sparse association rules important? Give any 2 practical examples.

20. What are constrained association rules? What are the popular constraint categories?

21. Describe negative association rules. What modifications are needed to mine for negative associations?

22. Distinguish between Pareto analysis and paired-comparison analysis.

23. What are temporal association rules? What is rule-expiry? How can you prune temporal association rules already found from a database?

24. Describe the *is-a* and *whole-part* hierarchies, and how they are related to association rule mining.

25. What is an outlier in categorical data? Can you find data outliers using unusual attribute combinations using an ARM algorithm?

26. Describe the fuzzy association rules and its applications.

27. Distinguish between periodic and aperiodic association rules and give two examples.

28. If support(A→C) > support(C→A), which of the items (A or C) occur more often in the set of all transactions?

29. If support(A→B, C) is 66%, is the support(A→C) greater, lesser or same?

30. If there are 3 items in the frequent itemset, how many confidence measures can you form? What are they?

31. If {Bread, Milk} is a large 2-itemset, what are the possible candidate itemsets that could be generated from a transaction {Bread, Butter, Jam, Milk}?

32. If L_2={ab, ac, bc} are the frequent 2-itemsets, find the candidate 3-itemsets.

33. If L_3={abc, abd, ace, bcd} are the frequent 3-itemsets from 5 items, find the candidate 4-itemsets.

34. Consider the milk hierarchy for cow and goat milk (in chapter 5). The cow milk is available in 3 varieties – full fat, half fat and skim milk. The goat milk is available only in full fat and half fat varieties. A manufacturer adds two new types of milk - chocolate milk (milk with chocolate added) and banana milk (milk with banana flavor added). Draw the new hierarchy and explain how multiple level association rules in whole-part hierarchy and is-a hierarchy are different from higher-level hierarchies.

35. If confidence($S \Rightarrow T$)=confidence($T \Rightarrow S$), prove that support(S) = support(T).

36. If the database of all transactions from which rules are obtained is continuously updated with new transactions, which of the following measures is more likely to change fast? A) Confidence B) Lift C) Support D) Interestingness

37. Suppose "Y follows X" occurs m times, and "Z follows Y, follows X" occurs n times out of a total of N data records (transactions). What is the confidence of the rule $Z \Rightarrow (X,Y)$?

38. Suppose at a store that sells a small number of different items (like electronics stores, optician stores, etc), all customers buy either 1, or 2 or at most 3 items only. In other words, the maximum number of items that any customer could purchase is 3. Device an ARM algorithm for this purpose. Which of the algorithms mentioned in the text will perform best in such a situation?

39. Association rules are generated in the confidence-support framework from frequent itemsets as follows:– (i) Find all frequent itemsets satisfying minSupp. (ii) For each subset X $\subset$ I, form A = I - X, where the negative sign denotes difference set (I - X contains all elements in I that are not in X). Output the rule $A \Rightarrow X$ if confidence(A,X) $\geq$ minConf. Since confidence(A,X) = support(A$\cap$X)/support(A)=support(I)/support(A), we need to find the support of each and every subset of I to check for support threshold. Design an algorithm for fast checking of this condition.

40. Table 7.8 gives the transactions at an online store that sells gifts. Find the frequent 1-itemsets for support 50%. Form the candidate itemsets generated by the AIS and Apriori algorithms. Enumerate all final frequent itemsets. Generate association rules for confidence 60%.

Table 7.8: Online store sales

Trans Id	Items sold
T1	Watches, Diamond, sports shoe
T2	Rings, necklace, sunglass, deluxe cap, table clock
T3	Parker pen, wall clock, sports shoe, trophy
T4	Money purse, key-chain, mini torch
T5	Parker pen, Wall clock, money purse
T6	Rings, trophy, key-chain, table clock
T7	Money purse, Parker pen, Diamond, key-chain
T8	Watches, sunglass, Parker pen, deluxe cap
T9	Wall clock, necklace, trophy
T10	Diamond, rings, key-chain, table clock,deluxe cap

41. Solve the above problem using the FP-tree algorithm

42. Distinguish between Pareto analysis and paired-comparison analysis.

43. What are temporal association rules? What is rule-expiry? How can you prune temporal association rules already found from a database?

44. Describe the *is-a* and *whole-part* hierarchies, and how they are related to association rule mining.

45. What is an outlier in categorical data? Can you find data outliers using unusual attribute combinations using an ARM algorithm?

46. Describe the fuzzy association rules and its applications.

47. Distinguish between periodic and aperiodic association rules and give two examples.

48. If the rule $X \Rightarrow Y$ has 20% support and 50% confidence, what is support$(Y \Rightarrow X)$?

49. If $X \Rightarrow Y$ is a rule obtained using minSupp and minConf, which of the following cannot be true? a) confidence$(Y \Rightarrow X) \geq$ minConf b) support$(X \Rightarrow Y) \geq$ minSupp c) confidence$(X \Rightarrow Y) \geq$ minConf d) support$(Y \Rightarrow X) \geq$ minSupp

50. Can a rule have 100% support? If an itemset has 100% confidence, describe how you can use this information to speed-up the ARM algorithm.

51. Assume that L is a large itemset. A rule is setup in the form R: $X \to$ L-X where - sign denotes set difference. If R satisfy the minimum confidence, should we check for rules of the form L-X$\to$ X?

52. Prove that support$(\neg X) = 1-$ support(X), where $\neg X$ denotes non-occurrence of itemset X.

53. Prove that support$(X \cup \neg Y) =$ support$(X)-$ support$(X \cup Y)$.

54. Prove that support$(\neg X \cup \neg Y) = 1-$support$(X)-$support$(Y)+$support$(X \cup Y)$.

7.14.0.1 References

[AA00] Agarwal, R. Aggarwal, C. and Prasad V.(2000). A tree projection algorithm for generation of frequent itemsets, *Journal of Parallel and Distributed Computing.*

[AI93] Agrawal, R., Imielinski, T., Swami, A.N. (1993). Mining association rules between sets of items in large databases, *Proceedings of the 12th ACM SIGMOD international conference on management of data*, 207-216, Washington, DC.

[AS94] Agrawal, R., Srikant, R.(1994). Fast algorithms for mining association rules, *Proc. 20th international conference on Very Large Data Bases (VLDB)*, Santiago, Chile, 487-499.

[AM96] Agrawal,R., Mannila, H., Srikant,R. Toivonen,H., Verkamo,A. I. (1996). Fast discovery of association rules, Chapter 12 in U. M. Fayyad, *et.al.* (eds), *Advances in Knowledge Discovery and Data Mining*, 307-328, AAAI Press.

[AS96] Agrawal, R., Shafer,J.C.(1996). Parallel mining of association rules, *IEEE Transactions on Knowledge and Data Engineering*, 8(6), 962-969.

[AT04] Ashrafi, M.Z., Taniar, D., Smith, K. (2004). ODAM: An optimised distributed association rule mining algorithm, *IEEE distributed systems online*, 5(3), (csdl2.computer.org/comp/mags/ds/2004/03/o3001.pdf).

[AS01] Avidan, B., Sonnenberg, A., Schnell, T.G., Sontag, S.J. (2001). Temporal associations between coughing or wheezing, and acid reflux in asthmatics, *Gut Online*, 49, 767-772, (gut.bmj.com/cgi/reprint/49/6/767).

[BMS97] Brin, S., Motwani, R., Silverstein, C. (1997). Beyond market-baskets: generalizing association rules to correlations, *ACM SIGMOD Record*, 265-276.

[BM97] Brin, S., Motwani, R., Ullman, J.D., Tsur, S. (1997). Dynamic itemset counting and implication rules for market-basket data, *SIGMOD Record*, 26, 255-264.

[CR06] Ceglar, A., Roddick, J.F. (2006). Association mining, *ACM computing surveys*, 38(2), July 06.

[CH96] Cheung,D.W., Han,J., Ng,V.,Fu, A.W., Fu,Y(1996). A fast distributed algorithm for mining association rules, *Fourth international conference on parallel and distributed information systems*, 31-42, Miami beach, FL.

[CH98] Cheung,D.W., Hu,K., Xia, S. (1998). Asynchronous parallel algorithm for mining Association Rules on a shared-memory multi-processor, *Proceedings of 10th annual ACM symposium on parallel algorithms and architectures*, Puerto Vallarta, Mexico, 279-288.

[FB02] Florez, G., Bridges, S.M., Vaughn, R.B. (2002). An improved algorithm for fuzzy data mining for intrusion detection, *Proceedings of fuzzy information processing society*, NAFIPS, 457-462.

[FL96] Fortin, S., Liu, L., Goebel, R. (1996). Multi-level association rule mining: An object-oriented approach based on dynamic hierarchies, Technical report TR 96-15, *Univ. of Alberta*, Edmonton, Canada.

[GW02] Guoqing, C., Wei, Q. (2002). Fuzzy association rules and the extended mining algorithms, *Information sciences*, 147(4), 201-228.

[GG03] Gyórödi, C., Gyórödi, R., Coffey, T., Holban, S. (2003). Mining association rules using dynamic FP-trees, *Proc. of Irish Signals and Systems conference*, Limerick, Ireland, 76-81.

[HL99] Han, J., Lakshmanan, L.V.S., Ng, R.T. (1999). Constrained-based multidimensional data mining, *Computer*, 32(8), 46-50.

[HP00] Han, J., Pei, J., Yin, Y.,Mao, R. (2000). Mining frequent patterns without candidate generation, *Proc. of 2000 ACM SIGMOD international conference on management of data*, Dallas, TX, 57-87.

[HM03] Hegland, M. (2003). Algorithms for Association Rules, *Lecture Notes in Computer Science*, Volume 2600, 226-234.

[HG00] Hipp, J., Güntzer,U., Nakhaeizadeh,G. (2000). Algorithms for association rule mining – a general survey and comparison, *ACM SIGKDD Explorations Newsletter*, 2(1), 58-64.

[HS95] Houtsma, M., Swami, A. (1995). Set-oriented mining for association rules in relational databases, *Proc. of the 11th IEEE International conference on data engineering*, 25-34, Taipei, Taiwan.

[JP97] Javeed, Z.M., Parthasarathy, S., Ogihara, M., Li,W. (1997). Parallel algorithms for discovery of association rules, *Data Mining and Knowledge Discovery*.

[LX98] Liang, Z., Xinming, T., Lin, L.I., Wenliang, J. (1998). Temporal association rule mining based on T-Apriori algorithm and its typical application, Proceedings of international symposium on spatio-temporal modeling, spatial reasoning, analysis, data mining and data fusion.

[LH01] Lu S., Hu H., Li F. (2001). Mining weighted association rules, *Intelligent data analysis*, 5(3), 211-225.

[MT94] Mannila, H., Toivonen, H., Verkamo, I.A. (1994). Efficient algorithms for discovering association rules, *AAAI Workshop on Knowledge Discovery in Databases* (KDD-94).

[MAA05] Metwally, A., Agrawal, D., Abbadi, A.E. (2005). Using association rules for fraud detection in web advertising networks, *Proc. of 31st VLDB conference*, Trondheim, Norway, 169-180.

[PS02] Parthasarathy, S.(2002). Efficient progressive sampling for Association Rules. ICDM 2002,354-361.

[SO05] Silvestri, C., Orlando, S. (2005). Distributed approximate mining of frequent patterns, *ACM symposium on applied computing*, Santa Fe, NM, 529-536.

[SC99] Singh, L., Chen, B., Haight, R., Scheuermann, P. (1999). An algorithm for constrained association rule mining in semi-structured data, *Pacific-Asia conference on knowledge discovery and data mining*, 1-11.

[SS06] Srivastava, S., Sural, S., Majumdar, A.K. (2006). Database intrusion detection using weighted sequence mining, *Journal of computers*, 1(4), 8-17.

[SL04] Su, J.H., Lin, W.Y. (2004). CBW: An efficient algorithm for Frequent itemset mining, *Proc. of the 37th Hawaii International Conference on system sciences*, 1-9.

[SM06] Szathmary, L., Maumus, S., Napoli, A. (2006). Mining rare association rules, (hal.inria.fr/ view_by_stamp.php?label=INRIA-LORRAINE&langue=fr&action_todo=view&id=inria-00102909&version=1)

[TM05] Tao,F., Murtagh, F., Farid, M. (2005). Weighted association rule mining using weighted support and significance framework, SIGKDD.

[TH96] Toivonen, H. (1996). Sampling large databases for association rules, The VLDB Journal, 134-145.

[TS99] Tseng, S.V. (1999). An efficient method for mining association rules with item constraints, Technical report, CSD-99-1089, *UC, Berkeley*, (www.eecs.berkeley.edu/Pubs/ TechRpts/1999/CSD-99-1089.pdf)

[WT04] Wang, C., Tjortjis, C.(2004). PRICES: An Efficient Algorithm for Mining Association Rules, *Lecture Notes in Computer Science*, Volume 3177, 352-358.

[WY00] Wang, W., Yang, J., Yu, P. (2000). Efficient mining of weighted association rules (WAR), *Proc of the ACM SIGKDD conference on knowledge discovery and data mining*, 270-274.

[WG04] Weiss, G. M. (2004). Mining with rarity: A unifying framework, *SIGKDD Explorations*, 6(1), 7-19. (www.acm.org/sigs/sigkdd/explorations/issue6-1/weiss.pdf)

[ZM00] Zaki,M.J.(2000). Scalable algorithms for association mining, *IEEE transactions on knowledge and data engineering*, 12(2), 372-390 (csdl.computer.org/comp/trans/tk/2000/03/k0372abs.htm).

8
Regression

CHAPTER OBJECTIVES

- Understand linear regression and its uses

- Distinguish regression and correlation

- Explain least squares principle

- Introduce sample covariance and its algorithms

- Describe weighted least squares

- Applications of regression

8.1 Regression Basics

Regression is a supervised learning model in which the functional relationship is learned from the training data. It is used for prediction, description and control[1]. A linear regression model expresses a dependent variable as a linear combination of one or more independent variables. As an example, the GPA of a student can be expressed as a function of the qualifying exam score, ownership of computers or tablets, and family income as independent variables:

GPA =(a * QUALEXAM + b * OWNSCOMP + c * F_INCOME) +ϵ

where a,b,c are constants and ϵ is the error term. The presence of an error term indicates that the model is not an *exact* mathematical relation. The unknowns in the above equation are called regression coefficients. They are estimated from sample data using the least squares principle (§8.1.3). Regression can be used for predicting the dependent variable when all of the independent variables are known. Hence the dependent variable is called *response variable* and independent variables are called *predictor variables*.

8.1.1 Scatterplots and Regression

Definition 8.1 Scatterplots are graphical methods to visualise possible relationships between quantitative variables.

---

[1] The Logistic Regression is used for classification as well.

The variable values (x_i, y_i), i=1,2,..,n (n>2) are plotted in the Cartesian plane. They can be standardised (by subtracting the mean and dividing by the standard deviation), resulting in a plot called the z-plot. It can provide valuable visual information regarding the underlying population distribution from which the sample was drawn. If the functional relationship is linear in the parameters, it is called a *linear model*. Otherwise it is called curvilinear or nonlinear model. The *polynomial regression* is a special case of multiple linear regression in which the functional dependency takes polynomial form in the independent variables, but is linear in the parameters. A linear regression model is assumed in the following discussion.

The dependent and independent variables in a regression model can be given descriptive names (eg: 'Family income'), labels (eg:F_INCOME), subscripted variables (eg: y_1, x_i), single letter variables (eg: u, v) etc. The subscripted variable notation is universally used due to its compact representation. In addition, the ANOVA decomposition of regression sums of squares gives rise to expressions akin to (unscaled) sample variances (which are usually written using subscripted variables). Each and every sample observation is used in the estimation of the unknowns in an OLS regression model.

An assumption in MLR is that the predictor variables are uncorrelated (this is rarely the case as there could exist little correlation among the predictors in many practical applications). This phenomenon is called multi-collinearity. Remedial measures are available in such situations.

8.1.1.1 Advantages of Scatter Plots

1. Provide insights on the form of relationships

   If most of the data points fall on or around a line or a curve, we can distinguish between linear and nonlinear regression models. In the nonlinear case, we could also fix the approximate regression model (polynomial, log-linear or piece-wise nonlinear etc) to be used. If the variables are uncorrelated, the points are scattered all over the place without adherence to a mathematical curve. In data mining applications with dozens of variables, the form of the relationship is more difficult to find.

2. Identify presence of outliers

   Outlier detection (in bivariate case) is easy and straightforward using a scatterplot. Analytical procedures to detect outliers require the computation of (squared or absolute) distances between each observation from the centroid (grand mean), which may not be suitable in some data mining applications [HD80].

3. Ascertain the type of transformation to be used

   Sometimes, a suitable transformation may result in a simplified regression model. For example, the log-linear regression model transforms the dependent variable using the logarithm (to an appropriate base, dependent upon the data) to give an approximate linear regression model. The most useful transformations are 1) variance stabilising transformation, 2) linearising the regression, 3) transformation to normality. The type of transformation to be used can be inferred either from the scatterplot or statistically.[2] If the functional form of the model is approximately known, we can also apply analytic transformation as follows: 1) for y=ax^b, use $y' = \log(y)$ and $x' = \log(x)$, 2) for y=ae^{bx}, use $y' = \log_e(y)$ etc.

4. Reveal discontinuity in data scatter
   Discontinuities are easy to catch if data are scatter plotted. If there is strong relationship in some data intervals, we can fit multiple regression equations. For instance, the piecewise linear regression is used to fit a linear regression model to each interval of the data range. Range of variables where the prediction is highly accurate can be understood using a scatterplot. This is also a trend that some data miners look for in time-dependent data.

5. Understand data range and scale
   The data range R=$(x_{max} - x_{min})$ is easy to find if x_{max} and x_{min} are known. Scatterplots not only give the range of X and Y values, but can also provide extra information regarding any scaling needed along the axes. This is useful in data mining models with many independent variables.

6. Identify clusters
   If there are distinct clusters of points in the sample (as evidenced by a scatterplot), more elaborate analysis can be based upon each of the clusters. This helps in analysing large samples by dividing it into more manageable sub-samples. In addition, it can reveal whether clusters are formed in the direction of particular variables (X and Y in bivariate case) or a combination of them. This is used in decision trees and binary SVM models.

7. Inference on variance of observations around the regression line
   Too little or too many sample sizes tend to create interpretation problems in scatterplots. If there is a reasonable number of observations, correct conclusions about the variance of observations around the regression line can be drawn. This is more convincing in the presence of correlation, which can be ascertained from a scatterplot.

8. Heteroscedasticity
   If the variability of error terms is not constant, it is called heteroscedasticity (which literally means *unequal scatter of variance*). Symbolically, we denote it as var(ϵ_i) = $c_i\sigma^2$. In statistical quality control and process control applications, this pattern reveals a process or device that is not *in trend*. In time dependent data mining applications, this indicates non-steady variation over time. This may in turn suggest remedial actions to be initiated to bring the process under control. The two most common remedies for heteroscedasticity are weighted regression models and variance-stabilising transformations.

---

[2]Compare the variance of the error terms with higher order powers of expected values of response variable as:-var(ϵ) $\sim$ E(Y) $\Rightarrow$ square root transformation, var(ϵ) $\sim$ E(Y)^2 $\Rightarrow$ logarithmic transformation etc.

The choice of the axes X and Y (in bivariate case) for scatter-plotting are arbitrary. By convention, the independent variable is chosen for the X-axis and the dependent variable for the Y-axis. However, in OLAP mining that involves dozens of variables, the axes can be chosen in any order. OLAP operations help the data miner in visualising the relationships in sub-dimensions. Scatterplotting is a first step to decide the model, as it reveals the structure of the data distribution, direction of strength (+ve or −ve) and discontinuities (clusters) present, if any. In addition, it can also shed light on the presence of outliers, necessity for data scaling, presence of homoscedasticity or heteroscedasticity of variances.

Scatterplots not only give us an idea about the strength of the relationship, but also the direction of a relationship. If the mean of the dependent variable (Y) for a conditioned value of independent variable (X=c) is closer to the mean of all Y values, rather than to the mean of all X values, it is called regression towards the mean. Such regression lines will have a slope less than 45 degrees, if independent variable is plotted along the X axis and dependent variable along the Y axis.

8.1.2   Simple Linear Regression

In simple linear regression we have two variables of interest, one of which is called the dependent (response) variable and is assumed to depend linearly on the independent (or predictor) variable.

Definition 8.2 The simple linear regression is a supervised statistical model that expresses a response variable as a weighted linear function of the predictor variable through a random error variation.

Statistically, we express it as $Y = aX + b + \epsilon$, which is called the stochastic form[3], where x is the independent predictor. There are many types of errors that can creep into the regression model – measurement errors on variables, sampling errors and random errors. The error term ϵ (which can be +ve or −ve) is added to compensate for all aggregated error sources. The constant 'a' is called the slope and 'b' is called the intercept, which are the interpretations obtained from a graphical representation of the equation in the Cartesian (X-Y) plane. If the intercept (b) is zero, we get regression *through the origin*. The *regression coefficients* 'a' and 'b' are unknown, and are estimated from the sample in an optimal way.

The X and Y are connected through an unknown random variation. If x_i and y_i are the observed values for a particular sample, the error in predicting Y from X can be expressed as $e_i = y_i - (ax_i + b)$. These error terms will vary from observation to observation in a sample and can take any positive or negative value or can be zero. In most statistical studies, we assume that the error terms are normally distributed with mean 0 and variance σ^2 (known), and that they are uncorrelated.

Regression can be used to predict the dependent variable when the independent variables are known, and to control the dependent variable at a specified level or between two values (called a regression band). Prediction of the unknown value is possible for both the dependent and independent variable(s). For instance, if the income of a person is known, we can have a point

---

[3]The corresponding *canonical form* using Mathematical Expectation is E(Y|x) = ax +b.

estimate or an interval estimate for the age using the regression model. If the relationship is perfect, the regression band collapses to a perfect mathematical equation (to a straight line in the case of linear regression) and the prediction of unknown values becomes exact.

8.1.3 Ordinary Least Squares (OLS)

A method for estimating the unknown parameters from available data utilises the 'least squares principle'. In this method, the squared deviations are simultaneously minimised to obtain the unknowns. Each of the 'n' observations give a deviation and hence we sum all of the squared deviations to get the *least squares criterion* as

$$Q = \sum_{i=1}^{n}(y_i - ax_i - b)^2. \tag{8.1}$$

Because 'a' and 'b' are the unknowns, we differentiate the above equation wrt 'a' and 'b' to get the following *normal equations*:

$$\partial Q/\partial a = \sum_{i=1}^{n} -2x_i(y_i - ax_i - b) \ \text{ and } \ \partial Q/\partial b = \sum_{i=1}^{n} -2(y_i - ax_i - b). \tag{8.2}$$

These are so called because they are obtained by differentiating the least squares criterion[4]. In the case of linear regression, the error terms are the vertical distances of observations from the fitted line of regression. Since our aim is to simultaneously minimise the *squared* error terms, we set the above derivatives to zero and solve for the unknowns. The second derivatives are obtained (from either of these equations) as $\partial^2 Q/(\partial a \partial b) = 2n\bar{x}$ (by using the fact that $\sum_i x_i = n\bar{x}$), $\partial^2 Q/\partial a^2 = 2\sum_i x_i^2$, and $\partial^2 Q/\partial b^2 = 2n$. The matrix of partial derivatives is positive definite as the determinant is

$$|M| = \begin{vmatrix} 2\sum_i x_i^2 & 2n\bar{x} \\ 2n\bar{x} & 2n \end{vmatrix} = 4n(\sum_{i=1}^{n} x_i^2 - n\bar{x}^2) = 4n \sum_{i=1}^{n}(x_i - \bar{x})^2$$

which is always ≥ 0. Expanding the equations (and canceling out common terms) we get: $\sum_{i=1}^{n} x_i y_i = a\sum_{i=1}^{n} x_i^2 + bn\bar{x}$ and $\bar{y} = a\bar{x} + b$.

Solution of these equations gives an estimate of the slope 'a' as

$$\hat{a} = \sum_{i=1}^{n}(x_i - \bar{x})(y_i - \bar{y})/[\sum_{i=1}^{n}(x_i - \bar{x})^2] = rs_y/s_x \tag{8.3}$$

which can be written as $\hat{a} = \sum_{i=1}^{n} w_i y_i$, where $w_i = (x_i - \bar{x})/\sum_{i=1}^{n}(x_i - \bar{x})^2$ and the intercept as $\hat{b} = \bar{y} - \hat{a}\,\bar{x}$. Thus the regression slope ('a') is a weighted linear combination of y_i values where the weights have the property that $\sum_{i=1}^{n} w_i = 0$ and $\sum_{i=1}^{n} w_i x_i = 1$ (see exercise).

Theorem 8.1 The simple linear regression slope is $\hat{a} = \sum_{i=1}^{n}(x_i - \bar{x})y_i/K = \sum_{i=1}^{n} x_i(y_i - \bar{y})/K$ where $K = [\sum_{i=1}^{n}(x_i - \bar{x})^2]$.

---

[4]They are not related to the Normal (Gaussian) distribution.

Proof. This follows easily by breaking the sum in the numerator of above equation into two parts, and using the fact that $\sum_i(x_i - \bar{x}) = \sum_i(y_i - \bar{y}) = \sum_i w_i = 0$.

Theorem 8.2 If the variable y is transformed as $y' = y/c$, the regression slope becomes $\hat{a}' = \hat{a}/c$. **Proof.** By putting $y_i = cy'_i$, and $\bar{y} = c\bar{y}'$ in 8.3, the constant c becomes a common factor which can be taken outside the summation. This proves the theorem.

Corollary 13 If the variable x is transformed as $x' = x/c$, the regression slope becomes $\hat{a}' = c\hat{a}$. **Proof** The substitution above gives a constant c in the numerator and a c^2 in the denominator. Canceling out the common term gives the result.

Corollary 14 If both variables are transformed as $x' = x/c_1$ and $y' = y/c_2$, the regression slope becomes $\hat{a}' = (c_1/c_2)\hat{a}$, which is unchanged if $c_1 = c_2$.

8.2 Sample Covariance

The sample covariance is a non-standardised measure of the strength of relationship between the variables involved. It was introduced by the British genetist Francis Galton (1822-1911)[5] in 1888. The sample covariance has various applications in statistics, econometrics, engineering and data mining. Let $(x_1, y_1), (x_2, y_2), \cdots, (x_n, y_n)$ be a sample from a bivariate population. Then we define the sample covariance as cov=$\sum_{i=1}^{n}(x_i - \bar{x})(y_i - \bar{y})/n$. Some authors use (n-1) in the denominator of the sample covariance to denote the loss of one *degree of freedom* due to the estimation of $\bar{x}$ and $\bar{y}$ from sample data. We will use (n-1) in the denominator only in those situations where the means are *estimated* from the sample (as in testing of hypothesis problems). The expression obtained without the scaling factor n (or n-1) in the denominator is called the *unscaled covariance*, and is denoted by capital letters COV.

The equation 8.3 shows that the slope of the estimated regression line is the scaled correlation between the variables with scaling factor s_y/s_x (or equivalently regression slope is the ratio of covariance over the variance of X). A regression slope near zero need not imply zero correlation because this could happen when s_y is too small when compared to s_x.

Theorem 8.3 Covariance can be computed using a short-cut formula: $cov = \frac{1}{n}\sum_{i=1}^{n} x_i\, y_i - \bar{x}\bar{y}$. **Proof.** This follows easily using cross-multiplication $\sum_{i=1}^{n}(x_i - \bar{x})(y_i - \bar{y}) = \sum_{i=1}^{n} x_i y_i - \sum_{i=1}^{n} x_i\bar{y} - \bar{x}\sum_{i=1}^{n} y_i + n\bar{x}\bar{y} = \sum_{i=1}^{n} x_i y_i - \bar{y}n\bar{x} - \bar{x}n\bar{y} + n\bar{x}\bar{y} = \sum_{i=1}^{n} x_i y_i - n\bar{x}\bar{y}$. Dividing both sides by n gives the desired result.

Corollary 15 The unscaled covariance can be represented in various ways as:

$$COV = \sum_{i=1}^{n} x_i\, y_i - \bar{x}\sum y_i = \sum_{i=1}^{n} x_i\, y_i - \bar{y}\sum x_i = \sum_{i=1}^{n} x_i \left[\sum_{j\neq i}(y_i - y_j)\right]. \qquad (8.4)$$

---

[5]Galton also introduced the concept of correlation in 1888, which was at that time called *co-relation* or *Galton coefficient*. See www.galton.org for his complete life history and works.

Theorem 8.4 If either of the variables are scaled (as $y'=y/c$), the covariance is correspondingly scaled (becomes cov(x,y)= c * cov(x,y')).
Proof. We work with unscaled covariance. Assume that only the Y variable is scaled by dividing each value by c. Then COV=$\sum_{i=1}^{n}(x_i-\bar{x})(y_i-\bar{y}) = \sum_{i=1}^{n}(x_i-\bar{x})c(y_i'-\bar{y}') = $ c*cov(x,y'). When both variables are scaled as $x' = x/c$ and $y'=y/d$, we get cov(x,y) = c*d cov(x',y').

Example 8.1 The height in cm and the weight in kg of 10 persons are given in table 8.1. Find the covariance.

Table 8.1: Height and weight of 10 adult persons

Height	150	155	145	166	165	172	164	159	153	152
Weight	54	63	54	70	63	69	62	60	78	69

Solution: We arbitrarily choose a=145, which is the minimum, to change the origin of X values (see theorem below), and divide Y values by 10 (c=10). The average scaled weight is 6.42. Unscaled covariance is $\sum_i x_i'y_i' - \bar{y}'\sum_i x_i'$=861.8 - 6.42*131 = 20.78. The covariance of

Table 8.2: Height (scaled) and weight of 10 adult persons

x'	5	10	0	21	20	27	19	14	8	7	131
$y'=y/10$	5.4	6.3	5.4	7.0	6.3	6.9	6.2	6.0	7.8	6.9	64.2
$x' * y'$	27	63	0	147	126	186.3	117.8	84	62.4	48.3	861.8

original data in this problem is also the same because the scaling factor for y (c=10) cancels out with the number of observations (n=10).

Data mining applications can have dozens of variables in which some are ordinal and assume integer values in a fixed range. Examples are the number of items purchased, number of trips by a customer to a supermarket or patient visits to a clinic (say in a week), various counts like returned items, items on reserve in a library or video store, various frequency counts, age of items or people as integers, number of injuries in an accident, total stitches in an operation, number of payment defaults by a patron etc. Below, we derive simple expressions for sample covariance in these situations.

Theorem 8.5 If $(x_1, y_1), (x_2, y_2), \cdots, (x_n, y_n)$ is a bivariate sample of size n, in which either of the variables is ordinal or higher scale (and assume consecutive integer values), the *unscaled* covariance can be simplified as

$$COV = \frac{(n-1)}{2}F(n) - \sum_{i=1}^{n-1}F(i) \tag{8.5}$$

where F(i) $=\sum_{j=1}^{i} y_j$ (if x_i's are integers), and F(i) $=\sum_{j=1}^{i} x_j$ (if y_i's are integers).

Proof. We assume without loss of generality that X is ordinal, and takes distinct values in [1,2,..,n] (see corollary below for any consecutive integer range). The first step is to re-arrange the (x_j, y_j) pairs such that $x'_j s$ are in natural order (this will not affect the computed value). Using the shortcut formula, the unscaled covariance becomes COV$=\sum_{i=1}^{n}(x_i - \bar{x})y_i$. As the x_i's are rearranged in natural order, $x_i = i$. Hence COV $=\sum_{i=1}^{n}(i-\bar{x})y_i = \sum_{i=1}^{n} iy_i - \bar{x}\sum_{i=1}^{n} y_i$. Now $\sum_{i=1}^{n} iy_i = \sum_{i=1}^{n}\sum_{j=i}^{n} y_j = nF(n) - \sum_{i=1}^{n-1} F(i)$. Also, $\bar{x}=n(n+1)/(2n) = (n+1)/2$. Substituting in the above we get COV $= nF(n) - \sum_{i=1}^{n-1} F(i) - ((n+1)/2)F(n)$. Combining first and third terms, and using $(n-(n+1)/2) = (n-1)/2$ we get COV$=\frac{(n-1)}{2}F(n) - \sum_{i=1}^{n-1} F(i)$.

Example 8.2 Compute the covariance for the following set of data [(2,3.5), (5,1.0), (3,3.8), (1,4.0), (4,1.1)].

Solution: We first rearrange the data as [(1,4.0),(2,3.5),(3,3.8),(4,1.1), (5,1.0)]. The F(j) values are given by F(1)=y_1=4, F(2) = $y_1 + y_2$ =7.5, F(3) =11.3, F(4) = 12.4, F(5) = 13.4. Hence $\sum_{i=1}^{n-1}$F(i) = 4+7.5+11.3+12.4 =35.2. From this we get the unscaled covariance as COV =2*13.4 - 35.2 = −8.4. The mean of x is $\bar{x}$=15/5=3. Thus by the shortcut method, we obtain COV=-2*4 -1*3.5 + 0 + 1*1.1 + 2*1 = −8.4. Dividing by n=5 gives the covariance as cov=-1.68.

Theorem 8.6 An arbitrary change of origin can be applied to one of the variables in covariance to ease the computation.

Proof. Assume without loss of generality that X values are large. We can write the covariance as $cov = \frac{1}{n}\sum_{i=1}^{n}(x_i - c)(y_i - \bar{y}) = \frac{1}{n}\sum_{i=1}^{n} x_i(y_i - \bar{y}) - \frac{c}{n}\sum_{i=1}^{n}(y_i - \bar{y})$. The second term is zero, giving the result. This theorem shows that the constant c can be any real number including mean, median, mode, minimum or maximum of the (x) observations, or it could be zero.

Example 8.3 Number of traffic accidents per year and number of injuries received for a sample of drivers are as follows. {(3,5),(2,3),(1,1),(4,2), (5,4).} Find the covariance.

Solution: We first rearrange the data in ascending order of the number of accidents as

$$(1, 1), (2, 3), (3, 5), (4, 2), (5, 4).$$

The cumulative sums of injuries are F(1) = 1, F(2) = 4, F(3) = 9, F(4) = 11. $\sum_{i=1}^{n-1} F(i) = 25$. The covariance is 24/4 - 25/5 = 1.

8.2.1   Recursive Algorithm For Covariance

In this section, we develop two useful recursive algorithms for covariance that have interesting implications in online computations. Sometimes, the entire data for computing the covariance may not be readily available, but could come over a wired or wireless connection one by one. In such situations, we could start computing the covariance with the already available data and update it using new data points that come afterward. The first theorem below is for the special case when one of the variables takes integer values, and the second one is for the general case.

Theorem 8.7 When the X variable takes consecutive integer values (in any range), the covariance can be recursively computed as $(1 + \frac{1}{n})\,\text{cov}_{n+1} = \text{cov}_n + \frac{1}{2}(y_{n+1} - \frac{1}{n}F(n))$, with initial values $\text{cov}_{n=1}=0$, or $\text{cov}_{n=2} = (y_2 - y_1)/4$, where $F(n)=\sum_{i=1}^{n} y_i$.

Proof. Without loss of generality, we assume that X takes values starting with 1. From the above theorem (where x is integer), the unscaled covariance is

$$COV_n = \frac{(n-1)}{2}F(n) - \sum_{i=1}^{n-1} F(i). \tag{8.6}$$

Replacing n by (n+1) and subtracting results in $COV_{n+1} - COV_n = \frac{n}{2}F(n+1) - \frac{(n-1)}{2}F(n) - F(n) = \frac{n}{2}F(n+1) - \frac{(n+1)}{2}F(n) = \frac{n}{2}y_{n+1} - \frac{1}{2}F(n)$. Thus $COV_{n+1} = COV_n + \frac{1}{2}(n\,y_{n+1} - F(n))$. Dividing both sides by n (and writing 1/n on the LHS as $\frac{1}{n+1}(1 + 1/n)$) gives the desired result. By expanding F(n), and replacing it by $\bar{y}_n$, we could also write the above expression as $(1 + \frac{1}{n})\,cov_{n+1} = \text{cov}_n + \frac{1}{2}(y_{n+1} - \bar{y}_n)$. When n=1, there is no covariance ($\text{cov}_{n=1}=0$). When n=2, let $(1,y_1)$, $(2,y_2)$ be the sample values. Then $F(1) = y_1$ and $F(2) = y_1 + y_2$. According to our formula, unscaled covariance $= (2\text{-}1)/2\ F(2) - F(1) = (y_1 + y_2)/2 - y_1 = (y_2 - y_1)/2$. By the short-cut method, unscaled $\text{cov}_{n=2} = \sum_{i=1}^{2} x_i y_i - 2(1 + 2)/2 * (y_1 + y_2)/2 = (y_1 + 2y_2) - 3(y_1 + y_2)/2 = (y_2 - y_1)/2$. Dividing by n=2 (sample size) gives $\text{cov}_{n=2} = (y_2 - y_1)/4$. This result can be extended to any consecutive range for X by a change of origin transformation. Due to the symmetry of X and Y variables, a similar result can be obtained when Y variable takes consecutive integer values, but X may not.

Example 8.4 Find the covariance for the data in example 8.3 (page 8-8) recursively.

Solution: As before, we first rearrange the data as $[(1,4.0),(2,3.5),(3,3.8),(4,1.1),(5,1.0)]$. We recursively find the unscaled covariance (obtained by multiplying throughout by n). The F(j) values are F(1)=4, F(2) = 7.5, F(3) =11.3, F(4) = 12.4. $\text{cov}_{n=1}=0$, $\text{cov}_{n=2}=\text{cov}_{n=1} + \frac{1}{2}(1 * y_2 - y_1) = (y_2 - y_1)/2 = (3.5\text{-}4.0)/2 =\text{-}.25$, $\text{cov}_{n=3}=\text{cov}_{n=2} + \frac{1}{2}(2\ y_3 - F(2)) = -.25 + \frac{1}{2}(2 * 3.8 - 7.5) = -.25 + .05 = -.20$, $\text{cov}_{n=4}=\text{cov}_{n=3} + \frac{1}{2}(3\ y_4 - F(3)) = -.20 + \frac{1}{2}(3 * 1.1 - 11.3) = -4.2$, $\text{cov}_{n=5}=\text{cov}_{n=4} + \frac{1}{2}(4\ y_5 - F(4)) = -4.2 + \frac{1}{2}(4 * 1 - 12.4) = -4.2 - 4.2 = -8.4$. Dividing by n=5 gives the covariance as -1.68.

Theorem 8.8 Let $(x_1, y_1), (x_2, y_2), \cdots , (x_n, y_n), (x_{n+1}, y_{n+1})$ be a bivariate sample of size (n+1). The sample covariance can be recursively computed as

$$\left(1 + \frac{1}{n}\right)\,cov_{n+1} = cov_n + \frac{1}{(n+1)}(x_{n+1} - \bar{x}_n)(y_{n+1} - \bar{y}_n), \tag{8.7}$$

with initial values $\text{cov}_{n=1}=0$, or $\text{cov}_{n=2} = (x_2 - x_1)(y_2 - y_1)/4$.

Proof. Using the short-cut formula we have $(n+1)\ cov_{n+1} = \sum_{i=1}^{n+1}(x_i - \bar{x}_{n+1})y_i$. Splitting the index of summation from 1 to n, and for (n+1) separately, this becomes

$$(n+1)cov_{n+1} = \sum_{i=1}^{n}(x_i - \bar{x}_{n+1})y_i + (x_{n+1} - \bar{x}_{n+1})y_{n+1}. \tag{8.8}$$

Substituting $\bar{x}_{n+1} = [\bar{x}_n + (x_{n+1} - \bar{x}_n)/(n+1)]$, the first expression above becomes

$$\sum_{i=1}^{n}(x_i - \bar{x}_n)y_i + \frac{1}{(n+1)}(\bar{x}_n - x_{n+1})\sum_{i=1}^{n} y_i. \qquad (8.9)$$

The first sum in (8.9) is easily seen to be n cov_n. The second expression when combined with $(x_{n+1} - \bar{x}_{n+1})y_{n+1}$ gives

$$\frac{n}{n+1}y_{n+1}(x_{n+1} - \bar{x}_n) + \frac{1}{(n+1)}(\bar{x}_n - x_{n+1})\sum_{i=1}^{n} y_i. \qquad (8.10)$$

Taking the common factor $(x_{n+1}-\bar{x}_n)/(n+1)$, this becomes $(x_{n+1}-\bar{x}_n)/(n+1)\left[ny_{n+1} - \sum_{i=1}^{n} y_i\right]$. Combine with (8.8) and divide both sides by n to get

$$\left(1 + \frac{1}{n}\right) cov_{n+1} = cov_n + \frac{1}{(n+1)}(x_{n+1} - \bar{x}_n)(y_{n+1} - \bar{y}_n). \qquad (8.11)$$

See [CR12] for an algorithmic implementation of covariance. When the X values are consecutive integers, $(x_{n+1} - \bar{x}_n) = (n+1) - [n(n+1)]/[2n] = (n+1)/2$. The n+1 in the numerator and denominator cancels out giving the result in theorem 8.7.

As the variance of a sample can be written as $s_n^2 = \sum_{i=1}^{n}(x_i - \bar{x})(x_i - \bar{x})/n$, we could get a recurrence relation for the sample variance as

$$\left(1 + \frac{1}{n}\right) s_{n+1}^2 = s_n^2 + (x_{n+1} - \bar{x}_n)^2/(n+1), \qquad (8.12)$$

with initial values $s_{n=1}^2 = 0$. See Ross (1987), page 143, for an alternate recurrence relation for sample variance. The analogous results when (n-1) is used as the scaling factor in the denominator can easily be derived in our case as

$$cov_{n+1} = \left(1 - \frac{1}{n}\right) cov_n + \frac{1}{(n+1)}(x_{n+1} - \bar{x}_n)(y_{n+1} - \bar{y}_n), \qquad (8.13)$$

from which the corresponding recurrence for variance (with n-1 as df) could be obtained as

$$s_{n+1}^2 = (1 - 1/n)s_n^2 + (x_{n+1} - \bar{x}_n)^2/(n+1). \qquad (8.14)$$

This expression has slight computational advantage over the expression given by Ross (1987) that requires the computation of two means. However, the equivalence of our expression, and that due to Ross can easily be established. Write $\bar{x}_{n+1} = (n/(n+1))\bar{x}_n + x_{n+1}/(n+1)$. Then

$$\bar{x}_{n+1} - \bar{x}_n = (n/(n+1))\bar{x}_n - \bar{x}_n + x_{n+1}/(n+1) = (x_{n+1} - \bar{x}_n)/(n+1). \qquad (8.15)$$

The updating term $(n+1)(\bar{x}_{n+1} - \bar{x}_n)^2$ of Ross simply reduces to $(n+1)(x_{n+1} - \bar{x}_n)^2/(n+1)^2 = (x_{n+1} - \bar{x}_n)^2/(n+1)$. This is our updating term, proving our conjecture. See [SC15] for a divide and conquer algorihm and some further results.

Table 8.3: Average market price of a brand of used cars (in '00)

Age	1	2	3	4	5	6	7	8	9	10
Price	8000	7200	6300	6000	5300	4200	2600	1900	1400	1200

Example 8.5 Table 8.3 gives the market price of a used car brand at an auto dealer. Fit the regression model using the short-cut method.

Solution: We will use the above corollary to simplify the calculation, by dividing the prices by 100. We form the cumulative prices of transformed data as: F(1)=80, F(2) = 152, F(3) = 215, F(4) = 275, F(5) = 328, F(6) = 370, F(7) = 396, F(8) = 415, F(9) = 429, F(10) = 441. By using the short-cut method, we get the slope as $\hat{a}$=(12/990)(4.5*441 - 2660)= -8.187878. Multiplication by 100 give $\hat{a}$=-818.7878. The $\hat{b}$ can be obtained as $\hat{b}$ = F(10)/10 - (-818.7878) $\bar{x}$= 4410+5.5 * 818.7878 = 8913.333

The fitted regression model is price = -818.7878 * Age + 8913.333 + ϵ. A scatterplot of the

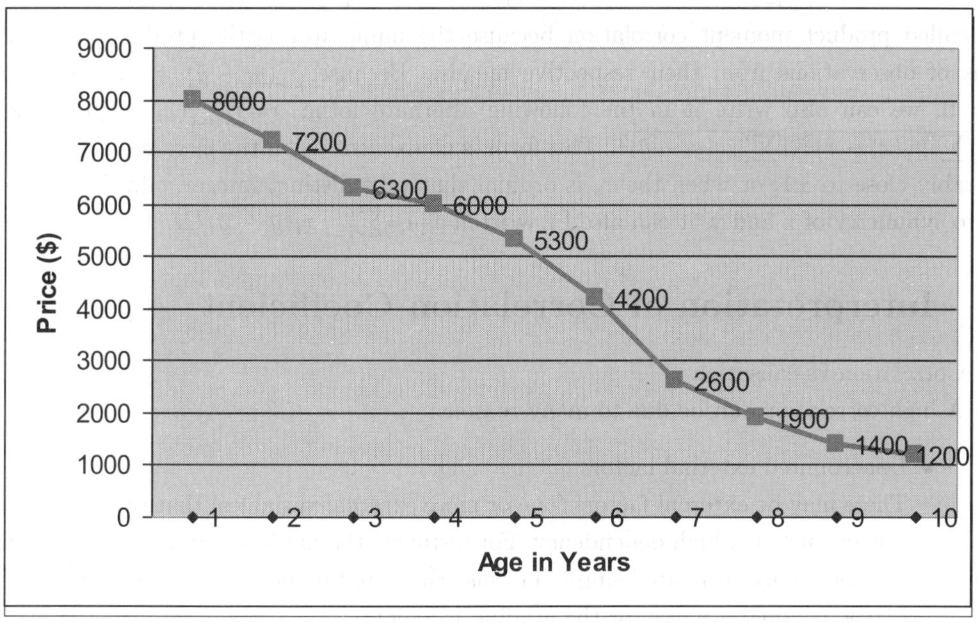

Figure 8.1: Market value of used cars.

data appear in figure 8.5.

8.2.2 Correlation Coefficient

The simple correlation coefficient is a standardised (unit-less) measure of linear relationships between two variables. It is also called Pearson's product moment correlation, or Galton coefficient. The sample correlation is denoted by 'r', and the population correlation is denoted by ρ. It is a ratio measure in NOIR typology that is always in the interval $[-1,+1]$ $(-1 \leq r \leq 1)$. Absolute value of r gives the strength of the linear relationship [YK65]. When $|r| \simeq 0$, and n>2, the linear regression model cannot be used effectively in prediction or classification. There are many data mining models that utilise the correlation coefficient. For instance, latent semantic indexing (LSI) uses correlation or cosine similarity to retrieve related documents in close proximity in the feature space. Web mining uses the cosine similarity metric, which is related to correlation coefficient. It is also used in cluster analysis and neural networks.

8.2.3   From Scatterplot to Correlation

The scatterplot only gives a 'visual estimate' of the strength of a relationship. To quantify the strength of the relationship, we need a well-defined unit-less measure that is symmetric in the variables involved. For a bivariate sample $(X,Y) = ((x_1, y_1), (x_2, y_2), \cdots, (x_n, y_n))$ it is defined as:

$$r = \sum_{i=1}^{n}(x_i - \bar{x})(y_i - \bar{y}) / \{[\sum_{i=1}^{n}(x_i - \bar{x})^2]^{1/2}[\sum_{i=1}^{n}(y_i - \bar{y})^2]^{1/2}\}. \tag{8.16}$$

It is called product moment correlation because the numerator is the product of the deviations of observations from their respective means. Because $\sum(y_i - \bar{y}) = 0$, and $\sum(x_i - \bar{x}) = 0$, we can also write it in the following alternate form: $r = \sum_{i=1}^{n}(x_i - \bar{x})y_i/D$ where $D = \sqrt{\sum_{i=1}^{n}(x_i - \bar{x})^2}\sqrt{\sum_{i=1}^{n}(y_i - \bar{y})^2}$. This form is convenient when the y_i values are small and preferably close to ± 1, or when the x_i is ordinal that take distinct integer values in [1,2,...,n]. Due to symmetry of x and y, it can also be written as $r = \sum_{i=1}^{n} x_i(y_i - \bar{y})/D$.

8.3   Interpretation of Correlation Coefficient

1. Correlation vs causation

   A high correlation can be due to many reasons:-

   - Unaccounted external factors

     There may be external factors (one or more external variables) that play a significant role in causing a high dependency. For instance, the marks of students in two courses may be highly correlated simply because they are taught by the same teacher, or the textbooks used are easy or the grading is lenient.

   - Wrong or too small sample

     If the sample is not a true representative of the population, the correlation coefficient may be spurious and not generalisable. A too small sample may result in high correlation. For instance, if there are just two sample values, the regression line is perfectly linear and the correlation is 1. A statistical significance of the correlation is based on the sample size and it must be consulted before drawing any meaningful conclusions.

   - Geometric interpretation

     Correlation is the cosine of the angle between the two vectors of samples (for centered

Table 8.4: Family income vs annual travel spending

X	40	44	53	58	66	71	80	86	93	102	120	135
Y	5.2	5.7	6.3	7.0	7.1	7.5	8.3	8.9	11.4	12.0	13.6	12.8

data). If (x_1, y_1), $(x_2, y_2), \cdots (x_n, y_n)$ is a sample of size n, take $X = (x_1, x_2, \cdots, x_n)$ as a single point in n-dimensional space and $Y = (y_1, y_2, \cdots, y_n)$ as another point. Then correlation is the cosine of the angle between X and Y. In other words, it is the angle between the residual vectors as $cos(\theta) = e_i' e_j / [\sqrt{e_i' e_i} \sqrt{e_j' e_j}]$ where $e_i = (X - \bar{X})$ and $e_j = (Y - \bar{Y})$.

Example 8.6 Table 8.4 give the family income (X) and total annual spending (Y) on travel (figures in 1000's) for a group of customers. Find the correlation coefficient.

Solution: A plot of the data indicates an approximate linear relationship. As the scaled spending are small numbers, we find the correlation using the short-cut method as $r = \sum_{i=1}^{n} (x_i - \bar{x}) y_i / \sqrt{\sum_{i=1}^{n} (x_i - \bar{x})^2} \sqrt{\sum_{i=1}^{n} (y_i - \bar{y})^2}$. $\bar{x} = 948/12 = 79$. $\sum_i (x_i - \bar{x}) y_i = 915.2$, $\sqrt{\sum_{i=1}^{n} (x_i - \bar{x})^2} = 9688$, $\sqrt{\sum_{i=1}^{n} (y_i - \bar{y})^2} = 92.93667$. Hence r = .964507

8.4 Multivariate Data

The data gathered by observing many variables on each observation are called multivariate data. Methods for analysing multivariate data are well-studied in statistics (known as multivariate analysis). Several data mining models like neural networks, SVM, LSI etc use multivariate data. For example, consider a series of measurements or observations made on a number of *subjects, customers, patients, transactions* or other *entities* of interest. This may be represented by a rectangular matrix X called a *multivariate data matrix* [or simply a *data matrix*] as

$$X = \begin{pmatrix} x_{11} & x_{12} & \cdots & x_{1p} \\ x_{21} & x_{22} & \cdots & x_{2p} \\ \vdots & \vdots & \ddots & \vdots \\ x_{n1} & x_{n2} & \cdots & x_{np} \end{pmatrix}$$

whose typical element x_{ij} represents the value of the jth variable for the ith individual. Here n denotes the number of individuals under investigation, and p, the number of measurements taken on each individual. In LSI, the term-by-document matrix is represented as a data matrix in which columns denote the documents and rows denote the terms.

8.5 Multiple Linear Regression (MLR)

In simple linear regression, we have one predictor and one response variable. In most of the data mining applications, we have many (> 2) predictors.

Definition 8.3 MLR is an extension of simple linear regression in which more than one predictor variables are related to a continuous response variable linearly through the parameters. We assume that the variables are quantitative and express the response variable as

$$Y = b_0 + b_1 X_1 + b_2 X_2 + \cdots + b_n X_n + \epsilon$$

where b_0, b_1, etc are unknowns to be estimated from the sample and ϵ is the random error variation. This equation (without the error term) represents a hyperplane in n-dimensions. Tilt of the hyperplane indicates the direction of relationship. When the predictor variables are correlated with the response variable, most of the data points will lie close to the hyperplane. Most often, OLAP miners look for this type of pattern (rather than for an absence of correlation) while mining multivariate data. The corresponding stochastic form is $E(Y|x_1, x_2, \cdots, x_n) = b_0 + b_1 x_1 + b_2 x_2 + \cdots + b_n x_n$. Multiple regression through the origin is a special case when the hyperplane passes through the origin ($b_0 = 0$).

The b_i's (multiple regression coefficients) measure the effect of a unit change in X_i on the response variable by conditioning all other predictor variables as constant. This is also called the first order MLR to distinguish it from higher order linear models that have variable interactions ($Y = b_0 + b_1 x_1 + b_2 x_2 + b_{12} x_1 x_2 + \cdots$). The total variability is measured using the statistic SSTO $= \sum_{i=1}^{n} (y_i - \bar{y})^2$ which can be broken down into the sum of squares due to regression (SSR) and error sum of squares (SSE) as $\sum_{i=1}^{n} (y_i - \bar{y})^2 = \sum_{i=1}^{n} (\hat{y}_i - \bar{y})^2 + \sum_{i=1}^{n} (y_i - \hat{y}_i)^2$, where $\hat{y}_i$ is the fitted value of y. SSE measures the deviations of observations from the fitted model, and the ratio $R^2 = SSR/SSTO$ is a measure of the goodness of linear fit [NK04].

8.6   Exercises

1. Mark as true or false:

    a) Regression when used for prediction is a supervised learning model

    b) A regression through the origin indicates zero correlation among the variables

    c) Every sample observation contributes to the regression coefficients

    d) Scatterplots are useless in providing strength of relationships

    e) Covariance of n data values is always positive

    f) Correlation is independent of change of origin and scale transformation

2. In OLS (8.1.3 in page 8-5) if the weights are defined as $w_i = (x_i - \bar{x}) / \sum_{k=1}^{n} (x_k - \bar{x})^2$, prove that $\sum_{i=1}^{n} w_i = 0$, and $\sum_{i=1}^{n} w_i x_i = 1$.

3. Compute the covariance of the following data using theorem 8.5 (page 8-7) and corollary 8.7 (page 8-9):– (i) [(6,25),(3,26),(1,18),(5,31),(2,27),(4,23)]

4. The BMI and systolic BP of 6 patients are as follows: (BMI,BP)=[(24,114), (25,122), (22,110), (27,151), (26,132), (23,129)]. Find the covariance using theorem 8.5 (page 8-7).

Table 8.5: Predicting sales price of used cars

CustId	Car age	Miles	Past service costs	Current Cost
10	4	50K	1520	250
20	4	43K	740	600
30	6	72K	1250	725
40	8	110K	2750	430
50	5	70K	1240	600
60	8	84K	3240	900
70	7	87K	1960	180

Table 8.6: Using regression model for prediction

Exp	Gender	Education	IQ	Income
5	1	2	?	55650
3	1	3	100	?
1	0	1	90	?
14	?	2	120	63540
8	1	2	?	55680
0	0	?	76	49920
7	0	1	104	?
?	1	3	92	62090
4	?	2	115	55040
20	1	3	85	?

5. What are the primary objectives in regression analysis and how is it useful in data mining?

6. The maintenance cost of an automobile is dependent on the age (in months), miles driven, past maintenance services and the make (Toyota, Ford, Chevrolet etc). Assuming that the maintenance cost during the first 3 years are negligible, data in table 8.5 give the maintenance cost for 5 subsequent years. Fit a multiple regression model, and estimate the costs for years 9 and 10 for a car with 100K miles and $1500 in past repair costs.

7. If the salary of skilled professionals is linearly dependent upon the work experience, education level, gender and IQ index as salary = 1210*experience + 4200*educlevel + 8600 * gender + 120 * IQ + 15000 where experience is the number of years of relevant work experience, educlevel={1=undergraduate, 2=graduate, 3=post graduate}, gender={0=Female, 1=Male}, and IQ=100 * (mental age / chronological age), predict the missing values (that are indicated by a ?) of the persons in table 8.6. Find minimum education level for a female with no experience and an IQ score of 70 to get a salary of 32000. How much experience is required for an undergraduate male with an IQ of 85 to get a salary of 44000?

Table 8.7: Average monthly electricity bill vs number of rooms

# Rooms	2	3	4	5	6	7	8
Bill amt.	20	24	31	38	43	46	51

8. What can you infer if the regression line passes through the origin?
   A) zero correlation B) negative correlation C) positive correlation D) none of these

9. If the regression line is parallel to the X-axis (independent variable), it indicates: A) zero correlation B) positive correlation C) negative correlation D) None of these.

10. In multiple regression there are: A) multiple predictor variables and a single response variable B) multiple response variables and a single predictor variable C) multiple predictor variables and multiple response variables D) single predictor and response variables

11. The number of rooms in a house and average monthly electricity amount are given in table 8.7. Use the short-cut method to compute the covariance and correlation coefficient (note: substitute bill amount=0 for rooms=1 column).

12. Prove that the covariance could be found using $\text{cov}_n = \frac{1}{n} \sum_{i=1}^{n-1} \sum_{j=i+1}^{n} (x_i - x_j)(y_i - y_j)$.

13. Prove that the correlation of a sample is always in the range $[-1, +1]$.

14. Prove that the matrix of partial derivatives in OLS regression is positive definite

8.6.0.1 References

[CR12] Chattamvelli, R. (2012). Statistical Algorithms, Narosa publishing house, New Delhi.

[GF88] Galton, F. (1888). Co-relations and their measurement, chiefly from anthropometric data, *Proceedings of Royal society*, London, 45, 135-145.

[HD80] Hawkins, D. (1980). *Identification of outliers*, Chapman and Hall.

[JA01] Joarder, A.H. (2001). *On some representations of sample variance*, Technical report series, TR269, King Fahd university of petroleum and minerals, Dhahran, Saudi Arabia (www.kfupm.edu.sa/math/TechReports_DATA/269.pdf).

[JM04] Jones, M.C. (2004). On some expressions for variance, covariance, skewness and L-moments, *Journal of statistical planning and inference*, 126, 97-106.

[NK04] Neter,J., Kutner,M., Nachtsheim,C.J., and Wasserman, W.(2004). Applied linear regression models, 4th ed.,Richard D. Irwin Inc.

[RS87] Ross, S.M. (1987), *Introduction to probability and statistics for engineers and scientists*, Wiley, NY.

[SC15] Shanmugam, R., Chattamvelli, R. (2015). *Statistics for Scientists and Engineers*, John Wiley, New York.

[YG98] Yatracos, Y.G. (1998). Variance and clustering, *Proc. of American Math. society*, 126, 1177-1179.

[YK65] Yule, G.U., Kendall, M.G. (1965). *An introduction to the theory of statistics*, Charles Griffin Co., London.

9
Cluster Analysis

Chapter objectives

- Understand clustering and cluster analysis

- Distinguish spatial and temporal clustering

- Discuss dissimilarity metrics (Manhattan, Minkowski, Mahalanobis, Chebychev metrics)

- Taxonomy of clustering algorithms

- Describe hierarchical and partitioning algorithms

- Briefly review other clustering algorithms

- Introduce cluster validity measures

- Applications of cluster analysis

9.1 Meaning of Clustering

Cluster analysis started during 1960's as a statistical technique to group large datasets into smaller subsets called clusters. An automatic clustering method or algorithm is used to extract the possible clusters from *unlabeled data*. Input to this algorithm is the entire dataset, and the output is the clusters found, if any. Many early clustering algorithms were developed by statisticians to summarise voluminous data or reduce data dimensionality. An example is the k-means algorithm that arbitrarily partitions the entire dataset into k (≥ 2) groups, and uses the arithmetic means of each group to iteratively re-assign elements to the nearest cluster. Unfortunately, some of the statistical algorithms are not scalable to very large and multi-dimensional data. This has rekindled an interest among computer scientists to invent scalable clustering algorithms. Clustering is useful in data mining that typically have millions of data records, and thousands of variables [ZJ82]. Data miners use clustering for dimensionality reduction, cluster sampling, and to look for patterns and trends in focused sub-dimensions. These sub-dimensions are the clusters, if any, found in the original data. In addition to the classical clustering methods, newer algorithms based upon SVM (chapter 11) and LSI (chapter 12) have recently became popular for extended data types (text, audio, video, etc). Clustering techniques are also being applied to spatial and temporal data to locate proximity points with respect to a frame of reference. For instance, cellular operators use spatial clustering to identify customer clusters with special characteristics like movie downloads, bulk MMS services, etc. This information can then be used to provide better services to those persons through enhanced bandwidth, expanded signal coverage, hardware and software upgrades, etc. As these could depend upon time, such information are extracted on a periodic basis.

Definition 9.1 A cluster is a logical group or collection of objects, attributes or properties of entities such that the elements within each collection are more alike than elements in different collections.

This definition has an implicit assumption that distinct groups or collections do exist in the data. Although rare, there are situations where no proper clustering is possible by some of the algorithms. Nevertheless, some of the recent clustering algorithms like density-based methods, fuzzy clustering methods etc can still extract some meaningful clusters where other algorithms fail. Most clustering algorithms produce more or less identical clusters when data do contain distinct clusters.

As clustering data on a single continuous variable is too trivial, we assume that there are two or more variables, some of which may be categorical (variables are quite often quantitative in statistical clustering applications, but there could be a mix of quantitative and categorical data in most data mining applications). The clusters, if any, are easy to visualise using a scatterplot when there are just two-variables. Multi-dimensional visualisation techniques may not always reveal distinct clusters when attributes are highly correlated.

9.1.1   Geometric Interpretation

Geometrically, data clusters are groups of data points in Euclidean space $\mathbb{R}^n$ such that the data within clusters are more similar (closer) among themselves, and more dissimilar (spread-out) from elements in other clusters with respect to a *distance* measure. Clustering algorithms use a dissimilarity measure computed from the data to decide how close (or dissimilar) the data points are. Distance-based measures are the most common dissimilarity measures for quantitative data (see below).

A change of origin and scale transformation does not alter the cluster distributions. As a change of origin transformation simply translates all data points by a fixed amount (or equivalently, it moves the coordinate axes by a fixed amount), it is useful when the data values are large. Similarly, a change of scale transformation simply compresses or enlarges the data scatter without affecting the geometry of cluster distributions. These are especially useful in hand-computations, and in density-based algorithms.

Sometimes distinct clusters may not be present in the data, in which case all data are grouped into a single cluster. Humans have a tendency to demarcate cluster boundaries using a curved surface or region. As computer algorithms use a single dissimilarity metric, the demarcating boundary very much depends upon the chosen metric (see below). The boundaries may also overlap with sparse data points. When at least two clusters are present, it has a geometric interpretation in terms of distances between the centroids of possible clusters in the n-dimensional space. This is the basis for geometrical shape based clustering techniques. When all variables are categorical, the distance metrics (see §9.3 in page 9-5) based techniques, and geometrical shape based techniques are inapplicable. Agglomerative categorical clustering with entropy criterion (ACE), entropy based clustering technique for categorical data (BkPlot) etc are applicable in such situations.

9.2 Cluster Display

The results of a cluster analysis can be presented in graphical form (2D and 3D scatterplots, bubble charts etc), as a dendrogram or in multi-way tables. Result of a hierarchical algorithm is presented as a dendrogram. Results of temporal clusters are displayed using one or more time-axes. Similarly spatial clusters are better displayed using an overlay chart (like the human body, interior of crafts, etc).

Definition 9.2 The dendrogram is a graphical (tree) representation of the order in which clusters were agglomerated at different steps of the algorithm, with the root representing the entire data collection.

Clustering has its origin in biological taxonomy, where it is used to represent species hierarchies. It partitions a data set X into nested hierarchies $H = H_1, H_2, \cdots, H_m$ such that $C_i \in H_k$ and $C_j \in H_l$ implies that $C_i \subseteq C_j$ if $i > j$ or $C_j \subseteq C_i$ if $i < j$ or $C_i \cap C_j = \phi$. In some applications, a partition of the data into hierarchies is more appropriate because the analyst can cut the dendrogram at suitable levels to obtain clusters of different granularities. Usually, dendrograms are drawn vertically. In this case the X-axis represents the data points, and the Y-axis is calibrated with the computed distances between the clusters represented by arms (division) of the dendrogram, so that the Y-axis calibrations occur at the split points only. A dendrogram can be stored in a compact form in computer memory, or on external storage devices. If the data gets updated, the clustering algorithm can be speeded up using the information present in the saved dendrogram.

A scatter-plot (chapter 2) can easily visualise clusters in low dimensional data. This method is inappropriate in data mining applications with hundreds of variables. This representation has the disadvantage that it is dependent upon different user's perception of geometric structures that demarcate the clusters. In addition, when there are thousands of data items, the boundaries that demarcate the clusters may get blurred. A solution (when the number of variables are small) is to use multiple clustering runs using much smaller random samples. For instance, if there are 4000 data items, we could draw 100 to 200 random samples, run the clustering algorithm, and repeat the procedure an appropriate number of times until the cluster distributions are clear.

Example 9.1 Text-based Internet search engines use keyword(s) submitted by users to retrieve all matching records from the web. The number of matches (matching records found) could be of the order of millions for common keywords. For instance, a search on 'monitor' will bring up records pertaining to computer monitor, medical device monitors, control station monitors, industrial monitors, as well as monitoring of various jobs, processes and activities. Browsing time can be tremendously reduced if the matches are automatically grouped using a clustering algorithm. If a user opts to view records for control station monitors, the browser can discard all other contexts (groups), and simply present relevant records in the focused group. Sometimes, sub-clustering can narrow down the search space further. As an example, diabetics and non-diabetics differ slightly in some physical and medical aspects. Separately clustering them can reveal patterns and trends exclusively among the groups.

9.2.0.1　Cluster Formation

Clustering variables may be spatial, temporal, or both. In temporal clustering, the primary variable is temporal (and others can all be spatial). In spatial clustering, we look for groups of data points spread across a geographical region (like cities, countries). For example, a 'star cluster' is a patch of the sky where the stars appear to be close together wrt the observer's viewing angle. Data for spatial clustering are usually not 'time-stamped'. It could have been collected over an unspecified period of time. Consider spatial clusters used to locate earthquake-prone regions anywhere on the earth (including sea floor). This data might have been collected over a period of decades, or even centuries. Similarly, disease prevalence in geographical regions, computer hard disk errors due to sector damages on the hard-disk platter, family income distribution in a country, market segmentation of customers etc are identified by spatial clustering [DL03]. Data are collected over a long period of time in all these examples. A voter cluster denotes geographic regions in which a candidate has favorable votes or majority of votes over their opponents, using data from prior elections (which may be collected over decades). In some of the above examples, the time may be important in a reverse chronological order. For the voter clusters, the recent results convey more information than those in the distant past. Similarly, disease prevalence and income distribution in a country can vary over time, so that decade-old data convey little meaning than recent ones. Spatial clusters over geographical regions are best displayed using an overlay chart or spatial histograms (chapter 2, §2.2).

Data for temporal clustering are collected over a time period as in the following example, and are hence time stamped[1]. Temporal clusters may contain spatial data, but time is the most important variable. Spatial variables simply revolve around the temporal axis, and reveal sub-clusters. See [NH94], [SC98] for spatial clustering in data mining.

Example 9.2 Consider customer orders at an e-commerce site. The number of orders, total order amounts, order cancellations, order returns etc will exhibit seasonal and non-seasonal variations. In addition, special incentives, discounts or promotional offers can result in increased sales among some customer categories. These can be obtained using a temporal clustering algorithm. Other examples are detecting unusual stock transactions using a time stamped log-file, intruder detections in computer networks, money laundering etc.

9.2.1 Cluster Analysis Step-by-Step

There are three distinct steps in cluster analysis – (i) data preparation, (ii) clustering algorithm execution, and (iii) result interpretation.

9.2.1.1 Data Preparation

Data preparation starts by identifying, cleansing and formatting the data. Data standardisation may also be done at this stage, as change of origin and scale does not alter the cluster distributions. This step may be helpful if some variables (like income, stock purchase amount, bank loan amount etc) are measured on a large scale. These variables can have a large influence on the computed similarity measures. Data may also be filtered to remove unwanted records. For instance, in a study to segment purchasing behavior of senior citizens, we may filter out all customers below the age of 60. Other examples are identifying customers who default on large loan amounts (in which small loan amounts are filtered out), clustering frequent flyers (low income passengers are filtered out). Most clustering algorithms accept data in comma or tab separated format. Note that the input format will depend on the data type.

Input to some of the clustering algorithms is either a data matrix, or similarity/dissimilarity matrix. In the case of image and multimedia clustering, a data matrix is derived from the

---

[1]time is considered a variable and is measured wrt a fixed epoch (days, week, month, year,decade)

available input. In the following discussion, X denotes a data matrix on 3 attributes for 5 subjects, and D denotes the proximity matrix:

$$X=\begin{pmatrix} x_{11} & x_{12} & x_{13} \\ x_{21} & x_{22} & x_{23} \\ x_{31} & x_{32} & x_{33} \\ x_{41} & x_{42} & x_{43} \\ x_{51} & x_{52} & x_{53} \end{pmatrix} \quad D=\begin{pmatrix} d_{11} & d_{12} & d_{13} & d_{14} & d_{15} \\ d_{12} & d_{22} & d_{23} & d_{24} & d_{25} \\ d_{13} & d_{23} & d_{33} & d_{34} & d_{35} \\ d_{14} & d_{24} & d_{34} & d_{44} & d_{45} \\ d_{15} & d_{25} & d_{35} & d_{45} & d_{55} \end{pmatrix}$$

where $(i, j)^{th}$ entry in D denotes how similar (or dissimilar) the corresponding subjects are wrt the chosen metric.

9.2.1.2 Clustering Algorithm Execution

There are many algorithms for data clustering. Running an algorithm on the chosen data is called an execution of the algorithm. Clusters present in the data can be found by any of the algorithms. Some algorithms have better performance than others in certain domains, and for certain shapes of the cluster boundaries. Grid-based algorithms are the preferred choice in image clustering, and density-based algorithms are preferred in spatial clustering [SM81]. Users may decide upon the similarity or dissimilarity metric (see below) to be used in the algorithm. Choice of an algorithm depends upon the speed (execution time), accuracy, and the quality of final results [LM00], prior knowledge about the demarcation boundaries (if any) and outliers, personal preferences, and software availability [MB05]. Some algorithms require the user to input one or more parameters. For example, the k-means family of algorithms (k-median, k-mode, c-means, etc) require a positive integer k$\geq$ 2 to be input.

9.2.1.3 Result Interpretation

The results of a clustering algorithm can be interpreted in various ways. If distinct clusters are detected, summary measures can be obtained for each of them. In the case of hierarchical clustering algorithm that outputs a dendrogram, clusters at various levels can reveal hidden structure in the data. Numeric measures obtained from the extracted clusters can be used to validate research hypotheses. Text and web clustering algorithms look for documents or web pages with content similarity.

9.3   Dissimilarity Metrics

Hierarchical clustering algorithms use a similarity or dissimilarity metric to segment the data. Using available data, most of the algorithms first constructs a dissimilarity matrix in which $(i,j)^{th}$ entry (D_{ij}) measures the dissimilarity between objects i and j. As this is a symmetric matrix, only the upper (or lower) diagonal portion need be stored in computer's memory. Attributes must preferably be standardised prior to computing the dissimilarity matrix. As most similarity and dissimilarity metrics work on quantitative data and return real numbers, the above matrix must be declared as real or double type. A metric defined on numeric data should satisfy the following conditions.

- $D_{ij} \geq 0$ (non-negativity condition)

- $D_{ii} = 0,$ and $D_{ij}{=}0$ iff i=j (Identity condition. This requirement is for dissimilarity metric. For similarity metrics $D_{ii} \geq \max_k d_{ik}$ for all i.)

- $D_{ij} = D_{ji}$ for all i,j (Symmetry). This condition ensures that the order in which data items are considered by an algorithm is unimportant.

- $D_{ij} \leq (D_{ik} + D_{kj})$ (Transitivity or triangle-inequality). This condition has the interpretation that the least similarity between i^{th} and j^{th} data pairs is always less than or equal to the sum of the similarities between $(i, k)^{th}$ and $(k, j)^{th}$ pairs.

Many metrics are available that satisfy the above criteria [RS98]. Most metrics produce more or less identical clusters when data are uncorrelated. Some metrics perform better in the presence of outliers. We describe the popular dissimilarity metrics in the following section, and discuss their advantages, and inter-relationships for particular parameter values. Note that all distance metrics are dissimilarity metrics because a smaller distance between clusters means that they are more similar.

9.3.1 Euclidean Distance Metric (L_2 Metric)

In the following discussion, x_i and x_j are assumed to be numeric data vectors of the same size in $\mathbb{R}^d$, and T denotes the transpose. Let $D_{ij}^2 = (x_i - x_j)^T(x_i - x_j) = \sum_{k=1}^d (x_{ik} - x_{jk})^2$, where x_{ik} is the kth attribute value of ith object be the squared Euclidean metric. This metric is so called because D_{ij}^2 is the squared Euclidean distance between x_i and x_j in $\mathbb{R}^d$. This is a dissimilarity metric because small values indicate that the entities or objects considered are similar and large values indicate that they are dissimilar.

Example 9.3 Let two data vectors be X_1=(7, 3, 5) and X_2=(3, 2, 6), the squared Euclidean distance is $(7 - 3)^2 + (3 - 2)^2 + (5 - 6)^2 = 16 + 1 + 1 = 18$, and the Euclidean distance= $+\sqrt{18}$=4.2426

An advantage of this metric is its ease of computation. It fully utilises all data values. If the statistical distribution of X is known, the distribution of Euclidean metric also is known, or can be derived. A disadvantage of this metric is that a large number of outliers, and few distinct outliers can misrepresent the computed value. The squared Euclidean metric makes the implicit assumption that the cluster boundaries are hyper-spherical.

9.3.2 Manhattan Metric (L_1 Metric)

It is a dissimilarity ratio-measure defined as $D_{ij} = \sum_{k=1}^d |x_{ik} - x_{jk}|$, where the vertical bars (|) denote absolute value of a number. Another name for it is 'taxicab metric' (or city-block metric) because it represents the distance between two points (in two dimensions) as a sum of connected path distances. An advantage of this metric is that it does not require the square-root calculation as in the case of Euclidean distance metric and Minkowski metric described below.

Example 9.4 Let X_1=(7, 3, 5) and X_2=(3, 2, 6) be 2 data vectors. The Manhattan metric gives the distance as $|7 - 3| + |3 - 2| + |5 - 6|$=4 +1+1=6.

Extreme observations do not cause a severe problem for Manhattan metric, as in the case of Euclidean metric. As the Euclidean metric geometrically represents the shortest distance between two points, $L_1 > L_2$ always.

9.3.3 Minkowski Metric

It is defined as $D_{ij}^p = ||x_i - x_j||_p = \left(\sum_{k=1}^{d} |(x_{ik} - x_{jk})|^p\right)^{1/p}$, x,y $\in R^d$, p$\in$ R., where d is the dimensionality of data, and p is a user chosen parameter, which is usually a small number. This becomes Manhattan metric for p=1. When p=2, we get the Euclidean metric because $|(x_{ik} - x_{jk})|^2 = (x_{ik} - x_{jk})^2$.

An advantage of this metric is that it is more flexible because the parameter p can take integer or non-integer values. When p is a fraction (need not be less than 1), the resulting metric is called fractional Minkowski metric (FMM). The FMM is used in robust regression estimation, and robust image matching and retrieval. This metric is more difficult to compute for fractional p values. If the distribution of parent data is known, it is not easy to derive the distribution of this metric (as in the case of Euclidean, Mahalanobis and Chebychev's metrics), except in some particular cases. Nevertheless, it is often used in many clustering algorithms, and cluster validation indices. For example, the Davies-Bouldin index (DB Index) uses this metric. See [HR05] for an application of fractional distance measures for content-based image retrieval.

Example 9.5 Let X_1=(7, 3, 5) and X_2=(3, 2, 6) be 2 data vectors. Compute the Minkowski metric for p=.5. Using the above formula we get the Minkowski distance as $[|7 - 3|^{.5} + |3 - 2|^{.5} + |5 - 6|^{.5}]^2$=(2+1+1)^2=16.

9.3.4 Mahalanobis' Distance Metric

One major weakness of above metrics is that any correlation among the features can distort the distance measure. The Mahalanobis' squared distance metric considers correlation among the variables, and is defined as: $D^2(x_i, x_j) = (x_i - x_j)'S^{-1}(x_i - x_j)$, where 'S' is the pooled sample variance-covariance matrix. This metric reduces to the (weighted) Euclidean metric when the variables are independent (S^{-1} is diagonal).

Inverting the sample variance-covariance matrix can be time consuming when the data size is large, and variables are correlated ([VY88], [HW89]). In addition, as $(x_i - x_j)'A(x_i - x_j) = c$ defines a hyper-ellipsoid in the Euclidean space $\mathbb{R}^n$, this metric makes the implicit assumption that the cluster boundaries are hyper-ellipsoidal.

9.3.5 Chebychev Metric (L_∞ Metric))

It is defined as $D_{ij} = \max_{k=1,2,...,d}(|x_{ik} - x_{jk}|)$. The difference between this metric and Manhattan metric is that instead of summing the absolute differences of the components, we simply take the maximum of the absolute differences. An outlier observation in one of the components can obviously influence the computed value. Hence $L_\infty \leq L_1$. Both of these metrics can be computed easily in one pass through the data (with (n-1) extra comparisons needed to update the maximum in the sequential method).

Example 9.6 Let two data vectors be X_1=(7, 3, 5) and X_2=(3, 2, 6), the Chebychev metric gives the distance as $\max(|7 - 3|, |3 - 2|, |5 - 6|) = 4$.

It is easy to update this metric when new data values are added. Missing observations can sometimes have a large impact on the computed measure.

9.3.6   Other Metrics

The Canberra metric is defined as $D_C(x, y, q) = \sum_{k=1}^{p} [|x_{ik} - y_{ik}|/(|x_{ik}| + |y_{ik}|)]^q$. Because the numerator of the ratio is always $\leq$ denominator, the ratio is always ≤ 1, and the sum is always bounded and small. For q=1, we get standard Canberra metric (A related metric is the Bray-Curtis coefficient $\sum_k |x_{ik} - y_{ik}|/\sum_k (x_{ik} + y_{ik})$). Fractional Canberra metric results when q$\neq$ 1. If q> 1, $[|x_{ik} - y_{ik}|/(|x_{ik}| + |y_{ik}|)]^q$ is less than $|x_{ik} - y_{ik}|/(|x_{ik}| + |y_{ik}|)$ (the opposite relationship holds for $0 < q < 1$). Hence the chosen value of q must be small (and preferably close to 1). It can also be generalised in the form of Minkowski metric as $[\sum_{k=1}^{p} (|x_{ik} - y_{ik}|/(|x_{ik}| + |y_{ik}|))^q]^{1/q}$. But as the ratios are all smaller than 1, this is of little use in clustering. Other choices are the (squared) Hellinger metric $d(x, y) = \sum_{k=1}^{p} (\sqrt{x_{ik}} - \sqrt{x_{jk}})^2$, and spherical versions of Euclidean and Manhattan metrics. Examples of similarity metrics include the cosine metric, correlation coefficients (chapter 3) etc. Because the correlation (and cosine metric) lies in the interval [-1,+1], it is either squared, or one of the following transformations is used:

i) d(x,y) = $(1 - r_{x,y})/2$

As $r_{x,y}$ has range in [-1,1], this transformation maps highly positively correlated values to numbers close to 0, highly negatively correlated values to number near 1, and nearly uncorrelated values to .5

ii) d(x,y)=1-$|r|$

This maps both positively and negatively correlated values to numbers close to zero. Thus it is a measure of dissimilarity. This metric need not satisfy the triangle inequality, and hence is called semi-metric. Another possibility is to use 1-r^2 as a metric. See [SJ99] for further discussion of similarity measures.

9.4   Clustering Algorithms

A clustering algorithm (also known as segmentation algorithm) is used to extract distinct clusters, if any, in input data. From the data miner's viewpoint, the actual clustering algorithm being used is often unimportant. All algorithms will extract more or less the same clusters when distinct boundaries exist in the data. But some of the algorithms could return spurious and unreliable clusters when distinct cluster boundaries are absent. This is more so in 2D and 3D image segmentations, when the dimensionality is large, or when there are too many categorical variables. Separate categorical clustering algorithms, and extensions of the k-means algorithm are available in the last case ([HZ98], [GR99], [CL05]).

There are dozens of algorithms available for data clustering ([JM99], [ZW05]). Most of them originated during the 1960's and 70's in Statistics. Some of these algorithms make multiple passes (multi-pass) through the data, and could perform poorly in large data mining applications that have millions of data records. A clustering algorithm is called 'scalable' if its running time grows polynomially (linearly as a special case) to the size of the data (if running time grows exponentially, it is non-scalable). The hierarchical algorithms induce a tree structure in the data, and have some interesting advantages as described below.

9.4.1   Hierarchical Clustering Algorithms (HCA)

The HCA partitions available data into sub-clusters iteratively using a dendrogram. A dendrogram represents a nested grouping of clusters from coarser to finer details, and is obtained as a bye-product of clustering. The root node of a dendrogram represents the entire data to be clustered. The internal nodes of a dendrogram are the clusters, and arcs represent the 'join' or 'split' of the nodes (depending upon its orientation). An advantage of HCA is that it will produce a dendrogram even if distinct clusters are nonexistent. A disadvantage of HCA is that once a subject has been assigned to a cluster, it can't be reassigned to another cluster. There are two popular strategies for HCA, called agglomerative (bottom-up) and divisive (top-down). CHAMELEON is a popular hierarchical clustering algorithm [KH99].

9.4.1.1 Agglomerative Algorithm

The literal meaning of agglomerate is to clump things together. We start with n different clusters, and assume that each data point is assigned to its own cluster. We then iteratively combine alike sub-clusters to produce bigger clusters. Merging of most similar sub-clusters is continued until either further merges become meaningless, or the desired number of clusters is extracted. The merging step should identify those sub-clusters that are most similar. This is found using one of the methods described below:

1. Single linkage (nearest neighbor)
   Two intermediate sub-clusters are considered alike if their *nearest* neighbors are close together. In other words, the minimum of the minimum-distances from a cluster to all other clusters is used. This is found using one of the distance metrics discussed in §9.3 (page 9-5).

2. Complete linkage (farthest neighbor)
   Two intermediate clusters are considered alike if their *farthest* neighbors are close together. In other words, we take the minimum of the *maximum distances* between pairs of data points in each sub-cluster.

3. Average linkage
   As the name implies, this method uses the minimum distance between means (averages) of the sub-cluster pairs as the merging criteria. This method is also known as the centroid method.

4. Ward's method
   This method uses the minimum sum of squared deviations of the points from their respective cluster centroids as the merging criteria.

Example 9.7 Table 9.1 gives the average insurance premium amount vs family income (in '000) of 25 insurance customers. Apply the hierarchical algorithm to cluster the data.
Solution: A scatterplot of the data (see page 9-13) reveals 3 clear clusters, in which one is a singleton cluster (for data (20,415) with serial number 22), and the other has 7 sample points. We will omit the calculation details here, and use one of the software packages mentioned in page 9-18. The dendrogram obtained by a hierarchical algorithm is given in figure 9.1 (with root at the top). This dendrogram can be cut at a desired level to get various clusters. Cutting it at top-level gives two clusters C1={22}, C2={All other data points}. By cutting it at level 2 gives three clusters C1={22}, C2={8,11}, C3={All other data points}. Cutting at level 3, and merging alike clusters gives C1={22}, C2={3,5,8,9,11,20,24}, C3={All other data points}. Finer level clusters can be obtained by cutting the dendrogram at subsequent levels.

Table 9.1: Family income and average insurance premium amounts data for 25 customers

ser. num.	inc- ome	amo- unt	ser. num.	inc- ome	amo- unt	ser. num.	inc- ome	amo- unt
1	31	250	10	38	310	18	38	98
2	25	187	11	63	513	19	45	212
3	57	354	12	23	86	20	50	430
4	40	300	13	29	112	21	48	290
5	52	475	14	35	170	22	20	415
6	27	208	15	40	260	23	51	287
7	33	262	16	36	225	24	62	413
8	72	470	17	27	170	25	50	255
9	56	390						

9.4.1.2 Divisive Algorithm

Whereas the agglomerative algorithm starts by assigning each data to its own cluster, the divisive algorithm starts in the reverse. Initially we assume that all data points belong to a single cluster. At each subsequent iteration, the data are appropriately partitioned into sub-clusters using a dissimilarity metric. This process is continued until further splits become meaningless.

This algorithm can be speeded up if approximate cluster boundaries are known (say from prior runs of the algorithm). It is preferred when the data reside entirely on external storage.

If the user was looking for 'process monitor', the search engine could discard all other categories, and find sub-clusters in the user chosen category. This is easy to find using a data driven hierarchical segmentation algorithm. If the user later decides to traverse up the hierarchy, and amend 'program monitor' to 'process monitor' to find combined matches, it is a simple matter of merging the results of 'program monitor' category in the dendrogram, and combining both results (this will work only if software implements the interactive mode. Otherwise a batch mode execution of 'program monitor', followed by a merging is required). Further discussions can be found in [MS80], [BG90], [MB04] and [KL04].

Example 9.8 In the example cited in page 9-3, a search engine could automatically cluster the returned records, and present major categories to the user. Assume that a search using 'monitor' as keyword returns millions of matches. A clustering algorithm finds {computer monitor, equipment monitor, process monitor, program monitor, industrial monitor} as the major categories.

9.4.2 Partitioning Algorithms

Partitioning algorithms (also called parametric clustering) use an optimisation function (similarity function) to partition the available data into successive clusters. The most popular criteria used is the squared error function, which minimises the sum of the squared errors (distances) of each data item in a cluster from its respective centroid[SW82]: $E^2 = \sum_{i=1}^{K} \sum_{j \in C_i} (x_j^i - c_i)^2$ where K is the number of clusters assumed, and c_i is the ith cluster's mean vector. An advantage of partitioning algorithms is that an element assigned to a cluster can be reassigned to another cluster as the algorithm progresses.

Figure 9.1: Dendrogram (single linkage) for data in table 9.1.

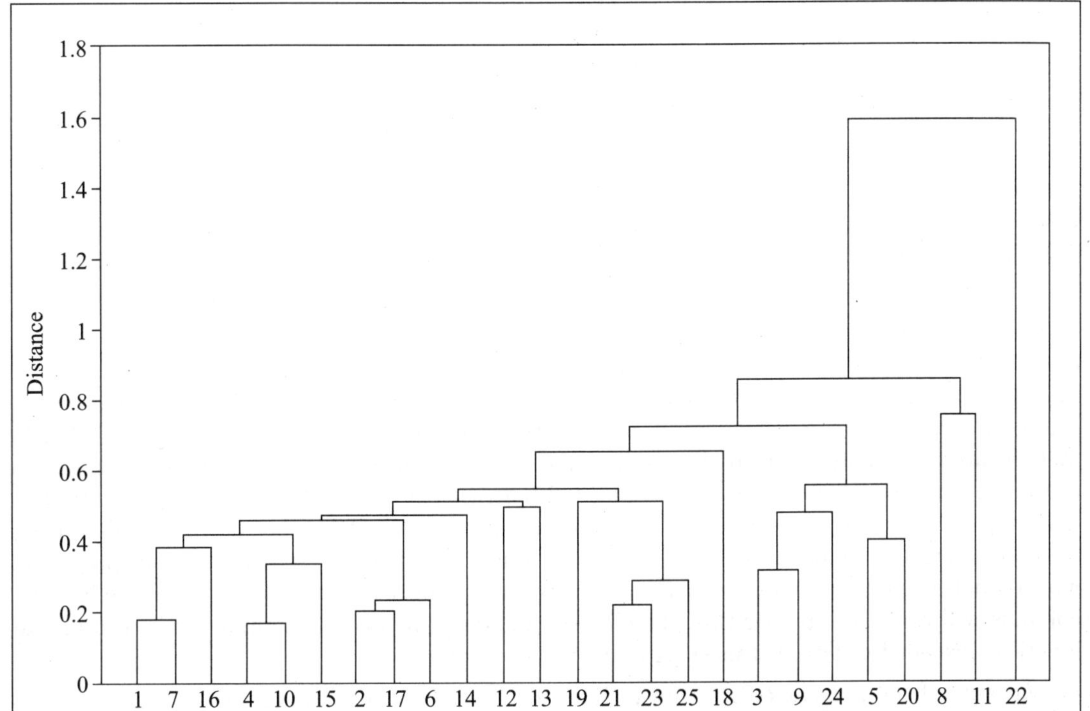

9.4.2.1   K-means Clustering Algorithm

This algorithm and most of its variants originated in Statistics [HW79]. It utilises one of the distance metrics to demarcate the clusters. A user should specify a positive integer k, that represents the best guess on the number of clusters present in the data.

We may not have an idea about the possible number of clusters for high dimensional data, and for data that are not scatter-plotted. Similarly, possible number of clusters is hidden or ambiguous in image, audio, video and multimedia clustering applications; and for text segmentation when data are scattered throughout the web. In such cases, we may try successive values of k starting with 2. This process is stopped when two consecutive k values produce *more-or-less* identical results. Another solution is to cluster analyse a small random sample of available data using one of the other algorithms to get a rough estimate of k, which is then used in the k-means algorithm.

Algorithm 3 K-means algorithm

```
step 1:
Input number of clusters k (>1), data matrix X.
step 2:
Initialise the means (centroids) of the k clusters
```

```
for each of the data instances do
   compute distance between current data instance and k cluster centroids
   assign the current instance to that cluster to which it is closest
end
Check the convergence criteria.
If at least one point has changed clusters during the last iteration
begin
 Update centroids using current partition
 repeat step 2.
end
Otherwise output the clusters
```

Example 9.9 Use k-means algorithm to cluster the data in table 9.1.
Solution A plot of the data indicate at least 3 major clusters, as shown in figure 9.9. The k-means algorithm with k=3, and HCA algorithm with Euclidean distance measure using several linkage metrics were run. Results obtained by the K-means with average linkage, and Ward's method are exactly identical in this example. Clusters found by K-means algorithm and HCA are entirely different, as can be seen in table 9.2 (entries denote the cluster to which a data item belongs). This is simply due to the opposite labeling used by HCA and K-means algorithms to data instances in clusters 1 and 2. Users have no way of telling (most) software programs, which labels should be used for which cluster. By relabeling 1 for 2 (and 2 for 1) in the last two columns will reveal the reader that the clusters found by the two algorithms are indeed almost identical (except for data instances 24 and 25).

Variants of the K-means algorithm include the K-medoids algorithm (that uses medoids instead of centroids), accelerated K- means[2] [EC03], k-harmonic means [ZB00], k-modes [CG01],X-means [PM00], fuzzy c-means ([BJ81],[OP07]), non-Euclidean c-means [KR03], and partial K-means [KM02]. Mean is influenced by outlier observations. This can be alleviated to an extent by using medoids, modes and harmonic means instead of centroids. CLARANS is a popular k-medoids algorithm [NH94]. The inflated K-means (also called adaptive K-means) algorithm starts with a higher number K' such that $(K < K' << n)$, and merges alike clusters at subsequent steps (thereby reducing K') to eventually produce exactly K clusters. For instance, if the initial guess is K=3, we may start with K=5, and reduce the number of clusters to K=3 by merging most similar clusters. See [BF88], [SC01], [HE02], [TA03], [TS06] for further details.

9.4.3   Density-based Methods

This method is more suited for data clustering of quantitative variables. It finds the clusters using local density information, under the assumption that data distribution follows a statistical law [SM81]. User can specify a minimum density and radius values to be used. Dense regions are then grouped together into a single cluster, with cluster boundaries running along low-density regions within the specified radius from the center of high-density cluster.

The DBSCAN (Density Based Spatial Clustering of Applications with Noise) [EK96] and its extension GDBSCAN [SE98] that works on arbitrary data types are density based clustering algorithms. Other approaches include grid-based methods (useful for image clustering), graph

---

[2]an efficient implementation of k-means

Figure 9.2: Scatterplot of data in table 9.1.

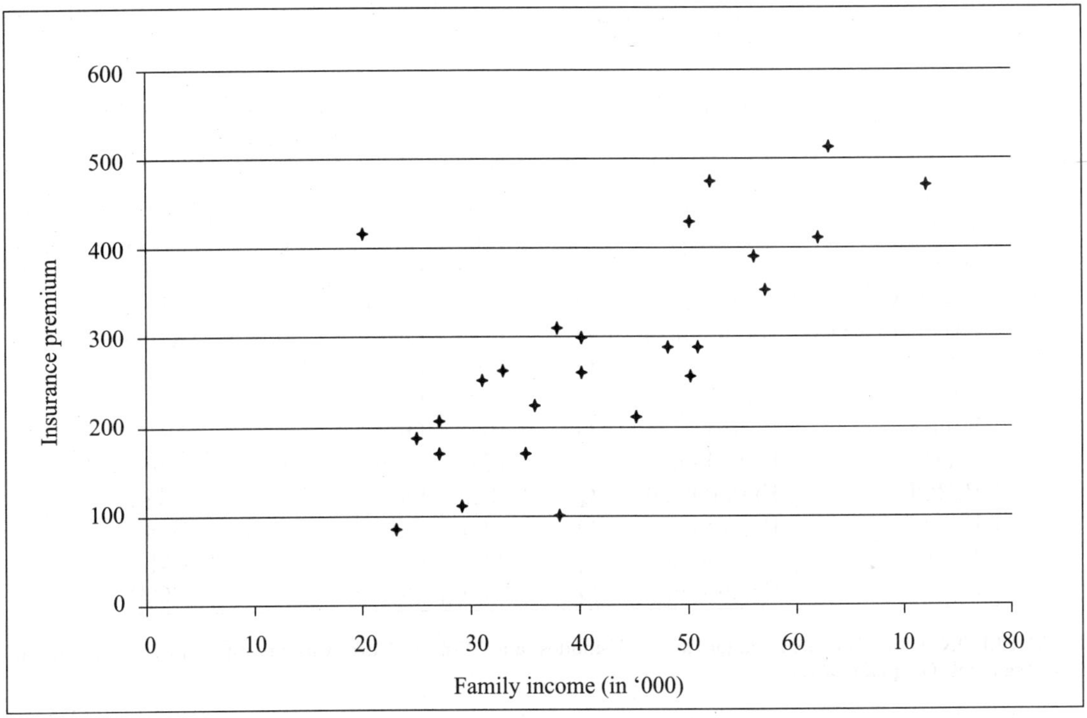

Table 9.2: Results of clustering using k-means algorithm (k=3) and Hierarchical clustering (with complete linkage, Ward's method)

Id.	Inc. (K)	prem ium	hca k=3	k-m k=3	k-m Wrd	Id.	Inc. (K)	prem ium	hca k=3	k-m k=3	k-m Wrd
1	31	250	2	1	1	14	35	170	1	2	2
2	25	187	1	2	2	15	40	260	2	1	1
3	57	354	3	3	3	16	36	225	2	1	1
4	40	300	2	1	1	17	27	170	1	2	2
5	52	475	3	3	3	18	38	98	1	2	2
6	27	208	1	2	2	19	45	212	2	1	1
7	33	262	2	1	1	20	50	430	3	3	3
8	72	470	3	3	3	21	48	290	2	1	1
9	56	390	3	3	3	22	20	415	2	1	1
10	38	310	2	1	1	23	51	287	2	1	1
11	63	513	3	3	3	24	62	413	2	3	3
12	23	86	1	2	2	25	50	255	3	1	1
13	29	112	1	2	2						

theoretic methods that embed a graph (or a minimum spanning tree) within the data, and fuzzy-clustering methods. A survey of clustering algorithms can be found in [JD88], [JM99], [KP04] and [ZW05]. Neural networks, SVM, and LSI can also be used for clustering. Specialised clustering algorithms are discussed in subsequent chapters.

Table 9.3: Comparison of Common Clustering algorithms

Algorithm	Category	Data	Year	Complexity	Refer
CURE	Partitional	Q	1998	$O(n^2 \log n)$	[GR98]
MCLUST	model-based	any	1998		[FR98]
ROCK	Hierarchical	C	1999	$O(n^2 \log n + mn^2)$	[GR99]
CLARANS	Partitional	Q	1994	$O(mn^2)$	[NH94]
CHAMELEON	Hierarchical	Q	1999	$O(n^2)$	[KH99]
MULTI	Hierarchical	Q	1999	$O(n^2 \log n)$	[KH99]
GDBSCAN	density-based	Q	1998	$O(n \log n)$	[SE98]
K-means	Partitional	Q	1957	$O(lmn)$	[LS57]
PAM	Partitional	Q	1957	$O(lm(n-m)^2)$	[MS80]
BIRCH	Hierarchical	Q	1982	$O(n)$	[ZR82]
Fuzzy C-means	Partitional	Q	1981	$O(n)$	[BJ81]
X-means	Partitional	Q	2000	$O(n^{dk+1} \log n)$	[PM00]
K-modes	Partitional	Q	2001	$O(lmn)$	[CG01]

CHAMELEON is a K-medoids algorithm. K-modes and X-means are variants of K-means algorithm. C=categorical, Q=quantitative

9.5   Cluster Validation Techniques (CVT)

A data clustering algorithm simply returns the distinct clusters present in the data. If the cluster boundaries are distinct, almost all of the algorithms will return more or less the same result. Cluster validation is a technique to quantitatively evaluate the correctness of an algorithm. Does the obtained clusters actually group data meaningfully, or is it an artifact of the algorithm? Does other parameter values result in better clustering?. Reasonable answers for all these can be found by CVI.

Definition 9.3 An automatic method to assess the validity of data clusters found using a clustering algorithm by a quantitative measure is called cluster validity index (CVI).

The CVI metrics are useful for checking departures of obtained results from an ideal clustering. There are many CVI measures for hard clustering, soft clustering and fuzzy clustering. For hard clustering, available indices include modified Hubert statistic, Davies-Bouldin index, Dunn's separation index [DJ74], and contingency coefficient [SV02]. For fuzzy clustering the popular measures are Bezdek's partition coefficient [BJ81], Gath-Geva fuzzy hyper-volume index [GG89], Duo-Xue-Duwu's fuzzy adaptive index [DX07], etc. The Hubert statistic, and its modified versions are goodness-of-fit measures for non-hierarchical cluster validation. See [BP98], [HB01], [MB02], [MB05], [KR05] for performance evaluation of validity indices.

9.6   Applications

Clustering has many data mining applications. For example, it is used to find group of customers with similar traits or patterns (spending, buying or ordering [SG00]), market segmenting, optimal pricing, etc. In web mining, it finds clusters of web documents accessed by groups of people on a regular basis, clusters of users with nearly identical buying habits etc. Clustering can reveal viewer categories, casting, and content of popular programs, etc. For example by cluster analysing viewers (adults, teenagers, kids) in various demographic groups can reveal distinct viewer characteristics. It can also be used as a modeling tool to classify unseen data instances using labels assigned to distinct clusters already found. A few of the applications are reviewed below.

9.6.1   Marketing

Market segmenting is one of the primary steps in building an effective marketing strategyThe objective is to partition the clients or objects into homogeneous subgroups using common (socio-demographic, geographic, product specific, psychographic, etc) traits. Among these, the socio-

Table 9.4: Data for customer segmentation (left) and segmentation obtained using hierarchical clustering (right)

Driver Age	Years of driving	Inc-ome	Insu amnt	Current car age	Car value (in '000)	Row Id.	Cluster Id.
23	3	30	125	5	4	1	1
64	9	70	90	6	5	2	2
35	12	42	60	1	10	3	3
30	8	38	75	2	11	4	3
28	9	26	100	1	9	5	1
21	1	20	115	5	6	6	1
70	13	50	80	12	2	7	2
53	29	46	78	7	4	8	2
27	6	31	95	2	10	9	1
22	2	20	105	1	12	10	1
49	15	52	65	5	5	11	2
50	7	61	80	8	3	12	2
34	8	43	60	4	5	13	3
21	1	22	85	2	9	14	1

demographic segmentation based upon demographic variables like gender, race, income brackets, education levels, age groups, employment status, personal habits (smoking, drinking) is the most popular. Geographic segmentation utilises location info or attributes (zip codes, regional codes, country codes, telephone area codes, etc). They are categorical or quantitative and usually produce clear segmentations due to non-overlapping boundaries. The psychographic segmentation deals with hobbies, interests, lifestyles, attitudes, and other personality traits and are often categorical (with many categories). They can identify eccentric customers, and changing fashions among sub-categories (eg: college students). They may give rise to spatial and temporal clusters. The challenges faced by marketing people are how to 1) minimise total marketing expenses, 2) optimally utilise available resources by meeting the needs of customers, 3) strengthen customer relationships, 4) prioritise market opportunities, 5) effectively target prospective customers, 6) new product or service development and marketing etc. The clustering techniques can be useful in all of these activities. In targeted marketing, a group of customers with similar attributes or interests are identified from a large clientele.

Table 9.4 gives the data on a sample of customers for an automobile market segmentation application. The income and car values are expressed in thousands. Driver age and car age are truncated to next smallest integer. The result obtained using a hierarchical clustering algorithm is given in the rightmost columns. For this training data, 'years of driving' is found to be a major deciding factor, which was verified by clustering on just the first two columns.

9.6.2   Insurance

Clustering is a powerful tool in the insurance domain. All types of insurance companies (life and health insurance, fire insurance, auto insurance, equipment and farm insurance, etc) use it for customer segmentation, attribute segmentation, for offering new services and products. For instance, attribute clustering can reveal groups of customers with similar characteristics or behaviors.

The insurance claims of policyholders vary widely among different segments. An objective is to identify persons or families with high (or low) average claim amounts. Clustering techniques can identify insurance fraud groups, discontinuing groups etc. Fraud detection in insurance claims is an expensive and time-consuming operation. This is tremendously eased using clustering techniques. Prior fraudulent claims are clustered to reveal structural patterns in fraud data. This information can then be utilised in narrowing down suspicious claims for further scrutiny.

9.6.3   Medical Sciences

Both data clustering and attribute clustering have various applications in medical sciences [PS70]. Clustering techniques are used to validate known findings as well as to explore new theories. For instance, clustering a group of patients can reveal critical, fatal, serious etc symptoms for heart ailments, genetic disorders etc [MW66]. Clustering can also provide information on outbreaks of epidemics, taxonomy in biological sciences (identification of bacterial strains, etc) [GN71], identifying effective drugs. Data clustering can reveal distinct groups of patients having a common illness with various treatment plans and drug combinations. In addition, variable clustering can give common causative factors and other dependencies. This information can be used for patient specific treatment plans, predicting relapse periods, finding harmful drug combinations for different patient categories, alternate options, and possible correlations with external factors (eg: smoking, family histories, eating habits etc).

Table 9.5 gives data on 10 patients in a diabetes clinic. The systolic and diastolic BP are averages obtained from multiple measurements. The BMI, height, weight, waist and hip sizes are measured to one decimal place. Clusters found using the K-means algorithm (3 classes) are given in second column. A plot of the entire data (on 26 different subjects) indicates more than 3 clusters in sub-dimensions with many outliers present.

9.6.4   Web Mining

Web search engines become less effective in grouping the results as millions of gigabytes are added annually to the web. A primary reason is the multiple contexts in which a word (search

Table 9.5: Data for patient segmentation and clusters obtained using k-means clustering

Row Id.	Cluster	Age (yr)	Height (cm)	Wght (kg)	BMI ratio	Waist (cm)	Hip (cm)	Systo. BP	Diasto. BP
1	2	40	152.2	54	23.3	75.2	99.1	146	68.5
2	1	40	155	63	26.2	81.4	91.5	116.5	74
3	1	38	145.2	54	25.6	78.6	93.6	111	69.5
4	3	47	166.8	69.8	25.1	92.9	90.1	154.5	85.5
5	2	44	165	63.4	23.3	73.9	97.6	108	68
6	2	59	172.5	69.2	23.3	87.6	87.9	136.5	84
7	3	45	164.2	62	23	84.4	92.6	146	65
8	3	53	159	60	23.7	81.4	94.7	142	84
9	2	41	153.4	77.6	33	93.9	103.8	104	69.5
10	1	52	152.4	69	29.7	91.5	104.3	127	80.5

key) can occur. Results ranking criteria may not satisfy the user needs completely. For example, the keyword 'mining' is used in geology (coal, ore etc mining), warfare (mine laying is called mining), data mining and so on. In addition, a search engine may also return company names (MiningCo), names of research institutions (Central Mining research institute), web site names, product names, and even human names (Mr. Mining) as 'matches'. The hits returned by a search engine can be clustered into various contexts and linguistic groups using additional keywords, links, and images present in the documents. If multiple non-overlapping clusters are identified, additional information can be requested from the user to unambiguously identify the correct context and display meaningful results.

Web mining often involves millions of records. There are three popular types of web mining as discussed in chapter 10 - web usage mining, web content mining and web structure mining. Clustering techniques can be applied to group various document types accessed by different users. In addition to clustering of visitor demographics, psychographics, etc, innovative techniques are being developed to mine characteristics of visitor's machines, user domains and cultures, browsers, operating systems, firewall attributes etc at client's end. These in turn provide information on the visitor's culture, languages, age-groups and socio-economic backgrounds. This may be used to offer client-specific services and online advertisements. Some servers maintain a time-stamped web-log of user activities, which may be cluster analysed to discover groups of similar access patterns, (URL) jumps, file requests etc. An example of clustering to web usage mining is discussed in [AL03]

Some of the clustering algorithms (especially in web mining) create clusters on the fly. To cope with the CPU load, shorter snippets of documents (eg: web pages or metatags) are analysed rather than entire documents. Scalable and efficient algorithms are the most appropriate in these situations.

9.6.5   Aviation

The black boxes (digital flight data recorders) are rugged storage devices that record cockpit activities in commercial jets. Since its inception in the 1970's, they have proven to be of immense value in aviation safety. They can identify crash causes, routinely analyse flight exceedances[3], record abnormal conditions. A flight signature is a similarity measure derived from recorded parameters for certain stages of a flight (eg: take-off or landing). An 'atypicality' score is then computed by comparing selected flight data with similar data for same stage of other (normal) flights. Software then cluster analyses flights data using flight signatures derived from recorded parameters. Other applications include identifying frequent flyers, profitable routes, cost overruns, ticket cancellation patterns, etc.

9.6.6 Miscellaneous Applications

Image clustering is a popular technique to cluster various types of images. Web pages may also be clustered using various types of images linked from it. Each pixel value is the unit to be used in image clustering applications. We divide a large image uniformly into sub-frames, and use the distance of each pixel from the centroid of each frame. These distances are averaged to get a 'score' for each frame. An iterative displacement method can be used to search for improvement in the cost function of current frame position.

Flat surface shaped computer storage devices (hard disks, CDs etc) develop sector damage over a long period of time. For example, most of the *copper platter*-based hard disk surfaces develop segment faults and data errors after a few years of use. These damaged sectors are randomly distributed over the disk surface. A disk repair program can sometimes fix these errors by reformatting those blocks in which the damaged sectors are present. Repairing (by fixing errors) each and every sector is tedious. The data points (damaged sectors) may be represented using either polar coordinates or by fixed imaginary axes through the spin point (in which case the data are located in all quadrants for each surface). A fast and efficient technique can be developed using cluster analysis, which finds clusters of maximum size (influential clusters) that have special shapes and orientations with reference to the spin point.

Clustering techniques have also been applied in many other disciplines. The root-cause analysis of failures of complex machinery, processes or complex softwares like operating systems utilise the clustering algorithms. For example, software crashes of Internet services, windows applications that utilise multiple shared libraries, DLL's, etc can be effectively carried out by automatic clustering ([GZ05], [OG03]). Attribute clustering find applications in sports and games in finding winning strategies, in grouping 'activities' (cultural, sports, etc), in establishing scoring profiles (accurate long throw, missed short throw, etc in basket ball). Clustering can also be used to classify businesses using technologies and skill sets of their employees, classifying service sector enterprises using quality of service, customer satisfiability, service offerings, hand-written character recognition [KS99] etc.

9.7 Software for Clustering

There are many clustering software programs that come bundled with statistical packages. Here we provide some of the free software packages for academic use. The site has a cluster utility for HCA. XLMINER is another excellent software for hierarchical and partitional clustering. Model-based clustering [BR93] programs are available at www.stat.washington.edu/raftery/Researc h/Mclust/mclust.html,

---

[3]an exceedance occurs when one or more parameters of an aircraft are beyond the allowed limits in its current state. It may be caused by pilot's prior actions, equipment malfunctions, or external conditions like wind, snow etc.

Table 9.6: Software for Cluster Analysis

URL	Name
www-users.cs.umn.edu/~karypis/software/index.html	CLUTO
bonsai.hgc.jp/~mdehoon/software/cluster/software.htm	package
faculty.fuqua.duke.edu/~kamakura/Downloads/	k-means
www.cs.umd.edu/~mount/Projects/KMeans/	k-means
rana.lbl.gov/eisen/?page_id=7	k-means
lib.stat.cmu.edu/	MULTI and MULTIV
www.stat.washington.edu/fraley/mclust	MCLUST
www.cs.utk.edu/~icat	ICAT
rana.lbl.gov/FuzzyK/software.html	Fuzzy
www.fuzzy-clustering.de	Fuzzy
pub/ai/stolcke/software/cluster-2.9.tar.Z	HCA
www.resample.com	XLMiner
cs.waikato.ac.nz/ml/Weka	Weka

ftp://ftp.math.waikato.ac.nz/pub/maj/ [FR98]. The site www.clustan.com has clustering programs for Windows and Unix.

9.8 Exercises

1. Mark as true or false:
   a) Most clustering algorithms make a single pass through the data
   b) K-means clustering works best for categorical data
   c) Extreme observations cause a severe problem for Manhattan metric
   d) If the variances of attributes are small, the Manhattan metric results in a small value than Euclidean metric
   e) Minkowski metric reduces to Hamming distance when the variables are binary
   f) The agglomerative algorithm starts by assigning the entire data to a single cluster
   g) Distance between intermediate clusters formed are found by linkage metrics
   h) Data instances near the leaves of a dendrogram more homogeneous than those near the root

2. If x_1=[15,20,9,14], x_2=[14,18,7,10] find the Euclidean and Manhattan distance between them.

3. Given the following pair of data values on 8 subjects, find the similarity using (i) Euclidean metric (ii) Manhattan metric (iii) L_{max} norm.
   X1: (8, 2), X2: (1, 5), X3: (9, 3), X4: (2, 7), X5: (0, 4), X6: (8, 6), X7: (4, 9), X8: (3, 5).

4. Discuss any three variants of the k-means algorithm, and state their advantages over the classic k-means algorithm.

5. Which of the following linkage methods use the minimum sum of squared deviations of data items from their respective cluster centroids as the merging criteria?

Table 9.7: Monthly income and total minutes online per week

ser. num.	inc- ome	min- utes	ser. num.	inc- ome	min- utes	ser. num.	inc- ome	min- utes
1	31	93	5	38	25	9	38	98
2	25	56	6	63	130	10	45	70
3	57	114	7	23	55	11	90	92
4	40	39	8	29	19	12	48	61

A) single linkage B) Ward's method C) Complete linkage D) Average linkage E) None of these

6. Can the k-means algorithm result in singleton clusters? Empty clusters? If so, under what conditions?

7. Discuss how data outliers influence the cluster formation in each of the following algorithms:– (i) k-means with Euclidean norm (ii) k-means with Manhattan norm (iii) k-medians with Manhattan norm (iv) agglomerative clustering with single and complete linkage (v) inflated k-means algorithm.

8. What does a horizontal cut of a dendrogram at a nontrivial level produce?

9. How many iterations are needed if an agglomerative algorithm starts with n distinct clusters (each containing one element) and always merges just two clusters together, until a single final cluster is reached?

10. Describe any 3 cluster validation indexes.

11. What are the popular linkage metrics? How are they different from similarity metrics?

12. Explain the boundary phenomena in clustering. What geometric boundary shapes are preferred by (i) Euclidean metric (ii) Mahalanobis metric (iii) k-means algorithm?

13. Describe density-based clustering technique. What are two popular algorithms in this category? Describe an application area where it is likely to outperform the k-means algorithm.

14. Compare clustering and classification. Does the classification accuracy depend upon the choice of a clustering algorithm?

15. Distinguish between hierarchical agglomerative and divisive algorithms. When data used for clustering resides on external storage devices, which one will you prefer?

16. Which of the clustering algorithms allow the re-assignment (to another cluster) of an already assigned data item to an intermediate cluster?

17. How is the k-means algorithm terminated? What are the disadvantages of k-means algorithm?

18. What calibrations are used on the X-axis and Y-axis of a dendrogram?

19. For the data matrix given in table 9.7, compute the similarity matrix and perform a complete-link hierarchical clustering.

9.8.0.1 References

[AL03] Arnoux, M., Lechevallier, Y., Tanasa, D., Trousse B.(2003). Automatic clustering for the web usage mining, Analele Univeritatii din Timisoara, vol XLI, *Seria Matematica-Informatica*.

[BR93] Banfield, J.D., Raftery, A.E. (1993). Model-based Gaussian and non- Gaussian clustering, *Biometrics*, 49, 803-821.

[BJ81] Bezdek, J.C.(1981). *Pattern recognition with fuzzy objective function algorithms*, Plenum, NY.

[BP98] Bezdek, J.C., Pal, N.R.(1998). Some new indexes of cluster validity, *IEEE transactions on systems, man, and cybernetics*, Part B 28(3), 301-315.

[BF88] Bradley, P.S., Fayyad, U.M. (1988). Refining initial points for K-means clustering, Proc. of 15th international conference on machine learning, 1-9.

[BG90] Brossier, G. (1990). Piecewise hierarchical clustering, *Journal of classification*, 7, 197-216.

[CG01] Chaturvedi, A. D., Green, P. E., Carroll, J. D. (2001). K-modes clustering, *Journal of classification*, 18, 35-56.

[CL05] Chen,K., Liu,L.(2005). The 'best K' for entropy-based categorical clustering, Proc of scientific and statistical database management (SSDBM05), Santa Barbara, CA, June 2005.

[DL01] Dolnicar, S., Leisch, F. (2001). Behavioral market segmentation of binary guest survey data with bagged clustering, Lecture notes in CS, 111-118.

[DL03] Dolnicar, S. (2003). Using cluster analysis for market segmentation - typical misconceptions, established methodological weaknesses and some recommendations for improvement, *Journal of market research*, 11(2), 5-12.

[DJ74] Dunn, J.C. (1974). Well separated clusters and optimal fuzzy partitions, *Journal of cybernetics*, 4, 95-104.

[DX07] Duo,C., Xue, L., Du-wu,C.(2007). An adaptive cluster validity index for the fuzzy c-means, IJCSNS *International journal of computer science and network security*, 7(2), 146-157.

[EC03] Elkan, C. (2003). Using the triangle inequality to accelerate k-means, Proc of the 20th international conference on machine learning (ICML-03), 1-7.

[EK96] Ester, M., Kriegel, H.P., Sander, J., Xu, X. (1996). A density based algorithm for discovering clusters in large spatial databases with noise, Proc. of second international conference on knowledge discovery and data mining, Portland, 226-231.

[FR98] Fraley,C., Raftery,A.E. (1998). MCLUST: software for model-based cluster and discriminant analysis, Dept of statistics, University of Washington, Tech report 342-98.

[GZ05] Ganapathi, A., Zimdars, A. (2005). Automated clustering of Win32 applications based on failure behavior, UC Berkeley.

[GG89] Gath,I., Geva,A.B. (1989). Unsupervised optimal fuzzy clustering, *IEEE transactions on Pattern analysis and machine intelligence*, 11, 773-781.

[GH93] Gnanadesikan, R., Harvey, J.W., Kettenring, J.R. (1993). Mahalanobis metrics for cluster analysis, *Sankhya - A*, 55, 494-505.

[GN71] Goodfellow, M. (1971). Numerical taxonomy of some nocardioform bacteria, *Journal of general microbiology*, 69(1), 33-80.

[GR98] Guha, S., Rastogi, R., Shim, K. (1998). CURE: An efficient algorithm for clustering large databases. In proc of SIGMOD 1998 International conference on management of data, Seattle.

[GR99] Guha,S., Restogi, R., Shim, K. (1999). ROCK: a robust clustering algorithm for categorical attributes, Proc of 15th international conference on data engineering.

[HW89] Hager, W.W. (1989). Updating the inverse of a matrix, *SIAM review*, 31(2), 221-239.

[HB01] Halkidi, M., Batistakis, Y., Vazirgiannis, M. (2001). On clustering validation techniques, *Journal of Intelligent information systems*, 17(2/3), 107-145.

[HE02] Hamerly, G., Elkan, C. (2002). Alternatives to the K-means algorithm that find better clusterings, Proc of the 11th international conference on information and knowledge management, McLean, Virginia, 600-607.

[HW79] Hartigan, J., Wong, M.A. (1979). A k-means clustering algorithm, *Applied statistics*, Royal Statistical Society, 28, 100-108.

[HR05] Howarth, P., Ruger, S. (2005). Fractional distance measures for content-based image retrieval, Lecture notes in CS 3408, ECIR 2005, 447-456.

[HZ98] Huang,Z. (1998). Extensions to the k-means algorithm for clustering large data sets with categorical values, *Data mining and knowledge discovery*, 2(3), 283-304.

[JD88] Jain A.K., Dubes,R.C. (1988). *Algorithms for clustering data*, Prentice Hall, Englewood Cliffs, NJ.

[JM99] Jain, A. K., Murty,M. N., Flynn, P. J. (1999). Data clustering: A review, *ACM computing surveys*, 31(3), 264-323.

[KR03] Karayiannis, N.B., Randolph M.M. (2003). Non-Euclidean c-means clustering algorithms, *Intelligent data analysis an international journal*, 7(5), 405-425.

[KG02] Karypis,G.(2002). CLUTO - A clustering toolkit, Technical Report 02-017, University of Minnesota, Department of Computer Science, Minneapolis.

[KH99] Karypis,G., Han,E.H., Kumar, V.(1999). CHAMELEON: A hierarchical clustering algorithm using dynamic modeling, *Computer*, 32, 68-75.

[KS99] Kato, N., Suzuki, M., Omachi, S. (1999). A handwritten character recognition system using directional element feature and asymmetric Mahalanobis distance, *IEEE Transactions on pattern analysis and machine intelligence*, 21(3), 258-262.

[KR05] Kim, M., Ramakrishna, R.S. (2005). New indices for cluster validity assessment, *Pattern recognition letters*, 26(15), 2353-2363.

[KP04] Kotsiantis,S., Pintelas,P. (2004). Recent advances in clustering: A brief survey, WSEAS Transactions on information science and applications, 1(1), 73-81.

[KL04] Kummamuru, K., Lotlikar, R., Roy, S., Singal, K., Krishnapuram, R. (2004). A hierarchical monothetic document clustering algorithm for summarization and browsing search results, Proceedings of the 13th international conference on World Wide Web, New York, 658-665.

[KM02] Kanungo, T., Mount, D.M., Netanyahu, N.S.,et.al (2002). An efficient k-means clustering algorithm: analysis and implementation, *IEEE transactions on pattern analysis and machine intelligence*, 24(7), 81-892.

[LM00] Laura,M. A.(2000). A linear algebra measure of cluster quality, *Journal of the American society for information science*, 51(7), 602-613.

[MW66] Manning, R.T., Watson, L. (1966). Signs, symptoms, and systematics, *Journal of American medical association*, 198, 1180-1184.

[MB02] Maulik, U., Bandyopadhyay, S.(2002). Performance evaluation of some clustering algorithms and validity indices, *IEEE Trans. pattern analysis and machine intelligence*, 24(12), 1650-1654.

[MS80] Mizoguchi, R., Shimura, M. (1980). A nonparametric algorithm for detecting clusters using hierarchical structure, *IEEE transactions on pattern analysis and machine intelligence*, PAMI-2, 292-300.

[MB04] Mirkin, B. (2004). *Clustering for data mining*, CRC Press, FL.

[MB05] Muata,K., Bryson,O. (2005). Assessing cluster quality using multiple measures - A Decision Tree based approach, in *The Next Wave in Computing, Optimization, and Decision Technologies* (Golden, B. et.al.), 371-384, Springer, US.

[NH94] Ng,R.T., Han, J.(1994). Efficient and effective clustering methods for spatial data mining, Proc. of 20th international conference on very large data bases, 144-155.

[OP07] Oliveira, V.J., Pedrycz, W. (eds.) (2007). *Advances in fuzzy clustering and its applications*, John Wiley, UK.

[OG03] Oppenheimer, D., Ganapathi, A., Patterson, D.A.(2003). Why do Internet services fail, and what can be done about it? 4th USENIX symposium on Internet technologies and systems, March 2003.

[PS70] Paykel, E.S.(1971). Classification of depressed patients - clusters analysis derived grouping, *British journal of psychiatry*, 118, 275-288.

[PM00] Pelleg, D., Moore, A. (2000). X-means: extending K-means with efficient estimation of the number of clusters, Proc. 17th international conference on machine learning, Stanford, CA, 727-734.

[RS98] Roussos,L.A., Stout,W.F., Marden, J.I. (1998). Using new proximity measures with hierarchical cluster analysis to detect multidimensionality, *Journal of educational measurement*, 35 (1), 130.

[SV02] Salazar,E.J.G., Velez, A.C., Parra, C.M. (2002). A cluster validity index for comparing non-hierarchical clustering methods,

[SE98] Sander, J., Ester, M., Kriegel, H., Xu, X. (1998). Density-based clustering in spatial databases: The algorithm GDBSCAN and its application, *Data mining and knowledge discovery*, 2(2), 169-194.

[SJ99] Santini,S., Jain, R.(1999). Similarity measures, *IEEE trans. pattern analysis machine intelligence*, 21(9), 871-883.

[SW82] Sarle, W.S. (1982). Cluster analysis by least squares, Proc. of the seventh annual SAS users group international conference, 651-653.

[SC98] Sheikholeslami,G., Chatterjee,S., Zhang,A.(1998). WaveCluster: A multi-resolution clustering approach for very large spatial databases, Proc international conference on very large data bases, 428-439, NY.

[SG00] Strehl, A., Ghosh, J. (2000). A scalable approach to balanced, high-dimensional clustering of market-baskets, HiPC-2000.

[SC01] Su,M.C., Chou,C.H. (2001). A modified version of the k-means algorithm with a distance based on cluster symmetry, *IEEE tran on pattern analysis and machine intelligence*, 23(6), 674-680.

[SM81] Symons, M.J.(1981). Clustering criteria and multivariate normal mixtures, *Biometrics*, 37, 35-43.

[TS06] Tan, P.N., Steinbach, M., Kumar, V. (2006). *Introduction to Data Mining*, Addison Wesley.

[TA03] Tarsitano, A. (2003). A computational study of several relocation methods for k-means algorithms, *Pattern recognition*, 36(12), 2955-2966.

[VY88] Van Hui, Y. (1988). Updating the inverse of the dispersion matrix, *Applied statistics*, Royal Statistical Society, 37(3), 474-477.

[ZR82] Zhang,T., Ramakrishnan,R., Livny, M.(1982) BIRCH: An efficient data clustering method for very large databases.

[ZB00] Zhang, B. (2000). Generalized k-harmonic means - dynamic weighting of data in unsupervised learning, Technical report HPL-2000-137, HP Labs.

[XQ05] Zu, R., Wunsch, D. (2005). Survey of clustering algorithms, *IEEE transactions on neural networks*, 16(3), 645-678.

[ZJ82] Zupan, J. (1982). *Clustering of large data sets*, Wiley, NY.

10
Genetic Algorithms

Chapter objectives

- Introduce genetic algorithms (GA)

- Explain genetic learning model

- Discuss advantages and disadvantages of GA

- Understand genetic operators

- Explain schema theorem

- Introduce parallel genetic algorithms

- Understand Genetic programming

- Applications of Genetic Algorithms

10.1   Introduction

This chapter discusses genetic learning models that are not as popular as other data mining models due to their applicability to a limited set of search and optimisation problems. Nevertheless, it has been successfully applied in a variety of fields like intrusion detection, detecting anomalies in financial transactions, combinatorial optimisation, etc.

Definition 10.1 Genetic algorithms (GA) are parallelisable heuristic search and optimisation techniques to solve complex problems based on the principle of natural selection in biological evolution, using a finite starting set of *candidate solutions*, each of which is uniquely encoded using a finite alphabet of distinct symbols, and operated upon by GA operators (selection, mutation, crossover) with an intention to weed out unfit solutions using a fitness criteria, and to iteratively evolve into an optimal set using an evaluation criteria.
It is used in optimisation, function evaluation, classification, and NP complete problems.

It was developed in 1960's by John Holland [HJ75] whose work originated in cellular automata. GAs are so called to keep an analogy with biological evolutions involving DNA strands. They are useful in modeling situations where the problem can be cast with selection and recombination operators playing prominent roles in progressively converging upon an optimal solution by starting with a *candidate solution set* (Other heuristic techniques use a single candidate (feasible) solution to start the iterations, whereas GAs work on a finite set of possible solutions). Unlike other models studied before, the GA's have a few operators (eg: mutation and cross-over) that manipulate the chromosomes (see §10.2 in page 10-13) by preserving the critical encoded information in its components to come up with better-fitting candidate solutions. They may be used as specialised heuristic search techniques for finding approximate (near-optimal) solutions to optimisation problems, or best decision boundaries in classification problems. When used for classification, it creates homogeneous sets of elements using a similarity or utility measure. To locate the optimum of complex multi-dimensional functions, the GA can serve as a preliminary step to locate the 'hills', and other faster traditional algorithms can start from the finishing point of GA to locate the true optimum fast.

All optimisation problems have one or more objectives that are usually expressed as a mathematical function (called objective function). If the variables are all discrete, it is called a discrete optimisation problem. If all variables are continuous, it is a continuous optimisation problem for which other methods (calculus, linear/nonlinear programming, etc.) are the preferred choice. Hybrid optimisation problems have a mixture of discrete and continuous variables (eg: mixed integer programming). Constrained optimisation problems have one or more constraints on individual variables or functions thereof (eg: constraints on linear combinations of variables). They may be equality or inequality constraints, that appear on the low-end, high-end or both (eg: $\sum_{i=1}^{n} c_i x_i \leq k$, and $k_1 \leq x_2 - x_1 \leq k_2$, etc). GAs are typically applied in discrete optimisation problems, with or without constraints.

Evolutionary algorithms (EA) are a collection of techniques based upon the theory of evolution. The most important members of this group are Genetic Algorithms, Evolutionary programming, classifier systems, genetic programming and evolutionary strategies. A unification of these techniques for fast solution of complex problems is called evolutionary computing.

GAs have been applied to many practical problems in construction of structures, local optimisation, online monitoring (of networks, plants, production facilities), load balancing, design of neural networks [HS91], dynamic process control, scheduling, circuit layout designs [KT95], telecommunications, feature selection, pattern recognition etc. See references [GO89], [WD94] for a comprehensive list of applications.

10.1.1 Searching for Optimality

The classical search techniques for finding optimal solutions are as follows:

1. Calculus-based methods
   These techniques have mathematically well-defined objectives in continuous variables, and use either derivative-based direct methods or indirect methods to find an optimal solution. The functional form and constraints, if any (as in Lagrangian relaxation), must be unambiguously known. Variables are quite often continuous.

2. Enumerative methods
   They make use of combinatorial enumeration (small size problems), branch and bound techniques or dynamic programming [DP] (medium size problems) to find optimal solutions by checking each candidate solution in an orderly way. DP is a multi-stage optimisation technique in which the stages are independent. Discrete dynamic programming is a branch-and-bound search technique that uses the DP principle.

3. Random search techniques
These techniques use heuristics to search a subset of the solution space to come up with near optimal solutions.

1) Simulated annealing
Simulated annealing is a probabilistic meta-algorithm that moves to an optimal point in the local neighbourhood by starting from a random point in the search space using a probability and step size. It works with one candidate solution at a time, and locates the global optima using an 'acceptance probability' either to replace a current solution with a random nearby solution, or to keep the current solution. Further information can be found in [KG83], [DL87], or at citeseer.ist.psu.edu/kirkpatrick83optimization.html,en.wikipedia.org/wiki/Simulated_annealing.

2) Tabu search, beam search, heuristic search techniques
These are neighbourhood search methods with data mining applications. The essential idea of tabu search is a set of *tabus* (taboos) that are marked as potential solutions in local 'cells' or 'regions'. This set is kept track of using data structures (one or more *tabu lists*) to preclude the possibility of re-searching in the same local regions, and to speedup the algorithm. It is a meta-heuristic technique that modifies unexplored neighbourhoods as the search progresses, using a search space S, neighbourhood structure(s) [N(i,k)], and search memory M^1 that stores information on the exploration process, and helps to forbid wrong moves and infinite loops (through cycles of length at most k). The moves in tabu search (*tabu moves*) are often reversible, and N(i,k) is defined for every potential solution at each iteration k. Subsequent potential solutions are searched for in these neighbourhoods, and its structure could get modified as well. This will prevent the possibility of skipping regions that could contain better solutions. Tabu search has been extended by many researchers to various domains (see [GL97], [PK00], [RH02], and the sites www.upt.pt/tabusearch, www.dei.unipd.it/~fisch/ricop/tabu_search_glover_laguna.pdf). It is related to scatter search, path re-linking, beam search etc. Beam search also limits the search space using a combination of BFS and best-n heuristic. It uses a tree (or forest), and always moves down the tree levels from the best-n nodes.

3) Evolutionary algorithms
Genetic algorithms fall under the general category of evolutionary algorithms that globally sample the space of alternative solutions using a candidate solution set with 2 more more elements. A fitness function is used to iteratively improve the 'utility' of candidate solutions, until an acceptable solution set is obtained. The members in the solution set can then be ranked on a utility score.

10.1.2 Genetic Learning Model

The genetic learning model randomly seeds an initial population of members that represent potential solutions, each of which can be encoded uniquely using a finite alphabet of symbols.

Definition 10.2 A *chromosome* is a template that comprises of ordered genes, and uses an encoding scheme to represent solutions to a problem.

Low-level representational details of each solution are hidden in the chromosome implementation, which uses data types or data structures [MZ99]. For instance, an array, a string, a

---

[1]Explicit memory stores elite solutions, and attributive or explorative memory stores guidance information.

linked list, a tree or any other linear or nonlinear data structure may be used to mathematically represent the chromosomes. Each chromosome in the candidate solution set is conceptually a string that encodes a potential solution to the given problem. Mathematically speaking, a chromosome is an ordered list of values that encode useful information to be used in optimising the problem. Encoding may be binary, integer, real, character, text or special symbols based [CD97]. Deciding upon the best encoding scheme is the very first step in solving a problem using the GA.

Example 10.1 Consider the problem of locating the roots of a polynomial in the interval $[0,15]$. A chromosome for this can be represented as $[b_1, b_2, b_3, b_4]$ where each of the b_i's is a bit $[0,1]$, because bit strings in the range '0000' to '1111' represent decimal values between 0 and 15. Thus $[1,0,1,1]$, $[0,1,0,1]$ are example chromosomes for this problem that represent the roots 11 and 5 respectively.

Gradient based algorithms are most often deterministic, and work on quantitative variables (exceptions exist as in interior point methods where categorical dummy variables can be used). GAs are always stochastic, and best suited when a large number of parameters are involved with *high nonlinear interactions* among them. Reasonable estimates of the unknown parameters are used at the start of GA. An advantage of GA is that it can accommodate quantitative and categorical variables. The variables or functions of them may be constrained or unconstrained. Optimising function in GAs is most often nonlinear in the variables. Due to the parameter interactions, each parameter cannot be tuned individually even in the most trivial genetic models. GA is the winner when the best performing solutions or best optimal solutions set to multi-modal optimisation problems are required.

Definition 10.3 The unit of encoding (usually a 'bit'[2]) is known as *gene*, and the encoding sequence is the chromosome (all possible combinations of chromosome values is the genome). Each gene has its own unique position (called locus) within a chromosome. The property (worthiness) of a chromosome is collectively determined by the locus of various genes.

Example 10.2 In the binary chromosome $[1,0,0,1]$, each of the bits is a gene. In the traveling salesman problem (TSP) for finding a tour through a finite number of cities, each unique city symbol is a gene. Integer encoding can be converted into binary encodings in which each digit occupies a fixed width in the chromosome. For instance, chromosome=11011 01001 01111 represents an encoding for $\{27,9,15\}$.

10.1.3   Population

All GAs work on a finite set of candidate solutions with (n>2) distinct elements. The choice of n depends upon the application as mentioned below.

Definition 10.4 A group of genomes in the current generation (see below) is called a population. It is also called a breeding-pool. Each member of the population has an attached ID and a fitness score.

A too small population may lack the genetic diversity to fully explore the search space. This

---

[2]A 'bit' in computer terminology denotes either a '0' or a '1'. This is different from the unit for entropy (chapter 3) called 'bits'.

may results in unexplored sub-spaces in the solution space leading to sub-optimal results. A large enough population can maintain the diversity uniformly. The population size can either be fixed (throughout the GA run) or can be varied adaptively. It is beneficial to sort (rearrange) the population in decreasing order of a utility index (defined below). This can be used to weed out non-performing and unfit solutions easily.

Definition 10.5 A member of the population which is one of the possible solutions to the problem at hand is called an *individual*. Species are individuals with one or more common properties or characteristics.

Example 10.3 For the TSP with n cities, the population is a set of n unique city combinations without repetition (that represent a tour). For the root-finding problem using binary GA, the population is a set of binary strings (all of the same length) of an appropriate size.

Definition 10.6 A generation is the set of genomes in existence at any iteration cycle. The transition from one population to another is called a new generation of the GA.

Definition 10.7 A fitness is a measure of the suitability or success of a chromosome. It measures the suitability of survival and reproduction of a genome. It is either found explicitly using a mathematical function, or implicitly obtained from the properties of a chromosome. Most often, it returns a real number in ordinal or higher scale.

Definition 10.8 An *allele set* is the set of discrete values which can be used to code a solution. The allele set for every binary GA is $\{0,1\}$.

Allele combination determines the trait of a chromosome. There are two types of allele – *dominant* and *recessive*. Recessive alleles survive for many generations without change, whereas dominant alleles expresse from genotype to phenotype. Most GAs have one dominant allele. As a biological analogue, if the dominant allele is for *left-handed* persons, the offspring will exhibit left handed-ness for many generations. Each trait is mapped to its own allele, and is independently inherited.

Definition 10.9 The *search space* (also called *state space*) is the set of all feasible solutions to the problem. It can be finite, countably infinite, or infinite. The (GA) population is an abstract representation of the candidate solution space.

Example 10.4 Binary encoding uses two symbols (chosen conveniently as 0,1), that can represent the presence or absence of a trait or characteristic. The binary genotype space is $\{0,1\}^L$ where L is the number of symbols (length of each chromosome).

In Gray encoding[3], each chromosome has exactly 3 neighbours that are unit Hamming distance away (see table 10.1 in page 10-6). For instance, in a size 3 binary encoding, 011 has neighbours $\{010, 001, 111\}$ and 000 has neighbours $\{001, 010, 100\}$. The n-bit Gray codes can be generated using a reflect and prefix method. For n=1, we simply keep 0 and 1 as a column (one below the other). For each higher value of n, we keep 2^{n-1} Gray codes as a column and then reflect them about an imaginary line below the last one to get $2^{n-1} + 2^{n-1} = 2^n$ codes. Then prepend a '0' to the first 2^{n-1} codes, and prepend a '1' to the rest of them.

---

[3]The Gray encoding was invented by Frank Gray of Bell Labs in 1947. It was then called a reflected binary code. It was renamed to 'Gray code' in his 1953 patent application.

Table 10.1: Binary vs Gray encoding

Binary	Dec	Gray	Binary	Dec	Gray	Binary	Dec	Gray
000	0	000	0000	0	0000	1000	8	1100
001	1	001	0001	1	0001	1001	9	1101
010	2	011	0010	2	0011	1010	10	1111
011	3	010	0011	3	0010	1011	11	1110
100	4	110	0100	4	0110	1100	12	1010
101	5	111	0101	5	0111	1101	13	1011
110	6	101	0110	6	0101	1110	14	1001
111	7	100	0111	7	0100	1111	15	1000

An advantage of Gray codes is that if the crossover point (in non-binary encoding) falls within a gene, the offspring values do not diverge. Let $chromosome_1 = \left[11011\,00111\,01111\right]$ and $chromosome_2 = \left[10010\,11001\,01010\right]$ be two chromosomes encoded as binary (called binary chromosomal strings). Corresponding decimal values are $\{27,7,15\}$ and $\{18,25,10\}$. If the crossover is between bit positions 7 and 8, the resulting values are $\left[11011\,00001\,01111\right]$ and $\left[10010\,11111\,01010\right]$ (which in decimal are $\{27,1,15\}$ and $\{18,31,10\}$). If Gray encoding is used, this divergence effect will be minimal.

Bit string based representations may not be suitable for some application domains. Hence other non-standard representations have been applied to particular problems (see [HW07], [MT99]).

Definition 10.10 Permutation encoding uses distinct numbers or symbols from a finite alphabet.

The encoding represents a concatenation of symbols and the fitness may or may not depend upon the position of a symbol in the sequence. Each gene is a number (or symbol) in a sequence. The order in which cities are visited in the TSP can be encoded as a string using a unique city identifier or label. They are also used for ordering problems, time table preparation etc.

Example 10.5 In the TSP, each number would represent a city to be visited. For instance, $\{5,2,8,1,3,4,6,9,10,7\}$ represents a tour through the cities numbered.

Definition 10.11 Value encoding uses arbitrary strings or symbols. They are used with large real or complex problems that are difficult to encode in binary. They use special GA operators.

Example 10.6 Consider a number scrambler (see page 10-22). We need to encode each possible number of a fixed width using letter(s) assigned to each digit. Hence an encoding will use character (assigned to each digit) concatenation. As an example, the number 608 21334 can be encoded as 'say hello' using the mapping defined in §10.4.1 (page 10-22).

Definition 10.12 Tree encoding use a tree of ordered values. It is used in Genetic programming, symbolic regression, etc.

Example 10.7 Prüfer tree encodings
This is a simple and straightforward O(n) time complexity method to encode an arbitrary tree
using numbers or symbols, where n is the number of nodes. Without loss of generality, we will
label each node of a tree by *unique* numbers (preferably consecutive integers in a desired range).
The method first selects that leaf node which has the least label (which is discarded from the
tree) and stores the label of the neighbouring node (which is kept in the tree). This process is
recursively applied to the reduced tree until we are left with a single arc. The algorithm stops at
this instant (leaving the largest labeled node as the root). It is easy to verify that if the labels
are known, the tree can be reconstructed from Prüfer sequence (although the orientations need
not be exact). As an example, consider a tree with 3 nodes labeled 1,2,3 with 2 as the root and
3 as left child. The Prüfer sequence is {2} because the least node (label 1) when deleted from
the tree results in a single edge (2,3). The reconstructed tree is either (1,2,3) or (3,2,1).

Example 10.8 Construct the Prüfer sequence of the trees shown in figure 10.1.
On the left we have a labeled tree with 5 nodes. The minimum leaf node label is 1, which is
discarded and its neighbour 3 is stored. Next minimum is 2, which is discarded and its neighbour
5 is stored. Discarding the next minimum leaf label 4 leaves us with a single edge. Hence the
desired sequence is {3,5,3}. On the right we have an undirected tree with 7 nodes. Follow
exactly the same procedure. Minimum leaf node is 4, which is discarded and its neighbour 2 is
kept. Next minimum is 5 which is discarded and and its neighbour 2 is kept, and so on. This
gives us the desired sequence {2,2,1,3,3}.

In some problems, a real number encoding with the integer part denoting one entity (or
trait), and the fractional part denoting another entity (trait) may be used in a compact way.
For instance, in scheduling problems, the integer part may represent an entity to be sched-
uled (processors, machines, etc) and the fractional part may represent entity to be assigned
(programs, jobs, etc) in a specific order.

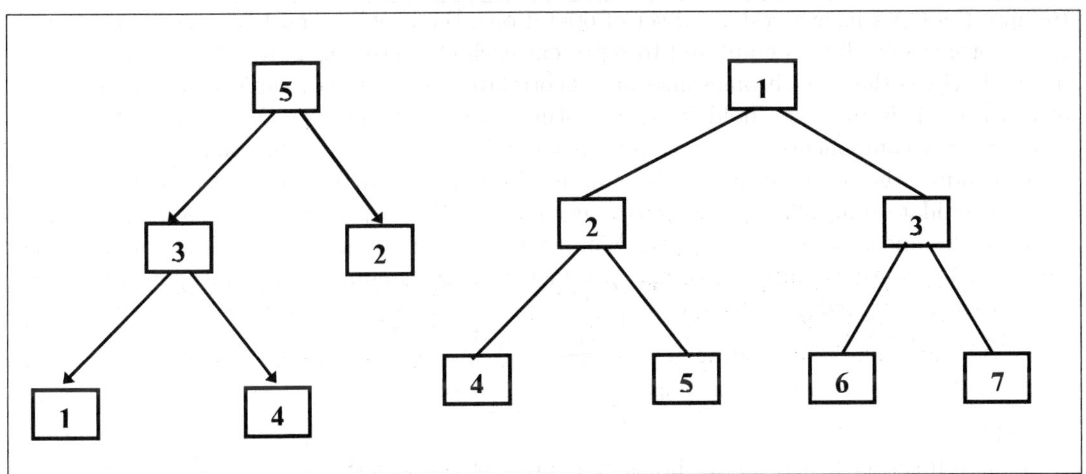

Figure 10.1: Trees to illustrate Prüfer encoding

Definition 10.13 A function (or mapping) with the search space as the domain, and the real
line as range is called the fitness function. It will determine how good each of the currently

maintained solutions are (instead of a positive fitness, we could also utilise a negative error function that need to be minimised). For some applications, the fitness function is obvious as shown below.

Example 10.9 In function evaluation problems, the fitness function is the functional value. For the TSP, the fitness function is the cost of the current tour found. In insurance claim processing applications, the fitness function is the claimed amount (or a function of it). In an application to optimally organise web pages, the fitness function is derived from the number of hits and inter-page links traversed, among other measures.

Table 10.2: Natural genetics vs Genetic algorithm

genetics	Genetic algorithm
gene	a feature of chromosome
chromosome	string
allele	feature value
locus	string position for operation
genotype	structure
phenotype	parameter set

In some problems, the fitness function may not be obvious without good domain knowledge. For instance, in structural optimisation problems like bridge design, the fitness function comprises of a combination of strength/weight ratios, maximum loads (under varying conditions of stress), spans, etc. It may also be based upon sampling methods like tournament selection.

Because the GA's have a restricted set of operations, the problem must be cast in such a way that a proper encoding is employed to represent each chromosome value. A basic assumption in simple GA is that the chromosomes are uncorrelated (so that they can evolve independently of others). This may not hold in some optimisation problems like structural optimisation where nearby components are correlated. A candidate set, a fitness function and parameters for concluding the termination are the minimal inputs. Cardinality of the candidate set (size of the candidate solutions[a]) is most often even (preferably a power of 2 to reduce redundant binary codes and facilitate some parallel implementations). In most data mining applications, we encounter continuous variables (eg: customer purchase amount, family income, etc). These variables are converted into discrete intervals using discretisation (or quantisation) with enough resolution (see chapter 2). Alternatively, the parametric encoding may be used.

[a]Size is the number of elements and length is the number of genes in each chromosome.

If GA is used for function optimisation (finding the maxima/minima of any type of real or complex functions), a properly bracketed range of the function can be mapped to a fixed length binary string. If the domain of the function is unknown, a multi stage GA (see §10.6.2 in page 10-30) may be employed. Hence the search range along with the desired precision determines the length of the chromosome in these problems. As an example, if a transcendental function is known to have an optima in [-512,512), a string of length 10 (2^{10}=1024) may be used[4] if accuracy to the nearest integer is sufficient. Accuracy can be increased by increasing

---

[4]We have 1024 points in the above interval with one open end to accommodate the 0 at the center.

the size of the encodings, or by using other encoding schemes (value encoding, tree encoding etc). Reducing the search space (by discarding regions that do not contain the optima) can speedup the convergence.

10.1.4 Advantages of Genetic Algorithms

There are many practical problems where other models either fail miserably, or are computationally intractable. For example, the TSP with more than 50 cities is hard to solve using conventional algorithms. But, the GA has been successfully applied to find an optimal solution to TSP with thousands of cities. GAs can identify most significant variables in a complex design, and incorporate new variables that might have been overlooked by other algorithms.

- Easy to understand
  The concept of genetic algorithms is easy to understand because it closely resembles natural selection and survival of the fittest principle[5] of Charles Darwin (1809-1882), found in biological evolution.

- Supports multi-objective optimisation
  GA works with discrete, continuous and mixed variable problems. It can simultaneously optimise multi-objective functions (to locate the unique optima, or 'k' best optima) in multi-modal search space (of multivariate functions).

- Easy to implement
  Genetic algorithms are easy to implement since it uses very few macro operations – selection, mutation, crossover, etc. In addition, it does not require derivative computation as in backpropagation learning of neural networks. It works with Monte Carlo data or experimental data. For example, in the number scrambling problem (page 10-22) we generate candidate solutions and try maximal matching words in the dictionary by comparing maximal prefixes of individuals, because of the lexicographic arrangement of words in dictionaries. Hence an exponential random number generator (which is biased to the left) may be used to quickly obtain maximal matching prefixes in the solution space.

- Easy to speed-up
  The convergence rate of GAs is easy to expedite as more domain knowledge is gained. Heuristics can be used to seed the initial candidate solutions to minimise the time spent by the algorithm in reaching the optimal solution. In addition, we can save the current optimal candidate solution set, and restart the algorithm at a later point in time. This feature is difficult to implement for neural networks and decision trees.

- Sequential/Parallel implementations
  Genetic algorithms can be implemented using sequential (steady-state or generational) or parallel (shared/centralised, distributed) paradigms and are ideal for parallel implementation. Parallel algorithms can be classified into data and control (instruction) parallelism. Data parallelism is scalable for large GAs [CE95]. Parallel GAs can use single-population (replicated) or distinct multi-population as training sets. Distributed GAs (DGA) can be implemented as master/slave model, where each slave performs genetic operations (mutation, crossover, selection, etc.) and fitness evaluations. The master node synchronises the solutions after each epoch, and communicates the current best solution set to each

---

[5]less worthy individuals are weeded out and fitter individuals are favored in each generation of evolution. This implies a maximisation procedure in the GA context.

slave. Operations can be parallelised with proper mutual exclusion principles to avoid two concurrent changes occurring simultaneously on any overlapping regions of a chromosome. This parallel nature guarantees that GAs will locate the global peak much faster than traditional optimisation techniques.

- Adaptive learning
  GAs can be programmed to learn adaptively. The parameters, operators and matching functions can be dynamically adapted, based on which we get a variety of GAs. It is especially useful in time-dependent optimisation problems, like portfolio return optimisation in stock markets. Multiple parameters, if any, can be separated into *time-dependent* and *time-independent* sets. Separate adaptation can then be employed for each set.

- Flexibility
  GAs can incorporate noisy objective functions (as in stochastic optimisation), composite objective functions (comprised of multiple functions), conflicting objectives (as in new highway constructions with minimum distance between cities and minimum construction cost, which may sometimes conflict because minimising the distance may necessitate construction of new bridges and flyovers which in turn increases the cost). In addition, the GA parameters (population size, encoding size, mutation and crossover rates, etc) can be tuned between generations. To reduce bias caused when some attributes assume unduly large values, neural networks normalise data to an appropriate range before the input layer captures it. In the case of GA, there is no such transformation required because the only operations are based upon the gene patterns of the input vectors (and not on the magnitudes of the vectors themselves).

- Hierarchical GAs
  The GA can be extended to select optimised feature subsets that serve as attribute values. This is known as hierarchical genetic algorithm (HGA).

10.1.5   Disadvantages

In spite of the impressive advantages mentioned above, the GAs have disadvantages too. For binary encodings, the loci for mutation and crossover are chosen as uniformly distributed random numbers in the interval $(1,L)$ or $(.5,L+.5)$ (see §3.10.4) where L is the length of each chromosome. But a uniform random number in the representation space does not necessarily translate into a uniform number in the solution space.

Example 10.10 Consider the solution space $(0,1,\cdots,7)$, which is encoded in binary using 3 bits ('000',$\cdots$,'111'). Thus binary '111'$=1*2^0 + 1*2^1 + 1*2^2 = 7$. For simplicity, consider the simple mutation operation that selects a gene position (a crossover operation selects the loci as an in-between gene position). A uniform (real) random number in $(1,3)$ has 33.33% chance of operating upon each of the genes (if the number is between 1 and 1.999, we may choose position 1 for mutation. Alternatively, we could generate random numbers in $(.5,3.5)$ and choose position 1 if the number is between .5 and 1.4999) etc). Mutation of the first bit (most significant bit (MSB)) from a 0 to a 1 will map to any number in $(4,5,6,7)$ in the solution space depending upon the other bits. For instance, a '011' upon mutating on MSB gives '111' $(=7)$ whereas '010' give '110' $(=6)$. But this range of numbers $(4,5,6,7)$ is 50% of the search space. Similarly, if the last bit is changed from a 0 to a 1, the possible numbers are $(1,3,4,5,7)$ as each of these decimal numbers has a 1 in their binary representations as the last digit. This represent $5/8 = 62.5\%$

of the search space. Hence the strides taken in the search space are non-uniform, although the effects cancel out when we change a 1 to a 0. This does not matter in most practical problems, as it could be advantageous in exploring the solution space fast. However, in some applications (eg: some game programs) the solution space may have to be explored uniformly for which binary encoding with loci chosen using a uniform random number is inefficient, and random numbers from other distributions may be better suited.

If the encoding alphabet is non-binary, the genetic operations can be more computationally intensive. As an example, assume that the alphabet is $\{0,1,2,3,4\}$. As there are five symbols, we need minimum 3 bits. We can encode this in binary as $\{000, 001, 010, 011, 100\}$. A crossover operation is then restricted[6] to the boundaries that start at $3L - 2$ (if indexing starts at 1). For instance, if two chromosomes are (1,3) and (2,4) which are encoded as (001 011) and (010 100) respectively, a crossover at locus[7] between position 4 and 5 will result in a valid offspring (001 000) and an invalid one (010 111). An inversion operation will have additional restriction on the ending bit position to be on $3L$ for integer $L \geq 1$. Hence the GA solution may not be too fast in some problems due to compliances, restrictions, and bound checks. In addition to these problems, there are a few other disadvantages discussed below:

- Trapping in the local optima.
  For large input strings, the GA can get trapped in one local optima due to short-term fitness employed. Modality is the number of local optima of the objective function in the fitness landscape. Some sub-regions of the data vector, or loops around the boundaries is repeatedly selected for crossover operations, and it never reaches the global optimal solution. Solutions to this problem include the imposition of a penalty function to get out of the local optima, fitness de-rating functions [BB93], etc.

- Non-applicability problems
  Some problems cannot be solved directly by GA (eg: if the only meaningful fitness measure is right/wrong decision).

- Overhead in handling strings
  If the input alphabet is non-binary, we may have to encode each input symbol as binary or use a mapping as done in §10.4 below. The string representation will have to selectively apply the bit operations on compatible pairs (for which a mask may be employed). This may in turn create an overhead, which is especially cumbersome if there are many input symbols.

- Evolutionary dead-end
  Biological life evolves towards better and better circumstances (for most life-forms), resulting in an evolutionary dead-end. The same thing can happen in GAs resulting in entrapments leading to approximate (rather than exact) solutions in some domains that lead to a dead-end.

10.1.6 Steps in GA

Various steps involved in solving a problem using a GA are different from that used in neural networks or decision trees. To solve a problem using GA,

---

[6] Without these restrictions, we could sometimes end up with infeasible child strings such as (110, 101, 111), which are not in our encoding.

[7] In GA, locus represents a bit position within the string representation

1. Identify the problem boundaries and the solution strategies
   The GA always starts with a finite number of candidate solutions that are successively improved in each iteration. Possible boundaries of the problem in each dimension should be approximately known for the solution to eventually converge. There are multiple solution strategies for solving some problems. The most suitable strategy for GA may not be the least computationally attractive.

2. Decide upon an encoding scheme to represent the entities (chromosomes) involved
   This step is crucial for fast convergence of the algorithm. The encoding scheme depends upon the problem, and is easy to find using search ranges, objectives, fitness measures etc. This step may be time consuming for complex and large-size problems.

3. Start with an initial population
   The quality of initial population can speedup the convergence in most cases. This may be known from domain knowledge, from prior runs, or obtained analytically (as in function optimisation). The size of the candidate solution set is problem dependent. Too big a size may slow down the convergence to optimality, and results in duplicate chromosomes in the optimal candidate set.

4. Fix the evaluation function
   The evaluation function must be easy to compute, and it must result in bounded real numbers in a fixed range (a finite score in ordinal or higher scale in NOIR typology). For binary encodings, a bit-fiddling function may be more appropriate. It can be a monotonically increasing or decreasing function. In root-finding applications, it is a decreasing function because roots represent those points where the function crosses the appropriate axis (vertical functional value near the root should become zero).

5. Identify the genetic operators (mutation, crossover, selection, shuffle, etc)
   There are a multitude of operators available in GA, some of which are discussed in §10.2. All of them may not be applicable to every problem. Those that are most suitable for the problem at hand should be identified.

6. Determine the termination criteria
   Convergence to the global optima should occur in a reasonable number of iterations. Termination can occur using optimality condition(s), or using iteration cut-offs (it exceeds a preset limit).

7. Decide upon other parameter settings, if any.
   Candidate solution set can be initialised either randomly or using a greedy method. Random initialisation may use Monte-Carlo methods (chapter 3). If parameters are assumed to follow statistical laws (eg: follows truncated statistical distributions), additional parameter tuning may be needed for the Monte-Carlo model to perform optimally.

The genetic model can continuously evolve into a perfect learning model over a sufficient number of iterations. When the genetic algorithm is applied for the first time, we start with a heterogeneous candidate solution (training) set, and evolve it iteratively into a homogeneous training set. If we already have a training set available from past runs or external sources, it could serve as the basis for a training model with necessary modifications made to the fitness function. Test instances are used to validate and modify the training model. Other variants of the GAs have been proposed, which includes a variety of parallel GAs (PGA), Hybrid GA (HYGA) that use hill-climbing heuristic to skip over local optima, Hierarchical GA (HGA) etc.

Figure 10.2.1 gives the general flow of GA. The convergence criteria can be based upon the number of generations (which depends upon the complexity of the problem and desired precision) or improvement in fitness of two subsequent populations, or a fixed desired fitness (as in structural designs).

10.1.7 A Notation for GA

The GA on a binary alphabet may be represented as a set GA$\{\{0,1\}, [m,x], n, s, f(y)\}$ where m, x are the operator symbols for mutation, crossover, etc., n is the size of each string, s is the population size and f(y) is the fitness function. This can also be represented as $GA_{n,s}^{[m,x]}(\{0,1\}, f(y))$. Additional parameters like probability of crossover, termination conditions etc can also be specified in this notation, which gives rise to the following general notation:

$$GA(S, O, n, L, f, M, I, R, p_1, p_2, \cdots)$$

where S is the alphabet set, O is the set of genetic operators, L is the string length, f is the fitness function, n is the size of the population, M is the selection mechanism, I is the initialisation function, R is the replacement policy, p1,p2 etc are the probabilities. This notation can be extended to the general alphabet with additional parameters.

10.2 Genetic Operators

There are three common genetic operations to induct an element into, or to modify elements of a training set – crossover, mutation and selection.[8] Most genetic operators take either one or two elements from the candidate solutions, and return a structurally altered result (it may also retain the chromosome if a probability or fitness condition is not met).

Definition 10.14 Copying of same information to the next generation is called mitosis.

In the mutation operation, the locus point is on a gene (bit position) whereas in the crossover operation, it is assumed to be in-between two genes (if it is on a gene, we have a choice of either including it or excluding it for crossover, as discussed below). Each iteration of the above operations on a population is called a generation. After each generation, the set of fitter chromosomes are kept track of. Subsequent iterations are biased towards most-fit individuals, so that the average population fitness increases gradually (cost is minimised).

Using a cutoff threshold, the iterations can eventually be terminated when the global optimum is reached (further improvements are not possible), or some run-time limits are exceeded (CPU time exceeds a limit, number of iterations are above a threshold, memory requirements cannot be satisfied etc).

10.2.1 Selection

Selection operation of GA is analogous to the natural selection mechanism that drives the *survival of the fittest* principle ([DL96],[VM99]). The current population for a problem is divided into two disjoint sets using a fitness measure as *well performers* (better-fit individuals) of size

---

[8]Crossover is also called recombination [BF92]. Other operations like double crossovers, uniform crossovers, shuffling, evolutionary cycle etc also have been proposed. The evolutionary variation resulting from these genetic operations mimic the Darwinian principle of fitter organisms that adapt better to their environment.

n_{keep}, and *poor performers* (unfit individuals) of size n_{drop} such that $n_{keep} + n_{drop} = n$. Selection operation is intended to weed out the poor performers by creating new offsprings from better-fit individuals. The n_{keep} should preferably be an even integer. It creates better performing individuals by applying genetic operators on all the n_{keep} elements as potential replacements for n_{drop} unfit individuals. At the end of each selection process, n_{keep} individuals will survive along with n_{drop} offsprings. There are no hard-and-fast rules regarding the relative sizes of n_{keep} and n_{drop} as it is problem dependent. *Selection pressure* is the ratio n_{drop}/n_{keep}. If it is too low, the rate of convergence is likely to be slow. A too high value indicates few better performing offspring in each generation. A plot of the entire fitness values in the current generation is the best way to decide on the size of n_{keep}. If the individuals are kept sorted on fitness (or error) measure, it is often easy to find the cutoff point. Instead of using a fixed cutoff value, this split point can also be determined adaptively. For example, we could start with $n_{keep} = n/2$, and as the algorithm progresses, increase it to $n_{keep} = 2n/3$, or to a higher level.

Selection operation often precedes other operations. It ensures that better fit chromosomes have more than average chance to be carried over to subsequent generations. There are many selection schemes:– (1) roulette wheel selection (2) ranked selection, (3) random selection (4) Boltzmann selection (5) tournament selection (6) steady-state selection. The popular selection schemes are discussed in the following section.

10.2.1.1 Roulette Wheel Selection

The basic idea of roulette wheel selection is to assign higher chances to better-fit individuals. Each individual is assigned a slice of an imaginary roulette wheel with sector size proportional to the fitness (if all sector areas are equal, the mapping is linear as each individual has the same probability of selection (called uniform selection)). Otherwise the probability of selection is proportional to the fitness of an individual. A starting point (preferably along the boundary of two adjacent slots) will determine the absolute origin (Z), and an orientation (clockwise or counter-clockwise) will drive the search. The probability of selection of an individual is $P(k) = F(k)/\sum_{j=1}^{n} F(j)$, where $F(k)$ denotes the fitness score (numeric) of the kth individual. The wheel is spun using a randomly generated number (in [0,1] that can be scaled to [0,360) (see §2.10.3) to select the chromosome using a relative value or angle wrt a fixed origin. If two (or more) chromosomes are to be selected, the above process is simply repeated.

10.2.1.2 Advantages of Roulette Wheel Selection

It is simple to implement, as it involves the generation of a random number in an appropriate interval, and a table search (a program loop or nested if). It is easy to understand and can easily be extended. The fitness of each individual (chromosome) could be determined using its rank (among all other individuals in the population), instead of its absolute fitness. This is called rank-based fitness measure.

10.2.1.3 Disadvantages of Roulette Wheel Selection

The biggest disadvantage is that fitter individuals are selected too often (elitist preserving selection). If the initial population contains too many individuals (n is very large), a batch selection (using sorted fitness values) may be required to reduce the redundant computations.

A generalisation of the above is the stochastic roulette wheel selection in which 'n' equidistant markers are placed around the wheel, and matching individuals are selected in each run. This is

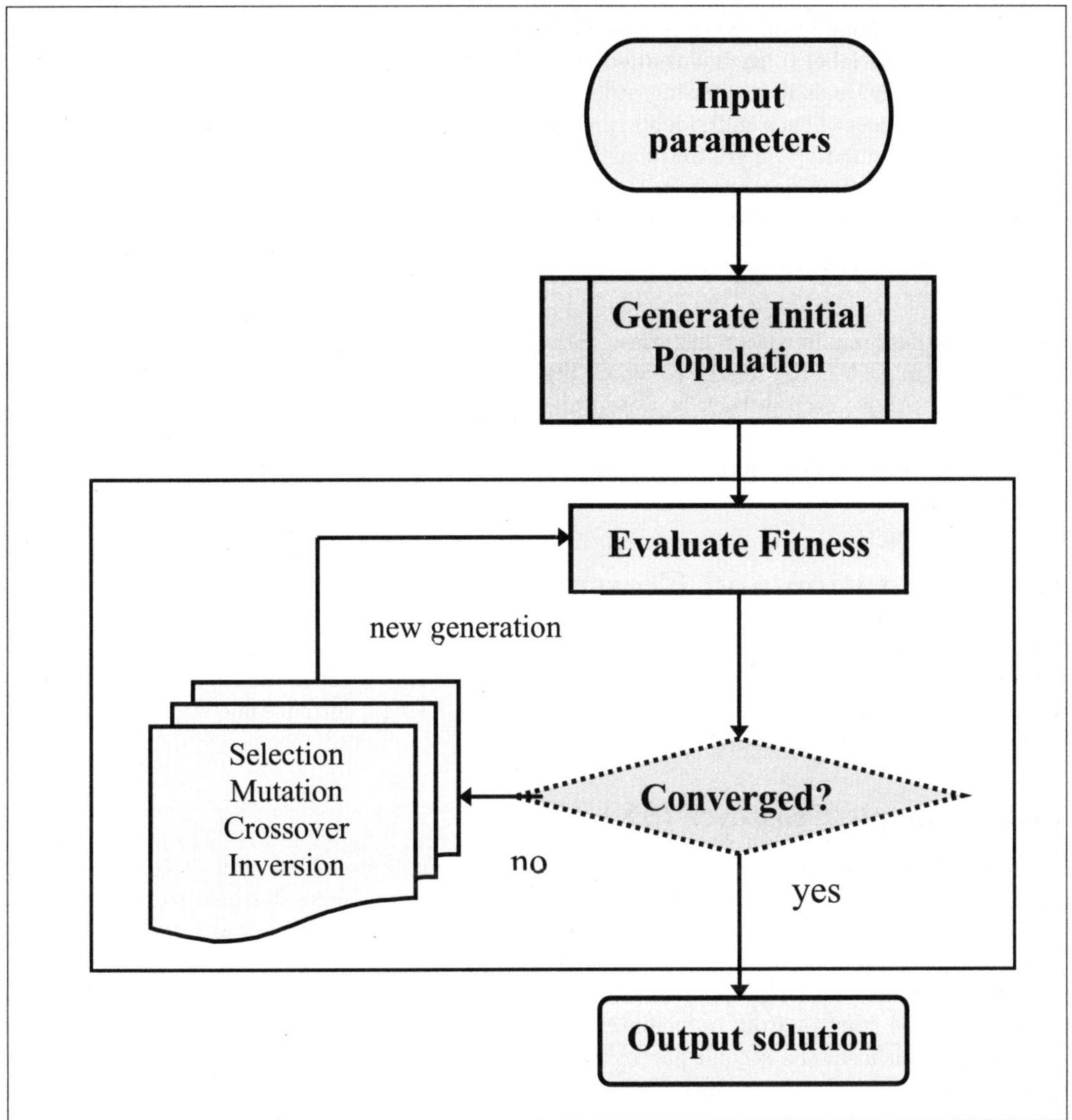

Figure 10.2: General flow of GA

implemented by generating 'n' unique random numbers, scaling each of them into the appropriate range, sorting them, and selecting the matching individuals as described above.

10.2.1.4 Tournament Selection

This is a two-phase algorithm. In the first phase, we choose k (tournament size) individuals in the population randomly. Fitness of each of the individuals are evaluated and ranked in descending

---

Algorithm 10.1 Roulette Wheel Selection Algorithm

1: Input initial population (S) of n chromosomes
2: Assign unique label to each chromosome in the current generation
3: Sort the individuals in decreasing order of fitness (optional step)
4: Evaluate fitness of each individual
5: Compute cumulative ranges and find sum of fitness FTOT
6: Generate a random number in [0,1] and scale it to [0,FTOT)
7: Select the individuals whose cumulative fitness range contains the scaled number
8: **return**

---

order (for fitness maximisation criteria). Individual(s) with highest fitness are selected as the best. The number k is the degree to which best individuals are favored, and is obviously much less than n. Above step is then repeated all over again n_{keep} times. Large tournament size precludes weak individuals from being selected.

A disadvantage of selection mechanism is that it is sensitive to multiple fitness values that are too small.

10.3   Mutation and Crossover

This section discusses the popular genetic operators crossover and mutation. It is assumed without loss of generality that the encoding is binary. But most of the following operations work on any alphabet. The mutation operator works on a single chromosome while the crossover works on a pair of chromosomes.

10.3.1   Simple Crossover (SX)

A crossover[9] is the simplest operator for faster global sampling of the search space. It works on chromosome pairs. Instances are randomly paired, and two instances destined to be eliminated are chosen as the candidates for forming new elements. This process called crossbreeding can be applied in parallel on disjoint pairs. Crossbred elements are more alike than dissimilar with the kept elements. In single-point crossover, the information from the first parent is copied (either from the start or from immediately previous crossover point) until the crossover point, and information from the second parent is copied from the crossover point (either to the end, or up to the next crossover point). It is an explorative operation because it makes big jumps into the unexplored genotype space. It builds upon the fitness of the past to explore new data points in the search space.

In the case of single crossover, the new chromosome inherits the 'head' of the first parent, and 'tail' of the second parent or *vice versa*. The features retained thus depend upon the locus of the crossover. In the special case when the crossover point is at the extremes (either in the beginning or at the end), the chromosomes will retain majority of features of the parents. Single crossover effect is maximal when the locus is at (or near) the middle of the chromosome. Other operations homogenise the results obtained by the crossover operation. A preset probability may be used to decide whether to apply the crossover or not. A uniform random number r_0 is generated in [0,1]. It is compared to the set probability (say p_0). If $r_0 < p_0$, crossover is

---

[9]crossover is a natural genetic operation to indicate meIosis of chromosomes by pairing of genes leading to slight variations in inherited characteristics.

Before crossover	After crossover
x1 x2 x3|x4 x5 x6 x7	x1 x2 x3|y4 y5 y6 y7
y1 y2 y3|y4 y5 y6 y7	y1 y2 y3|x4 x5 x6 x7

bye-passed, otherwise it is applied. The crossover locus is chosen randomly or adaptively. When chosen randomly, a real number in the open interval (0,n) determines the position of crossover. There are many options if this locus falls exactly on a gene (real number is an integer). In the first method, we may keep the aligning gene intact and apply crossover to the genes on either side of this point. In the second strategy, one may decide to include the aligning gene to the left (or right) portion (thereby moving the crossover point to its immediate right interval) if it is not on the rightmost (leftmost) bit. In the third method, we discard that random number and keep generating new numbers until an in-between bit position is obtained.

10.3.2   Uniform Crossover (UX)

A crossover mask of the same size as the chromosome may be used to produce the offsprings as illustrated below. If the mask bit is '1', the offspring inherits corresponding gene of the first parent and if it is '0', it inherits corresponding gene of second parent. In the special case when all mask bits are 0's, the UX favors the second parent, and if it is all 1's it favors the first parent. This technique, known as *uniform crossover*, can produce either a single offspring from two parents or two complementary offsprings (the second offspring being the one obtained by exchanging the order of the parents). The complement offspring of the above is y0 x1 y2 x3

Table 10.3: Illustration of uniform crossover

Mask	1	0	1	0	0	0	1	0	1	1
Parent 1	x0	x1	x2	x3	x4	x5	x6	x7	x8	x9
Offspring 1	x0	y1	x2	y3	y4	y5	x6	y7	x8	x9
Parent 2	y0	y1	y2	y3	y4	y5	y6	y7	y8	y9

x4 x5 y6 x7 y8 y9. As can easily be verified, this is obtained by exchanging the parent's roles. The crossover mask can be fixed or generated randomly. Fixed masks are helpful in favoring parent traits at fixed gene positions. There are many ways to generate the mask randomly. As $2^n - 1$ is the maximum magnitude of a binary number with n bits, we could generate a random integer in $[0,2^n$-1] and convert it to binary to get the mask. Alternatively, we could generate the individual genes or groups of them using random numbers. An already generated mask can also be manipulated (left or right rotations, adjacent bit shuffling, etc) to explore solution space in an ordered way. The mask could also be generated using random numbers from a statistical distribution.

10.3.2.1   Advantages of Uniform Crossover

The UX can be implemented using special hardware instructions. It is easy to parallelise as there is dependency only between matching bits in same loci in the parent chromosomes. In the case of sequential implementations, the mask may be kept entirely in memory and applied in a fixed order (eg: left to right). It is also possible to apply UX without storing the mask in memory by first generating the integer (that will be converted to the mask) and using the decimal to binary conversion process by repeated division by base (=2) method. This algorithm generates the binary digits of the decimal number from right to left (LSB to MSB) order. Hence UX also takes place in the same order. A comparison of various crossover operators can be found in [EC89].

10.3.3   Multi-Crossover (MX)

In multi-crossover operation, more than one locus point is chosen randomly. In generalised multi-crossover operation, each of the n/2 pairs of chromosomes chooses independent locus points. The crossover point tallying with bit position is more computationally involved in this situation, and is resolved individually using one of the strategies mentioned above. Alternative strategies may have to be employed if the new offset position (due to skipping an integer position) tallies with another crossover point. For example, if two random numbers generated are 6.0 and 6.9, the second strategy will result in a collision of two crossover points between bit positions 6 and 7. If the chromosome length is small, and number of crossover points is large, two (or more) generated random numbers can also collide between two bit positions (if the random numbers are 6.2 and 6.8, for example). As done in the third strategy above, we can discard such cases and continue regeneration of random numbers until collisions are avoided. If the size of the candidate solution set is not a power of 2, some of the intermediate encodings obtained could be redundant. In this case a remapping strategy may be used to map them back to valid encodings, which in turn imposes a performance penalty. If the number of redundant encodings is not a multiple of the number of legal encodings, a random remapping may be required or they are remapped into best performing chromosomes. Other possibilities are discarding redundant chromosomes, assigning them low fitness or probabilistic remapping.

The MX uses alternate crossover points (say odd numbered) for pivoting and other loci (even numbered) as boundary markers to preclude the possibility of two successive crossovers canceling each other (resulting in original parent retention). Simulated binary crossover uses a probability distribution to derive the offspring, and works directly on real parameters [DA95], [DB99]. Other types of crossovers include partially mixed, order-based and heuristic crossovers.

10.3.4   Mutation

Mutation[10] operation is used to modify or alter an element randomly. This alteration is dependent upon the attribute types, and how good an instance the process will create. It involves only a single parent, and ensures that none of the random points in the solution space have a zero expected value of being selected. Single mutation creates binary strings that are unit Hamming distance from its parent. Some adjustments can lead the instances to deviate from similarities of kept elements. However, mutation operation will in general create new matching elements if it is applied to n_{keep} elements as replacements for n_{drop} elements. It operates on a chromosome by simply flipping (change 0's to 1's and vice-versa in binary alphabet) randomly selected pairs of bits. It is applied with low probability than the other operators. A random *integer* in the range [0,n] is used as mutation locus, so that the likelihood of looping through a fixed set of symbols (positions) is minimised. Because the binary mutation operator simply flips the bits, the process can be speeded up by recomputing the new fitness measure from prior

---

[10]mutation is a genetic operation that indicates intra-alteration of chromosomes leading to slight variations in inherited characteristics

fitness value (with proper adjustment for mutated bit). The probability of mutation is problem dependent, but typically of the order .005 or less. Mutations apply unanticipated changes to a chromosome without reference to other chromosomes. It is an exploitative operation, since it induces small random diversions in the genotype space. Hence a 'lucky mutation' is sometimes needed to hit the unique optimum. Because it is less frequently applied, mutation is also called a 'background' operator.

Mutation can happen at many levels - *micro mutation* sets a random byte, where as fine grained mutation (*nano-mutation*) sets a randomly selected bit. In some applications, the worth of an attribute should continuously increase (or decrease) to reach an optimal solution. For instance, in an auto insurance application, the number of years of driving experience can be an indicator of accident proneness (As a driver gets more and more experienced (up to a certain age), the chances of getting involved in fatal accidents will, in general, come down). Hence a genetic mutation trying for a criterion to find optimal attribute combinations may be looking for higher experienced drivers. These can be tracked using schemata discussed below. For tree-structured encodings, this could happen at leaf level or at any of the internal nodes. The nano-mutation is less effective in exploring the search space and is seldom used. If a mutation changes a 0 to 1, and results in worse fitness, an *anti-mutation* can be applied to either restore the original or to improve upon it (using a new locus).

A *mutation only* genetic algorithm (MOGA) explores the search space using various mutation schemes described above. In some optimisation problems, a single unique solution may not exist. In such situations, we get either a finite number of optimal solutions or an infinite number of equally valid solutions for the current objective at hand. Such multiple solutions are known as *Pareto-optimal* set. In image pattern matching applications, one may be interested in finding a set of images (instead of a single image) that have maximal similarity with a target image. The MOGA has been successfully applied to solve such problems.

Accumulated stochastic errors through various generations could take the GA through a *genetic drift* effect in which one or more predominant genes are carried over in fixed locations. Single and multiple mutations can reduce the genetic drift. Other extensions to mutation include those with restricted positions, hierarchical mutation, etc.

10.3.5 Inversion

The simple inversion operation is easy to understand. This operator can be applied to any contiguous block of the chromosome. It simply reverses the bit string in the selected segment: if the selected region is [m,n], then newval[m+k] = oldval[n-k] for k=0,1,2..,n-m. If the inversion point is chosen randomly in-between two bit positions, boundary adjustments must be made properly.

To be meaningful, the size (number of bits in-between two inversion points) must be more than one, and preferably even. If two random inversion points bracket a single bit, we may either keep the corresponding bit intact, or preclude such a possibility by regenerating random numbers until all sizes are greater than one. This rule applies to both regular and nested inversions. Because it simply shuffles the bit positions without changing the number of bits of any type, the fitness of an individual remains the same after an inversion if the fitness measure is position independent. Disjoint inversions select multiple disjoint regions in the chromosome and applies the inversion to each segment, which can be parallelised. Nested inversions select portions of larger regions and applies the innermost inversions, followed by each successive outer inversions. Inversions could also be applied to overlapping boundaries, provided an order of

Table 10.4: An illustration of inversion operation

Before inversion	After inversion
x1 x2 x3|x4 x5 x6 |x7	x1 x2 x3|x6 x5 x4 | x7
y1 y2 | y3 y4 y5| y6 y7|y8	y1 y2|y5 y4 y3|y7 y6 | y8

application is maintained in the overlapping segments. It is useful when the fitness measure is dependent upon relative gene positions (as in asymmetric TSP).

10.3.6   Advanced Operators

Selection, mutation, crossover, and inversion operators are sufficient for solving most common problems using GA. But these are insufficient (or inefficient) in some specialised problems. A cyclic operator rotates the genes cyclically (clockwise or counter-clockwise) around the edges (as if each chromosome represented a circular buffer). This may be used to eliminate predominant gene positions discussed above. Another variant in which the empty bit position created by the acyclic rotation (block move to left or right) is always filled with a fixed bit (either 1 or 0) could also be employed (so as to change the fitness value). Other feasible operators are:– sliding window operators, knowledge derived operators, transposition operators etc. A creep operator works with floating point values by changing them by small random adjustments, and is considered a higher-level operator since it does not work at bit level. A decimation operator can be defined to get rid of low-fitness individuals, when the population accumulates too many of them. A threshold for the proportion of unfit individuals may be used to automatically activate the decimation process. A permutation operator involves re-ordering (permuting) the genes between two selected points. This also does not change the fitness value, unless the fitness measure is position dependent. Since there are multiple permutations possible when the enclosing genes are more than 2, a random number may be used to select the appropriate permutation indexed lexicographically. The editing operator is used to recursively apply a pre-defined set of editing rules to each chromosome.

Knowledge based operators are less robust, but could considerably improve performance. Domain knowledge can be used to eliminate unfit chromosomes, and identify regions of the search space that are likely to contain local improvements to current solution. Alternatively, regions that contain unfit chromosome can be identified, and a low probability associated to operate in those regions. The knowledge based multiple inversion (KBMI), and knowledge based neighbourhood swapping (KBNS) operators for TSP are discussed in [RB04].

An adaptive GA learns on the fly, and dynamically adapts the selection and rate of application of operators (by changing the corresponding probabilities). Adaptive mutation only GA (AMOGA) dynamically adapts mutation rate. The adaptive crossover only GA (ACOGA) adapts the crossover parameters, but the convergence to the optimum is not guaranteed. For example, if the optimum should have a '1' bit in position 5, but all generated candidate sets have a '0' in bit position 5, crossover alone may be insufficient to take us to the optimum fast. Since the inversion operation(s) does not (in general) improve the fitness measure, the adaptive inversion only GA is meaningless in most situations. A meaningful combination of the above methods can lead to a variety of hybrid GAs. Since the (single and multiple) crossover operation with proper locus points explores the search space rapidly, it is preferred over the other operations mentioned. Other variants of the GA operators include anti-symmetric operator in which crossover starts with high and mutation starts with low probabilities, and these are varied in opposite directions so that finally crossover ends up with low probability, and mutation with high probability. Other GA related acronyms include VEGA (Vector Evaluated GA [Schaffer 85]), MObGA(Multi Objective GA), etc.

10.3.7 Arithmetic Crossover (AX)

All of the binary crossovers discussed above are discrete. In some engineering and optimisation applications, we come across continuous chromosomes ([FZ97],[MC00]). The arithmetic crossover works on continuous genes by choosing a random number (r) in [0,1] and generating a pair (x', y') as x'=rx + (1-r) y and y'=(1-r) x + r y. As both r and (1-r) $\in$ [0,1], the ranges are preserved by AX. These are used in search and optimisation problems.

10.4 General Alphabet Set

The cardinality two binary alphabet is not directly applicable to problems involving a large number of continuous attributes. They can use any finite set of symbols space. This naturally leads to higher cardinality alphabets in which the symbols may either be integers, floating point numbers, or a set of distinct characters or labels. For example, consider all possible 3D protein representations with a desirable substructure. If each amino-acid is encoded by a separate symbol (say capital/lowercase English alphabet) the input vector will comprise of different combinations of these letters. All of the operators discussed for binary encoding are valid for general encoding too. The mutation, for example, can cyclically shift valid values, or utilise a mathematical function to map each element to its preceding and succeeding element. For numeric alphabets (numbers), we can use functions (statistic) like means (arithmetic, geometric, harmonic), symmetric functions or other group theoretic operators. Another problem involves pairing two sets of entities (applicants vs vacancies, jobs vs machines, salespersons vs regions etc). In such problems, we encode the less important entity as numbers and other (more important) as letters or symbols. For instance, if the number of applicants is less than 27, we could encode each by an English alphabet and encode vacancies by unique integers (say starting with 1). Then an encoding 'ALGORITHM' will imply that first applicant 'A' is assigned to vacancy 1, twelfth applicant 'L' is assigned to vacancy 2, seventh applicant to vacancy 3 and so on. In other words, the ordinal position from start of the string indicates the vacancy number to which that candidate is assigned. If the number of applicants is more than 26, we could use upper and lowercase letters, numbers and other ASCII or unicode characters to encode them as appropriate. A two-way table (a matrix) may be used to store the suitability measure. In the job scheduling example, this matrix will contain the cost or time involved in processing i^{th} job on j^{th} machine. For the salesperson assignment to geographic regions, this matrix will contain either the expected number of units sold or net income generated (in various regions). The worth of an encoding can then be computed by summing the corresponding matrix entries.

If the encoding uses real numbers, a Gaussian perturbation can be used to mutate gene values. For this purpose we use a zero mean Gaussian distribution with a small standard deviation $N(0,\sigma)$. A random number (r_i) is generated from $N(0,\sigma)$ (which can be positive or negative, and most often in the range $[-3\sigma, +3\sigma]$) and the ith gene is modified as $x' = x + r_i$, where r_i is i^{th} random number. Problems with solution set in the complex domain are tackled by representing the real and imaginary parts separately.

10.4.1 A Number Scrambler

Most people need to remember several numbers as part of their daily life – telephone numbers of friends and family members; business associates, travel agents, credit card numbers; numeric passwords and PINs; flight numbers; insurance numbers; number locks etc. Some of these (especially PIN numbers of cash cards) need to be protected to avoid fraudulent use, in case the card is stolen. In this section, we use genetic algorithms to develop a simple method that takes the burden out of your head to remember most of your important numbers.

In most countries, the telephone keypad is marked with letters in addition to numbers. It differs in land phones and various cell phone keypads. Whatever may be the labeling scheme, the letters can be used to advantage in remembering numbers. For our discussion, we will assume the labeling 2={'a','b','c'},3={'d','e','f'},4={'g','h','i'},5={'j','k','l'}, 6={'m','n','o'},7= {'p','q','r','s'},8={'t','u','v'},9={,'w','x','y','z'}, with the digits 0 and 1 as unlabeled. On some phone pads, {p,r,s} appear together on one digit, and the digit 0 is labeled q,z. Readers can simply recode the above as required, and substitute fixed symbols (eg: capital vowel, a blank or dash) for unmarked digits.

This problem is well-suited for GA due to the following reasons – (i) there could exist multiple meaningful strings for a single number, (ii) as the number of digits increases, the possible combinations also increase exponentially, (iii) generating all possible combinations is inefficient, as it involves multiple lookups in one or more dictionaries, (iv) The intermediate results can be saved (externalised) and the algorithm restarted at a later time if needed (v) a general alphabet GA can solve the problem using a fitness measure derived from matching prefixes (of chromosomes and dictionary words).

A recursive algorithm can generate all meaningful combinations of the letters, with the help of a dictionary lookup. For m=7, there are 2187 combinations (3^7), and for m=8 there are 6561 combinations. Hence this method is too slow and inefficient (because we have to look them up in dictionary) for m>5. To solve the problem using GA, we must decide upon the encoding scheme, population size, fitness measure, etc. The encoding scheme is obvious. As each digit is associated with a set of letters, we can represent the number using corresponding letter combinations. Each chromosome has the same length as the length of the number to be crypted (arbitrarily chosen as 7 in this example). The population size is arbitrary. If we just want 5 meaningful combinations, we could set $n_{keep} = 5$ and n_{drop} to a much higher value (say 10). In our notation (see page 10-13), this may be represented as GA({a,$\cdots$,z},[m,x],7,10,f(prefix match)), where (a,$\cdots$,z) is the encoding alphabet, operators used are mutation (m) and crossover (x), 7 is the length (of the number to be scrambled), 10 is the population size (arbitrary), and f() is the fitness function.

On occasion, a chromosome in the candidate solution pool may be near perfect during the nascent stage itself. If a generated string has an exact match in the dictionary, we remove it from the population as a matured chromosome that need not undergo further evolutions (see elitism below). Chances of perfect matches are however too small. Hence we will look for two or three substring matches in the dictionary. For instance, 4653 7424 (with length 8) can be encoded as GOLD RICH, GOLF RICH, HOLD RICI etc on using the above encoding. If a meaningful prefix has been found, we have the option to either keep it and grow the rest of the strings, or allow it to be changed with the hope of finding a better word. This will result in meaningful small word combinations (eg: 'say hello' for 729 43556). In the worst case, it could happen that none of the sub parts also turn out to be meaningful. Although such cases are rare (eg: 99999 1234), we could try to remember it from right to left, or use some other heuristic to reduce the number to more amenable form. A weakness of our implementation is that it could result in duplicate solutions, because each of the chromosomes evolves independently, without knowing other chromosome values.

A single MOGA may not take us to the global optimum fast. Suppose a chromosome obtained so far for 24862 is 'chvnb' with fitness score 40 (because there are many English words starting with 'ch'). As there are no words with chv as first 3 characters, we could try to mutate the third gene 'v' with 'w' or 'x', so as to keep the maximum prefix match. But neither of these mutated chromosomes give better fitness. So we need to mutate one of the previous genes (first or second) to escape out of local optimum. Even this may not take us always to the global optimum. This is more of a problem towards the end of the chromosome. As an example, suppose m=8 and we match 6 characters (either by a single word or multiple words). The last two characters may not even be a part of a valid word, in which case we will be seeking a better solution that is non-existent. On the contrary, sometimes our chromosome may be a proper subset of a larger size word. As an example, 736362 encodes into 'rememb' which is a stem of 'remember'. This is the reason why a GA works on a set of candidate solutions, some of which will never contribute to the global optima.

Many words (with 3 or more characters) in natural languages have an abundance of consonants with vowels in-between them for easy utterance[11]. This implies that the occurrence position of vowels can be predicted if the number of consonents before that position are known. This property of word formation can be utilized in our problem to accelerate the convergence. If the first character chosen is a consonant, the second letter should preferably be a vowel. This dependency among adjacent genes (letters) is called epistasis (see §10.5.2 in page 10-29), and could considerably speed-up our search. As there are a large number of English words in which the vowels repeat either alternatively, or after two consonants, and less frequently after 3 letter consonants (eg: 'sprinkle', 'strings'), and rarely after 4 or higher letter combinations as in 'strengthen'), it is a case of multi-epistasis.

Table 10.5: Initial population and results of GA after 5 and 10 iterations

chromosome	fitness	chromosome	fitness	chromosome	fitness
hdofuhc	1	heofuhc	4	heoftic	7
gdodugb	1	geodugb	6	geodugb	6
hdnesha	1	henesha	7	henesha	7
ifmduhc	2	ifmeuhc	4	ifmeuhc	4
gdoduha	1	geoduha	3	geoduia	3
hdmfsgc	1	hemfsgc	3	hemfugc	6
geoethb	3	geoethb	3	geoethb	3
hdoeugb	1	heoeugb	2	heoeugb	2
gfnfthb	1	genfthb	3	genetgb	5
idmfuhb	2	idmeuhb	5	idmeuhb	5

Table 10.5 gives the initial population, and results obtained after 5 and 10 iterations. After sufficiently large number of iterations, the 9th row will 'converge' to the code 'genetic'. If iterations are stopped before it, those codes for which the fitness score is maximum are chosen as the codes to remember. For example, choosing fitness=7 in table 10.5 gives us 'heoftic' (he+of+tic)

---

[11]It is more so in German and Russian than in English

and 'henesha' (hen+es+ha) as our codes. Sometimes it could happen that meaningful nouns (names of persons, places, movies, buzzwords, and words in other languages) that are not present in our primary lookup dictionary could emerge to show up with low fitness (because our fitness is purely calculated using prefix matches in dictionary words). Secondary dictionary lookups can avoid the algorithm from discarding such meaningful strings as non-optimal. Dictionaries may also be cross-linked to generate a variety of easy to remember word combinations comprising of names and words; or words in multiple languages intermingled (as in the above example where 'es' is a word in Spanish, German, etc). A high-level description of the algorithm (using mutation and crossover operations) is given below. Other operations (like inversion, multi-crossover etc) could also be used to improve it.

Algorithm 4 A genetic algorithm for scrambling numbers

```
Step-0:
    Get the number to be scrambled (number), initial population size n,
    maximum number of iterations to be performed (ITERMAX),
    maximum score desired (SCOREMAX) between 80 and 100 for elitism,
    maximum matches desired (MATCHMAX)
    (***    Array description: (D denotes any dictionary)
    mismatchPos[] stores position of mismatch of chromosome in dictionary word
    matchType[] stores 0 if complete chromosome is a (prefix of) word in D
    stores 1 if dictionary word (of smaller length) is a prefix of chromosome
    stores -1 if presently matched sub-word is kept, fresh search started
    stores 2 if only few characters match between a chromosome and a  word
    Initially, most of the matches will be of type 2 (no exact matches in D)
    fitness[] stores current fitness
    score[] stores global fitness ***)
    Let m = length of number (after removing white spaces)
    If m is less than 2, go to Exit
    Set iter=1, popsize = n, Elitists = 0
    Seed starting population of size n randomly with each chromosome of size m
    Score[i]=fitness[i]=mismatchPos[i]=matchType[i]=0, Head[i]=1 for i=1..n

Step-1:
    Open all dictionaries and read entries (just the words in sorted order)
    into character arrays (for binary searching a pattern in it).
    Save the number of words in each dictionary (upper limit for binsearch).
Step-2:
    Compute fitness score of current Population in fitness[] array using maximum
    matching prefix from Head[j] to m. As generations continue, the Head[]
    pointer will move towards the right end of each chromosome.
    Store mismatching prefix position of each chromosome in mismatchPos[] array,
    store match type (see above) in matchType[] array.
while (iter < ITERMAX) do begin
(* optional:-- Sort chromosomes in decreasing order of local fitness.
    Each chromosome has two fitnesses associated with it. Global fitness
    (score[]) is the worth from leftmost position to Head[j], and local
    fitness is the fitness from Head[j] to current locus (stored in fitness[]
    array). *)
```

```
(* divide current population into 3 distinct groups:--
   1) Elitists that do not undergo further evolutions
   2) Well-performers who have a valid match by a dictionary word
   3) Poor-performers who do not have an exact prefix match *)


   j = 1
   while there are more chromosomes in the current population do begin
     (* we will ignore the last character of a chromosome, if all others
     together is either a word, or a group of words concatenated together. *)
     if ((score[j]+fitness[j]>=SCOREMAX) or (head[j] >= m-1)) then
        Elitists = Elitists + 1
        remove it as an elitist from current population
     (* Elitists are matured chromosomes, carried over to next generation. *)
     (* Check current population size and exit if required matches are found. *)
      if (Elitists >= MATCHMAX) then
        go to Print
      else
        jump to nextWhileLoop
      end if
     end if

Crossover:
     (* Apply crossover operation on well-performing pairs. This may sometimes
        create a loop if last few characters never form a complete word or stem.
        To avoid looping, we will apply crossover with probability cp=.90 and
        mutation with probability mp=.10, which could improve the fitness score
        (if the locus point is halfway or less to the left, use cp=.95).
        Before applying the crossover, we will check if there can be an improvement
        in the fitness. If not, we will skip the crossover and either keep
        the subword(s) found, or apply a mutation at the mismatching
         position. This may sometimes miss more fit chromosomes *)

   if (matchType[j] = 1) then (* current chromosome is well-performer *)
   (* Check if there is another one that could be crossed over. If there is
      just one matchType[j] = 1 chromosome, or there is a left-over chromosome
      without a matching pair, we will keep the subword found, and advance
      afresh from the mismatching position looking for a new word.
      This case is distinguished by reversing the sign of matchType[j] as -1.
      Crossover may improve the fitness score when both mismatchPos[j], and
      mismatchPos[k] are either equal or are close-by (differ by 1  position).
      This is because a 5 letter prefix when crossed over with a two letter
      prefix at any position < 5 usually results in lesser fitness. If no such
      pairings are possible, proceed as above (mutation or fresh search) *)

       (* generate a probability (cp) between 0 and 1 and apply a crossover
          with a 'not-yet-selected' compatible chromosome (say k). *)
   if (cp <= .90) then
     (* Crossover can sometimes result in bigger or perfect matches. Hence
```

```
    we check the fitness score to see if solution has improved. If not, we
     update Head[] to speedup convergence (by partial word matches) *)
    (** If pairing is possible, we have 2 chromosomes at j and k **)


    minLocus = min(mismatchPos[j], mismatchPos[k])
    (* Assuming a reversible hypothetical crossover operation at locus, *)
    (* find newFitnessJ, newFitnessK & check if improvement is possible *)
    if (newFitnessJ >= fitness[j] || newFitnessK>=fitness[k]) then
   (* at least one fitness can be improved, so make crossover permanent *)
       Crossover (j, k, minLocus)
       jump to nextWhileLoop
     else (* crossover cannot improve either fitnesses *)
    (* We will keep the prefix matches found and start afresh from Head[] *)
       update Head[j] = mismatchPos[j], Head[k] = mismatchPos[k]
       matchType[j]=matchType[k]=-1 (* to distinguish afresh starts *)
     end if
     (* Else if pairing is not possible for current chromosome *)
     (* generate a probability ps to decide whether to mutate or keep *)
     if (ps<.99) then (* keep current prefix matches of chromosome j *)
       update Head[j] = mismatchPos[j]
       matchType[j] = -1
     else (* mutate at mismatchPos[j], hoping to improve fitness *)
       Mutation (j, mismatchPos[j]) (* mutate gene at mismatchPos[j] *)
     end if
   else (* cp > .90 *)
     Mutation (j, mismatchPos[j])
   end if
   jump to nextWhileLoop
Mutation:
   (* Apply mutation operation on poor-performing chromosomes *)
   if (matchType[j] = 2) then (* poor performer *)
    (* avoid the possibility of getting stuck in local optima
    Apply mutation at mismatchPos[j] with high probability (say .98)
    or at mismatchPos[j]-1 with low probability (say .01), and so on
    provided the index do not go beyond the lower limit. Initially, only the
    leftmost one or two letters will match with a dictionary word. *)
    (* generate mutation probability (mp) between 0 and 1 *)
    if (mp <= 0.98) then
      Mutation (j, mismatchPos[j]) (* mutate gene at mismatchPos[j] *)
      else if (mp <= 0.99) then (* 0.98 < mp <= 0.99 *)
        Mutation (j, mismatchPos[j]-1) (* and so on.. *)
    end if
   end if


nextWhileLoop:
   (* flag the chromosome(s) already operated upon *)
    j = next available index (or -1 if all chromosomes finished )
  end while (* there are more chromosomes in current population *)
```

```
(* Now we have a changed population, with better fit chromosomes *)

    for each chromosome j in the current population do in parallel
    (* update fitness of every non-elitist chromosome *)
      if (matchType[j]=-1) then
        (* accumulate global fitness of such chromosome(s) *)
          score[j] = score[j] + fitness[j];
      end if
      if (matchType[j]<>0) then (* non elitist *)
        (* extract chromosome suffix substrings starting with Head[j].
        (Head[j]-th to m-th characters). Could skip if Head[j]=m-1 *)
          compute fitness score in fitness[j] array using maximum matching
          prefix by dictionary lookup.
          store mismatching prefix position in mismatchPos[] array,
          store match type (see above) in matchType[] array
      end if
    end for
    iter = iter + 1
  end (* while *)

Print:
Sort current population on global fitness score (score[j]+fitness[j])
Print population with MATCHMAX largest fitness values
Exit:
End
```

10.5   Schema Theorem

Since candidate landscape has many possible bit patterns, the fitness landscape is a proper subset, out of which some are impossibilities for a solution. The impossible combinations are represented by a special symbol (eg: a '*' which denotes a "don't care").

Definition 10.15 The schema is a pattern of gene values in which a wild-card[12] symbol matches any symbol of the alphabet. Schemata[13] are templates that specify alignment of known symbols and wildcards at various positions.

Example 10.11 '**010*11*' is a schemata that matches {110100110, 100101110, 000100110, etc} in binary encodings. This schemata restricts positions 3,4,5, 7 and 8 to respectively contain 0,1,0,1,1 and allows other positions to be filled by any of the symbols in the alphabet. Hence whenever a schemata is specified, the corresponding alphabet used must also be specified.

Definition 10.16 The order of a schema is the number of defined (fixed) positions, and is denoted as $O(S)$.

---

[12]The wild-card is a concept that originated in playing cards and became popular during the punch-card era of mainframe computers. Literally it represents a placeholder for multiple characters or entities like a playing card that is a substitute for other cards. In operating systems, a '?' is a wildcard for a single character and a '*' is a wildcard for 0 or more characters, so that '**', etc are collapsed to a single '*'. In GA, we use a '*' or '#' as the equivalent of '?' and hence multiple occurrences cannot be collapsed.

[13]schemata is plural of schema – a plan, rule or visual map of a structure.

Table 10.6: Schema and possible binary encoding

Schema	Encodings	Schema	Encodings	Schema	Encodings
00$\star$	(000,001)	0$\star$0	(000, 010)	$\star$00	(000, 100)
01$\star$	(010,011)	0$\star$1	(001, 011)	$\star$01	(001,101)
10$\star$	(100, 101)	1$\star$0	(100, 110)	$\star$10	(010, 110)
11$\star$	(110, 111)	1$\star$1	(101, 111)	$\star$11	(011,111)

Definition 10.17 If $\Sigma = \{0, 1, *\}$ is the alphabet, a schema S over Σ is any combination of strings with at least one 0 or 1 (note that the trivial schema in which each character is '*' denotes the totality of all solutions).

Theorem 10.1 A schemata of size ν with 'm' wildcards at arbitrary positions defined on an alphabet with 'n' symbols represents a set of $\binom{\nu}{m}n^m$ encodings.

Proof. Without loss of generality, we assume that the wildcards occupy the first 'm' positions. Since each of these positions can be filled by any of the 'n' symbols, the total number of matches are n^m, by the counting rule. But the 'm' positions can be chosen out of ν positions in $\binom{\nu}{m}$ different ways. Thus the result. If the wildcard loci are fixed, the number of matching encodings are n^m.

Corollary 16 A schemata of size ν with 'm' wildcards on a binary alphabet matches $\binom{\nu}{m}2^m$ encodings.

The proof follows easily by putting 2 for 'n' in the above theorem.

If the encoding size (ν) is a power of 2, each encoding can be mapped to vertices of a ν-dimensional hypercube. A schema with a single wildcard will then represent an edge of the hypercube (so that the encodings are the vertices at the end of this edge), those with 2 wildcards will define a hyperplane (4 vertices at corners of the hypercube), and those with more than 2 wildcards will represent subcubes (encodings are all vertices of the subcube). Since there are $2^{\nu-1}\nu$ edges in a ν-dimensional hypercube, there are this many schematas possible. Table 10.6 lists all 12 single wildcard schematas for $\nu = 3$.

Because there are $\nu(\nu - 1)2^{\nu-3}$ hyperplanes (faces) for a ν-dimensional hypercube, there are this many two-wildcard schematas possible.

Theorem 10.2 Schema Theorem

This was mainly developed by Holland ([HJ75]) during 1970's. Let e_n^S denote the expected number of instances of a schema S at n^{th} generation, f(S) is the mean fitness value of individuals containing schema S at current generation, p_m and p_x denotes mutation and crossover probability, L(S) is the distance between the outermost fixed positions of schema. Let len denote the string length, and O(S) be the order of the schema S. Then $e_{n+1}^S \geq e_n^S f(S)(1 - p_x L(S)/(len - 1) - O(S)p_m)/f_{pop}$. When $p_m=0$, this reduces to $e_{n+1}^S \geq e_n^S f(S)(1 - p_x L(S)/(len - 1))/f_{pop}$.

An implication of the schemata is that solution to larger optimisation problems can be constructed from smaller 'building blocks' of solutions to sub-problems. The term $(1 - p_x L(S)/(len - 1))$ relates to the crossover operation, and $e_n^S f(S)/f_{pop}$ relates to selection operation. The inequality is due to the crossover operation, and it becomes an equality in the absence of crossover.

10.5.1 Elitism

Elitist strategy retains the current best to the next generation. The best performing chromosome(s) in the current population is evaluated using the fitness maximisation (or error minimisation) criteria, and copied to the next generation as such to prevent the loss of best found solutions till now. In the telephone number scrambling example discussed above, a scrambled number that has an exact match (or subword matches) in the dictionary can be copied as an elitist. Similarly, in a TSP, those routes with globally minimal costs aer always retained as there are no cheaper ways to traverse between these cities.

10.5.2 Epistasis

Definition 10.18 Inter-dependency among gene positions in a chromosome is called *epistasis*[14]. A gene at a particular position may suppress or enhance the effect on an associated gene in another position. These positions are usually contiguous in a DNA strand. In single epistasis, a pair of genes are associated, whereas in multi-epistasis either *two or more pairs* or *groups of 3 or more combinations* of genes are associated. The epistasis effect may be learned and utilised for fast convergence in some problems. As an example, if the cost between cities in the TSP has a unique minimum (say between cities U and V), the salesman will always traverse the path UV in either direction and the corresponding bits in the chromosome representations are in epistasy. Multiple epistasis can be used to advantage in reducing the dimensionality of the GA. Such genes can be identified using heuristic techniques, by solving sub-problems, or from prior domain knowledge. See [CH02] for further information.

10.6 Implementation of GA

A finite plausible solution set is used to start the GA algorithm. These are samples drawn from a population of solutions called the candidate solution set. New samples that have a better prospect of survival are generated in a progressively incremental way. This process is continued until an optimality criterion is satisfied. Our aim is to tune the parameters of the underlying problem in such a way that the output of the system is improved in each step so as to converge to the optimal (minimal or maximal) solution in a finite number of steps. A complete knowledge about the parameter space of a problem is often unknown. Reasonable estimates of the parameters are used as initial values. The initial solution set can be generated randomly (using a uniform random number generator to set each bit in the unstructured case), or using greedy methods. If a solution set is available (from previous runs of the program or from other methods), the initial set can also be chosen by picking best fitting chromosomes from such a list. If the representation space is fixed, the elements are called linear chromosomes. Although easier to work with, linear chromosomes are ineffective for representing domain knowledge of some problems. Many approaches have recently appeared to overcome this weakness that utilise complex data structures (tree-structured, multi-linked list etc).

Computations take place in time units called 'epochs' in which the population evolves according to one of the well-defined operations explained in §10.2. A fitness function based on all attribute interactions of the set is used to evaluate each element's worth in the set to decide either to keep it or drop it. The fitness function is dependent on the domain and the criteria, and provides a performance measure. It must be relatively fast to compute and the resulting value

---

[14]This phenomena is called epistasy.

(output) must be within reasonable range of the input values in the search space. The fitness of each individual in the candidate population is evaluated with respect to other members in the current generation. Since mutation operation changes the encoded string at a few places only, the function must preferably be able to update itself using such small changes. It may either be a mathematical function, an experiment or an operation.

For example, if each chromosome is represented as a bit vector, the sum of the bits can be used as the worth measure.[15] In each step, either one or two chromosomes are selected from the training set. The probability of selection being equal to the ratio of its worth over the worth of all elements in the pool (called relative worth of chromosome).

If only one chromosome has been selected, a mutation operation will change some of its bits from 1 to 0 or *vice versa* in the binary case. This may affect the worthiness of the chromosome. If it satisfies the fitness criteria (the fitness score is within a specified range) the instance is kept in the training set as unchanged. Otherwise, the instance is either dropped, and a new instance that is a better replacement with respect to the similarity measure is added to the training set, or is modified using other 'kept elements'. Unlike most of the other rule induction methods that consider one attribute at a time, fitness functions used in GAs take all attribute interactions into account. Fitness function returns a single score and are the most crucial aspect of GA, followed by the coding schemes and weights. An anti-fitness function can also be used if the aim is to minimise some nonfitness criteria.

A probability may be attached either to the individuals (so that more fit individuals are retained for subsequent iterations) or to the operators (so that some operators are more often applied than others).

10.6.1   Parallel GA (PGA)

One of the reasons for the popularity of the GAs is the inherent parallelism involved in its implementation [GB02]. It can be fine grain or coarse grain parallelism. Multiple mutations can be parallely implemented on a chromosome. Multiple inversions can also be applied as long as there are no overlapping regions. Multiple cross-over operations can be parallely implemented on disjoint chromosome pairs. Advantages of the PGA are the computational speedup achieved, and cost reduction. A discussion on parallel GA package (PGAPack) can be found in [MD99]. Details on an implementation of a parallel genetic algorithm in C that uses the Chameleon message-passing library is discussed in [DL96], [AH99]. PIKAIA is a parallel implementation of genetic algorithms that utilises the PVM message passing library.[16] DGenesis is a distributed implementation of a Parallel GA available from ftp.aic.nrl.navy.mil/pub/galist/src/ga/dgenesis-1.0.tar.Z.

10.6.2   Multi-Stage GA

The search space boundaries may be unknown in some applications like complex function optimisation. A high encoding size may have to be used to get an accurate solution, by sacrificing computational overhead. A solution is to use a multi-stage approach as follows. During the first stage, a large search interval is used to bracket the optima. This interval is properly chosen such that 1) the real optima is not missed, 2) the convergence is not too slow. Each subsequent

---

[15]Fast hardware register based bit manipulation instructions exist on most computer systems.

[16]www.whitedwarf.org/metcalfe/node14.htm

stage utilises the information obtained during the prior stage to narrow down the search interval further.

Messy GA is an adaptive version of GA in which the chromosome length is dynamic and position independent operators are used [GD89].

Example 10.12 Assume that a family possess m portfolios[17] where each portfolio has a maximum of n stocks. We will take the epoch as a trading day. At any time during the epoch, when the family feels that the market is reaching a maximum for that day, a decision is made to update the portfolio. The genetic operators correspond to the following:

mutation: A portfolio is chosen at random and then a stock is selected at random from that portfolio. This selected stock is then replaced by another stock from the stock population, such that the worth of the portfolio is increased.

crossover: two arbitrary portfolios are chosen and then two *nonidentical* stocks are randomly chosen from each portfolio and they are exchanged between the portfolios (shuffled), which in practice implies buying/selling of the stocks.

selection: The worth of each of the portfolios is determined using the current market price and at least one of the portfolios is kept untouched for the next day (next generation). If the market is down, none of the above operations may be able to give a positive yield on the portfolios. Hence we introduce a new operation called shuffle as follows:

shuffle: Choose a portfolio at random and identify two stocks in that portfolio such that by changing their share percentages will result in a positive yield.

In a computer simulation of the above problem, we may assign probabilities for each of the operations and iterate over the epochs trying to maximise the worth.

10.7   Neuro-Genetic Models (NGM)

Other supervised learning models can be used in sequence or in parallel with GAs. A good choice is a multi-layer neural network model. Neural networks and genetic algorithms have their own strengths and weaknesses. The NGM is a hybrid computational paradigm built using a combination of neural networks and genetic algorithms by optimally combining them to solve complex practical problems. As the GAs expect an encoded bit-string representation, the neural networks can be used as a front-end to produce the chromosomes with a wide variability in the genes, which can be used as the initial population of GA. It has been applied in bankruptcy prediction [SL04], [KB05], digital I/O buffer circuit design [UR05], etc. GAs can also be combined with fuzzy logic to produce intelligent multi-objective hybrid control systems. In addition, GAs can be used to construct neural networks ([AS94],[JM98]).

10.7.1   Genetic Programming

An automated method to create or select the most fit computer program from a starting set using artificial evolution is known as genetic programming (GP) [KR92]. It extends the character encoding of GA space to program structures (organised as a tree). An advantage of GP is that it is flexible enough to work in any specified languages. GP has been applied in symbolic regression (function identification), optimal control, game strategy development, sequence induction,

---

[17]A portfolio is a collection of shares of different firms. Three IBM shares, 2 Microsoft shares and 5 Adobe shares is an example portfolio.

creation of computer programs for automata, multi-agent systems, character recognition, cluster computing, automated synthesis of electrical circuits, controllers, antennas, etc.

A tree structure is used to represent the genetic programs efficiently. GP mutation works at the internal nodes or leaf, and involves replacing the sub-tree (or leaf) with a newly generated subtree (or leaf). The size of the tree can change dynamically (during run time). The crossover operation works at the internal nodes of the tree. In the case of selection, the starting set should contain complete programs for the problem to be tackled and a fitness measure that can be applied to each of the programs during execution. The ability of the program to solve a problem is measured by its fitness value (which is available either during or after successful completion of the program). A high level description of the requirements (what need to be solved and how) can be used as a starting set. GP can use arithmetic operators $(+, -, *, /,;\%)$, built-in functions (sin, cos, exp, tan, log), Boolean operators (AND, OR, XOR, NOT, NAND), arithmetic functions (max, min, range, avg, etc).

An important issue in GP is the space and time complexity. The space complexity (memory requirement) is proportional to the product of the population size and individual programs, while the time complexity is a function of the number of generations, size of data in addition to the above. GP has produced several human-competitive, patented or patentable inventions in various fields (see www.genetic-programming.com/humancompetitive.html for several human-competitive results produced by Genetic Programming).

10.8   Applications

The GA has been applied in a wide variety of fields, primarily due to the ease with which they can be implemented in complex situations. Some examples are classification [BM95] and clustering ([PS98],[EM98], [BM01], [MB00]), scheduling, data fitting, trend plotting and time series, combinatorial optimisation problems, pattern discovery [PG97], feature selection, medicine [MV99], and information retrieval.

10.8.0.1   Insurance

Insurance business use data mining for claims processing, customer profiling, modeling customer specific optimal premium rates, etc. Insurance claims data contain dozens of attributes, some of which are correlated. Available past data may have missing values, which makes the modeling more difficult. However, identifying groups of individuals in various categories can be facilitated using a well-trained genetic model. These models can then be used either to characterise new customers or to catch fraudulent insurance claims. Fraud patterns are analysed for data compromise. Consider the problem of identifying customer categories with high claims in auto insurance. We will use the attributes and encoding schemes in table 10.8.0.1.

We have left out some of the other important attributes like age, physical ailments, body stature, driving on highways, total traffic violations etc. Since all variables have been numerically encoded, we can use a fixed string representation to encode a chromosome. The mutation operation can be applied at micro or macro level. Because some of the attributes are encoded with a single bit, it is preferable to group them together and apply a single mutation for all of them and apply separate mutations for longer blocks of strings. For instance, if the education levels are from 0 (high school or less) to 10 (post-doctoral), we need 4 bits (3 bits are sufficient for up to 8 levels) to encode it. Table 10.7 gives a sample population of 5 chromosomes. We hypothesise that the fatal accidents=f(years of driving, smoking habits, drunken

Attribute	Encoding
Gender	[1=Male, 0=Female]
Education level	[0=HS, 1=diploma, 2= degree, etc]
Total years of driving	[0 to 15, truncated integer]
Total fatal accidents	[integer] (up to 7)
Smoking while driving	[0=No, 1=Yes]
Drunken driving	[0=No, 1=Yes]
Mobile use while driving	[0=No, 1=Yes]

Table 10.7: Population and training data for insurance application

Encoding	Fitness	Encoding for training data
0 010 0100 000 001	4.00	0 010 0101 010 101
1 011 0010 001 101	6.80	1 110 0010 001 011
1 110 0101 010 011	7.64	1 110 1101 100 010
0 100 0011 011 011	7.512	0 010 0100 011 001
1 110 0110 101 101	6.32768	1 100 0010 001 111

driving, mobile use) where years of driving is an integer (conveniently chosen between 0 and 15, but in practice could exceed it) and all other arguments are binary (see exercise 24 that incorporates gender and education level also). We have chosen the fitness function as f(n, s, d, m)=$((1 - p)^n + 2 * s + 3 * d + 4 * m)/10$, where p is a constant (say .2), n is the years of driving and all others (smoking (s), drinking (d) or mobile use (m) while driving) take values in $\{0,1\}$ (the multiplier constants here are chosen as arbitrary, but can be fine-tuned using past data or other models). If the bits for mutation are generated automatically, we will end-up with all 1's in highly favored loci. Hence we will use actual sampled data (available from prior insurance claims) rather than random data to identify risk categories.

10.8.0.2   Fraud Detection

Many fraud prevention and detection tools are used by financial institutions, telephone companies, stock-brokers and government agencies. Examples of fraud include stolen cards, overdrafts, counterfeit cards, bankruptcy fraud (where a customer transfers huge amounts and files for bankruptcy), falsified applications etc. Different types of frauds may require different detection algorithms. For instance, stolen card fraud can be detected using previous usage patterns (usual amount withdrawal range, time of day of withdrawals, day of week/month most often withdrawn, number of transactions per day). Some customers have a regular withdrawal pattern (on weekends or holidays, during lunch break etc) and definite withdrawal range. In addition, if the transaction takes place at an ATM, the number of attempts in keying pin code, total time taken per transaction, keying speed, how many successive attempts, etc may also be used to advantage.

Attribute	Encoding
Amount	integer $>$ min limit
Hour of day	1 to 24
Minute within hour	0 to 59
Weekday index	1 to 7
Total password attempts	0 to 7
Total key-in time	0 to 63 seconds

Fraudsters have a tendency to spend (or withdraw money) as much as possible without bothering about the actual spending (withdrawing) pattern of the true cardholder. Remote fraud occurs in e-commerce transactions where the cardholder is not physically present, but only the card details are submitted over the net. Supervised fraud detection techniques use a known fraudulent pattern available from past data to detect probable fraudulent transactions. A suspicious score or probability can be assigned to each transaction to categorise them for further analysis by other software tools or for manual processing. The attributes used could be the transaction amount (quantised in proper range), time of transaction, date of year, from/to account details, number of transactions and security clearance agencies (if any). Jam jarring is a technique used by some customers in restricting the use of a card type for a given service. For instance, if a user always pays for diesel/petrol by one type of card and expensive purchases at super markets by another and online e-commerce transactions by a third card, a misuse is easy to detect.

Fraud committed advertently by the customer itself is more difficult to catch. In falsified application form (FAF) type of fraud, the customer deliberately falsifies information or uses forged signatures in the application for an original or duplicate card. This type of fraud should be detected before the card is issued. We will consider the problem of detecting financial fraud at ATMs. Some banks offer different types of ATM cards – regular, premium and Gold cards – with different cash withdrawal and purchase limits. For simplicity, we have assumed only one type of card (see exercise 25). As the amount can be any integer (greater than the minimum amount that can be withdrawn, which is bank specific), we will use decimal encoding. We will not use random data in this example also, due to its tendency to optimise using loci with best-fit numbers. We will assume that we have a large database of prior transaction records that can be sampled.

A population of size 6 is arbitrarily initialised. An array of numbers is stored for each chromosome in the population, as explained below. For each member of this population, we will take a random sample (without replacement) from the available training data. The amount of transaction in the training data will be used to update three things:– the mean amount, the minimum and maximum amounts seen by this chromosome thus far. For example, if the mean in prior (t-1) iterations is $\bar{x}_{t-1}$ and the amount in the sampled training set is x_t, the new mean is $[(t-1)\bar{x}_{t-1} + x_t]/t$. The total password attempts and total key-in time (total seconds taken to enter required information before a transaction is validated) are updated as above. Because the card usage patterns differ among employed and unemployed customers (eg: housewives), it is usual to observe multiple peaks in the time of withdrawal. For instance, employed persons most often use the ATM cards between 7AM and 9AM, during lunch time (which is company dependent) and between 5PM and 9PM on working days. Thus the distribution of withdrawal time is multi-modal (see chapter 3). Hence we will keep a fixed number of modes (say 3) and use

Table 10.8: Training data for fraud application

Transact Amount	Hour of day	minute of hour	week index	password attempts	total time
3000	1	30	6	2	60
2300	05	00	7	3	55
900	08	40	2	1	30
1800	18	30	1	3	42
5500	22	00	1	5	64
3000	12	20	3	3	58
400	06	00	4	1	60
6000	23	40	7	4	58

Table 10.9: Test data for fraud detection

Transact Amount	Hour of day	minute of hour	week index	password attempts	total time	Fraud (Y|N)
2500	22	45	7	3	65	Y
3000	17	15	4	4	40	N
1400	07	30	2	1	70	Y
550	13	00	1	3	36	N
1000	22	50	6	5	55	Y
400	22	25	1	4	43	Y
3000	22	05	3	3	62	Y
1000	15	45	7	1	38	N

training data to update this information. The week index is counted as 1=Sunday, 2=Monday, etc. up to 7=Saturday, which is correlated with withdrawal amount and transaction time. We will use mode (instead of mean, which can result in fractional values) to keep track of week index. We can keep the transaction time either as two separate fields or as a single integer (in minutes between 0 and 1439). The pseudocode is as follows:

```
for iter=1 to numberOfIterations do begin
  select chromosome
  randomly select training instance
  update transaction amount (mean, min, max)
  update mode(s) of time of transaction
  update mode of week index
  update mean of password attempts, total key-in time
end
```

Table 10.8 gives the initial and final populations obtained after applying the above operations.

10.8.0.3 Miscellaneous Applications

The Set Partitioning Problem (SPP) is a combinatorial optimisation problem with many applications, the best-known of which is the airline crew scheduling (ACS) problem [LD94]. Given a finite set of solutions, a set of constraints, and a cost function, the ACS aims to find the schedule that satisfies the constraints at minimum cost. The GA alone may not be enough to find the optimal solution in a practical situation. Hence HGA is used to find near-feasible solutions using a hill-climbing heuristic to search for global optima. Other examples of SPP are tanker routing, assembly line balancing, location of off-shore drilling platforms, and switching circuit design.

Weather prediction systems are extremely complex and contain thousands of input attributes making it an excellent topic for GAs (see also Lanczos method in §12.3, page 369). GAs have been applied in building weather prediction models, modeling atmosphere dynamics etc[GB02]. Other applications include multiple fault diagnosis systems that involve a combination of non-linearly related faulty conditions, timetabling problem ([KL97], [MB03], [BN04]), parallel process scheduling [MM04], medical image segmentation ([GM06], [MH06]), ultrasonic tomography [DR98], clustering ([EM98], [BM02]) and grouping [FE98], [CL00] etc.

Table 10.10: Software for genetic algorithms

URL	Name	F/S
www.bitstar.com	ComputerAnts (Windows)	S
gaul.sourceforge.net/tutorial/simple.html	Gaul	F
geneura.ugr.es/GAGS/	GAGS	F
lancet.mit.edu/ga/dist/	GALib	F
jgap.sourceforge.net	JGAP	F
www.duane.com/~dduane/gannet	GANNET	F
GARAGe.cps.msu.edu/	GALOPPS	F
www.vipbg.vcu.edu/~edwin/lga972home.html	LGA972	F
ftp://ftp.eos.ncsu.edu/pub/simul/GAOT	GAOT (MATLAB)	F
fast.to/EO	Evolutionary Objects (C++)	F
lumpi.informatik.uni-dortmund.de/pub/GA/src/	GENEsYs	F
ftp.neuro.informatik.uni-kassel.de/pub/NeuralNets/	GENlib (GA-and-NN)	F
ftp.cs.ualberta.ca/pub/TechReports/	GIGA	F
www.aracnet.com/~wwir/software.html	Imogene	S
ftp.dai.ed.ac.uk/pub/pga/pga-3.1.tar.gz	PGA	F

Legend: C=Commercial, F=Free, S=Shareware;

10.9 Software for GA

There are dozens of software packages available for solving problems using genetic algorithms. GAUL and GAGS are libraries in C++ for GA, and GALib is a highly customisable C++ library. JGAP is a Java framework for GA and GP. BUGS (Better to Use Genetic Systems) is a Suntools, and X Windows based software written in C to demonstrate GA. It has the standard GA operators and can model genetic drift (www.aic.nrl.navy.mil/pub/galist/src/BUGS.tar.Z) and premature convergence The EM (Evolution Machine) is a software in turbo C and turbo C++

versions, available by FTP from ftp-bionik.fb10.tu-berlin.de/pub/software/Evolution-Machine/. lga972 is a java program that uses casual mutations to control false discovery rates, and to reduce genotyping [RE04]. Genetic Algorithm Optimisation Toolbox (GAOT) [HJ95] is written in MATLAB (version 5 or later). Other implementations include GA Workbench (simtel120 archives eg:wuarchive.wustl.edu), GAC (ftp.aic.nrl.navy.mil/ pub/galist/src/ga/GAC.shar.Z), GALab Optimization Toolbox, etc.

10.10   Exercises

1. Mark as true or false:
   (a) GAs can be used for minimisation and maximisation problems
   (b) The fitness measure of GA should always return a real number
   (c) Crossover operation is applied with higher probability than the mutation operator
   (d) Mutation operator works only for binary encoding
   (e) GAs can be built using mutation operations only
   (f) The inversion operation does not change the fitness of a chromosome
   (g) All chromosomes in the candidate solution set have the same length.

2. A random number generator returns a real number between 0 and 1. Describe how you will use it to generate a chromosome of size n (n bits of 0' and 1's).

3. What is a Genetic or Evolutionary Algorithm? How is genetic programming related to GA?

4. What does a *population* mean in GA? How is the goodness of a population evaluated? Discuss the merits and demerits of starting with a too large and too small initial population.

5. What are the most important factors that a GA designer has to consider in formulating a solution strategy? Discuss some advantages of parallel GA over sequential GA.

6. What factors determine the length of a chromosome for a given problem? What are the deciding factors for choosing initial population size?

7. Apply the inversion operator to the chromosome 1001011010 with (i) left=3 and right=6 (ii) left=5, right=9

8. If the set of symbols are 0,1,2,3 explain how you will use the string representation for a chromosome.

9. What is the solution set for the schemata 1***1, ****0

10. Explain how a coding scheme can be devised for the traveling salesman problem in which each city is identified by a label other than a number.

11. Can the crossover operation produce two identical twins? If so, explain how to tackle it.

12. How many simple 1-point crossover points (that fall in between genes) are there in a string of length n? How many meaningful inversions are there if the genes in inversions are contiguous?

13. Prove that the total number of schemata of length k with cardinality m is $(m+1)^k$.

14. What are the offsprings if parents are given by P1=(001010111), P2=(010111010) and the operations are a) mutation at position 3, b) crossover between positions 2 and 3, c) crossover at position 6 (with bit position kept intact).

15. A chromosome Y is unit Hamming distance away from chromosome X, and a chromosome Z is unit Hamming distance away from chromosome Y. What is the Hamming distance between X and Z?

16. Compute the selection pressure for a population in which the (n_{keep}, n_{drop}) pairs are respectively:– (i) (10,5) (ii) (8,22) (iii) (35,5) (iv) (6,100)
    What does too low and too high selection pressure values indicate?

17. What are the different types of selection mechanisms? Discuss the merits and demerits of each. Which ones do not require random numbers?

18. List any three neighbours of the following encodings that are unit Hamming distance away.
    (a) 1010 (b) 11111 (c) 110011 (d) 101010

19. Find the offspring in each of the following problems, if the two parents are $P_1 = [100001110100]$ and $P_2 = [011101100010]$, and the mask for uniform crossover is (i) [001 010 100 111], ii) [001 010 000 100]

20. How many chromosomes are there for the following schematas in which the alphabet is binary: (i) 1*0**1 (ii) *00*1**

21. Apply the inversion operation at specified positions and block lengths for the following encodings:
    (i) X=[01101011], (pos1=3, pos2=6), (pos1=1, pos2=5), (ii) X=[110001], (pos1=2, pos2=5).

22. Modify the number scrambler program to match words from right to left of the number.

23. Draw the undirected tree whose Prüfer sequence is given below:
    i) (3,3,1,6,6) (with labels 1 through 7) (ii) (1,2,2) (with labels 1 through 5)

24. Modify the sample problem in insurance application (table 10.7) by incorporating the gender (0=Female, 1=Male) and education level (numerically coded) to model the fatal accidents using GA.

25. Assume that a bank issues three types of ATM cards (regular, premium, Gold). Modify the ATM fraud detection application to incorporate the withdrawal limits on these types. Given an arbitrary transaction, apply a two level detection technique to identify if it is fraudulent. The first level compares transaction attributes with global parameters of all customers using that type of card and the second level then uses customer specific past transaction data to pinpoint fraudulent transaction.

10.10.0.4   References

[AH99] Allan, R.J., Hu, Y.F., Lockey, P. (1999) Parallel application software on high performance computers: survey of parallel numerical analysis software, CLRC Daresbury Laboratory, UK. (www.dl.ac.uk/TCSC/Subjects/ ~Parallel_Algorithms/lib_survey/).

[AS94] Angeline P.G., Saunders, G., Pollak, J.(1994) An evolutionary algorithm that constructs recurrent neural networks, *IEEE Transactions on neural networks*, 5(1), 54-65.

[BM95] Bandyopadhyay, S., Murthy, C.A. (1998) Pattern classification using genetic algorithms, *Pattern recognition letters*, 16, 801-808.

[BM01] Bandyopadhyay, S., Maulik, U. (2001) Nonparametric genetic clustering: comparison of validity indices, *IEEE transactions on Systems, Man, and cybernetics* -C, 31, 120-125.

[BM02] Bandyopadhyay, S., Maulik, U. (2002) Genetic clustering for automatic evolution of clusters and application to image classification, *Pattern recognition*, 35, 1197-1208.

[BB93] Beasley,D., Bull,D.R., Martin,R.R. (1993) A sequential niche technique for multimodal function optimization, *Evolutionary computation*, 1(2), 101-125.

[BF92] Bergman, A., Feldman, M. (1992) Recombination dynamics and the fitness landscape, *Physica-D*, 56, 57-67.

[BN04] Burke, E.K., Newall,P. (2004) Solving examination timetabling problems through adaptation of heuristic orderings, *Annals of operations research*, 129, 107-134.

[CE95] Cantu-paz, E. (1995) *A summary of research on parallel GA*, University of Illinois technical report, Urbana, IL, July 95.

[CH02] Cordell, H.J. (2002) Epistasis: what it means, what it doesn't mean, and statistical methods to detect it in humans, *Human molecular genetics*, 11(20), 2463-2468 (hmg.oxfordjournals.org/cgi/content/full/11/20/2463).

[CL00] Chambers, L.D. (2000) *The practical handbook of genetic algorithms: Applications*, 2nd Edition, Chapman & Hall/CRC.

[CD97] Coley, D.A. (1997) *An introduction to genetic algorithms for scientists and engineers*, Morgan Kaufman.

[DL87] Davis, L.(Ed) (1987) Genetic algorithms and simulated annealing, (Booker, L., *Improving search in genetic algorithms*, 61-73) Pittman, London.

[DL96] Davis, L. (1996) *A handbook of genetic algorithms*, International Thompson computer press, Boston.

[DA95] Deb, K., Agrawal, R.B. (1995) Simulated binary crossover for continuous search space, *Complex systems*, 9, 115-148.

[DB99] Deb, K., Beyer, H.G. (1999) *Self-adaptive genetic algorithms with simulated binary crossover*, Technical report CI-61/99, University of Dortmund, Germany.

[DR98] Delsanto,P. P., Romano,A., Scalerandi,M., Moldoveanu,F. (1998) Application of genetic algorithms to ultrasonic tomography, *Journal of the acoustical society of America*, 104(3), 1374-1381.

[EC89] Eshelman, L.J., Caruna, R., Schaffer, J.D. (1989) Biases in the crossover landscape, Proc of third international conference on genetic algorithms, Morgan Kaufman, 10-19.

[EM98] Estivill-Castro, V., Murray, A.T. (1998) Spatial clustering for data mining with genetic algorithms, International ICSC symposium on engineering of intelligent systems.

[FE98] Falkenauer, E. (1998) *Genetic algorithms and grouping problems*, Wiley, NY.

[FZ97] Fleming,P.J., Zalzala,A.M.(eds) (1997) *Genetic algorithms in engineering systems*, IN-SPEC, Incorporated/Institution of Electrical Engineers.

[GL97] Glover, F., Laguna,M. (1997) *Tabu Search*, Kluwer Academic Publishers, MA.

[GM06] Ghosh, P., Mitchell, M. (2006) Medical image segmentation with genetic algorithms, *Proceedings of the genetic and evolutionary computation conference*, GECCO-2006.

[GB02] Gibson, E., Burger, J., Knox, D (2002) Parallel genetic algorithms: An exploration of weather prediction through clustered computing, www.tcnj.edu/~crew/

[GO89] Goldberg, D.E. (1989) *Genetic algorithms in search, optimization, and machine learning*, Addison Wesley, USA.

[GD89] Goldberg, D.E., Deb, K., Korb, B. (1989) Messy genetic algorithms: motivation, analysis and results, *Complex systems*, 3, 493-530.

[HS91] Harp, S.A., Samad, T. (1991) Genetic synthesis of neural network architecture, *Handbook of genetic algorithms* (Davis, L. (ed)), chapter 15, Van Nostrand Reinhold, 202-221.

[HW07] Haupt,R.L., Werner,D.H. (2007) *Genetic algorithms in electromagnetics*, Wiley, NY.

[HJ75] Holland, J. (1975) *Adaptation in natural and artificial systems*, University of Michigan press, Ann Arbor, MI (MIT press, Cambridge, MA [1992]).

[HJ95] Houck,C., Joines,J., Kay,M. (1995) A genetic algorithm for function optimization: A Matlab implementation, NCSU-IE TR 95-09, North Carolina State University, Raleigh.

[JM98] Jain, L.C., Martin, N.M. (1998) Fusion of neural networks, fuzzy systems and genetic algorithms: industrial applications (International Series on Computational Intelligence).

[KT95] Karafyllidis, I., Thanailakis, A. (1995) An adaptive genetic algorithm for VLSI circuit partitioning, *International journal of electronics*, 79(2), 205-214.

[KB05] Kasabov, N., Benuskova, L. (2005) Theoretical and computational models for neuro-, genetic-, and neuro-genetic information processing, *Handbook of theoretical and computational nanotechnology* (Reith, M., Gennes, P-G., Schommers, W. (eds)), - vol X, chapter 41, American Scientific publishers, LA, 1-38.

[KG83] Kirkpatrick, S., Gelatt Jr. C.D. (1983) Optimization by simulated annealing, *Science*, 220, 671-680.

[KR92] Koza, J.R.(1992) *Genetic programming: on the programming of computers by means of natural selection*, MIT press, Cambridge, MA.

[KL97] Kragelund, L.V.(1997) Solving a timetabling problem using hybrid genetic algorithms, *Software practice and experience*, 27(10), 1121-1134.

[LD94] Levine,D. (1994) A parallel genetic algorithm for the set partitioning problem, MCS-P458-0894, Mathematics and Computer Science Division, Argonne National Lab, (ftp:// info.mcs.anl.gov/pub/tech_reports/reports/ANL9423.ps.Z)

[MT99] Man, K.F., Tang, K.S., Kwong, S. (1999) *Genetic Algorithms concepts and designs*, Springer, London.

[MD99] Margaritis, K.G., Digalakis, J.G. (1999) An experimental study of genetic algorithms using PGAPack, *Proceedings of 7th Hellenic conference on informatics*, University of Macedonia, Greece 54006, (medlab.cs.uoi.gr/hci99/applications7.htm).

[MV99] Martin-Bautista M.J., Vila M.A. (1999) A survey of genetic feature selection in mining issues, *Proceedings of Congress on Evolutionary Computation* (CEC-99), 1314-1321, Washington D.C.

[MB00] Maulik, U., Bandyopadhyay, S. (2000) Genetic algorithm based clustering techniques, *Pattern recognition*, 33, 1455-1465.

[MH06] McIntosh,C., Hamarneh,G. (2006) Genetic algorithm driven statistically deformed models for medical image segmentation, GECCO'06, Seattle, July 8-12, 1-8.

[MM04] Meijer, M. (2004) Scheduling parallel processes using genetic algorithms (www.science. uva.nl/research/~scs/papers/archive/Meijer2004a.pdf)

[MB03] Merlot, L.T.G., Boland,N.,Hughes, B.D. (2003) A hybrid algorithm for the examination timetabling problem, LNCS 2740, Springer, 207-231.

[MZ99] Michalewicz, Z. (1999) *Genetic Algorithms + Data Structures = Evolution Programs*, Springer-Verlag.

[MM98] Mitchell, M. (1998) An introduction to genetic algorithms, MIT Press.

[MC00] Mitsuo,G., Cheng,R. (2000) *Genetic algorithms and engineering optimization*, Wiley.

[PS98] Park Y., Song M. (1998) A genetic algorithm for clustering problems, Genetic Programming 1998: Proceeding of 3rd annual conference, Morgan Kaufmann, 568-575.

[PG97] Pei, M., Goodman, E.D., Punch, W.F. (1997) Pattern discovery from data using genetic algorithms, Proceeding of 1st Pacific-Asia conference - Knowledge discovery & data mining (PAKDD-97).

[PK00] Pham,D. T., Karaboga,D. (2000) *Intelligent optimization techniques: genetic algorithm, tabu search, simulated annealing and neural network*, Springer.

[RB04] Ray, S.S., Bandyopadhyay, S., Pal, S.K. (2004) New operators for genetic algorithms for traveling salesman problem, ICPR-04, 2, 497-500, Aug. 23-26, Cambridge, UK.

[RH02] Ribeiro, C.C., Hansen,P. (eds.) (2002) *Essays and surveys in meta-heuristics*, Kluwer Academic Publishers, Norwell, MA.

[RE04] Robles, J.R., Edwin, J.C.G (2004) lga972: a cross-platform application for optimizing LD studies using a genetic algorithm, *Bioinformatics*, 20(17), 3244-3245.

[SL04] Shin, K.S., Lee, K.J. (2004) Neuro-genetic approach for bankruptcy prediction modeling, LNCS 3214, *Knowledge based intelligent information and engineering systems*, Springer, 646-652.

[UR05] Ulrich,R.M., Ratcliffe, S., Mutnury,J., de Araujo,B., Cases, D.M. (2005) Neuro-Genetic models in modeling non-linear digital I/O buffer circuits, Electronic components and technology conference, 55(2), 1543-1548.

[VM99] Vose, M.D. (1999) *The simple Genetic Algorithm: Foundations and theory*, MIT Press, Cambridge, MA.

[WD94] Whitley, D. (1994) A genetic algorithm tutorial, *Statistics and computing*, 4, 65-85.

11
Neural Networks

CHAPTER OBJECTIVES

- To introduce neural networks and its inspiration

- Discuss the advantages of neural networks

- Understand the components of neural networks

- Discuss neural network topologies

- Review special types of ANNs

- Understand backpropagation learning

- See some applications of neural networks

11.1   Introduction to Neural Networks

Definition 11.1 A neural network is a biologically inspired nonlinear network computing model for information processing and exploratory data analysis, having a distinct ordering among the sets of neurons arranged as input and output layers with zero or more processing (hidden) layers that are interconnected by signal channels and fine-tuned by a training algorithm.

They are aptly called *neural networks* because they are biologically inspired models based upon the neural structure of the human brain, and its ability to learn from experience. They are also called Artificial Neural Networks (ANN) to distinguish them from 'natural neural networks' in the brain of living organisms. The nodes that process information in an ANN are also called *neurons* because it is the basic building block of the network computing model. They operate in parallel in a manner reminiscent of biological neural networks in the brain. They transform incoming data using activation and transfer functions and pass the results to each subsequent layer neurons up to the output layer.[1]

Neural networks as a computer science topic began with the seminal paper by McCulloch

---

[1] All neural networks do not use the layering concept. For example, the Hopfield network has multi-directional data flow among the nodes and is used for optimisation problems like TSP.

and Pitts [MP43], in which the authors explained the formal logic behind threshold switching circuits. In 1961, Rosenblatt [RF61] discovered an iterative learning procedure for single-layer perceptrons, which was extended by Minsky & Papert [MP69], Rumelhart [RH86], and many others. They have become major contenders to other classification, prediction and process control models due to their flexibility and ease of implementation in hardware [MF96], [MC02][2]. Since conventional computers use a cognitive algorithmic approach, and a *stored program* concept, they are more general in solving problems from a wide variety of disciplines. Neural networks complement conventional computers rather than compete with it.

> Neural cells of the brain are called neurons and were discovered in 1836. The first mathematical representation of an artificial neuron appeared in 1943. Natural neurons have 4 components called *axon*, *dendrites*, *soma*, and *synapses*. Dendrites are used to receive incoming signals to the neuron. Soma is like a CPU that process information nonlinearly. Transformed signals are passed to the axon and synapses to transmit it to neighboring neurons. Input to the natural neurons come as a stream of impulses from exterior sources (sight, sound, smell, taste, touching) or internal sources (hunger, happiness, fear, other emotions, feelings and body needs). Natural neurons often output various commands to different parts of the brain. They work in tandem with other neurons.

Neural networks are used as predictive or classification models in supervised, or semi-supervised modes. They can also be used as unsupervised models like clustering, image annotations and recognitions, etc. A few of the popular application areas are as follows:

1. Classification (medical procedures, medical images [ECG, EEG, X-rays], web sites, outcomes of applications (for admission, job, filing tax, loan etc), frauds, financial transactions or (unsuccessful) attempts to money transfer, email messages (as spam or non-spam).)

2. Forecasting and prediction (stock market, bond returns, future yields, apartment rents, real-estate prices, medical conditions (eg: diabetes onset [SE88], blood glucose levels ([PJ97],[PJ98]), convalescence periods, hospital stays, machinery failures and reliability, survival periods, etc.)

3. Clustering (attribute clustering, data clustering of web pages, grouping customers, students, industries, etc.)

4. Optimisation (discrete, continuous multivariate optimisations, 0-1 optimisation, n-bit parity problems [LH00], other combinatorial optimisation problems)

5. Function evaluation and approximation (used to approximate continuous functions as well as functions with a finite number of discontinuities [FS05])

6. Modeling and model evaluation (buildings, structures, plants, organic compounds, evolutionary models)

---

[2]Hardware implementation of neural network processors are called neural chips that have focused applications like signature analysis and verification, fingerprint matching, etc. An advantage of neural chips is that complex networks can be built by connecting them in series and parallel. Neuromorphic chips are miniaturised VLSI chips.

7. Monitoring (conditions of machinery, online financial transactions, automatic drug administration in hospitals, cruise control of vehicles).

8. Expert systems (A trained neural network acts like a "domain expert". It can be used to provide projections for new situations of interest[3]).

9. Miscellaneous (blind source separation [BG92], [GM99], [HL05], blind deconvolution, data compression [JJ99], image registration, pattern recognition and visualisation.)

11.1.0.1 Neural Network Inspiration

Study of neural networks was inspired by the following characteristics of human brain:

- Parallel and distributed nature of the brain

  The human brain is an amasing organ with about 10 billion neurons in adults, each of which is connected on the average with 10000 synapses. Hence it can be regarded as a massively parallel information-processing engine with input coming from either external sources (sight, sound, etc) or internal processes (hunger, internal pain, happiness). It is divided into distinct sections called cortex, mid-brain, brain-stem and cerebellum with their own functionality. The largest system in the brain is called visual cortex, which has distinct processing layers. The visual cortex processes visual data and can work in parallel with auditory and temporal cortices so that we can 'see', 'hear' and 'talk' simultaneously. In fact, each of the senses create parallel data channels using neuronal connectivities in respective regions. Each of these regions are further subdivided into areas according to anatomical structure (this is analogous to multi-dimensional neural networks). Synergy of a large number of neurons that work in parallel towards a common goal makes the brain act like a distributed computer system. Analogously, the ANNs are divided into multiple layers (input layer, output layer, processing layer(s)) with their own functionalities.

- Learning from experience

  The normal brain learns from experience. It stores information as multi-dimensional 'patterns' (which is why we can identify an invisible creature (a person, a bird, an animal, etc), or an object (a train, an aircraft etc) by 'pattern matching' the sound/noise coming from it with all familiar sound patterns in store. When our friends talk over the telephone, our brain can instantaneously identify them by comparing the incoming speaker signal with all familiar sounds we remember). A healthy brain can recognise complex patterns and distinguish between minor variations in frequencies, colors, intensities or color combinations of the same pattern. The self-learning nature of the brain has been trained from childhood to recognise visual stimuli from many sources. Neural network training is the corresponding analogue, which uses a *learning algorithm* (The collective minimisation of interconnection

---

[3]Expert systems are software programs that act as human experts and represent knowledge for reasoning, decision making or advising. Most expert systems have knowledge bases, which are accessed using a control program.

weights using the errors observed at each of the neurons is called 'learning'). Multidimensional neural networks are used in places where uni-dimensional networks either perform poorly or are inadequate.

- High connectivity, modifiable connectivity
  Neurons are densely interconnected with neighboring neurons through synapses. These connections are either static or dynamic. Power of the human mind comes from the dynamic nature of synaptic interconnections, and quickly modifiable electrochemical pathways, which allow the brain to retrieve vital information instantaneously during conscious state. This is analogous to dynamic neural networks, modifiable architectures, and weight modifications discussed below.

- Supervised learning mode
  Most of our childhood learning occurs in supervised mode. Similarly the neural networks are not pre-programmed, but rather obtained by 'training' it using empiric data. Most neural networks have high computational requirements during the training and validation phases than the running phase. A nonlinear function of input variables is minimised iteratively during the training phase by adjusting model parameters (weights or linking factors) so as to achieve a minimal error on the observed output.

- Fault tolerance
  The human brain has an amasing fault tolerance capability. If the brain is damaged (due to medical conditions or external impulses), it tries to adjust the consciousness in seconds. This technique is adopted in neural networks during training phase as well as in probabilistic and dynamic neural nets (see discussion below).

- Feedforward and feedback connections
  Our brain has feed-forward connections (from earlier processing stages) near sensory input and feedback connections near motor output. This idea is utilised in a variety of neural nets discussed below. Feed forward ANNs do not have loops, whereas feedback (recurrent) ANNs can have loops at any of the nodes (except perhaps the input layer). If all nodes have loops, it is called fully recurrent.

- Information storage
  The human brain stores patterns sequentially. This is especially true for text and audio data. This is easy to prove using a simple experiment. Ask any kid to utter the alphabet of their mothertongue from first to last letter (A to Z in English) as fast as possible, and note down the time taken. Then ask them to utter the alphabet in reverse (Z to A in English). Most kids will either abandon it after a few letters or take much more time than the forward order, unless they have specifically memorised it in reverse order as well (in which case it creates a new linked list). This is because the brain stores text patterns (including alphabets and songs) as a forward chain (analogous to forward linked lists) in one direction only (exceptions exist for some humans). This shows that neuronal

interconnections are either sparse or non-existent in the reverse direction. This idea is used in ANNs to recognize patterns in forward direction.

- Visualising neuronal firing
  Calcium channels are crucial for rapid neuronal functioning in the brain. Synapses exhibit sudden calcium ion concentrations in specific patterns while firing neuronal impulses in forward direction. Using dyes that emit fluorescence in response to increased calcium concentrations, researchers analyse pathway connectivities in dynamic synaptic interconnections as they participate in collective communications. This allows purely optical probing of brain signals using laser beams. As ANNs are built entirely in memory, displaying architectural diagrams is extremely useful during the training phase to visualise neuronal firing patterns, especially in hidden layers. As the neural processing occurs too rapidly, time dilation and synchronisation parameters may be used to slow the ANN computations down to any desired level to make the visualising easier.

11.1.1 Advantages of Neural Networks

There are many advantages to neural networks over other popular data mining models. Applications developed using the neural network model can work across geographical boundaries because localisation is possible during the training stage. They learn continuously from new data patterns and can outlive other models.

- No assumptions on data
  While statistical models make many assumptions on the nature or characteristics of the data, neural networks are more general and work on all types of data (numeric, categorical, binary, etc). Neural networks can accommodate noisy data, including outliers.

- Adaptive learning
  Using training data, the model learns how to do tasks for a complex combination of input values to be expected when the model is put to use. Validation instances are then input either to extract patterns or detect trends. Input instances should have maximum information for the model to apply what it has learned during the training. They are programmed to adapt continuously to new data by tuning the weights and to track changes dynamically. For example, fraud detection of handwritten bank checks use advanced adaptive decision technology with the help of a huge database containing digitised images of all bank checks written by customers of financial institutions that constantly acquire new customers, and terminate the accounts of existing customers.

- Self-organisation
  During the learning phase, the model is taught how to adjust itself or self-organise. Once the model is built, the redundant data transmission links can be deleted. Special types of ANNs like Kohonen's self-organising maps are used for classification, clustering and visualisation.

- Real-time operation

  ANN computations may be carried out in parallel, and special hardware devices are being designed and manufactured which take advantage of this capability. They may be used in batch mode or real-time mode. Timing restrictions between computation steps can be easily incorporated, which is analogous to the synaptic delay (which is half a millisecond in mammals) in the brain.

- Fault tolerance via redundant information coding

  Neural network models used in critical applications are expected to have high fault tolerance capabilities. Failure of nodes or communication links (or both) should have minimal impact on the output of the model. Partial destruction of a sub-network could lead to degradation of performance. However, some network capabilities may be retained even with random network damage.

- Can work as ensemble model

  As optimisation is done using backpropagation learning algorihm, it may be combined with prediction, classification or clustering simultaneously. See [AP92] for a simultaneous optimisation and segmentation model, and [DC98] for optimisation and prediction model.

- Scalable processing power

  The massive processing power of multiprocessors (machines with multiple CPUs) can be utilised to build neural networks by representing each node with a processor. Distributed network models are also possible if communication cost (to transmit data between nodes or layers) is negligible when compared to computation cost, but is uncommon due to the number of node interconnections involved. Processing power of the model is scalable to any complexity to suit computational demand of the problem at hand. Hence it is ideal for solving NP complete problems.

- Flexibility

  Classical statistical techniques like MLR (multiple linear regression), nonlinear regression etc have assumption on the structure of the underlying model or on the error terms. An advantage of neural networks is that it can model data for which the exact form of relationships are unknown. Different statistical models are used when the response variable is categorical (Logistic Regression (LR)) and continuous (MLR). But the neural networks can have categorical and continuous variables, can accommodate multiple dependent variables and can be used even in cases like multicollinearity, and nonlinear separability in pattern space. They can model linear as well as nonlinear relationships and have the ability to use a variety of transfer functions. This flexibility allows a wide variety of models development possible. In addition, they can be implemented purely in software on conventional computers, distributed computers or specialised devices, in hardware (using analog or digital circuits) [MF96], hybrid circuits or optical networks.

In spite of the above advantages, there are a few disadvantages as well. The major ones are (i) sensitivity to error propagation (ii) implementation cost (iii) local optima entrapment (iv) domain knowledge.

In contrast to the statistical models that explicitly express the dependence as a functional form with error terms, the neural networks neither reveal the functional form nor the strength of dependence among the variables. Neural networks implementation costs used to be high in the beginning. But with the availability of generic neural net software packages using which the topology and parameters are specified on the fly, the implementation costs (using software) have substantially come down. A good domain understanding is needed to decide about the proper topology of the neural network to be used in any application[4]. In addition, some mathematical skills are required to manually train (nontrivial) neural networks.

11.2 Components of Neural Networks

Most neural networks have a simple structure. They have an *input layer*, an *output layer* and zero or more *intermediate layers* (called *hidden* or *processing layers*). Input layer receives data from the outside world and pass it on to the processing layer (if it exists). The output layer represents decisions or control signals that are presented as the output of the neural network. A neural network is characterised by the network topology (layers, number of nodes in each layer and interconnections, feedforward or recurrent), number of input and output nodes, activation function, threshold function and learning algorithm to be used. Neural network structures involve five main concepts:

- Network architecture
  Each neural network has its own architecture to learn the relationships between input and output values. The topology of a network is determined by the number of layers, number of neurons in each layer and their inter-connectivity (see §11.4).

- Activation and transfer function
  Neuronal behavior is specified by two functions (i) the activation function and (ii) the output function. Activation function is used to transform the input values using synaptic strengths to a neuron. The transfer (threshold or output) function is used to output a value (called firing) from the neuron (see §11.3.1).

- Synaptic weights
  These are either completely pre-specified by the user, obtained from previous runs of the model, assumed by the developer of the model or generated randomly. Weights can change during the training phase, but are kept constant during the running phase. Input layer either applies a weight, or passes the values unchanged (all weights are one) to the hidden layer. If all outgoing synaptic weights are ≥ 1, the neuron magnifies the output values. If all weights are $0 < w_i < 1$, the output from that neuron is diminished. Hence at network initialisation time, all unknown synaptic weights should preferably be initialised to 1, instead of zeros.

---

[4]Hence the real question is not whether the results obtained by the model are right or wrong, but whether it was trained right or wrong

- Training algorithm

  Backpropagation(BP) is the most popular learning algorithm for neural networks. It uses the squared deviation between expected and observed outputs to incrementally update the current synaptic weights at successive levels from the rear to the front of the network. Thus the name *backpropagation*.

- Training and testing sets

  Input instances (in batch mode) are divided into a training set and a test set. A training algorithm builds "intelligence" necessary for the model to solve practical problems. Hence an untrained neural network model is almost useless.

11.2.1 Layering Concept

Nodes in a multi-layer neural network are arranged into distinct layers (physical or logical). Connections (also called signal channels) are established between nodes in adjacent layers. They are indicated by arrows for visual comprehension. Data transmission takes place only through these synaptic connections in the forward pass. One neural network is distinguished from another by the number of layers, grouping of neurons in each layer, interconnections between layers, the summing and transfer functions, in addition to the input and output nodes. A neuron at an intermediate layer may pass its output to all of the neurons below (to the next layer) or to a selected subset of them (if the synaptic strength is zero, the corresponding link is defunct).

11.2.1.1 Input/Output Layers

Input/output layers provide external interfaces to the real world. Results of processing are made available at the output layer (in the form of numbers, categories, bit settings, signals etc). For example, an ANN used to check if a complex electronic component has acceptable quality can input various signals from components and can output a value of 0 to indicate 'unacceptable' or a 1 to indicate 'acceptable', which can be interfaced with visual indicator lights (Green=acceptable, Red=Unacceptable).

Number of nodes in the input and output layers will vary according to the application. For example, in a function evaluation application, the number of input nodes is the same as the number of variables (plus network control parameters like thresholds) and the number of output nodes is usually one (function value). In clustering applications, there are as many input nodes as the number of variables and the number of output nodes may either be pre-specified (desired number of clusters to be extracted) or unspecified (in unknown number of clusters). In classification applications, the number of output nodes is the same as the number of classes. If the input layer has as many nodes as the number of features and the output layer has as many nodes as the number of classes, the ANN can be used as a total classifier.

Information is processed by the nodes in discrete time intervals (discrete-time neuron) or on a continuous scale (continuous-time neuron). For example, in process control applications that have multiple sensors (temperature, pressure, voltage etc sensors) the input comes continuously as multiple streams that are synchronised at the input layer. Automatic defect detection (using neural networks) in textile, plastic or paper rolls that have just been manufactured uses a window of appropriate size that depends on the physical size of scanning device. Each window is identified (to mark defect areas) and the roll is advanced after each step.

All input attributes to a neural network should preferably be numeric (interval or ratio scale). Any categorical data must be re-coded into numeric form before it is input. For example, if an input attribute is 'Gender' with values 'Male' and 'Female', we can recode these as Male=1 and Female=0 or vice versa. Which coding scheme is used does not matter really, because the model is built and adjusted using the training instances and validated using test instances. Hence the same numeric equivalents must be used in training and test instances. In other words, Male=1 and Female=0 coding is used in all phases. Neural networks are also being used to process other types of data like text, audio and image data. Either numeric equivalents are derived from such data or they are transformed to appropriate frequency domains.

An input instance is a set of input data values applied to the input layer nodes. It is preferable to standardise input values to a proper range (usually [0,1] or [-1,1], so that the firing functions can limit the output of each of the neurons to a strict range. Input neurons multicast their values to the neurons in the next layer. The Kohonen networks (without hidden layers) multicasts its output directly to all output layer neurons.

11.2.1.2 Data Transformation And Communication

Data communication is easy and straightforward in neural networks implemented entirely in software. If input data comes from sensors, data capture devices, or hardware components, a validity check may be required to catch missing values, out-of-bounds values and noise present at the input layer. Data standardisation (change of origin and scale), data transformations (eg: logarithmic, orthogonal etc), filtering (low-pass, high-pass, band-pass filters) may also be done at the input layer. The output layer similarly can perform transformations (eg:re-coding, scaling, etc) to present the output in an easily understandable form to the user. In an early warning neural network model to check for intrusion on runways in a busy airport, the numbers at output layer (2D/3D minimal separation distance between aircraft pairs on ground or in flight) may have to be re-coded as warning levels that are used to highlight the concerned aircraft position on the screen (continuously) or as text, voice or alarm messages to the air traffic controller (ATC) one or more times or sent as messages/commands to the respective pilots/autopilot (at intervals), or a combination of these may be triggered. Data at the output nodes may also be post-processed to relieve the output node(s) off its overload. If the separation distance is below a threshold (or is likely to fall below the threshold in the near future), remedial measures can be initiated immediately and the intensity of the alarm/message may be gradually increased to warn about an imminent action on the side of both the ATC and pilots and the process reversed when remedial actions have been initiated by the aircrafts. Because these involve interfacing with many other external systems, the output layer may be programmed to trigger appropriate external systems using different value combinations received at the output layer.

11.2.1.3 Training Phase

Neural network models have two stages during its lifetime, called *training* (or learning) phase and *test* (or running) phase.[5] Network connectivity and parameters are adapted during the training phase to minimise the error deviations between observed and estimated outputs.[6] Number of training instances needed depends upon the complexity of the underlying function to be approximated, time constraints and expenses involved in preparing data and chosen thresholds. All training instances should have the same number of variables. Missing data, if any, are replaced with reasonable estimates. Once built, the neural network can accept any number of input stimuli less than or equal to the maximum number of input nodes. Success rate is highest when a new input instance has the maximum number of input attributes. This is in sharp contrast with DT and other supervised models that require complete data values.

In the software implementation of neural networks, the initial weights are either chosen from prior domain knowledge[7] or generated randomly at run time. Weights may also be assigned empirically or analytically in the most trivial networks. This configuration is used to compute the output value(s). Weights are adjusted in the proper direction using the difference between observed and expected values (so that the errors in each of the output nodes are minimised). This is known as *supervised learning*.[8] Positive weights are *excitatory* and negative weights are *inhibitory* to the neuron response. If the training phase input instances are true representative samples of the population, the network will exhibit optimal computational behaviors for all types of input patterns during running phase.

11.3 Training Algorithms

Gradient-based methods are the most popular training algorithms for nonlinear multilayered networks. They are either first-order (steepest descent[9]) or second-order (conjugate gradient, quasi-Newton). They often require multiple passes (iterations) through the training data. A learning rate[10] (η) is used to adjust the weight threshold, which is dependent on the network topology and magnitude of input and output vectors.

Stochastic training methods use parallel sampling of the solution space to determine the best fitness weights [MD89]. Other training methods include genetic algorithms ([MJD89],[WS90]) simulated annealing, expectation maximisation [MF97], Gauss-Newton optimisation,[BK93] that

---

[5] If input variables are time dependent, there can also be a pruning or validation phase due to dismal performance over the passage of time.

[6] Some authors consider the validation phase to be distinct from training phase. Weights modification is the most frequent activity during this phase because deleting network connectivity is accomplished by setting the corresponding weight to zero. See pruning phase below.

[7] Once the model stabilises, the input/output variable orders and connection weights can be saved (to permanent storage or external memory) for future use.

[8] Mathematically, this is accomplished as the product of the weights and partial derivatives of the function to be minimised.

[9] French mathematician A.L.Cauchy (1789-1857) introduced the steepest descent algorithm in 1847.

[10] which assume large values during training phase to minimise training time.

involves the inverse of a Hessian matrix[11] and hence is not popular. Variants of the BP algorithm include adaptive BP [RB93], quick propagation [FS88], batch propagation, incremental BP, rational BP, and second-order BP (Levenberg-Marquardt algorithm [HM94], Pseudo-Newton) [TW03]. The evolutionary algorithms are much faster in general than gradient-based methods for training neural networks. A discussion of the hybrid neuro-genetic model can be found in [KW94], [CM00].

Learning phase need not stop after the initial epoch. Data distributions constantly change in some data mining applications. Hence an option to change the mode to 'training' or 'running' phase is usually provided, using which the model can be re-trained, if needed. If the weights are not tuned periodically, error propagation may result in wrong classifications or interpretations at the output nodes. The proportion of falsely rejected cases (in classification models) is called False Discovery Rates (FDR). Statistical measures (based on FDR, distance metrics or correlation coefficient) are available to decide if and when the model needs to be re-trained.

As done in the case of decision trees, neural networks can also be constructed as a self learning and adjusting model. A problem in this mode is the slow network convergence and related issues. A too slow convergence may be due to a trapping in the local optima, choice of a wrong model, or too small learning rate.

Figure 11.1: Linear (a), threshold (b) and sigmoid (c) output functions

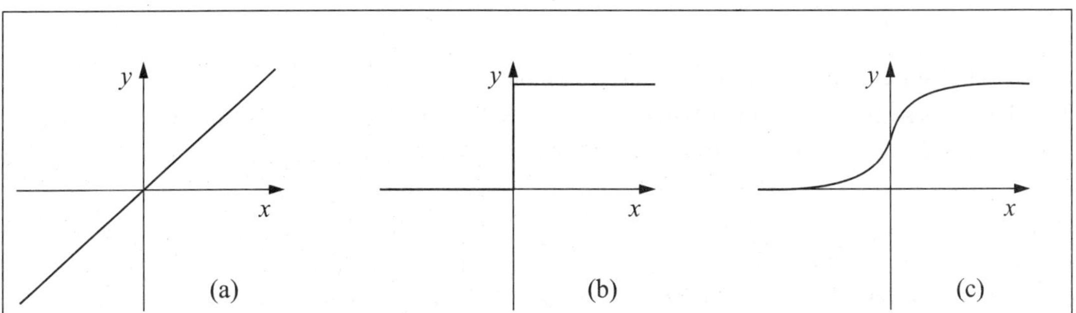

11.3.1 Activation Functions

In a FFN, each node accepts input from the nodes in the preceding layer, applies its own transformation using an activation (evaluation) function, and passes the result to the next level forward or outputs the results (at output layer). Depending upon the problem, the evaluation function can be any mathematical transformation with finite range as output. In other words, the neurons can apply any type of computation on the input. The most common operations

---

[11] A matrix obtained from the second derivative of the loss function wrt the weights between two layers is called Hessian matrix. It is used to measure the quality of a network in terms of the condition number (which is the ratio of largest and smallest eigen values).

are ORing, ANDing, XORing, finding the average, range, min, max, etc or applying a simple 'summing' function to the input:-

$$u = \sum_{j=1}^{n} w_j x_j + \theta \tag{11.1}$$

where w_j are the synaptic weights[12], θ is a threshold (or bias), and n is the number of input values to the neuron. Other popular choices for weighting functions include second order weighting schemes, density based weighting etc [TT95], [DK01]. Result of the weighing function is applied as argument to a threshold function for amplitude bracketing. It can have a variety of forms – linear, sine, hyperbolic tangent (with range -1 to 1), Gaussian[13] or sigmoid functions (f(x)=1/[1+exp($-\lambda$x)]) being the most popular. The sigmoid (with range 0 to 1) is the most preferred[14] choice, as it provides a smooth transition between low and high outputs. In the case of a two-layer network, the results are represented as a vector-matrix product. Let

$$W = \begin{pmatrix} w_{14} & w_{15} & w_{16} \\ w_{24} & w_{25} & w_{26} \\ w_{34} & w_{35} & w_{36} \end{pmatrix}$$

denote the weights associated with various nodes in two adjacent layers. When an input $X = [x_1, x_2, x_3]$ is applied to the input layer, the values get transformed as linear combinations (ie. vector-matrix product WX), so that the net input at node 4 becomes $W_{14}X_1 + W_{24}X_2 + W_{34}X_3$. Linear neurons are too simplistic to solve general problems and are used as building blocks for larger networks.

11.3.1.1 Sigmoid Function

A good choice for an output function is the sigmoid[15] function $y = f(x) = 1/(1 + e^{-\lambda x})$, ($\lambda > 0$). The derivatives are $y' = \lambda$ sig(x)[1-sig(x)] and $y'' = \lambda$ sig(x)[1-sig(x)][1-2 sig(x)]=$\lambda y'$[1-2 sig(x)]. When $\lambda = 1$ it is called standard sigmoid. It crosses the Y-axis at y=.50 irrespective of the value of λ. Output of the evaluation function close to 1 should indicate a sufficiently excited system. If there are 'p' input nodes and one hidden layer with 'q' processing nodes, the computations are carried out as follows: $a_j = 1/(1 + e^{-S_j})$ for j=1,2,$\cdots$, q where $S_j = \sum_{k=1}^{p} w_{k,j} I_k + \theta_j$, where I_k is the kth element of input vector and $w_{k,j}$ is the connection weight between nodes i (input) and j (hidden), and θ is the threshold.

11.3.2 Running Phase

Once properly trained, the neural network can be used in running mode to accomplish what it is trained to do. This phase is used with new or unseen instances. Thus the data and learning algorithms used at the training phase are the most crucial in determining the success of the

---

[12]The weights are real in most networks, but could also be complex numbers [IT01].

[13]The Gaussian function is called the normal distribution in Statistics and is named in honor of the German mathematician C.F. Gauss (1777-1857).

[14]Sigmoids are easy to implement, has continuity (and differentiability) and tractability over a wide range. It's derivative can be calculated from its output as $\partial sig(x)/\partial x = \lambda$ sig(x)[1-sig(x)]. A scaled sigmoid with argument (x/T) is sometimes used.

[15]Other choices of mathematical functions like hyperbolic tangent, etc also exist, but the sigmoid is preferred due to its simplicity

neural network model. Parameters (and node connection weights) are held fixed at this phase. If the training phase is immediately followed by the running phase, the network parameters are readily available. In some data mining applications, there can either be a delay between the two phases, or the training may be done on one machine and the network executed on multiple machines. In such situations, the parameters are externalised (saved to persistent storage) and restored prior to the running phase. Proper attribute types, orders and symmetry in the input layer that were used during the training phase may have to be maintained at this phase also (exception to this rule exist as in function evaluation).

Neural networks may be used in real-time mode (`continuous` basis) or in batch mode. An example of the first type is a neural network model to identify flight exceedences throughout the duration of a flight (from the takeoff point from origin airport to landing or taxiing position at destination airport). The flight data including crew commands are continuously fed-in as input. Other examples are process control networks, intrusion detection networks, video surveillance networks on busy intersections and roads, etc. Most of the neural network models used in data mining fall in the second category (batch-mode operation), where a finite number of test cases are fed into the model to produce a finite number of values (or signals) at the output layer. Examples are classification, clustering and prediction problems.

11.3.3   Pruning Phase

The pruning stage is used to remove redundant connections, if any. It simplifies the interconnectivity by deleting irrelevant neurons (and corresponding incoming, outgoing edges), replacing multiple synaptic connections by lesser ones, etc. The least useful connections are identified and weights are set to zero (if it falls below a small threshold). In case of a 'tie', the connections are removed one at a time if tied connections are between different layers because the deletion of one connection can make a 'tied' connection to exceed a threshold. If two tied connections are from (or to) a single neuron, they may be dropped simultaneously. Sparse and contiguous connectivity patterns (low weight spatial clusters) in large networks can also be identified during the pruning stage either to collapse the connectivities or to modify the architecture. A pruning algorithm that involves solving a linear system in least squares sense was suggested for second order recurrent networks in [CF06]. Alternative pruning algorithms have been suggested by many authors [MK00] and a survey can be found in [RR93]. Pruning with constraints utilise one or more conditioning on model parameters.

11.4   Network Topologies

Dozens of different types of neural networks can be classified as {weighted or unweighted}, {feed forward or recurrent}, {synchronous or asynchronous}, {with or without hidden layers}, and {with or without stored unit states}. Many popular topologies like unstructured networks, layered networks, modular and probabilistic networks etc also exist for neural networks. In the feed-forward neural network (FFN), signal transmissions take place in the forward direction.

The number of nodes to which a fixed neuron is connected is called its *degrees of freedom*. Too many degrees of freedom may lead to poor generalisation capabilities (exceptions are Kohonen and Hopfield networks). Thus the degrees of freedom must be maintained at an optimal level. A fully-connected FFN (FC-FFN) has each node at a layer fully connected to all other nodes in the adjacent layer.

In a Recurrent Neural Networks (RNN), the output of one layer is fed-back as input of previous layer neurons. Recurrent networks provide a much compact structure than an equivalent FFN, in analogy with IIR filters over FIR filters in digital signal processing. RNNs are suitable for representing dynamic nonlinear problems, and find applications in speech recognition [MV02], system identification etc [MC01]. As the gradient of the error wrt the connection weights are not easily solvable for recurrent multi-layer networks, the BP algorithm is suitable for only the most trivial or restricted feedback networks[16]. However, in simple feedback networks, the backward connections can be unfolded in time, and the network converted into a purely FFN and trained using BP as if it were of the standard form.

Neural network connectivity can be modeled as a connected directed graph (chapter 5). If the network is recurrent, the graph is cyclic. The RBF networks (§11.5.9) can be considered as directed bipartite graphs. In most graph theory applications, there exist one special node called the *start node* where the graph algorithms start executing. In the case of neural networks, the start nodes are those in the input layer. They need not be unique (in some problems with multiple input layer nodes, prepending another node may reduce it to a single start node). For example, in a neural network used to catch fraudulent transactions, the input nodes will capture transaction attributes (amount, date and time, number of attempts to key-in the pin code, etc). Date and time can be re-coded either to be numeric offsets after a fixed origin, using concatenation of sub-fields or as a linear function of their components. A special 'driver node' (called a virtual input node) added in front of the (real) input nodes can be used to capture the entire attributes, and communicate them to appropriate real nodes. In most neural networks applications, we start with a fully connected graph (unless prior domain knowledge about connection weights are available) and reduce it to partially connected network using the training and pruning phases.

In lateral neural networks, the output of a neuron in a layer can be fed as inputs of neurons in the same layer (if the input is fed to the same node, it is a matter of storing the previous output to be utilised in the next round of computations). The FFN with back-propagation (FFN-BP) is the most popular choice for supervisory learning, due to its ability to improve the likelihood of a correct output using a simple algorithm that adjusts the weights in the backward pass. In Feed Back Networks (FBN), data transmission can take place from an intermediate node to another node in one of the preceding layers.

11.4.1   FFN vs FBN

- FFN evolves and works in weight space while FBN works in state space (weights are prescribed)

---

[16]RNN can be trained by backpropagation through time (BPTT) algorithm, simulated annealing, or GA.

- FFN solutions are known whereas FBN solutions are unknown

- FFN is used for classification, prediction, function approximation, etc. FBN is used for optimisation, constraint satisfaction, and feature matching.

The topology of a layered neural network can be expressed in "m-n-p-o" format, where m is the number of neurons in the input layer, 'o' is the number of neurons in the output layer and 'n', 'p', etc, if present, are the number of neurons in hidden layers. Adjacent layers with the same number of neurons can also be combined to produce a more compact representation. For example, a 4-3-3-2 architecture can be abbreviated as $4\text{-}3^2\text{-}2$ etc. For Kohonen SOM, this simplifies to "m-n x p" if output layer is rectangular and "m-1..p" if output is triangular. The topology in figure 11.3 is 4-4-1 and in figure 11.6 is $6^2\text{-}4\text{-}5$. Neural networks with no hidden layers are obviously less flexible than multi-layer networks. These are most often bipartite graphs, but exceptions exist (eg: Hopfield networks, lateral networks). Variants of the architecture include Kohonen's Self Organising Maps (SOM) [KT97], Bayesian confidence propagation networks which use Bayes' law [OL00], recurrent networks, etc. Kohonen networks have many similarities and self-organising characteristics and are used to extract similar groups among heterogeneous populations, by examining the connection weights.

11.5 Special Types of ANNs

ANNs vary widely in terms of their flexibility, processing power and architectures. When used for classification, the simplest architecture results when data are linearly separable.

Definition 11.2 Two pattern sets P_1 and P_2 in n-dimensional space are linearly separable if at least one hyperplane HP_0 exist such that all points in P_1 are on one side of HP_0 and all points in P_2 are on the other side of HP_0.

11.5.1 Single Layer Perceptron (SLP)

The first generation neural networks called *perceptrons*, had only one hidden layer and could classify only linearly separable patterns. They transform the input $\text{X}=(x_1, x_2 \cdots, x_n)$ using a weighting function as $y = \sum_{j=1}^{n} w_j x_j - \theta$ to produce an output of $+1$ if $y > 0$ and 0 otherwise. Since $y - \sum_{j=1}^{n} w_j x_j = \pm\theta$ defines a hyperplane in multi-dimensions, the perceptron induces a decision boundary using the input vector and threshold. In the case of linearly separable classes, the Perceptron algorithm will converge in a finite number of iterations to the optimal solution.

11.5.2 ADALINE

Adaline is a special type of SLP in which the output function is not explicitly present. In other words, the transformed input $y = \sum_{j=1}^{n} w_j x_j \pm \theta$ is directly passed as output. Weight updating by the training algorithm is considerably simplified in this case as $w_i \leftarrow w_i + \eta(O_j - d_j)x_i$ (table 11.1, pp.11-23).

Figure 11.2: Connection weights for two adjacent layers of neural network

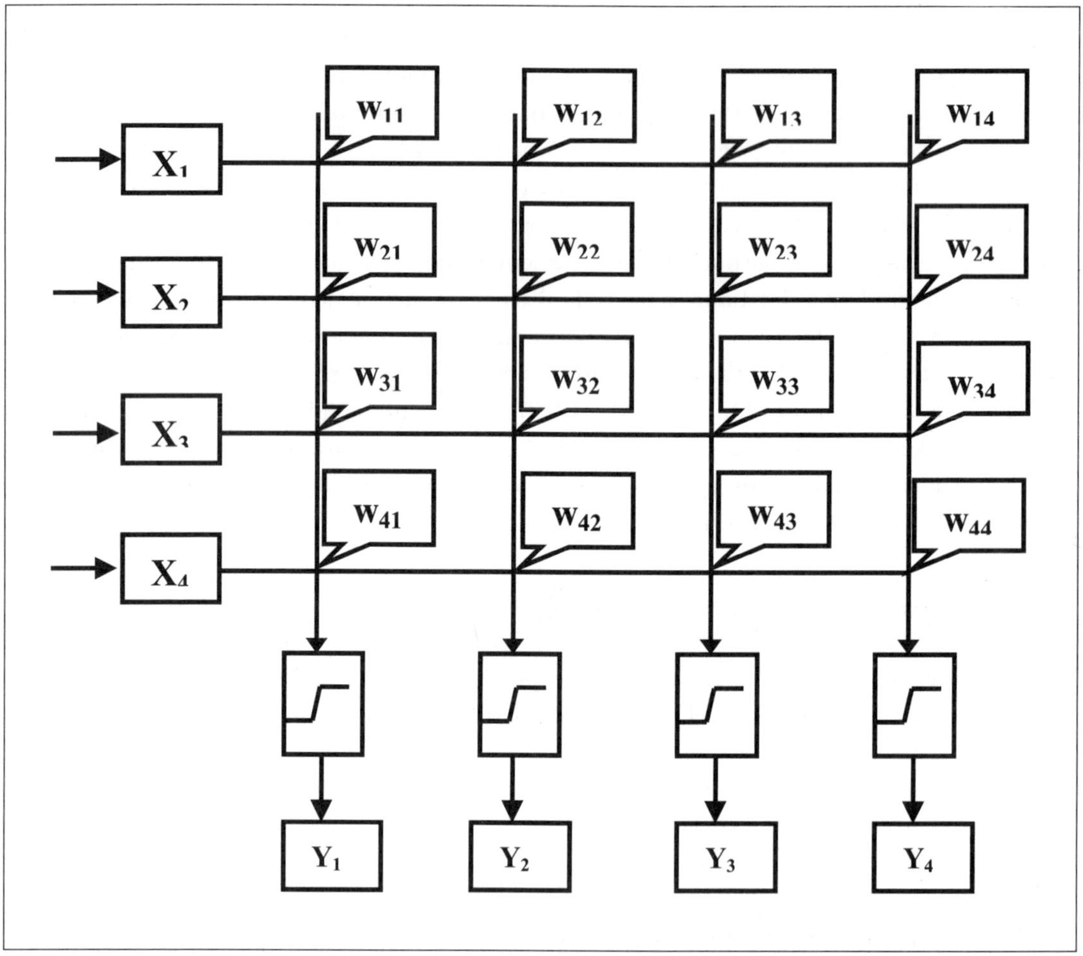

11.5.3   Multi-Layer Perceptron (MLP)

As the name implies, this type of ANNs have multiple processing or hidden layers. The output of the nodes at each layer (except the nodes in the input layer) is modeled such that the network can approximate any input-output combination. They are trained with static back-propagation algorithm, which is applicable to MLP with any number of hidden layers. With additional processing layers, any continuous function can be approximated reasonably, if there are enough hidden neurons in the layers [KN63]. In addition, MLPs with enough neurons in the hidden layer can correctly classify nonlinearly separable patterns. The output of a neuron can be expressed as $y = f(\sum_{j=1}^{n} w_j x_j \pm \theta)$ which in vector form can be written as y=f(W'x-θ), where f() is the output function. If firing function is the sigmoid, the output can be expressed as $y = 1/(1 + e^{-\lambda[W'x-\theta]})$ where $\lambda > 0$ is a parameter (table 11.1, pp.11-23).

Slow convergence of MLP occurs in some problems because the BP training algorithm of MLP is a generalised Least Mean Square(LMS) algorithm. Many techniques have been suggested to speedup the back-propagation. The most prominent of them are (i) normalisation of adaptation term to the derivative of the output activation function [PL93] (ii) pre-orthogonalisation (iii) Fahlman's [FS88] quick propagation algorithm (QPA).

The standard BPA updates the weights after each iteration if the error is significant at the output layer (If $(O_i - d_i) \simeq 0$ for all neurons in the output layer, the BP is skipped for that epoch). QPA updates the weights in batch by finding the average gradient of error surface during the past k epochs. Weight updates occur according to

$$W_{n+1} = W_n + \eta \Delta W_n = W_n + \eta \frac{S_{(n)}}{S_{(n-1)} - S_{(n)}} \Delta W_{n-1} \qquad (11.2)$$

where $S(n) = \partial E / \partial w_n$ is the error gradient wrt the weights in current iteration. Hence previous gradients must be stored by each of the neurons. The QPA makes two assumptions (i) the weight error curve can be approximated in a suitable small interval by a parabola (ii) the rate of change of error curve is not affected by other weights than the one under consideration simultaneously.

Multilayer Perceptron is the most used network model to mine time series data. Extensions to the MLP model include time lagged recurrent MLPs (used in temporal pattern classification), Jordan & Elman MLP with context units, etc. Output functions other than the sigmoid can also be used by the neurons. See appendix A for details.

11.5.4 Knowledge-based Networks

These incorporate domain knowledge into the model. The apriori domain knowledge must be maximised for a good network design. This will speed-up the search by reducing the dimensionality of feature space and are ideal for large data mining applications [AG99]. They find applications in artificial intelligence, robotics, expert systems etc.

11.5.5 Kohonen Networks

This type of ANNs have the ability to compare the output produced at every hidden layer to pick up the node that produced the optimum output. The Kohonen feature maps have only input and output layers (fully connected) without a hidden layer. It results in a self-reorganisation of the output units into a spatial map. For each training instance fed into the input layer, the output layer neurons compete to be the winner. The winning node as well as its neighbor(s) have their weights modified (belongs to winner takes all (WTA) category). This process is repeated for other training samples to adaptively update the weights. As the training progresses, distinct output units emerge with dominating weights (and weights in other units in its neighborhood tails off).

11.5.6   Self Organising Map (SOM)

SOMs are fully connected neural networks, used to detect structures in high-dimensional data [KT01]. It has applications in image and speech recognition, signal processing, clustering (especially text and multimedia files), diagnosis (faults, medical conditions)[LM04],[DR00] source separation, and visualisation.

Algorithm 11.1 Self Organising Map algorithm

1: Step 0: initialisation
2: Input number of patterns (attributes) n, number of neurons in the output layer (N), number of rows and columns in the output layer (p, q), threshold level ϵ.
Ensure: (n$\geq$2, N$\geq$1, (p,q)$\geq$2, 0$< \epsilon <$1)
3: Assign randomly generated weight vectors $m_j(t)$ of size n to each of the 1..N output nodes
    Input neighbor connectivity indicators for output layer neurons
4: **for** (each of the training instance 't' in parallel) **do**
5:        step 1: winner identification
6:        Construction:
7:        **for** (each of the output nodes j=1 to N) **do**
8:            Compute $m_c(t) = \min_j\{d^2(x(t) - m_j(t))\}$
            (* d is the distance measure (chap. 9) and c is the winning node identifier*)
9:        **end for**
10:
11:        step 2: Weight adaptation
12:        **for** (each of the cells j=1 to N) **do**
13:            **for** (each of the component k=1 to n of weight vector) **do**
14:                $m_{jk}(t + 1) = m_{jk}(t) + h(t)[x_k(t) - m_{jk}(t)]$
                where h(t) is gain term, $m_{jk}(t)$ denotes k-th component of j-th vector
15:            **end for**
16:        **end for**
17: **end for**
18: step 3: Output patterns found.
19: **for** (i=1 to p) **do**
20:        **for** (j=1 to q) **do**
21:            Compute r=i*p*j
22:            **if** $(m_r(t) > \epsilon)$ **then**
23:                Output information
24:            **end if**
25:        **end for**
26: **end for**
27: **return**

SOMs are unsupervised exploratory learning models used to nonlinearly map multi-dimensional data into a lower dimensional space (usually 2D rectangular or hexagonal grids) so as to preserve topological relations in feature space. They have a two-layer topology in which input layer has as many neurons as there are features (attributes) in the data. The output layer (also called mapping layer or topological map) is a two-dimensional grid structure, with each internal node having neighboring nodes (left, right, up and down). This is analogous to the connections in the

cerebral cortex where neuronal stimuli from various body parts are mapped to adjacent regions. Each input node is fully connected to the neurons in the output layer. Each output neuron is connected to adjacent neurons in the same layer as follows:
(i) those on the corners have either 2 or 3 neighbors {o(0,1),o(1,0)} or {o(0,1),o(1,0), o(1,1)} for top left corner, {o(0,n-2),o(1,n-1)} or {o(0,n-2),o(1,n-1),o(1,n-2)} for top right corner and so on.
(ii) other boundary nodes can have either 3 or 5 neighbors (o(0,j) has neighbors {o(0,j-1),o(0,j+1), o(1,j)} or {o(0,j-1),o(0,j+1),o(1,j-1),o(1,j),o(1,j+1)})
(iii) internal nodes can have either 4, 6 or 8 neighbors.

An extension of the fixed neighborhood is the circular neighborhood (cyclic maps), which uses a *wrap-around* such that the top neighbor of topmost row is the bottom row, and right neighbor of rightmost column is leftmost column etc, as in some of the algorithms for *magic-square* puzzle. The grid in this case becomes a torus. As training progresses, the number of neighbors may be decreased. During the initialisation step, each of the neurons in the output layer is assigned a randomly generated weight vector with the same size as the number of attributes in the input vector. All weights may be normalised to the same range (say [-1,1]) or scaled to have a norm in [0,1] range. During the training stage, these weight vectors get modified using an error minimising criteria discussed below. Those units with lowest error are declared as winners[17] and in a two step process, the weight vectors of the winner as well as those of its neighbors are updated. Each neuron in the output layer is uniquely identified either on an ordinal scale, or using the row and column numbers (j,k). If the output layer is rectangular with m rows and n columns, these pairs may be mapped to unique numbers as (j-1)*n+k if 'j' starts at 1 or as j*n+k if 'j' starts at 0.

In the standard SOM, the input neurons do no processing (other than simple data transformation, if any), since each input neuron knows only one of the m attributes. However, as each input neuron multicasts its output to each of the output layer neurons, all input attributes are known to each neuron in the output layer. Thus the processing takes place at the output layer by each neuron computing the distance measure and identifying the winner collectively. Neural networks in general, and SOM in particular are ideal for intrinsic parallelisation. It is used in data visualisation [KK97], clustering, transportation, web mining ([NS00], pharmaceutical design [AB00],). The step 2 above uses an error learning neighborhood function $h(t)$ (called gain term), which monotonically decreases slowly and has values less than 1 for fast convergence. For instance, the gain may be modeled using exponential decay $h(t) = c_0 e^{-t/\lambda}$ for suitable constants.

This rate may be adjusted to decay over time, and made proportional to the distance of a node from Best Matching Unit (BMU). This could be accomplished by expressing $h(t)$ as the product of $\Theta(d)h_1(t)$ where $\Theta(d)$ models the distance decay.

The complexity of SOM is $kO(n^2N)$ where n is the pattern size and k >1 depends upon the output layer connectivity. This can be improved by arranging the training instances with maximum overlap together (which is easy to do if prior training data are available, and the intermediate master setup is used). Previous iteration values can be used to update the distance measures

---

[17]Winners are also called Best Matching Units (BMU)

without re-computing the entire components. Other acceleration techniques are (i) heuristic search and match for next winner in old winner's neighborhood (ii) incremental output layers in which we start with a small set of nodes and grow it to larger set (iii) Random projections.[18]

> Nodes in the input layer must be synchronised with the nodes in the output layer during training and running phases. In software implementations, this can be accomplished either using a shared variable (mutex) or using a special master node, which can be used for time synchronisation, to keep the iteration tagged winner attributes and other relevant information like number of input records processed, number of clusters found, etc. If the running phase is not immediately followed by the training phase, we may have to save the 'trained' weights (and other parameters) externally, and later restore it prior to each running phase. This can also be done by the master node. The attributes of past 'k' winners may be kept (in a structure array) by the master and this information mined to reveal useful winning strategies. Since the spatial cluster structures found are buried in the local weight vectors of each output layer neuron, a master node can ease the task of externalising the clusters found in *nonvisual* environments.
>
> Weight updates in the topological layer are carried out in parallel, as the Kohonen networks are intrinsically parallel structures. Sequential updates are also possible at the expense of additional memory.

11.5.7 Fuzzy-Neural Networks (FNN)

Overlapping classes and incomplete input stimuli occur in many practical data mining applications (eg: a disease progression may be in {mild, moderate, severe, critical, fatal} stages which may vary as time progresses. FNN is a hybrid network incorporating the fuzziness (uses vague information). A fuzzy neuron accepts elements of a fuzzy set as input (synaptic weights, operations etc). Object fuzzification deals with neurons and weights. The output is represented as $y = \sum_{j=1}^{N} [w_j \; t \; x_j]$ where t is a norm rather than one of the operators (OR, AND, etc) for ANNs. A fuzzy neuron can have a non-fuzzy input and fuzzy synaptic weight or *vice versa*. Fuzzy arithmetic (fuzzy addition and multiplication) or fuzzy logic (fuzzy logical operations) are the frequent operations. FNN use the fuzzy BP algorithm for training.

11.5.8 Stochastic Neural Networks

This type of networks produces a stochastic function of the inputs and transfer functions. In other words, the output is a stochastic (rather than deterministic) function of its inputs and neuron interactions. Hence the BP learning algorithm is rather helpless and the simulated annealing algorithm that readily converges to the global optimum is the preferred choice.

11.5.9 Radial Basis Function (RBF) Networks

These are FFN that often have a single hidden layer. The input-output relationship is modeled as $y(x) = \sum_{j=1}^{N} w_j \phi(||x - x^j||)$ where $\phi()$ is an appropriately chosen function and x^j are learning

---

[18]See www.helsinki.fi/~niskanen/tk/koho.html for further details

data values. Whereas MLPs divide the pattern space using hyperplanes, the RBF uses hyperspheres (or circles in two dimensions) with a center and radius.

Figure 11.3: An RBF network

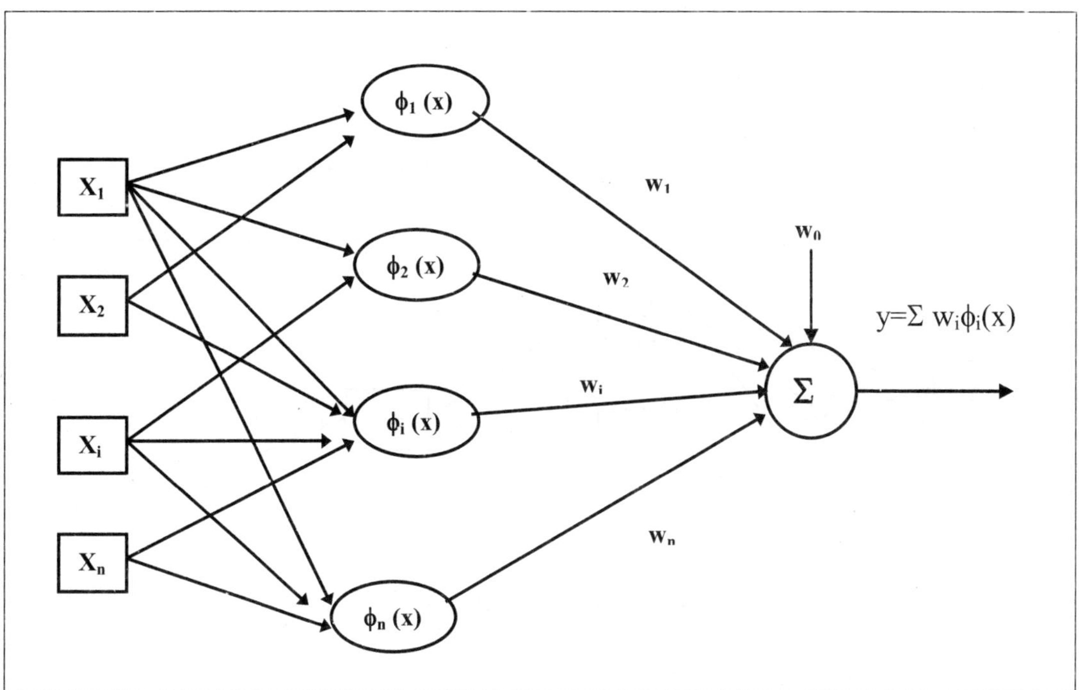

Hidden layer in an RBF network has more nodes than the input layer, and performs a nonlinear transformation of input space. The output layer is a linear regressor. A nonlinear transformation followed by a linear transformation at the output layer increases the likelihood of linear separability[19] and finds applications in nonlinear SVMs (chapter 13). The RBF output function is usually chosen to be a Gaussian $(\phi_\sigma(||x - t||) = \exp(-||x - t||^2/2\sigma^2))$, where t acts as the center of the Gaussian distribution. Thus the learning stage of an RBN NN with a given architecture should estimate the centers, spread and the network parameters (weights) to be used (from hidden to output nodes). The firing function can also be the square-root of a quadratic $(\phi(x) = (x^2 + c^2)^{1/2})$ or inverse quadratic $(\phi(x) = (x^2 + c^2)^{-1/2})$ for c>0. A hyperspheric RBF is

$$\phi(||x - t||) = \begin{cases} 1 & \text{for} \quad ||x - t|| \leq c; \\ 0 & \quad otherwise \end{cases}$$

An RBF is more flexible than that of sigmoid neural network, although weight updating (BP training) can be more cumbersome. See [MD89] for further discussions.

---

[19]This is called Cover's theorem and states that "a complex nonlinear pattern-classification problem in high-dimensional space is more likely to be linearly separable in low dimensional spaces".

11.5.10   Probabilistic Neural Networks (PNN)

PNNs[20], introduced in 1990 by Specht [SD90], perform classification into categories whereas general regression neural networks (GRNN) perform continuous variable regression. They have applications in classification schemes (medical procedures like EEG, ECG, X-rays, etc), clustering, etc. They have four categories of layers – *input layer*, one or more *hidden layer*, *pattern layer* and *decision layer*. There are as many nodes in the input layer as there are predictor variables.

Figure 11.4: A probabilistic neural network

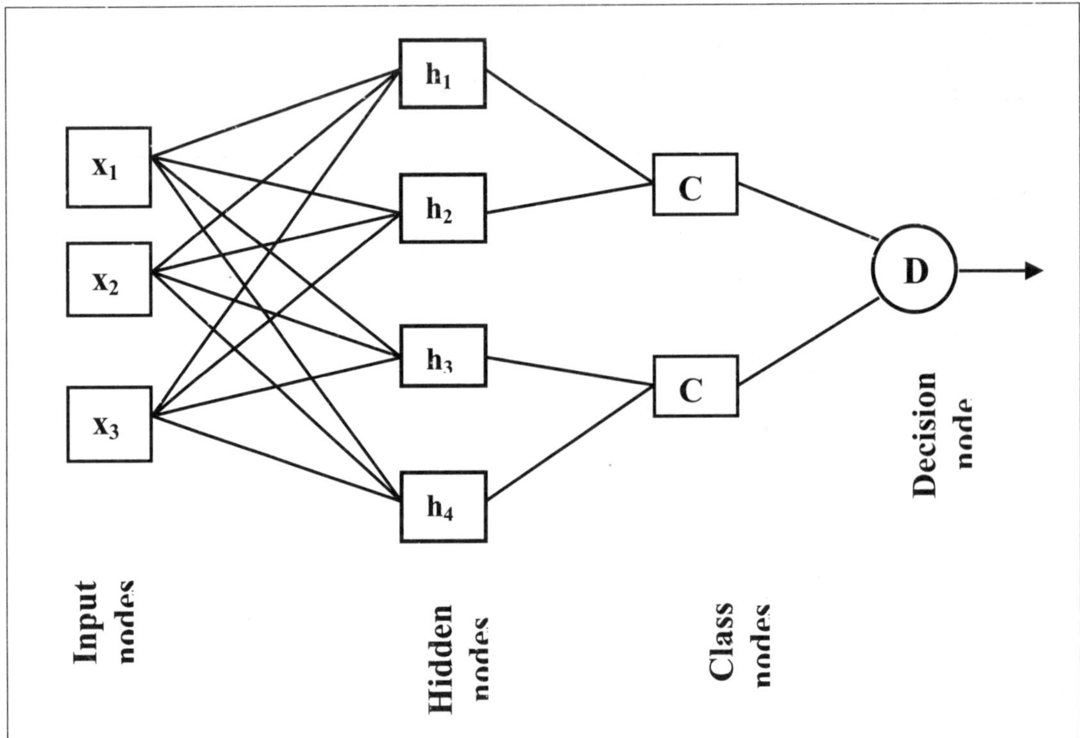

Hidden neurons apply the RBF kernel function to the summed input value. The pattern layer has as many neurons as there are categories of target variables. They are much faster to train than MLP, but the task of classifying new cases is slower. Due to the Bayes optimal classification approach, the space complexity is in general more than that of MLP. GRNN and PNN have many architectural similarities, but GRNN has two hidden layers and is used for regression and prediction. In spite of their training speed and other advantages, there are a few disadvantages too. Hidden layer(s) in PNN/GRNN have as many nodes as there are training

---

[20]PNN is also called kernel discriminant analysis. Kernels are also known as "Parzen windows".

instances. Hence they are unsuitable when a large number of training instances are needed to bring the network up to the desired accuracy. The network accuracy can be reinforced at any time using new data.

Table 11.1: Back Propagation Summary Table

Type	Activation function	Output function	Back Propagation
ADALINE	$y=\sum_{j=1}^{N} w_j x_j \pm \theta$	Nil	$\eta(O_j - d_j)x_i$
SLP	$y=\sum_{j=1}^{N} w_j x_j \pm \theta$	$1/[1 + \exp(-y)]$	$\eta(O_j - d_j)O_j(1 - O_j)x_i$
MLP	$y=\sum_{j=1}^{N} w_j x_j \pm \theta$	$1/[1 + \exp(-\lambda y)]$	$\eta\lambda(O_j - d_j)O_j(1 - O_j)x_i$
RBF	$y=\sum_{j=1}^{N} w_j\phi\|x - x_j\|$	$\phi_\sigma(\|y - t\|)$	$\eta(O_j - d_j)O_j\phi(x_i)$

Legend: SLP = Single Layer Perceptron, MLP= Multi-Layer Perceptron, RBF=Radial Basis Function. Note: The SLP uses standard sigmoid, and MLP uses general sigmoid.

11.5.11 Hopfield Networks (HN)

These are recurrent symmetric networks belonging to the minimum energy (attractor) networks family invented by the physicist John Hopfield. There are many graphical representations of the HN, some of which are shown in figure 11.5. The HNs do not use backpropagation algorithm, but use the Hebb's learning algorithm[21] which modifies the strengths of the connections between the nodes iteratively. It uses a symmetric weight matrix defined as $W = \frac{1}{n} \sum_{j=0}^{m} z'_j \, z_j$ where n is the size of each weight vector and m is the number of class patterns consisting of ± 1 elements. During each iteration, an input instance is used to continuously update the weights until the outputs are constant. Each neuron can be in active (firing) or inactive (not firing) state. Firing neurons activate positive weights to all other connected neurons, and others multiply their own output with the weight of active neuron connection. This product is summed and active neurons compare it to a threshold (which is neuron specific). If it is above the threshold, the active state is continued. Otherwise it becomes inactive and another neuron is selected for next iteration. Neuron selection and firing proceeds either using a fixed order or using a random permutation.

They are used for boundary detection [ZY97], classification problems with binary input patterns, and finding approximate solutions to TSP ([HT85], [WP88], [GL01]), facility location problem [GL00], and other combinatorial optimisation problems [SP98], [AG99], [AGM99]. Boltzmann machines are noisy versions of Hopfield networks to escape from local optima. Bidirectional Associative Memory (BAM) are networks that utilise underlying information and find applications in search engines, image retrievals etc.

---

[21]Hebb's learning rule was introduced by Donald Hebb in organisational behavior context [HD49]. It states that neuronal connection strengths are proportional to the frequency of neurons firing together and the weights between active neurons should be strengthened. A related rule called Hopfield rule states that 'if the desired outputs and inputs are both active or both inactive, weights must be incremented by learning rate".

Figure 11.5: Two views of a Hopfield neural network

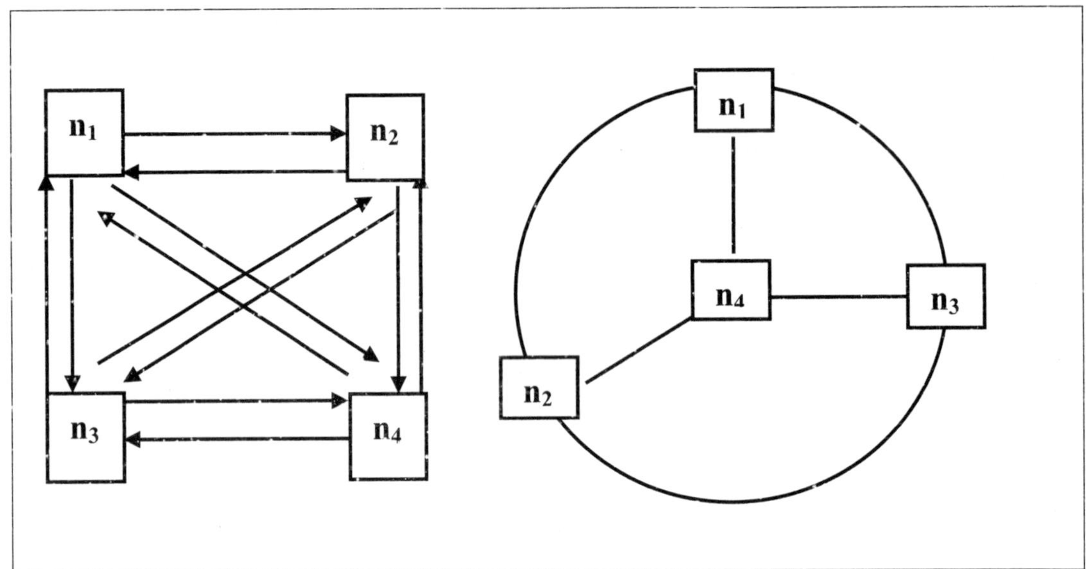

11.5.12   Miscellaneous Types

Generalised FFNs have multiple hidden layers and the connections can jump over one or more hidden layers (node connections can bye-pass layers, so that an input layer node can send its output directly to an output layer node). Modular networks are small logical sub-networks (all belonging to the same category (eg: MLP with sigmoid output)) that collectively work on high capacity problems. As they involve interactions between sub-networks and extensions to BP algorithm, they are more difficult to train. If the sub-networks are arranged hierarchically, it is called hierarchical modular networks. The subnets should preferably work in parallel to optimally recombine the results. Cellular neural nets (CNN) are massively parallel n-dimensional networks that are arranged as 3D array, torus, or other geometric structures (triangular, rectangular, hexagonal etc). Neurons in CNN are multiple input and single output nodes that operate in continuous or discrete time, and typically use $f(s) = (|s + 1| - |s - 1|)/2$ as output function. Complex networks have sub-networks embedded within them.

Self Organising Neural Networks (SONN) are used in "0-1" optimisation problems and have applications in sequencing, assignment and transportation problems, and other similar graph theoretic problems. Learning Vector Quantisation networks are variants of SOM with input layer, single Kohonen layer and an output layer. It uses supervised learning and is used for classification and clustering. Dynamic neural networks are those in which the connectivity varies over time and are used in traffic control and other massively parallel networks. Hybrid networks are those in which more than one of the above architectures are utilised. For example, a hybrid network may utilise unsupervised training at a lower layer and back-propagation at a higher layer. Multi-dimensional networks are generalisation to multiple dimensions (eg: time dimension, geographic dimension). Time delay neural networks [TDNN] are MLPs that operate in two dimensions with 'time' as one dimension with temporally nonaligned connection units.

11.6 Neural Networks vs MLR

The MLR model (chapter 8) is used for prediction or classification. Neural network modeling requires a paradigm shift from a MLR user's perspective. In MLR modeling, the user assumes an appropriate model (linear, curvilinear, piece-wise linear, nonlinear etc) using hints obtained either from graphing the data (scatterplots or extensions) or from prior domain knowledge. Data transformations can also be applied, if needed, before estimating the parameters of the model using the least squares principle. Neural network modeling is often an iterative process with many parameters (number of layers, number of nodes in each layer, their interconnectivity, error minimisation method, transfer function, thresholds, pre and post processing requirements for data, result monitoring methods etc). In MLR, we have only one continuous response variable, which is linearly related to the predictor variables (which are continuous or numerically coded categorical variables). In neural networks there can be one or more output variables (nodes at the output layer). If there is just one output node, the neural network model (used for prediction) resembles a MLR model - with few important differences.

- The weights (or regression coefficients) in classical MLR are chosen using ordinary 'least squares' criterion or its extensions. Weights in a neural network are chosen to maximise feature similarity or homogenisation, or minimise lack of fit using one of a variety of available criteria (see below).

- Neural networks make no assumptions on the data and can accommodate correlation among predictor variables. The MLR model make explicit assumptions on the data and error distributions. It assumes one response and 'n' predictor variables and that the error terms are normally distributed.

- Neural networks can accommodate multi-collinearity in the predictor variables, missing data during the running phase.

Neural network modeling also provides new capabilities for processing voluminous data over and above the MLR and other statistical models. Nonlinear partial least square models are often built using neural networks. Generalised Regression Neural Networks (GRNN) and its variants like adaptive GRNN (AGRNN) are used for linear and nonlinear regression modeling. They are used in nonstationary applications as well [RL04].

11.7   Learning Algorithms

An algorithm proposed by Widrow and Hoff [WH60] and popularised by Rumelhart, Hinton and Williams [RH86],[22] for adjusting the parameters of a neural network is called back-propagation (generalised delta rule). It is so called because it starts from the output layer and propagates the errors (squared magnitude between observed output and expected output) at each neuron by adjusting (increasing or decreasing) the weights collectively, so as to have a smaller expected error. BP algorithm is built upon the *Delta Rule* that reinforces the connectivity between two correlated neurons using first unit's output O_j and second units potential for error reduction relative to its target d_k as $\Delta w_{j,k} = \epsilon O_j (t_k - d_k)$. In a two layer network, adjusting the weights is known as *Perceptron Learning Rule*. The weight adjustment in a multi-layer network is more involved because of the dependence of individual weights on the overall structure of the network, and is called *backpropagation* training.[23] A hidden layer (and input layer) neuron can receive a vector of delta values from each of the connected neurons in the next layer. Thus the back-propagation is carried out layer-by-layer. If the sigmoid (or other continuously differentiable function) is used as the threshold function, the output activation that indicates how to modify the weight to optimally reduce the error can be learned using its mathematical properties (see below). Simplifications are possible in some applications by the use of a heavy-side function instead of standard activation functions discussed above [MF96].

Data for each instance in the training set is applied to the input layer and propagated through the network using the current parameter settings. Results obtained at the output layer are compared with the desired or expected output. Any discrepancies (positive or negative) between desired and computed output values are used to adjust the weights by back-propagation (if there are no discrepancies at each of the output nodes for one input instance, the back-propagation step is skipped for that instance). Error terms at the output layer are propagated through the network backwards. In the case of categorisation, the maximum among the output nodes decides the category of the corresponding input instance.

Problems of underfitting and overfitting could occur if the training data are noisy or impre-
cise. An estimate of node error is the percentage of misclassification at that node. The variance of the outputs of each of the neurons in the hidden layers can be kept track of to avoid over-fitting and accelerate the back-propagation [SD97]. Each node in the hidden layer collectively contributes to adjust the weights.

11.7.1   Backpropagation Algorithm (BPA)

The BPA is a first order training algorithm for neural networks with numeric responses (interval or ratio type) at the output layer. Variants of BP algorithms exist when output stimuli are categorical. The algorithm comprises of three steps:– (i) the input nodes are labeled, and data are fed into the layer in the correct order, which get propagated through the intermediate layers (ii) output produced at the output layer by the forward pass is noted down, and the error E is

---

[22]This algorithm was discovered independently by many others including Werbos [WP90], Parker [P82]
[23]Also called Hebb rule or Widrow-Hoff rule.

found at each of the nodes (iii) in a backward-pass, starting with output layer, the errors are propagated, and synaptic strengths are modified to minimise expected error.

Backpropagation training is used to adapt the synaptic weights [ST89]. For high cardinality hidden layer networks, the correlation between weight changes (from one iteration to previous in two passes of the backward iteration) can be used in early termination of weight adaptation. First partial derivative of the output function is proportional to the rate of change in the weights space, which when multiplied by a learning rate provides a criterion for weight adjustment.

> Gradient descent methods are optimisation techniques that utilise the direction of maximum change (along an error surface or multidimensional nonlinear function) to incrementally move to the global optima. It uses a step size chosen by the user (which is constant in most algorithms, but could be variable in adaptive algorithms). They can be first order (steepest descent) or second order (conjugate gradient). Gradient descent method as used in neural networks consistently decreases the output error of a neuron using small adjustments to the input weights (neuronal strengths). The gradient $\partial E/\partial w_{ij}$ gives the relative change in observed error for a unit change in the weight. Hence we have to move in the steepest direction in the weight space by simultaneously modifying the n input weights for a fixed output error at that neuron. A disadvantage of gradient descent method is that the derivative of the function should be available either in closed form (as an equation), or should be easy to compute numerically. In addition, it often converges to the local optima rather than the global optima. In neural network applications this is seldom known in closed form.
>
> The parameters tuned by a training algorithm must result in better prediction of output signal. Hence the network must adapt continuously as more and more training instances are seen. Time required by the network to fully learn the input-output relationship is called the learning (or convergence) time. An unreasonable convergence time or a fluctuation in convergence is a clear indication of the inadequacy of the network topology, or an implementation error (in the training algorithm, activation or threshold logic, data scaling or conditioning errors, encoding errors, fact selection errors etc). Over-training the network can create problems during the running phase. Hence a performance index may be used to stop the training at an appropriate level. The mean squared error (MSE) cost function $s^2 = (1/k)\sum e(k)^2 = 1/k \sum [y(k) - \hat{y}(k)]^2$, where $\hat{y}(k)$ is the predicted network output and y(k) is the expected output may be used as a stopping criterion. Because the weight updates occur layer-by-layer, the variance of the ratio $\frac{1}{k}\Delta w_{kj}/w_{kj}$ may also be used.

The network training is a mathematical exercise behind the scenes, but a user need not know much mathematics other than vector products and simple arithmetic. BP can be derived in many ways [RR96a,b]. An elegant derivation using differential calculus appears in Appendix-A.

In generalised FFNs (GFFN), connections are possible from one layer to distant layers by *bye-passing* neighbouring layers. This most often occurs either from the input layer to one of the hidden layers (by bye-passing immediate hidden layer(s)), or from one of the hidden layers to the output layer (by bye-passing immediate hidden layer(s)). Hence the nature of the network connections must be known before applying the BP algorithm for GFFN. A sliding 'time-window' may be used in time-series models. There are many popular back-propagation strategies in use :- standard procedure, backpropagation with variable step size (BPVS), Online Adaptive backpropagation (OABP), learning rate and momentum adaptive BP [MW90], resilient back propagation (RBP), scaled conjugate gradient backpropagation(SCBP), delta-bar-delta (DBD), ALECO (algorithm for learning efficiently with constrained optimisation), recursive least squares (RLS) BP etc. See [WP90], [WP93], [SD95] for details. An extension of the delta rule called p-Delta rule can be found in [AB02], and a parallel implementation of BPA in [FS97].

11.7.2   Ill-Conditioning

Overfitting and ill-conditioning[24] are two major problems to be tackled during training runs [SD97]. The gradient descent method uses a learning rate (η) that is set to high magnitude during initial training (for fast convergence) and adaptively scaled to lower magnitudes as the training progresses. A too low learning rate will result in slow convergence, but will eventually locate the true optima. A large value can lead to network divergence (without locating true optima). The ideal learning rate will depend upon weight distributions, and network topology. An ill-conditioned network has large deviations in the learning rates [SB93]. Some of the learning algorithms (Quickprop, RPROP, conjugate gradients, quasi-Newton, etc) are adaptable to ill-conditioned error surfaces. A sensitivity analysis step may be required in some applications to locate ill-conditioning.

11.7.2.1   Implementation Issues

Once the network topology, parameters (input and threshold), activation and output functions, and learning algorithms are known, the network can be built. In most of the software implementations of neural networks, the data resides in information systems, spreadsheets, databases, datamarts or datawarehouses[25]. Data can be extracted using macros, queries or using custom built programs. Synchronisation techniques (using timers, shared variables, lock/unlock mechanisms etc) can be used to properly time the operations. Proper (output) logging, execution tracing and debugging facilities can be incorporated during the training phase. In all of the networks discussed in this book, the activation and output functions are well-defined mathematical functions defined on the real line. Several extensions are possible, including complex functions, lookup tables (which is sometimes used when all inputs are binary), filters, functions in transformed (frequency) domains, hypercube mappings, etc.

Prior to building complex networks, a scaled-down prototype may be useful in some applications areas to validate a 'base model'. The parameters of a trained model may be saved to external data files with proper identification information and time stamps. This information is utilised in continuously evolving networks. Externalisation is also useful in debugging program errors, network failures, machine crashes, and other unanticipated errors. Neural networks deployable in various geographical regions may have to undergo additional localisation (in different Locales and time zones) testing, customisation to various hardware and software platforms etc.

---

[24]Ill-conditioned is the opposite of well-conditioned. It is also used in linear equations to indicate the sensitivity of the system to small changes in the coefficient matrix or rhs vector.

[25]As discussed before, the data may also come from devices, sensors, other systems etc. which is more common in hardware or hybrid implementations

Some of these options may also be left to the deployment stage, without affecting the model performance.

11.8   Applications

Neural networks have a large number of practical [LV00],[SG01], image analysis [SZ00],[SDL97]. In the following section we briefly describe a few applications of practical interest to data miners. A fairly extensive reference is provided for interested readers from other fields.

11.8.0.2   Advertising and Media Planning

Allocating advertisement budget to different media like newspapers, magazines, TV, radio, web sites (supporting ad banners) in different regions, languages or territories on an ongoing basis is crucial in optimising the sales revenue. Past data on advertisement costs in different media and sales revenue over different regions and time periods are used for this purpose. Assuming that sales depend upon product features, sales promotional offers, seasons, economic indicators, foreign exchange rates and nonlinearly on advertisement costs, a neural network can be built to predict optimal allotments for upcoming time periods or to predict the sales revenue using fixed input parameters (advertisement spending, promotional offers, discounts). Input data values are total ad budget, aproportioned past budget data into different media, past sales data, time of year, promotional offers and discounts, product features, length of advertisement (for Radio, TV, web), number of times shown (TV, Mobile, Web) or displayed (printed media) etc. The inverse prediction of the proportions of the total advertisement budgets to various media is more challenging than the sales revenue prediction. A simple network for this purpose is shown in figure 11.6. The success of this model very much depends on training data. Suppose the sales revenue for several pervious years are available, along with prior budgeted ad costs in different media. Then an empirical functional dependency exists between the amount spent in various media and the sales revenue generated. The network aims to estimate this empirical relationship.

Regions are numerically coded (country-wise and region-wise within countries), and year can be broken down into seasons, months or weeks and coded numerically. Alternately, the start and end indices (using Gregorian calendar) can be used for the date(s) of advertisements (this can create problems if the ad is released in December- end and continues into the new year). Discounts and promotional offers may not be deciding factors in advertisement budgeting in some product domains. But, as exposures in print media last longer than in other media, the rates and offers may be varied for various media advertisements (in other words, these can be vectors instead of a scalar). Web banner durations are more important for flash, movie clip or animation based ads than for static or frame based banners. Number of insertions indicate the actual number of times the ad is to be published (in print media) or to be shown on TV etc). Other relevant parameters for print-media include the ad size (quarter, half, or full page), ad location (cover, internal cover, or interior page), embellishments (colours, type faces, etc), special editions etc. The output layer contain predicted proportions of total budget under

Figure 11.6: A sample neural network for aproportioning ad-costs

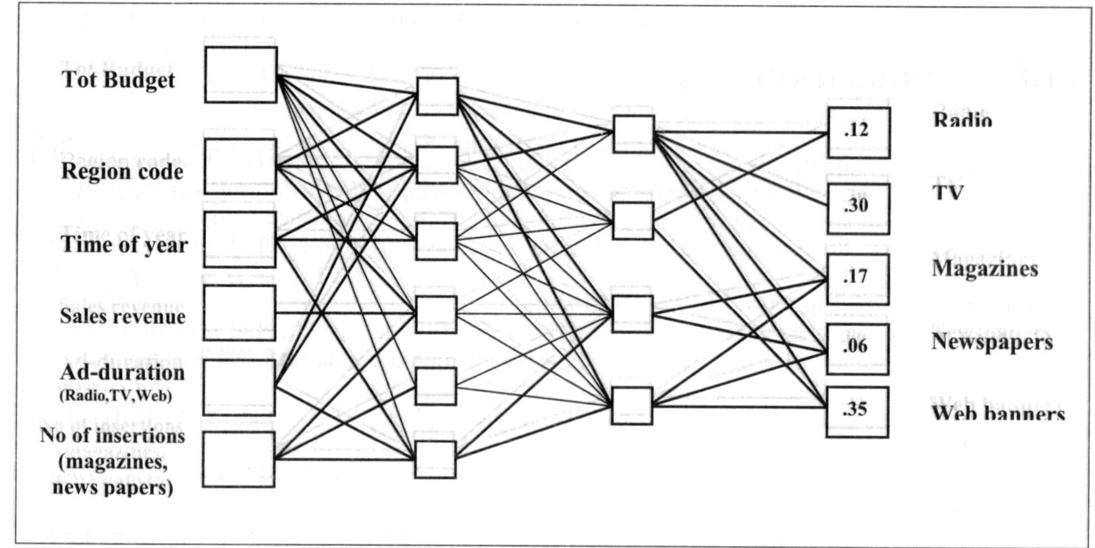

other constraints. Other applications include targeted advertising, interactive multi-language advertising (as web banners), tele-sales campaign, advertising catalog design etc.

11.8.0.3 Pattern Recognition

Neural networks serve as powerful classification and recognition models when mathematical or statistical models are incapable of providing reasonable solutions or when the distribution of objective instances are unknown. Moreover, due to its capability to incorporate fuzziness, neuro-fuzzy models have been developed when pattern boundaries are unclear. When the objective is to identify the membership of an instance's group using labeled training data, it becomes a supervised learning model. If a pattern is to be assigned to a hitherto undefined group, it becomes an unsupervised learning model. Hausdorff distance (§7.3) based pattern classifiers that match between a learned representation with a new input pattern for two and higher dimensions can also be implemented using neural networks [RR98]. In contrast to SVM that works best when variables are weak discriminators (ie. variables have too little discriminatory power), the ANN works in all situations.

Recognising characters and symbols using a computer into a known encoding scheme (like ASCII or unicode) is called character recognition. The characters and symbols may be hand written, typed text, scanned forms or of special forms (as in highway sign boards). If they are recognised using a scanner (from scanned digital images), it is called optical character recognition (OCR).[26] It is used in automatically recognising address labels, zip codes on letters, product

---

[26]Some authors call computer recognition of characters as OCR, though it originated from laser scanning. A

codes on packets, scan codes on inventory items, flight identification codes on baggage, codes on scanned forms (eg: patient data in hospitals), barcodes etc. Extracting typed text and numbers using a computer from images, videos, X-rays, MRI scans etc that are stored in digital form also involve recognition of characters in various fonts and sizes [SS90], [AP92]. Automatically recognising highway sign boards using a video camera installed in a speeding vehicle without human intervention require the capability to recognise numbers (speed limits), letters (Parking, stop symbols), warning symbols (no parking, no overtaking), arrow patterns (turning ahead, U-turn allowed, zig-zag road, two-way traffic) and even pictures (up-hill, pedestrian crossing, school zone) ([LS94], [ZK97]). Depending upon the speed of the vehicle, the size and angle of motion of these characters, shapes and symbols (with respect to the fixed camera which always points ahead at a steady angle) will vary. In addition, rain and snow can obscure some of the signboards. Another application is to recognise the license plate numbers of moving vehicles using traffic cameras installed at intersections and busy roads, and mining for useful information (eg: traffic violators, perpetrators) [KR94]. Similarly, robot controlled and remote controlled driving require a vision based guidance system to detect other nearby vehicles and obstacles, road surface (traffic lanes, humps etc), traffic sign detection and traffic light (current color and color transitions (eg: Yellow to Red)) detection. Intelligent driver-less vehicles are finding increasing applications in war-fields, toxic environments and disaster stricken regions [NF00], as also on main roads.

The neural network models have been successful in tackling the complexities involved in such applications [LF99]. To recognise symbol(s) from a single image (as in highway signs), the input image can be split into non-overlapping blocks (usually of size 2^k x 2^k pixels, where k=2,3, or 4) and each sub-image (and its coordinates or relative position information) assigned to a neuron. For zip-code, and address label recognition, each character image can be assigned to one or more neurons that work in tandem. The recognition process may be driven by table-lookups, using distance and similarity measures of probable shapes or a combination of these. For example, to recognise handwritten digits on postal envelopes and bank checks, each possible digit shapes (written by customers or letter mailers) are stored and each of the extracted digit images are compared with the stored digit patterns to find the most similar digit. Various distance metrics have been suggested for this purpose [RB96], [PB99]. In financial institutions with millions of customers, this could create problems due to the multifarious handwriting styles, different writing tools, non-uniform ink thickness, different font styles and shapes in addition to erasures and smudges on the symbols. Due to the heavy computational requirements in such situations, heuristics are utilised in fast matching. The stored patterns can be sorted in descending frequency of occurrence or distance similarity (most frequent shape or symbol is first matched to the input symbol, followed by next frequently used ones) using a neural network [TT99]. Due to the flexibility offered by neural networks during the training and running phases, such systems can be used across the globe, irrespective of the characters and symbols used in different countries. An application to automatic face recognition can be found in [TC01]. See [GS87], [HC00] for other applications.

---

better term is computer character recognition (CCR)

11.8.0.4   Classification

Classifying items using features like color, size, shape, weight, density, softness, elasticity etc or a combination of these is easy and straightforward in most cases because of the distinguishing boundary between the classes. In most data mining applications, the boundary between the classes are not well defined, or are unclear due to the number of attributes involved, and probable correlations among subsets of them. The process of assigning items uniquely into one among the many possible categories or classes using their inherent properties is called classification. Fuzzy classification is an extension in which an item has an associated membership score for each category because it can belong to more than one category simultaneously. The neural network model is a popular tool for classifying very large number of items.

Consider a large-scale (national or global) sample survey being conducted using a question-naire. The survey has many inter-related questions to be answered by the respondents. The data may be captured by any of the popular online techniques (eg: through web forms) or of-fline through e-mails, regular mails). A peculiarity of survey data is that many responses could have missing values (especially so, if the respondents are illiterate, have reading problems, are busy people or are not in good mood). Correlated questions could also result in missing values. For instance, some questions could be skipped depending upon the answer to prior questions. Our task is to classify the respondents into one of the known categories (which depends on the research objective). The captured data are first filtered to remove irrelevant information (eg: respondent's identity) and cleansed (see chapter 2). The input attributes are then identified and re-coded, if necessary. Survey data could have quantitative and categorical attributes. The numbers of output responses are selected as the number of classes (two output nodes for binary classifiers). If there are 'm' classes into which the respondent will be classified using 'n' instances, the model serves as an unsupervised map from n-dimensional space into a m-dimensional space. This model can be extended for other data types. For instance, by incorporating the responses in audio format (survey respondent's answers are stored as sound data) in online data collec-tion, we could analyse the audio data[27] and use in the classification model as one or more input stimuli. The number of output neurons is the same as the number of classes. An application to incomplete survey data can be found in [LL06], patient classifications ([LH97], [SS03]), and traffic classification [KR94].

11.8.0.5   Data Compression

The ANNs have profound applications in data compression due to the vast processing require-ments expected by some of the compression algorithms with the matching processing power available with neural nets. There are three stages in most compression algorithms - predictive coding, transform coding and quantisation. The predictive coding exploits correlation among neighboring pixels or voxels to predict current value of the pixel, using which the prediction

---

[27]Mining the audio data is more powerful than simple speech to text conversion, because it will reveal more information about the respondent's feelings, emotions, certainties etc., when the response is based upon a question that uses animation, video clip or multimedia.

error is coded. Most of the compression algorithms employ linear prediction, that assumes that the signal varies linearly in a short time interval. Neural networks can implement linear as well as nonlinear prediction models. Transform coding techniques map the data into a transform domain using one of the well-known algorithms (DCT, FFT etc) that are computationally intensive. Neural networks have been used to implement Hebbian transforms, FFT, DCT etc ([OE89]). A block transform partitions the data into non-overlapping sub-blocks that are coded separately. Neural networks have been successfully applied for image processing [SZ00] and image compression [JJ99], [MK00].

11.8.0.6   Speaker Identification

Audio collections on the web are constantly increasing at a fast pace. The speakers (orators, singers, etc) in these collections are usually identified by a separate field (eg: a text field). But there are situations where this information may not be available apriori. Examples are (i) a threat coming over a telephone line, (ii) a new customer using spoken commands to order items from an e-commerce site, (iii) a trapped VOIP telephone conversation that requires imminent action by law-enforcement personnel, (iv) an emergency 911-call etc. Speaker indexing of audio collections and databases is a quick and easy way to identify the speaker's characteristics (like language, dialect, gender, age group, communication medium, etc). Due to the multi-dimensionality nature of this problem, a neural network is a fit choice in such situations. The input neuron(s) capture the speech signals and output neurons report distinguishing attributes (like gender, age-group, ethnicity, matches found etc). Other applications include speech analysis, voice synthesis, speech-to-text and reverse conversions [AE00],[MV02] etc.

11.8.0.7   Web Mining

The web with its plethora of data of various formats, distributed throughout the world, makes a good test-bed for AI researchers. With hundreds of general purpose and specialised search engines, intelligent mining of the web for a combination of data types is a challenge. Automatic identification of web documents using their languages (English, German, Japanese etc) and multi-lingual text categorisation can be tackled by neural networks [CY05]. There are two popular categories of text-mining: entity extraction and attribute(s) extraction. Extracting entities of interest from large collection of text documents has many applications, especially in auto archiving, law enforcement, national security, hospitality industry etc. Attribute extraction filters relevant attributes of interest and their relationships, if any, from voluminous data. It is used in knobot crawling, web personalisation, web page clustering [SN03], and document classification. Clustering web documents, images etc into homogeneous groups involve minimal work at the processing layers. Hence Kohonen's SOM is an ideal choice in these situations [NS00]. Other applications include information retrieval, web services, integration of extracted information ([DR90], [BV98]), image and multimedia mining (chapter 10), word sense disambiguation.

11.8.0.8   Biometrics

Fingerprints are epidermal ridges on the outer layer of primate's inner finger skin, called the epidermis. They are formed (during 10th to 13th week, and finished by 17th week for humans) as a result of a buckling instability in the basal cell layer [KN04]. A finger-print is a unique bio-recognition mechanism to identify humans.[28] It is a biological signature certifying the person's identity. The shape of lines in the palm (on which palmistry is based) can also be used to identify persons, but is more difficult in the case of adults due to the large palm area. In most fingerprint systems, at least two fingerprints are stored per individual. Rarely, two (or more) humans can have identical fingerprints. For example, the fingerprints of twins can be almost identical[29] due to their genetic makeup. However, by storing the prints of two or more fingers can eliminate or reduce the possibility of duplicates. An advantage of fingerprint technology is that the fingerprint of a person vary only slightly according to age (from childhood to adulthood, to senility).[30]

Special pattern matching algorithms are available for searching fingerprint databases to retrieve matching records. These algorithms are based upon correlation, minutiae[31] or ridge shapes. Indexing of the databases using a numeric measure computed from each stored fingerprint is used in some of the searching algorithms. Another approach is to classify each fingerprint in a database into a finite number of categories using a template matching mechanism. This category name (or label) is also stored in fingerprint databases. When a new pattern is to be matched, the approximate category into which it falls is first determined using a neural network model and only those records matching the category label are further analysed. The categorisation may be extended to multiple levels (into a tree of categories) to speedup the searching. The desirable attributes are the demographic variables gender and ethnicity, hand identifier (left or right), finger identifier, age groups, body types, fingerprint pattern group (in spatial and frequency domains), number of minutiae, whorl, triradii[32] and other augmenting information like cut and wound patterns, bruises, burns, corrosion etc in addition to the digitised image itself. Gender and ethnicity are known to be uncorrelated with fingerprint patterns, although palm patterns have slight variation among males and females.

The fingerprint to be matched will be called the 'key'. An image of the key is input at the input layer neuron, along with known augmenting attributes, if any. The spatial patterns (minutiae information vector, pattern group) and transformed domain features are first extracted from the key. If the other attributes (age group, body type etc) are known, the query is submitted to retrieve all matching registered fingerprints. This can be done by output nodes. Visualisation

---

[28]Fingerprints are captured using optical sensors, semiconductor sensors, or a hybrid of these two or by the good old ink-based (pressure sensor) method.

[29]Twins have identical prints at macroscopic level, although at *microscopic* levels, they do differ slightly [JP02].

[30]Dirt, oily liquids, some of the skin diseases (atopic dermatitis etc) at the tip of fingerprint locations can create noise, resulting in partial/total obstructions to the print. Similarly, the finger tip of some of the senior citizens tend to be flat, making it difficult to capture.

[31]The literal meaning of minutiae is 'small detail'. In fingerprint theory (*dermatoglyphics*), it denotes peculiarities of a fingerprint like endings, sharp curving, loops, whorl, triradii, bifurcations, trifurcations etc.

[32]three ridges cuddling up locally or merging together (usually around a concave region) is called a triradius. They encapsulate different information than bifurcations. A whorl is a spiral or an 'open' loop.

helps the matching process tremendously, especially so for ridge-feature based methods. If there are large numbers of matches, the OLAP operations (chapter 5) can be used to narrow down on the matching registered fingerprints. Optimally matching fingerprints in each sub-dimensions (obtained by slicing on body type, age group etc) could be marked to obtain a potential matching set. The resulting set obtained by the output layer neurons after mining can further be analysed using other statistical pattern matching methods.

11.8.0.9 Miscellaneous Applications

A tremendous success of neural networks is in weather prediction, which involve thousands of correlated variables. Some of the data for weather prediction are collected from ground based radars, sensors, satellite images, and other devices. This generates huge amounts of data, some of which are irrelevant. As the neural network models can be scaled up to any complexity, they can be used for forecasting in different time brackets (say 12 hours ahead, 24 hours ahead etc).

Marketing new products, models or services into different geographical or political regions under constraints on production capacity, advertisement budgets, time to market, sales channels, etc so as to acquire more new customers than attract existing customers, or to have maximum market capitalisation involve many complex and correlated parameters on purchasing behaviors, demographic, cultural and socio-economic attributes, life styles etc. of customers, marketing strategies, sales discounts, servicing channels etc. A neural network model can be used for marketing and sales analysis, market segmentation and prediction ([ZK00]).

Traffic prediction and (re-)routing in transportation and communication networks is another area with dynamic behavior. Other applications include satellite communications [IC95], simulation of flows in rivers and waterfronts, simulation of biological systems, fast packet switching in data networks, electronic nose (that has applications in baggage sniffing, food industry, emission monitoring (factories, laboratories, mines, closed chambers), classifying emission (engine exhaust) tests as pass or fail, automatic defects detection in manufactured items (textile rolls, electronic components, circuit boards), automatic airbag deployment [TG00], e-commerce [GV99] etc. Other interesting applications include customer ranking and segmentation [BJ96], real-estate appraisals, software metrics estimation [SS99], medical imaging [MB92], [LN98], sports ([RW95],[BC97],[JK03]), science and engineering ([DB00], [HS98], [TL98], [HK00]).

11.9 Software For Neural Networks

There are many free and commercial software packages for neural networks [RR90]. NeuroSolutions is a graphical neural network development tool for Windows 95/98/Me/NT/2000. NetLAB is an extension to MATLAB that has neural network capabilities. NuMAP available from UT, Arlington has Multilayer Perceptron, Functional Link Networks, Piecewise Linear Networks, and SOM. RiskFinder (introduced in 1998) is a neural network based technology from Fair Isaac Inc to examine fraud patterns and the behavior of cardholders and merchants.

Table 11.2: Software for Neural Networks

URL	Name	comments	*
www.borgelt.net/software.html#neural	neural net	C++	F
heatonresearch.com/encog	Encog 2.4	Java	F
fann.sourceforge.net/projects/fann	Fast Ant		F
genn-team.github.io/genn	GeNN 2	Windows, Linux	F
brainsimulator.org	NN software	simulator	F
www.ra.cs.uni-tuebingen.de/software/javaNNS	java NNS	Java	F
sourceforge.net/projects/javanntrain	trainer	Java	F
wwww.justnn.com	Just NN		F
moose.sourceforge.net	Moose	Win, Unix, Linux	F
www.nengo.ca	Nengo	Win, Unix, Linux	F
neurojet.com	Neurojet	Windows, Linux	F
neuroph.sourceforge.net	Neuroph	Java	F
intelnics.com/neuraldesigner	NeuralDesigner		C
nemosim.sourceforge.net	NeMo	Win, Unix, Linux	F
ire.pw.edu.pl/~sulej/NetMaker	NetMaker	Win, Unix, Linux	S
www.neuron.yale.edu/neuron	Neuron	Win, Unix, Linux	F
www.nest-initiative.org	NEST 2.2	Unix, Linux	F
simtel.net	Neurak	UT Arlington	F
mathworks.com	NetLAB	MATLAB	C
opennn.cimne.com	Open source	C++	F
neuralensemble.org/PyNN	Python NN	meta language	F
simbrain.net	Simbrain 3	Java	F
www.ra.cs.uni-tuebingen.de/software/SNNS	Stuttgart NN	Windows	F
weber.ucsd.edu/~rtrippi/	ThinksPro		F
torch5.sourceforge.net	Torch a5	Win, Unix, Linux	F
www.vertexsimulator.org	Vertex	Win, Unix, Linux	C

Legend:* C=Commercial, F=Free, S=Shareware

11.10   Exercises

1. Mark as True or False

   a) In a recurrent network, all nodes (except input layer) have feedback loops

   b) Backpropagation algorithm updates the weights from front to the rear of the network

   c) Backpropagation learning is done during running phase

   d) Kohonen networks are intrinsically parallel structures

   e) If there are no discrepancies at each of the output nodes for one input instance, the back-propagation step is skipped for that instance

   f) All weights in a MLP should be initialised (at training startup)

   g) Supervised learning algorithms use a target output vector to which the output of a neural net is compared.

   h) GRNN has two hidden layers

Table 11.3: Attributes of a loan application

age (yrs)	gender (M,F)	educ-ation	income ('000)	empl. type	family debt	prev. credit	loan amt.	approval status
46	M	G	110	S	50	200	540	Yes
31	F	U	29	B	115	44	165	No
50	M	P	72	B	38	70	90	Yes
28	M	G	37	P	72	33	120	Yes
44	F	G	47	S	230	0	110	No
25	F	H	33	U	90	0	50	No
60	M	O	50	S	0	175	200	No
53	M	G	94	O	145	58	130	Yes
39	F	P	60	G	36	125	185	Yes
51	M	G	45	P	82	45	90	Yes

education: H=High school, U=Undergraduate, G=Graduate, P=Post-graduate, O=Other. Employment type: B=Business, G=Government, P=Private, S=Self-employed, U=Unemployed, O=Other

i) RBF networks always use the Gaussian transfer function

2. A pattern vector is a vector of–
   A) weights from start layer to hidden layer   B) outputs from nodes in a layer   C) measured features to input node   D) final error terms in a hidden layer.

3. The decision boundary between two classes for linearly separable pattern vectors (with n components) is
   A) a circle B) a hyperplane C) a hypersphere D) an ellipsoid

4. Should the input instances to a continuous neural network be synchronised? Should the output be synchronised?

5. Compare and contrast the neural network model and multiple linear regression model of statistics.

6. Does the backpropagation algorithm differ for classification, pattern recognition and function evaluation applications?.

7. How many weight values does a fully connected network with 5-4 neurons (no hidden layers, 5 input and 4 output neurons) have? How many weights for 4-3-3, and 4-2-2-3 fully connected networks?

8. What are the different layers used by a probabilistic neural network? What are some of its application areas?

9. Identify the input data attributes in each of the following neural networks –
   A) An advertisement budgeting system   B) An image classification system   C) Insurance

claims processing system   D) A cross-purchase rule mining system in a super-market   E) A fraud detection system   F) An e-mail spam filtering system

10. Should the nodes in input and output layers of a neural network be uniquely identified (say by numbers or labels)? If so, explain why? What about the nodes in hidden layer(s)?

11. Identify the output of the neural network in each of the following questions–
    i) A car insurance company's neural network model to predict accident proneness of drivers
    ii) A genetist's neural network to predict whether a stretch of a DNA sequence is a promoter   iii) An automatic emission testing system
    iv) A psychologist's neural network model to check learning disability in children

12. Can a neural network have the same number of input and output nodes? Can the input and output nodes be the same?

13. Explain how the date and time can be input to a neural network.

14. What are the building blocks of a neural network model?

15. What are the components of RBF network? Why is the sigmoid function not used in RBF? Distinguish RBF and Hopfield networks.

16. Explain difference between FFN and recurrent networks. Can the RBF network be recurrent?

17. If the input vector is X=[3.0,10.0,5.0,2.0] and weight vector is W=[1.2,3.8,-1.0,4.0], compute the output, if the output function is (i) sigmoid, (ii) hyperbolic tangent (iii) symmetric hard limit

$$y = \begin{cases} +1 & \text{for} \quad net >= 0; \\ -1 & \quad\quad otherwise \end{cases}$$

(iv) Saturating linear

$$y = \begin{cases} +1 & \text{for} \quad net > 1; \\ net & \text{for} \quad -1 < net <= 1 \\ 0 & \text{for} \quad net < -1 \end{cases}$$

18. How many node-node interconnections are there in each of the following fully connected feed forward neural network architectures?
    i) m layers with n nodes in each layer
    ii) m layers with n nodes in each layer except input and output layers, which have 1 node each,   iii) m layers with n nodes in input layer (i=1) and $\lfloor (n/2^{i-1}) \rfloor$ nodes in i^{th} layer, where $m \leq \lceil log_2 n \rceil$ iv) n layers with n-i+1 nodes in i^{th} layer.

19. Obtain a neural networks model to be used as an early warning system on airport runways/taxiways for the data in table 11.4. Explain how you will specify the direction of motion of an aircraft in 3-dimensions.

Table 11.4: Tracking objects in and around airports

Obj. id	body type	X coord	Y coord	Z coord	dir code	incl. code	speed (mph)	warning code
1	J	320	80	0	N	S	25	0
2	H	310	670	400	NW	D	60	3
3	J	350	-680	670	S	U	190	2
4	L	382	245	320	N	D	130	0
5	M	400	-448	480	S	U	175	2
6	L	310	820	560	N	D	265	3
7	M	320	1290	900	NW	D	300	1
8	M	390	-1082	860	SE	S	300	0
9	J	462	-664	1250	N	U	375	2
10	L	338	-585	540	SW	U	25	2
11	H	388	-770	200	S	U	75	0
12	L	370	1484	1440	N	D	365	0
13	M	392	900	690	S	D	150	0
14	J	892	700	1082	NE	S	280	0

Direction code: E=East, W=West, N=North, S=South, SE=South-East, etc. Inclination code: U=Up, D=Down, S=Steady (horizontal). Body type: H=Helicopter, J=Jumbo, T=Thin body, O=Other. Warning code: 0=none, 1=low, 2=medium, 3=severe, etc

20. Identify input attributes of an auto insurance company that wish to predict the annual premium of a new customer.

21. Explain the meaning of the following networks: (a) recurrent neural networks (b) Kohonen SOM (c) Cellular neural networks (d) Hopfield networks (e) probabilistic neural networks.

22. The firing function of an RBF network is

$$\phi(||x - t||) = \begin{cases} 1 & \text{for} \quad ||x - t|| \leq c; \\ 0 & otherwise \end{cases}$$

Find the output of a neuron for c=1 in each of the following:– (a) [x=2.5, t=.5] (b) [x=10, t=10] (c) [x=0, t= -1.5]

23. What types of fuzzifications are possible in a fuzzy neural network?

24. What is SOM? For which applications is it useful?

25. What are major implementation issues in neural networks?

11.10.0.10 References

[AP92] Amartur, S.C., Piraino,D., Takefuji, Y.(1992). Optimisation neural networks for the segmentation of Magnetic Resonance Images, *IEEE transactions on medical imaging*, 11, 215-220.

[AG99] Andrews, R., Geva, S. (1999). On the effects of initialising a neural network with prior knowledge, *Proc. of the international conference on neural information processing*, Perth, Australia, (Gedeon, T., *et.al.* eds.), 251-256.

[AB00] Arciniegas,F., Bennett,K., Breneman,C., Embrechts,M.J.(2000). Molecular database mining using Self-Organising Maps for the design of novel pharmaceuticals, in *Intelligent Engineering Systems through Artificial Neural Networks*: Smart Engineering System Design (C. H. Dagli et al., eds) Vol. 10, ASME Press, 477-482.

[AE00] Arciniegas, F., Embrechts, M.(2000). Text-to-Speech conversion with staged neural networks, in Intelligent Engineering Systems through Artificial Neural Networks: Smart Engineering System Design: Vol. 10, (C. H. Dagli *et. al.*, eds.), ASME Press, 733-738.

[AB02] Auer, P., Burgsteiner, H.M., Maass, W. (2002). *The p-Delta learning rule for parallel perceptrons,* Technische University of Graz, Austria.

[AGM99] August M., Gerald, W., Markus, S. (1999). *Applications of Hopfield networks,* Technical note SS99, University of Salzburg, Institute of Computer Science.

[BV98] Bansal, K., Vadhavkar, S., Gupta, A. (1998). Neural network based data mining applications for medical inventory problems, *International journal of agile manufacturing,* 1(2), 187-200.

[BC97] Bhandari, I., Colet, E., Parker, J., Pines, Z., Pratap, R., Ramanujam, K. (1997). Advanced Scout: Data mining and knowledge discovery in NBA data, *Data mining and knowledge discovery,* 1(1), 121-125.

[BJ96] Bigus, J.P. (1996). *Data mining with neural networks,* McGraw Hill, NY.

[BK93] Boese,K. D., Kahng, A. B. (1993). Simulated Annealing of Neural Networks: the 'Cooling' strategy revisited, Proc. *IEEE intl conference on circuits & systems,* 2572-2575.

[BG92] Burel, G. (1992). Blind separation of sources: A nonlinear neural algorithm, *Neural networks,* 5(6), 937-947.

[CF06] Castellano, G., Fanelli, A.M., Pelillo, M. (2006). Iterative pruning in second-order recurrent neural networks, *Neural processing letters,* 2(6), 5-8.

[CM00] Castillo, P.A., Merelo, J.J., Prieto,A., Rivas,V., Romero,G.(2000). G-prop: Global optimisation of multi-layer perceptrons using GAs, *Neurocomputing,* 35, 149-163.

[CY05] Chau, R., Yeh, C-H., Smith, K.A.(2005). A Neural Network model for hierarchical multilingual text categorisation, Proc. of the intl. symposium on neural networks, 2005.

[DB00] Dalgli, C.H., Buczak, A.L., Ghosh, J., Embrechts, M., Ersoy, O., Kercel, S.(2000). Intelligent engineering systems through artificial neural networks, *American society of mechanical engineering.*

[DC98] De Falco,I., Cioppa, A.D., Iazzetta, A., Natale, P,Tarantino,E. (1998). Optimizing neural networks for time series prediction, 3rd world conference on soft computing (WSC3).

[DK01] Dick, S., Kandel, A.(2001). Granular weights in a neural network, in Proceedings of the joint 9th IFSA World Congress and 20th NAFIPS international conference, Vancouver, B.C., Canada, July 25-28, 1708-1713.

[DR90] Doszkocs, T., Reggia, J., Lin, X. (1990). Connectionist models and information retrieval, *Annual review of information science and technology*, 25, 209-260.

[DR00] Dybowski, R. (2000). Neural computation in Medicine: perspective and prospects. In Malmgren, H., Borga, M., Niklasson, L. (eds.) Proceedings of the ANNIMAB-1 Conference (Artificial Neural Networks in Medicine and Biology), Goteborg, Springer, 26-36.

[FS88] Fahlman, S.E.(1988). Faster learning variations on Back Propagation: An Empirical Study: in Proc of the 1988 Connectionist Models Summer School, Morgan Kaufman, CA.

[FS05] Ferrari, S., Stengel, R.F. (2005). Smooth function approximation using neural networks, *IEEE trans on neural networks*, 16(1), 24-38.

[FS97] Foo, S.K., Saratchandran, P., Sundararajan, N. (1997). Parallel implementation of backpropagation neural networks on a heterogeneous array of transputers, IEEE trans Man and Cybernetics,-B, 27(1), 118-126.

[GS87] Gail C.A., Stephen,G.(1987). A massively parallel architecture for a self-organizing neural pattern recognition machine, Computer Vision, *Graphics & Image Processing*,37.

[GM99] Girolami M.(1999). *Self-Organising Neural Networks – independent component analysis and blind source separation*, Springer-Verlag.

[GL00] Guerrero F.,Lozano,S.,Smith,K.A.,Eguia,I.(2000). Facility location using neural networks, (in Y. Suzuki, S. Ovaska, T. Furuhashi, R. Roy, Y. Dote (eds.)), *Soft computing in industrial applications*, Springer-Verlag, London, 171-179.

[GL01] Guerrero, F., Lozano,S., Canca,D., Garcia,J.M., Smith,K.A.(2001). A new self-organising neural network for solving the traveling salesman problem, C. Dagli et al. (eds.), Smart Engineering System Design: Neural Networks, Fuzzy Logic, Evolutionary Programming, Data Mining, and Complex Systems, ASME Press, vol. 11, 865-870.

[GV99] Gupta, A., Vadhavkar, S., Au, S. (1999). Data mining for electronic commerce, *Electronic commerce advisor*, 4(2), 24-30.

[HM94] Hagan, M.T., Menhaj, M.B.(1994). Training feedforward networks with the Marquardt algorithm, *IEEE transactions on neural networks*, 5(6), 989-993.

[HK00] Ham, F.M., Kostanic, I.(2000). *Principles of neurocomputing for science & engineering*, McGraw Hill.

[HS98] Haykin, S. (1998). *Neural Networks – A comprehensive foundation*, Prentice Hall.

[HD49] Hebb, D.O. (1949). *The organisation of behavior*, John Wiley, NY.

[HT85] Hopfield, J.J., Tank, D.W. (1985). Neural computation of decisions in optimisation problems, *Biological cybernetics*, 52, 141-152.

[HC00] Hu, C.J. (2000). A novel, noniterative neural network and its application in pattern recognition, ANNIE, 713-720.

[HL05] Hu,S., Liu, D., Zhang,H.(2005). Gradient-based methods for simultaneous blind separation of mixed source signals, Proceedings of the IEEE International Symposium on Circuits and Systems, Kobe, Japan, 5690-5693.

[IC95] Ibnkahla, M., Castanie, F.(1995). Vector neural networks for digital satellite communications, *IEEE international conference on communications*, 3, 18-22, 1865-1869.

[IT01] Igelnik, B., Tabib-Azar,M., LeClair,S.R. (2001). A net with complex weights, *IEEE transactions on neural networks*, 12(2), 236-249.

[JP02] Jain, A., Probhakar,S., Pankanti, S. (2002). On the similarity of identical twin fingerprints, *Pattern Recognition*, 35, 2653-2663.

[JJ99] Jiang J. (1999). Image compression with neural networks - a survey, *Signal Processing: Image Communication*, 14, 737-760.

[JK03] Joshua, K. (2003). Neural Network Prediction of NFL Games, University of Wisconsin – Electrical and Computer Engineering Department, 2003.

[KW94] Kinnebrock, W. (1994). Accelerating the standard backpropagation method using a genetic approach, *Neurocomputing*, 6, 583-588.

[KR94] Kirschfink, H., Rehborn, H.(1994). Classification of traffic situations by using neural networks, ECAI.

[KK97] Kohonen,T., Kaski,S., Lappalainen,H. (1997). Self-organized formation of various invariant-feature filters in the Adaptive-Subspace SOM, *Neural Computation*, 9(6), 1321-1344.

[KT01] Kohonen, T. (2001) *Self-Organizing Maps*, Springer-Verlag, 3rd, extended ed.

[KN63] Kolmogorov, A.N. (1963). On the representation of continuous functions of many variables by superposition of continuous functions of one variable and addition, *Transactions of American Math. society*, 28, 55-59.

[KN04] Kcken,M., Newell,A.C.(2004). A model for fingerprint formation, *Europhysics Letters*, 68, 141-146.

[LN98] Lavrac, N.(1998). Data mining in medicine: selected techniques and applications, Proc of intelligent data analysis in medicine and pharmacology -IDAMAP98, Brighton, UK.

[LH97] Lim,C.P., Harrison,R.F., Kennedy,R.L.(1997). Application of autonomous neural network systems to medical pattern classification tasks, *Artificial Intelligence in medicine*, 11, 215-239.

[LM04] Lingxin, L, Mechefske, C., Weidong L (2004). Electric motor faults diagnosis using artificial neural networks. *Extenza*, 46(10), 616-621.

[LV00] Lisboa, P.J.G., Vellido, A., Edisbury, B. (2000). Business applications of neural networks, *Progress in neural processing*, vol 13, World Scientific, Singapore.

[LH00] Liu,D., Hohil, M., Stanley Smith (2000). A linear programming approach for solving the N-Bit parity problem using neural networks, Proceedings of the conference on artificial neural networks in engineering, St. Louis, MO, 203-208.

[LL06] Lu,C., Li,X. Pan,H. (2006). Application of extension neural network for classification with incomplete survey data, First International Conference on Innovative Computing, Information and Control - Volume III (ICICIC'06),190-193.

[LS94] Lu,S.W. (1994). Recognition of traffic signs – A multilayer neural network, Canadian conf. on Electrical and Computer Engineering, Halifax.

[LF99] Luo, F.L. (1999). *Applied neural networks for signal processing*, Cambridge University Press, Mass.

[MF97] Ma, S., Farmer, C.J. (1997). An efficient EM-based training algorithm for feedforward neural networks, *Neural networks*, 10(2), 243-256.

[MK00] Ma, L., Khorasani, K. (2000). New pruning techniques for constructive neural networks with application to image compression, Proceedings of SPIE, the International Society for Optical Engineering, 4052, 298-308.

[MC01] Mandic, D.P., Chambers, J.A. (2001). *Recurrent neural networks for prediction: Architectures, learning algorithms and stability*, Wiley, NY.

[MC02] Mathia,K., Clark,J. (2002). On neural network hardware and programming paradigms, Proc international joint conference on neural networks, May 12-17, Hawai.

[MP43] McCulloch, W.S., Pitts, W. (1943). A logical calculus of ideas immanent in nervous activity, *Bulletin of Mathematical biophysics*, 5, 115-133.

[MB92] Miller,A.S., Blott,B.H., Hames,T.K.(1992) Review of neural network applications in medical imaging and signal processing, *Medical and biological engg & comput.*,30,449-464.

[MW90] Minai, A.A., Williams, R.D.(1990). Acceleration of back-propagation through learning rate and momentum adaptation, International Joint conf. on Neural Networks, Vol. 1.

[MW90] Minsky, M., Pappert L. (1971). Progress report on artificial intelligence, web.media.mit.edu/~minsky/papers/PR1971.html

[MF96] Moerland, P., Fiesler, E. (1996). Neural network adaptations to hardware implementations, *Handbook of neural computation*, (Fiesler, E., Beale, R. eds), 1-13.

[MD89] Montana, D., Davis, L.(1989). Training feedforward neural networks using genetic algorithms, Proceedings of the 11th intrnl. joint conference on AI (IJCAI), 1, 762-767.

[MJD89] Moody, J.E., Darken, C.J. (1989). Fast learning in networks of locally-tuned processing units, *Neural computing*, 1, 28-294.

[MV02] Moonasar, V., Venayagamoorthy, G.K. (2002). Automatic speaker recognition using a committee of neural networks, ANNIE 2002, 835-840.

[NF00] Nakamiti, G., Freitas, R.(2000). Intelligent real-time traffic control, ANNIE 2000, 893-898.

[NS00] Ng, A., Smith, K. A.(2000). Web usage mining by a self-organizing map, C. Dagli et al. (Eds.), *Smart engineering system design: Neural networks, fuzzy logic, evolutionary programming, data mining, and complex systems*, ASME Press, vol. 10, 495-500.

[OE89] Oja, E. (1989). Neural networks principal components and subspaces, *International journal of neural systems*, 1, 61-68.

[OL00] Orre R, Lansner A, Bate A, Lindquist M. (2000) Bayesian neural networks with confidence estimations applied to data mining. *Comput. Stat. & Data Analysis*, 34,473-93.

[PJ97] Pender J.E. (1997). Modeling of blood glucose levels using artificial neural networks, Dissertation, University of Strathclyde, Scotland.

[PL93] Peng,M., Lev-Ari,H., Nikias,C.L., Proakis, J.G. (1993). Performance improvement of neural network equalizers, Proceedings of the 27th Annual asilomar conference on signals, systems, and computers, Pacific Grove, CA.

[PB99] Pfister, M., Behnke,S., Rojas,R.(2000). Recognition of handwritten ZIP codes in a realworld non-standard-letter sorting system, *J. of Applied Intelligence*, 12, 95-115.

[PJ98] Prank,K., Jurgens,C., Muhlen,A., Brabant,G. (1998). Predictive neural networks for learning the time course of blood glucose levels from the complex interaction of counter-regulatory hormones, *Neural Computation*, 10(4), 941-954.

[RR93] Reed, R. (1993). Pruning algorithms - a survey, *IEEE transactions on neural networks*, 4(5), 740-747.

[RB93] Reidmiller, M., Braun, H. (1993). A direct adaptive method for faster back-propagation learning: The RPROP algorithm, Proc IEEE international conference on neural networks, San Francisco, 586-591.

[RW95] Rick, W.(1995). Ranking college football teams: A neural network approach, *Interfaces*, 25, 44-59.

[RB96] Ripley, B. D. (1996). *Pattern recognition and neural networks*, Cambridge University Press, Cambridge.

[RR96a] Rojas, R (1996). *Neural Networks - A systematic introduction*, Springer-Verlag, NY.

[RR96b] Rojas, R.(1996). A graph labeling proof of the backpropagation algorithm, Communications of the ACM, 39(12), 202-206.

[RR98] Rosandich, R.G. (1998). HAVNET: A new neural network architecture for pattern recognition, *Neural networks*, 10(1), 139-151.

[RF61] Rosenblatt, F. (1961). *Principles of neurodynamics*, Spartan, New York.

[RH86] Rumelhart, D.E., Hinton, G.E., Williams, R.J. (1986). Learning internal representations by error propagation, *Parallel distributed processing 1*, MIT Press, 318-362.

[RR90] Russell E.,C., Roy, D.W.(1990). *Neural Network PC Tools: A practical guide*, Academic Press.

[RL04] Rutkowski, L. (2004). Generalized regression neural networks in time-varying environment, *IEEE tran on neural networks*, 15(3), 576-596.

[SB93] Saarinen,S., Bramley,R., Cybenko,G.(1993). Ill-conditioning in neural network training problems, *SIAM Journal of Scientific Computing*, 14, 693-714.

[ST89] Sanger, T.D. (1989). Optimal unsupervised learning in single-layer linear feedforward neural networks, *Neural networks*, 2, 459-473.

[SD95] Sarkar, D. (1995). Methods to speed-up error back-propagation learning algorithm, *ACM Computing Surveys*, 27(4),

[SS90] Scheff, K. Szu, H. (1990). Gram-Schmidt orthogonalisation neural networks for optical character recognition, *Journal of neural network computing*, Winter, 1990.

[SD97] Schittenkopf, C., Deco, G., Brauer, W. (1997). Two strategies to avoid overfitting in feedforward networks, *Neural networks*, 10(3), 505-516.

[SS03] Siew, E-G., Smith,K.A., Churilov,L. Wassertheil,J.(2003). A Comparison of patient classification using data mining in acute health care, in A. Abraham, K. Franke, M. Koppen (eds), *Intelligent Systems Design and Applications, Advances in Soft Computing*, 599-609, Springer.

[SE88] Smith,J.W., Everhart,J.E., Dickson, W.C., Knowler, W.C., Johannes, R.S. (1988). Using the ADAP learning algorithm to forecast the onset of diabetes mellitus, Proceedings of 12th Symposium on Computer Application in Medical Care (R. A. Greenes, Ed.), IEEE Computer Society Press, 261-265.

[SG01] Smith,K.A., Gupta, J.N.D.(eds.) (2001). Neural Networks in business: Techniques and applications for the Operations Researcher, *Computers & Operations Research*, Elsevier, 1023-1044

[SN03] Smith,K.A. Ng, A.(2003). Web page clustering using a self-organizing map of user navigation patterns, *Decision support systems journal*, special issue on web data mining, 35(2), 245-256.

[SP98] Smith, K., Palaniswami, M., Krishnamoorthy, M. (1998). Neural techniques for combinatorial optimisation with applications,*IEEE trans on neural networks*,9(6), 1301-1318.

[SS99] Smith, K.A., Siew, E.G., Milne,B. Luxford,K.(1999). Neural networks for software metrics estimation, in C. Dagli et al. (Eds.), *Smart engineering system design: Neural networks, fuzzy logic, evolutionary programming, data mining, and complex systems*, ASME Press, 9, 1073-1078.

[SD90] Specht, D.F. (1990). Probabilistic neural networks, *Neural networks*, 3, 110-118.

[SDL97] Su, M.C., Declaris, N., Liu, T.K. (1997). Application of neural networks in cluster analysis, Proc. *IEEE international conference on systems, man and cybernetics* (1), 1-6.

[SZ00] Swiatnicki Z.(2000). Application of Neural Networks for Image Analysis, ANNIE 2000, 1111-1116.

[TC01] Talukder, A., Casasent, D. (2001). Adaptive activation function neural net for face recognition, Proceedings. IJCNN'01 Intl joint conf on Neural Networks, vol 1, 549-552.

[TT99] Tambouratzis, T. (1999). A novel artificial neural network for sorting, *IEEE transactions on systems, man and cybernetics*, - B, 29(2), 271-275.

[TW03] Tang, Z., Wang, G., Tamura, H, Ishii, M. (2003). An algorithm of supervised learning for multilayer neural networks, *Neural computation*, 15(5), 1125-1142.

[TL98] Tarassenko, L. (1998). *A guide to neural computing applications*, Arnold, London.

[TG00] Tascillo,A., Gearhart,C. (2000). A neural network based airbag deployment algorithm, ANNIE 2000.

[TT95] Tresp, V., Taniguchi, M. (1995). Combining estimators using nonconstant weighting functions, Advances in neural info. proc. systems - 7 (Tesauro, G., et.al. eds), MIT Press.

[SH98] van der Smagt,P., Hirzinger,G. (1998). Solving the ill-conditioning in neural network learning, LNCS 1524, (Orr, G.B., Mueller, K.-R., eds.), 193-206.

[WP93] Wasserman, P.D. (1993). *Advanced methods in neural computing*, Van Nostrand.

[WP90] Werbos, P. (1990). Backpropagation through time, Proceedings of IEEE, 78(10), October, 1990, 1550-1560.

[WS90] Whitley, D., Starkweather, T., Bogart, C.(1990). Genetic algorithms and neural networks - optimizing connections and connectivity, *Parallel computing*, 14, 347-361.

[WH60] Widrow, B., Hoff, M.E. (1960). Adaptive switching circuits, New York Institute of Radio Engineers, IRE WESCON convention record, 96-104.

[WP88] Wilson, G.V., Pawley, G.S. (1988). On the stability of the TSP algorithm of Hopfield and Tank, *Biological cybernetics*, 58, 63-70.

[ZK97] Zadeh,M. M., Kasvand,T., Suen, C. Y. (1997). Localisation and recognition of traffic signs for automated vehicle control systems, In Conf. on Intelligent Transportation Systems, part of SPIE's Intelligent Systems & Automated Manufacturing, Pittsburgh, PA.

[ZK00] Zhao, J., Kulkarni, A.(2000). Market segmentation using self-organizing neural networks, ANNIE 2000, 929-934.

[ZY97] Zhu,Y., Yan,H.(1997). Computerized tumor boundary detection using a Hopfield neural network, *IEEE transactions on medical imaging*, 16, 55-67.

12
Web mining

Chapter objectives

- Introduce Web basics, Search engines

- Explore Web mining and its implications

- Introduce web content mining, web structure mining, web usage mining

- Describe the PageRank algorithm and its generalisations

- Explore Text mining and metrics for text mining

- Understand text categorisation

12.1   Introduction to the Web

The web is built upon the client/server technology[33]. Clients connect to the web using Hyper-Text Transfer Protocol (HTTP). A web server keeps track of all client connections using *session* objects. Each user session object has vital information about the client (eg: the time of connection, port number, idle time, etc). If a client is inactive for a specified time duration, the server may terminate the connection (session timeouts). Intranets are public networks accessible from within, and owned by an organisation. If partial accessibility from outside (for clients, telecommuters, managers on the move and workers from home) is permitted, it is known as extranet. The intraweb is a corporate web that is accessible from within an organisation.

12.1.1   Web Pages

A *web page* is a resource on the world wide web (WWW) that is created using one of the markup languages[34]. A web page can either display static information (like a poem or story or a newswire article) or the contents can be generated dynamically using scripting languages or other technologies (servlets, cgi programs, etc). Most of the web pages are created using a data markup language known as *Hyper Text Markup Language*(HTML) or its variants XHTML, WML (Wireless Markup Language), etc. They have the MIME content type 'text/html', and

---

[33]It is more aptly known as a distributed, heterogeneous client-server network due to the multifarious servers situated all over the world.

[34]It must be online and accessible by users on the Internet or intranet to be considered as a webpage.

is standardised by the world wide web consortium, and International Standards Association (ISO 8879/1986). They are stored on a web server and accessed simultaneously by multiple users from anywhere on the Internet. HTML files are saved with an '.html' extension (or .htm extension in some operating systems). A web page that uses secure connection (over secure socket layer (SSL)) will have '.shtml' extension. Other possible extensions as .asp, .aspx, .php, .cgi, etc. Hypertext documents use tags (like <HTML>, </HTML>, <BODY>,< /BODY>, etc) to demarcate each web page, and each section or portion of a web page. A web site can have any number of logically demarcated web pages. As mentioned in §12.8 (p.12-19), some of the Internet algorithms (eg: pageRank, HITS) work at the web-page level rather than at the web-site level.

Each web-page can contain *hyperlinks* that provide navigational links to (other sections of) the same page or other web pages. In other words, we can jump from any page of a document to any other page (intra-document links) of a multi-page document, or to another page anywhere on the Internet (inter-document links). These links called out-links are executed by clicking the mouse button (usually the left mouse button), using pointing devices or by keyboard shortcuts. The web server will then automatically invoke the contents of the new page and display it on the user screen. By looking at the page source (generated HTML behind a page), we could count the number of out-links. Number of users who visit a web page is called its in-link. This is more difficult to compute than the out-links. Page counters can be used to count incoming links during a specified time interval.

Users view a web page on their computer, cell phones, or other hand-held devices using a browser[35]. The most popular browsers are Internet Explorer, Navigator, Firefox, Chrome, etc. A web page can have any amount of data, so that a single web page may map to multiple physical pages on the user screen. A web page developer determines this mapping between a logical web page and physical devices screen size.

Timothy (Tim) Berners-Lee and Robert Cailliau at CERN, Switzerland are considered to be the originators of URL, HTTP specification and the *publishing language* HTML which is an extension of SGML (Standardised Generalised Markup Language) published by ISO in 1986. They proposed a global hypertext network project in 1989, which was started in October 1990. Original intension was the 'wise use' of it to share information among people with common interests. The "World Wide Web" became a reality at CERN in December 1990, and on the Internet in the summer of 1991. It became instantly popular due to the enormous funding by government agencies, and enthusiastic work done by researchers at many universities including Stanford, University of Illinois, and MIT during the initial stages. Netscape communications (started in April 1994 by Marc Andreesen, the creator of NCSA Mosaic browser), and Microsoft have also helped tremendously in subsequent years. Far beyond the imagination of its origina-tors, the web has already grown as a medium of communication with a multitude of uses for people in all age-groups and walks-of-life.

12.1.2   Web Sites

A *web site* is a collection of logically related web pages[36]. It usually has a single entry point (index.html) with hyperlinks to various other web pages. This is called the site address, and is

---

[35]Web pages can also be accessed using a user written program in Java or C++

[36]Trivially, you could also construct a personal web site with one web page. Unless otherwise specified, a 'page' will mean a 'web page' in the rest of the book.

published on the web for others to visit the site. Users connect to a web site by entering the site address in their browser's address bar. This address is known as the *Uniform Resource Locator* (URL) because of the fact that the same addressing scheme is used to access other resources like network news sites, file download sites, gopher sites etc. The *Uniform Resource Identifier* (URI) is an extension of the location identifier URL and includes services. An *IP address* of a site is the server's identity on the Internet. It can be specified either in *dotted* notation (198.116.144.49) or in human readable form (www.nasa.gov).

Most web sites are constructed as a hierarchy of web pages. Users first connect to a default page (default.html or index.html or a script/cgi file) which is the starting point for browsing (IP addresses are usually mapped to this page). Users navigate through a web site by clicking on hyperlinks, command buttons (Forward and Back buttons), or by clicking on portions of image maps (images that are mapped to different web pages or other resources). The lifetime of a web page varies between a few days (news pages, stock market listings, auction listings) to a few years (or even decades). For most global businesses, the key to success is their presence on the web with *up-to-date* (and sometimes *up-to-the-minute*) information available online. The web is also used for trade, communication, information dissemination and broadcasting, customer support, entertainment, marketing, advertising and various services (including employment services), among other things. Hence web pages frequently get revised and updated, which is especially rapid in e-commerce, and in forecasts (weather, stocks, etc). A few exceptions also exist in archival services, history pages, information repositories, FAQs, abstract services etc where pages are rarely modified or updated[37]. For example, some academic institutions still maintain old pages on course offerings, facilities, etc with minimal or no updates. Some web pages display the 'last updation date' at the bottom of the page (while some sites display both the creation date and last updation date). This information may sometimes be needed in web content mining applications. These dates could vary a lot for various pages at a site. For instance, the 'rules and regulations' page at e-com sites rarely varies, whereas the product catalogs, new offerings etc pages vary rapidly.

Electronic commerce sites (e-com sites for short) use web technology to sell items, services etc over the Internet. Most e-com sites have many pages for user navigation, including pages for displaying product catalogs, detailed item display, user registration, shopping cart viewing, changing items in an order, placing an order, entering credit card details, etc. Additional pages for viewing 'rules and regulations', refund policy, customer reviews etc are also common. Every user to a web site need not traverse these pages in the same order. The objective of web user mining is to find out hidden navigational trends of a single user in multiple sessions, or common trends among a group of visitors. These trends are then utilised to reorganise web pages, provide dynamic content, additional links in desired pages, remove redundant pages, identify preferred customers or even re-design an entire web site. Some e-com sites collect data on visitors such as total time spent, time of visit, originating URL, total pages viewed (per session), order amounts, etc. This is called visitor statistics, which is tagged per user-session. Summary information can be generated for each user by identifying each visitor session over a period of time (say per month, or from start of year to present). Many software packages exist for this purpose.

---

[37]See history of mathematics pages at www-groups.dcs.st-and.ac.uk/~history/Chronology/index.html

12.2 Search Engines

Information extraction from the web is a hot area of research. Most of the users connect to the web to view or extract information in various forms. These include textual information as presented by web pages, news contents, special interest groups etc. Information is most often extracted from the web by keywords or phrases. A search engine serves as an interim agent between the surfer (user) and the web. Users submit a query and the search engine brings back either the results found or indicates its inability to find any information.

12.2.1 Structure of Search Engines

A search engine has four main parts: (1) a query engine (2) a crawler (3) an indexer/sorter with navigation control and (4) a user interface component.

12.2.1.1 Query Engine

A query engine is used to process user queries that are submitted to the engine using a web form. It checks user submitted data for completeness (missing fields, incorrect syntax, codepage used, etc), and prepares a query to be executed.

12.2.1.2 The Crawler

The crawler takes a query as input and retrieves matching web pages as a collection. If the *hits* (matching records) are above a specified upper limit (say 10 million), the query engine may decide to abort the search with a proper warning issued to the user. Due to the explosive growth of online information, some of the search engines are becoming more intelligent in filtering desired content. As the web content, the markup languages, tools and technologies used to build them gradually changes over time, mining the web (by scanning the contents just once) presents technical challenges.

Some of the search engines (like Google[38]) utilise the inherent information present in hypertext and XML documents to filter out context specific results, and order them using "pagerank" objective measure. Intelligent web agents learn user preferences and utilise artificial intelligence (AI) criteria to analyse and filter relevant information, by totally eliminating or minimising irrelevant content [CB97]. All search engines work in two phases in query mode. In the first phase, user queries are parsed and matching records are retrieved. The second phase sorts the retrieved results using a relevance rank, which may be combined with other measures and heuristics. PageRank (PR) and HITS are two link-structure based metrics used by search engines in the second phase to relevance rank the retrieved records (in descending order of their popularity). Even narrowly focused user queries like "PageRank algorithm" or "Google Mathematics" result in millions of hits. Even if a user query results in millions of hits, the second phase will eliminate matches beyond a threshold because no one is going to view all of the million matches at a time. As discussed below, the relevance ranked matches are displayed on the user screen 'n at a time' where n is the number of matches displayed per screen. By default, Google, Yahoo etc use n=10 (This can be set to a higher number by the 'Advanced search' option).

---

[38]Google is derived from the word googol=10^{100}, which is a large number by current web standards. It had its origin with the NSF-funded Digital Library Initiative (DLI) project done by Sergey Brin and Lawrence (Larry) Page at Stanford during 1994. See the article by Mark Malseed at www.momentmag.com/Exclusive/2007/2007-02/200702-BrinFeature.html.

Search engines work at the web-page level, rather than at the site level. It is assumed that each web site is comprised of one or more web pages. Most web sites have hundreds to thousands of pages[39]. This is the reason why the page rank measure (defined below) is defined in terms of the ranks of other pages that point to it. Thus it is a global relative measure, the magnitude of which follows approximately the power-law $f(x)=Kx^{-(c+1)}$, where $x \geq c$ is continuous, K is the normalising constant, and c is a real number. It is already known [PR05] that the in-degree and out-degree of web pages follow the power law with respective exponents c=2.1 and c=2.7 for general keywords and phrases. As the rank of a page depends upon both the in-degrees (from quality pages) and out-degrees, it is a convolution of two power laws, resulting in a power law [CR11].

The rank of a page on the web has no meaning unless it is indexed by one of the search engines (like Google), because a web surfer won't be able to crawl to that page using a search engine. This has the hidden meaning that the popularity of a web page is in the hands of the search engine indexers. An out-of-date index entry for a page in the index databases of search engines will give it a poor score. This in turn pushes the page to the bottom of the search results matching a user query.

12.2.1.3 Indexers

If there are more than one hits, the query engine invokes the indexer to build the inverted index for ordering the hits on a criteria. The indexer will remove duplicate pages, if any, that are not removed by the crawler before the pages are sent to the user. Some search engines use additional criteria (like relevance of a page) to order the hits. Crawler wrapper based searches extract specific pieces of information from targeted domains. Web queries use extended database queries (eg:XML-GL) to search for information using multiple attributes [MM97]. For example, an advanced search option of a search engine can be used to filter out web pages based upon multiple keywords, time intervals (from date - to date), specific domains ("*.edu" only), or negation conditions etc. Image-based extraction retrieves documents using matching image content (images.google.com, images.search.yahoo.com)[40].

12.2.1.4 User Interface Component (UIC)

The user interface component is responsible for scrolling through the results, keeping 'session timeout' information, invoking converters to other formats, etc. For example, some search engines allow a user to translate the displayed contents into different natural languages. As this varies among various users, it is best done by the UIC.

12.2.2 Information Extraction

Search engines are used mainly for information retrieval (IR). Information extraction (IE) is one step beyond IR, in that it not only retrieves relevant information, but also analyses the results and presents extracted information or summary measures to the user. The IE may be supervised or unsupervised, and is more compute (and memory) intensive than IR. It may also be a repetitive process to narrow down on the sought information (see chapter 14). The main tasks of IE are named entity extraction, coreference resolution (CO) (inter-reference among entities through pronouns) template element/relation construction (TE/TR), and scenario template

---

[39]Some search engines that work at the directory level have also appeared, but are not quite popular.
[40]Two Stanford students, Yang Chih-Yuan (Jerry) and David Filo started Yahoo! in 1994.

production. The CO is used to resolve anaphorics ('I', 'we', etc) and pronouns in a sentence with the immediate preceding nouns, and are very much domain dependent [CH05]. Relationships between employees and employers, persons and families, members and clubs/organisations etc are resolved using TR, whereas TE associates descriptive tags with entities.

12.2.3   Linguistic Search Engines

These are special purpose engines that accept natural language questions like "What is a virtual world?" and retrieve matching hits. Because the structure of sentences in different languages is different, these search engines use linguistic processing specific to each language. Some search engines like Ask.com, askme.com, etc look up information in precompiled databases and use meta-searching and linguistic filtering to extract information [RQ01]. Linguistic pre-processing removes white spaces (space and tab characters, CR/LF) and punctuation (comma, semicolon, colon, exclamation, question marks). They are used to separate the lexical units (tokens) in the text. Parts of speech tagging associates syntactic categories (nouns, verbs, adverbs, pronouns, etc) to tokens (words). Some search engines (eg: answers.google.com, a9.com etc) also utilise user interaction. The linguistic search engines are more popular among students and novices.

12.3   Web Mining

Definition 12.3 Web mining is the application of data mining techniques to data originating from the web:- web pages, web data repositories, web logs, click-streams, web traffic, web links, etc with an intend to discover models of objects, processes, interactions and relationships.

In addition, external data from routers, channels, devices, and application data captured in any stage of web access can also be mined. Most of the data mining techniques discussed earlier (association rules, clustering, classification, decision trees etc) are utilised in web mining.

Web mining can also incorporate variables like time and client attributes. The number of customers to a web site at any time depends on many factors, including the geographical location of the business, variations in demand patterns, time of day, special seasons or occasions, etc. A web site that sells "pop music", for example, has more customers from western hemisphere than the east. An e-com site that sells regional movies has more customers during evening/night time (in that region) or on weekends, than during daytime. A *flowers-on-demand* site will be overburdened with customers during special occasions like Valentines day or new year's eve. Web mining tools look for patterns with respect to the above-mentioned parameters, in addition to customer specific variations.

Web mining can be categorised as hypertext mining, hypermedia mining, and metadata mining. Hypertext mining looks for patterns in text, hyperlinks, text markups, etc. It is the mining of complex web content created using a markup language (HTML, XML, etc). Classification of web pages is used for fast content retrieval from focused pages. In addition to keywords, a category hierarchy can also be used for retrieval. Clustering of web pages is very similar in that it creates distinct clusters of web pages which are more homogeneous (with respect to a clustering criteria). For example, they may be clustered as academic, industry or news sites; clustered using inherent criteria like ease of navigation, content presentation, user friendliness, or clustered using access patterns or other criteria. Multimedia data mining involves all types of media including audio, animation, video, streaming media and various image formats in addition to the standard text data. Multimodal indexing techniques are used for fast retrieval of multimedia content. Association rules, clustering, similarity partitioning, classification, etc are the techniques used in this case. Multimedia mining is applicable to digital libraries, information repositories, distance education sites etc. Data stream mining is the mining of patterns on data streams [AC07].

12.3.1 Advantages of Web Mining

- Ease of navigation
  Proper application of web mining can substantially reduce the browsing time of users. A regular customer to an e-com site could browse to a specific page, or place an order in a few clicks due to the familiarity with the organisation of various pages. But novice computer users or first time visitors may take substantially more time. This behavior is kept track of in providing dynamic pages, or visitor specific navigational capabilities in an intuitive way. For instance, if the pages are generated dynamically, additional links to various help files can be provided to novice users, and irrelevant pages or links suppressed for other users. In addition, a web-site map can be automatically built, which will in turn make the navigation through large sites much easier.

- Reduction in time
  Not only is the order of different web pages visited by a user, but the time spent by various visitors in each page, user specific or session specific page revisiting behaviors (discussed below) etc are also important in web mining. The problem is complicated due to the client-server architecture of the Internet with multiple simultaneous client connections to a web site and the fact that multiple pages are cached by the client machines in each request. By disabling the caching and adjusting for the transmission time, the time spent by users on each page can be estimated. A web site could be restructured by utilising the average time spent by users in each page. Web mining can substantially reduce browsing times by eliminating or separating seldom accessed content (flash movies, large images etc), by using alternate data formats (for images, audios) and by using psychographic attributes of clients. It can also improve the productivity dramatically in some situations. For example, automatically mining multiple real-time video data using video motion mining techniques is an immense time saver (in security and monitoring applications, driver-less vehicle maneuvering, etc).

- Increase in revenue
  There are a few busy customers who always prefer to go to a site where an e-shopping can be accomplished in a few clicks. Identifying these customer segments and providing them quick service can bring in additional revenues. Revenues may also be improved by increasing customer traffic, decreasing product returns and cancelations, acquiring new customer segments, reducing customer attritions, and providing more services and facilities for various customer segments.

- Improved security and redundancy
  Web mining can reveal security holes and weaknesses. For instance, trojan horses that continuously send page requests to a server can cause the server to go slow. These can be identified by mining the user logs. Spam mining is a common technique used by email sites in which links that generate spam (called spam links) are grouped together to classify received mail as regular or spam. Most popular email sites have options for automatic spam blocking by mining the email headers, subject lines, content and attachments. However, the spammers are inventing techniques to generate unsuspicious spam mails. In supervised

spam blocking, each user can identify links (URLs) that generated spams. This information is stored on the server for spam blocking. By mining this information of each user (who have identified spam sites), a server can come up with a global spam blocking list or separate lists for users in different domains or areas.

12.4 Implementing Web Mining

Web mining embodies a range of techniques. Traditionally these are grouped under web content, web structure and semantic web mining. We give a more detailed categorisation in the following section.

12.4.1 Web Content Mining (WCM)

Web content comprises of data of different types and formats:- text, images (of various formats like .bmp, .gif, .jpg, etc), tables, audio, video, multimedia, hyperlinks, embedded objects, etc delivered to the user. Web content mining involves pattern discovery from a subset of the above types. Text mining is the most frequently employed technique, which extracts text data in the web content (HTML, XML, doc, rtf, txt, ppt, etc files or extract text from images, tables, etc) to detect patterns and trends. The web content may be static or dynamic (over a time period). Even though more and more web sites are moving towards dynamic content generation, there are still millions of web sites that have static web content. Web content miners look for patterns and trends in unstructured data residing in, or originating from web documents[41]. Formatting information could also be incorporated into the mining process. For instance, if a web page has a few terms in bold type or bigger fonts, it may indicate an emphasis point or important information. These terms may then be given more weights than other regular terms. Similarly HTML has various tags ($< b >, < u >, < i >, < em >, < strong >, < tt >$) and headers ($< h1 >$ to $< h6 >$). The contents within them may be assigned different weights appropriately. Nouns (and pronoun counts[42] also known as coreference counts) may represent document content better than verbs and prepositions. This is especially useful in entity extraction techniques. Content summaries (or abstracts), if any, can also be given priority over regular content, as this may have condensed information. Conditional content filtering works with one or more conditions. For instance, contents that appear in reference sections, header lines, within special tags like $< code >, < samp >, < cite >$ may be ignored for some data mining applications. The WCM also explores non-overlapping content in related sites in order to provide better services to visitors.

Intelligent agents are goal programmed software entities that are trained to perform a task efficiently and autonomously. They can return the results in desired formats to a user who deploys them or logs the results. They usually work with online (and often dynamic) data. They find applications in process control, network management, intrusion detection, personal assistance, etc. Knobots are specialised agents that have a narrower scope and are run intermittently to collect information (eg: filtering web news, web auctions, online tenders etc or interoperate multiple agents.)

Web citation mining looks for patterns in bibliographic citation links on the web ([MN02]). Instead of a set of terms, we start with a set of 'n' author identifiers (or affiliations) to obtain

---

[41]A web document is a unit for web mining, which is synonymous with a record in database terminology.

[42]A pronoun is a reference to another noun (it, he, his, him, she, her, hers, who, whose, they, theirs etc)

author by document matrix. If the analysis is carried out on large domains in which multiple authors share the same name, other identifiers may be needed for disambiguation. Agent-based systems (intelligent search engines/agents, smart web-crawlers and spiders, evolutionary knobots) can be incorporated into web content mining for automated pattern discovery.

12.4.2   Web Usage Mining (WUM)

Definition 12.4 Web usage mining (also called sequence mining, web-log mining) looks for navigation patterns of visitors to a site (page click order, time spent in each page, revisits of pages, path to order[43] in e-com sites), and is derived from the server logs.

Web server logs are text files of requests made by visitors and responses sent by the Web server. Depending upon the log file format and selected configuration options on the Web server, various kinds of information can be logged. Web sites with too much traffic keep server logs on a daily basis. Because the user access patterns differ significantly, usage mining is usually performed at the site level. Traffic analysers are programs that mine for visitor traffic and detect trends and variations in visit patterns. From the maintenance point of view, we may be interested in most/least frequently visited pages, least frequented hyperlinks, most revisited pages per session, most downloaded data files, exit pages, and other navigation strategies. Sequential pattern mining extracts access patterns of each individual user or group of users from a domain. For instance, a brand awareness campaign may aim to introduce customers to as many newer brands as possible in a short session. The order of visits of pages can be mined to categorise customers, fine-tune page contents, restructure pages and generate user specific dynamic content or advertisement banners. Some data (like ad banners) are being pushed (without being requested) to the users by web servers. The requested data (pull data) are usually common to all users who submit same queries or clicks on same links (as in news sites). This difference between push and pull data can either be constant or vary according to user characteristics. A proper choice of dynamic (user and time based) push data may in turn increase the revenues.

Temporal patterns are similar in that time dependent visit order is analysed and useful information extracted. Web usage miners discover common as well as eccentric usage patterns in web accesses. Web usage mining can be used in e-business (B2B, B2C) e-CRM, e-Services, etc. Distributed WUM utilise multiple (one or more) web logs to discover recurring trends and patterns among the users. Data for web usage mining can be captured either at the server side or on proxy servers. Statistical techniques (coupled with visualisation), AI techniques or neural networks are usually employed for web usage mining. Web usage mining is used to redesign a web site and to improve accessibility, performance, content and structure in addition to predicting or modeling user behaviors and temporal variations in web traffic. It is also applied in e-Learning, and distance education programs to mine for student activity levels and other traits with an aim to student modeling, in web caching, load balancing, or data distribution into multiple pages.

12.4.3   Web User Quality Mining (WUQM)

Web users differ significantly in their abilities, expertise, age groups and intelligence. Most web sites do not keep track of visitor qualifications, gender, age-groups etc due to the large number of visitors from various walks of life. The user quality has to be mined using metrics like the time spent in a webpage (TSW), total browsing time per session (TBTS), number of clicks per page/session (NCP/NCS), click intervals (CI), ad-banner clicks (AC), click-through rates

---

[43]also called *path to purchase* which is a list of ordered page IDs from home page to firm order page

(CTR), frequency of visits (FOV), idle time per page/session (ITP), etc [SS05]. Since the visitor commands (like scrolling a web page or clicking on a link) are executed on the client side, the server has no direct way to measure TSW and ITP of each visitor (these can be obtained by executing scripts on the client side and returning the results with the next HTTP request from the client). However, the time intervals between consecutive page requests when compared with the average viewing time for those pages can give a hint on the idle time. Similarly, the time interval between two successive page requests can be regarded as the TSW for the preceding page. Some of these attributes can also be combined to get other metrics like path to order (PTO) which is the path from home page to a firm order page. The PTO will be more or less similar for experienced users. It may contain loops, back and forth jumps etc for others, which are removed and the edges are appropriately weighted to get a weighted path for them. Since most users may not view the pages sequentially, the length of the SNP can provide additional information about first time visitors, and help to detect anomalies. The visitor quality may be an important factor in targeted advertising, creating customer specific click through content, group specific content expansion, age-group and gender based dynamic content creation etc in addition to building online business intelligence.

12.5   Web Structure Mining (WSM)

Web structure mining examines the dynamic link structures of pages within a site or among various sites in order to improve navigation. This involves hyperlink analysis, schema analysis etc to classify web pages, groups of pages or even entire sites into different categories. Examples include hyperlink structure mining, link based classification and clustering, link based event detection, link-type mining, and domain connectivity analysis. Web structure miners generate summaries of inherent structures present in one or more web sites, and domains in the form of tables, graphs or other visualisation aids. Web structure mining overlaps with web retrieval, because we are more interested in the dynamics of hyperlink connectivities.

12.5.1   Link Mining

Hyperlinks pointing to web pages indicate the popularity of the page[44]. Links coming out of a page generally indicate the variety of topics covered in that page. It could also be due to content condensation or a large size of the page (eg: hyperlink enabled books). Analysing links, link trees (and link graphs) using data mining techniques to reveal link dependencies and click patterns is called link mining. A link[45] denotes any type of jump (hyperlinks created using one of the anchor tags, image maps, etc) that relate abstract connections between two (or more) resources or 'objects' on the web. One-way links (uni-directional) may exist either to a resource that does not have hyperlink facility (a plain text file created in a simple text editor), or to a document that does not require further linking (a written song rendered as a text file, an audio clip, an entity at the bottom of a hierarchy). Links can be one-to-one, one-to-many or many-to-one. Entities and resources participating in a link have sets of properties (called property vector or property bag[46] if they are unordered). Some of these properties may be unknown to a client until the resource is accessed. Hyperlinks to different sections of the same document are

---

[44]Exceptions do exist as in browser plug-ins, online birthday and greeting card services, or various browser download pages.

[45]A link in database terminology is a connection between records in related tables.

[46]The bag is a terminology in data structures, where it denotes an unordered collection.

also common on the web. Link based ranking (LBR) is used to order or prioritise the links in dynamic web access.

Static link mining is used to summarise cross-linked pages and can be depicted as a directed or undirected graph (with self loops). Edges of this graph can also be weighted to indicate the strength (count) of links between pages. Dynamic link mining is more useful in identifying dead links, infrequently visited links, heavily used links etc. It is only when users access these links that they come alive and provide accessibility patterns. The link graph in this case will consist of only those edges that are active during a specified time interval (the graph may degenerate to a 'forest' (a disconnected graph, see chapter 5) for short intervals). Some of the links in documents may either be seldom used or can even be hidden. If the web content is generated dynamically, some of the links can be suppressed from teens and users who exhibit fraudulent activity or dishonesty. A taxonomy of link mining tasks can be found in [GD05].

12.5.2   The Page Views

Search engines report millions of matching records (hits) for most search words of random surfers. Majority of web surfers scroll through only a few of the most relevant screenfuls of pages displayed by the search engines (most search engines display 10 matches at a time by default on the user screen). They then click on the hyperlinks and navigate away to other web pages. We call the number of screenfuls of information returned by search engines to the user as the page count. The page count is zero if there are no matches for the search word(s) on the web (eg: a search for lexicronym results in no hits. If the number of matches are $> m$, the page count is pc=$\lceil n/m \rceil$, where n is the total number of matching documents found, m (usually 10) is the total documents displayed per page (screen) and $\lceil n/m \rceil$ is the least integer greater than n/m (due to the fact that the last page may not have the full m documents). Page views are the total number of pages viewed by the user. These need not be in sequential order. Number of page views is in general more when the search word is a person name[47] (like a researcher, historian, or artist) or place name (like a tourist spot, city name, etc) than when it is a general keyword. Statistical distribution of the number of page views of search engine displayed pages by a random surfer resembles a highly skewed exponential distribution (majority of web surfers click on the links displayed in the first page only, and a small percentage of them (especially researchers and academicians) may view a few of the subsequent pages). If the number of views are restricted to screenfuls, the *number of pages viewed* will be a discrete random variable (which takes values in $[1, \lceil n/m \rceil]$. In this case the number of pages viewed by a *random surfer* (for an arbitrary query) is approximately distributed as a zero-truncated Poisson law,[48] which is the discrete analogue of the exponential law. This can also be approximated by the right-truncated Zipf distribution

$$f(x) = Kx^{-(c+1)}/\zeta(c+1), \; x = 1, 2, ..., \lceil n/m \rceil$$

where $\zeta()$ is Riemann zeta function, and K, c are constants. In the non-truncated case, this is the discrete analogue of the power-law f(x)=$Kx^{-(c+1)}$, where x is continuous, and $x \geq c$. The

---

[47]This is due to the fact that lot many persons have identical names, and there is a tendency to reuse old names. Similarly, there are lots of places with exactly identical names, which is more so in the Western hemisphere than in the East. For instance, there are 46 places or cities called London in 6 continents. The *numero uno* in duplicate names is "Richmond" with more than 100 places or cities around the world (http://en.wikipedia.org/wiki/Richmond).

[48]The Poisson distribution takes values $x \geq 0$, whereas in our case the number of views are $1 \leq x \leq \lceil n/m \rceil$ pages because the first page is displayed by default. The distribution is approximate because it is doubly-truncated. Moreover, the intentions of surfers vary widely (some users jump from first page to 10th or last page, then to 20th or last page, and so on, and a few users could jump back and forth between various pages).

Zipf law for large c values tails off much faster than the Poisson law for small λ values (<1). Thus the Zipf distribution provides the *best fit* for random surfers (with a mode at x=1), while the doubly-truncated Poisson distribution provides the *best fit* for specialized surfers.

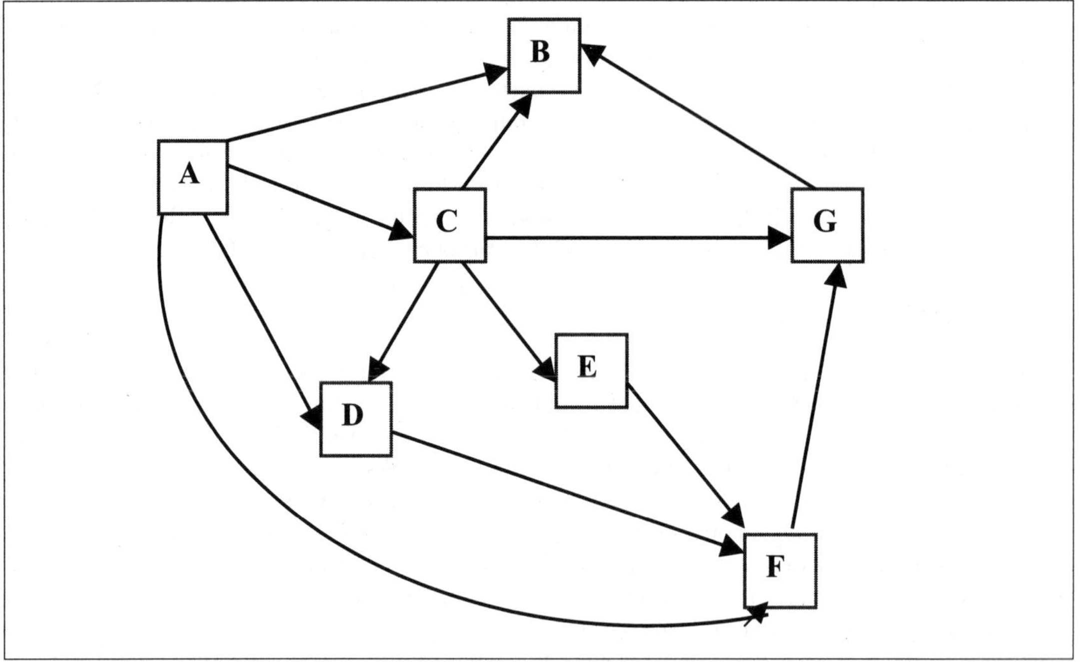

Figure 12.1: Hubs and Authorities computation example for web pages (example:12.1)

12.6   Measures for Web Structure Mining

PageRank and HITS are the most popular measures for web structure mining. Various search engines use these measures, based upon LBR. The following sections discuss the standard PageRank algorithm, and then generalises it in various ways.

12.6.1   Heuristic Ranking Algorithms

Popular search-engines order (or prioritise) the retrieved pages using a utility index. This indicates the popularity of the pages on the web at the time of retrieval. Some search-engines also incorporate other heuristics in addition to the popularity score. As a simple example, suppose that a major natural calamity like a devastating tsunami or earthquake has just occurred somewhere. There could hardly be any related pages or documents on the web yet. But if thousands of user requests come immediately, search engines may want to promote any page containing a piece of the information about such happening (in TV channels, radio stations and online news sites) to the top. This can be done either by bye-passing the normal page ranks, or by adding a fictitious constant K to all such new pages. As more and more pages become available on this happening, this promotion will automatically rank all new pages in high-to-low priority order. The constant K can gradually (eg: exponentially) be decreased over time to demote such pages as time goes by, so that they will eventually blend-in with regular pages.

12.6.2 Hubs and Authorities

A similar crawling strategy is called Hub and Authorities, that distinguishes two types of pages [KJ99]. Hubs are connection points of important pages that point to important authorities. Authority is a page pointed to from many important hubs. Associating a hub-score H(x) and authority-score A(x) with a page x, they are recursively updated as

$$H_{i+1}(x) = \sum_{x,y} A_i(y) \text{ and } A_{i+1}(y) = \sum_{x,y} H_i(x) \tag{12.1}$$

where the summation index (x,y) denotes that pages x and y are hyperlinked (at least once).

Algorithm 12.1 Algorithm for Hubs and Authorities

1: Input a set S of web pages with information on out-links to other pages (this can be captured in a 2D adjacency matrix with Boolean weights).
2: Special checks and Initialisation:
begin
If S is empty, return error code
If S has a single page, or at least one outlink in S is to a page not in S, return error code
If S has a cycle, break the cycle or return error code
end
Let x_p denote the Hub weight and y_p denote the Authority weight of page p
Count n = number of distinct logical pages in the input
Initialise all Hubs and Authority scores to $1.0/\sqrt{n}$ each $(A_0(x) = H_0(x) = 1.0/\sqrt{n})$.
3: Construction:
4: **repeat**
5:      Update the value of Hub weight x_p as the sum of the values of Y_q for all pages q with outlinks to page p using eq 12.1
6:      Update the value of Authority weight y_p as the sum of the values of X_q for all pages q with inlinks from page p using eq 12.1
7:      Find the sum of the squared Hub weights as H_w, and divide each of the Hub weights by $\sqrt{H_w}$ to get $X_p = x_p/\sqrt{H_w}$.
8:      Find the sum of the squared Authority weights as A_w, and divide each of the Authority weights by A_w to get $Y_p = y_p/\sqrt{A_w}$.
9: **until** (there are no more improvements)
10: Sort the Hub weights X_p's, and Authority weights Y_p's in descending order
11: Determine the cutoff to decide Hubs and Authorities
12: **return**

Example 12.1 Find the Hubs and Authorities for the following list of web pages in figure 12.1.
Page A:$\Rightarrow$ links to pages B, C, D, F
Page C:$\Rightarrow$ links to pages D, E, G
Page D:$\Rightarrow$ links to page F
Page E:$\Rightarrow$ links to page F

Page F:$\Rightarrow$ links to page G
Page G:$\Rightarrow$ links to page B.
Solution: There are 7 pages (n=7). We initialise $A_0(x) = H_0(x) = 1/\sqrt{7}=1/2.6458=0.3780$. These are the first 2 rows of table 12.1 (the last 2 entries are the sum of squares of previous elements, and it's positive square-root). Third row is the Hub score computed using equation 12.1. For example, the page B has incoming links from A, C, and G. Hence $H_1(B) = Au_0(A) + Au_0(C) + Au_0(G) = X_0(A) + X_0(C) + X_0(G) = 3*0.3780 = 1.1339$, where we have indicated Authority by Au to avoid confusion with the page labeled 'A'. Similarly, the page G has incoming links from C, F and outgoing link to B. Hence $H_1(G) = Au_0(C) + Au_0(F) = X_0(C) + X_0(F) = 0.7559$. We have indicated un-normalised Hub-scores by x_i, and normalised scores by X_i (by

Table 12.1: Determine the Hubs and Authorities of web pages

	A	B	C	D	E	F	G	SSQ	SQRT
X_0	0.3780	0.3780	0.3780	0.3780	0.3780	0.3780	0.3780	1.0000	1.0000
Y_0	0.3780	0.3780	0.3780	0.3780	0.3780	0.3780	0.3780	1.0000	1.0000
x_1	0.0000	1.1339	0.3780	0.7559	0.3780	1.1339	0.7559	4.0000	2.0000
X_1	0.0000	0.5669	0.1890	0.3780	0.1890	0.5669	0.3780	1.0000	
y_1	1.5119	0.0000	1.5119	0.3780	0.3780	0.3780	0.3780	5.1429	2.2678
Y_1	0.6667	0.0000	0.6667	0.1667	0.1667	0.1667	0.1667	1.0000	
x_2	0.0000	1.5000	0.6667	1.3333	0.6667	1.0000	0.8333	6.6111	2.5712
X_2	0.0000	0.5834	0.2593	0.5186	0.2593	0.3889	0.3241		
y_2	1.7008	0.0000	1.5119	0.5669	0.5669	0.3780	0.5669	6.2857	2.5071
Y_2	0.6784	0.0000	0.6030	0.2261	0.2261	0.1508	0.2261		
x_3	0.0000	1.5076	0.6784	1.2814	0.6030	1.1307	0.7538	6.5852	2.5662
X_3	0.0000	0.5875	0.2644	0.4994	0.2350	0.4406	0.2937	1.0000	
y_3	1.7502	0.0000	1.6853	0.3889	0.3889	0.3241	0.5834	6.6513	2.5790
Y_3	0.6786	0.0000	0.6535	0.1508	0.1508	0.1257	0.2262	1.0000	

dividing un-normalised score by the square-root value in last column). Similarly, the Authority-scores are indicated by y_i and Y_i respectively. As 'A' has outgoing links to B,C,D,F; $Au_1(A) = H_0(B)+H_0(C)+H_0(D)+H_0(F)=4*0.3780=1.5119$, and $Au_1(G) = H_0(B)=0.3780$. Fourth row (normalised) is obtained by dividing each value in third row by the square root column value 2.0, and sixth row is normalised by dividing each element in fifth row by 2.2678. Iterations are continued until normalised scores stabilise to reasonable accuracy. The Hub-scores can now be sorted in descending order to get [0.5875, 0.4994, 0.4406, 0.2937, 0.2644, 0.2350]. The Authority-scores are [0.6786, 0.6535, 0.2262, 0.1508, 0.1508, 0.1257, 0.0]. Choosing .23 as the cutoff, we get the Hubs as {B, D, F, G} and Authorities as {A, C, E}.

A disadvantage of H & A Algorithm is that it may not be able to distinguish between Hubs and Authorities when there is a cycle in the graph (A$\Rightarrow$B, B$\Rightarrow$C, and C$\Rightarrow$A is a cycle of length 3). If there are lot many other links along with a cycle of length k, the algorithm will converge in k-1 iterations. A solution is to either remove, or break the cycle (by removing one of the links). Another problem is choosing the cut-off threshold (for the normalised scores). In addition, many pages could serve as both Hubs and Authorities. For example, the page 'F' has scores .4406 and .1257, both of which are not at the extremes. Thus it may be considered as both a Hub and an Authority. This may not always be apparent. When there exist thousands of pages, one may be interested only in super-hubs and super-authorities only, which are easy to spot. We could speed-up the algorithm by a simple technique. The normalised scores (X_i, Y_i) either decreases, or increases with slight fluctuation during the first few iterations. For example, the hub-scores of B steadily increases as $X_i(B){=}.378{\rightarrow}.5669{\rightarrow}.5834{\rightarrow}.5875$ whereas that of C fluctuates and decreases as $X_i(C){=}.378{\rightarrow}.189{\rightarrow}.2593{\rightarrow}.2644$. Thus after the first few iterations, we could form two sets as potential hubs and others. Thereafter the updates of X_i's and Y_i's (rows of table 12.2) occur only in their respective sets. This can reduce the computations by almost half. Many researchers have extended the HITS algorithm. See [SB07], [AT08] and the references therein for other algorithms and a comparison of HITS and its variants.

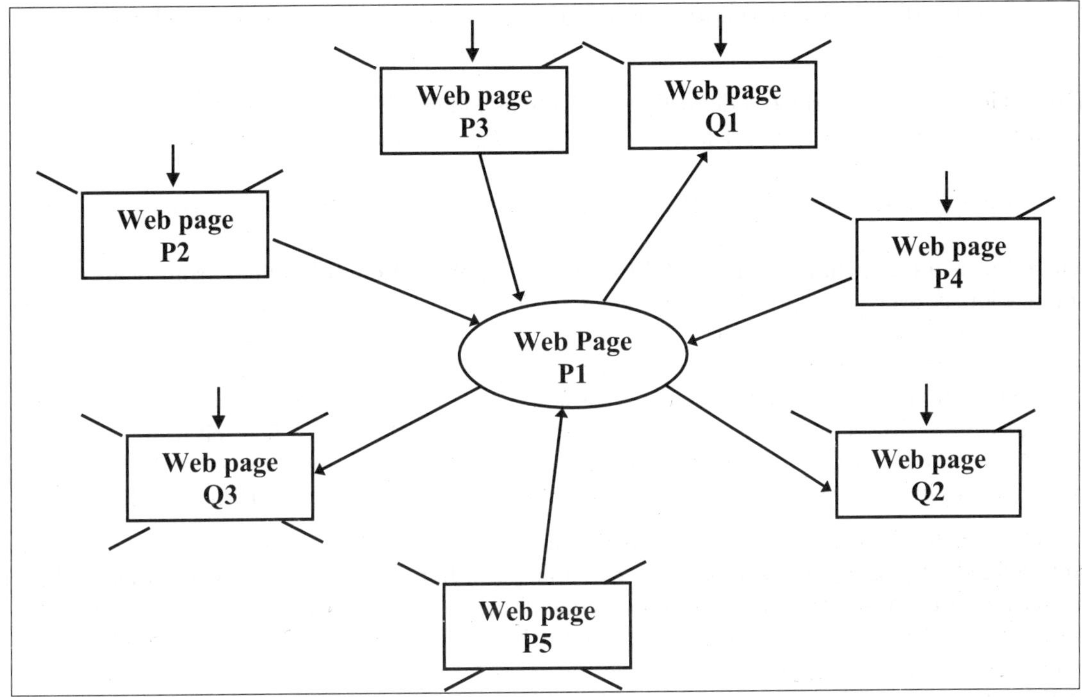

Figure 12.2: Incoming and outgoing links to web pages

12.7   PageRank Algorithm (PRA)

The page rank measure introduced by Brin & Page (1998) is used by some search engines to relevance rank retrieved document collections on a high-to-low priority order. This ordering indicates the current (at the time of query) page popularity of retrieved pages on the web. There are many ways to quantify the popularity of a web page. One of the widely accepted methods is the Page-Rank, which is a global property of a page irrespective of its content or the encoding language used (links could exist from English web pages to pages developed in other languages like Japanese or German). It is a soft-link based ranking method to relevance rank the retrieved pages. This enables the user to browse through the most relevant documents that match their query in the high-to-low priority (popularity) order. Since more meaningful documents are filtered out, it can tremendously save user's time and effort. In addition, search engines can reduce the execution time of their internal query engines by discarding all pages whose ranks fall below a cutoff threshold that is query specific. Page ranks are assigned during the pre-processing stage in a user, browser, language and query independent manner. In the OPRA, all out-going links like hypertext links, image, and multimedia links etc are equally weighted. Some search engines allow the user to query the web using images (Google and Yahoo both allow image querying), audio clips, video and multimedia clips, etc. In section 2, we generalise the OPRA by considering these types of hyperlinks. The computed page ranks are kept in inverted index files on search engine sites, and are incrementally updated on a periodic basis (say, once a month). As very low page ranks (near 0) indicate least popular pages on the web, corresponding index entries are not stored. This results in huge storage savings.

Definition 12.5 Page rank is defined recursively as follows[49]:
Let a page P_1 be linked to other pages P_2, P_3, $\cdots$, P_n as shown in figure 12.2.

$$\mathrm{Rank}(P_1) = (1 - m)\left[\mathrm{Rank}(P_2)/\mathrm{OL}(P_2) + \cdots + \mathrm{Rank}(P_n)/\mathrm{OL}(P_n)\right] + (m/n) \qquad (12.2)$$

where 'm' is a dampening constant, $\mathrm{OL}(P_i)$ is the number of outgoing links from page P_i and summation is over all pages P_i hyperlinked to P_1. This dampening constant is application specific. For example, in simple web surfing we choose m=.15 by assuming that a random surfer will click around 6.66 hyperlinks on the average (the average can be any real number, not necessarily integer, so that the leakage probability is 1/6.6666=.15) before the surfer moves over to a new site or terminate the current session. It is chosen as .5 in citation searching. The decay factor can be interpreted as an aproportioned constant score for each of the n pages participating in the island of inter-linking. More details can be found in [BP98], [CH07] etc.

The PageRank is a numeric metric on the ratio scale in NOIR typology, as the zero point is well defined. It is used by not only web search engines, but also by intranet and specialised information retrieval systems that contain document citations (documents in the collection should be inter-linked for the PR to work. Also note that it is used only to relevance rank the retrieved documents, and not to the information retrieval part as such). As used by web search engines, the PR indicates the popularity of a web page in relevance to other pages. Raw page ranks are positive real numbers (≥ 0). Most of the crawlers and search engines use a logarithmic transformation (to the base 10) to map the computed page rank to the range 0 to 10 (so that a raw page rank of 10^{10} will become $\log_{10}(10^{10}) = 10$). Even the most popular web pages in sites like yahoo, Bing and Google have page ranks well below this limit at present. We note that this transformation is done under the false assumption that page ranks are always >1. But in practice, there could exist many web pages whose page ranks are ≤ 1. The logarithmic transformation maps such ranks to negative numbers (as $\log(1)=0$ and $\log(0)=-\infty$, whatever may be the base of the logarithm). For example, suppose there are two newly created web pages P1 and P2 (with several links in between them) that have been hosted on the web recently. Even if there are hundreds of outgoing links between these pages, there could hardly exist any incoming links to such pages. In such a case, the page rank could be less than 1. A solution is to use a cutoff threshold to discard all web pages whose popularity is $\leq$ k, where k $>$ 1 is a real number, making logarithmic scaling meaningful.

---

[49]The multipliers used here were different in the original paper of Brin & Page [BP98] ($\mathrm{Rank}(P_1) = (1 - m) + m\left[\mathrm{Rank}(P_2)/\mathrm{OL}(P_2) + \cdots + \mathrm{Rank}(P_n)\mathrm{OL}(P_n)\right]$)

12.7.1   The Original Page Rank Algorithm (OPRA)

The page rank measure introduced by Brin & Page (1998) is used by some search engines to relevance rank retrieved document collections on a high-to-low priority order. This ordering indicates the current (at the time of query) page popularity of retrieved pages on the web. There are many ways to quantify the popularity of a web page. One of the widely accepted methods is the Page-Rank, which is a global property of a page irrespective of its content or the encoding language used (links could exist from English web pages to pages developed in other languages like Japanese or German). It is a soft-link based ranking method to relevance rank the retrieved pages so as to enable the user to browse through the most relevant documents that match their query in the high-to-low priority (popularity) order. As this filters out more meaningful documents, it can tremendously save time and effort. In addition, search engines can reduce the execution time of their internal query engines by discarding all pages whose ranks fall below a cutoff threshold that is query specific.

12.7.2   Power-law Distribution

It is known that the distribution of PageRank follows a power law ($f(x)=cx^{-a}$, where c is a normalizing constant, and a is any real number > 0) for general keywords [CR11]. As the pageRank is directly proportional to in-links from quality pages, and inversely proportional to out-links, we could first derive the distribution of X/Y where X and Y are both power-law distributed with pdf $f(x)=c1\ x^{-a1}$ and $f(y)=c2\ y^{-a2}$. Make the transformation U=X/Y and V=Y, so that the Jacobian becomes $|J|=v$. Under our assumption of independence of X and Y, the pdf of U is obtained as $f(u)=c1c2 \int_v (uv)^{-a1}v^{-a2}v\ dv = c1c2 \int_v u^{-a1}v^{-(a1+a2-1)}dv$. Upon simplification, this becomes $f(u) = c1c2(a1+a2-1)u^{-a1}/[(a1+a2)-2]$. This is the pdf of a power-law. Hence each of the terms on the RHS of the above equation (12.2) is power law distributed. To find the distribution of pageRanks, it suffices to establish that the sum of two independent power-law distributed variables is power-law distributed. Let U=X+Y and V=Y, so that the Jacobian is $|J|=1$. Proceeding exactly as above, we find the distribution of U as $f(u) = c1c2(a1+a2)\ u^{-a1}/[(a1+a2)-1]$. As this is a power-law, it follows that the (n-1) terms in the sum on the RHS of equation (12.2) are also power-law distributed. This proves our result.

The power-law can take different shapes depending upon the parameters. It is a special case of the Pareto distribution. For narrowly focused keywords in specific areas, the power law comes close to an exponential law ($f(x)=c\ e^{-cx}$) . For instance, a search on "global warming" brings up news items, scientific papers, public debates and chat-room discussion documents on the topic, whereas a search on "ozone layer depletion", which is more narrowly focused, brings up scientific articles more than news items, and others. Adding more and more keywords narrows down the filtered documents further and further, thereby eliminating 'noise' and increasing precision. Here noise refers to irrelevant documents retrieved due to synonymy, polysemy, polycronymy, etc.

Theoretically, the raw page rank is a real number that can vary in the range $[0,\infty)$. If two pages have the same page rank, it only tells us that both pages are equally popular. But it tells us neither about among whom those pages are popular, nor the content similarity of those pages. This is a weakness of the pageRank measure. Due to the heterogeneity of web pages, there are very few web pages that have very high page rank. Majority of web pages have low page ranks, which could vary dynamically over time. This fluctuation in popularity score is more for less-popular pages, than otherwise. Thus the page rank distribution (for a fixed epoch) is highly skewed to the right. This skewness will of course decrease over time, as more and more web pages vie for a piece of the elite pie (simultaneously the page ranks also will increase globally, but there could be many new entrants into the elite page rank groups). Hence the distribution is expected to remain a power law.

12.7.3 Quality of Inlinks

The pageRank of a page is directly proportional to the number of quality incoming links and inversely proportional to the number of outgoing links. In other words, pages with high pageRanks are either being conferred more importance by few other highly ranked pages, or have too many inlinks from not-so-popular pages. Thus it is not the quantity (number) of in-links, but the quality that counts more in determining its pageRank. This suggests a weighted average of inlink counts as the proper choice in the numerator. Although page developers have a 'say' over limiting the number of out-links from a page, they in general do not have much control over the in-links. These links are quite often unknown for new pages due to the fact that lot many users come to a page through links in other pages scattered over the billions of web pages.

Algorithm 12.2 Algorithm for Computing the PageRank of Interlinked Pages

---

1: Input a set S of web pages with information on out-links to other pages, damping constant m (0<m<1)
2: Special checks and Initialisation:
   begin
   If S is empty, return error code
   If S has a single page, or do not have interlinked pages (all links are to external pages), return error code
   end
   Count n = number of distinct logical pages in the input, Compute K = m/n, C= 1.0-m Initialise a 1D array of ROLINK[] scores (reciprocals of outgoing links) to $1.0/OL(P_j)$ for j=1,2,$\cdots$,n, where $OL(P_j) \neq 0$ is the number of outgoing links from page P_j. If there are "sink pages" (with no outgoing links), the corresponding ROLINK[] values are set as zeros. Each such sink page will reduce the number of equations by 1 (if there are k sink pages in S, we may end up solving a system of m (<n) equations in n unknowns where m=n-k).
3: Form the coefficient matrix 'A' as A[i,i] = 1.0, for i=1,2,$\cdots$ n;
4: **for** (i=1,2,$\cdots$,n) **do**
5:      **for** (j=1,2,$\cdots$,n) **do**
6:          **if** (i $\neq$ j) **then**
7:              **if** (P_i has incoming links from P_j) **then**
8:                  A[i,j] = $-$C*ROLINK[j]
9:              **else**
10:                 A[i,j] = 0
11:             **end if**
12:         **end if**
13:      **end for**
14: **end for**
15: Form the RHS vector b[] as b[i] = K, for i=1,2,$\cdots$ n
16: Solve the system of linear equations A*X = b for the unknown X
17: Assign the pageRank scores X[j] to pages $OL(P_j)$ in the same order
18: **return**

Visitors can also come to a page through search engine displayed results. We call such visitors as transient visitors, so these links (called transient links) are not permanent (If we count transient links also to compute the pageRank, we could increase the rank of any web page by repeatedly searching for it using various search engines and clicking on those links displayed by the search engine. Similar argument holds for incoming links from blogs, social networks and soft links coming from application programs (like java programs, applets or servlets). This is the reason why most of the popular search engines prepend their URL to a user query. This in turn enables web sites to distinguish visitors coming through search engines from regular visitors coming through web pages. But anyone can write their own search engine, and suppress their URL. In this case, the visited page has no way of telling whether the request is coming from a search engine displayed page. A solution is for every HTTP server to store the URLs of known search engines and distinguish HTTP requests coming from non-search engines. Due to the large number of existing search engines that are likely to grow in the coming years, this may not be a permanent remedy. Most search engines compile the index file using static web pages downloaded from the web due to this reason.

12.7.4   Drawback of PageRank Algorithm

The pageRank algorithm has many drawbacks. The most important one is that it totally ignores to where the outgoing links point to. In other words, the entire sets of web pages are considered as equally likely to have outgoing links. Thus the outgoing link weights are equally proportioned in the algorithm, giving equal importance to each of the pointing pages. As mentioned above, this number is in the hands of content developers or web site administrators.

12.8   Generalised PageRank Algorithms

This section discusses several generalisations to the OPRA that are reported in [CR10]. These generalisations are intended to improve the quality of search results, allowing a user to find the most relevant information among the displayed results.

12.8.1   PageRank Linear Equations

If the constant factor (m/n) is removed from the above definition, we see that $\text{Rank}(P_1)$ is a weighted average of other page ranks with respective weights (1-m)/OL(P_j) for j=2,3,.., n. As the definition is recursive, this gives rise to a set of linear equations in $x_i = \text{Rank}(P_i)$ as the unknowns. This can be written in matrix form as R = AR where R is a column vector of page ranks, and A is a matrix with entries $a_{i,i}$=0, and $a_{i,j} = $ (1-m)/OL(Pj), and $a_{j,i}=$ (1-m)/OL(P_i), ∀ i and j. Rewriting this as (A-I) R = **0**, where **0** is a column vector, we see that this gives rise to a set of n homogeneous linear equations. It has a unique solution when its determinant |A-I|=0. This is in general not possible. That is the reason why the constants m/n's are added to each equation to make it inhomogeneous. The corresponding linear equations then become R = AR + b where b is a column vector of all (m/n)'s. This can be written in matrix form as $(I_n$-A)R = b. The condition for a unique solution then becomes $|A - I_n| > 0$, or equivalently the inverse $(I_n - A)^{-1}$ exists (so that R = $(I_n - A)^{-1}*$b). It can hence be found using one of the algorithms for solving linear equations. But this may not always be practical for all pages on the web. Some matured web pages could have millions of incoming links, which results in a linear equation in million variables. Iterative algorithms, D & C algorithms, and cluster-based decomposition techniques are available to find the page ranks fast [DP05], [IK06].

12.8.1.1   Page Ranking Using Topological Sort

In some applications, we may be simply interested in ranking the web-pages among themselves rather than explicitly obtaining the page-ranks. Consider for example, the set of pages in figure 12.1 in page 12-12. Page B has 3 incoming links (sink node), whereas page A has 4 outgoing links (anti-sink or source node). As there are no incoming links to A, we could immediately guess that it is either a newly hosted page on the web or a web-spam page. As there are 3 incoming links to B and F, we expect these pages to have a higher rank than others. We could use the principle of topological sorting to remove the pages one-by-one, and order the pages among themselves. There are two approaches known as 'source-first' and 'sink-first'. Consider the source-first algorithm. Removing page marked 'A', and all its outgoing edges results in node 'C' to become a source (as C now has no in-links). Removing C and its outlinks results in D, E to become sources. Proceeding similarly, we get F as the next source, followed by G and B. Hence we could immediately order the pages as A<C<{D,E}<F<G<B. This means that page A has the lowest page rank and B has the highest. Next consider the sink-first algorithm. Start with page B and discard all incoming links. Now G becomes a sink. Proceed recursively to get the rest of the pages in high-to-low page rank order.

The linear equation method is the most straightforward way to compute the pageRank. We exemplify it for two sets of pages.

Example 12.2 Find the page-ranks of web pages linked as shown in figure 12.3.
Solution: We first consider the left figure marked (a). The numbers next to each node indicate the incoming and outgoing links. For example the page P_1 has 2 incoming and 2 outgoing links (because the $\Leftrightarrow$ indicates that there is a link from P_1 to P_2 and reverse, and P_1 to P_3 and reverse). Take damping constant m=.9, and number of pages n=3 (m/n=0.3) and denote rank(P_1)=X, rank(P_2)=Y, and rank(P_3)=Z. We get the following system of linear equations:– X = .1*(Y + Z/2) + 0.3, Y = .1*(X/2+Z/2) + 0.3, Z = 0.1*(X/2) + 0.3 (as P_3 has incoming link only from P_1). Eliminate Z between the first two equations to get 21*X=22*Y, or X=(22/21)*Y. Substitute in the third equation to get 10*Z = X/2 + 3 = 11/21*Y+3. Solution of these equations give X=0.349206, Y=1/3, Z=0.317460. These are the required pageranks, that add up to 1. Notice that there is a cycle in the links:– P_3 to P_2, P_2 to P_1, P_1 to P_3. If this cycle is removed, there are just two links remaining (P_3 to P_1, P_1 to P_2). This shows that page P_1 is more important than others and it points to P_2. This justifies our answer that P_1 has the highest score, followed by P_2. Had we taken the damping constant as m=.6, the results would have been X=0.388889, Y=1/3, Z=0.277778. Next consider the figure marked (b). Take m=0.6. Denote the pageRanks of P_1 to P_5 by V, W, X, Y and Z. The system of equations are V = .4[X/2]+0.12, W = .4[V/2]+0.12, X = .4[Y]+0.12, Y = .4[W/2+ Z]+0.12, and Z = .4[V/2+W/2+X/2]+0.12. This can be written in matrix form as

$$\begin{pmatrix} -1 & .2 & 0.0 & 0.0 & 0.0 \\ .2 & -1 & 0.0 & 0.0 & 0.0 \\ 0.0 & 0.0 & -1 & .4 & 0.0 \\ 0.0 & .2 & 0.0 & -1 & .4 \\ .2 & .2 & .2 & 0.0 & -1 \end{pmatrix} \begin{pmatrix} V \\ W \\ X \\ Y \\ Z \end{pmatrix} = \begin{pmatrix} -0.12 \\ -0.12 \\ -0.12 \\ -0.12 \\ -0.12 \end{pmatrix}.$$

This has the solution V=0.163290, W=0.152658, X=0.216449, Y=0.241123, Z=0.226479. Had we chosen m=.9, the solution is: V=0.19005, W=0.1895, X=0.201038, Y=0.210378, Z=0.20903.

We get the following system of linear equations, by taking n=6 (excluding A) and m=0.9. Denote rank of a page by the page label itself. For page B, we have B = .1[A/4+C/4+G] +

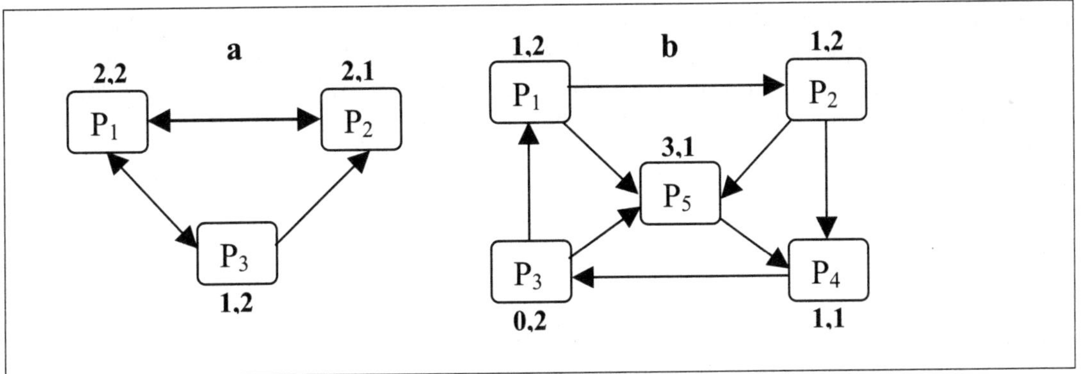

Figure 12.3: Linked web pages for example 12.2

0.9/6.0, Similarly, C = .1[A/4]+0.9/6.0, D = .1[A/4+C/4]+0.9/6.0, E = .1[C/4]+0.9/6.0, F = .1[A/4+D+E]+0.9/6.0, G=.1[C/4+F]+0.9/6.0. These can be put in matrix form as 6 equations in 7 unknowns.

$$\begin{pmatrix} .025 & -1.0 & .025 & 0.0 & 0.0 & 0.0 & 0.1 \\ .025 & 0.0 & -1.0 & 0.0 & 0.0 & 0.0 & 0.0 \\ .025 & 0.0 & .025 & -1.0 & 0.0 & 0.0 & 0.0 \\ 0.0 & 0.0 & .025 & 0.0 & -1.0 & 0.0 & 0.0 \\ .025 & 0.0 & 0.0 & 0.1 & 0.1 & -1.0 & 0.0 \\ 0.0 & 0.0 & 0.025 & 0.0 & 0.0 & 0.1 & -1.0 \end{pmatrix} \begin{pmatrix} U \\ V \\ W \\ X \\ Y \\ Z \end{pmatrix} = \begin{pmatrix} -0.15 \\ -0.15 \\ -0.15 \\ -0.15 \\ -0.15 \\ -0.15 \end{pmatrix}.$$

Assume the rank of sink node to be a small number (say 0). Solve for the rest of the variables to obtain the pageRanks B=0.325575, C=0.15, D=E=0.15375, F=0.18075, G=0.171825. Arranging them in ascending order, we see that A<C<{D,E}<G<F<B. This is almost what we got using topological sort, except for the order of F and G. This is because F has 3 in-links and one outlink, whereas G has 2 in-links and one outlink node.

The following section introduces many novel and relatively efficient modifications to the well-known PageRank metric used by some search engines. These modifications are intended to filter-out noise resulting from spurious links, irrelevant links and spam links. The proposed methods are effective in filtering out the most relevant content from very large number of documents. This can tremendously reduce the browsing time of surfers and researchers.

12.8.2 Noise-Removed Page Rank Algorithm (NoRPRA)

The first modification that we consider is the noise removal. Even trivial pages on the web use counters (like page counters that accumulate the total number of visitors to a page, or visitors to a web site after an administrator set epoch date), and toolbars that are borrowed from elsewhere (like Google toolbar). These in turn increase outgoing links, thereby reducing the pageRank. A simple solution is to eliminate such secondary links, and use genuine out-links as $GOL(P_i)$. This is done by subtracting the secondary link count $SL(P_i)$, resulting in NoRPRA equation as

$$\text{Rank}(P_1) = (1-m)[\text{Rank}(P_2)/\text{GOL}(P_2)+\text{Rank}(P_3)/\text{GOL}(P_3)+..+\text{Rank}(P_n)/\text{GOL}(P_n)]+\left(\frac{m}{n}\right)$$

where $GOL(P_j)=[OL(P_j)-SL(P_j)]$ are the number of *Genuine Outgoing Links*. All links to web counters, online advertisement sites, character animations, cgi-bin directories, popup applets, animation buttons, movie clips, news-feed bars, etc may be considered as secondary links. On occasion, it could happen that this subtraction results in $GOL(P_j)$ becoming zero (for example, if the only outgoing links from a site are to cgi-bin programs). In this case, the page becomes a sink, and we reset its outgoing link count to 1; or it is dropped from linear equations (resulting in one less equations than variables; see algorithm 12.2). In most cases, this simple modification can improve the quality of pageRank to a great extent.

12.8.3   Alpha Page Rank Algorithm (APRA)

The OPRA uses the number of outgoing links as the divisor. This implies that those pages with a large number of out-links, but with the same content as another page with lesser number of out-links will be adversely lowering their pageRanks. The extra out-links in a page may be due to advertisements, data validations, third-party controls, etc being used in a page. The NoRPRA may not always be practical for the entire web, as it must distinguish between genuine links and noisy links for each page, which could change over time. Hence a modification of the denominator as $[OL(P_j)]^\alpha$ where $0< \alpha <1$ (preferably in [.5,.99] range) is a real number will favorably lower the penalty of pages with too many outgoing links. A proper choice is $\alpha =.5$, in which case we get the APRA equations as

$$\text{Rank}(P_1) = (1-m)[\text{Rank}(P_2)/\sqrt{GOL(P_2)}+\text{Rank}(P_3)/\sqrt{GOL(P_3)}+..+\text{Rank}(P_n)/\sqrt{GOL(P_n)}]$$

$+(m/n)$ where all square-roots are positive. This is structurally identical to the above system of linear equations, except that the coefficient matrix entries are bigger because we have used the square-root of outgoing links in the denominator. It can be solved using any of the exact or approximate methods for linear equations. The average number of outgoing links in a web page will double every 10 to 12 years. Hence this modification has great practical utility. As the denominator terms are smaller than OPRA, the sum of the page ranks won't add up to 1.

12.8.4   Weighted Page Rank Algorithm (WePRA)

The OPRA considers all outgoing links to be equal. But links could exist from pure hypertext, from images, or graphics. There are also other types of links. These links must be differentially weighted to make them more meaningful. We distinguish between hypertext links to hypertext documents (H2H), hypertext links to images (H2I), image links to hypertext documents (I2H), image links to other images (I2I), multimedia links, graphics links etc. If the differential weights assigned to these links are known, we could modify the OPRA to get the WePRA equation as

$$\text{Rank}(P_1) = (1 - m)\{\text{Rank}(P_2)/[\sum_i w_i * GOL(P_{2i})] + \text{Rank}(P_3)/[\sum_i w_i * GOL(P_{3i})] + \cdots + \quad (12.3)$$

$$\text{Rank}(P_n)/[\sum_i w_i * GOL(P_{ni})]\} + (m/n)$$

This modification can tremendously improve the quality of PR using a properly chosen weighting scheme. If a link of a particular type does not exist from a page, the corresponding weight is set to zero.

12.8.5 Filtered Page Rank Algorithm (FiPRA)

As noted above, inlinks from search engine displayed pages, blogs and social networks should not be considered as genuine, and must be discarded from pageRank calculations. This is because creating unfaithful links from search engines, blogs and social networking sites, etc can inadvertently alter the page rank. Most search engines download entire static web pages to local servers to create their indexes, which are updated on a periodic basis. Nevertheless, anyone can write a bare wrapper for several search engine retrieved set of pages, and post it on their web site as if it were a genuine page. This type of spoofing may result in wrong entries for such pages to be created in search engine index files resulting in spurious pageRanks. A solution is to adjust the numerator of the above equation to discard such dynamic links. As the numerator contains unknown numbers (pageRanks), we need to use a clever trick. Assume that all pages in the pageRank equation are getting hits from search engine displayed pages. Number of such clicks coming from various search engines can be obtained by a URL analysis (we assume that search engines prepend their address when the user request is redirected to the actual page). We assign various weights to each search engines based upon this observation. For example, if there are 100 visitors coming through Google, 65 through Yahoo and 35 through Bing, the respective weights for transient links are .5, .325 and .175. But transient links must be differently weighted from static links. Hence the combined transient weights are differentially weighted wrt static link weights to obtain the actual weighting in the numerator. To get the actual incoming links, we need to subtract the weighted secondary dynamic links. Instead, we add these numbers to the m/n on the RHS to get the modified set of linear equations as:

$$\text{Rank}(P_1) = (1-m)[\text{Rank}(P_2)/\text{GOL}(P_2)+\text{Rank}(P_3)/\text{GOL}(P_3)+...+\text{Rank}(P_n)/\text{GOL}(P_n)]+(\text{M}) \tag{12.4}$$

where M=(m/n)+ (1-m) [TW(P_2)/ OL(P_2) +$\cdots$+ TW(P_n)/ OL(P_n)] where TW(P_i) are the transient weights for page P_i due to the incoming links from various search engines, advertisements, discussion groups and chat-rooms, etc. This is again a set of linear equations with a changed RHS vector. It can be solved using one of the methods described earlier.

12.8.6 Hybrid Page Rank Algorithm (HyPRA)

We could combine the above modified PRA's suitably to get a variety of useful measures. For instance NoPRA and APRA may be combined to get the HyPRA equation as:

$$\text{Rank}(P_1) = (1-m)[\text{Rank}(P_2)/\text{GOL}(P_2)^\alpha + \text{Rank}(P_3)/\text{GOL}(P_3)^\alpha +...+\text{Rank}(P_n)/\text{GOL}(P_n)^\alpha] \tag{12.5}$$

$+(\frac{m}{n})$ where GOL(P_j)=[OL(P_j)-SL(P_j)] as before. This modification can tremendously improve the quality of pageRanks. As in the case of APRA, the denominator terms are much smaller, and the sum of the page ranks won't add up to 1. Many other authors have also generalised the pageRank algorithm. See [DP05] for a divide-and-conquer algorithm, [LM06] for a reordering algorithm, [IK06] for a convergence analysis, and [BP05], [BM06], [SS10] for various surveys.[50]

12.8.7 TrustRank Algorithm

The trustRank algorithm developed by researchers at Yahoo and Stanford university is used to identify spam sites from others. It is a generalised version of pageRank algorithm. It uses a

---

[50]Several sites offer online pageRank computing facility. See for example www.checkpagerank.net, www.prchecker.info, www.liveprcheck.com, and www.smartpagerank.com

biased version of OPRA to propagate trust to descendant (linked) web-sites. Gyöngyi, Garcia-Molina & Pedersen [GG04] used it to combat web-spam[51]. Spam sites attempt to achieve higher web ranking than they deserve using various techniques. Thus the algorithm keeps a source vector of positive pages and negative pages, which are updated continuously as described below.

TrustRank algorithm starts with a set T of *trusted pages* (also called *seed pages*) that are either manually compiled or filtered for trustworthiness. All the members in T are assumed to have high trust-score. These pages can transfer or disseminate similar trust levels to other pages by linking to them. An underlying implication of this linking is that good pages on the web seldom links with spam sites or other rogue sites (but the converse is not true). It then creates a transition matrix T, as $T_{i,i}=0$ $\forall i$, $T_{i,j}=0$ $\forall j$ if there are no in-links to page i, and $T_{i,j} = 1.0/OL(P_j)$ if inlink exist from page P_j to page P_i, and $OL(P_j)$ is the number of outlinks from page P_j. It then iteratively updates a score distribution vector until user specified number of iterations. Details can be found in [GG04]. See www.trustrank.com, trustrankcheckers.com, etc for checking the trustRank of your home page.

Static hyperlink relationships can also be summarised in tabular form, but is of limited utility. Consider two web pages P1 and P2. Hyperlinks may exist from P1 to P2 (a 'vote' by page P1 to P2), with or without links in the reverse direction (from P2 to P1). These are called incoming and outgoing links. This two-way (bi-directional) relationship is called co-citation. Unidirectional 'voting' among a group of pages can be represented by a directed graph. The Pagerank algorithm looks at not only the votes received by a page (incoming links), but the number of important voters (pages that have cast their votes) too.

Co-citation may be between pages on the same site (intra-site co-citation) or inter-site (between two sites). Inter-site co-citation is more interesting from WSM point of view because it can reveal high coupling between web pages. These are used in B2B e-commerce.

12.8.8   Link Categorisation

Links may be categorised as homogeneous or heterogeneous. Homogeneous (or standard) links are the most common and they are from web pages (html, asp, php, jsp etc) to (same or) other web pages of similar type. Markup languages allow links to be created from images, thumbnails, buttons etc to other resources (text documents, web pages, pdf files, other application files, audios, videos and multimedia files) on the web. These type of links are called extended or heterogeneous links. Linking a web page to application files (pdf, word, postscript, etc) may result in automatic launching of these applications on the client machine/device by the user's browser. Hence a variety of object types may be interlinked to provide rich online content. Different link mining techniques either ignore target type or explicitly consider it. An analysis of the variety of target types that are linked to a web page is useful in time reduction, page restructuring, resource zipping etc. The table 12.2 summarises heterogeneous links in web pages, which can also be represented as a multi histogram.

Standard links can further be broken down into local (intra-document links to other sections[52] or inter-document links at the same site) and remote (links to markup files elsewhere). There are special types of links that can appear in markup languages. As an example, <a

---

[51]Web-spam is different from email spam. While the former is a web-page created with bogus keywords or links to fake the web indexers to rank it high; as if it is a genuine page, the latter is a fake email message that is unsolicited. Web-spam (also called URL spamming) usually have a dishonest business intention to invite web-users to their site, spam emails may or may not have a dishonest intention.

[52]The sections are given names and the # character used in anchors to jump to it.

Table 12.2: Heterogeneous links in web pages

Sr no.	Page label	Standard links	Image links	Audio links	Video links	Multimedia links	Other links
1	P0	12	3	0	1	0	2
2	P1	8	0	1	0	1	0
..	..	. . .	. . .	. . .	. . .	. . .	. . .
n	Pn	. . .	. . .	. . .	. . .	. . .	. . .

href="mailto:administrator@yoursite.com> indicates a special link that generates an email message (by invoking the local mailer program). Similarly, links may exist to other protocols (ftp, news etc). For instance, <A HREF="ftp://anonymous:anonymous@somesite.com"> indicates a link to an ftp site with anonymous login,[53] and <A HREF="news:comp.lang.java"> is a link to a news site. Other possible links are to WAIS, gopher, image maps (<A HREF=/cgi-bin/imagemap/label>), multimedia files (<A MREF="multimediaresource">Description</A>) etc.

Since the execution time for viewing a resource is the sum of the application loading time (on client machine) and data downloading time (from server to client), this data can also be mined to provide valuable information about client machine, network latencies, memory and cache, processor speed, etc). Static link mining cannot provide information on broken (non-existent) links. This information is useful in maintenance of web pages. Link mining has applications in identifying duplicate or overlapping content, link prediction, bibliographic citation and coupling analysis, and event based ranking.

12.8.9 Link Stepping

Definition 12.6 Link stepping is a technique to automatically analyse trustworthiness of hyperlinked sites and is started from a parent site.

It involves stepping from the parent page (eg: home page of parent site) to each of the child links (web pages that are one click away) and repeating the process with each child page (to a specified depth)[54] looking for unwanted or banned sites or those that have restrictions or viewer caution. When a loop (which represents a cycle in the graph) is detected, the corresponding thread is terminated. As an example, if a student home page that is part of an academic site has links to rogue sites or illegal organisations, the link stepping can identify them. It may either be based upon lookup tables (that store URL of illegal sites) or can be dynamic (links from rogue sites are often to other rogue sites. If the pages are identified by unique labels and URLs, the relationships can be captured into a table. Additional columns can be used to identify the depth from a specified page, degree of trustworthiness of linked page, etc. These rogue links can be used to update lookup tables, which must be updated on a periodic basis due to the possibility of links becoming non-existent over time).

Link stepping can be implemented recursively by starting the search at a specified page (usually the homepage). A maximum depth from the start page may be used to stop the recursive descent. The output of link stepping is a list of pages that contain links to rogue

---

[53]You could also add a resource identifier after the site address.

[54]Higher order links to rogue sites are most often less harmful than those at a few clicks away.

Table 12.3: Sample data for link stepping

Label	URL	Parent	Description
P0	www.univ.edu	www	University web site
P1	www.univ.edu/John	P0	John's homepage
P2	www.banned.com	P1	A banned site
P3	www.rogue.com	P2	A rogue site
..	. . .	. . .	Links from rogues sites

sites in (*page id, depth, rogue link*)[55] format or a message indicating that no rogue links are found. Link stepping can be restricted to local pages alone, or to pages in pre-specified domains (eg: .org, .int, .net) using an appropriate filter (which discards unwanted links). It can be implemented by growing a dynamic tree with the start page as the node to skip links to prior (already considered) pages. If a page does not have new links (other than those to parent pages of the tree), or if the depth of a node exceeds maximum specified depth, the corresponding branch will stop growing.

12.8.10   Link Analysis

Link analysis[56] is a term used in law enforcement, intelligence gathering, homeland security, online fraud detection, and other monitoring and control systems.

Definition 12.7 Link analysis is a semi-supervised or unsupervised technique that involves a *connecting-the-dots* metaphor by domain experts using computer generated visual models [ST03]. It relies heavily on link mining, and provides hints on a combination of rules among entities, activities and inter-correlations over a period of time.

Data for link analysis need not be entirely web based. Links may be weighted such that more recent ones receive higher weights than those in distant past. Various weighting patterns (exponentially decaying (into past), uniform, half-normal etc) can also be used to evaluate different scenarios. A discussion of link analysis ranking algorithms can be found in [NZ01], [BR05], [DB03].

12.9   Web Query Mining (WQM)

Most web sites are designed to interact with users. Each query encapsulates the information needs of a user. Web and DW queries are IR requests (for data from the server to the client) whereas DB queries can retrieve or update data on a server. Most of them are targeted at text documents (HTML, XML, etc type). But queries may also be used to retrieve images, audio and multimedia files. A query can contain wild cards to search for documents with unspecified or partially matching search keys.

The WQM is used for targeted advertising and generating dynamic content. Another application is to mine for query execution time (the time elapsed between query submission and

---

[55] Alternatively (*page id, depth from root, rogue link count*).

[56] Link analysis has different meanings in other disciplines like accounting, AI, neural networks.

results delivery). Because generic queries result in millions of hits, it takes longer to execute. However, some search engines have distributed crawlers that collaboratively filter out desired information. Query execution time may depend upon the number of keys and phrases (keywords, images, etc) used, wildcards (if any) present, domains searched (English documents or multi-lingual documents), time interval specified (eg: all documents with a time stamp between 1/1/2014 and 12/3/2014), and so on. Temporal query mining can be used in organising indexes, installing new crawlers, etc. Incremental query mining looks for patterns and trends in refined queries submitted by the same user. Web queries may be stored (in databases as pure text or markup language format) for later use and the stored queries mined to get generic patterns of user queries.

12.9.1 Query Performance Measures

Precision (P) and recall (R) are the popular measures of query performance. Precision is a measure of the preciseness of a search, and is defined as the ratio of the number of relevant documents retrieved to the total number of documents retrieved. Irrelevant document retrieval may be due to ambiguous words (as in 'java' which appears in dozens of contexts, 'bank' that appears in financial contexts, as side of waterways, in blood bank etc). It may also be due to insufficient information, wildcards in a query, common words and phrases in multiple languages,[57] and unspecified time intervals. Higher precision means lesser irrelevant documents. Precision can be improved either using several disambiguated keywords and phrases or by re-querying. If P=0, all retrieved documents are irrelevant (which is rare). If P=1, every retrieved document is relevant. Due to the large number of hits returned by today's search engines, it is quite often impossible to get a numerical value of precision. However, most search engines rank the hits in the order of relevance. Hence users need not bother about the numeric value of precision, so long as they find the desired information from the first few hits displayed.

Recall measures the completeness of a search. Higher recall values indicate lesser missing documents. It is defined as the following ratio:
Recall = number of relevant documents retrieved/entire relevant documents in the collection (say on the web). Recall and Precision does not depend upon each other. Recall depends more on the ability of the search engine in finding and filtering requested documents than on user query. Since the number of relevant documents on the web is most often unknown, the recall can only be estimated.

12.10 Semantic Web Mining

Semantic web mining is a term coined from 'semantic web' and 'web mining'. The semantic web is a conceptualised view of the web as comprising of different layers of content with an aim to enrich the structure of web documents and improve machine interpretability. This results in layered search engines to filter out proper information, thereby increasing precision and recall. The architect of the Web, Tim Berners-Lee has suggested a multi-layer view of it as:- (1) XML syntax layer, (2) RDF - Data layer (3) Ontology layer (4) Logic layer (5) The Proof layer, (6) Unicode/URI and trust layers. This will in turn result in the next generation web. The XML layer can be considered as the 'syntactic layer' for semantic web. The Resource Description Framework (RDF) is a foundation for processing metadata providing interoperability for web

---

[57]Lot many English words have different meanings in German. As examples :– 'war' means 'was', 'was' means 'what', 'kind' (pronounced kin-d) means 'child' and 'hose' means 'pants'.

based applications, and is considered as the 'vocabulary layer'. RDF has 3 types of entities called resources, properties and statements. Resources are objects of any type (web page, video clip etc) on the web that can be accessed by URI.

It can either be used to build a semantic web from the information obtained by web mining or an already built semantic web model can be used to improve the web mining process [BH02]. A standardisation effort by the W3C includes the Web Ontology Language (WOL, which is also abbreviated as OWL), which is designed to process the semantic content of web data. The OWL has several sub-languages called OWL *Lite*, OWL *DL* and OWL *Full*, suited for different user tasks. The ISO standard *Topic map* is the analogue of the W3 standard RDF/OWL. A discussion on an automated tool for semantic tagging of large corpora (SemTag) can be found in [DE03].

12.10.1   Metadata Mining

Metadata are summary data about other data[58]. Metadata can contain physical attributes like location and size information, version numbers, creation dates[59], resource dependencies, identifiers, language(s), formats, publisher, source(s), relations, and special features (like calling conventions of programs/scripts, plug-in requirements (if any), automation info[60]) or derived attributes. Metadata are also used to store information about various software components (eg: DLLs, COM components under Microsoft platforms). The attributes of metadata may be *obvious*, *shadowed* or *hidden*. Obvious attributes are those obtainable without accessing the metadata content. The file size, file type, creation dates, etc are examples. The version numbers and derived attributes are shadowed attributes whereas security restrictions, encryption passwords (if any) are hidden attributes. Another categorisation is *content dependent* and *independent* metadata. Content dependent metadata attributes are derived from the actual content (file size, formats used, etc), whereas the content independent attributes do not depend on actual content (file name, creation date, security access control attributes). Other subdivisions of metadata can be found in [KV97].

Irrespective of the data type that they represent (text, structured, image, audio, video, multimedia, or various application data), the metadata are often kept as text files, either in flat form or in structured form (in XML format). Since they are created once and modified rarely, they may be kept as 'in-memory' files, on file systems (under various OS), in databases, in datawarehouses or as embedded files. One XML document can store the metadata of multiple files. They are more structured than the data they represent and thus facilitate pattern discovery from large data. In addition, they summarise the features of the data implicitly, thereby guiding the mining process and reducing dimensionality of search space.

Metadata mining extracts unknown patterns and trends from metadata files. They are most often utilised for ontology extraction, summarisation, classification, and clustering. Metadatawarehouses are centralised repositories of metadata. They facilitate data extraction and mining processes into different layers. In the absence of such datawarehouses, they are extracted on the fly and could provide useful information about the parent documents. For instance, metadata for text documents contain the type, file size, creation date, language used and other content dependent attributes and identifying information of the document. Metadata mining applications include data warehousing, multimedia data mining, focused searching, con-

---

[58]It represents the schema in databases.

[59]This may indicate the creation, access and modification dates of data files or of metadata file or both.

[60]The ability of a program to know the properties/methods of an object at runtime is called automation.

tent and version management etc. Since datawarehouses contain data from multiple sources, a metadata mining step can be performed for fast information retrieval and pattern discovery. They are well suited for web content mining, especially so with distributed collaborative data mining, and in multilingual web mining. Metadata creation tools may be automatic [AB05] or semi-automatic.

12.10.2 Multilingual Web Mining

At present, roughly one-third of the web content is non-English. This ratio is expected to grow as internet access is spreading fast all over the world. Mining the multilingual web pages is more challenging due to the interaction between pages in various languages, and the presence of multiple languages (like Chinese and English) in the same web page. Multilingual queries may originate from native languages (searching for Japanese government documents using Japanese language) or foreign languages (searching for Japanese documents using English). This is called cross-language retrieval. Cross-language mining is an extension that involves further analysis of the retrieved data. A two dimensional matrix can be used to capture the cross-language relationships. A discussion on cross language mining can be found in [TZ05].

12.10.3 Web Personalisers

Web personalising is a technique to present user specific content or services in an optimal way. User profiles are created using web personalisers and logging programs. These typically are multi-threaded programs that run on an HTTP server in supervisor mode to capture all user accesses. Because HTTP is a stateless protocol, the state of each user must be kept track of using session variables, client-side cookies, URL rewriting or some other popular technique. The profiling techniques used are association rules or cluster detection. Since a user may have multiple association rules, a probability is associated with each rule to indicate its frequency of occurrence. Clusters, if any, found by the personalisers may be indicative of the extra hyperlinks to be fused into the pages to make them more coherent. Behaviors of most users change over time. Hence the mined information should be incorporated into site redesign on a periodic basis. User profiles could also vary according to religions, cultures, geographic location of the user, education levels, or even the user's age. This extraneous information may also be saved in coded form in addition to the profiling data.

12.11 Image Mining

Video and image data in multifarious formats are flooding the web due to the availability of highly efficient compression algorithms. These are rich information sources that have hidden visual knowledge in the content.

Definition 12.8 Image mining is the extraction of hidden patterns and trends in image documents using one or more images or fractions (sub-image blocks) thereof. It is a convergence of content based retrieval, pattern matching, computer vision and statistical techniques.

Any types of documents that have an image link or embedded image can be mined. Some images (like surveillance scans of highways/runways) can also have embedded text, audio data and other multimedia data in it. All these can be utilised in the mining process, with various weights associated to each component or using different algorithms for each component. For instance, embedded text can be separated out and a text mining algorithm applied on it. The pure video data can then be mined using SVM. It is an *image-driven* discovery of hitherto unknown patterns in large data containing images. It can also be applied to corporate intrawebs.

Document retrieval using an image as a key is different from image mining. In image-based retrieval, we are interested only in the spatial features of an image key that matches (either exactly or approximately) one of the images stored or linked to a document like a weg page. In image mining, we mine image content, image features, and other information that are not explicitly stored in an image. For instance, automatic classification of a person into lean, athletic, normal, or heavy category using his or her image, automatic recognition of fighter jets from radar images of aircrafts, etc are image mining examples. Image association rule mining involves at least one image along with other attributes that generate a consequent.

12.11.1   Issues in Image Mining

Images are retrieved from the web using one or more keys and kept in local cache memory or secondary storage. Each image is processed a minimum number of times or transformed into another domain (say using DCT or DWT) and the features are extracted for further analysis. Images with matching features may be indexed and subjected to further 'in-memory' analysis. This may be implemented in multiple stages. The first step eliminates irrelevant images and identifies outlier images. The summary data extracted from resulting samples is used for feature matching, similarity matching, and clustering. A representative sample of *training images* may be used during learning stage to build up models. The built model can be used for future classifications, labeling and quality testing. As the information conveyed in images are visual rather than verbal (some images have limited text information in their metadata), fast and efficient algorithms are employed for image mining. Results of image mining augmented with patterns and hints obtained from other components (text, audio, animation) can be used for concept extraction and classification.

12.11.2   Multimedia Mining

Ever increasing web server storage capacities, efficient compression techniques, and higher network bandwidths have all contributed to the popularity of multimedia content on the web. Size of digital multimedia data on the web is growing fast and could overtake text data in the near future. Multimedia is retrieved using *query-by-example (QBE)* [LH00], or by content-based retrieval [WB96], [YI99]. In QBE, the user specifies a sample segment and the system uses pattern extraction and (exact or approximate) matching techniques to retrieve relevant documents.

Main challenges in multimedia data mining are the dimensionalities involved. A multimedia file may contain audio, video, images, animation, multimedia objects in addition to hypertext data. Each of them will be called an element. The set of elements will vary dynamically over time, and can be mixed or interspersed. Hence pattern discovery involves intra-mining of element features as well as mining for inter-relationships between objects (cross-dimensional patterns) in spatial and temporal dimensions. Feature extraction is carried in cross dimensions, as well as in 'regions of interest' within a single dimension (uni-modal patterns). Strong correlations across media are used to collect summary statistics and video shots to be used in queries. This stage has high computational complexity due to the heterogeneity of media, variability among element content, volume of data, various compression standards, special tools required, in addition to probable correlation between different types of media. In multimedia mining, users may have to specify the number of features to be extracted (as in clustering), number of hidden variables and thresholds, or the models to be assumed. See [CS03], [VZ04] for further details.

12.12   Table Mining

Tables are convenient tools to display relational information (with rows and columns) in structured format (see chapter 2). Tables can be nested (without overlaps) to any depth. Most business sites utilise tables to display summary data. Presence or absence of a table in a web page is not evident without physically examining the contents (markup language), or metadata. Tables can also be generated dynamically using scripting languages, servlets, and cgi programs. Presence of the $< table >$ tags need not indicate valid tabular data in a page [CT00]. Some web designers use the table mechanism to preserve space by arranging information in multiple columns. Table mining may be used either to categorise data using table contents or to explicitly mine tabular data. The data in online tables may be of categorical, numeric, character, or text forms or could contain links to another resource. It can also be a combination of the above in different columns of a table.

A preprocessing step can greatly simplify table mining. Multiple tables can have different dimensionalities and overlapping contents. These issues are resolved during the pre-processing stage using table and column headings, tabular metadata etc. Text data in tables are mined by accessing corresponding columns as well as immediate links pointing to text resources. Tabular link mining looks for patterns in soft (URL) links or hard links (tables in some softwares like MS Word can contain links to any type of resource like images, pictures, and objects).

12.13   Data Stream Mining (DSM)

A data stream is an unbounded data source of individual data items (a binary stream is the simplest). Mathematically, a single stream can be represented as a sequence $x_0, x_1, \cdots, x_t, ..$ For example, data captured by sensors (traffic sensors (road, rail, air etc or [computer, telephone, wireless, ATM] networks), geographical sensors, sonar sensors, industrial sensors etc) can result in a continuous data stream. Many applications exist for the stream model and new applications are emerging due to the size and cost reduction of sensors. Each sensor is designed to collect specific types of data. Real world data are often encoded and compressed before it is transmitted as a stream.

Definition 12.9 Analysing the data streams to extract (unknown) knowledge or detect anomalies and deviations rapidly is termed data stream mining.

Efficient mining of these streams for patterns and trends is done on an ongoing basis. Most stream mining algorithms scan the stream just once, so that they can run in real-time. The unit for stream mining may be individual stream sources or a collection of sources. Distributed stream mining applications employ hundreds of sensors that collect and transmit stream data over wired or wireless networks. Most stream mining occurs at central or distributed servers. The stream sources may be dependent in some applications so that data streams are correlated. For instance, temperature and pressure sensors in an industrial setting may have their own individual thresholds as well as dependencies. Incoming requests and outgoing responses in a web server may have separate sensors to detect possible trojan horses and weird programs that waste server's CPU cycles. Popular stream mining algorithms process the data stream once or utilise a sliding window of suitable size over recent data. A tilted window technique stores past summaries at various critical points and processes current data in a window of appropriate size. A weighting scheme can also be used to emphasise trends and drifts at various time intervals. An approximate and distributed algorithm for stream data as well as a logarithmic tilted-time window are described in [OP06], [LC06]. Online stream mining algorithm for semi-structured data can be found in [AA02].

12.14   Applications

Mining data that originate from the web has a large number of applications. Web mining is a general term used in mining web content, web structure or web usage [MJ97],[KM02]. Text mining is a special type of web mining in which data are restricted to plain text format (in unstructured or derived from structured files). Image mining is used for pattern extraction from large heterogeneous image collections. Multimedia mining has potential uses in automatic image captioning, automatic classifications (at various levels from a frame to entire file), event and anomaly detection, video surveillance, annotations, trend analysis, pattern extraction, content recommendations, etc. In the following sections, we briefly review some of the applications of the above techniques.

12.14.0.1   Web-page Clustering

Clustering the semi-structured web pages using keywords, images, links and other media present in them can provide quick insight and visual information on the vast number of web pages into a small number of groups (distinct clusters). Incorporating the hyperlink attributes will in addition provide co-citation, page coupling and cross-linking information and reduce the dimensionality in some instances. Webpage clustering works by first selecting a feature vector for each document, and applying one of the clustering techniques discussed in chapter 7 to extract distinct features to form homogeneous group of documents with common traits. This technique can be applied to summarise the multitude of documents on a large web site to find distinct and meaningful topic groups. The K-means algorithm is not quite applicable for text clustering because the number of clusters is unknown. Since the data used for web clustering is most often online, rerunning the algorithm with different K-values is costly. A simple solution is to download the target pages to a local system and rerun the algorithm locally. This may not always be feasible due to copyright restrictions, large number of images or videos present in the pages, and dynamic nature of the content. The X-means algorithm estimates the optimal number of clusters using a range of values and is better suited for text clustering [MS00]. Semisupervised clustering techniques can also be applied by first extracting phrases using methods like KEA [WP99], CNC [FA00], TFIDF etc and invoking the clustering algorithm [ZZ05]. The KEA (Keyword Extraction Automatically) has a training phase and extraction phase. It utilises the TFIDF and first occurrence measures for each candidate phrase.

12.14.0.2   Web Marketing

Web-based marketing has revolutionised the 'marketing concept' beyond imagination. Due to the continuous (24x7) presence of the web, location specific marketing strategies can be built at different times of the day. For instance, marketing effort needed to reach a desired result using the web is much less compared to conventional marketing. In addition, the cost of information is lower, and quality is higher than conventional marketing. Web marketers strategically place themselves in an environment where demographic and psychographic attributes and their interactions among various visitors provide crucial information needed for an extended marketing concept [BM98].

Table 12.4: Software for Web Mining

URL	Name	C/F
angoss.com	Angoss knowledge studio	C
estard.com/poducts	Estard DM	C
kdknuggets.com/software/web-analytics.html	Web & socialmedia	F
programurl.com/mediaminer.htm	MediaMiner	F
megaputer.com/site/polyanalyst.php	PolyAnalyst	C
miner3D.com	DM suite	C
solver.com/xlminer-data-mining	XLMiner	C
smartertools.com/smarterstats/website-data-mining.aspx	DM stuite	C
www.softpedia.com	Ez Web Miner	F
www.3d2f.com	WebFax miner	F
www.cs.waikato.ac.nz/ml/weka/	Weka	F
wizsoft.com/web-content-mining.html	WizSoft	C

Legend: C=Commercial, F=Free, S=Shareware

12.14.0.3 Miscellaneous Applications

There are numerous other applications for web based data mining. Here we briefly summarise them.

12.14.0.4 Website rankings

Automatically ranking web sites can tremendously reduce browsing time and decrease congestion. As an example, web servers that host downloadable public domain software programs are scattered throughout the world. Users choose one of these servers due to geographic proximity or personal preferences. By mining the past data, separate rankings can be provided to users from various parts of the world. Similarly, arbitrarily ranking a set of pre-specified web sites based upon their content richness, artistic presentation or ease of navigation involve a multitude of techniques derived from past user accesses and click-through rates. Other applications can be found in [CS03], [CH05].

12.15 Software for Web

Ez Web Miner is a very powerful automated tool to search and mine web data. It has a multi-threaded mining-engine with custom filters, export facilities etc. MediaMiner is used to mine news sites, stock reports, e-com sites, forums and blogs etc with export facility to XML, HTML and PDF formats. Webfax Miner is used to extract targeted fax numbers from websites and search engines (3d2f.com/programs/0-221-webfax-miner-download.shtml). Other software include, Sawmill5 (flowerfire.com), Vivismo clustering software (clusty.com), Leximancer textminer (www.leximancer.com), Temis Insight (www.temis-group.com),etc. Commercial software include SAS text miner (www.sas.com) and SPSS Clementine (www.spss.com), IBM speed tracer, etc.

12.16   Exercises

1. True or false:
   (a) A synonym for web usage mining is sequence mining
   (b) Software version numbers are shadowed attributes
   (c) Web usage mining is performed at individual sites
   (d) The Pagerank measure is used for web query mining
   (e) Link mining uses only remote links in web pages
   (f) The CNC measure is used for image classification

2. Explain the following terms:
   (i) conditional content filtering (ii) web citation mining (iii) web server logs (iv) push data and pull data (v) Hubs and authorities

3. Describe any three advantages of web mining.

4. Explain coreference resolution using an example. What is a coreference count?

5. Describe any three criteria used for web-page clustering. Which clustering algorithm is better suited for text clustering? Why?

6. Explain the following terms:
   (i) co-citation (ii) Page rank (iii) coupling and cohesion (iv) Link stepping, (v) "path to purchase".

7. What is the range of values that the PageRank can take?

8. List 3 advantages of link mining. Compare link mining and link stepping. What are one-way links? How are they useful in web mining?

9. If a web page P1 is linked to two other pages P2 and P3 with 2 links from P2 to P1, and 3 links from P3 to P1, compute the page rank (m= .85) if Rank(P2) = 12, Rank(P3) = 15.

10. If two pages P1 and P2 have bidirectional links with 4 links from P1 to P2 and 3 links from P2 to P1, find the pagerank of each page (m= .85).

11. Identify each of the following attributes as standard, hidden or shadowed (ii) content dependent or independent.
    (a) Last modified date (b) version number (c) video formats in the document (d) plugins required for a video (e) security certificate.

12. What are the commonly used query performance measures? How are they related?

13. Describe two popular measures for web structure mining.

14. Explain any three uses of web user quality mining.

15. Where is metadata stored? What are the categories of metadata? List any three advantages of metadata mining.

16. If the average viewing time (in seconds) for the pages are respectively given as P0 = 16, P1 = 40, P2 = 35, P3 = 73, P4 = 54, P5 = 60, P6 = 29, P7 = 18, P8 = 43, P9 = 71, P10 = 19, find the total time spent by 20 users in pages 7 and 10. Find the average time for an arbitrary user to reach page P10 from home page P0.

17. What is datastream mining? Where does the data for it come from? What are some uses of it?

18. Describe table mining. What are some of its uses?

12.16.0.5 References

[AC07] Aggrawal, C.C. (2007). *Data Streams: Models and algorithms*, Kluwer Academic, London.

[AP98] Allan, J., Papka, R., Lavrenko, V. (1998). Online news event detection and tracking, SIGIR 21, Melbourne, Aug 1998, 1-9.

[AA02] Asai, T., Arimura, H., Abe, K., Arikawa, S. (2002). Online algorithms for mining semi-structured data streams, Proc. IEEE international conference on data mining (ICDM02), Maebashi, 27-34.

[AT08] Asano, Y., Tenuka, Y, Nishizeki, T. (2008) Improvements of HITS algorithms for spam links, *IEICE transactions on info systems*, E91-D, 2, 200-208

[AB05] Atanas, K., Borislav, P., Ivan, T., Dimitar, M., Damyan, O. (2005). Semantic annotation, indexing and retrieval, *Journal of web semantics*, 2(1), 1-39.

[BH02] Berendt, B., Hotho, A., Stumme, G. (2002). Towards semantic web mining, LNCS 2342, (Horrocks, I., Hendler, J. (eds)), 264-278, Springer Berlin.

[BR05] Borodin, A., Roberts, G.O., Rosenthal, J.S., Tsaparas, P. (2005). Link analysis ranking algorithms, theory and experiments, *ACM Transactions on internet technology*, 5(1), 231-297.

[BP98] Brin, S., Page, L. (1998). The Anatomy of a large-scale hypertextual web search engine, *Computer networks and ISDN systems*, 30(1), 107-117.

[BM98] Buchner, A., Mulvenna, M. D.(1998). Discovering Internet marketing intelligence through online analytical web usage mining, SIGMOD Record, 27(4), 54-61.

[CS03] Chakrabarti, S. (2003). *Mining the Web — Discovering knowledge from hypertext data*, Morgan Kaufmann.

[CR10] Chattamvelli, Rajan (2010). *Some generalisations of the pagerank metric*, National conference on current trends in advanced computing, CTAC'10, Kristu Jayanti College of Management and Technology, Kothanur, Bangalore 560077, India.

[CR11] Chattamvelli, Rajan (2011). Power of the power-laws and an application to the PageRank metric, *PMU Journal of Computing Science and Engineering*, Vallam, Thanjavur 613403, Tamil Nadu, India.

[CT00] Chen, H.H., Tsai, S.C., Tsai, J.H. (2000). Mining tables from large scale HTML texts, 18th international conference on computational linguistics, Saarbruecken, Germany.

[CH07] Chen, P., Xie, H., Maslov, S., Redner, S. (2007). Finding scientific gems with Google's PageRank algorithm, Journal of Informetrics, 1, 8-15, Elsevier (www.sciencedirect.com).

[CZ06] Chundi, P., Zhang, R., Castellanos, M. (2006). Entropy based measure functions for analyzing time stamped documents, Proc of 4th SIAM text mining workshop, Bethesda.

[CB97] Cooley, R., Mobasher, B., Srivastava, J. (1999). Web mining: Information and pattern discovery on the world wide web, Technical report TR 97-027, University of Minnesota.

[CH05] Cunningham, H. (2005). Information extraction, automatic, *Encyclopedia of languages and linguistics*, 2nd edition, Springer.

[DB03] Davison, B.D. (2003). Unifying text and link analysis, Proceedings of the IJCAI-03 workshop on text-mining and link-analysis (TextLink), Acapulco, Mexico.

[DE03] Dill,S., Eiron, N., *et.al* (2003). SemTag and Seeker: Bootstrapping the semantic web via automated semantic annotation, 12th international world wide web conference, Budapest.

[FA00] Frantzi, K., Ananiadou, S., Mima, H. (2000). Automatic recognition of multiword terms: the C-value/NC-value method, *International journal on digital libraries*, 3(2), 115-130.

[GD05] Getoor, L., Diehl, C.P. (2005). Link mining: a survey, *SIGKDD explorations*, 7(2), 3-12.

[KV97] Kashyap, V. (1997). Information brokering over heterogeneous digital data: a metadata based approach, CS department, Rutgers University, NJ.

[KJ99] Kleinberg, J.M. (1999). Authoritative sources in a hyperlinked environment, *Journal of ACM*, 46(5), 604-632.

[KM02] Kohavi, R., Masand, B., Spilipoulou, M., Srivastava, J (2002). Web mining, *Data mining and knowledge discovery*, 6, 5-8.

[KL04] Kummamuru, K., Lotlikar, R., Roy, S., Singal, K., Krishnapuram, R. (2004). A hierarchical monolithic document clustering algorithm for summarization and browsing search results, Proc of WWW 04, 658-665.

[LJ07] Ling, X., Jiang, J., He, X., Mei, Q., Zhai, C., Schatz, B. (2007). Generating gene summaries from biomedical literature: a study of semi-structured summarization, *Information processing and management*, 43(6), 1777-1791.

[LH00] Liu, Z., Huang, Q. (2000). Content-based indexing and retrieval-by-example in audio, ICME 2000, vol 2, New York, 877-880.

[LC06] Liu, Y., Choudhary, A., Zhou, J., Khokhar, A. (2006). A scalable distributed stream mining system for highway traffic data, Knowledge discovery in databases (PKDD), *Lecture notes in AI* 4213, Springer, 309-321.

[MZ05] Mei, Q., Zhai, C. (2005). Discovering evolutionary theme patterns from text - an exploration of temporal text mining, KDD-05, Chicago, IL.

[MM97] Mendelzon, A., Mihaila, G., Milo,, T (1997). Querying the world wide web, *Journal of digital libraries*, 1(1), 68-88.

[MN02] Miki, T., Nomura, S., Ishida, T. (2002). Semantic web link analysis to discover social relationships in academic communities, Proceedings of symposium on applications of internet, 38-45. (citeseer.ist.psu.edu)

[MJ97] Mobasher, B., Jain, N., Han, J., Srivastava, J. (1997). Web Mining: Pattern discovery from World Wide Web transaction. In international conference on tools with artificial intelligence, 558-567, Newport Beach.

[MS00] Modha, D.S., Spangler, W.S. (2000). Clustering hypertext with applications to web mining, Proc of 11th ACM symposium on hypertext and hypermedia, New York, 143-152.

[MS05] Mullins, I.M., Siadaty, M.S., Lyman, J., *et. al.* (2005). Data mining and clinical data repositories: Insights from a 667 000 patient data set, *Computers in Biology and Medicine*, 36(12), 1351-1377.

[NZ01] Ng, A.Y., Zheng, A.X., Jordan, M.I. (2001). Stable algorithms for link analysis, Proc. of 24th annual international ACM SIGIR conference on research and development in information retrieval, New York, 258-266.

[OP06] Orlando, S., Perego, R., Silvestri, C.*et.al.* (2006). Algorithms and frameworks for stream mining and knowledge discovery on Grids, CoreGRID technical report TR-0046 (coregrid.net), 1-28.

[PR05]Pandurangan, G. Raghavan,P. et.al. (2005). Using PageRank to characterize web search, *Internet Mathematics*, 2(1), 1-20.

[PB04] Perkio, J., Buntine, W., Perttu, S. (2004). Exploring independent trends in a topic based search engine, Proc. of international conference on web intelligence, 664-668.

[RQ01] Radev, D.R., Qi, H., Zheng, Z. Blair-Goldensohn, S., Fan, Z., Prager, J.M. (2001). Mining the web for answers to natural language questions, Proc of international conference on knowledge management, 143-150.

[RK03] Rogelj, P., Kovacic, S. (2003). Point similarity measure based on mutual information, LNCS 2717, Springer-Berlin, 112-121.

[RG02] Roy, S., Gevry, D., Pottenger, W.M. (2002). Methodologies for trend detection in textual data mining, Textmine '02 workshop, SIAM international conference on data mining, 2002.

[SB07] Senellart, P., Blondel, V.D. (2007). Automatic discovery of similar words, in *Survey of text mining: clustering, classification, and retrieval*, 2 ed., Berry, M.W., Castellanos, M (eds.), chapter 2, 25-44.

[ST03] Senator, T.E. (2003). Link mining applications: progress and challenges, *SIGKDD explorations*, 7(2), 76-83.

[SS05] Sun, J., Shen, D., Zeng, H., Yang, Q., Lu, Y., Cheni, Z. (2005). Webpage summarization using clickthrough data, Proc of 28th annual international ACM SIGIR conference on research and development in information retrieval, 194-201.

[TZ05] Tao, T., Zhai, C. (2005). Mining comparable bilingual text corpora for cross-language information integration, Proceedings of KDD 05, Chicago, IL., 1-6.

[VZ04] Velivelli, A., Zhai, C., Huang, T.S. (2004). Audio segment retrieval using a short duration example query, IEEE international conference on multimedia (ICME), Taipei, 1603-1606.

[WI05] Weiss S.M., Indurkhya, N., Zhang, T., Damerau, F.J. (2005). *Text mining: predictive methods for analyzing unstructured information*, Morgan Kaufman.

[WP99] Witten, I., Paynter, G., Frank, E., Gutwin, C., Nevill-Manning, C. (1999). KEA: Practical automatic keyphrase extraction, Proc of 4th ACM conference on digital libraries, Berkeley, CA, 254-255.

[WB96] Wold, E., Blum, T., Keislar, D., Wheaton, J. (1996). Content based classification, search and retrieval of audio, *IEEE multimedia*, 3(3), 27-36.

[YP97] Yang, Y., Pedersen, J.O.(1997). A comparative study on feature selection in text categorization, Proc of 14th international conference on machine learning, Nashville, TN, 412-420.

[YI99] Yoshitaka, A., Ichakawa, T. (1999). A survey on content-based retrieval for multimedia databases, *IEEE transactions on knowledge and data engineering*, 11, 81-93.

[YZ03] Yu, H., Zhai, C., Han, J. (2003). Text classification from positive and unlabeled documents, CIKM 03, Nov 3-8, New Orleans, LA.

[ZH04] Zeng, H., He, Q., Chen, Z., Ma, W., Ma, J. (2004). Learning to cluster web search results, Proceedings of SIGIR 2004.

[ZZ05] Zhang, Y., Zincir-Heywood, N., Milios, E. (2005). Narrative text classification for automatic key phrase extraction in web document corpora, Proc of 7th ACM international workshop on web information and data management, 51-58, Bremen, Germany.

13
Support Vector Machines

Chapter objectives

- Introduce support vector machines and hyperplane classifiers

- Discuss advantages and disadvantages of SVM

- Introduce multiclass SVMs, and its solution techniques

- Distinguish primal and dual forms of SVMs, and corresponding classifiers

- Summarise various forms of SVM (SSVM, MC-SVM, ν−SVM, LP-SVM, LS-SVM, SVR)

- Discuss Sequential Minimal Optimisation (SMO)

- Understand Kernels, nonlinear SVM, and Incremental SVM

- Discuss Support Vector Regression

- Compare Statistical Classifiers and SVM; and review variable selection methods

- Review some applications of SVM

13.1 Introduction

There are many classifiers that originated in statistics. Examples are the naive Bayes classifier, maximum entropy classifier, Fisher's discriminant classifier, partial least squares classifier, and Mahalanobis-distance based classifier. In addition, multiple (linear and nonlinear) logistic regression models can be used as classifiers. Some of these classical models for pattern classification and prediction have assumptions on the data distributions. For instance, logistic regression models assume that the error terms are normally distributed with zero mean, and that the independent variables are uncorrelated. Similarly, normality is assumed in discriminant analysis, canonical correlation, etc. The Support Vector Machine (SVM) is a supervised classification model without any assumptions on the data distribution, or error terms. It filters out distinct data points in the input space towards a maximal margin region by ignoring a large number of data points away from the margin. This information is not exactly available in least-squares based classifiers. Vapnik & Lerner [VL63]; and Vapnik & Chervonenkis ([VC64],[VC91]) developed the SVM as an extension of generalised portrait algorithm. Another name for SVM is

Kernel machines (as nonlinear SVM uses a kernel mapping, see page 13-29). A machine learning algorithm tries to learn the relationship (X→y) from the training data X to the known class or category labels y, so that it can be used to classify new data instances. It is used for pattern recognition (eg: face, retina, fingerprint and other images, handwriting and speech recognition), classification (eg: astronomical classification), clustering (web page and image clustering), regression (SVR), and function approximation.

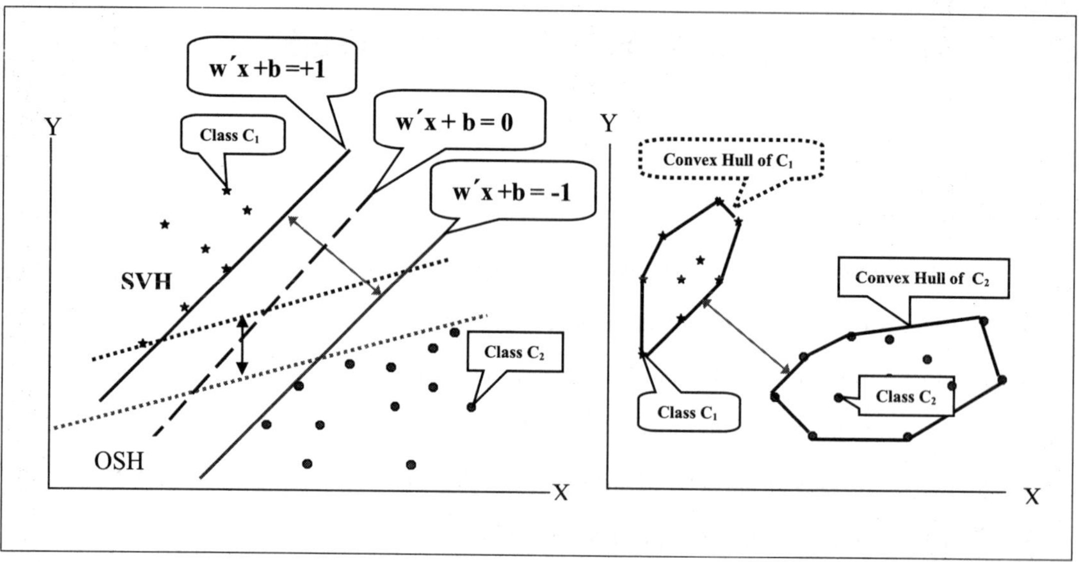

Figure 13.1: There could exist multiple separating hyperplanes when the number of data points is larger than the dimensionality, and the data are linearly separable.

One reason for the popularity of SVM is that it is easily scalable to very large data sets as is common in many data mining applications. It has a small number of user-tunable parameters (eg: C in SM-SVM, LS-SVM, LP-SVM and SVR; and ν, ρ in ν-SVM), and the same model can be generalised to nonlinear, overlapping and fuzzy models. It has good generalisation capabilities depending upon the data distributions (separability margin in input or feature space). A classifier is well-generalisable if it can predict the correct class of unseen data with minimal error. In addition, many stable SVM software packages are already available in the public domain (see §13.15 in page 13-37), and the list is continuously growing. All these factors have instigated researchers from many fields to apply SVM and its variants to various classification, regression and prediction problems.

The classical primal SVM is solved as a quadratic programming optimisation problem with as many variables as the dimension of training data [VV95],[BC98]. In other words, if data (X) are multivariate with dimensionality n, the primal SVM seeks a vector w of size n, and a scalar constant 'b'. The dual SVM is also a quadratic program with as many variables as data instances (N) in the training set, but with just one constraint. The Least Squares SVM (LS-SVM) is always solved in the dual space as a set of (N+1) linear equations in (N+1) unknowns. A necessary condition for solving a problem using SVM is that N > n (the SVM is trivial when n is less than or equal to the data dimensionality N). This is explained in the following section.

13.1.1 Structural Risk Minimisation Principle

Most of the classical models mentioned above use the empirical risk minimisation (ERM) principle by minimising a distance norm, which is usually L_2 or L_1 norm (Least squares regression typically uses the squared L_2 norm). SVM utilises the structural risk minimisation (SRM) principle of statistical learning theory, introduced by Vapnik & Chervonenkis ([CV95],[VV98]). The SRM principle minimises an upper bound on the expected risk. This in turn results in a least generalisation error bound on the optimal surface (which is a hyperplane in linear SVM). Using labeled training data $\{(X_i, y_i)\}_{i=1}^{N} \in \{\mathbb{R}^n \times (C_1, C_2, \cdots, C_m)\}$, SVM estimates an empirical function f(X,y):$\mathbb{R}^n \to \{C_1, C_2, \cdots, C_m\}$ so as to generalise well to unseen data in $\mathbb{R}^n$. It uses the concept of VC dimensions. For linear SVM, it states that *the VC dimension of hyperplanes in $\mathbb{R}^n$ is (n+1)*.

13.1.2 Solution Techniques

SVM is an optimisation theory based classifier technique that finds a maximal margin surface (linear or nonlinear in the input space), which separates the data into distinct classes. Basic knowledge in differential calculus (derivatives and Lagrangian relaxation), algebraic equations, matrix theory, n-dimensional geometry and Kernel mapping (§13.9,p.13-24) are needed for a good grasp of SVM. Other data mining models like decision trees, neural networks etc need more extensive theoretical preparation. All of the algorithms mentioned below use the SRM principle. Classical SVM is formulated as a quadratic programming problem with linear inequalities. This can alternatively be formulated as a linear programming (LP-SVM) problem using the L_1 norm [TA06]. Another method is to formulate it as the solution of linear equations (LS-SVM) by minimising a least-squares criterion [SV99], [SG02]. An efficient method called Sequential Minimal Optimisation (SMO) introduced by Platt utilises the Karush-Kuhn-Tucker (KKT) conditions to iteratively reach an optimal solution in the dual-space using a simple linear equation solver for 2 variables ([PJ99], [KS03]). The incremental SVM (I-SVM) introduced by Cauwenberghs & Poggio [CP01], and extended by Fung & Mangasarian [FM02] is another approach, in which dual variables can be added or dropped during any intermediate solution step. Both the SMO and I-SVM work with the dual formulation, while the others work in both the primal and dual spaces (see table 13.3, pp.13-31). In addition, approximate algorithms [WS06], hierarchical-SVM [CH04] etc are available for solving large SVM problems. As the number of constraints in a primal problem is the same as the number of data points, all SVM problems (except for the trivial ones) are solved as dual-SVM, and the primal parameters are then found.

13.1.3 Linear Separability

Real-world data are assumed to be in the Euclidean *input space* $\mathbb{R}^n$. Training data are linearly separable if there exist at least one hyperplane $w'x + b = 0$ such that $w'x_k + b > 0$ for $y_k \in C_1$ and $w'x_k + b < 0$ for $y_k \in C_2$, or *vice-versa*. In other words, if we were to place a hyperplane between the two classes, a viewer anywhere on the boundary of class C_1 will see all points belonging to class C_2 on the other side of the hyperplane and *vice-versa*. If no such hyperplane exists, the data are called non-linearly separable. In this case they are mapped to a higher-dimensional space called the *feature space* (where it becomes linearly separable). Before proceeding to solve a problem using SVM, a linear separability test may be carried out to choose the appropriate algorithm to be used [EL07]. If the data are linearly separable, the simple SVM or its dual form is enough to obtain a classifier. Real-world data are seldom linearly separable in the input space.

The well-known solution techniques in this case are soft-margin SVM (page 13-8), LS-SVM (page 13-30), and kernel maps (page 13-24).

The most often used distance measures in SVM are the L_2 norm and L_1 norm. The L_2 norm of a hyperplane $w'x + b = 0$ is $w'w = \sum_{i=1}^{n} w_i^2$, and the L_1 norm is $\sum_{i=1}^{n} |w_i|$. All unknown parameters of a model are estimated by minimising the norm (called the loss function, that could also be a function of the norm), which is a function of the data points implicitly through the constraints (see below). Most formulations of the SVM try to locate a minimum-norm hyperplane (in the input space if data are linearly separable, and in the feature space if they are inseparable) that separate the training data using one of the optimisation techniques.

13.1.4   Hyperplane Classifiers

A hyperplane is an n-dimensional generalisation of a straight line in 2D. It can be visualised as a plane surface in 3D, but is not easy to visualise when dimensionality is greater than 3. The Euclidean equation of a hyperplane in $\mathbb{R}^n$ is $w_1x_1 + w_2x_2 + \cdots + w_nx_n = b$, where w_i's are real numbers and $b \in \mathbb{R}$ is a real constant called the intercept, which can be positive or negative. When b=0, the hyperplane passes through the origin (irrespective of the dimensionality). Without regard to the sign of b, we will write it either in summation form as $\sum_{i=1}^{n} w_ix_i + b = 0$, or in vector form as $w'x + b = 0$. There are many classifiers (called Linear Classifiers) that utilise the concept of a *separating hyperplane*. All of them assume that the data are linearly separable in the input space (there exist at least one hyperplane which separates the positive from the negative data points. Note that this is hypothesised only for the training data). The SVM is one such classifier that originated in statistical learning theory.

Consider the training data $\{(X_i, y_i)\}_{i=1}^{N}$, where $y_i \in \{-1, +1\}$ are the class labels. The choice of which class should be labeled as +1 or −1 is arbitrary. If the hyperplane intersects all of the positive axes, those points towards the origin belong to the negative class. This information is unknown until we solve the problem. Irrespective of the labels assigned to the classes, the final solution obtained by SVM will remain the same, as shown below. In linearly separable models, we seek a hyperplane $w'x + b$=0 such that all data points with class label y=+1 are on one side, and all data points with class label y=−1 are on the other side of the hyperplane. Coefficient vector w (of the same size as the dimensionality of data vector X) determines the orientation of the hyperplane, and the scalar b is proportional to the offset of this hyperplane from the origin. If both w and b are scaled (up or down) by dividing by a nonzero constant, we get the same hyperplane. This implies that we can have an infinite number of solutions using various scaling factors, all of them *geometrically* representing the same hyperplane. To avoid such confusion, we make w and b unique by the additional constraint that $w'x + b = \pm 1$ for data points on the boundary of each class. (As $w'x + b = 0$ represents a hyperplane with offset b (intercept on axes proportional to b), the equations $w'x + b = \pm 1$ represent two parallel hyperplanes. The resulting hyperplane $w_0'x + b_0 = 0$ is called the *canonical form* of the optimal separating hyperplane (OSH). These are mathematically expressed as

$$w'x + b \geq +1 \forall x_i \in C_1, \quad \text{and} \quad w'x + b \leq -1 \forall x_i \in C_2, \tag{13.1}$$

where C_1 and C_2 are the two classes. These two equations can be combined, and written as a single equation as f(w,b) $= y_i(w'x_i + b) \geq 1$, because the inequality simply reverses when both sides of second equation are multiplied by −1. The class labels have been chosen as $\{-1, +1\}$, rather than (0, 1), due to this reason (see also the discussion in page 13-26 on LS-SVM). Differentiating f(w,b)-1 wrt w, we get $\frac{\partial}{\partial w}$f(w,b) $= y_ix_i$, where $y_i \in \{-1, +1\}$. Hence the gradient of the constraining surface is in the increasing direction for positively labeled class ($y_i=+1$), and in the opposite direction for negative class. As shown below, the primal classifier is independent of the class labels y_i's (see table 13.4, pp.13-31). It totally depends on the weight vector w, intercept term b, and the new data values to be classified. This is the reason why we mentioned above that the choice of which class should be labeled as {-1,+1} is quite arbitrary. As mentioned below, the same class labels must be used during training and classification stages.

The offset b can be so chosen that these two hyperplanes touch the nearest data points from the OSH. Such data points on the boundary are known as support vectors.

Definition 13.1 Support vectors (SV) are boundary data points (in each class) which are closest (in the geometric sense) to the optimal separating hyperplane (in linear SVM) or hypersurface (in non-linear SVM).

They are easy to visualise in 2-D, but may not be obvious in higher dimensions. In most practical problems, the support vectors form a small fraction of the entire training data set. Number of support vectors found depends on the solution technique (LS-SVM gives more support vectors), and the kernel mapping used (in NL-SVM). RBF kernels (§13.9,p.13-24) in general obtain more support vectors than linear and polynomial kernels. Most SVM software programs report not only the OSH, but the support vectors as well. If the support vectors, weight vector w, and the offset b are given, we could easily compute the OSH equation or the classifier. But the converse is not in general possible (if we are given the classifier, we cannot find the SV's because they can lie anywhere on the SVH (see below); and there could be any number of them parallely aligned along the SVH). Data scaling affects only the margin and not the support vectors. The change of origin transformation can force the offset to become zero, so that the OSH hyperplane passes through the origin. These points being on the margin have a special status in SVM.

Those parallel hyperplanes that touch the support vectors, and are parallel to $w'x+b=0$ are called *Support Vector Hyperplanes* (SVH). See §13.6, page 13-21 for ≥ 3 classes. They are easy to find if one of the classes C_1 or C_2 has just one data point. This is the driving principle behind the incremental SVM and SMO, both of which works in the dual-space (as discussed below, the primal constraints map to dual-space variables). Ideally, we expect a sufficiently large number of data points in each class for the SVM to optimally align the hyperplanes. The solution is greatly simplified if the support vectors in one of the classes are known (either from prior SVM runs, or by other means). Most algorithms can be speeded up if one class has substantially less number of points than the other class.

13.2   Binary SVM

Definition 13.2 An SVM with exactly 2 classes is called a binary SVM.

The primal binary SVM tries to minimise $\frac{1}{2}||w||^2$ with the constraints $y_i(w'x_i+b) \geq 1$ where the weight vector w and the scalar b are the unknowns. Binary SVM is most often encountered in medical sciences, where the result of a test or procedure can be positive or negative. Other examples are for spam mail classification (as spam or non-spam), experimental trials (success or failure, acceptable or unacceptable), insurance (insured or uninsured), online marketing (likely to purchase, unlikely) etc. Hierarchical classifiers can be built using several binary SVMs. When data are linearly separable, the binary SVM obtains a single hyperplane ($w'x+b=0$) that maximally separates the positive and negative classes. This is called the optimal separating hyperplane (OSH). The surface $D(x) = w'x + b$ is called the decision function, or discriminating boundary of linear SVM in the primal space.

13.2.1   Binary SVM Classifier

The binary classifier can be represented as $f(x) = \theta(w'x + b)$, where θ is the sign function,[1] w and b are the known values obtained by a training algorithm. If $\theta(w'x + b) \leq -1$, the corresponding data instance is classified as belonging to the negative class. If $(\theta(w'x + b) \geq +1)$ it belongs to the positive class. The region $-1 < \theta(w'x + b) < +1$ is known as the generalisation region. As a special case, if $\theta(w'x + b) \simeq 0$, the new data instance falls almost on the OSH, and it cannot be classified. This classifier is obtained as a weighted sum of a subset of training data (ie. support vectors), and satisfies $(w'x_i + b)y_i = 1$ for SV points, where y_i's are the (numeric) class labels of the training set. This is why the SVM class labels are always chosen as numeric; usually as integers, rather than real numbers.

Binary classifier of the dual-SVM also has a compact representation as a linear combination of dot products of support vectors and the new data vector x as $f(x) = \theta \left(\sum_{i=1}^{|SV|} \alpha_i y_i (x.x_i) + b \right)$ (see §13.3.1 in page 13-13). Support vectors being points on the borderline, data points falling in-between the support vector hyperplanes are more difficult to classify (see 13.4.0.2 in page 13-17).

13.2.2   Binary SVM Margin

Definition 13.3 The *margin* $(\mu(w, b))$ of a linearly separable SVM is the smallest distance between support vectors in different classes, which is the same as the shortest distance between the convex hulls of the classes (see Fig. 11.1(b)).

It does not directly depend upon the intercept term 'b', but it is customary to include the b in the definition[2] (as they are related). In the two-class case, it is the sum of the smallest Euclidean distances (L_2 norm) between the optimal decision boundary, and pairs of points (one in each class) on either side. This is called the *distance* between the classes. This could be symbolically defined for the two-class (binary) case as $\mu(w, b) = \min_{x \in C_1} d(w, b; x) + \min_{x \in C_2} d(w, b; x)$, where d(w,b;x) denotes the perpendicular distance from x to the OSH (see §13.4 in page 13-25).

Let x_1 and x_2 be two arbitrary data points (vectors) on the OSH. Then $w'x_1 + b = 0 = w'x_2 + b$. Subtracting these two equations gives $w'(x_1 - x_2) = 0$. As x_1 and x_2 are vectors on the OSH, the difference vector $x_1 - x_2$ also lies on the OSH. Therefore the vector w is orthogonal to the decision hyperplane (because it is orthogonal to any vector on the surface of the hyperplane). Since $y_i(w'x_i + b) \geq 1$, maximising the margin (equivalently minimising $||w||$ or $||w||^2$) will automatically align the parallel hyperplanes (SVH) at a distance $1/||w||$ from the OSH, so as to touch the nearest points in each class. When there are more than two classes, we define the margin as the minimum among the pair-wise margins.

---

[1]There are two types of sign functions. The hard sign function returns either a $+1$ or a -1 only. The soft sign function returns any real value. We assume a soft sign function, due to the possibility of new data points (to be classified) falling in the generalisation region (in between the SVH's) where it takes values in $(-1, +1)$.

[2]The intercepts on the respective axes are $-b/w_i$.

13.2.2.1 Confidence in Classification

The confidence in classification depends on both the margin and the value returned by our soft sign-function. In other words, the confidence in classification is high if (i) the margin is large, (ii) $|\theta(w'x+b)|$ is large (where the vertical bars denote absolute value). We can combine both these criteria, and obtain a single confidence measure as $2|\theta(w'x+b)|/||w||$. This is a ratio-measure that can take any value ≥ 0. The denominator term $||w||$ is very large when the margin is too narrow, so that even points that fall reasonably away from the SVH have low confidence. In fact, the margin width should be weighted more than the value returned by the soft sign function. To understand this better, consider two classifiers C_1 with a large margin D and C_2 with a narrow margin d. Should a new data point that lies close to the margin of C_1 have more confidence in classifying than data points that lie at least at a distance D from the margin of C_2?. Intuitively, a large margin reveals more discriminatory power, than otherwise. Hence we could obtain better confidence measures by using different weighting schemes to the margin width and $|\theta(w'x+b)|$. For example, a convex combination $2\beta/||w|| + (1-\beta)|\theta(w'x+b)|$ where $\beta \in (.5, 1)$ returns a better ratio-measure. The β can either be fixed (say .7) for both the classes, or they can be different for each class (say .7 for C_1 and .6 for C_2) depending upon the number of training data instances in respective classes. A simple choice is $\beta_1=.5[1+m/(m+n)]$ and $\beta_2=.5[1+n/(m+n)]$ where m and n are the number of data points in respective classes. As $||w||$ is known, the first term $2\beta/||w||$ need be computed only once when multiple data are to be classified.

Another choice is to adaptively vary β at run-time using the total number of points classified up to now into each class (and those falling into the unclassifiable region) using a proper choice as starting value. Separate β values may also be adaptively varied for each class using two different starting values as discussed above, resulting in an adaptive learning machine. Data points falling inside the margin have $-1 < \theta(w'x+b) < 1$ (or equivalently $a_i=C$ where a_i's are the dual variables). Some of these points can still be classified using a proper cut-off value of the confidence measure (see 13.4.0.2 in page 13-17).

Example 13.1 Prove that if the support vector hyperplanes are given by $w'x + a = 0$ and $w'x + b = 0$, then the OSH in the unweighted case is given by $w'x + (a+b)/2 = 0$.

Solution: We know that the OSH and SVH are parallel. Hence the slope of OSH is the same as that of SVH, implying that all corresponding coefficients w_i's of OSH and SVH are equal. Assume that 'a' and 'b' are either both positive, or both negative and $|b| > |a|$. As the OSH is halfway between the SVH's the distance $|(a+b)/2-a| = |b-(a+b)/2| = |(b-a)|/2$. If $|a| > |b|$, a similar reasoning gives $|(a+b)/2-b| = |a-(a+b)/2| = |(a-b)|/2$. If one of them is positive and the other is negative (say 'b'), make a simple substitution $b' = -b$, and apply the above reasoning to get $|(a+b')/2-b'| = |(a-b')|/2 = (a+b)/2 = |a-(a+b')/2| = |(a-b')|/2$ as half the margin. Hence the intercept of the OSH is $(a+b)/2$, giving the OSH equation $w'x + (a+b)/2 = 0$. This can also be proved using the distance of SVH from the origin as given in next page. Consider any point on an arbitrary hyperplane $P=w'x + (a+b)/2 = 0$. To prove that it is an OSH, it is sufficient to prove that it is equidistant from the SVH. Let α be an arbitrary point on P. As α lies on P, we have $w'\alpha + (a+b)/2 = 0$. Rearranging gives $w'\alpha=-(a+b)/2$. Distance from α to the plane $w'x + a = 0$ is $(w'\alpha + a)/||w||$. Similarly, the distance from α to the plane $w'x + b = 0$ is $(w'\alpha + b)/||w||$. Substitute $w'\alpha=-(a+b)/2$ to get the distances as $\frac{1}{||w||}(a - (a+b)/2)$ and $\frac{1}{||w||}(b - (a+b)/2)$. This upon simplification reduces to $(a-b)/(2||w||)$ and $(b-a)/(2||w||)$, which are equidistant from the OSH (the -ve sign simply indicates that the hyperplanes are on both sides of the OSH).

Consider the parallel lines H_1 and H_2 in figure 13.2 with equations $w_1x_1 + w_2x_2 - b_i = 0$ for i=1,2. To find the perpendicular distance between them, we draw a right-angled triangle ABC as shown in page 13-9 with hypotenuse AB. Being parallel, the slope of both lines is $\tan(\theta) = -w_1/w_2$. Taking *sine* of the angle at B, we get $\sin(180 - \theta) = AC/AB$, from which we get the perpendicular distance AC as AB * $\sin(\theta)$, by using $\sin(\pi - \theta) = \sin(\theta)$. From $\sin^2(\theta) = 1/(1 + \cot^2(\theta))$, by taking the positive square root, and substituting $\cot(\theta) = -w_2/w_1$, we get $\sin(\theta) = w_1/\sqrt{w_1^2 + w_2^2}$. The coordinates of the points A and B on the X_1 axis are obtained by putting $x_2 = 0$ in the above equations as $(b_1/w_1,0)$ and $(b_2/w_1,0)$, so that the magnitude of the hypotenuse AB is $|b_1/w_1 - b_2/w_1| = |b_1 - b_2|/w_1$. Hence AC is given as AB*$\sin(\theta) = |b_1 - b_2|/\sqrt{w_1^2 + w_2^2}$. Because the numerator contains only the constants b_i's, this can be generalised to find the distance between two parallel hyperplanes in n-dimensions as $|b_1 - b_2|/\sqrt{w_1^2 + w_2^2 + \cdots + w_n^2} = |b_1 - b_2|/||w||$. By a simple scaling, it is always possible to bring the hyperplane equations to the standard form $w_1x_1 + w_2x_2 - b = \pm 1$ (for those points in each class nearest to the optimal hyperplane), whence the margin becomes $2/||w||$. For example, consider the hyperplanes (lines in 2D) $w_1x_1 + w_2x_2 - 5 = 0$ and $w_1x_1 + w_2x_2 - 25 = 0$. Dividing throughout by $(25-5)/2 = 10$, and denoting $w_i' = w_i/10$ (i=1,2), we get $w_1'x_1 + w_2'x_2 - .5 = 0$ and $w_1'x_1 + w_2'x_2 - 2.5 = 0$. This could be expressed as $w_1'x_1 + w_2'x_2 - b = \pm 1$ where b=1.5. This can easily be generalised to the n-dimensional case as $w'x - b = \pm 1$. If the last element of **w** vector is chosen as 'b' (or $-$b), and the last element of **x** is chosen as '± 1', we can get rid of the constant 'b' in this equation.

The above result could also be derived using the formula (from analytic geometry) for finding the perpendicular ($\perp$r) distance of a point from a plane. Consider an arbitrary point u= $(u_1, u_2, \cdots, u_n)$. The $\perp$r distance from u to the hyperplane $w'x + b = 0$ is $|w'u + b|/||w||$. Hence if u is taken as the origin (O), and the equation of the parallel bounding hyperplanes are taken as $w'x + b = \pm 1$, the $\perp$r distance between them is the same as the difference between the $\perp$r distances of O from the first plane, and O from the second plane. By combining ± 1 on the RHS with 'b' on the LHS, the distance between support vector hyperplanes become $|b+1|/||w|| - |b-1|/||w|| = 2/||w||$, which is independent of the intercept term 'b'. Alternatively, choose any point on the first hyperplane and find the $\perp$r distance to the second hyperplane from it. If u=$(u_1, u_2, \cdots, u_n)$ is an arbitrary point on w' x+b_1=0, then w' u+b_1=0 or equivalently w' u $= -b_1$. The $\perp$r distance from u to the second hyperplane is $|w'u + b_2|/||w||$. Substitute w' u $= -b_1$ to get $\perp$r distance as $|b_2 - b_1|/||w|| = |b_1 - b_2|/||w||$.

13.2.3   Simple SVM (SSVM)

SSVM is a binary classifier where data are assumed to be linearly separable in the input space. Another name for it is hard-margin SVM (HM-SVM). In the binary case, it is assumed that a hyperplane exists such that all points belonging to one class (C_1) are on one side, and all points belonging to the other class (C_2) are on the other side of the hyperplane. In the unweighted case, we can always consolidate on that hyperplane which is equally distant from the margin points of C_1 and C_2 (it is halfway between boundary points of C_1 and C_2). Geometrically, this is equivalent to locating that hyperplane which is halfway between the convex hulls of the classes.

A remarkable property of SVM is that if all training data except the support vectors are removed, and the problem is re-solved, we get the same OSH and SVH as the original. All training data can also be support vectors. This is especially true when data dimensionality is high and training data size is low, in LS-SVM and in NL-SVM using RBF kernels.

SSVM aims to minimise the classification error and maximise the margin (other versions of SVM like ν–SVM and SM-SVM have additional objectives, as explained below). Optimal margin is the sum of the $\perp$r distances from the support vectors to the OSH $w'x + b = 0$. We know from analytic geometry that the distance between the optimal hyperplane and a point x not on it is $d(w, b; x) = |w'x + b|/||w||$. Thus the margin for binary classifier is $\mu(w, b) = \frac{1}{||w||} \left(\min_{x \in C_1} |w'x + b| + \min_{x \in C_2} |w'x + b| \right)$ where b is the bias parameter. We can choose w and b such that $\min_{x \in C_i} |w'x + b| = 1$. Hence the margin becomes $\mu(w, b) = 2/||w||$.

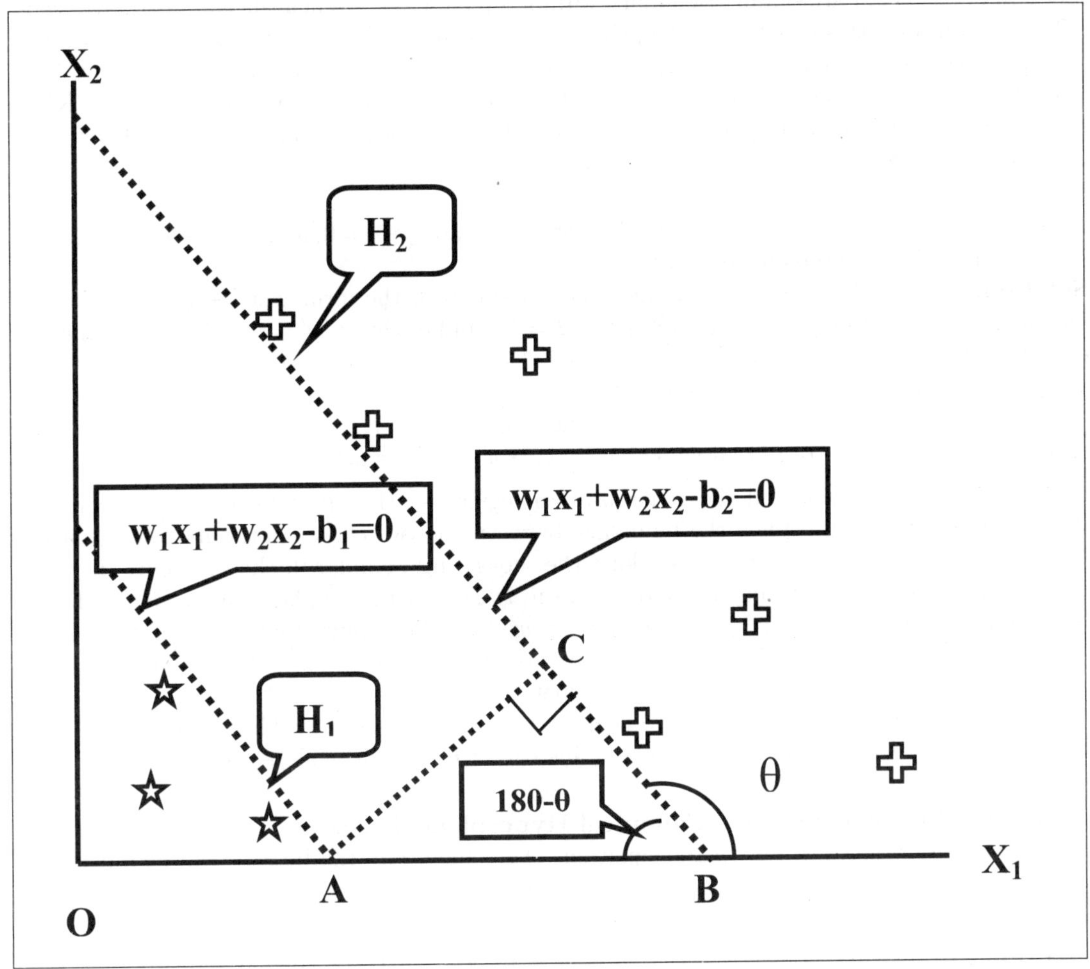

Figure 13.2: Distance between parallel hyperplanes

The margin $2/||w||$ is maximum when the denominator $||w||$ is minimum (because $||w||$ is always >0), or equivalently the squared-norm $||w||^2 = \mathbf{w}'\mathbf{w} = \sum_{i=1}^{n} w_i^2$ is minimum. All primal SVMs (except LP-SVM, LS-SVM) minimise $\frac{1}{2}||w||^2$, where the constant $(1/2)$ is added for mathematical convenience. The SM-SVM, LS-SVM and ν-SVM (all defined below) have additional expressions in the objective function.

Example 13.2 Find the SSVM margin, if the OSH equation is $3.2x_1 - 2.4x_2 + 5 = 0$.
Solution: We know that the margin is independent of intercept term 'b', and is given by $2/||w||$. Here $||w|| = \sqrt{3.2^2 + 2.4^2} = \sqrt{10.24 + 5.76} = 4$, giving the margin as 2/4=0.5

The SSVM can be expressed mathematically as minimise $\frac{1}{2}||w||^2$, subject to the constraint $w'x_i + b \geq 1 \, \forall x_i \in C_1$ and $w'x_i + b \leq -1 \, \forall x_i \in C_2$. As shown in §13.1.4 in page 13-4, these two constraints can be combined into a single constraint $(w'x_i + b)y_i \geq 1 \, \forall x_i \in C_1 \cup C_2$, where $y_i \in \{-1, +1\}$ are the class labels. Since the inequality ($\geq$) remains the same irrespective of the class labels y_i's, we can label either of the classes as +1 or -1. In other words, the class labels in binary SVM are symmetric binary variables (chapter 1). When the training data size is small, we can solve this problem in the primal weight space to find the optimal w and b values. As the objective function ($||w||$ or $\frac{1}{2}||w||^2$) is purely a function of the weight vector w, it is *explicitly* obtained by the training algorithm. The scalar b is *implicitly* obtained using the KKT conditions. But in LS-SVM (§13.10(p.13-26)), the 'b' is also obtained along with dual variables explicitly as the solution to a set of linear equations.

Example 13.3 SVH equations in an SSVM are $w_1x_1 + w_2x_2 + 8 = 0$, and $w_1x_1 + w_2x_2 - 22 = 0$. What is the value of intercept term 'b'?
Solution: To find 'b', we need to reduce the equations to the form $w_1x_1 + w_2x_2 - b = \pm 1$. Dividing throughout by $|b_1 - b_2|/2 = (22-(-8))/2 = 15$, and denoting $w'_i = w_i/15$, and adding -1 on both sides of equality we get $w'_1x_1 + w'_2x_2 + 8/15 - 1 = -1$, which reduces to $w'_1x_1 + w'_2x_2 - 7/15 = -1$. Similarly, dividing throughout by 15, adding +1 on both sides, the other equation is $w'_1x_1 + w'_2x_2 - (22/15) + 1 = +1$, which reduces to $w'_1x_1 + w'_2x_2 - 7/15 = +1$. Hence b=7/15 (note: If the SVH equations are assumed as $w_1x_1 + w_2x_2 + b = \pm 1$, we get b = $-$ 7/15).

In section 13.2, we defined the generalisation region as $-1 < \theta(w'x + b) < +1$ where $\theta()$ is the soft sign function. When the boundary between the two classes is too narrow (negative and positive classes almost *criss-cross* along the separating boundary), the difference $|b_1 - b_2|$ is too small. As the margin in the canonical form is $2/||w||$, this implies that w values should be large (remember that $||w|| = (w_1^2 + w_2^2 + \cdots + w_n^2)^{1/2}$). This shows that the intercept term b is implicitly dependent on the magnitude of the margin. In their original development of SVM, Vapnik & Chervonenkis [VC64] scaled the hyperplane equations such that $|w'x + b|$=1. The following section gives a new theorem to achieve it in an alternate way. This theorem paves the way for choosing the class labels as $\{-k,+k\}$ instead of the traditional $\{-1,+1\}$.

Theorem 13.1 Chattamvelli's Canonical Hyperplane Theorem

Any two arbitrary parallel hyperplane equations $w'x + b_i = 0$, i=1,2; $b_1 \neq b_2$ can be put in the canonical form as w'x+b=$\pm$ k, where k is a real number; and b=$(b_1 + b_2)/2$ if $|b_1 - b_2| \geq 2$, and b = k * $(b_1 + b_2) / |b_1 - b_2|$ otherwise.

Proof. We first consider the case where $|b_1 - b_2| \geq 2$. Let the canonical form be w' x + b = $\pm$ k. Then either $b_1 = $ b - k and $b_2 = $ b + k or *vice versa*. Adding these cancels out the k giving $b_1 + b_2 = $ 2*b, so that b = $(b_1 + b_2) / 2$. In the second case, we need to do a scaling before it is reduced to the canonical form. We have seen in page 13-8 that the perpendicular distance between two parallel hyperplanes $w'x + b_i = 0$, i=1,2 is D=$|b_1 - b_2|/||w||$. Write this as D=$(|b_1 - b_2|/(2k)) * (2k/||w||)$. Therefore, when $b_1 \neq b_2$, we could make the perpendicular distance as $2k/||w||$ by dividing throughout by $|b_1 - b_2|/(2k)$. Denote $v_i = 2kw_i/(|b_1 - b_2|)$. Then the above equations become $v'x + 2kb_1/(|b_1 - b_2|) = 0$ and $v'x + 2kb_2/(|b_1 - b_2|) = 0$. To put this in the form $v'x + $ b=$\pm$ k, we must either have b+k = $2kb_1/(|b_1 - b_2|)$ and b-k = $2kb_2/(|b_1 - b_2|)$ or *vice-versa*. In both cases, adding these equations gives 2b = 2k * $(b_1 + b_2)$

/ $|b_1 - b_2|$. Divide throughout by 2 to get b = k * $(b_1 + b_2)$ / $|b_1 - b_2|$. In the particular case when k=1, we get the canonical form as v' x + b=$\pm$ 1, where b=$(b_1 + b_2)/|b_1 - b_2|$. Here the

---

Algorithm 13.1 Canonical Hyperplane Algorithm

1: Input k, b_1 and b_2
2: Let y = $|b_1 - b_2|$
3: If y > 2, then b = $(b_1 + b_2)$ / 2.
4: else b = k * $(b_1 + b_2)$ / $|b_1 - b_2|$
   Return the hyperplane equations as w'x+b=$\pm$ k

---

sign of b_i's are unrestricted. They can be positive or negative. The above result can also be extended to the nonlinearly separable case by finding such a hyperplane in the feature space.

Example 13.4 The SVH equations in an SSVM are $w_1x_1 + w_2x_2 + 3.112 = 0$, and $w_1x_1 + w_2x_2 + 3.115 = 0$. Bring it in the canonical form and obtain the intercept term 'b'.

Solution: At first sight, it may look that it is impossible to obtain the canonical form in this case because the b values are too close. Here $|b_1 - b_2|$=3.115-3.112=.003 <2, so that the divisor by the above theorem becomes .0015. Dividing throughout by .0015, we get $v_1x_1 + v_2x_2 + 3112 = 0$, and $v_1x_1 + v_2x_2 + 3115 = 0$ where $v_i = (10000/15) * w_i$. From this the canonical form can be easily obtained as $v_1x_1 + v_2x_2 + b = \pm1$, where b=3113.50

In fact, when b_1 and b_2 are too close (and $|b_1 - b_2|$ <1), we could directly obtain the OSH using example 13.1 in page 13-7. We could also express the canonical form of the OSH as $w'x + b = \pm\epsilon$, where b=$(b_1 + b_2)/2$. In the separable case (linear SVM) this has the interpretation that the classes are barely linearly separable. In the nonlinear case it means that the classes are separable by a smooth curve, so that a simple kernel (§13.9,p.13-24) could easily result in linearly separable classes. Put k = ρ in theorem 13.1 to get the ρ-SVM formulation given below.

13.2.4 ρ-SVM

The classical Vapnik's SVM uses a margin width of $2/||w||$. A direct extension in the separable case is to parameterise the margin with a new variable (say ρ) so that the margin width is $2\rho/||w||$. Then the bounding hyperplane (SVH) equations become $w'x + b = \pm\rho$, which are $\rho/||w||$ units away from the OSH (on opposite sides). The ρ may be known from prior runs of the problem, during the pruning stage of SVM, or using a genetic algorithm to explicitly find an approximate margin. We could use the same formulation as above in these situations. The classifier in this case becomes: $\theta(|w'x + b| - \rho)$ implying that if $\theta(w'x_i + b) \geq \rho$, classify the item into C_1. Otherwise $(\theta(w'x_i + b) \leq -\rho)$, classify the item into C_2 (the region $-\rho \leq \theta(w'x_i + b) \leq +\rho$ is the generalisation region). As ρ is unrestricted, this formulation can avoid the data scaling mentioned above. The ρ is seldom exactly known in practice. Therefore, we need to parameterise it and include it in the objective function. As the margin is $2\rho/||w||$, it can be optimised by maximising ρ (or minimising $-\rho$) in the numerator, and minimising $||w||$ (or $||w||^2$) in the denominator *simultaneously*. Combining these two objectives gives rise to the ρ-SVM:- Minimise $\sum_{i=1}^{n} w_i^2 - \rho$ subject to the constraint $y_i(w'x_i + b) \geq \rho$, i=1,2,$\cdots$,N and $y_i \in \{-1, +1\}$ are the class labels, and $\rho > 0$. This can also be formulated as LP-SVM (see pp.13-23) by minimising $||w||_1 - \rho$ (here $||w||_1$ is the L_1 norm. See §13.3.2.1,pp.13-15 for dual formulation).

13.3   Lagrangian Formulation

Lagrangian relaxation is a popular differential calculus-based technique for constrained optimisation[3]. The above primal SVM is not easy to solve because there are as many constraints as the training data size. It is possible to convert it into an equivalent form with a single constraint. With this aim, we formulate it using Lagrangians $\Lambda = (\lambda_1, \lambda_2, \cdots, \lambda_N)'$, a vector of continuous variables as follows:

$$L(w, b, \Lambda) = \frac{1}{2}||w||^2 - \sum_{i=1}^{N} \lambda_i \left[(w'x_i + b)y_i - 1 \right], \tag{13.2}$$

where the negative sign (before the summation) implies maximisation wrt λ_i, and minimisation wrt w and b. The optimal solution will correspond to a saddle point of the Lagrangian.

Differentiating (13.2) wrt w and b, and equating to zero results in:

$$w = \sum_{i=1}^{N} \lambda_i y_i x_i, \tag{13.3}$$

$$\text{and } \sum_{i=1}^{N} \lambda_i y_i = 0 \tag{13.4}$$

(here we have used the fact that $\frac{1}{2}\frac{\partial}{\partial w}||w||^2 = \frac{1}{2}\frac{\partial}{\partial w}(w'w) = w$. This is why we kept a multiplier $\frac{1}{2}$ in the objective function). Substitute for w from 13.3 to get the binary classification problem as:– Find decision function

$$f(\lambda_i, b) = (\sum_{i=1}^{N} \lambda_i y_i x_i'x) + b \text{ such that } (\sum_{j} \lambda_j y_j x_j.x_i + b)y_i \geq 1 \forall i \tag{13.5}$$

The constraint 13.4 has an interesting interpretation for the separable binary SVM. As our class labels (y_i's) in this case are ± 1, the weighted sum (or equivalently the weighted average) of the class labels with λ_i's as the weights must be zero. Suppose that in a medical study to distinguish patients falling in positive and negative categories, there is just one patient in the positive category yet. This means that $y_i=+1$ for just one data point, but $y_i = -1$ for many data points. In such cases where the number of data points in one class is too few, all but one of the λ_i's are too small. This is apparent from 13.4 by rewriting it as $\sum_{i:y_i=+1} \lambda_i y_i = \sum_{i:y_i=-1} \lambda_i y_i$, where C_1 (with $y_i=+1$) and C_2 (with $y_i=-1$) denote the separate classes. This can be used to speed-up the SVM training.

If the hyperplane passes through the origin, the equality constraint (13.4) is not needed. In addition, the KKT orthogonality condition[4] must be satisfied at the optimal point. This gives us $y_i(w'x_i + b - 1) = 0 \forall i$. Substitute these values into both the objective function and constraints of (13.2) and simplify to get

$$L(\Lambda) = \sum_{i=1}^{N} \lambda_i - \frac{1}{2}\sum_{i,j=1}^{N} \lambda_i \lambda_j y_i y_j (x_i.x_j), \tag{13.6}$$

---

[3]It was invented by Italian born, French mathematician Joseph Louis Lagrange (1736-1813).

[4]The KKT complementarity condition states that the product of dual variables and primal constraints should be zero at the optimal point.

subject to the constraint $\sum_i \lambda_i y_i = 0$, $\lambda_i \geq 0$. This can be written in vector form as $L(\Lambda) = \mathbf{e}'\Lambda - \frac{1}{2}\Lambda'M\Lambda$, where $\mathbf{e}$ is a vector of all 1's, and M is a symmetric matrix with elements $m_{i,j} = y_i y_i x_i' x_j$. In the rest of the chapter, we will use a_i in place of λ_i as a notational convenience.

13.3.1 Dual SVM Formulation

All versions of dual SVMs *maximise* the objective function by convention (see below), which has as many variables as the training data points. In other words, there is a Lagrange multiplier a_i for every data point in the training set. The dual SVM can be obtained from the above as follows:– Given the training data $\{(X_i, y_i)\}_{i=1}^N$, find $\{a_i\}, i = 1, 2, \cdots, N$ that maximise

$$L(a) = \sum_{i=1}^N a_i - \frac{1}{2}\sum_{i,j=1}^N a_i a_j y_i y_j (x_i.x_j) \tag{13.7}$$

such that $a_i \geq 0$, $\sum_i a_i y_i = 0, \forall i$. (As maximising f(x) is equivalent to minimising $-$f(x), we could also write the objective function as *minimise* $L(a) = \frac{1}{2}\sum_{i,j=1}^N a_i a_j y_i y_j (x_i.x_j) - \sum_i a_i)$. Input data appear in the above equation as dot products $(x_i.x_j) = (x_i' x_j)$. The a_i's can be scaled (up or down) by dividing by a nonzero constant. Hence we make the solution unique by the restrictions that $a_i \in [0,1]$, and $\sum a_i = 1$.

SSVM solves a convex quadratic programming problem with equality constraints in the dual space, from which the solution to the primal problem (that involves inequality constraints) is obtained as described below. As $a_i \geq 0 \forall i$, $y_i, y_j = \pm 1$, and $\sum_i a_i = k$ represents a hyperplane in the dual space that cuts equal amounts on the coordinate axes, the objective of maximising the RHS of equation 13.3.1 always yields a finite solution (if it exists)[5]. The classifier in dual-SVM is $D(x)=\theta(\sum_{i=1}^{|SV|} a_i y_i (x.x_i)+b)$, where $|SV|$=number of support vectors, and b is a constant to be estimated. As the number of support vectors in large data mining applications is usually much smaller than the training data size, it becomes easy to classify new data instances using this classifier (it takes $|SV|$ cross-products, multiplications and additions. This can be slightly speeded up if the y_i's are rearranged such that all $y_i=+1$ are together in the beginning, followed by all $y_i = -1$).

13.3.1.1 Estimating The Intercept Term

Let x_r and x_s denote support vectors in the two separate classes C_1 and C_2. From the SVH, we get $\hat{b} = 1 - \hat{w}'x_r \rightarrow (1)$ and $\hat{b} = -1 - \hat{w}' x_s \rightarrow (2)$. Adding these two equations gives $2\hat{b} = -\hat{w}'(x_r + x_s)$. An estimate of the bias term b can be obtained as $\hat{b} = -\frac{1}{2}\hat{w}'(x_r + x_s)$. The estimated weight vector $\hat{w}$ is of the same size as data dimensionality n. As the support vectors x_r and x_s are boundary data points in separate classes, they are also of the same dimensionality, so that the above reduces to a vector by vector multiplication resulting in a scalar. The scalar $\hat{b}$ could then be written as a dot product $\hat{b} = -\frac{1}{2}\hat{w}'.(x_r + x_s)$. If all support vectors are known, we could improve the accuracy of this estimate by averaging over all such cases. Substitute the value of b in primal classifier equation, and simplify to get $f(x)=\theta(w'x - \frac{1}{2}\hat{w}'(x_r + x_s)) = \theta(w'[x - \frac{1}{2}(x_r + x_s)])$ where x is the new data to be classified.

It was mentioned above that x_r and x_s are support vectors in the two separate classes. If we are given two arbitrary support vectors, how do we decide whether they belong to the same

---

[5]This is unlike the LPP where the solution can be unbounded.

class or not? As SV's are boundary points, $w'x_r + b$ and $w'x_s + b$ must both be ± 1. If both of them belong to the same class, the signs are exactly the same. If they belong to different classes, the signs are different. Thus we can easily distinguish between the classes C_1 and C_2.

Those x_i's with nonzero a_i's are the support vectors. By splitting the index of summation over SV and non-SV (NSV) data instances of the training set, we get[6] $\sum_{i:x_i \in SV} a_i [(w'x_i + b)y_i - 1] + \sum_{i:x_i \in NSV} a_i [(w'x_i + b)y_i - 1] = 0$. An interpretation of this result is that for non-SV points $a_i = 0$ and $(w'x_i + b)y_i \geq 1$, and for SV points $a_i \neq 0$ and $(w'x_i + b)y_i = 1$ (or equivalently $(w'x_i + b) = \pm 1$). Hence the a_i's (Lagrangian multipliers) represent the information content of x_i's in optimally separating the groups. Thus the SVM works in the *data space* (it can tell us which primal training data points (SVs) contribute to classification), whereas (most) statistical classifiers work in the *attribute space* (they can tell us which variables or combinations thereof contribute to classification). Any inherent correlation among the variables is totally ignored by SVM in its search through the data values for a maximal margin. See also §13.13 in page 13-34.

13.3.2   Properties of Dual SVM

The dual SVM formulation has many interesting properties. Several popular sequential and parallel algorithms have appeared in the literature based on the dual SVM formulation. We list some of the important properties below:

1. Sparseness of solution vector
   As the number of variables in the dual formulation is the same as the training data size, many of the resulting $a_i's$ are zeros, resulting in sparse solution vector. Classifier in this case is $f(x) = \theta[\sum_{i:x_i \in SV}^{|SV|} \alpha_i y_i x_i'x + b]$, where α_i's are the a_i's for SV points, and x is the new data to be classified. The dual classifier indeed utilises the class labels ($y_i's$ $= \pm 1$ for SSVM) of support vector points, but our primal classifier does not utilise any of them (see §13.2.1,pp.13-6). The Lagrangian multipliers ($a_i's$) obtained as the dual solution are called *support values*. All support values are non-negative ($a_i \geq 0$) (for all SVMs except LS-SVM; see page 13-27, and I-SVM. These can be sorted in decreasing order of magnitude to reveal the relative importance of primal data points in the training set. Very small support values indicate that the contribution of corresponding primal data point is negligible for discrimination purposes. Only problem is that the one-to-one correspondence between support values and primal data points is not readily available (we can know this information if we solve the problem manually. Computer programs may rearrange training data points on the fly, which in turn results in rearranged $a_i's$). As the same class labels are used in the primal and dual versions, they can disambiguate the training data. We could discard all corresponding primal training data points by using a cutoff threshold for support values in the dual space.

2. Solution is global
   The coefficient matrix of dual objective function is symmetric, which can be positive definite (all eigen values are >0), or positive semi-definite (all eigen values are ≥ 0). In the first case, the solution is global and unique; else the solution need not be unique.

3. Classifier involves intercept term
   The dual formulation (13.3.1) does not contain the intercept term b, but involves only the a_i's, whereas the dual classifier contains b. One way to estimate the b is to find just two of

---

[6]The summation expression '$i : x_i \in SV$' is read as 'i such that x_i is a support vector'

the support vectors (one in each class C_1 and C_2), and use the technique discussed 13.3.1.1, pp.13-13. Another way is to find the mean of the minimum of $\sum_{i:x_i \in SV}^{|SV|} a_i y_i(x.x_i)$ for the positive class, and maximum for the negative class (labeled -1) and flip the sign.

Example 13.5 If the optimal weight vector $\hat{w}=[1.2, .2, 0.9]'$, and two support vectors are $x'_r = [1,0,-1]$, and $x'_s=[1,1,1]$, estimate the bias term 'b' in the OSH hyperplane equation $w'x + b = 0$, and obtain the classifier.

Solution: We know that the bias term can be estimated as $\hat{b} = -\frac{1}{2}\hat{w}'(x_r + x_s)$. Here $(x_r + x_s)' = [2, 1, 0]$, and $w'(x_r + x_s) = [1.2, .2, 0.9].[2, 1, 0]' = 2.4 + .2 = 2.6$. Thus $\hat{b} = -\frac{1}{2}(2.6) = $ -1.3, from which the classifier is obtained as $f(x) = \theta(1.2x_1 + .2x_2 + .9x_3 - 1.3)$.

Example 13.6 Use the above classifier to classify the new data points [-2, -1, 3] and [2,4,-1]. Are any of these new data points support vectors?

Solution: Substituting [-2, -1, 3] in the classifier $\theta(1.2x_1 + .2x_2 + .9x_3 - 1.3)$, we get $\theta(.1 - 1.3) = \theta(-1.2) = -1.2$. Hence [-2, -1, 3] belongs to the negative class. Similarly substituting [2,4,-1] in the classifier gives $\theta(2.3 - 1.3) = +1$. Hence this data point is a support vector, and it belongs to the positive class.

13.3.2.1   Dual of ρ–SVM

In section 13.2.4, we extended the classical SVM to have a margin of $2\rho/||w||$. An exactly similar procedure as above can be used to obtain the dual of this SVM. Let $\Lambda = (\lambda_1, \lambda_2, \cdots, \lambda_N)'$ be a vector of continuous variables. Then we form the Lagrangian as:

$$L(w, b, \Lambda, \rho) = \frac{1}{2}||w||^2 - \sum_{i=1}^{N} \lambda_i [(w'x_i + b)y_i - \rho] \tag{13.8}$$

where the negative sign (before the summation) implies maximisation wrt λ_i and minimisation wrt w and b.

Differentiate (13.8) wrt w, b, and ρ and force the derivatives wrt the variables vanish to zero to get: $w = \sum \lambda_i y_i x_i$, $\sum \lambda_i y_i = 0$ and $\sum \lambda_i = 0$. The last condition shows that the λ_i's in this case can be negative. In addition, the KKT orthogonality condition[7] must be satisfied at the optimal point. This gives us $y_i(w'x_i + b - \rho) = 0 \forall i$ (see below). Substitute these values and simplify to get $L(\Lambda) = \sum_{i=1}^{N} \lambda_i - \frac{1}{2}\sum_{i,j=1}^{N} \lambda_i \lambda_j y_i y_j (x_i.x_j)$. This can be written in vector form as $L(\Lambda) = e'\Lambda - \frac{1}{2}\Lambda'M\Lambda$, where e is a vector of all 1's, and M is a symmetric matrix with elements $m_{i,j} = y_i y_j x'_i x_j$.

Example 13.7 If a_i's are the solution to the dual-SVM, prove that $||w||^2 = \sum_{i:x_i \in SV} a_i$ in the linear separable case.

Solution: We know that the primal and dual solutions are connected as $\hat{w} = \sum_{i=1}^{N} a_i y_i x_i$, where a_i's are the weights obtained by dual training algorithm, y_i's are the class labels, and x_i's are the primal data points. Splitting the index of summation over SV and NSV points results in $\hat{w} = \sum_{i:x_i \in SV} a_i y_i x_i$, where we have restricted the summation to support vector points only. We have also omitted the NSV term because the a_i's for them are zeros. In addition, the dual KKT condition states that $y_j \left(\sum_{i:x_i \in SV} a_i y_i (x_i.x_j) + b \right) = 1$, or equivalently $y_j \left(\sum_{i:x_i \in SV} a_i y_i (x_i.x_j) \right) =$

---

[7]The KKT complementarity condition states that the product of dual variables and primal constraints should be zero at the optimal point.

$1 - by_j \rightarrow$ (A). This implies the primal-dual relationship:– if $a_i > 0 \rightarrow y_i(w'x_i + b) = +1 \rightarrow x_i$ is a support vector. Moreover $\sum_i a_i y_i = 0$. Hence

$$||w||^2 = w'w = \sum_{i:x_i \in SV} \sum_{j:x_j \in SV} a_i a_j \, y_i y_j (x_i.x_j) = \sum_{j:x_j \in SV} a_j \, [y_j \sum_{i:x_i \in SV} a_i y_i (x_i.x_j)]. \quad (13.9)$$

Substitute for the second expression from (A) above to get $||w||^2 = \sum_{j:x_j \in SV} a_j(1 - by_j) = \sum_{j:x_j \in SV} a_j - b[\sum_{j:x_j \in SV} a_j y_j]$. Now use $\sum_i a_i y_i = \sum_{j:x_j \in SV} a_j y_j = 0$ so that the second expression vanishes, giving the result. This shows that minimising the squared norm (or equivalently, maximising the margin) is equivalent to minimising the sum of the Lagrangian multipliers in the dual problem. This is the guiding principle of dual LP-SVM. This result holds good even in the case of kernel mapping, as each dot product $(x_i.x_j)$ gets replaced by their kernel product $K(x_i, x_j) = \phi(x_i).\phi(x_j)$.

Example 13.8 An insurance company keeps track of the delay in filing a claim, and the amount claimed for each of its customers with an aim to catch any fraud. A sample data set is given in table 13.1, where 'F' indicates Fraud and 'G' indicates a genuine case. Formulate it as an SSVM and obtain the classifier using one of the softwares (§13.15 in page 13-37).

Solution: A plot of the data indicates that it is linearly separable. Hence we can solve it using SSVM. The primal SVM aims to minimise $\frac{1}{2}||w||^2$, subject to the constraints $y_i(w'x_i + b) \geq 1$. We label genuine as +1 and fraud as −1. The respective constraints are $(6\ w_1 + 480\ w_2 + b) \geq 1$, $-(6\ w_1 + 1200\ w_2 + b) \geq 1$, $(2\ w_1 + 250\ w_2 + b) \geq 1$, $(4\ w_1 + 475\ w_2 + b) \geq 1$, $-(13\ w_1 + 700w_2 + b) \geq 1$, $-(9w_1 + 800\ w_2 + b) \geq 1$, $(w_1 + 375\ w_2 + b) \geq 1$, $-(7\ w_1 + 900\ w_2 + b) \geq 1$, $-(5\ w_1 + 1050\ w_2 + b) \geq 1$, $(3\ w_1 + 600\ w_2 + b) \geq 1$. This can now be solved using any of the software for SVM.

Table 13.1: Delay in insurance claim filing (X-axis) vs amount claimed (Y-axis)

Type	Delay (days)	Amount	Type	Delay (days)	Amount
G	6	480	F	6	1200
G	2	250	G	4	475
F	13	700	F	9	800
G	1	375	F	7	900
F	5	1050	G	3	600

13.4   Weighted SVM (W-SVM)

Data instances in positive and negative classes are equally weighted (all weights are 1) in SSVM. Thus the optimal hyperplane is halfway between the convex hulls of these classes. In practice, there could be a higher penalty in classifying positives as negatives, and *vice-versa* in the binary-case [ZS06]. For example, in a medical diagnosis problem or a lab test, there could be a large number of negatives with low risk, and a few positives with high risk. If both cases are equally weighted, the negatives are at an advantage because the OSH is equally distant from both classes. As another example, in an SVM used to predict if a new customer would buy an insurance policy or not, the OSH can be biased from a profit maximisation viewpoint (if it is unbiased, some customers who are likely to buy a policy will be wrongly classified).

There are many ways to use weights in W-SVM – class weighting, instance weighting and attribute weighting. In the simplest case, all positive data instances are given one weight, and all negative instances are given another weight. The class weighting (using properly chosen different weights[8] or vice versa.) can improve the separability margin. This is useful in support vector clustering. Choosing different class weights when data are inseparable can sometimes result in separable data, or at least decrease the non-separability (this is not of much use for classification purposes when data are linearly separable). We have to use a clever trick for classifying new items. Assume that class C_1 is weighted by k_1 and class C_2 is weighted by k_2 such that the new data are linearly separable. We can fit an SVM and obtain the classifier. Now if we wish to classify a new data instance (say x_{new}) how do we know whether to use constant k_1 or k_2 as the multiplier for x_{new}?. A solution is to compute both $y_1 = k_1 x_{new}$ and $y_2 = k_2 x_{new}$, and use the above classifier to classify both y_1 and y_2. There are many possibilities to consider. If both of them are classified to C_1 (respectively to C_2) we could unambiguously classify x_{new} to C_1 (respectively to C_2). If y_1 is classified as belonging to C_1 ($\theta(w'x + b) \geq 1$) and y_2 falls in the generalisation region ($-1 < \theta(w'x + b) < +1$), we could classify x_{new} to C_1 with a confidence ($p = .5^*[1 + \theta(w'x + b)]$); and classify x_{new} to C_2 with confidence 1-p. Similar argument holds when y_1 is classified as belonging to C_2 and y_2 falls in the generalisation region. If both y_1 and y_2 falls in the generalisation region, x_{new} cannot be classified. But the sign of $(\theta(w'y_1 + b) + \theta(w'y_2 + b))$ can be used to classify it to either of the classes with probability $p = .5^*(|\theta(w'y_1 + b) + \theta(w'y_2 + b)|)$. There is another possibility – y_1 is classified as belonging to C_1 and y_2 to C_2 or *vice-versa*. Although rare, the soft sign function value can be used to disambiguate this possibility.

Attribute weighting assigns a fixed weight to each attribute (component) of the vector x. In other words, it computes a linear combination of the variable values. The Fisher's Linear Discriminant Analysis (FiLDA) method is an attribute weighting technique to map higher dimensional data to possibly lower dimensional subspaces so as to yield the largest mean difference between the original classes. When data are nonlinearly separable, we could use the kernel mapping technique (§13.9,pp.13-24) to obtain linearly separable data in the feature space. This technique is called kernel-FiLDA.

13.4.0.2   Classifying the Unclassifiable

How do we classify data that fall exactly on (or in the very vicinity of) the optimal hyperplane? These data are unclassifiable. This is more of a problem when the margin is large, than otherwise. A smart solution to overcome this dilemma is to maintain two classifiers – a largest margin classifier and a next-largest margin classifier. As mentioned in §13.10(p.13-26), the LS-SVM can be used to find the largest margin classifier by choosing the tuning parameter C to be very small (near zero). Increasing C successively shifts the optimal margin from its current position to another one with lesser margin. Of course the tilt of the corresponding hyperplanes will be different. Hence data points that fall on or close by the OSH can still be classified using the second classifier. This method is extended to multi-class problems by weighting class-i instances with weight w_i. In the second approach, each data instance in a class may be weighted individually. This is called instance weighting, and is useful when a few outliers are present in a class. A statistical distribution may be used for 'instance weighting' if there is enough data spread in $\mathbb{R}^n$. A third approach is to use hierarchical weights using a distance metric from a fixed point. An advantage of W-SVM is that it can minimise the penalty associated with classifying

---

[8]The weights must be so chosen that the classes have a receding effect from each other, so as to increase the separability. A simple choice is to use one of the weights as $w_1 > 1$ and the other as $0 < w_2 < 1$

high-risk instances into other classes. This is crucial in medical diagnosis where a patient can wrongly get classified into the positive class.

13.4.1   Overlapping Classes

When classes are not linearly separable, it is called inseparable or overlapping case. SVM takes care of overlapping classes using slack variables (SM-SVM), kernel methods (nonlinear SVM), or using Fuzzy-SVM. If the dimensionality is large (> 3), it is not easy to *visualise* whether the data are linearly separable or not. Nevertheless, analytical separability tests can reveal the nature of overlap. If the inseparability in the input space is due to a small fraction of training data, or if the depth of overlap along the boundary is shallow, the soft-margin SVM is the preferred choice. Slack variables (ξ_i) are associated with each data instance that lies in the other class, and the sum of the magnitudes of their distances from the opposite SVH is simultaneously minimised as described below. Majority of these slack variables will turn out to be zeros in the solution found. In the NL-SVM, a suitable kernel (§13.9,p.13-24) is used to map the data into a linearly separable higher dimensional feature space [CS00]. Computations are carried out directly in the input space as cross products in NL-SVM as described in §13.11 (page 13-29).

13.5   Soft-Margin SVM (SM-SVM)

In some applications, the data are not strictly linearly separable. This may happen due to faulty data collection mechanisms, various types of errors, data outliers etc. Overlaps may also occur due to natural or genuine reasons. If the number of training data instances violating linear separability is small, we can still seek a linear separating hyperplane by penalising the violating points. In SSVM, the objective is purely a function of the unknown weights w's (and implicitly on b's through the constraints). In the inseparable case, we introduce positive slack variables ξ_i's to minimise the guaranteed risk bound. Let (X_i, y_i),i=1,2,..N where $x_i \in \mathbb{R}^n$ be the training data. Let $\xi_i \geq 0$ be slack variables which should ideally be very close to zero (if $\xi_i = 0, \forall i$, it becomes linearly separable and HM-SVM is applicable). The objective is to simultaneously minimise $||w||^2$ and $\sum_{i=1}^{N} \xi_i^k$. Hence the primal SM-SVM can be expressed as

$$J(w, b, \xi) : \underset{w, b, \xi_i}{\text{Min}} \left(\frac{1}{2}||w||^2 + C\sum_{i=1}^{N} \xi_i^k \right) \tag{13.10}$$

where $k > 0 \in \mathbb{R}$ such that $(w'x_i + b)y_i \geq 1 - \xi_i$ and $\xi_i \geq 0$ for i=1, 2, $\cdots$, m, and C is a constant to compromise between the size of the margin large and magnitude of the ξ's small. Theoretically, the ξ's are needed only for those points that violate the margin. Unfortunately, this information is often unknown. If the dimensionality is 2, we could plot the points and locate the overlapping points. But this method is ambiguous in nonlinear SVM and higher dimensional data. This is the reason why we have included the positive slack variables ξ's in the objective function.

Most popular choices of k are 1 and 2. An advantage of this case is that the dual formulation does not contain the ξ_i terms. The ξ_i values indicate where x_i lies wrt the optimal hyperplane, and its count (for nonzero ξ_i) is a measure of the number of points that violate a margin of $1/||w||$. If $0 < \xi_i < 1$, then x_i is classified correctly, but it lies inside the margin. If $\xi_i \geq 1$ then x_i is misclassified with the magnitude of ξ_i indicating the confidence in misclassification (the larger the ξ_i, the higher the confidence in misclassification). If $\xi_i=0$, then x_i is classified correctly.

As before, we take the Lagrangian to find the dual form (for k=1 in 13.10) to get:

$$Q(w,b,\xi,\Lambda,\beta) = \frac{1}{2}||w||^2 + C\sum_{i=1}^{N}\xi_i - \sum_{i=1}^{N}\lambda_i\left[(w'x_i + b)y_i - 1 + \xi_i\right] - \sum_i \beta_i\xi_i. \tag{13.11}$$

The negative sign is used in this expression because we intend to maximise wrt λ_i, β_i and minimise wrt w, ξ_i and b (as $y_i(w'x_i + b) - 1 + \xi_i \geq 0$). Differentiating (13.11) wrt w, ξ_i and b, and imposing stationarity (and replacing λ_i by a_i) we get the following equations: $w = \sum a_iy_ix_i$, $a_i + \beta_i = C$, and $\sum a_iy_i = 0$. Substituting these values in (13.11), combining like terms and simplifying using the above equations result in the dual SM-SVM:– Maximise

$$L(A) = -\frac{1}{2}\sum_{i,j}a_ia_jy_iy_j(x_i.x_j) + \sum_i a_i, \tag{13.12}$$

with the constraints $\sum_i a_iy_i = 0$, and $0 \leq a_i \leq C$. This formulation is similar to the separable case, with the exception of a refined upper bound on the a_i's. The a_i's can be associated with data labels of positive and negative classes. Hence we can write it in the alternative form $a_iy_i \in [0, C]$ for y_i=+1, and $a_iy_i \in [-C, 0]$ for y_i=-1. Those SVs with their multipliers at the upper bound C are called bounded SVs, and those for which $0 < a_i < C$ are called unbounded SVs. The KKT conditions in this case become $a_i[(w'x_i + b)y_i - 1 + \xi_i] = 0, \forall i$. An interpretation of this condition is that:– $a_i = 0 \Rightarrow (w'x_i + b)y_i \geq 1$ and $\xi_i = 0$, for non-support vectors, $a_i = C \Rightarrow (w'x_i + b)y_i \leq 1$ and $\xi_i \geq 0$, for bounded support vectors, and $0 < a_i < C \Rightarrow (w'x_i + b)y_i = 1$ and $\xi_i = 0$ for margin support vectors (MSV). This shows that the solution is sparse (in the a_i's). Smaller C values result in larger SVs (both bounded and unbounded). The bias term b is found by averaging over those sample data points for which $0 < a_i \leq C$. The condition $(w'x_i + b)y_i - 1 \geq 0$ is called feasibility condition, and $a_i \geq 0$ is called non-negativity condition. New data instances are classified using a dual classifier

$$D(x) = \theta\left(\sum_{SV's} \hat{a}_iy_i(x_i, x) + \hat{b}\right) \text{ where } \hat{b} = -\frac{1}{2}\sum_{MSV's}\hat{a}_iy_i\left[(x_r, x_i) + (x_s, x_i)\right]. \tag{13.13}$$

Primal and dual variables are connected as w=$\sum_i a_iy_ix_i$, where the summation is over support vector points only.

Convergence speed and efficiency of SM-SVM depend on the initial choice of the parameters. Parameter C must be specified before we start the algorithm. Many methods have been proposed for its estimation. The popular parameter estimation techniques are (i) empirical estimation (ii) grid search methods (iii) stochastic heuristics method (simulated annealing, etc). Details can be found in [AS05b], [SS02].

13.5.1 Weighted Soft Margin SVM (WSM-SVM)

Weighted SVM (W-SVM) uses a biased margin obtained by weighting data instances or attributes in each class separately (when all weights are 1, it reduces to SSVM). For example, in an SVM model to classify an ATM transaction as fraud or genuine, the variables involved (eg: transaction amount, time of transaction, number of PIN attempts) may be weighted differently as they have unequal importance in catching fraud. Each transaction may also be weighted differently depending upon the time of transaction. Once a classifier has been obtained from training data, unlabeled data instances can be classified using a simple mathematical evaluation

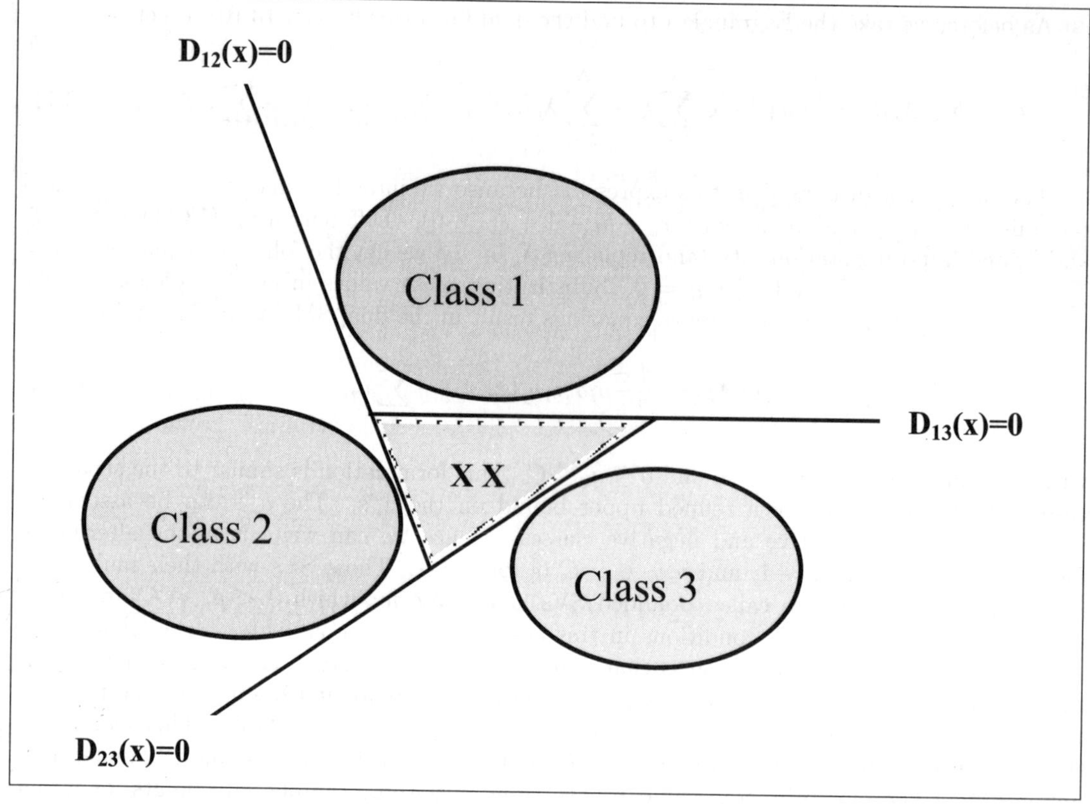

Figure 13.3: Unclassifiable regions in pair-wise SVM

and a sign test. As noted before, asymmetric classification problems are best tackled by assigning different weights to data instances in various classes. The weighting may be class, instance or attribute specific. A weighting scheme that attenuates white noise (random data errors) and outliers will contribute positively than a random weighting scheme. Let $\{(X_i, y_i, s_i)\}_{i=1}^{N} \in \{\mathbb{R}^n$ x $(C_1, C_2, \cdots, C_m)$ x $(0,1]\}$ be the input data, where s_i are the weights normalised to the range $(0,1]$. The primal WSM-SVM is:– Minimise

$$\frac{1}{2}||w||^2 + C\sum_{i=1}^{N} s_i\xi_i \tag{13.14}$$

such that $[(w'x_i + b)y_i] \geq 1 - \xi_i,$

where $\xi_i \geq 0$ and $s_i \in (0,1]$ determine the trade-off between classification error and margin separation. The dual is obtained from the Lagrangian formulation as Maximise

$$\sum_i a_i - \frac{1}{2}\sum_{i,j} a_i a_j \, y_i y_j \, (x_i.x_j), \tag{13.15}$$

such that $\sum_i a_i y_i = 0$ and $0 \leq a_i \leq s_i C$. See [SB99], [WY04] for further details.

13.6 Multi-class SVM (MC-SVM)

The MC-SVM is a natural generalisation of SSVM to the multiple class case. We denote the classes by $C_1, C_2, \cdots, C_m$. The classes may either be all pair-wise linearly separable, or with overlaps (which is more frequent). Thus there are three cases to consider — (i) all of the classes are pair-wise linearly separable, (ii) some of the classes are pair-wise linearly separable (iii) none of the classes are pair-wise linearly separable. In case (i), we could extend the SSVM principle to each pair. This may not be apparent in higher dimensions without an explicit linear separability test. In case (ii), we build linear SVMs for pairwise linearly separable classes; and nonlinear SVM for others. Case (iii) is best handled using nonlinear SVM.

There are many ways to obtain the classifier in the case of multi-SVM. In one approach, we obtain one classifier for each of the class pairs. In another approach called one-against-all, we obtain separate classifiers for "one class against all other classes combined". These are described below.

13.6.1 Pair-wise SSVM (One-versus-One [OVO])

In pair-wise SVM, we find an OSH for each pair of classes, from which the corresponding classifier can be obtained. For n classes, theoretically there are $\binom{n}{2}$ classifiers possible, some of which may be redundant. In figure 13.4 (right) there are 4 classes, and 6 OVO SVMs (they are denoted by i-j notation). Unclassifiable regions are possible in multi-class SVM (see figure 13.3, page 13-20). Classifying an unseen data instance to its correct class is more difficult in this case. We

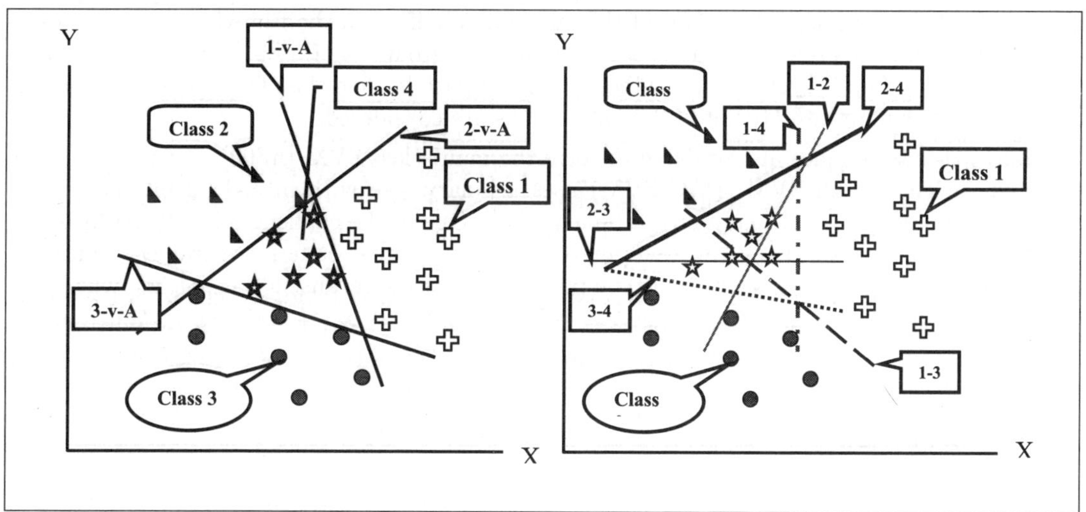

Figure 13.4: Binary one-versus-all and one-versus-one SVM

test each of the classifiers in turn and note down the counts assigned to each class. If these counts have a unique maximum, the new data instance is unambiguously assigned to that class. Otherwise, heuristic techniques (based on distance from the OSH, number of data instances in each class etc) are used to choose the correct class. This method is not useful for large number of classes, as the computational complexity of this approach dramatically increases as the number of classes becomes large.

If there are more than 2 pairwise SVMs, we could arrange the classifiers in high-to-low priority order using the margin width (in decreasing order). This is under the assumption that the higher the margin width, the better is its discriminatory power. As an example, if the margins of 3 pairwise SVMs are $\mu(C_1, C_2)=10$, $\mu(C_1, C_3)=16$, and $\mu(C_2, C_3)=20$, a new data instance to be classified will be put through the third classifier first, followed by second and first. This could improve classification speed when a large number of classifiers are to be tested.

13.6.2   One-versus-All (OVA) SVM

In this method (also called one-versus-rest (OVR), or one-against-rest), we fix one class (say C_i) and collapse the data for all other classes into a single set. The singled out class should systematically be labeled as either $+1$ or -1 in each pass. As in SSVM, one binary classifier is found for these two sets of data. The process is cyclically repeated for each of the other classes. When the loop is complete, we have m classifiers. Let $D_k(x) = \theta(w_k'x + b_k)$ be the classifier for class 'k' against all others. An unseen data instance is put through each of the classifiers. The correct class can be obtained in (m-1) iterations, and the m-th iteration can be used as a validation step. Mathematically, this could be represented as

$$J(w^i, b^i, \xi^i) : \mathop{Min}_{w^i, b^i, \xi_i} \frac{1}{2}||w^i||^2 + C\sum_{i=1}^{N} \xi_j^i \qquad (13.16)$$

such that $((w^i)'x_i + b^i) \geq 1 - \xi_j^i$ if $y_j = i$, $((w^i)'x_i + b^i) \leq -1 + \xi_j^i$ if $y_j \neq i$ and $\xi_j^i \geq 0$ for $i=1, 2, \cdots, m$, where C is a constant to compromise between the size of the margin large and magnitude of the ξ's small. A variant of this method is half-versus-half in which $\lceil m/2 \rceil$ classes are collapsed into one group G_1, and $\lfloor m/2 \rfloor$ into another group G_2. If m is even, we evenly split the classes. This step is cyclically repeated by moving data for one class from G_1 to G_2, and an original G_2 class data into G_1 until all combinations are considered.

Many other techniques are available to solve the multi-class SVM [AS03]. For instance, the Directed Acyclic Graph SVM (DAG-SVM) uses a binary tree with m(m-1)/2 internal nodes and m leaves, each of which represents a binary SVM. One of the challenges in MC-SVM is the selection of class labels. They can take any convenient positive or negative values, which we denote by a_i for i^{th} class. They may also be pair-wise labeled. Because $\hat{b} = a_i - \hat{w}'x_{c_i}$ where x_{c_i} is any support vector in class C_i, we get $\hat{b} = \frac{1}{m}[\sum a_i - \hat{w}'(x_{c_1} + x_{c_2} + \cdots + x_{c_m})]$. When m=2, this reduces to $\hat{b} = -\frac{1}{2}\hat{w}'(x_{c_1} + x_{c_2})$ as $a_1 + a_2 = -1 + (+1) = 0$.

13.7 ν-SVM

A major disadvantage of SM-SVM is its inability to bind the fraction of training data that violate the soft margin, which is determined by a user chosen constant C. The ν-SVM remedies this weakness to an extent by using a parameter ν that bounds the fraction of data instances violating the margin (fraction of unbounded SVs) ([SS00], [CL01b], [CL03]). The primal ν-SVM is formulated as follows:

$$\text{Min} \ \frac{1}{2}||w||^2 - \rho\nu + \frac{1}{N}\sum_{i=1}^{N} \xi_i \ \text{ such that } \ [(w'x_i + b)y_i \geq \rho - \xi_i], \qquad (13.17)$$

where $\nu \in [0,1]$, $\rho \geq 0$, $\xi_i \geq 0$. The last term in 13.17 is the arithmetic mean of margin violations. Thus the objective is to maximise the sum of the margin and the mean of violations minus a quantity $\rho\nu$. This can be reparameterised by scaling both the objective function and constraints. Divide the objective function by $\nu^2/2$, constraints by ν, and let $\mu = 2/(\nu N)$. Also divide each of w, b, ρ and ξ_i by ν (ie. denote $w'/\nu = w$, $b'/\nu = b$, etc) to get Min $||w||^2 - 2\rho + \mu\sum_{i=1}^{N}\xi_i$ with the same constraints, but scaled coefficients. It is called the $\mu-$SVM to distinguish it from the original $\nu-$SVM.

The corresponding dual form is Max $\frac{1}{2}\alpha'Q\alpha$ subject to $0 \leq \alpha_i \leq 1/N$, $y'\alpha = \sum_i \alpha_i y_i = 0$, $e'\alpha \geq \nu$, where $Q_{i,j} = y_i y_j K(x_i, x_j)$, and e is a vector with all 1's (so that $e'\alpha = \sum_i \alpha_i$). (equivalent constraints in $\mu-$SVM are $y'\alpha = 0$, $e'\alpha = \sum_i \alpha_i = 2$, and $0 \leq \alpha_i \leq \mu$). KKT conditions for $\mu-$SVM are $(\mu - \alpha_i)\xi_i = 0$, and $\alpha_i(y_i[w'x_i + b] - \rho + \xi_i) = 0$. The dual formulation shares many similarities with the dual SM-SVM with two notable differences – (i) objective function does not contain the $\sum_i \alpha_i$ expression and (ii) there is an additional constraint $\sum_i \alpha_i \geq \nu$. Dual classifier is independent of ν, and is given by f(x) = $\theta(\sum_i \alpha_i y_i(x_i.x) + b)$. An instance-weighted $\nu-$SVM is obtained from (13.17) by replacing ξ_i with $\beta_i\xi_i$ in the objective function. Resulting dual optimisation problem is exactly identical, except for the refined constraint $0 \leq \alpha_i \leq \beta_i/N$. An advantage of ν-SVM is that it constrains the fraction of bounded support vectors (because nonzero a_i correspond to support vectors). See [SS00], [CL01b], etc for further information.

13.8   LP-SVM

We have seen (pp.13-8) that the distance between two parallel hyperplanes $w'x+b_1 = 0$ and $w'x+b_2 = 0$ is $|b_1 - b_2|/||w||$, where $||w|| = \sqrt{w_1^2 + w_2^2 + \cdots + w_n^2}$. Hence the margin can be maximised by minimising the denominator $||w||$. As minimising $||w||$ is equivalent to minimising $||w||^2$, the standard formulation of SVM minimises the L_2-norm, and solves a quadratic programming problem. This method is not easily scalable to very large training data sizes. The LP formulation of SVM uses the L_1-norm $\sum_{i=1}^{n} |w_i|$. As this is non-differentiable, the classical optimisation techniques are not applicable. The classical linear programming optimizes $c'x$ subject to Ax$\leq$ b where the coefficient matrix A, RHS vector b and cost coefficients c are known, and x's (plus any slack variables introduced) are the unknowns. To cast the above as a standard LPP, we write the objective function as Min $\mathbf{1'w} + \mathbf{0b}$ subject to $[X,\mathbf{1}][w,b]' \geq y_i$, where $\mathbf{1'}$ is a vector of all 1's, the weight vector w, scalar b (and any slack variables added) are the unknowns, and X is the training data (which are known; here X is analogous to the coefficient matrix A in the LPP). This gives the following LP SVM formulation:

$$\text{Minimise } L(w,\xi) = \sum_{i=1}^{n} |w_i| + c\sum_{i=1}^{N} \xi_i \quad \text{subject to} \quad [w'x_i + b]y_i \geq 1 - \xi_i, i = 1, 2, ..N. \quad (13.18)$$

When the data are highly nonlinearly separable, we could do the kernel map ($\S$13.9,p.13-24), and obtain an LPP formulation in a higher dimensional space where the data become separable as $L(w,\xi) = \sum_{i=1}^{n} |w_i| + c\sum_i \xi_i$ subject to $[w'\phi(x_i) + b]y_i \geq 1 - \xi_i$, i=1,2,..N where $\phi(x)$ is the kernel map. LP algorithms can solve this by introducing slack variables. The decision function is f=θ(g(x)) where g(x)=$(\sum_{i=1}^{l} a_i y_i(x.x_i) + b)$. Thus w=$\sum_{i=1}^{N} a_i y_i x_i$. Substituting this value of w in above equation gives the decision boundary as f(x)=$(\sum_{i=1}^{N} a_i y_i x'x_i + b)$ in which x_i are the support vectors and $a_i > 0$ (so that the summation need be carried out over support vector points).

An advantage of LP-SVM is that it is easily scalable to voluminous data. The decomposition principle of LP, or one of the interior point algorithms can be used to solve large problems [TA06]. An error analysis of LP SVM and a bound on misclassification error can be found in [WS06].

13.9   Kernels

Data overlap is a common phenomenon in practical applications. A scatter plot can reveal the extent of overlap in low dimensions. When the dimensionality is large (say > 5), it is not easy to ascertain either the presence or the extent of overlap. Kernel method is a popular technique to transform nonlinearly separable data to linearly separable form. They map the input data into a higher dimensional feature space using one of the kernel functions. The Gaussian kernel is useful in classification of inseparable data. Because e^{-x} has an infinite series expansion, Gaussian and Laplacian kernels map the input space into an infinite dimensional feature space, where data are guaranteed to be linearly separable.

Definition 13.4 A kernel K(x,y) is a real valued function (in one or more variables) defined on $\mathbb{R}$ such that there is another function $\phi : X \to Z$ such that K(x,y)=$\phi(x)'\phi(y)$, where $\phi(x)$ is a nonlinear mapping function. Symbolically we write $\phi : \mathbb{R}^m \to \mathbb{R}^n$: $K(x_i, x_j) = \phi(x_i).\phi(x_j)$,

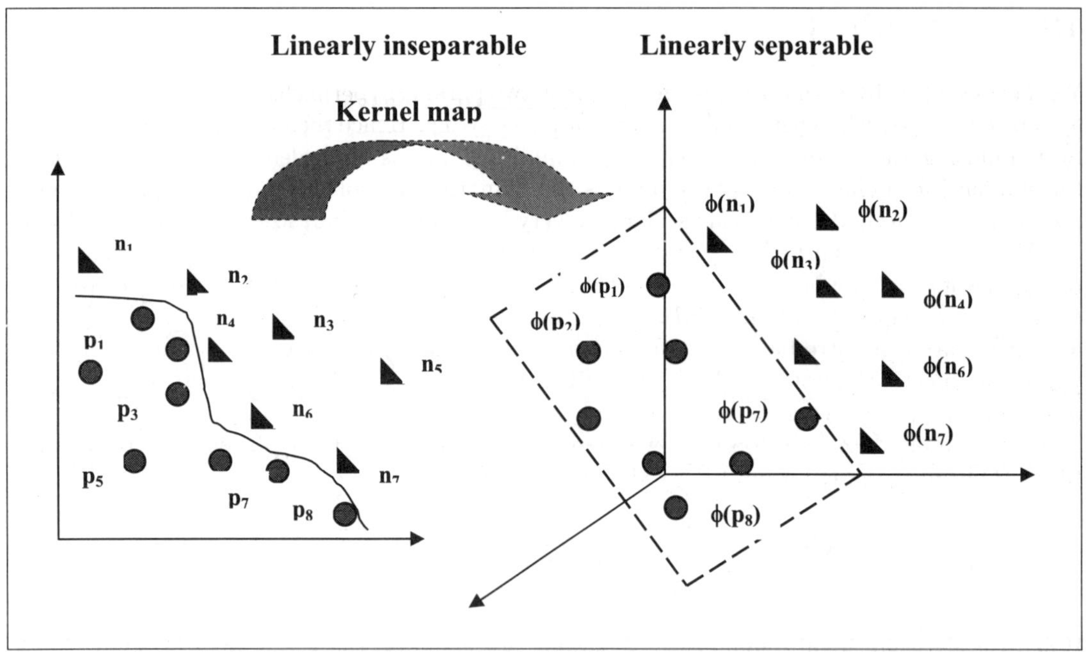

Figure 13.5: Kernel mapping of highly nonlinearly separable data to linearly separable form

where $n > m$. As data mining applications typically involve thousands of data instances, an explicit mapping of each data point in the input space to a feature space is highly computationally intensive. Instead, the dot products $(\phi(x_i).\phi(x_j))$ in feature space are replaced by the dot products $(x_i.x_j)$ in the input space. The scalar $\hat{b}$ is found as kernel-induced inner product

Table 13.2: Some commonly used kernels in SVM

Kernel name	Representation
Linear	$< x, x' >= x'x$
Polynomial	$(x'x + m)^d$
Laplacian	$\exp(-\lambda\|x - x'\|)$
Gaussian (RBF)	$\exp(-C\|x - x'\|^2/(2\sigma^2))$
Sigmoid	$\tanh(k< x.x' >+c)$
Mahalanobis	$\exp(-(x_1 - x_2)'A(x_1 - x_2))$

$\hat{b} = -\frac{1}{2}\hat{w}'.(\phi(x_r)+\phi(x_s))$. As shown in table 13.2, Kernels can be linear or nonlinear, the former being much simpler to work with. It finds applications not only in SVM, but with many other

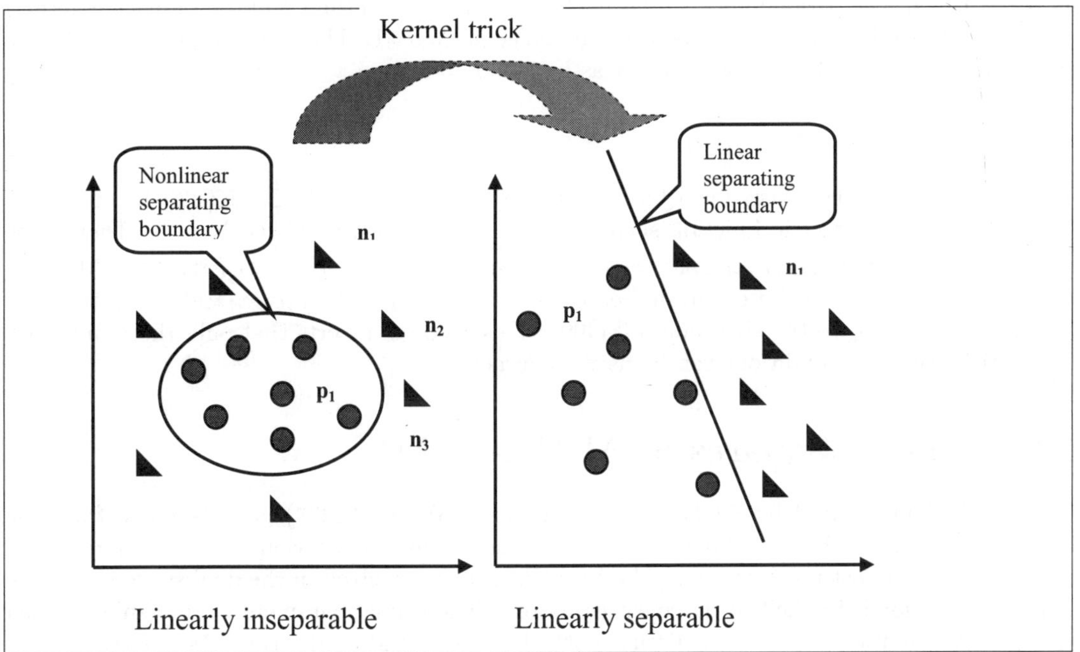

Figure 13.6: Kernel trick to map nonlinear data into linear separable form.

classification and prediction models as well (see §13.11.1).

Example 13.9 Quadratic kernels in 2D
Consider the kernel $K(x,y) = (x.y)^2 = x_1^2y_1^2 + 2x_1x_2y_1y_2 + x_2^2y_2^2$ where $x = (x_1, x_2)$ and $y = (y_1, y_2)$. There are many mappings ϕ from $\mathbb{R}^2 \to \mathbb{R}^3$ such that $(x.y)^2 = \phi(x).\phi(y)$. Some examples are $\phi(x) = \left[x_1^2, \sqrt{2}x_1x_2, x_2^2\right]'$, and $\phi(x) = \left[x_1^2 - x_2^2, 2x_1x_2, x_1^2 + x_2^2\right]'$.

13.9.1 Properties of Kernels

If K_1 and K_2 are kernels, then (i) $K_1 + c$, (ii) aK_1, (iii) $aK_1 + bK_2$, (iv) $K_1.K_2$ are kernels, where a,b,c $\in \mathbb{R}^+$ [HR02].

In addition to new kernels obtained by operating them arithmetically, we could also apply reparameterisations, compositions and nonlinear transformations that satisfy Mercer's condition (see below) to get new kernels. Because linear combinations of Gaussian variables are Gaussian distributed, weighted linear combinations of Gaussian kernels can also be employed. Many other kernels are also being used for specialised applications (eg: Cauchy kernel, hyperbolic kernels, Fourier kernels, etc). See Schoelkopf,B., Burges,C.J.C. & Smola [SB99], Shawe-Taylor & Cristianini [SC04], Abe, S.[AS05b], etc.

Polynomial kernels can result in large dot products when the data are unnormalised. As the maximum value of the polynomial $(x.x' + m)^d$ is $(m + 1)^d$, the normalised polynomial kernel becomes $K(x, x') = (x.x'+m)^d/(m+1)^d$. Gaussian (RBF) kernel produces more support vectors in general than the other kernels mentioned in table 13.2. When the number of features are not very large, numerical stability problems and error propagation are minimal with RBF kernels, as $K_{ij} \leq 1 \forall i, j$. The parameters C and σ in the Gaussian kernel (and λ in the Laplacian kernel) should be chosen by the user. A good initial choice is C=σ=1. The constants (k and c) in the sigmoidal kernels are chosen conveniently as k=2 and c=1.

13.9.2 Mercer's Theorem

This theorem forms the basis for a function to be used as a kernel. Let K(x,y) be any semi-positive definite symmetric function such that $\int_{\forall f \in L_2} K(x, y)$ f(x)f(y) dx dy ≥ 0. Then there exist ϕ_k such that $\int K(x, y)\phi_k(x)dx = a_k\phi_k(y)$ and $K(x, y) = \sum_k a_k\phi_k(x)\phi_k(y)$. Mercer's Kernels are obtained as the eigenvalue expansion $K(x_1, x_2) = \sum_i a_i\phi_i(x_1)\phi_i(x_2)$ so that the features get mapped to the eigenvalues [MJ09]. This theorem asserts that only those functions that satisfy Mercer's conditions can be used as kernels.

13.10 Least Squares SVM (LS-SVM)

The LS-SVM introduced by Suykens & Vandewalle ([SV99], [SL99]) gets its name from the fact that the primal objective function contains a least squares criterion (sum of squared-error cost-function) for cost minimisation. The LS-SVM is always solved in the dual space where the objective function is implicit in the linear equations. The primal parameters are then estimated using KKT conditions. Below we discuss both the primal and dual forms. As it uses equality constraints

$$[w'\phi(x_i) + b]y_i = 1 - \xi_i, \quad \text{for} \quad i = 1, 2, \cdots, N \tag{13.19}$$

the quadratic programming problem reduces to solving a set of (N+1) linear equations in (N+1) unknowns if b$\neq$0 and NxN equations if b=0. In [MR01] and [KS03] the authors use the constraint as

$$y_i - [w'\phi(x_i) + b] = \xi_i. \tag{13.20}$$

Expressions 13.19 and 13.20 are equivalent for binary SVM. Write 13.19 as $1-[w'\phi(x_i)+b]y_i = \xi_i$. Next multiply both sides by y_i, to get $y_i - [w'\phi(x_i) + b]y_i^2 = y_i\xi_i$. Now use $y_i^2 = (\pm 1)^2$=1 in the second term on LHS, to get the LHS of 13.20. Put $\xi_i' = y_i\xi_i$ on the new RHS, so that $\xi_i = \xi_i'/y_i = \pm\xi_i'$. As shown below, ξ_i's appear in the objective function as a square, so that

$\xi_i'^2 = \xi_i^2$. In equation 13.20, ξ_i represents the error in predicting the correct class label y_i from the SVM estimated value. It can be positive or negative. By the KKT condition, this error term $\in [-1, +1]$. Hence we may substitute $y_i \xi_i$ by ξ_i' to get an equivalent problem.

13.10.1   LS-SVM Formulation

The objective function in LS-SVM is Minimise $\frac{1}{2}||w||^2 + C \sum \xi_i^2$. The constant multiplier (C) is associated with the least squares criterion due to historical reasons. There is nothing wrong if we consider the objective function as $\frac{C}{2}||w||^2 + \sum \xi_i^2$. In this case, we have better control over the maximum margin. The C values now cause the margin to behave like a step function. Large C values (say $\geq k$) favor the max margin classifier, whereas smaller C values favor other margins by giving up the maximum margin forever. For historical reasons, we write the primal objective function as Minimise $J(w, b, \xi_i) = \frac{1}{2}||w||^2 + \frac{C}{2}\sum \xi_i^2$ with equality constraints $[w'\phi(x_i) + b]y_i = 1$ - ξ_i for i=1,2,$\cdots$N (here b is implicitly minimised). This formulation has better control over margin violating points in each class. As the objective is to simultaneously minimise the sums of squares, we use 13.20 instead of 13.19, and write the Lagrangian formulation as

$$L(w, b, a_i, \xi_i) = \frac{1}{2}||w||^2 + \frac{C}{2}\sum_i \xi_i^2 + \sum_i a_i \left[y_i - w'\phi(x_i) - b - \xi_i\right]. \qquad (13.21)$$

Differentiating 13.21 wrt w, b, ξ_i and a_i; and equating to zero gives

$$\text{KKT conditions:} \begin{cases} \frac{\partial L}{\partial w} = 0 & \Rightarrow \quad w = \sum_i a_i \phi(x_i); \\ \frac{\partial L}{\partial b} = 0 & \Rightarrow \quad \sum_i a_i = 0; \\ \frac{\partial L}{\partial \xi_i} = 0 & \Rightarrow \quad a_i = C\xi_i; \\ \frac{\partial L}{\partial a_i} = 0 & \Rightarrow \quad y_i - w'\phi(x_i) - b - \xi_i = 0. \end{cases}$$

Inclusion of the bias term b results in a constraint $\sum_i a_i = 0$, forcing some of the a_i's to be negative (this is what was mentioned in 13.3.2,pp.13-14 that all support values are non-negative ($a_i \geq 0$) for all SVMs except LS-SVM). Hence we can drop the bias term from our model, and cast the optimisation problem as a set of linear equations. To this end, we put b=0, substitute for w from the first equation, and put $\xi_i = a_i/C$ from the third equation into the fourth equation to get

$$\phi(x_i). \sum_j a_j \phi(x_j) + a_i/C = y_i, \text{ for } i = 1, 2, .., N. \qquad (13.22)$$

This can be written in compact form as

$$(K + \frac{1}{C}I_N)\alpha = Y, \qquad (13.23)$$

where K is the N×N kernel matrix with $(i, j)^{th}$ entry $k_{i,j} = \phi(x_i).\phi(x_j)$, Y $= [y_1, y_2, \cdots, y_N]'$ is the vector of known class labels, I_N is the identity matrix, and $\alpha = [a_1, a_2, \cdots, a_N]'$ is the solution vector that we seek.

Multiply both sides of equation 13.22 by y_i and use $y_i^2 = 1$ on the RHS to get

$$y_i\phi(x_i). \sum_j a_j \phi(x_j) + a_i y_i/C = 1. \qquad (13.24)$$

This can be written in matrix form as a set of NxN linear equations $\left[K' + \text{Diag}(\frac{y_i}{C})\right]\alpha = e_N$, where $k_{i,j}' = y_i\phi(x_i).\phi(x_j)$, $\text{Diag}(\frac{y_i}{C})$ is a diagonal matrix of the same size as K$'$ with $(i,i)^{th}$

element$=\frac{y_i}{C}$, e_N is a column vector of all 1's, $\alpha = [a_1, a_2, \cdots , a_N]'$. As the RHS contains all 1's, this formulation is slightly better from a computational viewpoint. As the solution to a set of linear equations is unchanged upon rearrangement of the rows, or columns we could rearrange it in such a way that all $y_i = +1$ comes first as a block (starting with the first row), followed by all $y_i = -1$ values (at subsequent rows). This simplifies the kernel matrix computation. In addition, as the RHS is a vector of all ones, matrix decomposition methods (like Strassen's method) result in identical sub-problems. Note that the size of the problem must be at least 3 to have any meaning (N$\geq$3). In this trivial case, there is just one hyperplane that separates the data (two points belong to one class and the third belongs to the other class). As the training data size is much larger in practical problems, the sparseness can be exploited to find the solution fast.

Example 13.10 Obtain the LS-SVM solution to the example 13.8 in page 13-16.
Solution: As before, we denote Genuine by +1 and Fraud by -1. As the data are separable, no kernel mapping is necessary. The K matrix is easier to compute if the claimed amounts are scaled. The scaling of variables will not affect the placement of the OSH, but may affect the intercept term b, and the margin width. We arbitrarily divide the claimed amount by 100 to get the encoded input patterns: x_1=(6,4.8;+1), x_2=(6,12;-1), x_3=(2,2.5;+1), x_4=(4,4.75;+1), x_5=(13,7;-1), x_6=(9,8;-1), x_7=(1,3.75;+1), x_8=(7,9;-1), x_9=(5,10.5;-1), and $x_{10} = $ (3,6;+1). The Y vector (of class labels) is [1,-1,1,1,-1,-1,1,-1,-1,1]. Linear equations are

$$\begin{bmatrix} 0 & Y'_N \\ Y_N & \Omega + \frac{1}{C}I_N \end{bmatrix} \begin{bmatrix} b \\ \alpha_N \end{bmatrix} = \begin{bmatrix} 0 \\ e_N \end{bmatrix} \quad \text{where} \quad \Omega_N =$$

$$\begin{pmatrix}
59.04 & 93.6 & 24.0 & 46.80 & 111.6 & 92.4 & 24.00 & 85.2 & 80.4 & 46.8 \\
93.6 & 180.0 & 42.0 & 81.00 & 162.0 & 150.0 & 51.00 & 150.0 & 156.0 & 90.0 \\
24.0 & 42.0 & 10.25 & 19.875 & 43.5 & 38.0 & 11.375 & 36.5 & 36.25 & 21.0 \\
46.8 & 81.0 & 19.875 & 38.5625 & 85.25 & 74.0 & 21.8125 & 70.75 & 69.875 & 40.5 \\
111.6 & 162.0 & 43.5 & 85.25 & 218.0 & 173.0 & 39.25 & 154.0 & 138.5 & 81.0 \\
92.4 & 150.0 & 38.0 & 74.00 & 173.0 & 145.0 & 39.00 & 135.0 & 129.0 & 75.0 \\
24.0 & 51.0 & 11.375 & 21.8125 & 39.25 & 39.0 & 15.0625 & 40.75 & 44.375 & 25.5 \\
85.2 & 150.0 & 36.5 & 70.75 & 154.0 & 135.0 & 40.75 & 130.0 & 129.5 & 75.0 \\
80.4 & 156.0 & 36.25 & 69.875 & 138.5 & 129.0 & 44.375 & 129.5 & 135.25 & 78.0 \\
46.8 & 90.0 & 21.0 & 40.50 & 81.0 & 75.0 & 25.5 & 75.00 & 78.0 & 45.0
\end{pmatrix} .$$

The solution vector for C=0.2 is b=0.371685, α_N={-0.032977 , -0.026799, 0.054982, -0.010827, -0.002094, 0.020544, 0.040841, 0.021967, 0.013495, -0.024905}

Next we formulate the problem by including the bias term b, and using equation 13.19. The Lagrangian formulation in this case is easily obtained as

$$L(w,b,a_i,\xi_i) = \frac{1}{2}||w||^2 + \frac{C}{2}\sum_{i=1}^{N}\xi_i^2 - \sum_{i=1}^{N} a_i\left[y_i(w'\phi(x_i) + b) - 1 + \xi_i\right] \tag{13.25}$$

where the negative sign indicates maximisation wrt a_i and minimisation wrt w and b.

Differentiating 13.25 wrt w, b, ξ_i and a_i; and equating to zero gives:

$$\text{KKT conditions:} \begin{cases} \frac{\partial L}{\partial w} = 0 & \Rightarrow \quad w = \sum_i a_i y_i \phi(x_i); \\ \frac{\partial L}{\partial b} = 0 & \Rightarrow \quad \sum_i a_i y_i = 0; \\ \frac{\partial L}{\partial \xi_i} = 0 & \Rightarrow \quad a_i = C\xi_i; \\ \frac{\partial L}{\partial a_i} = 0 & \Rightarrow \quad y_i\left[w'\phi(x_i) + b\right] = 1 - \xi_i. \end{cases}$$

Substitute for w and ξ_i from the first and third equations into the fourth equation. Then write the second equation as $0.b + \sum_i y_i a_i = 0$, followed by $y_i\, b + y_i[\phi(x_i).\sum_j a_j y_j \phi(x_j)] + a_i/C = 1$. Thus the quadratic programming problem again reduces to solving a set of (N+1)x(N+1) linear equations:

$$\begin{bmatrix} 0 & Y'_N \\ Y_N & \Omega + \frac{1}{C}I_N \end{bmatrix} \begin{bmatrix} b \\ \alpha_N \end{bmatrix} = \begin{bmatrix} 0 \\ e_N \end{bmatrix}$$

where $Y_n = [y_1, y_2, \cdots, y_N]'$ is the vector of class labels, $e_N = [1, 1, \cdots, 1]'$, b is a scalar, α_N is a column vector of Lagrangians (unknowns), and $\Omega_{i,j} = y_i y_j \phi(x_i).\phi(x_j) = y_i y_j K(x_i, x_j)$. As $Y'_N K Y_N = \sum_{i,j} y_i y_j K(x_i, x_j) = \sum_{i,j} y_i y_j \phi(x_i).\phi(x_j) = ||y_i \phi(x_i)||^2 \geq 0$, the solution is unique. Multiplying throughout by y_i gives $b + [\phi(x_i).\sum_j a_j y_j \phi(x_j)] + a_i y_i/C = y_i$, which can be written in matrix form as

$$\begin{bmatrix} Y'_N \\ \Omega' + \mathrm{Diag}(\frac{y_i}{C}) \end{bmatrix} \begin{bmatrix} \alpha_N \\ 0 \end{bmatrix} = \begin{bmatrix} 0 \\ Y_N - e_N b \end{bmatrix}.$$

As b which appears on the RHS is an unknown, we take it to the LHS and rewrite the above system as

$$\begin{bmatrix} Y'_N & 0 \\ \Omega' + \mathrm{Diag}(\frac{y_i}{C}) & e_N \end{bmatrix} \begin{bmatrix} \alpha_N \\ b \end{bmatrix} = \begin{bmatrix} 0 \\ Y_N \end{bmatrix}.$$

The C is a user-chosen tuning parameter. Very small C values favor the maximum margin of SSVM, but larger C values forgo the maximum margin to settle upon a lesser margin hyperplane. This is because the cost function consists of two criteria – namely margin maximisation, and bias minimisation. The bias-minimisation criterion is weighted more when C is very large. Only those primal data points for which a_i's are nonzero contribute to the LS-classifier.

An inherent weakness of both of the above formulations is that they weight each ξ_i values equally. Intuitively, those points close to the SVH should influence the placement of the hyperplane more than those points away from it. This precludes the possibility of bias due to outliers that are too far away from the SVH. We could use a simple trick to modify the criterion as follows: We consider the objective function as $\frac{C}{2}||w||^2 + \sum 1/\xi_i^2$. Then large ξ_i values are weakened in favor of small ones. To bring it to the LS-SVM form, we simply do a substitution $\psi_i = 1/\xi_i$ so that the objective function reverts back to the original form $\frac{C}{2}||w||^2 + \sum \psi_i^2$ and the constraints become $[w'\phi(x_i) + b]y_i = (\psi_i - 1)/\psi_i$. Multiplying throughout by ψ_i this becomes $[w'\phi(x_i) + b]y_i \psi_i = (\psi_i - 1)$. This is now in the standard form discussed above, and can be solved in the dual space.

The classifier for LS-SVM is $f(x) = \theta(\sum_i \hat{a}_i y_i K(x, x_i) + \hat{b}))$, where $K(x, x_i)$ is one of the kernel mappings described in table 13.2. As noted before, the summation is carried out over support vector points only. LS SVM is applicable to linear as well as nonlinear classification and regression problems. Kernel mapping is not necessary when data are linearly separable. Due to the computational advantage of LS-SVM over others for large-scale classification problems, it has been extended to many other situations. Examples are multi-class LS-SVM (pair-wise, OVO, OVA, etc)[AS05b], weighted LS-SVM [SB02], recurrent LS-SVM etc [SG02]. LS-SVMLab is a MATLAB/C toolbox available from www.esat.kuleuven.ac.be/sista/lssvmlab/

13.11 Nonlinear SVM (NL-SVM)

Nonlinear SVM is a supervised learning model to predict class labels of overlapping data. Decision boundary in nonlinear SVM can be any continuous surface that may be either closed or open. Most of the practical data encountered in classification and regression problems are non-linearly separable. The degree of overlap is a deciding factor for a proper choice of the kernel. There are many ways to solve the nonlinearly separable SVM. The SMO algorithm can also use the kernel mapping prior to finding the Lagrangians. This is achieved by replacing each of the dot products $x_i.x_j$ by $K(x_i, x_j)$. When data have overlap, we have to use one of the methods discussed below.

1. Using kernel maps

   As mentioned above, the kernel maps transform the data into a higher-dimensional linearly separable feature space. The primal HM-SVM problem can then be written as Minimise $Q(w, b) = \frac{1}{2}||w||^2$ subject to the constraints $[w'\phi(x_i) + b]\, y_i \geq 1$. In figure 13.5, data in the input space are nonlinearly separable by an open curve. Transformed data in the feature space are linearly separable.

   Points (support vectors) closest to the optimal separating margin (OSM) satisfy $w'\phi(x) + b = \pm 1$, which represents a tube shaped region of size 2ϵ, where ϵ is the radius of the tube. The primal classifier is $\theta(\hat{w}'\phi(x) + \hat{b})$, showing that each new data instance to be classified must undergo the same kernel map used in the training algorithm [BE05].

2. Using NL-SM-SVM (NS-SVM)

   A nonlinear kernel mapping described above is not easy to obtain without software. However, if a kernel map improves the separability in the feature space, we could combine the SM-SVM on top of the kernel map to obtain the NS-SVM. Relaxing separability constraints by penalising with a cost parameter, the objective is minimising either $\frac{1}{2}||w||^2 + C\sum_i \xi_i$ if L_1-norm is used, or $\frac{1}{2}||w||^2 + C\sum_i \xi_i^2$, if L_2-norm is used. In the former case, we minimise $Q(w, b, \xi) = \frac{1}{2}||w||^2 + C\sum_{i=1}^N \xi_i$, subject to the constraints $[w'\phi(x_i) + b]\, y_i \geq 1 - \xi_i, \forall i$. Here C is a user-chosen constant. Small values of C seek large classification margins. This could often result in more support vectors. Introducing the Lagrangians a_i we get $Q(w, b, \xi, a) = \frac{1}{2}||w||^2 + c\sum_{i=1}^N \xi_i - \sum a_i([w'\phi(x) + b]y_i - 1 + \xi_i)$. Differentiating wrt w,b,ξ_i, and substituting the resulting equations in the above, we get the dual formulation as Maximise $Q(a, b) = \sum_i a_i - \frac{1}{2}\sum_{i,j} a_i a_j y_i y_j K(x_i, x_j)$. The primal-dual relationship in this case is $w = \sum_i a_i y_i \phi(x_i)$, and the classifier is $\theta(w'\phi(x) + b) = \theta(\sum_i \hat{a}_i y_i K(x_i, x) + \hat{b})$ where the summation is over support vector points only. In NS-SVM (using kernels) there are many parameters to choose that include the model parameters and kernel parameters. An optimal parameter set can be obtained using the grid-search method ([SS04],[KR06]).

3. Other SVMs

   Fuzzy-SVM provides an alternative to solve nonlinearly separable cases by associating fuzziness with data instance labels, or model parameters [TY03], [TA03], [AS04]. It has been extended to the MC-SVM also [AS05a], [AS05b].

Example 13.11 Predicting payment defaulting customers.
Table 13.5 gives some training data on customers who have taken a loan. The first column is the outstanding family debt, and second column is their annual income. Fourth column is -1 if the loan payment was not defaulted as indicated by third column and a $+1$ otherwise.
Solution: To apply the SVM, we should scale the data (preferably into the [-1,+1] range). Some of the SVM software programs discussed in 13.15 (page 13-37) in addition accepts the data in [attr:value] (subsequent pairs separated by blanks) format where attr is an integer that

Table 13.3: SVM summary table

Type	Criteria	Objective function	Constraint(s)
HM-Primal	Min	$\frac{1}{2}\|\|w\|\|^2$	$(w'x_i + b)y_i \geq 1 \; \forall \; i$
HM-Dual	Max	$\sum_i a_i - \frac{1}{2}\sum_{i,j} a_i a_j y_i y_j (x_i.x_j)$	$\sum_i a_i y_i = 0, \; a_i \geq 0 \; \forall i$
ρ-SVM	Min	$\sum_{i=1}^{n} w_i^2 - \rho$	$(w'x_i + b)y_i \geq \rho \; \forall i$
Dual ρ-SVM	Max	$\sum_{i=1}^{N} a_i - \frac{1}{2}\sum_{i,j}^{N} a_i a_j y_i y_j (x_i.x_j)$	$\sum a_i y_i = 0, \sum a_i = 0$
SM - Primal	Min	$\frac{1}{2}\|\|w\|\|^2 + C\sum_i \xi_i$	$(w'x_i + b)y_i \geq 1 - \xi_i \; \& \; \xi_i \geq 0$
SM - Dual	Max	$\sum a_i - \frac{1}{2}\sum_{i,j} a_i a_j y_i y_j (x_i.x_j)$	$\sum_i a_i y_i = 0, \; \& \; 0 \leq a_i \leq C \, \forall i$
ν-SVM	Min	$\frac{1}{2}\|\|w\|\|^2 - \nu\rho + \frac{1}{N}\sum_i \xi_i$	$(w'x_i + b)y_i \geq \rho - \xi_i, \; \& \; \xi_i, \rho \geq 0$
Dual ν-SVM	Max	$\frac{1}{2}\sum a_i a_j y_i y_j (x_i.x_j)$	$0 \leq a_i \leq \frac{1}{N}, \sum_i a_i \geq \nu, \sum_i a_i y_i = 0$
LP - Primal	Min	$\frac{1}{2}\|\|w\|\|_1 + C\sum_i \xi_i$	$(w'x_i + b)y_i \geq 1 - \xi_i \; \& \; \xi_i \geq 0$
LP - Dual	Max	$\sum a_i , \; 0 \leq a_i \leq C \, \forall i$	$-1 \leq \sum_i a_i y_i x_i \leq 1, \; \sum_i a_i y_i = 0$
NL-Primal	Min	$\frac{1}{2}\|\|w\|\|^2 + C\sum_i \xi_i$	$[w'\phi(x) + b]y_i \geq 1$
NL -Dual	Max	$\sum_i a_i - \frac{1}{2}\sum_{i,j} a_i a_j y_i y_j K(x_i, x_j)$	$\sum_i a_i y_i = 0, 0 \leq a_i \leq C \, \forall i$
LS-Primal	Min	$\frac{1}{2}\|\|w\|\|^2 + C\sum \xi_i^2$	$[w'\phi(x_i) + b]y_i = 1 - \xi_i$
LS -Dual	Max	$\sum_{i,j}[a_i y_i K(x_i, x_j) + b]y_j + \frac{a_j}{C} = 1$	$\sum_i a_i y_i = 0, a_i = C\xi_i \, \forall i$
SVR	Min	$\frac{1}{2}\|\|w\|\|^2 + C\sum_i(\xi_i^+ + \xi_i^-)$	$y_i - (w'x_i+b) \leq \epsilon + \xi_i^+,$ $(w'x_i+b) - y_i \leq \epsilon + \xi_i^-$
SVR-Dual	Max	$\sum_{i=1}^{N} y_i(a_i^+ - a_i^-) - \epsilon \sum_{i=1}^{N}(a_i^+ + a_i^-)$ $- \frac{1}{2}\sum_{i,j=1}^{N}(a_i^+ - a_i^-)(a_j^+ - a_j^-)K(x_i,x_j)$	$\sum_i(a_i^+ - a_i^-) = 0, \; a_i^+, a_i^- \in [0,C]$

Legend: HM = Hard-Margin, SM = Soft-Margin, LP = Linear Programming, NL = Nonlinear,
LS = Least Squares, SVR = Support Vector Regression, Min=Minimise, Max=Maximise.
Note: The ξ_i's can be positive or negative in LS-SVM primal, and the a_i's can be positive or
negative in LS-SVM dual. The LS-SVM results in a set of linear equations for various j values.

Table 13.4: Table of SVM classifiers

Type	Classifier	Comment(s)
SSVM-Primal	$\theta(w'x + b) = \theta(w'[x - \frac{1}{2}(x_r + x_s)])$	independent of class labels
SSVM-Dual,SM-SVM,	$\theta(\sum_{i=1}^{|SV|} a_i y_i (x.x_i) + b)$	depends on class labels
$\nu-$SVM, NL-SVM	$\theta[\sum_{i=1}^{N} \alpha_i y_i K(x, x_i) + b]$	K(x,x_i) is a kernel
LS-SVM	$\theta[\sum_{i=1}^{N} \alpha_i y_i \Phi(x_i)' \Phi(x) + b]$	$\Phi(x)$ is a kernel
$\rho-$SVM	$\theta(|w'x + b| - \rho)$	$\rho > 0$ is a parameter

Note: b values in column 2 denote estimated values.

identify the attribute, and value is a real number. These should be arranged in increasing order
of attribute numbers. The same scaling is used during training and testing stages.

When trained with SVMLEARN of Joachims[JT99] we got the VC-Dim as < 3.4316 using
polynomial kernel, number of SV=7, norm of weight vector $|w|$=1.51279, from which the margin
is obtained as 1.322. The intercept term b=-0.16759691, and a[i]*y[i] values for support vectors
are as given in table 13.6, where the column index denotes row-id given in table 13.5, and y[i]

Table 13.5: Which customers fault on loan repayment?

Sr.	Family Debt	Annual Income	Payment fault	class label
1	120000	32000	No	-1
2	200000	45000	Yes	1
3	75000	23000	No	-1
4	0	52000	No	-1
5	50000	30000	No	-1
6	170000	70000	Yes	+1
7	0	28000	No	-1
8	29000	35500	No	-1
9	78000	39000	Yes	+1
10	133000	46500	Yes	+1
11	96000	27500	Yes	+1
12	41000	32800	No	-1

are the class labels $\pm$ 1 (LIBSVM also gave the number of SVs as 7 (with 5 bounded SVs) in cross-validation mode).

Table 13.6: Results of SVM training

1	3	4	5	9	10	11
-1.365081	-1.365081	-0.976093	-0.388988	1.365081	1.365081	1.365081

13.11.1   Other Kernel Algorithms

There are many statistical procedures known as generalised eigen problems. These include kernel least squares, kernel discriminants, kernel principal components, kernel canonical correlation analysis [KG03]. These methods work in a higher-dimensional feature space using kernel functions that map input space data vectors. The kernel-based k-means algorithm minimises the distance of each point in the transformed pattern space to the respective cluster centroid using $\sum_{i=1}^{k} \sum ||\phi(x_i) - \mu_c||^2$ where μ_c is the mean of cluster c in the transformed space. Other kernel based methods include kernel ridge regression, kernel Mahalanobis distance and kernel SOMs, kernelised time-series analysis, kernelised LSA etc (see Abe[AS05b], and references [CS00], [CV05], [VV98]).

13.12   Support Vector Regression (SVR)

SVM is used for classification and prediction when class labels are categorical (and numerically coded). SVR is used to predict a quantitative variable. It is applied directly in input space when the data are linearly separable. Otherwise, we apply SVR in the feature space using kernel maps.

Let the training data $(X, y) = (X_1, y_1), (X_2, y_2), \cdots, (X_N, y_N) \in \mathbb{R}^n$ x $\mathbb{R}$ be inseparable. Consider the classifier for NL-SVM f(x,w,b)=$w'\phi(x)+b$ (where we have dropped the sign function

$\theta()$ because y_i's are real numbers). SVR seeks that function f(x) that has at most ϵ numeric deviation from actual targets y_i for all the training instances. Hence errors less than ϵ in magnitude are ignored, and more than ϵ are penalised.

Definition 13.5 SVR is a supervised learning method for empirical data fitting using the ϵ-insensitive loss function principle that minimises an empirical risk
Min $\frac{1}{N}\sum_{i=1}^{N} e(y_i, f(x_i, w, b)) + \mu||f||^2$ where the loss function is:

$$e(y_i, f(x_i, w, b)) = \begin{cases} 0 & \text{if } |y_i - f(x_i, w, b)| \leq \epsilon; \\ |y_i - f(x_i, w, b)| & \text{otherwise} \end{cases}$$

for $i = 1, 2, \cdots, N$. The empirical risk of SVR is the arithmetic mean of $e(y_i, f(x_i, w, b))$ values, and is given by $R_{emp}(w, b) = \frac{1}{N}\sum_{i=1}^{N} e(y_i, f(x_i, w, b))$, which reduces to $R_{emp}(w, b) = \frac{1}{N}\sum_{i=1}^{N}(y_i - f(x_i, w, b))^2$ when the least-squares criterion is used. Because $||f||^2 = ||w||^2$ where $w = \sum_i a_i \phi(x_i)$, and $\phi()$ is the kernel function, the above problem can be stated in alternate form as $Min \frac{1}{N}\sum_{i=1}^{N}(|w'\phi(x_i) + b - y_i| - \epsilon) + \mu||w||^2$

The tube parameter ϵ is not easy to choose for a given problem, because the margin of error in the fit may be unknown. A variant called $\nu-$SVR adjusts the ϵ automatically [SS00].

Above hard-margin SVR (HM-SVR) assumes that such a solution is always feasible. When the classes overlap, the solution to HM-SVR may be infeasible. As done in the case of SM-SVM, we introduce two slack variables ξ_i and ξ_i^* that represent the under- and over- estimation errors (Here we have used different symbols for positive and negative slacks as a notational convenience). Then the primal SVR can be expressed as

$$\text{Minimise } \frac{1}{2}||w||^2 + C\sum_{i=1}^{N}(\xi_i + \xi_i^*), \tag{13.26}$$

such that $y_i - w'\phi(x_i) - b \leq \epsilon + \xi_i$, $w'\phi(x_i) + b - y_i \leq \epsilon + \xi_i^*$, $\xi_i, \xi_i^* \geq 0$. The user-chosen parameter C>0 determines the trade-off between flatness of f(x) and ϵ-tolerability.

Duality principle (as in SVM) can be utilised to reduce the computational complexity [OG99]. Dual of the SVR is given in table §13.3. Objective function for the dual formulation is given by $-$ Minimise $\sum_{i=1}^{N}(a_i + \hat{a}_i) + C\sum_{i=1}^{N}(\xi_i + \xi_i^*)$ subject to the constraints given above. The KKT conditions in SVR reduces to $a_i^+(\epsilon + \xi_i - y_i + (w'x) + b) = 0$, $a_i^-(\epsilon + \xi_i^* + y_i - (w'x) - b) = 0$ and $(C - a_i^+)\xi_i = 0$, $(C - a_i^-)\xi_i^* = 0$. In addition, $a_i^+ * a_i^- = 0$ implying that one of the weights is zero for each data point. The primal-dual relationship is $w = \sum_i(a_i^+ - a_i^-)y_i x_i$, which is called the support vector expansion. See [CB01], [SS04], [MM06] for further details.

Decision function for SVR becomes $f(x) = \theta(\sum_{i:x_i \in SV}(a_i^+ - a_i^-)K(x_i, x) + b)$ where $0 \leq a_i^+, a_i^- \leq C$. SVR can be used to predict new dependent values as $f(z) = \sum_i(a_i - a_i^*)K(x_i, z) + b$, where the parameters are estimated from training data [AS05b], [VV99]. Platt's SMO algorithm has been extended to SVR by Flake & Lawrence[FL02]. Called Sequential Minimal Optimising Regression (SMOR)[9], it seeks two Lagrangian multipliers for each data point on the boundary (support vectors) that represent the width of the ϵ-tube in the primal data space. As the support vectors are unknown until we solve the SVR, the SMOR algorithm iterates through all data points as done by the SMO.

---

[9]Pronounced S-MOR; it uses three user-tunable parameters (C, tube-width ϵ, and tolerance δ).

LP-SVR is an extension of LP-SVM that formulates the SVR as an LP problem. The fuzzy SVR seeks a fuzzy hyperplane with fuzzy coefficients [WC07] or fuzzy membership functions for support vectors. See Smola & Schoelkopf [SS04] for a beginner's tutorial on SVR, and Chen, Lin & Schoelkopf [CL03] for other examples.

13.13 SVM vs Statistical Classifiers

It was mentioned in section §13.3.1 that the SVM works in the *data space*. This section further explores the differences between the SVM and popular statistical classifiers, and clarifies some myths about SVM. As noted in §13.3.1 in page 13-14, the Lagrangian multipliers $(a_i's)$ are nonzero only for the support vectors (which are data points on the boundary of each class). This is evident from the discriminating boundary (classifier) $D(x)=\theta(\sum_{i=1}^{|SV|} a_i y_i(x.x_i)+\hat{b})$, where $|SV|$=number of support vectors, x_i's are the support vectors, x is the new data instance to be classified, and a_i,b are the unknowns obtained by solving the dual-SVM. The OSH is a weighted sum of support vectors. Coefficients of variables in the OSH indicate the tilt of the hyperplane. It does not carry any information on the discriminatory power of the variables. This implies that only data points on the boundary contribute to the classifier (and all other data points are totally ignored). In the extreme case when there are protruding data points in each class towards the OSH (called the degenerate case), the SVM simply ignores the majority of data points, and comes up bluntly with a classifier which is a linear combination of those protruding points (as the classifier given above is simply a linear combination of the support vectors), unless a pruning step removes such points. Therefore, the coefficients in the SVM classifier do not carry discriminatory power of the variables, except in particular cases discussed below.

13.13.1 Boundary Dense and Boundary Sparse Problems

For simplicity, consider the binary SVM. The boundary of the classes could be dense in some classification problems. In other words, there are too many data points concentrated along the boundary (highly dense), and too few data points away from the boundary. We call such problems as *boundary-dense*. The SVM classifier obtained in such problems could come close to some of the statistical classifiers[10]. Practical data distributions are seldom boundary-dense. If the number of data points on the boundary are sparse (data points away from the boundary may or may not be dense), we call it as *boundary-sparse* problem. A great majority of practical classification problems are of this type. Even a single data point on the boundary can result in a major tilt of the OSH in this case. This is good enough evidence that the coefficients in the hyper-surface equation of SVM do not carry any discriminatory power of variables. On the other hand, statistical classifiers do not give due weightage to boundary points alone. The discriminatory power of all data points are statistically *blended in* in an optimal way by using all the variables, or linear combinations of a subset of them to come up with the respective weightages to be given to each of them. Deleting some of the boundary points does not have a major impact on the weights of the variables. Of course, in boundary-dense problems, a large number of data points have a major impact on the weightages of variables. This is why both SVM and statistical classifiers produce similar hyper-surface coefficients.

This does not mean that SVM is inferior to statistical classifiers. There are many problems where SVM is *the preferred choice*. First of all, statistical classifiers are computationally more

---

[10]This is why the SVM and statistical classifiers reported by Moguerza & Muñoz[MM06] are almost identical

intensive, and not easily scalable to very large problems (as some of them involve matrix inversions). One of the advantages of SVM is that it is easily scalable to large problems. But SVM being a supervised machine learning model could always be trained by a small random subset of the available data called training data. Secondly, support vectors encapsulate a combination of essential features (with high discriminatory power) of each class in some pattern classification problems that involve text, images, audios etc. They serve as threshold fences for all other elements of respective classes. In such cases, the SVM will work fine to classify new data instances. Thirdly, there are a large number of classification problems where variables (elements of X) are either meaningless, or unknown, or have too little discriminatory power (ie. they are weak discriminators). As an example, we could use external attributes like file type (or extension), file name and size, or intrinsic attributes like character encoding, compression or encryption information, file metadata and schemas, header information, etc in text-classification. Due to high overlap and flexibility of file naming and storage options in various operating systems, most of the above variables may not carry much discriminatory power. In such cases the statistical classifier performs poorer than the SVM. However, most people who use SVM for text, audio, image and multimedia classifications and pattern recognition (eg.handwriting recognition) are least bothered about the variables that contribute to classification; but are more interested in the end-result of identifying to which class their datum belongs to. Variables are important in education, medical sciences, process-control, psychology, robotics, etc. This is because either follow-up research studies are often carried out to further investigate variable interactions and dependencies that contributed to the classification, or remedial actions (or error corrections) are undertaken to correct a fault or condition (as in classifying a patient to a positive class due to a disease; which must immediately be treated to move the patient back to the negative class). Variables (like a patient's age, disease onset duration, other medical conditions, etc) can turn out to be immensely useful in these fields.

Some SVM formulations (like LS-SVM) can be forced to catch-up with statistical classifiers using a simple trick. In the soft-margin SVM (SM-SVM), ν−SVM, LS-SVM, SMO and other models, we have a user tunable parameter (C). Tuning this parameter results in more and more data points (away from the boundary) to be included in the classifier. Thus we could tune the parameter(s) to such an extent that a great majority of points from each class (or all points from one class as in rare positive class for a medical condition like H1N1 virus) are included in the classifier. This will certainly improve the variable contribution than otherwise. As the number of soft-margin support vector points (beyond the SVH) are increased due to the tuning of C, we will reach a point where the OSH (say of an LS-SVM) almost orients towards the hyperplane obtained by some statistical classifiers. But increasing C of course comes with a price – an increase in the computing time. We may not know what value of C to use when we run the SVM for the first time. It can be obtained by trial-and-error, grid-search method, etc. In this case the SVM loses the luster of a "maximum-margin classifier".

The Sequential Minimal Optimisation (SMO) algorithm always work in the dual space and solves several two variable equations for the Lagrangian multipliers. A detailed discussion of SMO and other SVM algorithms appear in [CR12], and variable selection methods in [SA11].

13.14   Applications of SVM

SVM and related methods have been applied in a variety of classification techniques. The popular ones are in hand-writing recognition (especially in handwritten digit recognition), detection of objects in videos, images etc, digitised text (captions) and features in images, image based retrievals, intrusion detection, text and web classification, color based image classification, face and retina recognition, signature identification, speech recognition [RV05], diagnosis from biomedical signals, target identifications (using video, radar, sonar), remote sensing etc.[11] [PM05].

13.14.1   Medical Application

Predicting onset of illnesses is a time consuming task for medical professionals. This task can be tremendously eased using SVM and its variants. Blood pressure (BP) data of patients to a diabetes clinic are used in this application to classify visitors to the clinic as high or low risk. Variables considered are the BMI, waist size over hip size ratio, systolic and diastolic BP, and serum cholesterol. Data for 20 patients are given in table 13.7, where the second column indicates the risk (high risk =+1, low risk =-1).

Table 13.7: Data for high BP and low BP risks

Sr no.	class labels	BMI index	W/H ratio	Sys BP	Dia BP	Chol mg.
1	-1	23.3	0.76	146	68.5	133
2	-1	26.2	0.89	116.5	74	205
3	-1	25.6	0.84	111	69.5	200
4	-1	25.1	1.03	154.5	85.5	218
5	-1	23.3	0.76	108	68	176
6	-1	23.3	1	136.5	84	184
7	-1	23	0.91	146	65	265
8	1	23.7	0.86	142	84	247
9	-1	33	0.9	104	69.5	170
10	-1	29.7	0.88	127	80.5	201
11	1	28.5	0.99	138	84	148
12	1	25.2	0.23	106	62.5	187
13	-1	22.1	0.81	144.5	85.5	238
14	1	28.4	1.01	152.5	87.5	184
15	-1	32.2	1.03	117.5	87	187
16	1	33.6	0.89	126	80	193
17	-1	27.4	0.93	122	80.5	210
18	1	24.4	0.93	144.5	89	245
19	-1	24.4	0.85	125.5	80.5	221
20	1	24	0.98	148	92	239

Legend: +1=high BP, -1=low BP; sys BP=systolic blood pressure; dia BP=diastolic blood pressure; Chol= serum cholesterol.

Several SVM and SVR models were run on the data. The results are summarised in table 13.8. Third column gives the width of ϵ-tube (for SVR), and fourth column gives the kernel used. Sixth column gives the support vectors at their upper bound. The RBF kernel in general gives more support vectors (but less bounded support vectors) than polynomial or linear kernels.

---

[11] see www.clopinet.com/isabelle/Projects/SVM/applist.html for an expanding list

SVM_LEARN outputs the product of class labels and weight vectors, using which it is easy to construct a classifier. This classifier can be used to predict the BP risk of unseen patients.

Table 13.8: Results of SVM training on patient data

Sr no.	SVM/ SVR	wt	kernel type	SV	SV (UB)	L1 norm	b	VCDim <
1	SVM	—	linear	16	12	13.996	1.02985	1.0042
2	SVM	—	poly	17	13	13.9738	1.06388	1.097
3	SVM	—	rbf	20	7	8.61926	0.73132	6.38462
4	SVR	.005	poly	18	13	13.91754	.725876	—
5	SVR	.5	rbf	20	7	1.61926	.231323	—
6	SVR	.8	rbf	20	7	.00001	.060144	—

13.15   SVM Software

There are many publicly available softwares for SVM in popular operating systems.

Table 13.9: Software for SVM

URL	Name	comments
svmlight.joachims.org	SVMLight	small size
www.ee.unimelb.edu.au/staff/apsh/svm/	SVMHeavy	see [SA01]
creativecommons.org/licenses/by-nc/2.0/	ISDA	regression also
www.lsi.upc.edu/~nlp/SVMTool/	SVMTool	
chasen.org/~taku/software/TinySVM/	TinySVM	
www.cenparmi.concordia.ca/~people/jdong/	HeroSvm	
www.biology.ucsd.edu/ ~gert/svm/incremental/	Incremental SVM	MATLAB
www.esat.kuleuven.ac.be/sista/lssvmlab/	LS-SVM	MATLAB/C
hwanjoyu.org/idis/svm-java/	SMO	in Java
www.esat.kuleuven.ac.be/sista/lssvmlab/	LS-SVMLab	MATLAB/C

The svmTorch II is a fast implementation of Vapnik's SVM that works on Windows, Solaris and Linux platforms [CB01]. It can be used for classification and regression problems. The iterative single data algorithm (ISDA) can be used for large-scale multiclass classification [AS03] and regression tasks. When used for binary classification, ISDA uses class labels '1' and '2' instead of '-1' and '+1'. Several software libraries for SVM are also available in C, C++, MATLAB etc (LIBSVM [CL01], MATLAB SVM Toolbox (www.isis.ecs.soton.ac.uk/resources/svminfo/) are just a few of these. Software programs (in C++) for incremental SVM are described in (www.cs.virginia.edu/~xj3a/research/SmartSketchpad/readme_SVM_INC.pdf)[XW00].

jSVM is a java wrapper for SVMlight. Java applets for SVM and SVR at svm.dcs.rhbnc.ac.uk/. See www.support-vector.net/software.html, www.support-vector-machines.org/SVM_soft.html for an up-to-date list, and www.svms.org/software.html for a ranking of various softwares.

Table 13.10: Emission test results for autos

Auto age	Mileage	Emission test
8	85000	Pass
11	104000	Fail
7	58000	Pass
9	975000	Fail
16	152000	Fail
8	56500	Pass
10	120000	Fail
13	129000	Fail
14	151000	Fail
10	79000	Pass
15	135000	Fail
12	75000	Pass

13.16 Exercises

1. Mark as True or False
   (a) SM-SVM is well suited when data set has slight overlap
   (b) All training data for SVM should be labeled
   (c) Kernel techniques are not needed when the data are linearly separable
   (d) SVM is not applicable when classes have a hierarchical tree structure
   (e) Data distributions or correlations are not considered by SVM
   (f) The ν in ν-SVM is an estimate of the fraction of support vectors that violate the margin
   (g) If the constant C in SM-SVM is very large, the model discriminates all training data without errors
   (h) Mercer's theorem gives conditions for a function to be a kernel
   (i) OSH and SVH have the same slope.
   (j) In general, a linear SVM has more support vectors than a nonlinear SVM.
   (k) The SVR works only with labeled data
   (l) The SMO algorithm works in the primal space
   (m) Each iteration of SMO (except first), we choose a KKT condition-violating data point.

2. Discuss the advantages of SVM over neural networks (NN). Are there any advantages for NN over SVM?

3. Prove that the XOR problem is separable using the polynomial kernel $K(x, y) = (x.y+1)^2$.

4. Which SVM model is preferred when the training data are not linearly separable, and has high overlap among the classes?

5. What is the relation between the primal weight vector w, and the dual solution a_i's?

6. Table 13.10 contain data on 12 automobiles tested for emission control. X values are the miles driven and age in months, and the Y values indicate emission passed (y=+1) or failed (y=-1). Fit an SVM and obtain the support vectors and margin.

Table 13.11: Annual repair cost of automobiles

Repair cost	Car Age (yrs)	Mileage
3900	12	98000
2500	8	80000
5700	10	102000
1100	3	24000
6200	9	108000
2350	8	64000
1600	5	67500
4300	14	90000
1500	4	43000
2900	6	75000
6500	10	110000

7. Express the primal classifier for a binary SVM in terms of support vectors, if the dual solution is $(a_1 = 2, a_2 = -1, a_3 = 1)$ and $(y_1 = +1, y_2 = +1, y_3 = -1)$.

8. What is the perpendicular distance from the origin to each of the following OSH? Does the magnitude of these distances indicate any discriminatory power of SVM? (a) $w'x + b = 0$ b) $w'x + b = +1$ c) $w'x + b = -1$

9. What does the dual SM-SVM training return? What are the differences between dual SM-SVM and dual HM-SVM?

10. Distinguish between bounded and unbounded support vectors. What are the support values for each of them? Prove that averaging 'b' over those data points for which $0 \le a_i \le C$ in SM-SVM results in an improved estimate.

11. What are the advantages of solving an SVM in its dual form? What is the classifier expression of dual SM-SVM?

12. Prove that the intercept term in the inseparable case can be expressed as $b^* = 1 - \sum_{j=1}^{|SV|} \alpha_j^* K(x_i, x_j)$, where x_i's are those points for which $\alpha_i > 0$.

13. For the kernel function $K(x, y) = (1 + xy)^2$, verify whether $\phi(x_1, x_2) = (1, \sqrt{2}x_1, \sqrt{2}x_2, x_1^2, x_2^2, \sqrt{2}x_1 x_2)$ is an applicable mapping.

14. Consider the quadratic kernel $K(x_i, x_j) = (x_i . x_j)^2$. Prove that it is equivalent to a mapping to 3 dimensional space induced by $\Phi(x) = (x_1^2, x_2^2, \sqrt{2}x_1 x_2)$. Derive the corresponding mapping for the quadratic $K(x_i, x_j) = (x_i . x_j + c)^2$ where c is nonzero constant $\ne \pm 1$.

15. If the weight vector $w = \frac{1}{5}[4, 3, 7, -1]'$ and two support vectors are $x_r = [0,1,-1,2]'$, and $x_s = [1,3,1,4]'$, estimate the bias term 'b' in the hyperplane equation $w'x + b = \pm 1$. What is the classifier of this SVM?

16. Plot the data in table 13.3 and answer the questions that follow:

(i) Are the data linearly separable? (ii) What is the optimal hyperplane equation? (iii) What is the margin? (iv) How many support vectors are there?

17. Give two practical examples that result in multi-class SVM with i) 3 classes ii) 4 classes

18. Describe the asymmetric classification problem. Which SVM is the most appropriate for it? How is it different from unbalanced classification problem?. Give two examples of it.

19. If a few classes of a MC-SVM are linearly separable and others are not, discuss how you will utilise the linear separability to speedup classification of new data.

20. Find the margin width for each of the following problems where the support vector hyperplane equations are given.
    (i) $4x_1 + 3x_2 - 3.5 = 0, 4x_1 + 3x_2 - 1.5 = 0$ (ii) $1.2x_1 + 1.6x_2 + 5 = \pm 1$

21. Write down the classifier equation in each of the following SVMs: (i) S-SVM - Primal (ii) S-SVM Dual (iii) SM-SVM primal (iv) LS-SVM primal (v) ν-SVM primal

22. How many OVO and OVA-SVM (see section §13.6) are there in a classification problem involving (a) 5 separate classes? (b) n separable classes?

23. Describe any three ways in which the MC-SVM can be solved.

24. Auto-emission test results for cars aged 7 or more years appear in table 13.10. Mileage column gives the speedometer readings, and the age is rounded to the nearest integer. Recode the emission test result column as [-1,+1] and obtain a classifier using RBF kernels, with a proper choice of the parameters.

25. Table 13.11 gives the annual maintenance cost of some automobiles. Use SVR to build a regression model and use it to predict the expected repair cost using the age and mileage of cars (Hint: svmtrain of LIBSVM uses -s 3 as the command line flag for ϵ-regression, and SVM_LEARN uses -z r flag for regression).

26. Maximising the margin $2/||w||$ of HM-SVM is equivalent to minimising $||w||^2$. But if $||w||$ is less than 1, the square of it is less than itself (eg: $.5^2 = .25$). Discuss how the SVM reaches optimal solution irrespective of the magnitude of $||w||$.

27. What class label will you give to new data that fall in the generalisation region? What will you do when many new data points lie exactly on the OSH.

28. Prove that the solution to the dual SVM uniquely determines the primal solution. If the dual-SVM is solved by multiple algorithms like an LPP, an LS-SVM and SMO, is the primal solution unique?.

29. Prove that when X is transformed by the change of scale transformation (chapter 2) x'=x/c, where c is a nonzero constant; results in an equivalent SVM. What is the new margin? What is the classifier?

30. Prove that the bias term b can be computed from the dual SVM solution as $\hat{b} = 1/y_i - \sum_{j=1}^{N} a_j y_j K(x_i, x_j) - a_i/[C * y_i]$ for any j for which a_j is a support value (ie. $a_j \neq 0$).

31. Suppose a measure of misclassification is defined for a test set of size n as $\sum_{i=1}^{n}(1/\theta(w'x_i + b)^2)$, where $\theta()$ is the soft sign function. What is its range and when is it maximum? What does a high value of this measure indicate?

32. Explain why the SMO algorithm always optimizes two Lagrange multipliers simultaneously. What happens when the number of primal points is odd?

33. Prove that $||w||^2 = \sum_{i \in \text{SV}} \alpha_i^*$ in the nonlinearly separable case as well, where the summation runs over support values only. Show that irrespective of separability, the maximum geometric margin is given by $1/\left(\sum_{i \in \text{SV}} a_i\right)^{1/2} = 1/\left(\sum_{i=1}^n a_i\right)^{1/2}$.

34. If $w'x + b = \pm k$ is the canonical form of parallel hyperplanes $w'x + b_1 = 0$ and $w'x + b_2 = 0$, prove that the sign of the intercept term b depends on $b_1 + b_2$. Prove also that the hyperplane passes through the origin when b_1 and b_2 are equal but of opposite signs. Prove that this result can be extended to the hypersurface in the nonlinear case.

35. Prove that the constraints $[w'\phi(x_i) + b]y_i = 1 - \xi_i$ and $y_i - [w'\phi(x_i) + b] = \xi_i$ are equivalent for the binary classification problem with class labels ∓ 1.

36. Let (v,y) be vectors with n+1 components in which v_i are orthogonal vectors (so that $v_i'v_i = 1$), and y is the class label. Prove that a binary labellings as $y_i = +1$ if $v_i \in C_1$ and $y_i = -1$ if $v_i \in C_2$ results in a linear classifier. If v_i's are of the form $(0,0,..,0,1,0,..0)$ with just a 1 at i^{th} position, prove that the margin is at least $2/\sqrt{n}$.

37. For what value of the tuning parameter C does the soft-margin SVM become HM-SVM? What values can C take in SMO?

38. Find the second derivative of $||w||^2$ wrt w, and prove that the objective function is strictly convex and minimising $\frac{1}{2}||w||^2$ will indeed produce a unique separating hyperplane (find the maximum margin).

39. If Sam labels class C_1 as +1 and class C_2 as -1, and Seetha labels the classes the other way (C_1 as -1 and C_2 as +1), will they both get the same solution to the SVM? Will the class labels assigned to new data instances be the same for both?

40. The SMO algorithm obtains pairs of dual variables a_i's sequentially. What if the data size is odd (so that we need to find an odd number of a_i's)? If the number of data points in one of the classes is too small, the corresponding a_i's are large (due to the sum constraint). Can you use this as a heuristic in choosing the first a_i's to optimise?

41. Suppose you need to solve an SVM as minimise $\frac{1}{2}||w||^2 + C\sum_{i=1}^N \xi_i^2$. As the objective function contains ξ_i^2, this can be solved either using SM-SVM or LS-SVM (as a set of linear equations). Unfortunately, you do not have a program for SM-SVM, or LS-SVM but only one for hard-margin SVM that minimises $\frac{1}{2}||w||^2$. Is it possible to express the given problem as an HM-SVM?

42. The LS-SVM solves a set of linear equations Ax=b. But the solution is non-existent when the determinant $|A|=0$. In addition, there exist multiple solutions if the number of constraints is less than the number of variables. Does the LS-SVM always guarantee a unique solution?.

43. Compare and contrast the SMO and I-SVM training algorithms. In what aspects do they differ? How do you select the initial data points to start the training?

44. What is the interpretation of the tuning parameter C in SM-SVM, SMO and I-SVM? What values can C take?

45. What are the conditions for updating the weight matrix in I-SVM for each added/dropped data point? How is the intercept term b updated?

46. What is meant by vector-migration in I-SVM? How do you keep track of migrations?

13.16.0.1   References

[AS03] Abe, S. (2003) Analysis of Multiclass Support Vector Machines, *International conference on Computational Intelligence for Modeling Control and Automation* (CIMCA), 385-396.

[AS04] Abe, S. (2004) Fuzzy LP-SVMs for multiclass problems, ESANN'2004 proceedings - *European symposium on artificial neural networks*, Bruges, Belgium, 429-434.

[AS05a] Abe, S. (2005) Variants of Support Vector Machines, in *Advances in pattern recognition: Support Vector Machines for pattern classification*, 129-154, Springer, London.

[AS05b] Abe, S. (2005) *Support vector machines for pattern classification*, Springer, NY.

[BY00] Bengio, Y. (2000) Gradient-based otpimisation of hyperparameters, *Neural Computation*, 12, 1889-1900.

[BE05] Bordes, A., Ertekin,S., Weston, J., Bottou, L. (2005) Fast kernel classifiers with online and active learning, *Journal of machine learning research*, 6, 1579-1619.

[BC98] Burges, C.J.C.(1998) A tutorial on support vector machines for pattern recognition, *Data mining and knowledge discovery*, 2, 121-167. citeseer.ist.psu.edu/burges98tutorial.html

[CH04] Cai, L., Hofmann, T. (2004) Hierarchical document categorization with support vector machines, CIKM'04, Nov 8-13, 1-10.

[CV05] Camastra, F., Verri, A.(2005) A novel kernel method for clustering, *IEEE trans. pattern anal. machine intelligence*, 27(5),801-805(ftp.disi.unige.it/person/CamastraF/wirn05.pdf).

[CP01] Cauwenberghs,G.,Poggio,T.(2001) Incremental and decremental support vector machine learning,*Advances in Neural Information Processing Systems*,12,409-415,MIT press.

[CL01] Chang,C.C.,Lin,CJ(2001) Libsvm:a library for svm. www.csie.ntu.edu.tw/~cjlin/libsvm

[CL01b] Chang,C.C., Lin,C.J.(2001) Training $\nu-$support vector classifiers: Theory and algorithms, *Neural Computation*, 13(9), 2119-2147.

[CV02] Chapelle, O., Vapnik, V., Bousquet, O., Mukherjee, S. (2002) Choosing multiple parameters for support vector machines, *Machine learning*, 46, 131-159 (oliver.chapelle.cc/).

[CV02] Chattamvelli, Rajan (2012) Data mining algorithms, Narosa, ND; Alpha science international, Oxford, UK.

[CL03] Chen, P., Lin,C.J., Schoelkopf,B.(2003) A tutorial on $\nu-$support vector machines. www.csie.ntu.edu.tw/~cjlin/papers/CheLinSch.pdf

[CB01] Collobert, R., Bengio, S. (2001) Support vector machines for large-scale regression problems, *Journal of machine learning research*, 1, 143-160.

[CV95] Cortes, C., Vapnik, V.(1995) Support vector networks, *Machine Learning*, 20, 273-297.

[CS00] Cristianini, N., Shawe-Taylor, J. (2000) *An introduction to support vector machines and other kernel-based learning methods*, Cambridge University press.

[DK03] Dong, J-X, Krzyzak, A., Suen, C.Y. (2002) A practical SMO algorithm, *Proc. of int. conference on pattern recognition* (www.cenparmi.concordia.ca/~jdong/)

[EL07] Elizondo,D., Lobato, J.J., Birkenhead, R.(2007) *A novel and efficient method for testing nonlinear separability*, LNCS 4668 (Artificial neural networks - ICANN 2007), 737-746, Springer.

[FC05] Fan, R., Chen, P., Lin, C. (2005) Working set selection using second order information for training support vector machines, *Journal of machine learning research*, 6, 1889-1918.

[FL02] Flake, G.W., Lawrence, S. (2002) Efficient SVM regression training with SMO, *Machine Learning*, 46, 271-290.

[FM02] Fung, G., Mangasarian, O.L. (2002) Incremental support vector machine classification, *Proceedings of 2nd SIAM international conference on data mining*, (Grossman, R.,*et.al.*(eds)), 247-260.

[GW00] Guyon, I., Weston, J., Barnhill, S., Vapnik, V. (2000) Gene selection for cancer classification using support vector machines, *Machine learning research*, 46, 389-422 (cbcl.mit.edu/publications/ps/optimal.pdf).

[GE03] Guyon, I., Elisseeff, A. (2003) An introduction to variable and feature selection, *Journal of machine learning research*, 3, 1157-1182.

[HR02] Herbrich, R. (2002) *Learning kernel classifiers, theory and algorithms*, MIT Press.

[HR05] Hongjian, F.,Ramamohana rao, K. (2005) A weighting scheme based on emerging patterns for Weighted Support Vector Machines, *Proceedings of IEEE international conference on granular computing*, 2, 435-440.

[HL02] Hsu, C.-W., Lin,C.-J. (2002) A comparison of methods for multi-class support vector machines, *IEEE Transactions on Neural Networks*, 13, 415-425.

[II06] Iria,J., Ireson, N., Ciravegna, F.(2006) An experimental study on boundary classification algorithms for information extraction using SVM, *Proceeding of the 11th conference of the European chapter of the association for computational linguistics*, April 2006.

[JT99] Joachims, T. (1999) Making large-scale SVM learning practical, in *Advances in Kernel methods*, Schoelkopf, B., *et.al.* (eds)(see [SB99]), MIT Press.

[JT02] Joachims, T. (2002) *Learning to classify text using support vector machines: Methods, theory and algorithms*, Kluwer.

[KR06] Kãrnà, T., Rossi,F., Lendasse,A. (2006) LS-SVM functional network for time series prediction, *European symposium on artificial neural networks*, ESANN'06, Belgium, 473-478.

[KS00] Keerthi,S.S, Shevade,S.K., Bhattacharyya,C., Murthy,K.R.K (2000) Improvements to Platt's SMO Algorithm for SVM Classifier Design, Technical Report CD-00-01, Dept. of Mechanical & Production Engineering, National University of Singapore (www.keerthis.com).

[KS03] Keerthi, S.S., Shevade, S.K. (2003) SMO algorithm for least squares SVM formulations, *Neural computation*, 15(2), 487-507 (www.keerthis.com).

[KG03] Kuss, M., Graepel, T.(2003) *The geometry of kernel canonical correlation analysis*, Technical report 108, Max Plank institute for biological cybernetics, (ftp://ftp.kyb.tuebingen.mpg.de/pub/mpi-memos/pdf/TR-108.pdf)

[LP02] Laskov, P. (2002) Feasible direction decomposition algorithm for training support vector machines, *Machine Learning*, 46, 315-349.

[LR94] Lewis, D.D., Ringuette, M. (1994) A comparison of two learning algorithms for text categorization, *Proceedings of the third annual symposium on document analysis and information retrieval*, 81-93.

[MJ09] Mercer, J.(1909) Functions of positive and negative type and their connection with the theory of integral equations, *Phil. trans. of the Royal society*-A, London, 415-446.

[MR01] Mika, S., Rätsch, G., Müller, K.R. (2001) A mathematical programming approach to the kernel Fisher algorithm, *Advances in neural information processing systems*, (Leen, T.K., Dietterich, T.G., Tresp, V. (eds)), MIT press, 591-597.

[MM06] Moguerza,J.M., Muñoz, A. (2006) Support vector machines with applications, *Statistical science*, 21(3), 322-336.

[NA07] Nagatani, T., Abe, S. (2007) Backward variable selection of support vector regressors by block deletion *IJCNN conf on neural networks*, 2117-2122 (www.lib.kobe-u.ac.jp/repository/90000478.pdf)

[OS08] Ojeda, F., Suykens, J.A.K, De Moor, B. (2008) Low rank updated LS-SVM classifiers for fast variable selection, *Neural networks*, 21, 437-449.

[OF97] Osuna, E.,Freund, R., Girosi, F. (1997) An improved training algorithm for support vector machines, *Proceedings of the IEEE NNSP* 97, 276-285.

[OG99] Osuna, E.E., Girosi, F.(1999) Reducing the run-time complexity in support vector machines, *Advances in Kernel methods*, 271-284.

[PM05] Pal, M., Mather, P. M. (2005) Support vector machines for classification in remote sensing, *International journal of remote sensing*, 26(5), 1007-1011.

[PJ99] Platt, J.C. (1998) Sequential minimal optimization: A fast algorithm for training support vector machines, Tech report 98-14, Microsoft Research, citeseer.nj.nec.comp/platt98sequential.html

[RA03] Rakotomamonjy, A. (2003) Variable selection using SVM-based criteria, *Journal of machine learning research*, 3, 1357-1370 (asi.insa-rouen.fr/enseignants/~arakotom/publications.html).

[RB07] Romero, E., Barrio, I., Belanche,L. (2007) Incremental and decremental learning for linear support vector machines, ICANN2007, *LNCS* 4668, 209-218.

[RO04] Ronneberger, O. (2004) LibSvmTL: A support vector machine template library, http://lmb.informatik.uni-freiburg.de/lmbsoft/libsvmtl/index.en.html

[RV05] Rossi,F., Villa, N. (2006) Support vector machine for functional data classification, *Neuro computing*, 69(7/9), 730-742, Elsevier.

[SB99] Schoelkopf,B., Burges,C.J.C., Smola,A.J. (1999) *Advances in kernel methods - support vector learning*, MIT Press, Cambridge, MA.

[SS00] Schoelkopf,B., Smola,A.J., Williamson, R.C., Bartlett,P.L.(2000) New support vector algorithms, *Neural computation*, 12, 1207-1245.

[SS02] Schoelkopf, B., Smola, A., (2002) *Learning with Kernels: Support Vector Machines, Regularization, Optimization, and Beyond*, MIT Press.

[SC04] Shawe-Taylor,J., Cristianini, N.(2004) *Kernel Methods for Pattern Analysis*, Cambridge University Press.

[SA01] Shilton, A. (2001) SVMHeavy: a support vector machine optimiser.

[SL07] Shilton, A., Lai, D.T.H. (2007) Iterative fuzzy support vector machine classification, *IEEE Fuzzy Systems Conference*, FUZZ-IEEE, 23-26 July 2007, 1-6.

[SP05] Shilton,A., Palaniswami,M., Ralph,D., Tsoi,A.C.(2005) Incremental training of support vector machines, *IEEE transactions on neural networks*, 16(1), 114-131.

[SS04] Smola, A.J., Schoelkopf, B. (2004) A tutorial on support vector regression, *Statistics and computing*, Springer, 14(3), 199-222.

[SB99] Smola,A.J. Bartlett,P. Schoelkopf,B., Schuurmans,C.(1999) *Advances in large margin classifiers*, MIT Press, Cambridge, MA.

[SP02] Sollich,P. (2002) Bayesian methods for support vector machines: evidence and predictive class probabilities, *Machine learning*, 46, 21-52.

[SA11] Statnikov, A.,Aliferis, C.F., Hardin, D.P., Guyon, I. (2011) *A gentle introduction to support vector machines in biomedicine*, World scientific., Singapore.

[SV99] Suykens, J.A.K., Vandewalle, J.(1999) Least squares support vector machine classifiers, *Neural processing letters*, 9(3), 293-300.

[SG02] Suykens, J., Gestel, T., de Brabanter, J., de Moor, B., Vandewalle, J., (2002) *Least Squares Support Vector Machines*, World Scientific, Singapore (ISBN: 9812381511).

[SB02] Suykens,J.A.K., Brabanter, J.D., Lukas,L., Vandewalle,J.(2002) Weighted least squares support vector machines: robustness and sparse approximation, *Neuro Computing*, 48, 85-105.

[SL99] Suykens,J.A.K., Lukas,L.,Van Dooran,P., De Moor,B. Vandewalle,J.(1999) Least squares support vector classifiers: A large scale algorithm, *Euro conference on circuit theory and design*, Aug/Sep 99, 839-842.

[SV02] Suykens,J.A.K., Van Gestel, T., Brabanter, J.D., De Moor, B., Vandewalle,J.(2002) Least squares support vector machines, *World Scientific*, Singapore.

[TA06] Torii, Y., Abe,S.(2006) Fast training of Linear Programming Support Vector Machines using decomposition techniques, *Proc. Second IAPR TC3 Workshop on Artificial Neural Networks in Pattern Recognition* (ANNPR 2006), 165-176, Gunzburg, Germany, August/September 2006, (LNCS 4087).

[TY03] Tsang, E.C.C., Yeung, D.S., Chan, P.P.K.(2003) Fuzzy support vector machines for solving two-class problems, *International Conference on Machine Learning and Cybernetics*, 1080- 1083.

[TA03] Tsujinishi, D., Abe, S. (2003) Fuzzy least squares support vector machines, *Proceedings of the international joint conference on neural networks*, 2, 1599-1604.

[VL63] Vapnik, V., Lerner,A.(1963) Pattern recognition using generalized portrait method, *Automation and Remote Control*, v24.

[VC64] Vapnik, V., Chervonenkis, A.Y. (1964) A note on a class of perceptron, *Automation and Remote Control*, 25, 103-109.

[VC91] Vapnik, V., Chervonenkis, A.Y. (1991) The necessary and sufficient conditions for consistency in the empirical risk minimisation method, *Pattern recognition and image analysis*, 1(3), 283-305.

[VV95] Vapnik, V.(1995) *The Nature of Statistical Learning Theory*, Springer Verlag.

[VV98] Vapnik, V.(1998) *Statistical Learning Theory*, Wiley, NY.

[VV99] Vapnik, V. (1999) Three remarks on the support vector method of function estimation, in *Advances in Kernel methods: support vector learning*, (Schoelkopf, B., Burges, C.J.C., Smola, A.J. (eds)), MIT Press, 25-41.

[VV99b] Vapnik, V. (1999) An overview of statistical learning theory, *IEEE transactions on neural networks*, 10(5), 988-999.

[WC07] Wang, T.Y., Chang, H.M.(2007). Fuzzy support vector machine for multi-class text categorization, *Information Processing and Management: an International Journal archive*, 43(4),914-929.

[WS06] Wang,L.,Sun,S., Zhang,K.(2006) A fast approximate algorithm for training L1-SVMs in primal space, *Neurocomputing*, 70, 7-9, 1554-1560.

[WY04] Wang, M., Yang, J. *et.al.* (2004) Weighted-support vector machines for predicting membrane protein types based on pseudo amino acid composition, *Protein engineering, design and selection*, Oxford University press, 1-28.

[WZ08] Wu, S., Zou, H. (2008) Structured variable selection in support vector machines, *Electronic journal of statistics*, 2, 103-117 (www.stat.umn.edu/~hzou/Papers/EJS-svm.pdf).

[XW00] Xiao,R., Wang,J., Zhang,F.(2000) An approach to incremental SVM learning algorithm, *Proc of 12th international conference on tools with artificial intelligence*, 268-273.

[ZS06] Zhang, X.F., Shen, L.S.(2006) Application of weighted SVM on the classification and recognition of tongue images, *Chinese journal of biomedical engineering*, 25(2), 230-233.

[ZZ07] Zhu, J., Zou, H. (2007) Variable selection for linear support vector machines, *Studies in computational intelligence - Trends in neural computing*, Springer, 35, 35-39.

14
Latent Semantic Indexing

Chapter objectives

- Introduce Vector-Space Models (VSM)

- Distinguish between VSM and LSI

- Explain geometric interpretations of LSI

- Discuss advantages and disadvantages of LSI

- Introduce singular value decomposition and its applications

- Introduce LSI query and pseudo-documents

- Discuss latent semantic clustering

- Applications of LSI

14.1 Vector Space Models

Vector spaces are used in various fields of applied sciences. It finds extensive applications in mathematics, physics, engineering and multivariate statistics [HS85]. Each vector has a direction and a magnitude with respect to a set of fixed coordinate axes. Hence each vector is denoted by a set of as many components (also called coordinates) as there are number of axes. For instance, a vector in 2D has 2 components, and one in n-dimensions (n-D) has n components. Some of the components of a vector could also be zeros. If the second component is zero, it denotes a vector parallel to the X-axis. Thus (5,0) denotes a row vector in the direction of X-axis and (0,8) denotes a row vector along the Y-axis. Similarly, a vector in 3D with the third component as zero lies in the XY plane, and with the second component as zero lies in the XZ plane. Thus (0,5,10) is a vector in the YZ plane, and (0,0,6) is a vector parallel to the Z-axis. In the above examples we have represented them as row vectors, but they are regarded as *column vectors* in information Retrieval (IR) and related applications. Vectors used in data mining and IR typically have a large number of components (of the order of thousands, or even millions), many of which could be zeros. For example, document collections are represented as Euclidean vectors in the term space (with as many components as there are distinct terms in the collection). Vectors in IR and related applications are all located in the positive quadrant due to containment relationships. Queries are user requests to retrieve matching documents from a large collection. Each user query gets mapped to the term-space (see below) as a vector. All vectors that are in the neighbourhood (in the geometric sense) of the mapped query can then be retrieved easily using a similarity metric. Document similarities (document with document), term similarities (term with term), and containment relationships (term within document) can all be compared using the VSM, which uses the *bag-of-words* representation. The containments of terms in text documents are captured in a term-by-document (TD) matrix described below.

14.1.1   Term-by-Document Matrix (TD matrix)

Information content of document collections are captured as real values (Boolean occurrence flags, counts, TF-IDF etc) in the TD matrix (say W). It is so called because it originated in text retrieval. But the same name is used in audio, video, image, and multimedia (AVIM) retrieval systems. The simplest possible mapping results in a Boolean matrix in which a 1 at $(i,j)^{th}$ position indicates the presence, and a 0 indicates the absence of ith term in j^{th} document (thus document vectors could contain many zeros if many terms are absent in it). By convention, terms are indexed along the rows, and documents along the columns as shown in figure 14.1. Thus w_{ij} summarises the occurrence of term i in document j. Two almost identical document vectors (columns) indicate that the corresponding documents have many things in common. In other words, correlation among the column vectors will be high if the document contents have large overlaps, and low otherwise. As an example, in table 14.1, Corr(Doc1,Doc3)=0 because $\sum x_i y_i$=1x0+0x2+2x0+0x1=0 showing that documents 1 and 3 have too little content similarity. This is evident from the bipartite representation in figure 14.1.

Two almost identical rows indicate affinity of co-occurrence of terms in these documents. For AVIM data, the columns represent files and rows represent features, correlation for a neighborhood (spatial or temporal), sub-images, embedded text, etc. Unless otherwise stated, the discussion is confined to text documents in the rest of the chapter. For text data, we will remove common words like 'a', 'the', 'is', 'are', 'with', etc that occur in almost all documents because these can unnecessarily increase the size of our TD matrix. Similarly, numbers present in documents are judiciously chosen as they may or may not convey meaning[1]. This process is called *noise removal.*

This matrix is quite large and sparse in typical IR applications that have hundreds of thousands of documents, and thousands of terms. It is seldom a square matrix. Working with such a large dimensionality is inefficient and cumbersome (the memory storage of such large matrices could be reduced by utilising sparsity of the TD matrix). The information contained in it can be captured into a bipartite graph (T,D) where T_i is connected to D_j if term i is present in document j. This graph need not be connected (in which case it is called a bipartite forest). But the truncated Singular Value Decomposition (SVD) of the TD matrix comes to the rescue, as it maps the information content of documents to a feature space of much smaller dimensionality (see §14.3 in page 14-9).

The Boolean matrix representation discussed above just checks the presence or absence of a keyword (for text data) or a feature (for AVIM data). Individual terms and features can be assigned weights depending upon their occurrence counts (number of times they occur in a document), as this carries more information. For example, the element at row 1, column 2 $(w_{1,2})$ is 3, which means that the term-1 occurs 3 times in document-2. If our aim is just to retrieve all documents containing a term, it does not matter whether we use a Boolean matrix

---

[1]See [BC85] for a list of around 500 most frequent non-discriminating words in English.

Table 14.1: Structure of a TD matrix where rows are indexed by terms and columns are indexed by documents. Entries denote occurrence of terms in documents

	Doc_1	Doc_2	Doc_3
$Term_1$	1	3	0
$Term_2$	0	1	2
$Term_3$	2	1	0
$Term_4$	0	0	1

Table 14.2: Bipartite representation

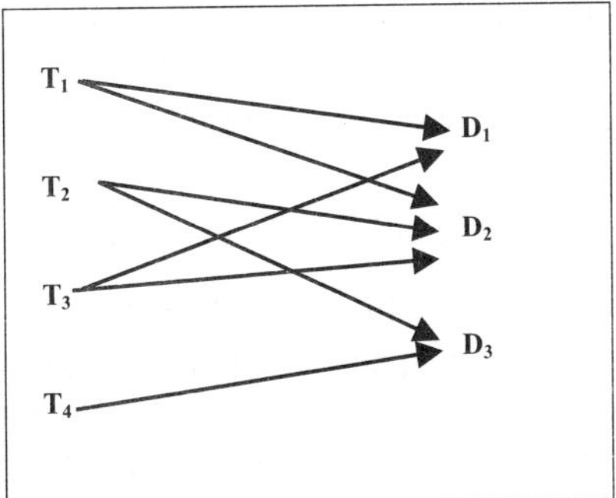

or occurrence count matrix. The occurrence count is useful when we need to rank the retrieved documents in a high-to-low relevance order. Similarly, the count is important and useful in clustering applications, data de-duplication, and image-steganography as it contributes more to similarity or dissimilarity metrics.

14.1.2   Textual IR

Information Retrieval (IR) has been in use for centuries. Catalogs and other indexing systems were used by libraries, hospitals and educational institutions even before the advent of computers. Text based IR is like searching a library catalog using subject keywords, author names, publisher name or words in the title. One can also search a library catalog using the call numbers (LCC) which are structured as a tree.[2] For example "QA 76.9 D343" could represent a data mining category. Each prefix of this code (from left to right) moves us towards more and more focused (narrow) category of topics.

IR requests often originate from user queries. Users can use one of the textual keywords, or one of the image/multimedia clips for IR. Some common keywords may produce a multitude of matching records (depending upon the size of the database). For example, a 'subject keyword' (eg:physics) or 'title name' (eg:database) used as keys for IR in search engines can result in millions of matches. Multiple keywords may be used to eliminate unwanted records and filter more narrowly focused content. For instance, the name of an actor and the year of release of a movie can be used together to find an unknown movie name (which may not be unique if an actor stars in two or more movies per year). Similarly, a patient name, symptom details, and approximate date of visit may be used together by a hospital to retrieve a past medical record.

Multimedia IR systems use images and sub-images, parts of audio/video data, and clips of animation sequences or other features (eg: color of an image, occlusion information in graphics) to retrieve matches. One notable difference in image-based IR is that the search keys (in user queries) are either exact sub-images or reduced-size images (like thumbnail of an image). Em-

---

[2]The Library of Congress Classification (LCC) is a coding scheme developed by the US Library of Congress to classify books, journals, media files, bound volumes etc. In LCC, Q=Science, QA=Mathematics, QA75, QA76=Computer Science.

bedded text (in images) and imaging location (eg: internal organ, brain) may be used together to retrieve medical images (like MRI, Doppler and CAT scans).

Document locations, data formats, access mechanisms, communication protocols etc are usually hidden from users (humans, crawlers, knobots, etc). As most of the IR requests are text-based, we will confine ourselves to keyword based IR in the rest of the chapter. See Letsche & Berry [LB97], Grossman & Frieder [GF04], Jessup & Martin [JM05] for further discussions.

14.1.3 Geometric Interpretation

Vector space models, introduced by G. Salton[3] [SW75], are built upon an intuitive notion of vectors in Euclidean space. Each document in the input space is mapped as a vector in the term-space (where each axis represents an independent term). Generalised VSM models use a linear combination of input terms as feature axes, thereby improving concept matching. LSA model

Figure 14.1: Adjacent vectors in semantic space, and a projected query for 'medical database'

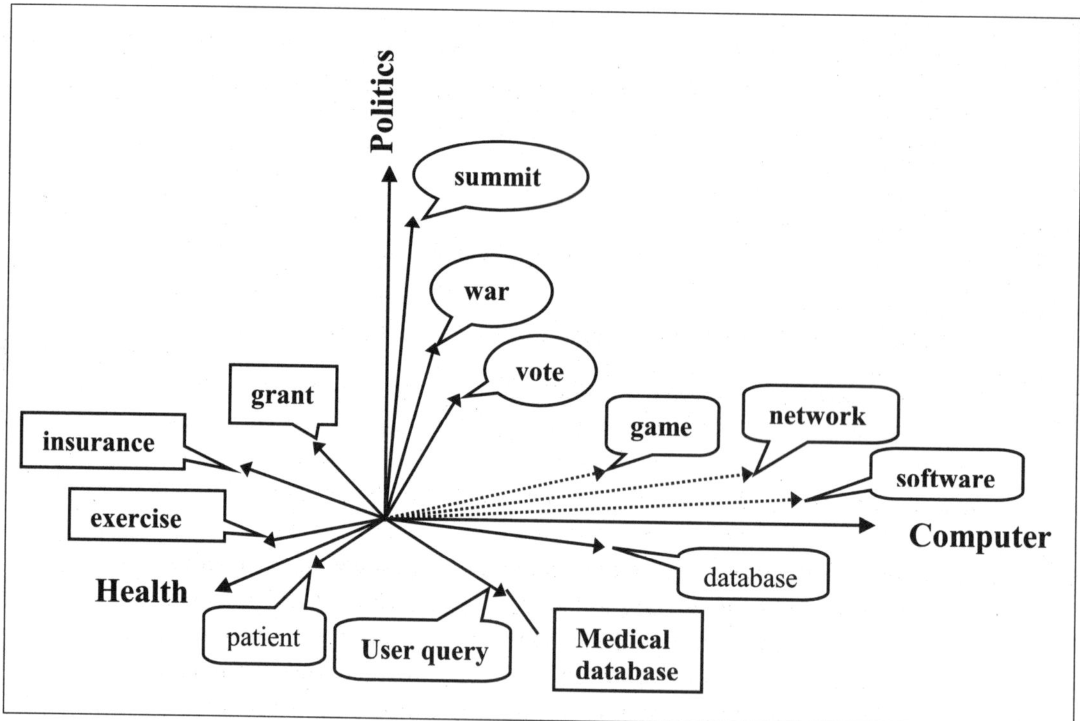

extends this idea by mapping input documents into a low dimensional feature space using reduced rank matrix factorisation methods, where each axis is a conceptual keyword. LSA removes high-frequency words (in text data), patterns, common backgrounds, etc (in AVIM data) with intent to improve the discriminatory power of the rest of the words or patterns. Similarly, those words or patterns that occur in at-most one document are also removed, as they can discriminate at most one document. This step is called downsizing the input space.

---

[3]Salton *et.al.* introduced it in citation searching. It was known earlier than Salton in abstract services [LH58].

Many other factorisation techniques like LU, QR, semi-discrete decomposition (SDD), Centroid and Generalised Centroid Decompositions (CD and GCD) [CF02] etc exist in Mathematics and numerical computing. But the SVD is the recommended choice in IR because it decomposes the available information into document-space and term-space in the transformed domain where each axis is weighted according to the relative importance of terms as evidenced by the diagonal elements of the singular value matrix (S) of SVD. In other words, two very similar documents in the input space are mapped to two vectors in close proximity in the feature space by the SVD.

14.2 Latent Semantic Analysis

Data in text format on the Web are increasing at an enormous rate. Majority of text data on the web are in HTML and XML formats. Search engines play a prominent role in retrieving relevance ranked documents from voluminous data to web users using a *query-results* paradigm. Most users query (request) for documents using abbreviations (acronyms), keywords or phrases in various languages. Search engines parse the keywords (some of them even remove duplicate words and fixes upper/lower case mismatches) and use it in an optimal way (using any user-supplied Boolean operators) to filter out the most precise documents from the web in a high-to-low relevance order. This method is called *lexical IR*. It is important to understand that the LSA is used by the document retrieval engine, and the relevance ranking of retrieved documents are done by the page-rank algorithm or HITS algorithm.

The Latent Semantic Analysis (LSA) works by establishing a many-to-one mapping from heterogeneous document collections in the input space, to feature vectors in a reduced rank *latent* semantic space (also called LSI space). This is accomplished by truncating the high to low ordered singular values of the TD matrix using a threshold (see below). This method called truncated SVD maps the input space to a semantic space of much less dimensionality. As we are interested in locating entities (documents, streams, images etc) that are approximately similar in content to key entities (extracted from a query), a proper truncation will remove 'noise' and 'impurities', and retain the essential semantic content of the data [JM05]. It has been shown that LSA even retrieves documents that have no lexical words in common with the query keywords. This is one of the strengths of *content similarity matching* provided by the reduced rank method. This aspect of LSA has been exploited in duplicate documents identification, alias documents (like copied images, papers, audio, video, etc) that have almost identical contents.

LSA has its origin in text data (where it is called LSI)[4]. It has been extended to other types of data, including image, audio, video, animation and multimedia data. The TD-matrix for each type of media is different. As mentioned in §14.1.1(pp.14-2), it is easy to form for text data. It has applications in automatic document indexing, topic segmentation, relevance ranking, uni-language and cross-language IR [LL90], inter-language translations, focused crawling [AK05], document ownership discovery, image annotations, data de-duplication, speaker verification, relevance scoring in electronic communications, and clustering documents using similarity measures [CT03],[KA04].

14.2.1 Steps in LSA

There are three main steps in LSA, the first two of which are executed only once for static document corpora. They are as follows:– (i) Form the TD matrix from cleansed input data (ii)

---

[4]This is why the information matrix is called Term-by-Document (TD) matrix

SVD factorise the TD matrix (iii) Map the user query to the LSI space, and filter all matching records in its neighbourhood using a similarity metric.

The TD matrix is formed from the document corpus. In the simplest case, a Boolean flag is used to keep track of the occurrence or non-occurrence of a term in a document. The second step is discussed in §14.3, and it need be carried out only once (if the document corpus remains static). The third step involves query mapping and retrieval using a cosine similarity as described in §14.4.

14.2.2 Characteristics of LSA

1. Solid mathematical footing
   The LSI algorithm uses truncated SVD to map the input space to a reduced rank feature space. Although many SVD factorisations are possible, the restrictions that S matrix should have diagonal elements in non-increasing[5] order, and off-diagonal elements as zeros, and number of diagonal elements of S is less than the rank of W matrix make the factorisation unique. Query matching involves inner products of query vector with feature vectors. Approximate LSI algorithm works on 'largest eigen value' generation algorithms (eg: Lanczos methods[6]). Hence LSI is built upon solid mathematical principles.

2. Simple run-time computations
   Most of the computations take place during the preprocessing stage. Results returned by the SVD algorithm (U, S, V matrices, §14.3) are stored for IR tasks. At run-time, each user query is first converted into a vector (using the same weighting scheme used in forming the TD matrix) in the input space. This vector is mapped into the semantic space (see 14.4 in page 14-12) to get the latent query vector, also called a pseudo-document vector. Most similar documents are easily found using simple algebraic operations at run time. Hence thousands of user queries can be parallely executed in real-time mode.

3. Reduced dimensionality
   As the feature space is much low-dimensional than the input space, the space complexity is substantially reduced. For instance, if the original TD matrix is n x m (m documents and maximum n terms), the total space complexity is $O(mn)$. In VSM (that uses a normal SVD) this is n * r + r * r + r * m = r*(m+n+r), where r is the rank of the TD matrix (which is $\leq \min(m,n)$). In LSI, we approximate the solution using a variance cutoff threshold (truncated SVD), and store only the r' nonzero diagonal elements of S matrix as a *vector*. This is called rank reduction, and it results in a reduced space complexity of $O(r'(m + n + 1))$ (because S is stored as a vector). Rank reduction techniques are useful to speedup the query execution, and to reduce 'noise' and 'redundancy' in the data.

14.2.3 Advantages of LSA

1. Better performance
   Larger the number and size of the keywords, better is the performance of LSI. This is in sharp contrast to the conventional (keyword-based) IR techniques that take more execution time, and result in less accuracy when a large number of keywords are present.

---

[5]decreasing is different from non-increasing because a non-increasing sequence can have duplicate elements. eg: 10,5,2,1 is a strictly decreasing sequence, but 5, 4, 4, 3, 2, 2, 1 is non-increasing.

[6]The Lanczos algorithm is available for download from www.netlib.org/lanczos/index.html

2. Reuses factored matrices
   LSI uses a mathematical technique called low rank orthogonal decomposition of the TD matrix. Once the TD matrix has been factored using SVD, all subsequent queries can be answered using factored matrices, as long as the document contents remain static. Several updating algorithms are available when document contents change over time, so that an expensive re-factorisation of the TD matrix can be avoided. This is an important issue in web information retrieval because most of the web documents continuously evolve over time.

3. Generalisation capability
   LSI induces a partial structure in unstructured data. It has better precision and recall than other vector space models (a major contender of LSI is the principal component analysis based covariance matrix analysis [KA04] which has advantages in distributed IR). In natural language data, it tries to remedy the *synonymy* (§14.3) in a better way.

4. Query execution speed
   LSI facilitates relevance feedback searching and hierarchical searching. The factored matrices returned by SVD can be saved and reused in query sessions, as long as the document corpora is static. LSI queries can be considerably speeded up using special techniques.

5. Ability to parallelise
   LSI can be parallelised in multiple ways. First of all, the SVD factorisation can be obtained using a parallel algorithm ([SI10]). When the input matrix is large and sparse, Rajamanickam [RS09] shows how to make it more efficient. We could pre-compute the needed matrices in parallel (see below). Multiple queries are then parallely executed, using matrix-vector multiplications. This ability is extremely essential in search engines that cater to the information requests of millions of users simultaneously.

6. Exploits labeled data
   LSA can fully exploit both data labels and document labels. Labeled training data are helpful to automatically assign cluster labels in latent semantic clustering (LSC) (pp.14-19), and in gene expression analysis. As an example, consider an LSA based system to automatically assign reviewers to a large number of papers received for an international conference, where each paper is tagged with a set of (>2) keywords and phrases. Each reviewer is labeled with a set of narrowly focused keywords that match his/her area of interest. There could exist multiple persons working in the same field (like 'green computing', 'latent semantic clustering', etc). Here the TD matrix rows are the reviewer's fields of interest (tagged by the reviewer name), and columns are the keywords and phrases in the documents received. An assumption here is that research areas of reviewers (rows) are present in at least one paper received; and that the keywords and phrases present in the documents semantically match with at least one reviewer field. LSI can still match the papers with multiple reviewers using semantic similarity even when some of the keywords (or abbreviations) do not have exact matches. Document labels are useful to detect and delete duplicate documents in large document corpora using the content, and to bookmark prior information retrieval queries.

7. Useful for clustering
   LSA can be used for document clustering, term clustering as also feature visualisation in subspaces. Document clusters are collection of documents that are grouped together using their semantic content. Term clusters on the other hand group together terms that have

semantic similarity. LSI thus becomes a sub-dimensional pattern discovery tool; rather than the conventional goal-oriented searching method.

8. Miscellaneous uses
   LSI based methods are useful for 'noise reduction' in text, audio, video and multimedia data. In addition, if data are sparse, LSI can be used for data compression, dimensionality reduction and outlier detection. This is achieved by using an appropriate cutoff threshold for truncated SVD. For dense data, a kernelised LSA may be used to impose a limited amount of sparsity in the feature sub-spaces.

14.2.4   Disadvantages of LSA

1. Computational complexity
   The TD matrix used in LSI is often very large in size. SVD takes the TD matrix as input and produces two orthogonal matrices (U, V), and a diagonal matrix (S) as output. The vectorised user query should be projected to the feature space (see below), and matched to other feature vectors to identify nearby vectors using a similarity metric (like cosine similarity), which involves additional computations.

2. The polysemy problem
   Words with multiple meanings are quite common in natural languages (more so in German than in English). In addition, *acronym polysemy* (polycronym)[7] results from using the same abbreviation for multiple phrases (eg: LSI means 'Latent Semantic Indexing', 'Large Scale Integrated' (as in VLSI), 'Large Scale Integration' (in manufacturing); NASA can mean 'National Aeronautics and Space Administration', 'National Auto Sports Association', 'National Association of Screening Agencies' etc). When the abbreviated letters itself is a valid word in a natural language, we call it acronym-word (acronym-noun, acronym-verb, etc) polysemy (lexicronym[8]). As examples, CAT (Computer Assisted Tomography), MAN (Metropolitan Area Network), MAGIC (Map And Geographic Information Center, see www.abbreviations.com/abbreviations/M/26), READ (Rapid Engineering Application Development), MARS (Machine Assisted Reference Services (www.abbreviations.com/abbreviations/M/43, www.definition-of.com/MARS), etc are lexicronyms in English. Similar examples abound in other languages like German. Lexicronyms can be uni-lingual or multi-lingual. If the abbreviated word is a valid word in multiple natural languages, we call it multi-lingual lexicronym (eg: WAR (which is a valid word in English, German, etc), PAN (English, Spanish), etc which are also acronyms). Multi-lingual lexicronyms (multilexicronym) are the most problematic because those when used as keyword(s) could retrieve documents from multiple languages, which includes dictionary words and acronyms from each such language. A related word is lexichronym, which is derived from 'chrono' which means 'order'. The abbreviated letters could also be a noun:– CLARA (CLairemont Amateur Radio Association), LISA (Laser Interferometer Space Antenna), ANN (Artificial Neural Networks) etc. called acronym-noun (or acronym-name) polysemy, which can be disambiguated to an extent using capitalisation, if any, supplied by the user. These are difficult IR issues to be tackled when user query contains one (or more) polysems.

---

[7]See the article by Jon Newman (www.verbatimmag.com/WordWords.html) for an alternate definition.

[8]People have a tendency to weave acronyms around natural words in various languages for ease of remembrance. This results in lexicronyms. A related word is lexichronym (with stem chrono) which are words formed by concatenating alternate letters of a bigger word. For instance, sweats=set+was is an English lexichronym.

3. Scalability to voluminous data
   As the queries are transformed into pseudo-document vectors using the pre-multiplier matrix U, and the (inverse) diagonal matrix S (see §14.4), these values are needed to execute each query. The vector space models do not scale up well to large document collections, and are often restricted to vertical search engines, portals and digital libraries that utilise a hierarchical retrieval, or uses relevance feedback to filter out matches in multiple steps.

4. Dynamic data updates
   Special updating algorithms are needed to avoid a new SVD factorisation when the document contents change over time. This is more of a problem in web-based retrievals, and in refined weather forecasting models where some of the atmospheric conditions vary over time continuously. Faster algorithms for SVD updating can be found in Tougas & Stern [TS06].

5. Phonetic differences of words
   LSI does not consider phonetic interpretation of words. This is more of a limitation than a disadvantage. But it is important in multilingual searches where the pronunciation of a word can disambiguate it wrt the languages. As examples, the word 'kind' is pronounced in English as 'kaind' whereas it is pronounced in German as 'kin-d' (as in kindergarten) and means 'child' (kin-der=children). Similarly, the word 'once' is pronounced in English as 'won-s' and in Spanish as 'on-se'. These problems can be alleviated using heuristic techniques that automatically identify the target language by analysing other keywords (if any) supplied by the user.

6. Miscellaneous
   For image, audio, video, animation and multimedia data, the TD matrix formation may not be fully automatic due to feature overlaps, spatial or temporal discontinuities in some features.

In addition, extracting some features may require domain transformations (eg: from time domain to frequency domain). These involve additional computational overhead (eg: FFT, DCT, DWT, etc in audio, image and multimedia data).

14.3   Singular Value Decomposition (SVD)

The SVD is a popular technique in Mathematics, invented independently by Beltrami (1873) [BE73], and Jordan (1875) [JC75]for square matrices, and extended by Eckart & Young (1936)[EY36] for rectangular matrices. It has applications in statistics, information retrieval, image processing, VLSI design, audio watermarking, steganography, gene expression analysis, and weather prediction, among many other fields [AM04]. Orthogonal factorisations preserve affinity information and norms, which are desirable properties in IR and topic segmentations. In numerical weather prediction using Lanczos method, one looks for quickly changing perturbation subspaces over time series frames of atmospheric data. The analogue of TD matrix has atmospheric variables as rows, and time frames as columns. Because the singular values can be obtained in monotonically decreasing order, largest singular vectors corresponding to a few large singular values are further processed through nonlinear models (eg: neural networks) to obtain ensemble weather forecasts. Statistical applications of SVD include principal component regression, multi-collinearity detection, etc [HS85].

Table 14.3: Summary table of some English words ending in -onym

Name	Meaning	Example
antonym	words of opposite meaning	(small x large), (left x right)
eponym	discovery known by discoverer's name	Alzheimer's disease, Parkinson's disease
hyperonym	*is-a* relationship	car→automobile → land vehicle
homonym	words spelt alike	right (correct, antonym of left)
meronym	*whole-part* relationship	word → line → paragraph → document
metonym	colloquial phrase	Bollywood, Silicon Valley
pseudonym	alias, duplicate human name	Student, Shakesphere
synonym	words of same meaning	petrol, gasoline, benzene

Bollywood stands for Indian film industry, and *Silicon valley* stands for California high-tech companies. Other examples are 'Washington' (US govt), 'Kremlin' (Russian govt), 'Wall street' (US financial market), etc

It was introduced for information retrieval during the 1990's by Deerwester,Dumais, *et.al.* [DD90], Berry, Dumais *et.al.* , [BD95], Landauer & Littman[LL90] among others [HS01], [JM05]. Let W be a nontrivial[9] rectangular (real or complex) matrix. The SVD is a factorisation of W into the form $W_{n \times m} = U_{n \times m} \ S_{m \times m} \ V_{m \times n} \Rightarrow$ (1) (fig 14.2), where the columns of U and V are orthonormal (U U' = I_n and V V' = I_m), and S is a diagonal matrix with *singular values* (SV) of W along the main diagonal (diagonal elements of S are the positive square roots of (real) eigen values of W'W. If the matrix W is positive semi-definite, the eigen values are all non-negative). These values contain hidden dimensionality and weighting information about the documents and terms. Symbolically we represent it as W=

$$
\begin{pmatrix}
u_{11} & u_{12} & \cdots & u_{1m} \\
u_{21} & u_{22} & \cdots & u_{2m} \\
| & | & \ddots & | \\
u_{n1} & u_{n2} & \cdots & u_{nm}
\end{pmatrix}
\begin{pmatrix}
s_{11} & 0 & \cdots & 0 \\
0 & s_{22} & \cdots & 0 \\
\vdots & \vdots & \ddots & \vdots \\
0 & 0 & \cdots & s_{mm}
\end{pmatrix}
\begin{pmatrix}
v_{11} & v_{12} & -- & v_{1m} \\
v_{21} & v_{22} & -- & v_{2m} \\
-- & -- & \ddots & -- \\
v_{m1} & v_{m2} & \cdots & v_{mm}
\end{pmatrix}
$$

$$\text{(left singular vectors)} \quad \text{(singular values)} \quad \text{(right singular vectors)}$$

This is called the *full-SVD* because the original dimensions are maintained. The columns of U (respectively V) are called left (right) *singular vectors*, and satisfy $u_i' u_j = 0 = v_i' v_j$. As the order of the documents (columns of TD matrix) and the order of the terms (rows of TD matrix) are unimportant, it is always possible to get the SVD with diagonal elements of S in decreasing order (see figure 14.2). Symbolically, $s_{11} \geq s_{22} \geq \cdots \geq s_{mm}$.

---

[9]By nontrivial, we mean that it is neither null, nor identity matrix and it has at least 2 rows and columns

The English suffix *onym* is derived from the Greek word *onoma*, which means *name*. There are many words in English that have onym as the suffix. Some of them are acronym, antonym, hyperonym (hyponym), meronym, pseudonym and synonym. Words or phrases in natural languages with the same meaning are called *synonyms* (*syn* = (together, same)).[a] Examples of synonym words in English are (student, pupil), (plane, aircraft, jet), (petrol, gasoline, benzene), (drug, medicine, medication). Synonyms can be further sub-categorised as noun, verb, adjective, adverb synonyms, etc. Above examples are noun-synonyms. Examples of verb-synonyms are – (walk, stroll), (write, scribe), (surf, browse); adjective-synonyms are (pretty,beautiful), (smart,clever); and adverb-synonyms are (quick, fast), (hard, difficult, arduous) etc. The state of being a synonym is called *synonymy* (adj is synonymous). But 'pupil' also means different things in other contexts – It is the central black or blue circular part of the eye, in music it is a Filipino rock-band, and it is used in Optics as the 'Entrance pupil' and 'Exit pupil'. Similarly, 'plane' has different context specific meanings – a leveled surface (eg: inclined plane), a flat figure in geometry, a carpenter tool to level-off edges, and a kind of tree. Quasi-synonyms are not always synonyms. This is because one of the synonyms may be a polysem (see below). The 'plane' in the above example is a polysem. Hence you cannot substitute an occurrence of 'plane' by an 'aircraft' everywhere it occurs in a text. Other examples of quasi-synonyms are 'drug', 'pupil'– because these are polysems. Synonyms can also be approximate in meaning as in (big, large), (angry, annoyed, irate, irritated), which are important issues in IR queries. Noun synonyms are more problematic in IR than the others discussed above, because the rest of them seldom appear in queries of experienced surfers.

Words that have multiple meanings are called *polysems*, which is derived from the Greek word *poly*, which means 'many'. The state of being a polysem is called *polysemy*. As examples in English, 'mouse' is a pointing device in computing, and a rodent elsewhere. Polysems can also be categorised as noun-polysems, verb-polysems, etc. The mouse-example above is a noun-polysem. There are many words in English that assume different meanings when used as nouns or verbs/adverbs. These are called noun-verb polysems. For instance, a 'well' when used as a noun means a water-well or oil-well, but when used as a verb/adverb means 'good' or 'OK'. Other examples of noun-verb polysems are 'drink', 'catch', 'cut', 'second', 'watch'.

Pseudonyms (pseudo=false) are alternate names for any object including humans used to hide the actual identity. Actors, stage artists, authors (usually of literary work), performers, etc use it. Homophones are cause for concern in voice-based IR. Sometimes, users also mistakenly use the wrong homophone (especially people with non-English mother-tongues, and English learners) because they are unsure of the correct spelling. This in turn results in semantically incorrect document retrievals. The same argument holds in other languages.

Synonymy and polysemy are intra-language issues. Synonymy results in poor recall and polysemy leads to poor precision. A synonym list (per language) can be used to alleviate the synonymy problem. In addition, various world languages have common words. For example, 'tag' in German (pronounced taa'g) means day, and 'kind' (kin'd) means child, 'Mar' in Greek means king, and 'pan' in Spanish means bread. In IR context, synonymy and polysemy are more important than the others mentioned in table 14.3. Whereas the previous generation pattern-matching, and indexing techniques perform poorly in these situations, the new generation LSI and covariance matrix analysis [KA04] methods are quite successful in tackling it.

---

[a]see the English synonym dictionary: http://dico.isc.cnrs.fr/dico/en/search

Dimensionality is of the order of thousands to millions in most data mining applications. For instance, search engines (like Google, yahoo) work with millions of records to filter out documents matching a user query. But the singular values (diagonal elements of S) will tail off rapidly in magnitude (they will eventually become negligible). Typical truncation sizes when dimensionality is very large are between 100 and 300 (it could be as small as 1.2% to 4% of the size in applications with $1000<n\leq10000$ [1.2% when n is towards its upper limit 10000, and 4% when n is near 1000 so that when the size is 1200, we will retain around 48 largest eigen values, and when the size is 9600, we will retain around $\lceil115.2\rceil=116$ largest eigen values], 4% to 10% of the size in applications with $100<n\leq1000$, 10% to 20% of the size in small sized applications with $40<n\leq100$, and 20% to 33% when $n\leq40$ where fractional numbers are scaled to the next highest integer). Alternatively retain the largest k eigen values such that $s_{kk}/\sum_{j=1}^{k} s_{jj}$ is less than a threshold.

They represent the relative importance of terms in the LSI space as larger singular values capture the essential features. Hence we can discard the singular values that are below a predefined threshold, and drop the corresponding number of (last) rows (columns) of U (V). This is called the *reduced rank method*. Notice the analogy with support values (nonzero Lagrangian multipliers) in SVM, where we ignored all data points whose support values are zeros (those that lie beyond the SVH on both sides in binary SVM). In LSA, we could ignore all projected vectors in the feature space whose singular values are zero or negligible, as they represent the *noise* in the input data. High-level pseudocode given below combines both full and reduced rank SVD. We get the full SVD if we ignore all entries in square brackets (eg: ignoring '[largest p]' tells us to 'Find the positive square roots of eigen values', etc)

14.4   LSI Query

Text-based IR uses one or more keywords. More keywords can improve the retrieval precision. For instance, 'planet mercury' will restrict the search for 'mercury' in the celestial domain (as there are companies, gadgets, elements etc called mercury). A user who sits in front of a terminal and submits a query is unaware of the intricate mathematical techniques used behind the scenes. It is the *retrieval engine* that interprets the user query as lexical or LSI-based, and applies different techniques in each case.

14.4.1   Query Processing

A text-based query is first parsed for keywords and connectives (some IR systems allow the user to use + (AND) operator to combine keywords, | (OR) operator to choose one among many keywords etc). The extracted keywords are converted into a query vector (q) using the same order of terms chosen for the TD matrix (same attribute space) and same weighting scheme used, if any (see below). As the order of terms (rows) in the TD matrix is unimportant, we could rearrange the rows such that the terms are in alphabetical order. As the order of keywords (without connective operators like '+', '|', etc) in a query is also unimportant, we could rearrange them in alphabetical order. This rearrangement may not work when order is important. As examples, "object oriented" and "oriented object" retrieve semantically different documents, as does "insurance premium" and "premium insurance". Hence the rearrangement must be done prudently by preserving any quote marks supplied by the user. The numeric query vector q (of size n) is initialised to all zeros. First keyword in the query is *binary-searched* (with low=1, and high=n) in the list of (n) keywords to get the index position (say j_1) where it occurs (here we

Algorithm 14.1 Pseudocode of the SVD Algorithm

1: Input: The dimensions (m,n) and the TD matrix W of size n x m
2: Form the symmetric matrix A = W'W of size m x m
    (* check if the matrix A is nonsingular (|A| not zero) *)
    Step 2:
3: Find the eigen values of A
4: Sort the eigen values in non-increasing order
    (* discard the small magnitude eigen values below a cutoff point p *)
5: Find the positive square roots of [largest p] eigen values
6: Form a diagonal matrix S using these [p] values along the main diagonal
7: Find the inverse of the diagonal matrix S as T (also diagonal)
8: **for all** ([p] singular values) **do**
9:       Find the corresponding eigen vector of 'A' matrix
10:     Form a column of V matrix using the eigen vector
11: **end for**
12: Compute U = W V T [dropping unwanted rows/columns beyond p]
13: Output: Orthogonal matrices U, V and diagonal matrix S (or T)
14: **return**

have the implicit assumption that all keywords in a query are present in the term-space), and j_1^{th} bit of q is set to 1 (in Boolean weighting). Binary-searches for subsequent keywords (if any), use low = $(j_{i-1} + 1)$ and high=n, for i=2,3,$\cdots$, p, where p is the number of distinct keywords in the user query.

Table 14.4: Full vs Truncated SVD

Name	LHS	RHS	comments
Full SVD	$W_{n\times m}$	$U_{n\times m}\, S_{m\times m}\, V_{m\times m}$	some s_{ii} can be zeros
Truncated SVD	$W_{n\times m}$	$U_{n\times k}\, S_{k\times k}\, V_{k\times m}$	s_{ii} negligible for i>k

Each query execution involves inner products and vector-matrix multiplications. The inner product of q' and U_j is found next for j=1,2,$\cdots$,k. This projects the user query into the document space. The inner product is next multiplied by the inverse of the diagonal matrix S (which is simply a diagonal matrix with elements that are reciprocals of the nonzero diagonal elements of S) to obtain

$$\hat{q}_k = q'U_kS_k^{-1} = \sum_{i=1}^{k}(1/s_i)q_iu_{ik} \tag{14.1}$$

where $s_i \neq 0$. This represents a pseudo-document in LSI space. The purpose of post-multiplying by S_k^{-1} is to tilt the pseudo-document to that similarity sub-space in which there is maximum feature overlap as evidenced by the decreasing singular values along the diagonal of S. Our aim in IR is to retrieve from the LSI space all documents that have maximum similarity to the pseudo-document. So, a similarity measure should be used to decide which are the vectors in the neighborhood of the pseudo-document vector $\hat{q}$. A good choice is the cosine similarity metric that works on normalised vectors (another choice is the Tanimoto coefficient $S(x,y) = $

$x'y/(x'x + y'y - x'y))$. We have multiplied the query vector by $(1/s_i)$ for differential weighting (the singular values capture the importance of each semantic vector). Once the query vector is so scaled, the similarity between the query vector and document vectors can be computed using the cosine metric. Because the cosine metric considers the angle between the vectors rather than their magnitudes, it is an ideal choice to locate similar documents using a cutoff threshold (alternatively, the total number of documents to be retrieved can be preset and similar documents retrieved until the count exceeds the total desired). This is accomplished by finding the maximum among

$$\sum_{i=1}^{k} \hat{q}_i v_{ij} / [\sqrt{\sum_i \hat{q}_i^2} \sqrt{\sum_i v_{ij}^2}] \tag{14.2}$$

for all j vectors in the immediate vicinity of $\hat{q}$. This can also be represented in inner-product form as $\frac{1}{||\hat{q}||} * \frac{\hat{q}.V_k}{||V_k||}$, because $\hat{q}$ is a constant vector for a user query. We have noted in the

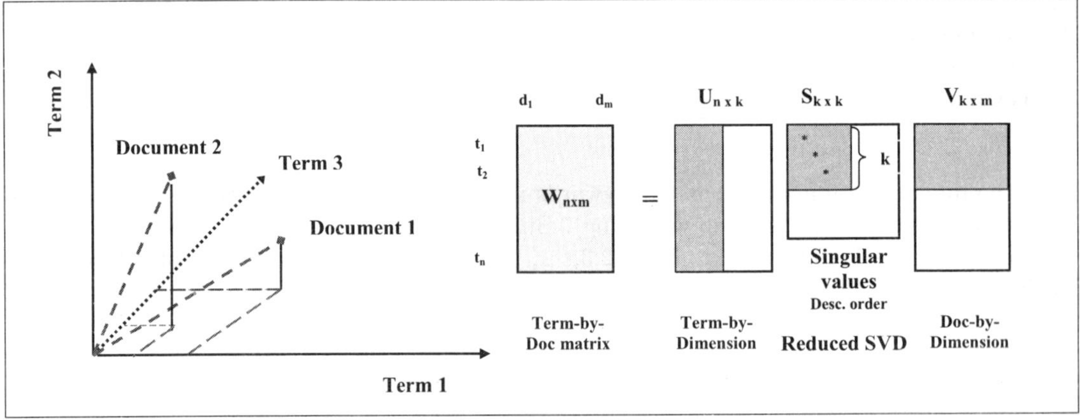

Figure 14.2: Pictorial representation of documents in term space (left) and reduced rank SVD (right). The shaded region indicates the values retained in computing rank k approximation.

above discussion that the same weighting scheme is to be used in forming the query vector from user-supplied keywords. But most of the IR requests use one or two keywords only (the average Google query is two words long). If a query uses just one valid keyword only (after eliminating common words like 'best', 'the', etc), the weighting scheme used for forming the query vector is immaterial because the cosine metric is unaltered when variables are scaled by a nonzero constant (this relationship does not hold for Tanimoto coefficient, unless both variables are scaled by the same constant. In other words, it is *scale variant*, whereas cosine similarity is scale-invariant). Just a Boolean weighting is enough in this case because the query vector will have a 1 in the corresponding position (say j), and zeros elsewhere. As mentioned in the introduction (page 14-1), this represents a vector parallel to the jth axis (in input space). Hence $\hat{q} = q'U_k S_k^{-1}$ is just the jth row of the matrix $U_k S_k^{-1}$ (this pseudo-document may not be parallel to the jth axis of feature space). A term weighting of the query will affect only the magnitude (and not the direction or orientation) of this vector. This is especially useful in incremental querying in which a user incrementally adds one keyword at a time to refine an already submitted query. There are many practical situations where some keywords (or nouns) *almost always* occur together, as in scientific inventions (eg: Kolmogorov-Smirnov test, Vapnik-

Chervonenkis dimension). Examples abound in medical sciences as eponyms[10] (eg: Kaposis sarcoma, Alzheimer's disease[11], Parkinson's disease, and various syndrome names like Albright's syndrome, Wiskott-Aldrich syndrome. See healthdirectorymoz.com/Conditions_and_Diseases/ for a complete list). When the relative merits of two keywords (or phrases) in a language are the same, it is reasonable to assume that their weighting is also the same, as in the above examples. A Boolean weighting is enough in such situations too, if these are the only two keywords present in a query. A convex weighting scheme with proper weights α and $1 - \alpha$ (where $0 < \alpha < 1$) may also be employed if the two keywords are known to have respective merits.

Above discussion considers the user queries one at a time (user submits a query and gets back a set of matching records). This is the *request-response* paradigm of design patterns. Multiple persons often query large data repositories simultaneously. Examples are library catalogs, medical databases, patent records, insurance databases etc. Popular search engines support millions of users who simultaneously search the web. All of these involve multiple executions of user queries. To speed them up, we compute the matrix $V_k^* = V_k/||V_k||$ once, and keep it in computer's memory (the U_k matrix is not needed at this stage). Normalised query vectors of each user are then multiplied by V_k^* to get the similarity vector as $\frac{1}{||\hat{q}||}\hat{q}.V_k^*$. Similarity of multiple query vectors can be exploited in speeding-up the search for similar documents. As an example, multiple persons search the web using very similar keywords after natural calamities like tsunamis and earthquakes, major accidents and political developments. Query vectors in these cases can be quite similar. Another technique is called relevance feedback querying that

Table 14.5: TD matrix for university documents

keywords	Doc1	Doc2	Doc3	Doc4
admission	2	0	1	0
car	0	1	0	0
campus	1	0	1	0
hospital	0	3	0	1
library	1	1	1	0
park	0	1	1	0
restaurant	1	0	0	1
street	2	1	1	0
university	3	1	1	0

incrementally filters out most-relevant documents. Users can relevance rank the returned results to indicate those subsets (records) that are most relevant to what is sought. This can eliminate synonymy and polysemy to a great extent. The refined pseudo-document in this case becomes $\hat{q}_{upd} = q'U_kS_k^{-1} + d'V_k$ where d is a vector of user filtered documents to be added to the query [LB97]. This results in a tilt of the pseudo-document vector in the feature space (in the direction of the resultant vector) towards similarity sub-spaces suggested by the user. These tilts can be nonrandom (for experienced users searching for focused information), or random (for novices

---

[10] Discovery named after one or more person(s) who first discovered or reported it are called eponyms. See www.doctorslounge.com/studlounge/downdirty/eponym.html for more information.

[11] Alzheimer's disease was first reported by a German physician Alois Alzheimer (1864-1915) at a talk in Nov 1906, after analysing the brain of one of his patients whose autopsy revealed neuritic plaques around brain nerve cells. It was his colleague Emil Kraepelin, who conjoined the term 'Alzheimer's disease' in 1910. see www.alz.org

and casual surfers). Nonrandom incremental tilts can be kept track of to adaptively retrieve most-matching records in the immediate neighbourhood of $\hat{q}_{upd}$.

14.5 Applications of LSI

LSI has been successfully applied in a variety of IR and related fields ([BC08], [PS10]). Examples are feature-based information retrieval, relevance feedback in IR, automatic document clustering, image based indexing, tracking conference submissions, computer-based question and answer systems, and social networking.

14.5.1 Performance based online examinations (PBOE)

In PBOE systems, an arbitrary question of an appropriate (low) difficulty level is first displayed online to the examinees. Without loss of generality, we assume that the questions are either True/False (T/F) type or multiple choice (MC) type (so that the answers can be validated immediately online using stored correct answers). If the examinee answers it correctly, a more difficult question is generated using the keywords, names, acronyms, phrases and images present in the current question.

If the answer to the current question is wrong (or the time limit for that question has exceeded), another question that does not include the leading keywords and features present in the current question can be generated with LSI using combination criteria (this can be done without LSI, if the question difficulty level is the only criterion to be used). This will eliminate question bias in certain fields, the terminology of which may be unfamiliar or uninteresting to the examinees. As an example, a student who is proficient in databases and programming may be least interested in theory of computation or neural networks. Thus the answer (or the un-answerability within the stipulated time limit) to the current question acts as a positive filter or a negative filter to subsequent questions generated. This can be extended to have LSI with memory, in which the common features present in all previous incorrect or unanswered questions are accumulated, and used as a filter to generate subsequent questions. One intuitive weakness of the above PBOE (with T/F or MC questions) is the "answer guessing" attempts by an examinee. If all answers are guessed, the correctly marked answers to difficult questions will exhibit a statistical law (For n questions, each with m choices, the expected number of correct answers by a random guessing examinee is n/m, which need not be an integer as it is an expected value). Similar applications can be found in AI.

14.5.2 LSI in Information Retrieval

Search engines that use keyword based retrieval techniques (without LSI) look for exact matches in the vast document corpora using a set of keywords and phrases. Each document is examined in turn and the occurrence information is captured into a ranked result set that drops all documents that have 0 ranks (absence of keywords in it). This method called lexical matching has the disadvantage that it looks only for exact spellings.

LSI aims to locate matching documents using semantic content. We can also use the LSI to find similarity of terms using known documents, and similarity among document collections using semantically similar terms present in them (see double clustering discussed below). Establishing document similarity using known terms is a problem of automatic clustering [DD90]. It even retrieves documents that do not have word-by-word exact matches. In addition, inter-language

Table 14.6: The symmetric matrix W W$'$

5	0	3	0	3	1	2	5	7
x	1	0	3	1	1	0	1	1
x	x	2	0	2	1	1	3	4
x	x	x	10	3	3	1	3	3
x	x	x	x	3	2	1	4	5
x	x	x	x	x	2	0	2	2
x	x	x	x	x	x	2	2	3
x	x	x	x	x	x	x	6	8
x	x	x	x	x	x	x	x	11

polysemy can be reduced to a great extent by LSI. As examples, the words 'war', 'links', etc have totally different meanings in English and German. Supplying additional keywords often disambiguates the multi-language polysemy. This is one of the reasons for the popularity of LSI in IR. See [SL96] for an LSI based neural network model, [IT02] for text segmentation, [LM07] for further discussions.

Example 14.1 Table 14.4.1 gives the TD matrix for 4 documents chosen arbitrarily from the document collection at a university. Entries denote the total number of occurrences of the term (indexed along the rows) within the document. Obtain a truncated SVD for k=3.

Solution 14.1 Singular values are found from the real eigen values of WW' matrix. This matrix is given in table 14.6 in which an 'x' indicates that the entry need not be computed as it is a symmetric matrix. The full SVD matrices (U, S, V) are first obtained for comparison (in practice, we need not carry out the full-SVD) as S=Diag(28.2265, 10.9740, 2.0765, 0.7229, 1.747E-15, 5.9357E-16, 3.9192E-16, 2.1222E-16, 7.478E-19),
U=

left singular vectors

−0.372	0.315	−0.0003	−0.128	−0.419	0.275	−0.223	−0.020	0.667
−0.087	−0.256	−0.069	0.277	−0.257	−0.557	−0.648	0.220	−0.016
−0.224	0.178	−0.240	−0.407	0.170	0.296	−0.305	0.580	−0.387
−0.277	−0.839	0.228	−0.039	0.003	0.376	0.040	0.114	0.099
−0.311	−0.078	−0.309	−0.130	0.236	−0.485	0.486	0.358	0.360
−0.163	−0.215	−0.549	−0.409	−0.357	−0.092	0.045	−0.506	−0.254
−0.164	0.066	0.674	−0.590	−0.003	−0.376	−0.040	−0.114	−0.099
−0.459	0.059	−0.069	0.149	0.674	0.004	−0.315	−0.451	0.048
−0.607	0.196	0.171	0.429	−0.304	0.003	0.312	0.039	−0.435

$V=$

$$
\begin{pmatrix}
\multicolumn{9}{c}{\textit{right singular vectors}} \\
-0.372 & -0.087 & -0.224 & -0.277 & -0.311 & -0.163 & -0.164 & -0.459 & -0.607 \\
0.315 & -0.256 & 0.178 & -0.840 & -0.078 & -0.215 & 0.066 & 0.059 & 0.196 \\
-0.0003 & -0.069 & -0.240 & 0.228 & -0.309 & -0.549 & 0.674 & -0.069 & 0.171 \\
-0.128 & 0.277 & -0.407 & -0.039 & -0.130 & -0.410 & -0.590 & 0.149 & 0.429 \\
-0.854 & -0.169 & 0.229 & -0.151 & 0.030 & 0.070 & 0.151 & 0.278 & 0.247 \\
0.071 & -0.667 & 0.253 & 0.330 & -0.377 & 0.028 & -0.330 & -0.238 & 0.263 \\
-0.028 & -0.415 & -0.034 & 0.118 & 0.714 & -0.498 & -0.118 & 0.041 & -0.197 \\
0.030 & 0.301 & 0.680 & 0.137 & -0.239 & -0.438 & -0.137 & 0.265 & -0.299 \\
-0.104 & 0.329 & 0.338 & -0.032 & 0.274 & -0.114 & 0.032 & -0.744 & 0.351
\end{pmatrix}
$$

In this example, there are more negative elements than positives. We could flip the sign of all elements in both the matrices (U and V) without affecting the final result (because $-U*-V = U*V$, and $\hat{q} = q'U_kS_k^{-1}$ will reverse its direction only). This is especially useful in hand-computation. If two vectors are in close proximity, the cosine of the angle between them is close to one (ie. $\cos(0) = 1$). After sign-flipping, the angles may be close to 180. As $\cos(\pi) = -1$, we need to take the absolute value in this case to retrieve matching records. Most of the entries in the query vector are zeros, and those nonzero entries are *comparatively larger* than the elements in the U and V matrices. Hence too small entries in the U and V matrices (perhaps with 6 or more leading zeros) can be set to zeros to induce sparsity in them. This is especially useful in LSC.

Largest four eigen values are [28.2265, 10.9740, 2.0765, 0.7229], and the trace of the matrix $W W'$ is 42. Computed eigen values sum to 41.9999 (due to truncation error). Truncation thresholds when 4 and 3 eigen values are retained are $\epsilon(4) = .00000238$ and $\epsilon(3) = .0172$. Thus choosing k=3 gives good enough results in this case. The factored sub-matrices with truncation at k=3 are as follows:

$$
\begin{pmatrix}
\multicolumn{3}{c}{\textit{left singular vectors}} \\
-0.3716 & 0.3152 & -0.0003 \\
-0.0870 & -0.2563 & -0.0689 \\
-0.2236 & 0.1783 & -0.2403 \\
-0.2765 & -0.8395 & 0.2275 \\
-0.3106 & -0.0780 & -0.3092 \\
-0.1626 & -0.2149 & -0.5492 \\
-0.1636 & 0.0664 & 0.6743 \\
-0.4586 & 0.0588 & -0.0692 \\
-0.6066 & 0.1957 & 0.1708
\end{pmatrix}
\begin{pmatrix}
\multicolumn{3}{c}{\textit{singular values}} \\
5.31286 & 0.0000 & 0.0000 \\
0.0000 & 3.3127 & 0.0000 \\
0.0000 & 0.0000 & 1.4410
\end{pmatrix}
$$

$$
\begin{pmatrix}
\multicolumn{9}{c}{\textit{right singular vectors}} \\
-0.372 & -0.087 & -0.224 & -0.277 & -0.311 & -0.163 & -0.164 & -0.459 & -0.607 \\
0.315 & -0.256 & 0.178 & -0.839 & -0.078 & -0.215 & 0.066 & 0.059 & 0.196 \\
-0.0003 & -0.069 & -0.240 & 0.228 & -0.309 & -0.549 & 0.674 & -0.069 & 0.171
\end{pmatrix}
$$

The reconstructed matrix from approximants has elements close to the original matrix.

$$\begin{pmatrix}
4.988 & 0.026 & 2.962 & -0.004 & 2.988 & 0.962 & 1.946 & 5.014 & 7.040 \\
0.026 & 0.945 & 0.082 & 3.007 & 1.026 & 1.082 & 0.118 & 0.970 & 0.914 \\
2.962 & 0.082 & 1.880 & -0.011 & 1.962 & 0.879 & 0.826 & 3.044 & 4.126 \\
-0.004 & 3.008 & -0.011 & 9.999 & 2.996 & 2.989 & 0.984 & 3.004 & 3.012 \\
2.988 & 1.026 & 1.962 & 2.996 & 2.988 & 1.962 & 0.945 & 4.014 & 5.040 \\
0.962 & 1.082 & 0.879 & 2.989 & 1.962 & 1.879 & -0.175 & 2.044 & 2.127 \\
1.946 & 0.118 & 0.826 & 0.984 & 0.944 & -0.175 & 1.748 & 2.064 & 3.183 \\
5.014 & 0.970 & 3.044 & 3.004 & 4.014 & 2.044 & 2.064 & 5.984 & 7.954 \\
7.0396 & 0.914 & 4.126 & 3.012 & 5.040 & 2.127 & 3.183 & 7.954 & 10.867
\end{pmatrix}$$

As discussed before, a user query is converted into a vector using the same order of keywords. Consider a query '(park automobile on campus)', which is converted into a query vector $q'=[0,1,1,0,0,1,0,0,0]$, in the unweighted (Boolean weighting) case.

The factorisation of WW' is a mapping to the term-space. This can be used for topic segmentation and attribute clustering.

14.5.3 Latent Semantic Clustering (LSC)

It was mentioned in §14.1.1(p.14-2) that two almost identical document vectors (columns) d_1 and d_2 in the TD-matrix indicate that the corresponding documents have many things in common. In other words, the inner product $d_1.d_2$ is low when documents have few similarities, and high otherwise. We have no way of knowing what is the threshold for low and high until we find all possible dot products (which could be cumbersome because there are $\binom{m}{2}$ of them). A solution is to normalise each column vector so that all inner products will lie in $[0,1]$. If two documents contain synonyms, the similarity could still be small. The LSC aims to remedy this weakness in the input space by clustering in the transformed latent semantic space.

There are two categorisations of LSC. The first one uses document similarities as evinced by the SVD matrices to find similar document clusters. The second approach is smarter in that it first finds a set of term-clusters using the U matrix of SVD (in the term-space). If distinct clusters do exist (which is often the case in large document corpora, heterogeneous image sources and multimedia files), only documents containing the terms in these clusters are used to form the document cluster. Each term cluster can be replaced by a centroid or medoid vector, which best represents the clusters. This is meaningful due to the fact that each term vector (a column of U matrix) encapsulates information content of one or more keywords in the input space. Analogously, each document vector (a column of V matrix) encapsulates information content of one or more documents in the input space. A cluster containing just one element is called a singleton cluster. Even if some of the term clusters are singletons, we cannot ignore them as it could be present in multiple documents. Similarly, singleton document clusters can be representatives of multiple input space documents.

In [CV06] the authors call this approach as double clustering. An advantage of double clustering is that in those situations where majority of the terms indeed forms well-separated clusters, the effective dimensionality is reduced. This means that if a group of words tend to co-occur together (they form distinct clusters), there probably exist documents that contain words semantically similar to the words in the cluster. But this will not always produce meaningful clusters, because of the assumption that term clusters and document clusters are independent. Most of the web data are heterogeneous. Hence the term-cluster patterns need not be reflected in document clusters. If the term clusters and document clusters are highly correlated, the document clusters found will have large overlaps. As the Mahalanobis distance metric exploits the correlations, it is the best choice in this case. A fuzzy clustering approach is also meaningful in such situations. A kernel mapping to a higher dimensional space could improve the chances of obtaining crisp clusters. Assume that TC_1 and TC_2 are two disjoint term clusters. Even if there exist documents that contain terms exclusively from TC_1 and from TC_2, there could exist a large number of documents that contain terms from both. An example is classifying newswire articles by topic (like politics, sports, science and technology, disasters, obituaries etc). Similarly, small training sizes complicate the document clustering, and small number of keywords complicates attribute clustering.

Most of the clustering algorithms work on a dissimilarity matrix. The term-space matrix returned by the SVD is in fact a similarity matrix. Several methods exist to convert a similarity matrix to a dissimilarity matrix. Hence it is a simple matter to utilise the U matrix to cluster similar terms (say by using the k-means algorithm). This gives us term clusters. But the k-means algorithm requires the user to input an integer k that represents the best possible number of clusters. This may not be known apriori. Hence a choice is to try a cluster validity index (CVI) on the clusters obtained with different k values, and settle upon the best clustering possible (say m clusters where m<k). Alternatively, a hierarchical clustering algorithm is employed to get a dendrogram first. This dendrogram can be cut at an appropriate level to get the clusters at the correct granularity. As the clustering algorithms are unsupervised learning methods, and so is LSC. But if the data are labeled, as in text classification, one could utilise this extra information to come up with more meaningful clusters (the cluster to which labeled data belong is known apriori), as well as cluster labels (names) described below. Hand-labeled data are difficult or expensive to get in some domains, as this may require some domain knowledge and expertise. See [SI10], [PS10] for further details.

Example 14.2 Cluster the terms in table 14.4.1(pp.14-15) using LSC.

Solution 14.2 We will first do an SVD factorisation to get the term matrix (U). We could form a similarity matrix from the rows of the U matrix by taking dot products (some of the clustering algorithms instead use a dissimilarity matrix). This gives the following matrix, where the elements in the first column are the dot products of first row vector with the rest of the row vectors, and so on. These values are proportional to the cosine of the angle between the row vectors. V=

		right singular vectors					
0.0002287							
$2.2E-5$	0.000611						
0.0006586	-0.000279	0.00127					
0.0001777	$-9.9E-5$	-0.0005	0.000447				
0.0003777	0.000386	-0.00036	-0.000378	$-7.6E-5$			
0.0001398	$6.3E-5$	-0.00056	0.000954	$5E-5$	0.000519		
0.0002567	-0.000451	0.000918	0.000363	$-4.6E-5$	$-7.5E-5$	0.00049	
0.0002807	-0.000512	0.000226	$2.9E-5$	0.000675	0.000509	0.000551	0.00067

Note that the cosine of the angle between two dissimilar vectors is zero. As we use a similarity matrix, we could simply pick out the maximum similarity pairs. Ten of the maximum elements and their corresponding terms are as follows:– .001272={campus, hospital}, .000954={hospital, restaurant}, .000918={campus, street}, .00067={street, university}, .000675={library, university}, .000659={admission, hospital}, .000611={car, campus}, .00056={campus, restaurant}, .000551={restaurant, university}, .000519={park, restaurant}. Here 'park' is a polysem. It can mean either a public park or vehicles park. Continuing this way, we get the clusters C_1={campus, hospital}, C_2={hospital, restaurant}, C_3={campus, street}, C_4={street, university}, C_5={library, university}, etc. We could merge C_1 and C_2 as they contain common items and proceed to higher levels to get less granular clusters. Similarly, we could find all document clusters by first finding the dot products of the columns of the V matrix and proceeding exactly as above (see exercise in page 14-24).

Algorithm 14.2 Algorithm for term (resp. document) clustering using LSC

1: Input the data matrix X, and decompose it using truncated SVD as X = U*S*V
2: Form a similarity matrix using dot products of rows (resp columns) of U (resp V) matrix
3: Pick out largest values and identify corresponding terms (documents) in the input space
4: Build a dendrogram using the above step
5: Merge adjacent clusters in the dendrogram using commonalities.
   Return the clusters found.

14.5.3.1 Labeling Semantic Clusters Found

Most of the statistical clustering algorithms either report the clusters found by grouping the data points in respective clusters, or by plotting the cluster boundaries. The hierarchical clustering algorithms report the clusters found as a dendrogram (which can be considered as a visualisation of the cluster formations from high to low granularity order). As the LSC carry supplementary information, we could automatically generate cluster labels. This is called unsupervised cluster labeling. Consider the term-clustering scenario first. Suppose that our LSC algorithm finds distinct clusters in the term space. We divide the clusters into two groups :– (i) those that have 2 or more data points, (ii) those that are singleton clusters. In case (i), we compute the mean vector (called the centroid term vector) of the cluster vectors in the term space. This centroid vector need not coincide exactly with a term vector. Hence we could find the medoid vector (that vector which is closest to the centroid vector in the term-space). We then map either the term-centroid or term-medoid vector to the input space to get a list of closest terms. If the mapped vector coincides exactly with a vector in input-space (which is rare), we use the corresponding term as the cluster label. In case of a tie, or in the presence of multiple term vectors in close proximity of the mapped centroid vector, we simply concatenate the corresponding terms by an appropriate connective ($-$, &, etc), and use the resulting string as cluster label. In case (ii), finding the centroid vector step is bye-passed. As mentioned earlier, a singleton cluster in term-space could represent a distinct cluster (with two or more elements) in the input-space due to the many-to-one mapping induced by the truncated SVD. Hence we proceed as above in the input space. Cluster labels are found in document clustering in an exactly similar fashion, except that rows (terms) are replaced by columns (documents). As synonymy and polysemy could sometimes result in ambiguous labels, a supervised cluster labeling method can also be employed with the help of a domain expert. In this case, each of the clusters found in the transformed space is mapped to the corresponding terms or documents and displayed for a human expert to decide the most appropriate cluster label.

When a cluster is formed by two or more keywords, we may have to use a heuristic to label it meaningfully. A simple rule is as follows. Assume that there are two keywords that belong to the same cluster. If one of them is a noun, and the other is a verb, adverb, pronoun, or gerund; we concatenate the noun at the end of the other term. Examples are teachScience, shoppingMall, academicUniversity, internationalConference etc. If both terms are nouns, we check for any possible *whole-part* relation among them. Then the *part* term is concatenated at the end of *whole* term. For instance, if the terms are {library, university}, we form the label as UniversityLibrary rather than LibraryUniversity because library is a part of university. Other types of relationships like cause-and-effect, is-a, has-a, etc may also be helpful in forming meaningful cluster labels. If both of them are verbs, adverbs, or gerunds, we form a label using a connective word or letter (like AND, N, _, &, etc). Examples are training_driving, Hire&Fire, serve&eat, etc. This method can easily be extended to more than two keywords. If at least one element in a cluster has a user-defined label, it may be used as a prefix if there are other unlabeled elements. When more than one element has user-assigned label, the above heuristic can provide more meaningful labels. When a cluster contains too many elements, an appropriate number of alternating sequences of verbs and nouns can be used to label it. As in our university-example in page 14-15, most or all of the keywords could also be nouns in some document collections.

14.6 Software for LSI

There are many free and commercial software available for a variety of tasks related to LSA. GTP (General Text Parser) is a software environment (written in C++ and Java) which has capabilities to parse text, generate TD matrix using a variety of weighting schemes, decompose it using SVD, and do the matching.

Table 14.7: Software for LSI

URL	Name	C/F
www.sourceforge.net/projects/tml-java/	Text Mining library for LSA	F
www.cs.utk.edu/~berry/projects/	General Text Parser	F
www.netlib.org/svdpack/index.html	SVDPACKC (Ansi C)	F
kt.ijs.si/Dunja/textgarden/	LSI based software	F
ailab.si/jure/pade/	Orange	F
alias-i.com/lingpipe/	SVD for LSA	F
tcc.itc.it/research/textec/tools-resources/jlsi.html	Java LSI (jLSI)	F
www.sas.com	SAS Textminer	C
spss.com	SPSS clementine	C

Legend: C=Commercial, F=Free

SVDPACK (in FORTRAN 77), SVDPACKC (ANSI-C) suite comprises of 4 numerical algorithms for singular value decomposition. See also webscripts.softpedia.com/scriptDownload/ Incremental-SVD-Download-34276.html (incremental SVD programs), www.download32.com/ matrix-tcl-d2065.html, jLSI) is an open source Java tool for LSI (see table).

14.7 Exercises

1. Mark as True or False
   (a) The order of occurrence of query keywords is important in LSI
   (b) LSI cannot be used for image classification
   (c) LSI uses a reduced rank approximation of TD matrix
   (d) LSI can be applied when documents have a hierarchical tree structure
   (e) SVD does not preserve sparsity in the TD matrix
   (f) Latent semantic clustering works for image and video data
   (g) Synonymy and polysemy are partially solved by LSI
   (h) Vector space models suffer from the curse of dimensionality
   (i) Singular-values of a TD matrix can be negative
   (j) Vectors used in text mining are all located in the positive quadrant
   (k) Polysemy results in irrelevant documents in IR, thereby reducing precision
   (l) Synonymy misses relevant documents thereby reducing recall ratio
   (m) Two almost identical columns of a TD matrix indicate that the corresponding documents are very similar
   (n) Class labels cannot be found by LSC algorithms
   (o) LSC algorithm finds clusters in an iterative loop using a threshold to terminate
   (p) LSC is a supervised learning algorithm.

2. A *die* has a different meaning when used as noun or verb. To which category does this word belongs to?   (a) synonym (b) meronym (c) metonym (d) polysem

3. What are the advantages of SVD over other factorisation techniques (LU, QR) in LSI?

4. What are some application areas of LSI? Can it be applied to non-text data?

5. List any three advantages of LSI over lexical (keyword-based) information retrieval.

6. Can a user query get mapped to align exactly with a coordinate axis in feature space? If so when?

7. Explain how a query vector is formed, how it is mapped to a pseudo-document vector, and how the query is executed in LSI?

8. Explain synonymy and polysemy with examples. What kind of polysem is each of the following? a) 'stand', b) 'board', c) 'make', d) 'drink'

9. Under what conditions are the dimensionality reduction helpful in IR? How does the LSI achieve reduced dimension? What criteria are used to decide the reduced dimension? How much is the savings in memory due to the dimensionality reduction?

10. How does LSI facilitate relevance feedback searching?

11. What are few situations where a Boolean weighting is applicable to form a query, even if the TD matrix was formed using more complex measures (like TF-IDF).

12. A pseudo-document vector corresponding to a user query is $\hat{q}=[.1, .5, .3]$, and table below gives the V^* matrix (§14.4, page 14-12).

$$\begin{pmatrix} \underline{term/doc} & Doc_1 & Doc_2 & Doc_3 \\ Term_1 & .8 & .3 & .25 \\ Term_2 & 0 & .1 & .9 \\ Term_3 & .5 & 0 & .6 \end{pmatrix} \quad \begin{pmatrix} \underline{keywords} & Doc1 & Doc2 & Doc3 \\ breakfast & 2 & 0 & 0 \\ bread & 1 & 0 & 0 \\ computer & 1 & 0 & 0 \\ jam & 0 & 3 & 1 \\ juice & 1 & 1 & 0 \\ park & 0 & 1 & 0 \\ restaurant & 0 & 0 & 1 \\ school & 1 & 1 & 0 \\ street & 2 & 1 & 1 \end{pmatrix}$$

Which document is semantically most similar?

13. For the TD matrix in table above, find the SVD and obtain matching vectors for the queries (i) [breakfast, jam, juice], (ii) [park, street], (iii) [bread, restaurant, school], (iv) [computer]

14. If two singular values of the TD matrix are exactly identical, can you speedup the LSI algorithm? If two rows and columns of the TD matrix are rearranged (swapped among themselves) does it affect the solution?

15. What is the cosine similarity of each of the following:- (i) two parallel vectors (ii) two perpendicular vectors

16. Create the term by document matrix using the Boolean weighting and TF-IDF weighting from the following terms for:- (i) the first two paragraphs in §14.2 in page 14-5, where (terms={data, document, feature, eigen values, HTML, latent, lexical, LSA, query, rank, semantic space, SVD, VSM, web}), and (ii) Four paragraphs of the box in §14.3 in page 14-11 where (terms={antonym, context, English, Greek, homophone, information, IR, polysem, precision, query, recall, synonym}). Consider each paragraph as a separate document (so that (i) gives a 14 x 2 TD matrix, and (ii) gives a 12 x 4 matrix).

17. Identify which of the following are lexicronym, polycronym. What are the solutions to increase the precision in IR in such situations? (i) ART (ii) AIR (iii) PAN (iv) UNO (v) TAN

18. Describe how you will extend the PBOE system so that LSI remembers each and every wrongly answered questions, and use it to generate a new question.

19. What is the TD matrix equivalent in image based IR. How will you form it?

20. Describe the latent semantic clustering problem. Why is it inefficient to do the clustering in the input space? What are the advantages of clustering in the feature space?

21. Some documents that are semantically similar could have well-separated clusters of keywords present in them. Describe how the LSC finds them.

22. Can the semantic space contain outliers? What are its implications in the input space of documents?

23. Cluster the documents in table above using LSC.

24. Describe a method to label the term-space and document-space clusters found by LSC
automatically in both the presence and absence of a few labeled training data.

14.7.0.2 References

[AM04] Akritas, A.G., Malaschonok, G.I.(2004) Applications of singular-value decomposition
(SVD), *Mathematics and computers in simulation*, Applications of computer algebra in
science, engineering, simulation and special software, 67,15-31 (sciencedirect.com)

[AK05] Almpanidis,G., Kotropoulos,C., Pitas,I. (2005) Focused crawling using latent semantic
indexing - An application for vertical search engines, LNCS 3652, 402-413, in *Research
and advanced technology for digital libraries*, Springer verlag, (citeseer.ist.psu.edu).

[BJ08] Bellegarda, J.R. (2008) *Latent semantic mapping: principles and applications*, Morgan
Claypool.

[BE73] Beltrami, E. (1873) Sulle funzioni bilineari (English translation by D. Boley at University of Minnesota, Department of CS, (www.cs.umn.edu) Technical report 90-37).

[BD95] Berry, M. W., Dumais, S. T., O'Brien, G.W.(1995) Using linear algebra for intelligent
information retrieval, *SIAM Review*, 37(4), 573-595.

[BM00] Berry, M.W.(ed) (2000) *Computational information retrieval*, SIAM, Philadelphia, PA.

[BC08] Berry, M.W., Castellano, M. (2008) *Survey of text mining - II: Clustering, classification
and retrieval*, SIAM, Philadelphia, PA.

[BC85] Buckley, C. (1985) Implementation of the SMART information retrieval system, Tech
report TR85-686, Cornell University, Ithaka, NY.

[CT03] Castelli, V., Thomasian, A., Li,C.S. (2003) CSVD: Clustering and singular value decomposition for approximate similarity search in high-dimensional spaces, *IEEE transactions
on knowledge and data engineering*, 15(3), 671-685.

[CF02] Chu,M.T., Funderlic, R.E.(2002) The centroid decomposition: relationships between
discrete variational decompositions and SVDs, *SIAM journal on matrix analysis and applications*, 23(4), 1025-1044.

[CV06] Csorba, K., Vajk, I. (2006). Double clustering in latent semantic indexing, 4th *Slovakian-
Hungarian joint symposium on Application of Machine Intelligence (SAMI)*,
(www.bmf.hu/conferences/SAMI2006/CSorba.pdf).

[DD90] Deerwester,S., Dumais,S., *et.al.* (1990) Indexing by latent semantic analysis, *Journal
of the American society for information science*, 41(6), 391-407.

[EY36] Eckart, C., Young, G.(1936) The approximation of one matrix by another of lower rank,
Psychometrika, 1, 211-218, https://ccrma.stanford.edu/~dattorro/eckart%26young.1936.pdf.

[GR01] Gee K. Randall (2001) *Text clustering using LSI*, MS Thesis, Dept. of Computer
Science, University of Texas, Arlington, TX.

[GF04] Grossman,D.A., Frieder,O. (2004) *Information retrieval: algorithms and heuristics -
2nd ed*, Springer, NY.

[HS85] Hammerling, S. (1985) The singular value decomposition in multivariate statistics, *ACM SIGNUM newsletter*, 20(3), 2-25.

[HS01] Husbands, P., Simon, H., Ding, C.H.Q. (2001) On the use of the singular value decomposition for text retrieval, *The Oxford handbook of computational linguistics*, Mitkov, R. (ed), 145-154 (citeseer.ist.psu.edu).

[IT02] Ishioka,T.(2002) Text segmentation by Latent Semantic Indexing, New development in psychometrics, Proceedings of the international meeting of the psychometric society IMPS-2001, Tokyo, 689-696, Springer, www.rd.dnc.ac.jp/~tunenori/doc/impsTxtsegLSI.pdf

[JM05] Jessup,E.R.,Martin,J.H.(2005) Taking a new look at the latent semantic analysis approach to information retrieval, www.cs.colorado.edu/~jessup/SUBPAGES/PS/martin.ps.gz

[JC75] Jordan, C. (1875) Essai sur la geometrie á n dimensions, *Bulletin de la societe Mathematique*, 3, 103-174.

[KA04] Kobayashi, M., Aono, M. (2004) Vector space models for search and cluster mining, chapter 5 in *Survey of text mining*, Berry, M.W. (ed), 103-122, Springer, NY.

[LL90] Landauer, T. K., Littman, M. L. (1990) Fully automatic cross-language document retrieval using latent semantic indexing. In Proceedings of the sixth annual conference of the UW centre for the New Oxford English dictionary and text research, 31-38. UW centre for the New OED and Text Research, Waterloo, Ontario.

[LM07] Landauer, T.K., McNamara, D.S., Dennis, S., Kintsch, W.(eds)(2007) *Handbook of Latent Semantic Analysis*, Lawrence Erlbaum.

[LB97] Letsche, T.A., Berry,M.W. (1997) Large-scale information retrieval with latent semantic indexing, *Information Sciences*, 100(1-4), 105-137 (sciencedirect.com).

[LH58] Luhn, H.P. (1958) The automatic creation of literature abstracts, *IBM journal of research and development*, 2, 159-165.

[PS10] Park, S.C. (2010) Latent semantic analysis for vector-space expansion and fuzzy logic-based genetic clustering, *Knowledge and information systems*, 22(3), 347-369.

[RS09] Rajamanickam, S. (2009) *Efficient algorithms for sparse singular value decomposition*, Ph.D. thesis, University of Florida, Gainsville, FL, USA (www.cise.ufl.edu/~srajaman/Rajamanickam_S.pdf).

[SW75] Salton, G., Wong, A., Yang, C.S. (1975) A vector space model for automatic indexing, *Communications of the ACM*, 18(11), 613-620.

[SI10] Seshadri, V., Iyer, V.K. (2010) Parallelization of a dynamic SVD clustering algorithm and its applications in information retrieval, *Software practice and experience*, 40(10), 883-896.

[SL96] Syu, I., Lang, S.D., Deo, N.(1996) A neural network model for information retrieval using latent semantic indexing. ICNN 96, *IEEE international conference on neural networks*, 1318-1323 vol.2.

[TS06] Tougas, J.E.B., Stern, H.(2006) Updating the partial SVD: Making LSI run faster, *Computational statistics and data analysis*, 52(1), 174-183 (src.acm.org/subpages/papers/Grand %20Finals%20 2005/Jane%20Tougas.pdf).

15
Text Mining

Chapter objectives:

- Describe text mining

- Distinguish between text-statistics and text mining

- Understand Term-by-Document matrix

- Describe text pre-processing

- Describe metrics for text mining

- Explain the facet identification algorithm

- Describe text-classification

- Explain text-clustering

- Real-world applications of text mining

15.1   Introduction

Data mining models can be built on a wide variety of data including quantitative and categorical data, image and multimedia data, spatial and temporal data, text data and other types of data. Text mining is an inter-disciplinary field with focus on text data in various forms. It draws upon information retrieval (IR), information extraction (IE), data visualisation, pattern recognition, machine learning, statistics, Natural Language Processing (NLP, or NaLP) and computational linguistics. It is also known as intelligent text analysis, text-data mining or Knowledge-Discovery in Text (KDT, or KDiT). The aim is to extract interesting and non-trivial information and knowledge from unstructured (free form) text. Text mining searches for valuable trends and patterns in document corpora containing text data (also called textual data) - Unicode characters in any language, binary strings representing specialised data, application files like spreadsheets, word documents etc or even numbers interpreted as text. Depending upon the goal, the Text Mining Pipeline (TMP) may include a combination work-flow of techniques from these fields [FS06]. Consider a 'Question Answer System (QAS)', say for medical diagnosis.

A patient accesses the system (say over a mobile internet connection) and answers the displayed questions either using cell-phone keypad or orally. The system uses pattern recognition, machine learning, IR, speech-to-text conversions etc to capture the answers to eventually narrow down the probable disease or condition of the patient to suggest the treatment options or procedures to be followed. A summarisation system[1] may use IR, visualisation, etc and can be used for structured, semi-structured or unstructured documents. Rule-based text mining utilises expert-derived rules. They use "rulebases" that are special databases which store rules and data in an optimal way, with added capabilities to draw deductive consequences. New rules can also be deduced and existing rules pruned during the process. Most common techniques for text mining include classification, link analysis, and clustering.

15.1.1 Document

Document has a different meaning in text mining than its literal meaning. A unit of discrete textual data that is accessible (readable or fetchable) and is a part of a collection is called a document in text mining. This means that each document is identifiable using a name, a URI (uniform resource identifier) or can be demarcated using an index (like time series and stream data). The name of a document (or the file extension) is irrelevant in text mining, as the name is used only to distinguish between documents. A document in a collection is indicated by d_i in this chapter. Text mining algorithms do not modify the document contents, but simply read (fetch) the contents of it. Each document belongs to a homogeneous set in majority of applications. Such a homogeneous set is called a *document corpora*. They can also be related as is-a or whole-part hierarchies.

Definition 15.1 Text mining is the process of extracting interesting and nontrivial information and knowledge from (unstructured or structured) text documents (called a corpus) using data mining techniques.

Novelty detection mines for novel associations in text data (using two or more terms, some of which may be pre-specified). As an example, if 'hazardous chemical' occurs in a document, 'safe storage', 'handling precautions' etc may also occur within a specified distance (say within a paragraph or within he next 10 words). From this, an association rule can be obtained as 'hazardous chemical'$\Rightarrow$'safe handling'(10%, 2%). Due to the increasing number of text corpora appearing on the web, text mining will continue to attract users in various fields :- biologists, medical and legal professionals, academicians, information technologists, and researchers from many other fields. A striking similarity among classical data mining and text mining processes is the commonality in the respective architectures. Both have a pre-processing stage, pattern discovery stage and presentation layers. The pre-processing stage is more important when text mining is non-repetitive (data are mined for the first time and are scattered in unstructured document corpora called unprepared text mining). Pre-processing stage may be nonexistent when data are taken from text-warehouses or from specially indexed text corpora. Pattern discovery stage is similar to that in classical data mining. Here we apply an appropriate data mining model to dig into the data. Presentation layer (visualisation stage or post-processing) is used to display the mined information in a meaningful way. They may or may not have navigational capabilities to explore complex data relationships online. Most of the state-of-the-art text mining software systems include such a capability for semi-supervised text mining

---

[1]Summarisation is a compact description of data into meaningful numbers, tables or graphs that succinctly characterise the essential content of parent data. Automatic Content Summarisation (ACS) is used to discover knowledge in text with the help of humans or machines who interpret the summary information.

tasks that may include selections and constraints on data or attributes. Some systems allow refinement constraints to be applied repeatedly to narrow down the search space. This can reveal patterns at finer granularity levels.

15.1.2 Text Statistics vs Text Mining

Text statistics is different from text mining. The former refers to the extrinsic or summary of features extracted from text. Examples are the length of minimum and maximum of the words, occurrence frequencies, positional characteristics (eg: the minimum distance between two occurrences of a word in a document, minimum or maximum distance of a word from the beginning or end of the document), most common words (mode), foreign words in text (eg: vice-versa, per-se), etc. These can all be obtained by simple string or character processing primitives (like length(string), index(), strstr(), etc functions) in programming languages, or by table lookups. These are not considered as text mining, but are called text statistics. They

Table 15.1: Occurrence Frequency of Words

a (2)	and (2)	called (1)	corpus (1)	data (1)	documents(1)
extracting(1)	from (1)	information(1)	interesting(1)	is (1)	knowledge(1)
mining (2)	non-trivial(1)	of (1)	or (1)	process(1)	structured(1)
techniques(1)	text (2)	the (1)	unstructured 1	using (1)	

Numbers in brackets denote the number of times the word occurs in the sentence.

are presented in tabular or graphical formats to the user, or stored as metadata. These are utilised by some of the text mining algorithms. As an example, consider the definition of 'text mining' given above. There are 26 words (excluding special characters like brackets, comma and fullstop) with respective frequencies given in table 15.1. The minimum length is 1 (for 'a') and maximum is 12 (for 'unstructured'). As the majority of occurrence frequencies are 1, the minimum distance from the beginning and end are the actual distances. For instance, the word 'is' has positional count 3, as it is the third word from the beginning. The word 'and' has distance 8 to the beginning (there are 8 words before the first occurrence of 'and'), and distance 14 to the end of the passage. The minimum distance between the two occurrences is 2 (as there are two words 'nontrivial information' in-between them). There is no unique mode, as there are 4 words with occurrence frequency 2. Many text summarisation measures have appeared in the literature:– precision, recall, relative utility, etc. These are described below.

15.1.3 Data for Text Mining

The web is undoubtedly the largest source of information at present. Approximately 90% of the data on the web are in text format in various languages. It contains text data in unstructured, semi-structured and structured formats, along with other types of data (like audio, video, animation and multimedia (AVAM), images, spatial and temporal data). In addition, email messages, chat-room discussion data, tweets and messages in twitter like sites, embedded text in images and videos (called anchor text), text quotes (as in stock markets), text quotations, and text lotteries generate copious amounts of text data. Other sources of text mining are medical and legal transcriptions, various archives, writeups (like essays, customer complaints), text extracted from scanned-documents (like fax, registration forms), "callouts" in text documents, spreadsheets and graphics etc. Data for text mining may also come from document collections (called document-corpora) which may be static or dynamic. Data for text mining can also come from web pages, XML files, wire services, fax messages, e-mails, Lotus Notes databases, call centre files, and patent and digital libraries.

Most common data sources include repositories maintained by government agencies, blog repositories and archives, medical and legal transcriptions, various log files (server logs, call center data, mail-order data, etc.) and streams. This data can be extracted from structured documents like HTML and XML files, product reviews, chat room discussions, medical and legal transcriptions, etc[2]. Partial information about attribute types, sizes, character codes, and languages may be available when this document is derived from spreadsheets, or extracted from a database using a query, or captured from web forms. For example, a numerically coded categorical variable may be output as a text (in quote marks) to distinguish between a regular number and variable value. Categorical variables containing numeric codes, and numeric/alphanumeric attributes can also be distinguished. This information is hard to catch when data are derived from dictations, and appraisals. A pre-processing step may be needed when data are extracted afresh from raw sources. The outcome of pre-processing is a partially structured text document in a proper format (comma separated, tab separated etc). If the data itself contain separators (eg: comma may be present in insurance claims, patient histories, etc. Also, some European countries swap the meaning of comma and '.' in decimal numbers so that 62.55 is written as 62,55), a special character that is not present in the text is to be used as a separator. Extracted data can conveniently be stored in text datawarehouses or special databases for later text mining sessions. A sequential order among the words and phrases has to be maintained in most extracted data. Examples are dictations, songs, essays, data in web pages, newswire articles, etc. But there are also instances where such an order is unimportant or irrelevant. Examples are stock quotes, auction bids[3], search keywords used by searchers at job-sites, auction sites etc. In the later case, each document can be represented by the "bag-of-words" method.

Unlike the web mining techniques discussed previously, that utilise the dynamic web pages, the text mining most often uses static text documents. But the data for text mining applications may result from dynamic events like tsunamis, natural disasters, etc. The Dec 2004 Asian tsunami appeal generated 726,875 text replies each costing around 3 dollars to the Disasters Emergency Committee's online fund raiser, and the star studded Live 8 text lottery on July 2, 2005 had 2,060,285 text messages each costing \$3 to win a pair of tickets to London Hyde Park concert, setting a new world record.

15.1.4 Textification

Converting speech, song, dictations etc into pure text formats for computer storage or manipulation is called *textification*. In the case of speech, songs, instructions and lectures, the textified document will contain meaningful words (exceptions are common in songs and dictations). However, similar-sounding words could create incorrect textified document entries. Examples are (born, Borne, Bonn), (break, brake), (road, rod, rode), (right, write, Wright) etc.[4] Other types of sounds or music could also be textified using specially built 'vocabulary'. For example, by assigning lowercase and uppercase letters and special symbols to each pitch interval of a musical

---

[2]Many programs exist to convert html files to pure text files. See www.jafsoft.com/detagger/, www.softinterface.com/Convert-Doc/Convert-Doc.htm, www.jetman.dircon.co.uk/software/web2text.html etc
[3]The highest bid is important in auctions as the auctioneer wishes to sell the item to the highest bidder.
[4]see www.frostburg.edu/clife/writingcenter/handouts/homonym.htm for a complete list.

Table 15.2: Data retrieval vs Data Mining

Data type	Goal-oriented Search	Opportunistic Discovery
Structured	Data retrieval	Data mining
Semi-structured	Information/Knowledge retrieval	Stream mining
Unstructured	Information retrieval	Text mining

Both search and discovery in unstructured data (text) can make use of metadata.

instrument's available pitch range, these may be textified and stored in pure text form. Textification is an optional step as it is unnecessary when data are already in pure text form. Textified data are stored in text datawarehouses, buffers, text fields or clipboards.

15.2 Text Mining Workflow

Text mining starts with raw unstructured text scattered over thousands or even millions of documents or entire data contained in a single document like millions of transactions recorded into a file. Text data are converted into numbers (or as vectors) using transformation techniques. Thus as a prelude to the text mining process, a term-by-document matrix (described below) is output by the first phase. The documents may be in various formats as in proprietary formats from Microsoft or Adobe, html and xml formats or in plain text files. Feature dimensionality (total possible combination of features) in text mining is more elaborate than in classical data mining.

15.2.1 Text Preprocessing

A text pre-processing step includes text cleansing and filtering (removal of stop-words, unwanted characters, punctuations etc)[5], tokenisation[6], NLP parsing, boundary detection[7], noise removal (mainly spelling errors, capitalisations (Cat, CAT, cat, etc are all converted into a single form)), parsing abbreviations, acronyms, spelling anomalies (British and US spellings), and pruning.[8] Words with very low occurrence frequencies in the corpus are removed during the pruning step. A filtering step removes stop words (the, of, an, and, to, at, above, below, bottom, far, many, etc in English), stems, and formatting information (from word processing files) as these are worthless in most (if not all) analyses[9]. Collocation identification may be needed in some documents.

---

[5]Multiple delimiters (eg: multiple spaces and tabs), special characters and punctuation marks, if any, that does not hold any discriminative power can be filtered during the filtering stage.

[6]Tokenisation splits text data into individual tokens, which are either words, numbers or logically related phrases.

[7]Boundary detection tries to identify logically related groupings. Text mining uses a pipeline of NLP tools that perform a sequence of tasks like sentence splitting, tokenisation, part-of-speech tagging, lemmatisation, etc. term extraction, For instance, boundary markers are the carriage return/linefeed in log files, end tags in HTML/XML documents and dot (.) in word documents/ HTML paragraphs with more than one sentence. Word processing documents may also contain hard/soft return characters.

[8]dozens of English words are spelt differently in various parts of the world. Examples are [color, colour], [neighbor, neighbour], use of 's' and 'z' in many words like civilisation/civilization, specialised/specialized, etc

[9]See [BC85] for a list of around 500 most frequent non-discriminating words in English.

Collocations are phrase combinations that make sense only when used in a proper order. The meaning of a collocation is totally different than the individual parts (eg: object oriented' and 'oriented object'). Collocation discovery is simple and straightforward using pattern matching and dictionary lookup techniques (see www.gutenberg.net for Webster's dictionary files).

15.2.1.1   Stemming

Stemming (also called lemmatising) reduces a word to its root (or stem) in the language with an intention to improve the scope for similarity matching and dimensionality reduction (as the number of terms in a document is often decreased). A major problem with stemming is the polysemy problem. Another is that stemming may not always be reversible (because a stem may appear in many other contexts) This of course is dependent on the language. Majority of stemming in English is suffix based, in which case it uses a part-of-speech tagger because different parts of speech may have to be lemmatised differently. But there are also negation stemming that starts by analysing the prefix of a word. Examples are prefixes like 'in' (as in incomplete, inaccurate, inequality), 'dis' (dislocate, disagree, disparity), 'un' (undo, unwell), etc. Due to the very large number of ways in which negation exist in English, it is not easy to do pre-stemming in English. Nevertheless, algorithms like Lovins stemming, Porter stemming etc does a good job.

Definition 15.2 Stemming is the process of replacing numerous permutations of a word wrt the grammatical considerations of the language.

This process is different for different natural languages. The discussion here is confined to English alone. English verbs can appear in past, present, future and gerund forms. Noun-stemming involves singular and plural fixing, pronoun and synonym replacements, etc. Terms are reduced to their stem or root variant. For instance, *computable, computation, computational, computed, computing, computerised*, etc are reduced to their stem *compute*[10]. Each root-word has an ambit – the extend of expandability of the root. For example, 'computation' in the above list need not involve a digital computer, as it could be hand computation, theory of computation, or analog computation. Hence syntactic loss of meaning may occur if every occurrence of 'computation' is reduced to the stem *compute*. This process is called *noise removal*.

Data-specific filters may be used to extract relevant text when data come from pre-formatted documents (eg: Excel like spreadsheets, XML files with metadata). Hence this step can be implemented as a pipeline[11], in which each stage performs a particular task. In addition, parsers[12], tokenisers, regular expression and semantic analysers, linguistic filters etc may also be required. For example, a linguistic pre-processing may involve parts of speech identification, word disambiguation (does 'java' mean the programming language, the island, or some other java? Is CAT an abbreviation, a word, or part of another word like catfish, catalyse, catch, etc) to make subsequent analysis easier[13]. Most software have builtin functions or commands to parse and tokenise the input. One example is the PERL *split* function that accepts delimiters (space, tab, comma, etc) as the first argument. Relevant data are extracted from these documents and captured into

---

[10]The mean or median modified Hamming distance between words and their stems is used to identify root words automatically.

[11]A pipeline is an architectural style that comprises of two basic components called pipes and filters. Two or more of these are sequentially connected, such that the pipes serve as conduits for data and instructions to subsequent filters.

[12]In compilers, a parser denotes a program that extracts tokens from the source file. In text mining, a parser extracts the words, phrases and other entities (like numbers, labels, etc) from the input data.

[13]There are many types of ambiguity — word ambiguity, semantic ambiguity, multi-lingual ambiguity, etc

a text-warehouse where the data are a stream of characters separated by predefined delimiters. A data dictionary or index may help in accessing this information faster. A visualisation tool may be used to view knowledge maps, structural relationships, summary measures, etc. Specialised text mining may utilise resources like grammars, lexicons, thesaurus, and annotated corpora. See [SB07] for an automatic synonym extraction method in a monolingual dictionary.

Linguistic concepts and grammar lie at the core of NLP. This is of course language dependent because the grammatical structure of sentences vary a lot among world languages. Linguistic features can be extracted in a domain-independent manner using NLP algorithms. Major tasks in NLP are POS tagging and parsing.

15.2.1.2 Parsing

Parsing is the process of identifying white-space, punctuation and other special characters in the data and extracting words from the data. This is easy to do in structured files like XML or VRML file. If the input file is in proprietary formats (like Adobe PDF, MS Word, etc), vendor supplied API's may be needed to quickly parse the data. This can be implemented as pipes-and-filters architecture as the inputs and outputs are well-defined. Lookup tables may be needed in some cases (as in parsing source programs or music files textified using specially built vocabulary).

15.2.2   Goals of Text Mining

A text miner's goal may be to extract entities, attributes, classes, features, parts-of-speech, ontologies[14], etc. Entity extraction is used to mine named entities (people, locations, organisation names, drug names, etc) while attribute extraction is used to mine for patterns and trends in attributes (drug combinations, payment defaults, etc). Concept extraction aims to extract key concepts and themes from document collections so as to glean meaningful information in natural language documents. Three properties of ontology are exhaustivity, granularity and specificity. Exhaustivity measures the breadth of coverage while specificity measures the depth of coverage. Granularity is a measure of the level of detail. Text-clustering techniques can be used to reduce data dimensionality substantially using semantic mapping of text corpora. Concept and term-clustering tries to automatically identify related groups of terms in large text corpora. Co-clustering is an extension that simultaneously clusters rows (documents) and columns (terms), thereby exploiting the duality between documents and terms. This often gives better results. Text mining can obtain partial structure among unstructured document corpora. It has also been applied to obtain reasonable estimates of missing values by extracting patterns from similar document corpora. Incremental algorithms allow data to be processed incrementally, and to save intermediate results for use in subsequent processing.

Attribute extraction aims to filter a subset of the original attributes that best describes the data, or satisfies certain criteria. Filter approach and Wrapper approach are two popular attribute selection methods. Filter approach evaluates attribute relevance purely based on the data, whereas wrapper approach evaluates attribute relevance using data and induction algorithms employed. The Filter approach is the simpler of the two. It uses a criterion to evaluate the quality of attributes. Popular numeric criteria are mutual information, entropy, gain ratio, correlation, etc (§15.5). Nonmetric criteria also exist for attribute selection (eg: Promise

---

[14]An ontology in computer science denotes a set of concepts and their inter-relationships (http://en.wikipedia.org/wiki/Ontology_computer_science) and can mean generic, domain, application, or representational ontology

criterion). Filter approach can be used for dimensionality reduction when a large number of variables are present in text corpora. Wrapper approach on the contrary searches the input space using heuristics or search techniques like evolutionary algorithms, ant-colony optimisation, DFS and BFS searches, Tabu search, beam search, etc. It can extract better attribute subsets than obtainable by filter approach as it uses subsets of attributes rather than individual attributes. Co-clustering of documents and terms can help to reduce data dimensionality substantially.

Text mining is being used in medical sciences to improve the potential to discover new uses for drugs, reveal dangerous drug interactions, for rule discovery [MS05] etc. It is used in biological sciences and genetics for disease gene identifications. It can be used as a first step to extract patterns or summary information, which are either subsequently mined with more robust models like decision trees, association rules etc or used as a basis for applications and services.

> Principal Component Analysis (PCA) is a statistical technique for dimensionality reduction. It works purely with quantitative data. It first locates the direction of a set of orthogonal vectors in the original space that accounts for maximum variance, and then projects the entire data onto the space spanned by these orthogonal vectors as coordinate axes, so as to retain the essential features of the original variables. This is especially useful to text data that typically have large dimensionality. Discriminant analysis is a similar technique to combine the original data features to most effectively discriminate between the classes. It can also be extended to reduce data dimensionality of a TD matrix by preserving any implicit cluster structures inherent in it. This is usually done in two stages [BM03]. See also cacm.acm.org/magazines/2010/2/69359-faster-dimension-reduction/fulltext.

Text mining improves information access by enabling semantic querying and can be used for event detection from text corpora (newswire articles, sales histories, e-commerce transactions, free-form text notes and transcriptions). A goal oriented query engine can be coupled to effectively mine useful patterns. Consider the purchases of books by an online shopper. A data mining model can be built by mining the subject combinations (eg: Fiction and Sports; Physics and Astronomy, etc), or topic combination within each subject (eg: Java programming and databases, data structures and C++ programming, etc) that most customers buy together. Thus the buying habits of past customers can be used to suggest related books, or offer discount on combination items for new customers. See [AP98], [WI05] for details.

15.2.2.1   Text Mining vs Data Mining

There are several differences between the two, both in terms of the data they deal with and the processes involved. Data for text mining can come from text-datawarehouses (that contain cleansed data), text extracted from elsewhere (say from the web), textified collections (say speech, conversations, lectures, songs) or free form text (as in dictations, textified documents). Data mining aims to extract previously unknown and potentially useful information from numeric historic data, while text mining extracts information which is either explicitly stated in the text or can be inferred using linguistic transformations. Text mining is more challenging because it depends on the language of the document, whether the data are time-stamped, and the domain. Computational requirements are maximal during the initial stages of text mining (pre-processing, filtering, linguistic analysis, etc) whereas it is comparatively easier in data mining.

15.2.2.2 Document Facets

Some of the words in a document may carry more weight than others. For example, the words "patient", "symptom", "duration", "diagnosis", and "medicine" may weigh more than all other words in a patient record or medical transcription. Those words that best represent or capture the key concepts, ideas or information in a document are called the *facet* of the document. There are many algorithms available to identify facets, most of them use the occurrence frequencies of words.

> Facets filtering is analogous to extracting factors in the statistical technique called factor analysis. The key differences are that (i) factor analysis works with quantitative data, whereas facets works with text labels (which are in a broad sense categorical data) (ii) facets do not uilise variance or correlation, whereas factor analysis uses the inherent variability in the data to the maximum, and comes up with independent linear combinations of attribute values (called factors) that account for maximum variability in original data.

A problem in facet identification is synonymy, which are multiple words with the same meaning. This is more so in large passages and documents than small ones. Moreover, scientific documents may contain "technical words" that are non-dictionary words that have a special meaning in a subject. Examples are "mining" which has a different meaning in data mining than its literal meaning, "recall" which is a measure used in text retrieval. If a list of technical words that are likely to appear in a document are known apriori, it can be used to compute the importance score in the algorithm.

Algorithm 15.1 Facet Identification Algorithm

1: Read Input Document D {* N is the total number of words in D *}
Ensure: (N$\geq$ 2)
2: E ={} {* E contains the list of Facets *}
3: Extract the list of keywords and their occurrence counts by pre-processing
4: Sort words on descending order of occurrence counts
5: Compute an importance score (real number) for each keyword or phrase iteratively using the document frequency, distinctness, etc.
6: Temporarily select the keywords with the highest scores {* m= total number of Facets *}
7: Prune the facet using KL-divergence measure
8: Compute pair-wise similarity between the active keywords in the document (the words in current Facet) using the Jaccard similarity measure.
9: Identify keywords and phrases with high average Jaccard coefficient as prominent facets
10: Form E = List of prominent facets
11: **return** E

Sometimes an expansion of predefined acronyms may be needed to extract matching records. As an example, consider Artificial PaceMakers (APaM) used to control abnormal heartbeats. They were invented by a Canadian electrical engineer named John Hopps in 1950. The first generation APaMs were all Battery-Operated Devices (BOD) installed externally with electrical cables running into the heart. Second generation APaMs with miniaturised batteries are Surgically Implanted Devices (SID) in close proximity of the heart. An ensemble naming convention is adopted to categorise APaMs based on which heart-chamber is paced, or which intrinsic heart-rhythm is modified. Third generation APaMs with miniaturised wireless transmitters that can be charged remotely are being developed.

Theoretically there exists 4^4 different possible combinations, some of which (eg: OOOO,

Table 15.3: An Artificial Pacemaker Nomenclature

Position	I	II	III	IV
Category	Chamber paced	Chamber sensed	Response	Rate response
Code	O = none A = Atrium V = Ventricle D=Dual (A+V)	O = none A = Atrium V = Ventricle D=Dual	O=None, T=Triggered I = Inhibited D=Dual (T+I)	O=None, R=Rate modulation A=Atrium,V=Ventricle D=Dual (A+V)

Multisite pacers are also in use that intelligently senses multiple sites. Other nomenclatures also exist.

which represents a dead pacemaker) are meaningless. The following are valid codes for the pacemakers:– AVIO, DATO, VATA, ADOR. Here AVIO describes a pacemaker[15] that senses the Ventricle (V), paces the Atrium (A) with Inhibitance (I), and with No rate response(O). Assume that a medical researcher is interested in correlating the type of pacemaker used by a group of patients with a fatal heart condition. If past data are available, a text query can retrieve all such patient records by a simple expansion of each abbreviated functionality. Similarly, another query might be the opportunistic illnesses in all patients having a specified type of pacemaker (say DOOR). Some of the important document facets in these cases are already known. Such examples exist in organic chemistry, genomics, proteomics, bioinformatics and many other fields.

15.3   Term-by-Document Matrix (TD-Matrix)

Some text mining operations may need to convert unstructured text data into semi-structured or fully-structured formats. One popular technique is the TD-Matrix (chapter 14, page 14-3). It converts raw data into a more tractable 2D-representation. It is assumed in the following discussion that the rows are labeled with terms[16] and columns are labeled with documents. This matrix will contain the frequency (count, percentage or fraction) of occurrence of various terms in the document, or a binary indicator of the presence or absence coded using 1 and 0 (the binary coding does not carry weightage and is not popular). The above matrix has numeric or categorical entries and hence lends itself to data mining process. Latent semantic indexing (LSI) (chapter 14) is a text mining technique that uses a *truncated SVD* of the term-by-document matrix. It provides a low rank approximation to the original data in a subspace thereby providing a dense approximation. It is typical to have thousands of terms and millions of documents in open (unrestricted) text mining applications, making it difficult to manage the often sparse TD matrix. The popular subspace reduction techniques are the LSI and Covariance matrix analysis discussed in chapter 14. These can save memory and computations tremendously. An example of biomedical text mining for generating gene summaries appears in [LJ07].

15.4   Text Classification

Majority of public domain documents on the web are of text form. As millions of documents are made available online each year, this collection is gradually growing. This has created an interest for fast text categorisation algorithms.

---

[15]By changing the order of the columns in table 15.3, this could also be abbreviated in other forms like VAIO.

[16]Terms are either single words, or logically related group of two or more words.

Table 15.4: Data Mining vs Text Mining

	Finding Patterns	Finding Nuggets	
		Novel	Standard
Non-textual data	General data mining	Exploratory Data Analysis	Database Queries
Textual data	Computational Linguistics	Text Mining	Information Retrieval

Definition 15.3 Automatically assigning (two or more) semantic categories to natural language text documents using text features and commonalities is known as text classification or categorisation.

In other words, applying one of a finite number of distinct labels to text documents using properties of text in labeled documents is called text classification. It is a confluence of information retrieval/extraction, computational linguistics, pattern recognition, machine learning, visualisation and statistics. It can be supervised or unsupervised. In supervised text categorisation, the classes into which documents get binned (as well as the similarity or dissimilarity measure(s) to be used) are provided by the data miner (during training stage). Semi-supervised text categorisation algorithms are also available for large-scale classification problems [JT02]. Automatically detecting the natural language in which a web-page content is developed is a simple example of categorisation. The input in this case is either a single web page or a set of pages from a web site and the output is the language of the content. This task is quite trivial as most of the HTML pages tag the main language in which the content is developed, but could require lookup tables or dictionaries when content is developed in multiple languages.

A predefined numeric metric on the document space is used to perform unsupervised categorisation or text clustering. The raw data for text categorisation are individual documents - web pages in the case of markup documents, application documents or logical units (in the case of dictations, transcriptions etc). They can be used as unsupervised tools or can supplement human efforts in discovering desired patterns (classification (syntactic or semantic), clustering (entity or attribute), associations (lexical, contextual, syntactic) etc). On the contrary, supervised text classification pre-computes the metric using the training set of documents that are assigned to target categories.

The instances in text categorisation are text documents or logical units of text. Manual classification is often time consuming and ineffective even for small document collections, due to the high dimensionality of feature space. This can be done without human intervention by softwares that employ cognitive intelligence, feature space mapping etc. Consider a large number of online survey responses. A text miners aim may be to extract unknown facts, truths, trends, anomalies or even casual associations or other information among either documents or entities that generate the data. Text categorisation aims to automatically label documents into distinct predefined groups. These are useful for search indexing and cataloging, document clustering (eg: spam filtering), data summarisation, selective archiving, etc. A binary classifier is the simplest, in which each category has two classes. Information extraction by filtering documents from large collections using user queries can be considered as a simple binary classifier (matching records are retrieved [class 1] and others discarded [class 0]). An unbalanced classifier is one in which disparity between the two categories are significant.

The categorisation process may involve several phases due to the large number of phrases of varying lengths. In the filtering phase, common features are extracted, non-informative features are discarded and document clusters are identified using commonalities, if any. See [JK11] for a survey of document clustering. Similar to outliers in random samples, there could be documents that do not have features in common with others, or that tend to be oddly different from others in the features sought. Hence hierarchical techniques in which each document belongs to a distinct group initially are employed. The groups are then combined using an appropriate similarity or dissimilarity measure. The above algorithms assume that the text documents are labeled. Classification becomes more challenging when the collection contains labeled and unlabeled documents [NK00]. In this case, the training data must comprise of a small number of labeled documents, and a significantly large number of unlabeled documents (noisy negatives). A support vector machine (chapter 13) (SVM) approach for this purpose can be found in [YZ03].

Text classification may also involve classifying entities based upon text data originated from them. Examples are – classifying customers using customer reviews, classifying students by analysing essay contents, insurance claimants using claims data, patients using medical history, teachers using student evaluations, respondents using survey data, tourists using guestbook suggestions or messages, and pilots using cabin conversations[MS05]. Similarly, software professionals can be classified using the program code and comments that appear in the source they develop. It is implicitly assumed that the data are uni-lingual. When data in multiple languages are present, the classification algorithm is applied to each language document, and the results are finally combined (eg: using multi-class SVM).

As mentioned above, the unit for classification can be documents of various types (ms word, pdf, ps, excel etc), web pages, email messages, chat-room discussions, news articles and so on. In the case of stream data (as in chat-room discussions), each user session can be considered as a unit. The classes into which a unit can fall must be provided during the training phase. For example, company documents can fall into various departments (accounts, business, HR, marketing, purchase, etc), subject areas, correspondence types (business, personal), line of business (e-commerce, direct marketing, etc), news types (calamities, entertainment, politics, science & tech, sports), etc based on their content. Each document to be classified must be mapped to a vector of numeric values [LR94],[II06]. In other words, each document is mapped to a numeric vector with weights normalised in the range $[0,1]$ as $D_j = (w_{1j}, w_{2j}, \cdots, w_{kj})$ where k is the number of features chosen. An interpretation of w_{ij} is that it represents the relative contribution (in the semantic sense) of ith term in j^{th} document. The simplest weighting scheme is the binary weighting in which $w_{ij} = 1$ if $term_i \in Doc_j$, and 0 otherwise. (Documents are mapped to a binary vector using terms, such that each vector element represents the presence (=1) or absence (=0) of a term in the document). The natural generalisation of this scheme is the term frequency (TF) weighting in which the weights represent the frequency (number of times) of term occurrence in a document. Measures that are more complex use term frequency (TF) or weighted versions of it (TFIDF) to reflect the relative importance of each term in the document. The TF measure could result in big numbers (at least for some terms) in large documents (eg: books, news articles and large reports). Scaled versions of this measure include reciprocal, exponential and logarithmic transforms of TF. Examples are log(1+TF), 1/(1+TF), exp(-TF), etc in which log is to the base 10 or e (a 1 is added (in the first 2 measures) to avoid the possibility of TF=0 causing overflow problems ($\log(0) = -\infty$)). As TF increases, log(1+TF) also increases. Other two measures are monotonically decreasing functions of TF (hence they are called inverse term frequency (ITF) measure). Above measures are usually weighted by simple weights (that have a scaling effect)[HR05], or by inverse document frequencies (IDF) that add feature discriminatory

Table 15.5: Weight matrix for text categorisation

Sr.	label	1	2	3	4	5	6	7	8	9	10	11	12	13	14	15
1	-1	1	0	1	1	1	1	0	0	1	0	0	0	0	0	0
2	+1	0	1	1	0	0	0	1	1	0	0	0	0	1	0	0
3	+1	1	0	0	0	0	0	0	1	0	0	1	0	1	0	1
4	-1	0	1	0	1	1	1	1	0	1	0	1	0	1	0	0
5	+1	1	0	0	1	0	0	1	1	0	1	0	0	0	0	0
6	+1	0	0	1	0	0	1	0	1	1	0	0	0	0	1	1
7	+1	0	0	1	0	1	1	0	0	1	0	1	0	0	0	1
8	-1	1	1	0	1	0	0	0	0	0	0	1	0	0	1	1
9	+1	1	1	0	0	1	0	1	1	0	0	1	0	0	0	1
10	-1	0	1	1	0	1	1	1	0	1	0	0	0	0	1	1

Legend: +1=newswire article, -1=private email message.

Table 15.6: Results of SVM training on text categorisation

Sr no.	SVM/ SVR	wt	kernel type	SV	SV (UB)	Ll norm	b	VCDim <
1	SVM	—	poly	10	2	1.49	-.779179	12.24
2	SVM	—	rbf	10	4	4.6636	-.66573	4.347
3	SVR	.5	poly	10	0	0	-.32382	—
4	SVR	.05	rbf	10	2	1.3247	-.732278	—
5	SVR	.005	rbf	10	4	4.6236	-.66073	—

power relative to other terms. Each data vector should preferably be normalised to the unit range. For example, the TF measure may be scaled by dividing it by the total number of terms in all documents combined as RTF=$n_i/\sum_j n_j$. Text data may also have parts with unequal weights. For instance, data about a book may include author names, title, reader reviews etc with unequal importance. Similarly, a journal article has a title, abstract, keywords and phrases, main body, and usually a summary/conclusions section. An insurance claim data may contain various sections (claiming reason, injury description, incurred loss etc) with unequal importance. Feature reduction using cutoffs (low frequency terms are ignored to speed-up the process using a cutoff frequency), term size (eg: 'a', 'I', 'an', etc in English), term relevance (irrelevant high-frequency terms are ignored), sparsity (terms that appear in one document are dropped) and other techniques can be used to speed-up the classification. The bag-of-words representations convert each document into a vector of values using the map $\phi_i(x) = TF_i * log(IDF_i)/k$ where k is the number of categories. A pre-processing step eliminates stop words, and does stemming. Several free softwares are available for this purpose (see below).

Example 15.1 Table 15.5 presents the term vectors extracted from an email correspondence (label -1) and a newswire article (label $+1$) using the keywords {assure,construction,dear,dinner, email,family, final,fund,hope,impounded,power,protest,secretary,social,technology}. Develop a classifier to automatically label a new document into the above categories.

Solution 15.1 The problem was solved using SVM and SVR with various parameters. Results are summarised in table 15.6. The weight vector of OSH for the last case is given by [.5,.33333,.33333,.5,.33333,.33333,.33333,.5,.33333,.5] with corresponding labels [-1,1,1,-1,1,1,1,-1,1,-1]. The classifier can be obtained using the intercept term 'b' as f(x) = $\theta(\sum_i \hat{a}_i y_i K(x, x_i) + \hat{b})$) (with some softwares, the intercept term b can be slightly sensitive to the scaled data range. As an example, if the scaled range is [0,1] we get b=-.66073, and if it is scaled to [-1,1] we get b=-.661666 using SVM_LEARN).

15.4.1   Temporal Text Mining (TTM)

Mining time tagged text data provides insight into temporal patterns and trends. Most of the web documents (like HTML, XML files) seldom carry time information (except last updation date). There are many specialised applications where the time of an activity is important. For instance in repeat visits of a patient to a clinic, or a client to a lawyer involve time-tagged materials in text form. Similarly, a customer service center may carry time tagged data about the services offered to the customer, reviews received from the customer, items shipped to the customer etc. Online newspapers, discussion forums, and email archives also tag each item with a time stamp (otherwise it could have disastrous consequences). Ignoring the time-stamp can sometimes lead to wrong results or conclusions. More than one time-tags may also appear in some applications. For example, an email message has a date of receipt, date first read, date replied and date last accessed, (and sometimes dates forwarded, date archived etc). News articles or queries on discussion groups similarly have an evolutionary pattern with a beginning date, a progression period, an impact period and (usually) an ending date [MM97]. Pattern similarities among recurrent or repeated events can be obtained using an ensemble of text mining and other data mining tools. Trend analysis can be temporal too. In this case the trends found in two time intervals (that may or may not overlap), usually of fixed duration, are compared for commonalities. A related concept is change and deviation detection in text. This usually applies to text data (or images that are tagged with text or have embedded text) that change over time.

Temporal documents are processed in chronological or reverse-chronological order. Extraction of historical data will require a time interval (starting and ending time), one (or both) end(s) of which can be open (unspecified). For instance, in mining for information about all 'car recalls' by automobile manufacturers to fix a design defect, one can keep the ending interval as unspecified. In this case, all cars recalled from specified starting date up to the present time is taken into consideration. To mine for patterns in relief efforts in natural disasters (like tsunamis, earthquakes, tornadoes) using data in the public domain, one may keep both time intervals open because chances are that online information about such disasters that have occurred before 1990 is rare or scanty. But as time progresses, this may create problems due to the large number of such documents being added online. A temporal clustering may be carried out as a prelude to filter out important events and the TTM performed afterwards.

Let D=$\{d_{t_1}, d_{t_2}, \cdots, d_{t_n}\}$ denote a time-stamped document collection arranged chronologically. A temporal query can specify the time interval, a theme (topic or subtopic) and a weightage. The results may be presented in tabular format, as a graph, flow diagram or by other methods. In [MZ05], the authors use a 'theme evolution graph', which is a special case of an edge weighted graph in time lanes. The time span need not be restricted to a single interval. If the approximate time spans of multiple occurrences are known, we could also specify multiple time spans. For example, to mine for relief efforts that span 100 days after devastating tsunamis

Figure 15.1: Architecture of a text classifier

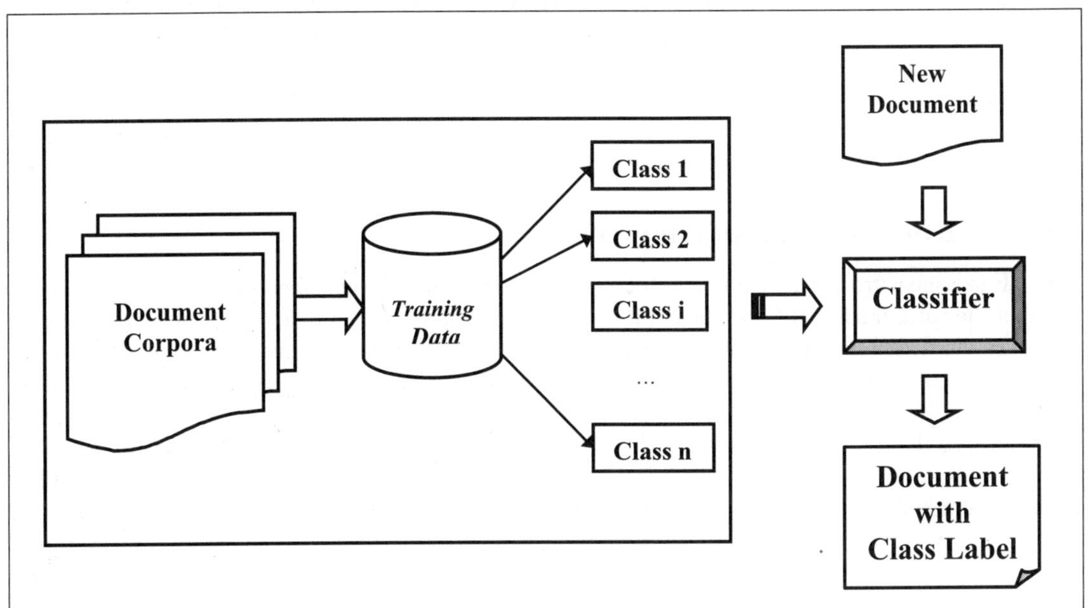

from 1998 to present, we can specify multiple time intervals as [17 July 1998, 25 Oct 1998][17] and [26 Dec 2004, 5 Apr 2005]. Since the time intervals are used as filters, the data are first filtered to remove irrelevant documents. Resulting data are then ordered in temporal dimension and analysed using text mining tools or specialised techniques (see [PB04] for an application of multinomial PCA for theme extraction). Similarly mining the complete works of well-known authors may require time windowing, author disambiguations, language selection etc (the complete works of Shakespeare contains 31500 different English words of which 14400 appear only once). A threshold based text mining approach for multiple time intervals was suggested in [CZ06]. More information about TTM can be found in [RG02], [PB04], [MZ05].

15.4.2 Distributed Text Mining (DTM)

Distributed text mining methods have emerged due to the computational complexity of mining a large number of documents spread spatially all over the web. One of the challenges involved in DTM is identifying the best data partition boundaries. Since the servers participating in DTM may be dispersed over geographically different regions, the data may either be mutually exclusively partitioned or stored on Internet Backbone Protocol (IBP) servers or depots. This is especially useful for mining of bioinformatics and genomics data. Collaborative filtering[18] is a special type of social filtering in which opinions of user groups spread across different geographic regions or interest groups are filtered using multiple servers in a distributed fashion.

---

[17] The 1998 tsunami of Papua New Guinea killed 2182 people.

[18] This term is also used in search engines that distribute the workload among multiple servers, and combine the results to be delivered to users.

15.5   Metrics for Text Mining

Because of the popularity of text mining algorithms in various disciplines, many metrics have been suggested for it [FA00]. Each document can be considered as a point in the feature space in which the terms are the axes, and the coordinates (along each dimension) are proportional to the number of times the corresponding term occurs in the document. Documents that tend to cluster together are more or less homogeneous, or have slight overlaps. In the following section, we summarise a few of the most popular metrics that are useful in later stages of text mining.

15.5.1   Document Frequency (DF)

The DF is a measure of the occurrence of a word-stem in document collections. The $DF(term) \in (0,m]$ is the total number of documents in which the *word-stem* occurs at least once (0-frequency cases are discarded in the analysis). If there are n non-overlapping terms and m documents, the DF can be conveniently arranged as an n x m matrix (with columns representing frequencies of terms). Low frequency columns will then indicate rare terms that are non-informative in unstructured documents or represent 'noise' terms (terms that are too frequent (eg: an, and, be, for, in, or, the, etc) may also be non-informative). In simple structured documents (like research papers, technical reports, theses), phrases may appear in the title, abstract, keywords, main text, and summary sections or in footnotes. Similarly in semi-structured web pages, the phrases may appear in headers, meta sections, tables, or in various parts of main text. Hence the frequency of occurrence may be weighted (with different weights for various sections) to get a weighted document frequency (WDF) measure. The weights can be normalised in the interval [0,1]. For example, the keywords section (which is usually prepared by the author(s) and captures all essential content in a few words) may be given highest weights followed by the title, abstract, summary, main text and footnotes. It is often used in clustering, classification and anomaly detection. A document D_1 is said to subsume another document D_2 if every term in D_1 is also in D_2. This information is useful in dimensionality reduction, common feature extraction, clustering and similarity matching. The DF measure can be used to check for subsumption $(d_i^1 \leq d_i^2 \forall i)$.

15.5.2   Term Variance

(TV) A document can contain 0 or more occurrences of a term. If a non-trivial term occurs many times in a document, it could contribute more to the measure than low frequency occurrences. This metric gives the *unnormalised* (or unscaled) variance of the frequency of occurrence of a term in various documents (with DF>0). $TV(term) = \sum_{i=1}^{m} O_i^2(term) - \frac{1}{m}(\sum_{i=1}^{m} O_i(term))^2$, where $O_i(term)$ is the number of occurrences of term in i^{th} document. This metric does not consider the relative size (number of words) of the document.

Table 15.7: Occurrence frequency of terms in documents

i	Doc_1	Doc_2	Doc_3	Doc_4	Doc_5	Doc_6
the	1	0	4	3	6	2
boy	5	2	1	0	3	4
Total	6	2	5	3	9	6

	present i		
1	p	q	p+q
0	r	s	r+s
sum	p+r	q+s	p+q+r+s

Example 15.2 Compute the term variance for the data in table 15.7.

Solution 15.2 TV(the)=$1^2+4^2+3^2+6^2+2^2$-$16^2/6$=66-42.66=23.44. Similarly, TV(boy)=5^2+ $2^2 + 1^2 + 3^2 + 4^2$-$15^2/6$=55-37.5=17.50.

15.5.3 Relative Term Variance (RTV)

This metric measures the variance of term occurrences, normalised by the total number of meaningful tokens (after removal of duplicates and stop words). Let T(i) denote the total number of nontrivial phrases in i^{th} document. Then RTV(t) = $\sum_{i=1}^{m}(O_i(t)^2/\text{T}(i))-(1/m)(\sum_{i=1}^{m} O_i(t)/\text{T}(i))^2$.

Example 15.3 Compute the relative term variance for the data in table 15.7.

Solution 15.3 Here T(1)=6, T(2)=2, T(3)=5, T(4)=3, T(5)=9, T(6)=6. RTV(the)=$1^2/6 +$ $0^2/6 + 4^2/5 + 3^2/3 + 6^2/9 + 2^2/6$-$16^2/6$=66-42.66=23.44. Similarly, RTV(boy)=$5^2 + 2^2 + 1^2 +$ $3^2 + 4^2$-$15^2/6$=55-37.5=17.50.

15.5.4 Information Gain (IG)

The IG is an entropy measure that utilises the presence or absence of terms in various documents [YP97]. It is defined as IG(w) =

$$-\sum_{i=1}^{m} P_r(c_i) \ln(P_r(c_i)) + P_r(t) \sum_{i=1}^{m} P_r(c_i|t) \ln(P_r(c_i|t)) + P_r(\bar{t}) \sum_{i=1}^{m} P_r(c_i|\bar{t}) \ln(P_r(c_i|\bar{t})) \quad (15.1)$$

where $c_i, i = 1, \cdots, m$ denotes the target space categories, $P_r(c_i)$ denotes the probability (fraction) of documents with topic c_i, which is easy to obtain by first counting the number of documents without its occurrence (say $P_0(c_i)$ and using $P_r(c_i) = (n - P_0(c_i))/m$ (This count is easy to obtain from the matrix representation) and $P_r(t) = \text{DF}(t)/m$. The $P_r(c_i|t) = m/\text{DF}(t)$ denotes conditional probabilities and $\bar{t}$ denotes the non-occurrence of term t.

15.5.5 χ^2 statistic

A contingency table is a popular statistical model for analysing multi-way categorical data. As the name implies, the data are captured into tabular form in which rows and columns are labeled with different attributes, and cells at intersections contain frequencies of corresponding occurrences. The Pearson's χ^2 statistic is used to test the significance of computed values with tabulated value. It is a normalised value used to measure the lack of independence between a term and a topic. It gets its name from the fact that the numerical measure is compared with a χ^2 distribution with (r-1)x(c-1) = 1 degrees of freedom to check for significance. In the following table, 'p' is the number of times term and category co-occur, and 's' is the number of times they do not occur, and so on. A term goodness measure is then defined as

Table 15.8: A list of words in several documents

Document#	List of words
1	{breakfast, bread, juice, milk, omlette, restaurant, street}
2	{bread, butter, coffee, omlette, school}
3	{breakfast, milk, park, restaurant, street}
4	{coffee, restaurant, school, uniform}

$$\chi^2(t,c) = \frac{n(ps - qr)^2}{(p+r)(q+s)(p+q)(r+s)} \tag{15.2}$$

If terms and categories are independent, the above statistic is zero.

15.5.6 Term Strength (TS)

This measure is defined over two (or more) documents that are distinct, but related. Let y denote a document and d_i denote distinct, but related documents. Then the simple term strength is defined as $TS_y^{x_i}(t) = P_r(t \in y | t \in x_i)$, which by the Bayes rule can be evaluated as $TS_y^{x_i}(t) = P_r(t \in y \cap t \in x_i)/P_r(t \in x_i)$. This measure can be extended to any number of documents as follows: $TS_y^{x_1, x_2, \cdots, x_n}(t) = P_r(t \in y | t \in x_1, x_2, \cdots, x_n) = P_r(t \in y \cap t \in x_1 \cap t \in x_2 \cdots \cap t \in x_n)/\prod_{i=1}^{n} P_r(t \in x_i)$

15.5.7 The TFIDF measure

Term frequency is the number of occurrences of a term in a document (If the term is a noun, it may be referred to in the document multiple times by pronouns like he, she, it, his, her, etc).

TF(t,d) = freq of occurrences of term t in document d. The relative TF is RTF(t,d)=TF(t,d) /tcard(d), where tcard() denotes term cardinality (total number of unique terms in document d, not counting pronouns). DF(t) is the total number of documents in which term t occurs (see 15.5.1). IDF(t) = log(card(D)/DF(t)) where card(D) is the total number of documents under consideration. Because the minimum value of card(D)/DF(t) is 1, the IDF(t) ≥ 0 and monotonically decreases as more and more documents contain the term. Hence the discriminating power is low if a term is too common (it occurs in many documents). Rare terms (occurring in few documents) provide higher discriminating power. The TF and IDF can be conjoined in determining the relevance, giving the TFIDF measure as TFIDF(t,d) = TF(t,d)*IDF(t). These weights can be normalised to the interval (0,1) using an appropriate data transformation.

15.5.8 The CNC measure

The CNC method involves 'to and fro' hyperlinks, which is useful for link analysis. It is more robust due to the fact that both linguistic analysis (parts of speech tagging, stop list, linguistic filters, etc) and statistical analysis (frequency counts, C/NC-values) are carried out and resulting phrases are ranked by NC-values. The TFIDF method uses frequency counts of words in text files (TF) and number of documents where it occurs (DF). Terms that appear in many documents get a low score over the terms that appear in a few documents. A document clustering technique for

Table 15.9: Document Frequency and Inverse Document Frequency

#	Word	DF	IDF
1	breakfast	2	0.301029996
2	bread	2	0.301029996
3	butter	1	0.602059991
4	coffee	2	0.301029996
5	juice	1	0.602059991
6	milk	2	0.301029996
7	omlette	2	0.301029996
8	park	1	0.602059991
9	restaurant	3	0.124938737
10	street	2	0.301029996
11	school	2	0.301029996
12	uniform	1	0.602059991

browsing search results can be found in [KL04]. An algorithm to find clusters using key phrase labels can be found in [ZH04]. See [MS00], [GK99] for further results on text mining.

15.5.9   Cosine metric

The cosine of the angle between two normalised vectors is a measure of similarity in terms of relative distributions of components. In other words, the more the value of cosine metric, the greater is the similarity (as $\cos(0) = 1$). This measure does not consider the relative magnitude of the components. Thus if one document is several megabytes in size and another is just a few kilobytes, the angle between them could be very small if they are very similar. Thus the relative magnitudes of documents do not influence the value of cosine measure. When used for text-retrieval or clustering, a threshold must be used to decide how many close-by documents need to be extracted.

15.5.10   Edit distance

Edit distance (ED) is a binary measure defined on two non-trivial string arguments. If S1 and S2 are two strings, ED(S1,S2) is minimum number of operations required to convert either of them to the other using substitution, insertion, deletion or juxtaposition operations. Normalised Edit Distance (NED) is the ED divided by the maximum distance between S1 and S2. NED=ED(S1,S2)/maxDist(S1,S2). Consider S1='text categorisation', S2='text categories'. As the first 13 characters are exactly identical, the strings differ starting with 14-th character. Thus ED(S1,S2)=diff('sation','es')=6 because if we delete the last 5 characters 'ation' in S1 and insert an 'e' before s, we get S2. ED is zero if the arguments are equal (ED(S1,S1)=0). The strings need not be of equal length.

Example 15.4 Table 15.7 gives the occurrence frequency of terms in documents. Compute the TV.

Table 15.10: TD-Matrix for list of words in several documents

D #	break-fast	bread	but ter	cof fee	jui ce	milk	oml-ette	pa rk	restau-rant	str-eet	sch-ool	uni-form
1	1	1	0	0	1	1	1	0	1	1	0	0
2	0	1	1	1	0	0	1	0	0	0	1	0
3	1	0	0	0	0	1	0	1	1	1	0	0
4	0	0	0	1	0	0	0	0	1	0	1	1

Table 15.11: TF-IDF measure for list of words in several documents

D #	break-fast	bread	but ter	cof fee	jui ce	milk	oml-ette	pa rk	restau-rant	str-eet	sch-ool	uni-form
1	.301	.301	.000	.000	.602	.301	.301	.000	.1249	.301	.000	.000
2	.000	.301	.602	.301	.000	.000	.301	.000	.0000	.000	.301	.000
3	.301	.000	.000	.000	.000	.301	.000	.602	.1249	.301	.000	.000
4	.301	.000	.000	.301	.000	.000	.000	.000	.1249	.000	.301	.602

Solution 15.4 For term 'the' $\sum_{i=1}^{n} O_i(t)=16$, $\sum_{i=1}^{n}(O_i(t))^2=(1+0+16+9+36+4)=66$, so that TV(the) = 66 - 16*16/6 =66-42.6667=23.3333. For 'boy', $\sum_{i=1}^{n} O_i(t)=5+2+1+3+4=15$, and $\sum_{i=1}^{n}(O_i(t))^2=25+4+1+9+16=55$, so that TV(boy) = 55-15*15/6=55-37.5=17.50.

Example 15.5 Table 15.8 gives a list of words that appear in 4 documents. Compute the TF and IDF measures.

Solution 15.5 The IDF is defined for each word as IDF(word(i)) =log(total documents/number of documents containing word(i)). If a term occurs in every document, the IDF returns log(n/n)=0. If it occurs in a single document, IDF returns log(n/1) = log(n). All other occurrence counts are mapped to real numbers in the above range. Calculations are shown in table 15.9.

Example 15.6 Represent each document in example 15.5 as a weighted vector using the TF-IDF measure.

Solution 15.6 Computing the TF-IDF measure in this example is easy as each term occurs in a document at most once. Hence the TF-IDF measure can be obtained directly by multiplying entries in the TD-matrix by the corresponding IDF(term) obtained in table 15.9.

15.5.10.1   F-score

Precision and Recall are the most popular metrics for text mining (also called scoring metrics). Text mining and document clustering applications use a combined score based upon P and R called an F- score.

Definition 15.4 The F- score is a ratio-measure defined as F= 2PR/(P+R) where P is the Precision and R is the recall.

This is the harmonic mean of P and R as seen by dividing numerator and denominator by PR. The F-score symmetrically combines the precision and recall. Because P and R vary between 0 and 1, their product will always be between 0 and 1. Thus the F-score will also vary in the same range. If distinct clusters are identified in the target, the precision and recall can be calculated for each cluster, resulting in a vector of F-scores. These scores can be combined to produce a unique score as $F_{combined} = \sum_j card(C_j)F(j)/card(S)$ where $S = \cup_j C_j$ is the entire retrieved document collection, $C'_j s$ are sub-clusters, and card(S) denotes cardinality of set S. This is a weighted average of each individual F-scores with relative frequencies of clusters as weights. The F score coincides with P when P=R. Let 'a', 'b', 'c' and 'd' denote the true-positives (retrieved and relevant), true-negatives (retrieved and irrelevant), false-positives (relevant not retrieved) and false-negatives (irrelevant and not retrieved). These are conveniently arranged in table 15.12:

Table 15.12: Summary count of document retrieval

	Relevant	Irrelevant	Total
Retrieved	a	b	a+b
Not retrieved	c	d	c+d
Total	a+c	b+d	a+b+c+d

It is easy to see that P = a/(a+b) and R = a/(a+c). Substituting these values give F = 2a/(2a+b+c). From this expression, it is evident that F approaches 1 when b and c approach zero, and F approaches 0 when 'a' is small compared to 'b' and 'c'.

15.5.10.2 Generalised F-score

The F-score gives equal importance to Precision and Recall, but can be generalised as

$$F_\beta = (\beta + 1)PR/[P + \beta R]. \tag{15.3}$$

If $\beta < 1, F_\beta$ is biased towards Recall. Substitute the value of P and R to get $F_\beta = a(\beta + 1)/[\beta(a + b) + (a + c)]$. Dividing both the numerator and denominator by $(\beta + 1)$, this reduces to the form a/[a+K] where K=(b+c β)/[($\beta + 1$)]. When $\beta \to 0$, K tends to b so that $F_\beta \to$P. Due to the symmetry of precision and recall, another measure could also be obtained as

$$F_\beta = (\beta + 1)PR/[\beta P + R]. \tag{15.4}$$

In this case $F_\beta \to$R as $\beta \to$0. When $\beta = 1$, we get the standard F-score.

15.6 Applications of Text Mining

Text Mining is a hot research area due to the plethora of text documents available on the web. Text mining is being used in medical sciences to improve the potential to discover new uses for drugs, reveal dangerous drug interactions, for rule discovery [MS05], disease gene identifications etc. It may be used as a first step in extracting patterns which are either subsequently mined with more robust models like decision trees, association rules etc or used as a basis for applications and services. Automatic topic detection and tracking (ATDT) systems are finding increasing use in detecting *new events* and tracking *known* (recently occurred) *events* in news broadcasts, lectures, speech etc. to identify topic boundaries [AP98].

15.6.1   Spam-Mail Classification

Spam mail is the junk mail that comes regularly into your mailbox from unknown senders with strange or unsuspecting subject headings. Most of the spam mails are harmless, and are found in web based email systems. With a large number of email id's at hand, it is a matter of time for spammers to generate bulk email messages to these users on a regular basis. This not only create unwanted spam mail, but also results in wastage of time for the users to check (and sometimes read) and delete the messages from their mailboxes. Many attempts have been made to either filter the spam mail into separate folders other than the users mailbox, or to send it back to the sender. Email messages can be of pure text format, RTF format, HTML format etc. Most of the spam filtering techniques first extract the contents of the message into a memory buffer in text form, and apply content filtering techniques to categorise the mail as spam or not-spam. Presence of keywords or combinations thereof like "click here", "click on the link", "win prize", "claim your prize", "fast cash", "instant money", "teens", "site of the day", etc in the body of the message can help to easily identify spam mail[19]. Some spam mails come with attachments of different types - images, audios or videos of various formats, power point files, MS Word documents and PDFs being the most common ones. Attachment filtering techniques utilise the properties of attachments to classify mail as spam or not.

A table-driven algorithm for spam filtering can be developed with a list of keywords ranked according to their probability of occurrence in a spam email body and header fields. Multiple tables can also be used to drive our algorithm - one containing words that are frequently found in spam mail, another containing improbable keywords in the user's regular mails, another containing suspicious URLs (of mail sender), etc. A table-driven link analysis technique can be used to analyse all URLs in your email message to see if they are pointing to known[20] spam sites (black-list filtering). A second table will contain the list of URLs from where you regularly get mail (white-list filtering). But this type of checking is time consuming and may not be practical for some users who frequently communicate through emails. In this case, a probability based searching can be employed to identify spam mail fast with a certain confidence level. The occurrence of a combination of words from your list will improve the prediction probability and reduce false positives. Identifying false positives is very important for many business users. The probability of identifying false positives can however be improved using a training model, which continuously learns from previous false positive emails. As spam filters get better and better, the false positive issues can be resolved almost completely, making it only a storage burden in spam-mail folders. These spam mails could also be mined to identify the psychography and ethnography of spam mailers.

One of the widely used techniques is a Bayesian filter that assigns probabilities to words appearing in an email message. Bayesian filters not only filter all spams, but also may result in zero false positives (mistakenly filtering away non-spam mail messages as spam). The subject line and sender id can also provide clues for classification.

15.6.2   Miscellaneous Applications

There are numerous other applications for text data mining. Here we briefly summarise them.

a) Chat-room discussion

Chat-rooms is a synonym for online discussion groups in public domain email systems and

---

[19]Some other high probability words has been deliberately omitted from the list due to reader discretion
[20]Some spammers use IP spoofing to hide their identity.

servers. Each chat-room is initially identified by a topic. Users join and leave chat-rooms as time goes by. As discussions progress, the topics may drift dynamically. Identifying current topic(s) need to be done on the fly by mining the text messages generated by the users (with more weightage for recent data). A list of likely topics is identified and probabilities are assigned to each of them to emphasise their importance.

b) Knobotic retrieval

Knobots are automated software programs launched by users to retrieve information that is narrowly focused or of general interest. For example, knobots for news retrieval filters out news documents on the web that have an overlap with user specified keywords and phrases. This involve IR, document categorisations and clustering, link mining and image/multimedia mining.

c) Customer segmentation

As discussed in chapter x, the objective of segmentation is to partition the visitors into homogeneous subgroups using common (socio-demographic, geographic, product specific, psychographic, etc) traits. The socio-demographic segmentation is based upon demographic variables like gender, race, income brackets, education levels, age groups, employment status, personal habits (smoking, drinking) etc. These traits are usually captured by providing a web form to the visitor on their first visit. Geographic segmentation is based upon attributes like zip codes, regional codes, country codes, telephone area codes, etc that are categorical or quantitative. The psychographic segmentation uses hobbies, interests, lifestyles, attitudes, and other personality traits and are often categorical. Technographics traits include information about the visitors domain, computer, networks, and other hardware components, firewalls etc. For instance, the OS, browser etc used by the user can be obtained from client requests. The IP address can provide domain information while the to and from data transmission rate can give a hint on modem speed or type of connectivity.

Table 15.13: Software for Text Mining

URL	Name	comments	C/F/S
cgi.csc.liv.ac.uk/~frans/KDD/Software	TextMiningDemo		F
leximancer.com	Leximancer textminer		F
opennlp.sourceforge.net	OpenNLP		F
co.umist.ac.uk	TextMiner		F
www.megaputer.com/products/ta/	TextAnalyst	Megaputer	S
www.temis-group.com	Temis Insight		C
www.sas.com	SAS text miner		C
www.spss.com	SPSS Clementine		C
www.ibm.com	IBM speed tracer		C

Legend: C=Commercial, F=Free, S=Shareware

15.7 Software for Text Mining

OpenNLP is an open source software that can tokenise, POS tag and chunk input text. FACT (Finding Associations in Collections of Text) to search and mine association rules in text data. It has a multi-threaded mining-engine with custom filters, export facilities etc (www.softpedia.com/ progDownload/Ez-Web-Miner-Download-56139.html). MediaMiner is used to mine news sites, stock reports, e-com sites, forums and blogs etc with export facility to XML, HTML and PDF formats (programurl.com/mediaminer.htm). Webfax Miner is used to extract targeted fax numbers from websites and search engines (3d2f.com/programs/0-221-webfax-miner-download.shtml). The Bow toolkit (www-2.cs.cmu.edu/~mccallum/bow/) is a text preprocessing tool that can output TD-matrix. Other software include), R tm library (of the programming language called R), Angoss knowledge studio, TextMiner (a text-mining tool designed and developed at UMIST-Univ. of Manchester, co.umist.ac.uk). See also www.textmininglab.net and [GJ00] for a review of 71 text mining tools available on the internet.

15.8   Exercises

1. True or false:
   (a) Creation of the term-by-doc matrix is a pre-processing step in information retrieval.
   (b) A synonym for web usage mining is sequence mining
   (c) Software version numbers are shadowed attributes
   (d) Web usage mining is performed at individual sites
   (e) The Pagerank measure is used for web query mining
   (f) Link mining uses only remote links in web pages
   (g) The F-score (§15.5.10.2) approaches 1 when 'a' is small compared to 'b' and 'c'
   (h) The CNC measure is used for image classification
   (i) In supervised text categorisation, each document is labeled uniquely.

2. Explain the following terms:
   (i) lemmatising (ii) text citation mining (iii) Document facets

3. What is meant by 'linguistic pre-processing'? How does it help to speed-up web mining?

4. Explain how you will use the term-by-document matrix to identify duplicate documents. Can you identify documents with overlapping content using the term-by-document matrix.

5. What are the different types of text mining? Can you use XML documents for this? List any 3 metrics used in text mining.

6. Describe any three criteria used for text clustering. Which clustering algorithm is better suited for text clustering? Why?

7. Describe hypertext mining and its relation with text mining.

8. Explain the following terms:
   (i) edit distance (ii) coupling and cohesion

9. What are the commonly used query performance measures? How are they related?

10. Compute the precision (P), recall (R) and F-score from the following table

	Relevant	Irrelevant	Total
Retrieved	250	80	330
Not retrieved	520	100000	—
Total	770	—	—

11. In the table 15.12, when 'a' is the mean of 'b' and 'c', prove that the F-score is one-half. If a new measure is defined as G-score=$a/\sqrt{(a+b)(a+c)}$, what is its range? Compute its value using the data in the above table.

12. Prove that the F-score is the harmonic mean of Precision and Recall. If $F_{\beta_1} = (\beta + 1)PR/[\beta P + R]$ and $F_{\beta_2} = (\beta+1)PR/[\beta R + P]$, prove that $(F_{\beta_1}/F_{\beta_2}) - 1$ is proportional to the difference between P and R.

13. Where is metadata stored? What are the categories of metadata? List any three advantages of metadata mining.

14. If the average viewing time (in seconds) for the pages are respectively given as P0 = 16, P1 = 40, P2 = 35, P3 = 73, P4 = 54, P5 = 60, P6 = 29, P7 = 18, P8 = 43, P9 = 71, P10 = 19, find the total time spent by 20 users in pages 7 and 10. Find the average time for an arbitrary user to reach page P10 from home page P0.

15. Which of the following can be the source for text mining?
    A) Online news papers B) Online research articles in PDF format C) newsgroup postings D) birthday e-cards E) Patent documents F) Email messages on mail servers G) video clips

16. Give examples of text data in unstructured, semi-structured and structured formats on the web.

17. What distinguishes text from other data? Why can't we use classical data mining models for text data?

18. What is the most appropriate data type that characterise nontrivial text data?
    A) it is one-dimensional B) it is two-dimensional C) it is unstructured multi-dimensional D) it is structured multi-dimensional

19. The goal of text mining is to
    A) extract new knowledge from free-form text data B) arrange text documents in easily accessible form C) reduce dimensionality of text corpora D) Delete unwanted text documents from databases.

20. Describe text classification and its applications.

21. What is stemming? How can you speedup stemming in each document?

22. Write a program to generate the Term-by-Document matrix using simple Boolean weights and TF-IDF metric. How can you speedup the process if all terms that are present in the documents are known apriori.

23. When is pre-processing stage for text mining nonexistent?

24. Distinguish between descriptive and predictive text mining.

25. What are the three important stages in the text-mining pipeline?

26. Explain how you will use the term-by-document matrix to identify duplicate text documents. Can you identify documents with overlapping content using the term-by-document matrix?.

27. Which of the following techniques can be used to select the most relevant attributes in a text corpora?
A) PCA B) boosting methods C) LSI D) Discriminant analysis

28. What is the maximum possible value of edit distance between 2 strings of size n characters?

29. Describe how the wrapper tags (like HTML <ul>, <b> etc) in markup languages like HTML and XML documents can be used to improve text mining?

15.8.0.1   References

[BR13] Banchs, Rafael E(2013) *Text mining with MATLAB*, Springer (www.textmininglab.net).

[BM03] Berry, M.W. (2003). *Survey of text mining: clustering, classification, and retrieval,* Springer, NY.

[BK10] Berry, M.W., Kogan, J. (2010) *Text mining: Applications and theory*, John Wiley, UK.

[C03] Chakrabarti, (2003). *Mining the Web — Discovering knowledge from hypertext data,* Morgan Kaufmann, (ISBN:1-55860-754-4)

[CB97] Cooley, R., Mobasher, B., Srivastava, J. (1999). Web mining: Information and pattern discovery on the world wide web, Technical report TR 97-027, *University of Minnesota.*

[DB03] Davison, B.D. (2003). Unifying text and link analysis,

[FS06] Feldman, R., Sanger, J. (2006). *The Text Mining Handbook: Advanced Approaches in Analyzing Unstructured Data,* Cambridge University Press.

[FA00] Frantzi, K., Ananiadou, S., Mima, H. (2000). Automatic recognition of multiword terms: the C-value/NC-value method, *International journal on digital libraries,* 3(2), 115-130.

[GJ00] van Gemert, J. (2000) Text mining tools on the Internet: An overview, Dept. of Computer Science, University of Amsterdam, Netherlands (www.science.uva.nl/~jvgemert)

[GK99] Goldstein, J., Kantrowitz, M., Mittal, V., Carbonnel, J. (1999). Summarizing text documents: sentence selection and evaluation metrics, SIGIR 99 (22nd international conf), 121-128.

[HM99] Hearst, Marti. A.(1999) Untangling Text Data Mining, *Proceedings of the 37th annual meeting of the Association for Computational Linguistics,* www.sims.berkeley.edu/~hearst/papers/acl99/acl99-tdm.html

[IV90] Ide,N., Vronis,J.(1990). Very large neural networks for word sense disambiguation, ECAI.

[JK11] Jayabharathy, J., Kanmani, S., Ayeshaa Parveen, A. (2011) A survey of document clustering algorithms with topic discovery, *Journal of Computing,* 3(2), 21-27 (www.journalofcomputing.org/volume-3-issue-2-february-2011/).

[JM09] Janecek, M.A. (2009)*Efficient feature reduction and classification methods*, Ph.D. dissertation, Univ. of Wien (www.univie.ac.at/andreas.janecek/).

[KL04] Kummamuru, K., Lotlikar, R., Roy, S., Singal, K., Krishnapuram, R. (2004). A hierarchical monolithic document clustering algorithm for summarization and browsing search results, Proc of WWW 04, 658-665.

[LR14] Leskovec, J., Rajaraman, A., Ullman, J.D. (2014). *Mining of large datasets*, Cambridge university press.

[LH00] Liu, Z., Huang, Q. (2000). Content-based indexing and retrieval-by-example in audio, ICME 2000, vol 2, New York, 877-880.

[MZ05] Mei, Q., Zhai, C. (2005) Discovering evolutionary theme patterns from text - an exploration of temporal text mining, KDD-05, Chicago, IL.

[MM97] Mendelzon, A., Mihaila, G., Milo,, T (1997). Querying the World Wide Web, *Journal of Digital Libraries*, 1(1), 68-88.

[MN02] Miki, T., Nomura, S., Ishida, T. (2002). Semantic web link analysis to discover social relationships in academic communities.

[MS00] Modha, D.S., Spangler, W.S. (2000). Clustering hypertext with applications to web mining, *Proc of 11th ACM symposium on hypertext and hypermedia*, New York, 143-152.

[MS05] Mullins, I.M., Siadaty, M.S., Lyman, J., *et. al.* (2005). Data mining and clinical data repositories: Insights from a 667,000 patient data set, *Computers in Biology and Medicine*,

[NK00] Nigam, K. (2000). Text classification from labeled and unlabeled documents using EM, *Machine learning*, 39, 103-134.

[PB04] Perkio, J., Buntine, W., Perttu, S. (2004). Exploring independent trends in a topic based search engine, *Proc. of international conference on web intelligence*, 664-668.

[RG02] Roy, S., Gevry, D., Pottenger, W.M. (2002). Methodologies for trend detection in textual data mining, Textmine '02 workshop, *SIAM international conference on data mining*, 2002.

[SB07] Senellart, P., Blondel, V.D. (2007). Automatic discovery of similar words, in *Survey of text mining: clustering, classification, and retrieval*, 2 ed., Berry, M.W., Castellanos, M (eds.), chapter 2, 25-44.

[SS05] Sun, J., Shen, D., Zeng, H., Yang, Q., Lu, Y., Cheni, Z. (2005). Webpage summarization using clickthrough data, Proc of 28-th annual international ACM SIGIR conf on research and development in information retrieval, 194-201.

[TZ05] Tao, T., Zhai, C. (2005). Mining comparable bilingual text corpora for cross-language information integration, Proc. of KDD 05, Chicago, IL., 1-6.

[WI05] Weiss S.M., Indurkhya, N., Zhang, T., Damerau, F.J. (2005) *Text mining: predictive methods for analyzing unstructured information*, Morgan Kaufman.

[WP99] Witten, I., Paynter, G., Frank, E., Gutwin, C., Nevill-Manning, C. (1999). KEA: Practical automatic keyphrase extraction, Proc of 4th ACM conf on digital libraries, Berkeley, CA, 254-255.

[WI05] Witten, I. H. (2005). Text mining, In *Practical handbook of internet computing*, M.P. Singh (ed), ch:14, 1-22, Chapman & Hall/CRC Press, Boca Raton.

[YP97] Yang, Y., Pedersen, J.O.(1997). A comparative study on feature selection in text categorization, *Proc of 14-th international conference on machine learning*, Nashville, TN, 412-420.

[YZ03] Yu, H., Zhai, C., Han, J. (2003) Text classification from positive and unlabeled documents, CIKM 03, Nov 3-8, New Orleans, LA.

[ZZ05] Zhang, Y., Zincir-Heywood, N., Milios, E. (2005). Narrative text classification for automatic key phrase extraction in web document corpora, *Proc of 7-th ACM international workshop on web information and data management*, 51-58, Bremen, Germany.

Appendix-A

The Backpropagation algorithm

First consider a simplified two-layer network that works with real numbers in interval or ratio scale. Denoting observed output at node j by O_j and expected output by d_j, the squared error at an output node j is $E_j = (O_j - d_j)^2$. The weight adjustments by the gradient descent algorithm minimizes

$$\Delta w_{ji} = -\eta \partial E_j / \partial w_{ji} \tag{0.1}$$

where η is a constant called learning rate, and negative sign indicates minimisation wrt w by starting with all w_i=1. Since the incoming weights are transformed (usually summed) by the activation function and passed through the threshold function, the error term is a function of the (linear) function of the weights. Hence we have to use the chain rule of differentiation[21]. To simplify the derivation, first differentiate the squared error by O_j to get $\partial E_j / \partial O_j = 2(O_j - d_j)$. Next find the rate of change of the output wrt the threshold input. If the threshold is the standard sigmoid function, the differentiation is considerably simplified to give $\partial O_j / \partial T_j = O_j(1 - O_j)$. If the activation function is linear (weighted summing function), the rate of change wrt the weights are constant coefficients (which in our case are the inputs x_j, which are assumed to be constants as unknown weights are the variables) giving $\partial T_j / \partial w_j = x_j$. Apply the chain rule twice to get

$$\partial E_j / \partial w_i = (\partial E_j / \partial O_j) * (\partial O_j / \partial T_j) * (\partial T_j / \partial w_i) = 2(O_j - d_j) * O_j * (1 - O_j) * x_i \tag{0.2}$$

where O_j and d_j being observed and expected values are often scalars, and x_j can be a scalar, vector or even a matrix in some applications.

Using eq.(0.1), the weight adjustments are obtained as $\Delta w_{ji} = -2\eta(O_j - d_j) \, O_j(1 - O_j) \, x_i$, where we have dropped the multiplier symbol *. As the standard sigmoid varies between 0 and 1, the $(1 - O_j)$ term is always positive. The RHS expression $(O_j - d_j)O_j(1 - O_j)$ when equated to zero gives a cubic in O_j with roots 0,1 and d_j. If O_j values are either close to 0 or 1, the corresponding neuron is quite excited (or inhibited). This also may be used as a criteria to skip the back-propagation. As the region of rapid rise of the sigmoid is in the middle (see figure

---

[21]The chain rule of differential calculus states that $\frac{\partial}{\partial x} f(g(x)) = (\frac{\partial}{\partial g(x)} f(g(x))) * (\frac{\partial}{\partial x} g(x))$ where g(x) is a continuous differentiable function.

11.1 (c), pp.11-11), faster adaptation is to be expected when output is in the neighborhood of 0.5. The output layer error can also be expressed as $(O_j - d_j)f'(x_k)$ where $f'(x)$ denotes the first derivative wrt x of the threshold function. The speed of adaptation also depends upon the user-chosen adaptation rate η and observed error $(O_j - d_j)$.

For hidden layers, we do not have direct values of 'expected' outputs d_j. Hence weight adaptation proceeds through hidden neurons using the computed errors at succeeding layer nodes using the following two steps. For each of the hidden layer neuron find $Err(j) = (\sum_k Err(k)w_{kj})f'(x_j)$ where $Err(k)$ is the computed output error at neuron k and w_{kj}'s are the connection strengths from neuron k to neuron j in succeeding layer. In the second step, the weights are updated using $w_{kj}(new) = w_{kj}(curr) + \Delta w_{kj}$. This step is repeated until one of the terminating criteria discussed above is satisfied.

Solutions to Selected Exercises

Chapter 1

Q1. (a) Nominal, {Hindu, Christian, Muslim, Sikh, Jain, Buddhist} (b) Ordinal, {mild, severe, high}, (c) Nominal (possibly binary) {No, Yes}, (d) Ordinal, {A, B, C, D, F} (it could also be an 'I' for incomplete grade, which cannot be compared with others), (e) Ratio, (f) Nominal {list of news papers}, (g) Nominal (although TV channel numbers can be ordered, it does not carry much meaning as far as TV programs are concerned. Watching a higher numbered channel is in no way better than watching channel 1), (h) Nominal {list of languages}, (i) Ratio (a real number in a proper range, usually between 0 to 4, or 0 to 5), (j) Nominal {list of languages} (although mother-tongue is unique for a person, different people have different mother-tongues), (k) Ratio (pair of positive or negative real numbers for left and right eye, power 0 means plain glass), (l) Ratio (0 to max pages, where 0 is well defined, fractional values possible), (m) Ratio (a positive integer in some range, depending upon language, literacy level, typing practice etc).

Q2. (a) Ordinal, (b) Ratio, (c) Ordinal (integer), (d) Ratio, (e) Nominal, (f) cyclic Ordinal, (g) Ratio, (h) Ratio, (i) Nominal, (j) Ratio, (k) Ratio (integer), (l) Ordinal

Q3. (a) Nominal (B, R, G, W), (b) Nominal (V, M, B), (c) Ordinal (U, R, V), (d) Ordinal (M,N,P,C), (e) Nominal, (f) Nominal, (g) Nominal

Q4. It is nominal (A)

Q5. 1) Interval or higher without zeroes 2) Ordinal or higher 3) Interval or higher 4) Interval or higher 5) Ordinal or higher

Q6. Insured=0, uninsured=1

Q7. Whether a variable is symmetric or asymmetric depends upon the research hypothesis. a) asymmetric (for a study to correlate typing errors of specific letter keys or key combinations, for accident proneness etc), symmetric for most of the other studies, b) symmetric, c) symmetric, d) asymmetric, e) asymmetric.

Q8. Mode

Q9. Yes, when the number of categories is large and number of records are small. (eg: international country calling codes)

Q10. (1) continuous, (2) continuous, (3) discrete, (4) continuous, (5) discrete, (6) discrete, (7) discrete, (8) discrete, (9) count, (10) continuous.

Q11. (a) binary, (b), (c) are nominal, (d) is binary, (e) is continuous (if measured to decimal places), (f) is continuous

Q12. (a) No. It is a ratio variable with 3 components (that together is used to identify a location on or near the Earth), (b) No (a patient may consult for multiple options like Ear and Nose) (c) No (d) Yes

Q13. All of them except (3) and (4)

Q14. New pattern discovery, cost/expense/time minimisation, efficient resource utilisation, etc

Q15. Databases, machine learning, statistics, visualisation, Q16. Access, count(), Q17. All of them, Q18. a) Unsupervised b) Unsupervised c) Supervised d) Supervised e) Supervised f) Unsupervised

Q16. count(), recoding 17. all of them 18. (a) and f) are unsupervised, b), d), e) supervised, c) can be both 19. previous val, next val, total choices,

Chapter 2

Q1. (a) F, (b) T, (c) T, (d) F, (e) F, (f) F, (g) T, (h) T, (i) T, (j) T, Q2. C

Q3. A bar chart in which the bars are segmented using a categorical variable.

Q4. Conditional histograms partition the entire data set into two or more groups using a nominal variable, and plots each set separately using a histogram.

Q5. Charts in which the primary variable of interest is time. Stock prices, average temperatures in a locality, expenses, natural disasters etc over a period of time.

Q6. When the base spikes of a histogram can be aligned with a spatial 2D- or 3D-map. Disease prevalence in various geographical regions, country-wise e-commerce orders over the entire world.

Q7. Pareto diagrams are overlay charts that work on the Pareto principle to prioritise activities on importance scale. It orders the root causes in decreasing order of relative importance. Processing incomplete applications, identifying causes for loan defaulting.

Q8. A, Q9. Charts used to visually compare different entities on various qualities. Compare prices of articles of various brands, compare job applicants with various majors (CS, EE, ECE, etc.)

Q10. They are used to summarise numeric data into a leading stem followed by leafs. Stems are chosen from most significant digits, and leafs are the leftovers.

Q11. (1) Data should all be of the same size. For example, if there are 2-digit and 3-digit numbers, a leading zero must be prepended to 2-digit numbers to make them 3 digits. (2) Not very suitable for large amounts of data, (3) Not convenient when data are positives and negatives

Q12. A structured diagram to analyse the effects resulting from a single cause. They can depict, identify and sort-out various causes of a problem. Also called cause-and-effect diagram or Ishikawa diagram. 4M approach and 4P approach.

Q13. Bars generally do not touch in barcharts, while adjacent bars can touch in histogram. Histogram can be used with continuous data (falling in various intervals), while bar charts often work with categorical data.

Q14. The Q-Q (Quantile-Quantile)-plot is a graphical tool for checking the normality assumption. Used to check the normality assumption of the parent population, to check if two samples come from identical populations.

Q15. (i) Sample sizes need not be equal, (2) can easily be extended to multiple dimensions and many distributional aspects can be tested simultaneously. Normal probability plots use expected value of kth order statistic along X-axis.

Q16. C, Q17. See example 2.1 in page 2-12.

Q18. Medical sciences (ultrasound, Doppler, magnetic bubble/resonance imaging), data mining (OLAP), engineering design, banking (anomaly detection).

Q19. See text.

Q20. If all spike sizes are not multiples of picture sizes, top (or bottom) picture may have to be cut (or scaled). This is more of a problem in picto-pie charts. They are visually more convincing, even to illiterate people.

Q21. A plot to represent a small set of numbers, text or strings that have some characteristic in common. In the case of numbers, the leading digit (MSB) is used as stem to group related items together. Rest of the digits form the leaf.

Q22. We can use separate Pareto charts, overlaid Pareto charts, or radar charts. See respective sections.

Q23. See §2.6.7 in page 2-22.

Q24. Temporal histogram has at least one time-dependent variable, which is usually chosen for the X-axis. A spatial histogram has an aligning map. This base map need not be a geographical map. Examples are sky maps, circuit board maps, maps of the human body or parts thereof, interior of crafts etc.

Q25. A multivariate scatter plot in which one or more variables are conditioned. The conditioning variable is preferably nominal type with as few categories as possible.

Q26. I/O exceptions ={File not found, File permissions wrong, File creation error, File corrupted, etc}, Null pointer exceptions ={Pointer not initialised, Wrong pointer assignment, Pointer out-of-scope etc}, Hardware exception={Faulty equipment, faulty/incompatible interfaces, disk read error, etc}, Data exception={Incorrect formats, wrong types, missing or out-of-range values}, Memory exception={Memory overflow/underflow, insufficient memory, corrupt memory, array out-of-bounds, divide by zero, etc}, Other={wrong parameters, library function errors, DLL errors, SQL exceptions, etc} Using this list, it is easy to construct a fishbone diagram with the stem labeled as Exceptions and leafs labeled using the above categories.

Q27. Linear mapping of brightness, Use different colours, Use a bubble chart, Use higher-dimensional scatterplots

Q28. (1) Chernoff plots do not specify actual data values, which is a limitation of it in data visualisation. (2) All possible states are difficult to remember, thereby limiting its use for heterogeneous data display.

Chapter 3

Q1. a) F, b) F, c) T, d) T, e) T, f) F g) F, h) T

Q2. Axiomatic approach, Geometric approach, Bayesian approach, Frequency approach

Q3. see section §3.2.1 in page 3-3

Q4. (1) Any writings by inexperienced or beginning learners, (2) Typewritten materials by left-handed and right-handed people, (3) All money transactions at the bank

Q5. standard deviation=0, mean = c, mode exists and is also c

Q6. Use y = 10*(x-65)/28, z = (3/28)*(2x - 158) = 3*(x-79)/14

Q7. When each observation is the same constant. Q9. see §3.4 in page 3-10.

Q9. Use P(A∪B)=P(A)+P(B)-P(A∩B)=47/95

Q10. $x_1 + x_2 = $ (p+q)/2+(p-q)/2 =p, from which $(x_1 + x_2)/2$ is p/2. Now $(x_1 - \overline{x}) = $ (p+q)/2-p/2 = q/2 and $(x_2 - \overline{x}) = $ (p-q)/2-p/2 = -q/2. Hence variance = $[(x_1 - \overline{x})^2 + (x_2 - \overline{x})^2]/2 = [q^2/4 + q^2/4]/2 = q^2/4$ if n is the scalling factor, $q^2/2$ otherwise.

Q11. (i) $26^3 * 10^4$, (ii) $26^3 * 10 * 9 * 8 * 7$, (iii) $26 * 25 * 24 * 10 * 9 * 8 * 7$

Q12. (i) (15*54+70)/16, (ii) (15*54 − 38)/14, (iii) (15*54+3*56)/18

Q13. (i) Rearrange the data in ascending order as X=[2,2,3,5,5,5]. Median is the arithmetic mean of middle values = (3+5)/2 = 4. Mode = 5. (ii) Proceed as above

Q14. Divide each number by 100 to get X'=[.50, .65, .72, .84, .9, .98], the GM of X' is $(0.17336592)^{1/6}$=0.74673. Multiplying by 100 gives the GM of original data as 74.673 For part (ii), divide by 10 000 and proceed as above.

Q15. Mean of above data is 459/6=76.5, from which the absolute deviations are Y=[26.5,11.5,4.5, 7.5,13.5,21.5]. Thus the mean absolute deviation is 85/6=14.1667, [Q18.] a) F, b) T, [Q19.] A) only

Q21. Let the numbers be p and q. We are given that AM=GM so that (p+q)/2=$\sqrt{pq}$. Square both sides and cross multiply to get $(p+q)^2$=4pq. Take 4pq to the LHS to get $p^2 + q^2 - 2pq$=0 or equivalently $(p-q)^2$=0. If the numbers are different this hold only if p=q. For [Q22.], [Q23.] proceed exactly as above. [Q24.] see text

Chapter 4

Q1. (a) T (b) F, (c) F, (d) T (e) T

Q2. , [Q3.] See text. [Q4.] 0, 1 [Q5.] see text

Q6. global summary, summary without duplicates [Q7.] see §4.4.4

Q8. see page 4-17.

Q9. see figure §4.4.1, page 4-16

Q10. see section §4.2 in page 4-15.

Q11. data impurity is called noise. It may indicate nonconformance, missing values, incompatible values or out of range values.

Q12. Metadata is summary or location data about other data. It is used for faster access, enforcing access restrictions and security policies, incremental updates, provide insight into the nature of the data, and for support operations. Metadata may be managed by utility programs, ETL tools, or manually by administrators using DW tools

Q13. Operational reports, tactical reports, and strategic reports. See §4.1.2 in page 4-5

Q14. see text

Q15. There are two ways to populate or update data in a DW/DM from ODS. In the *pull approach*, applications running on DW server pulls the data from ODS, using ODS schemas. In the *push approach*, applications on the ODS server pushes the data to the DW/DM using the DW structures.

Q16. Datamarts can be categorised using the data they contain (atomic datamarts and aggregated datamarts). Atomic datamarts contain multidimensional data at the lowest level of detail, usually in a star schema. Aggregated datamarts contain aggregated data derived from atomic datamarts, parent datawarehouse or created directly during data-staging.

Q17. Small size, focused functionality, finer security restrictions, faster data updates, easy scalability, ease of maintenance of indices, metadata.

Q18. Bulk update mode, incremental update mode, trickle mode

Q19. The smoothing process rounds each element according to $x_{new} = \lfloor x \rfloor$ if fractional part is less than .5, and $x_{new} = \lceil x \rceil = \lfloor x \rfloor + 1$ otherwise to give $X'=\{15, 62, 40, 44, 23\}$.

Q20. High scalability, replication and mirroring capabilities, integrated dataflows, workload balancing, all-time availability.

Q21. see 4.5.1, page 4-22

Q22. There are four types of datawarehouses – virtual DW, centralised DW, distributed DW and hybrid DW.

Q23. ODS consolidates operational data from multiple sources, and has limited update capabilities (add, change, delete). It has two purposes :— 1) to integrate data spread across a multitude of base files into a single and current repository, (2) to serve as a single platform for feeding DW/DM. But the DW is used by knowledge workers (executives, managers, business and financial analysts), support users, and clients for managerial decision making, and for ad-hoc reporting, with most of the operations as read-only.

Q24. Balancing of processor overloads so that users get quick responses.

Q24. Special DW to store spatial data. Spatial datawarehouses are optimised for spatial storage and operations (contains, contained in, adjacent, overlaps etc).

Chapter 5

Q1. (a) F (it is an analysis tool) (b) T, (c) T, (d) F, (e) T

Q2. See page 119.

Q3. C, [Q4.] a), e) & f).

Q5. summary measures, yes using drill down we can narrow down to cell level.

Q6. yes, empty cells indicate missing or incompatible combinations.

Q7. (Time, product, location, customers, sales persons)

Q8. see text. time can be century, decade, year, quarter, month, week, day, etc.

Q9. see text, [Q10.] ROLAP uses the std relational data store while MOLAP uses multidim data cube with aggregations and precalculations. ROLAP uses SQL while MOLAP uses specialised tools

Q11. 2 dimensional with (m+1)x(n+1) categories (including ALL category).

Q12. Data values sparse (nonzero). There are no valid products in Africa

Q13. see text. [Q14.] all types of digital data can be used. Categorical data can be recoded, interval and ratio data can be transformed, text data cleansed (ch 15), audio, video etc transformed to other domains using DFT, DWT etc.

Q15. A) roll-up   B) drill-down   C) slicing D) dicing

Q15. see text

Q16. Addition of new products and services, new features to existing products, new functionality requirement, compliance to new standards etc could all necessitate structural changes in commercial datamarts. Similarly introduction of new medications and procedures, discovery of new therapies, new diseases etc could result in changes in the structure of medical datamarts.

Q17. disease prevalence modeling in a geographic region, tax collection modeling, transportation modeling.

Q18. Sales amount is additive, Monthly balance and VAT are semi-additive, The customer's age, height, weight, blood pressure, blood sugar level, etc are nonadditive. Similarly temperature of a city, atmospheric pressure and humidity, physical prop- erties of materials like density, illumination, etc are also nonadditive.

Q19. ,[Q20.] see text

Q30. Third-party malicious users, data theft, data damage, other types of attacks. Most of them originate within the intranet. Security can be implemented in hardware level, OS level, middleware level, application level, cube level, cell level etc

Chapter 6

Q1. (a) F (b) T, (c) T, (d) T, (e) T, (f) F, (g) T (h) F, (i) T, (j) F, (k) F

Q2. (i) n, (ii)$\geq$n

Q3. A maximum gain split

Q4. Gini(S) is maximum when each of the probabilities are equal (1/m). The maximum is $1-\sum_i(1/m^2) = 1-(1/m)$

Q5. see section §6.3.2 in page 6-9

Q6. see section §6.3.3 in page 6-11

Q7. When the DT is put to use, we can form a frequency count of various assigned classes to test data. Those classes with low frequencies are then identified. Those attributes towards the leaf of the DT for these classes have low contribution.

Q8. The computational complexity to build a DT is O(mn) where m is the number of attributes and n is the size of the training set

Q9. Yes. Each class has an associated set of attribute constraints as evidenced by the path from the root to that class (or given by the production rule). Starting from the root attribute, we generate random numbers until we get a number corresponding to the sought path. This process is repeated for each of the subsequent nodes.

Q10. Gini's diversity measure, entropy, χ^2 measure, minimum classification error measure, Bhattacharya distance, Kolmogorov-Smirnoff distance.

Q11. see section §6.1 in page 6-17

Q12. 3 trees. For simplicity assume that the root split is for variable={A,B,C}. Then we can have a binary tree with the following branches at the root :–{AB,C}, {A, BC}, {AC,B}, where each of the two letter combinations are expanded to a binary tree.

Q13. Maximum when each of the probabilities are equal, minimum when one of the p_i's is 1, and all others are zero. This means that all instances belong to the same class.

Q14. Yes. The tree need not be unique, but could be isomorphous.

Q15. E, Q16. D

Table 0.14: Answers to exercises: Support table for ex 6.2

Item	count
U	33.33%
V	33.33%
W	60%
X	50%
Y	50%
Z	66.66%

Table 0.15: Answers to exercises: Support table 2

Item	count
W, X	16.66%
W, Y	16.66%
W, Z	16.66%
X, Y	16.66%
X, Z	50%
Y, Z	33.33%

Q17. Yes

Q19. a) Gini index $\in [0,1]$, b) entropy $\in [0,\infty)$, c) χ^2 statistic $\in [0,\infty)$

Q20. pre-pruning and post-pruning. It will generally reduce.

Chapter 7

Q1. (a) F (b) F, (c) T, (d) T, (e) F, (f) T, (g) F, (h) F, (i) F, (j) T, (k) F, (l) T

Q2. The support table for 1-item transactions are given in table 0.14.

Use $\text{conf}(X \to Z) = n(X \cup Z)/n(X)$ to get conf(X, Z)=3/4=75%, etc. Lift(X,Y) = supp(X, Y)/[supp(X) *supp(Y)] = (1/6)/[1/2*1/2]=2/3 = 66.66%, etc

Q3. Supp(A→B) = $n(A \cup B)/n$, and[22] Conf(A→B) = $n(A \cup B)/n(A)$, where n is the total number of transactions and n(A) is the number of transactions containing A. Obviously, n is $\geq$ n(A) [the max n(A) can take is n, if all transactions contain an 'A' item]. Thus the denominator of Supp(A→B) is $\geq$ that of Conf(A→B). Hence Support $\leq$ Confidence.

Q6. WARM assigns different weights to attributes or transactions. As weights are variable, they can reflect more realistic dependency relations based upon the utility or importance of attributes or transactions.

Q7. (a) support high [purchase of shoes in shoe-shop] (b) support low [swim glasses in optician store] (c) confidence high [printer or multimedia speakers with computer] (d) confidence low [purchase of accounting software with computer]. As mentioned in §6.2.1 these values depend upon where the respective items are bought.

Q12. Production rules are multi-level dependencies with one consequent. Association rules can have multiple consequents.

Q13. Using an extra variable like cost, utility, durability, etc.

Q14. For better comprehension.

Chapter 8

Q1. (a) F (b) F, (c) T, (d) F, (e) F, (f) F, (g) T, (h) T

Q2. D=$\sqrt{(15-14)^2 + (20-18)^2 + (9-7)^2 + (14-10)^2} = \sqrt{25} = 5$, Manhattan=9

---

[22]Here the $\cup$ notation has a different interpretation than given in chapter 3.

Q3. See section §9.3 in page 9-5

Q4. Inflated k-means, accelerated k-means, k-medoids

Q5. B

Q6. K-means algorithm can result in singleton clusters, especially when outliers are present. If the initial values of k are randomly chosen, the very first pass of k-means algorithm can have empty clusters.

Q8. It produces clusters of different granularities.

Q9. For hard clustering, available indices include modified Hubert statistic, Davies-Bouldin index, Dunn's separation index, and contingency coefficient. For fuzzy clustering the popular measures are Bezdek's partition coefficient, Gath-Geva fuzzy hyper-volume index, Duo-Xue-Duwu's fuzzy adaptive index, etc. The Hubert statistic, and its modified versions are goodness-of-fit measures for non-hierarchical cluster validation

Q10. Single linkage, complete linkage, average linkage. Linkage metrics are used to determine the similarity of sub-clusters.

Q11. Different clustering algorithms tend to carve different geometric shapes around data points in R^n. This depends upon the similarity metrics, weights, density functions (in density based clustering) etc. Euclidean metric carves hyperspheres, Mahalanobis squared distance tend to prefer hyper ellipsoids and Manhattan metric favors parallelopipeds. The preference of k-means algorithm depends upon the distance metric chosen.

Q12. Density-based clustering finds the clusters using local density information, under the assumption that data distribution follows a statistical law. DBSCAN, DENCLU, GDB-SCAN

Q13. Clustering is an unsupervised learning algorithm. If distinct clusters are absent in the data, a clustering algorithm may return either 1 or n clusters. Classification algorithms are mainly supervised learning algorithms. Using a model built from training data, these algorithms tries to assign a label to unlabeled data instances. Clustering can be used as a classification model if a few distinct clusters are identified in the data. Each such distinct cluster is assigned a class label. Each new data point is assigned to that cluster (and that class label) to which it is the closest. Ties are resolved by statistical measures, heuristics or fuzzy measures. For instance, if a new data item is equally distant from clusters C_i and C_j with respective sizes $|C_i| = k_i$ and $|C_j| = k_j$, the new instance belongs to C_i with probability $p_i = k_i/(k_i + k_j)$ and belongs to C_j with probability $1 - p_i$. The updated sizes are used in subsequent iterations.

Q14. Agglomerative algorithms builds clusters 'bottom-up', by starting with n distinct clusters. The divisive algorithms work in the opposite manner. It starts with a single cluster containing all the data and iteratively splits it into sub-clusters. The divisive algorithm is preferred when the data are on external storage devices.

Q15. K-means and its variants

Q16. Terminated when none of the data items changes clusters during 2 consecutive iterations. Before starting k-means algorithm, a best-guess on k must be known. In addition, it works only for quantitative data (ordinal or higher scale).

Chapter 9

Q1. (a) T (b) F, (c) T, (d) F e) F f) T

Q2. $\sum_{i=1}^{n} w_i = \sum_{i=1}^{n}(x_i - \bar{x})/\sum_{k=1}^{n}(x_k - \bar{x})^2 = 0$ because the numerator is zero. $\sum_{i=1}^{n} w_i x_i = \sum_{i=1}^{n} x_i(x_i - \bar{x})/\sum_{k=1}^{n}(x_k - \bar{x})^2 = \sum_{i=1}^{n}(x_i - \bar{x})(x_i - \bar{x})/\sum_{k=1}^{n}(x_k - \bar{x})^2 = 1$

Q5. see text. [Q8.] D, [Q9.] D, [Q10.] A

Q14.

$$|M| = \begin{vmatrix} 2\sum_i x_i^2 & 2n\bar{x} \\ 2n\bar{x} & 2n \end{vmatrix} = 4n(\sum_{i=1}^{n} x_i^2 - n\bar{x}^2) = 4n\sum_{i=1}^{n}(x_i - \bar{x})^2$$

which evaluates to $4n(\sum_i x_i^2 - n\bar{x}^2) = 4n\sum_i(x_i - \bar{x})^2 = 4n^2 s^2$. Hence the matrix of partial derivatives is positive definite.

Chapter 10

Q1. (a) T (b) F, (c) T, (d) F, (e) T, (f) F, (g) T

Q2. Use a cutoff point (say .5) to map each generated random number in [0,1] range to either a zero or a one, and the process repeated n times. For example, if the number generated is .4, corresponding bit is set as 0 because .4 is less than cutoff=.5

Q3. See §10.1 in page 10-1, and 10.7.1 in page 10-31.

Q4. An initial population of members represent potential solutions, each of which can be encoded uniquely using a finite alphabet of distinct symbols. Goodness is evaluated using a fitness measure. If the initial population is too large, the convergence is slow and there could be multiple duplicates in the final solution. If the initial population is too small, the speed of exploring the search space may be slow.

Q5. Chromosome encoding scheme, fitness function, population size (n_{keep}, n_{drop}), genetic operators to be used, termination conditions, probability (for cutoffs, to decide to operate on a chromosome or not, etc). Advantages of PGA are the computational speedup achieved, and cost reduction. In addition, they are less likely to get stuck at a local optima.

Q6. A variety of factors like (i) size of the problem (eg: in the number scrambling example, it was the length of the number or number of digits in it. In TSP, it is the number of cities = length of a tour, if each city is visited exactly once), (ii) accuracy desired (eg: in root finding problems where longer chromosomes can give more precise results for the fractional part), (iii) number of attributes to be included (eg: in a fraud detection problem, each and every variable that could contribute to fraud may be included)

Q7. (i) 1010101010 (ii) 1001100110

Q8. A concatenation of these symbols can be used to represent a chromosome.

Q9. (i) {10001, 10011, 10101, 10111, 11001, 11011, 11101, 11111}, (ii) all combinations of 0's and 1's of length 4 followed by a single 0.

Q10. See §10.10 in page 10-6

Q11. Crossover twins can result only from identical parent chromosomes. If two parent chromosomes are duplicates, we can discard that pair for the crossover operation. This applies to single and multiple crossovers. A solution is to apply one of the other operations (eg: mutation, inversion) at different positions to each parent.

Q12. (i) n-1, (ii) there are (n-1) inversions of length 2, (n-2) inversions of length 3, and so on 2 inversions of length n-1 and one inversion of length n (which simply reflects the entire chromosome). Hence total inversions are (n-1) + (n-2) + $\cdots$ +2+1 = n(n-1)/2 = $\binom{n}{2}$.

Q13. Each gene position in a schemata can be filled by a '*' symbol or by one of the possible m symbols. Thus there are (m+1) possible symbols for each of the k positions. By the product rule (chapter 3), the total possible number of schemata are $(m + 1)^k$.

Q14. a) P1=(000010111), P2=(011111010) b) P1=(000111010), P2=(011010111)
c) P1=(001010010), P2=(010111111) (note: bit position 6 kept)

Q15. It can be zero or two. Assume that X and Y differ in position i (so that the Hamming distance=1). If Y and Z also differ in position i, then Z must be identical with X. If Y and Z differ in position j ($\neq$i), then there are two positions in which X and Z differ. In this case the Hamming distance=2.

Q16. (i) .5, (ii) 2.75 (iii) 1/7 (iv) 100/6. Too low selection pressure indicates slow evolution of the population, which may result in slow convergence to the optimum. For example, if n_{drop}=2 and n_{keep}=20, the selection pressure is .1, and the maximum number of chromosomes that can be weeded out to induct better performing ones is just 2 at each iteration. Large selection pressure indicates faster evolution, few better performing offspring in each generation, and is ideal for parallel implementations of GA.

Q17. Roulette wheel selection, random selection, Boltzmann selection, tournament selection, steady-state selection. See §10.2.1 in page 10-13.

Q18. (a) {1110,1011,0010}, (b) {11110,01111,10111}, (c) {110010,110111,010011}, (d) {101011, 111010, 101110}

Q19. (i) 010101100100 (if the mask bit selects P_1 for bit 0 and P_2 for bit 1) (ii) proceed as before

Q20. $2^3 = 8$, $2^4 = 16$

Q21. proceed as in Q7.

Q22. This can be done by simply reversing the digits of the number to be scrambled, and looking for matches using this mirror image.

Chapter 11

Q1. (a) T (b) F, (c) F, (d) T, (e) T, (f) T, (g) T, (h) T, i)F

Q2. C, Q3. B, Q4. Yes, preferably yes

Q5. MLR is a linear model in the parameters. It assumes that the errors are normally distributed with zero mean. Standard MLR uses the 'least squares' criterion for error minimisation. The neural network has no linearity assumption – it works even when the parameters are nonlinearly related. It has no assumptions on the error distributions. It uses an activation function and an output function at each of the intermediate nodes.

Q6. The values propagated backwards are real numbers in pattern recognition and function evaluation. Neural network can be programmed such that classification takes place at the output nodes using a non-overlapping range of values or some other categorisation of a vector of values. Hence the values propagated backwards can still be real numbers.

Q7. 20, 21, 18. Q8. They have four categories of layers – input layer, one or more hidden layer, pattern layer and decision layer. They have applications in classification schemes (medical procedures like EEG, ECG, X-rays, etc), clustering, etc.

Q9. A) media types and rates, duration (for online media, TV, Radio), geographic locations, size and location in print (for print media), number of times to be advertised, total budget etc; B) Image type (format), boundary parameters, block size(s) and number of blocks, internal attributes like resolution, shade, intensity etc.; C) Insured info, accident info, injury info, expenses, prior accidents and claims, claimed amount; D) Purchase info, article info, locational details (where items are shelved); E) amounts involved, time and date of transaction, number of password attempts, total time taken for transaction; F) email text, URLs, sender details, subject line, file attachments and their types

Q10. It depends on the application, but in general if number of nodes > 1, they must be uniquely identified in input and output layers.

Q11. i) accident proneness={low, medium, high, extremely high} ii) promoter={yes, no}, iii) emission test={pass, fail}, iv) learning disability={present, absent}

Q12. Yes. The neural network architecture represents the input and output nodes as distinct.

Q13. Date and time can be re-coded either to be numeric offsets after a fixed origin, using concatenation of sub-fields or as a linear function of their components. If date is used for computations (for date comparisons, to find elapsed dates, overdue dates etc) it is preferable to input it as Gregorian format or as an offset after a fixed epoch date. But if used for other purposes (eg: date stamping super market or stock purchase transactions) it can be input as mm/dd/yy or similar components. Similar argument applies to time. If time differences are critical in a computation, it is input as sub-seconds (milli, micro seconds etc).

Q14. Network architecture, input and output formats, activation and transfer functions, initial weights, training method to be used.

Q15. See section §11.5.9 in page 11-20

Q16. In FFN networks, signal transmissions take place in the forward direction, whereas in recurrent network it could happen in both directions. No

Q17. $W'X = \sum_i w_i x_i = 44.6$

Q18. (i) $(m-1)n^2$, (ii) $2n + (m-3)n^2$ (iii) $n*(n/2) + (n/2)*(n/4) + \cdots + 1 = n^2/2(1+1/4 + 1/4^2 + 1/4^3 + \cdots +1/4^k)$ where k such that $\lfloor n/2^k \rfloor = 1$. This is a finite geometric progression, which can be simplified for fixed n., (iv) $n*(n-1) + (n-1)*(n-2) + (n-2)*(n-3) + \cdots +2*1$

Q19. As a velocity vector

Q20. Age, gender, income, education, years of driving experience, prior accidents, traffic violation points, physical and vision problems, body stature, drunken-driving habits, mobile use while driving, seat belt usage, vehicle details.

Q21. see respective sections

Q22. (a) $||x - t|| = 2$, hence $\phi(||x - t||) = 0$, (b) $||x - t|| = 0$, hence $\phi(||x - t||) = 1$, (c) $||x - t|| = 1.5$, hence $\phi(||x - t||) = 0$.

Q23. A fuzzy neuron accepts elements of a fuzzy set as input (synaptic weights, operations etc). Fuzzification is possible with neurons and weights. See §11.5.7 in page 11-20 for details.

Chapter 12

Q1. (a) F (b) T, (c) T, (d) F, (e) F, (f) F

Q2. (i) Conditional content filtering works with one or more conditions. For instance, content that appear in reference sections, header lines, within special tags like < code > , < samp >, < cite > may be ignored for some data mining applications. (ii) Web citation mining looks for patterns in bibliographic citation links on the web, (iii) Some data (like ad banners) are being pushed (without being requested) to the users by web servers. These are called push data in the content. The requested data (pull data) is usually common to all users who submit same queries or clicks on same links (as in news sites). This difference between push and pull data can either be constant or vary according to user characteristics., (iv) Hub and Authorities are crawling strategies that distinguishes two types of pages. Hubs are connection points of important pages that point to important authorities. Authority is a page pointed to from many important hubs.

Q3. Ease of navigation, reduced browsing time, increase in revenue, improved security and redundancy.

Q4. Coreference resolution (CR) (inter-reference among entities through pronouns, nouns and pronoun counts are called coreference counts. It may represent document content better than verbs and prepositions. This is especially useful in entity extraction techniques.

Q5. Web pages may be clustered using inherent criteria like ease of navigation, content presentation, user friendliness, or clustered using access patterns or other criteria. X-means algorithm. The x-means algorithm efficiently estimates the number of clusters, and is better suited for text clustering. The pages used for clustering may be spread across the internet, and all pages may not be downloadable to a central site due to copyright restrictions.

Q6. (i) Two-way (bi-directional) relationship between pages is called co-citation. It may be between pages on the same site (intra-site co-citation) or inter-site, (ii) The PageRank algorithm ranks a page high, if many other highly ranked pages link to it (iii) Coupling is a measure of inter-dependency between two pages. Cohesion is a measure of the similarity of elements *within* a group, (iv) Link stepping is a technique to automatically analyse the trustworthiness of hyperlinked sites by passing through hyperlinks in a depth first manner It is started from a specified parent site. (v) Path to purchase (also known as path to order) is the path from the main home page to a firm order page. It will be more or less similar for experienced users, but could vary widely (with cycles and loops) for novices.

Q7. The PageRank can be any positive real number. As the dampening constant is always between 0 and 1, the minimum possible value is $1/n$ (as 1-m=0 when m=1). When m is close to 0, the constant term m/n is negligible and the first term can increase to any large value (depending on the popularity of the page). It may also be noted that the PageRank is a logarithmic scale, rather than a linear one. The values obtained by PageRank as defined originally by Brin & Page differ only slightly from that given in the text.

Q8. Static link mining is used to summarise cross-linked pages. Dynamic link mining is used in identifying dead links, infrequently visited links, heavily used links etc. Link based ranking (LBR) is used to order or prioritise the links in dynamic web access. Link mining is used to summarise cross-linked pages, identify dead links, infrequently visited links, heavily used links etc or prioritise the links in dynamic web access. Link stepping is used to check the trustworthiness of a web page. It could change over time as directly linked pages modify their link structure. One-way links (uni-directional) may exist either to a resource that does not have hyperlink facility (a plain text file created in a simple text editor), or to a document that does not require further linking (a written song rendered as a text file, an audio clip, an entity at the bottom of a hierarchy). They represent outliers in a link mining algorithm. They may be separated out and mined separately, to speedup the link mining process.

Q9. As Rank(P2) = 12, Rank(P2)/OL(P2) = 6. Similarly as Rank(P3) = 15, Rank(P3)/

$OL(P3) = 5$. Now $Rank(P_1) = (1-m)+m\,[Rank(P_2)/OL(P_2) + \cdots + Rank(P_n)/OL(P_n)]$. Substituting values, we get Rank(P1) = (1-.85) + .85 *(6 + 5) = 9.5.

Q10. Let x and y be the pageranks. Then x=0.15+0.85(y/3), y=0.15+0.85(x/4). Solve as a set of equations to get x=0.204949083, y=0.19393794.

Q11. Obvious (standard) attributes are those that can be obtained without accessing metadata content. Hence last modified date is standard. Version number is a shadowed attribute, while (c) and (d) are hidden. Content dependent attributes are derived from the actual content (file size, formats used, etc) whereas the content independent attributes do not depend on actual content (file name, creation date, security access control attributes).

Q12. Precision (P) and recall (R) are the popular measures of query performance. Recall measures the completeness of a search. Higher recall values indicate lesser missing documents. It is defined as the following ratio:– Recall = number of relevant documents retrieved/entire relevant documents on the web. Recall and Precision does not depend upon each other.

Q13. The PageRank and HITS are the most popular measures for web structure mining.

Q14. The visitor quality is useful in targeted advertising, creating customer specific click through content, group specific content expansion, age-group and gender based dynamic content creation etc in addition to building online business intelligence.

Q15. Metadata can be stored internally in an application, externally in separate files (as text file, either in flat form or in structured form (eg: XML format)) or in metadata catalogs/metadata marts. One xml document can store the metadata of multiple files. Metadata mining extracts unknown patterns and trends from metadata files. They are most often utilised for ontology extraction, summarisation, classification, and clustering. There are many categorisations of metadata (eg: Content dependent and independent metadata). They are used for web content mining, especially so with distributed collaborative data mining

Q16. Since the average time spent by a user in page 7 is 18, total time spent by 20 users is 18*20 = 360 seconds or 6 minutes approximately. Similarly, total time spent by 20 users in page 10 is 19*20 = 380 seconds or 6 minutes and 20 seconds approx. The average time for an arbitrary user to reach page P10 from home page P0 cannot be found from the information given.

Q17. Analysing the data streams to extract (unknown) knowledge or detect anomalies and deviations rapidly is termed data stream mining. Data captured by sensors (traffic sensors (road, rail, air etc or [computer, telephone, wireless, ATM] networks), geographical sensors, sonar sensors, industrial sensors etc) can result in a continuous data stream. Data stream mining can automatically detect anomalies, overloads and exceedances, process deviations etc.

Q18. Table mining can be used to classify web pages and markup files using tabular data. In supervised mining, tabular data are extracted using a (user specified) key combination, and a model is built. Multiple tables can have different dimensionalities and overlapping contents. These issues are resolved during the pre-processing stage using table and column headings, tabular metadata etc. Table mining may be used either to categorise data using table contents or to explicitly mine tabular data. The data in online tables may be of categorical, numeric, character, or text forms or could contain links to another resource. It can also be a combination of the above in different columns of a table.

Chapter 13

Q1. (a) T, (b) T, (c) T, (d) F, (e) T, (f) T, (g) F, (h) T

Q2. No backpropagation training, no initialisation needed, no activation and output functions, No hidden layer architectures. Yes. SVM can be used only for classification, regression and clustering while neural network is much more general and can be used for function evaluation, process monitoring (conditions of machinery, online financial transactions, automatic drug administration in hospitals, cruise control of vehicles), Optimisation (discrete, continuous multivariate optimisations, 0-1 optimisation, n-bit parity problems, other combinatorial optimisation problems), expert systems etc.

Q3. In the XOR problem, we have inputs 1 and 0 and outputs 1 or 0 [(0 XOR 0) = 0, (1 XOR 1) = 0, (0 XOR 1) = (1 XOR 0) = 1]. To cast it as an SVM, we recode 0 as -1 so that the inputs are [-1,-1], [-1,+1], [+1,-1] and [+1,+1] with respective outputs -1, +1, +1, -1. Hence data instances[-1,-1] and [+1,+1] belong to class C_1, and the other two to class C_2. In SVM terminology, this can be represented as training samples ([-1,-1;-1], [-1,+1;+1],[+1,-1;+1],[+1,+1;-1]) where the class labels follow the semicolon (some SVM software expect the class labels at the beginning, followed by data). The best a linear classifier can classify (without kernels) is 75%. Let $[x_1, x_2]$ and $[y_1, y_2]$ be two points. Then $(x.y)=x_1y_1 + x_2y_2$, and $(x.y + 1)^2 = x_1^2y_1^2 + x_2^2y_2^2 + 2x_1x_2y_1y_2 + 2x_1y_1 + 2x_2y_2 + 1 = [x_1^2, x_2^2, \sqrt{2}x_1x_2, \sqrt{2}x_1, \sqrt{2}x_2, 1][y_1^2, y_2^2, \sqrt{2}y_1y_2, \sqrt{2}y_1, \sqrt{2}y_2, 1]$, giving $\phi(x) = [1\ x_1^2\ x_2^2\ \sqrt{2}x_1\ \sqrt{2}x_2\ \sqrt{2}x_1x_2]'$ (we moved the constant 1 to the beginning). The $\Phi()$ matrix can be written as

$$\begin{pmatrix} 1 & 1 & 1 & -\sqrt{2} & -\sqrt{2} & \sqrt{2} \\ 1 & 1 & 1 & -\sqrt{2} & \sqrt{2} & -\sqrt{2} \\ 1 & 1 & 1 & \sqrt{2} & -\sqrt{2} & -\sqrt{2} \\ 1 & 1 & 1 & \sqrt{2} & \sqrt{2} & \sqrt{2} \end{pmatrix}$$

The kernel matrix is obtained from the above as $K(x, y) = \Phi\Phi' =$

$$\begin{pmatrix} 9 & 1 & 1 & 1 \\ 1 & 9 & 1 & 1 \\ 1 & 1 & 9 & 1 \\ 1 & 1 & 1 & 9 \end{pmatrix}$$

The Lagrange weights are $(1/8)[1, 1, 1, 1]'$. The primal classifier is $(1/8)(-1)\phi(x_1) + (1/8)(+1)\phi(x_2) + (1/8)(+1)\phi(x_3) + (1/8)(-1)\phi(x_4)$, and the separating surface is $y = w'\phi(x) = -x_1x_2$. It is easy to verify that in the higher dimensional space, the points are linearly separable using the transformation $z = -x_1x_2$.

Q4. Kernel based Nonlinear SVM

Q5. $w=\sum_i a_iy_ix_i$, where a_i's are the dual weights, y_i are class labels, and the summation is over support vector points only.

Q7. $w=\sum_i a_iy_ix_i$ where x_i's are the support vector points. Here a_iy_i's are (2, -1, -3/2). Hence $w=2x_1 - x_2 - (3/2)x_3$, where at least one SV belongs to each class.

Q8. $\frac{b}{||w||}$, $\frac{b-1}{||w||}$, $\frac{b+1}{||w||}$. No, it has nothing to do with the discriminatory power of SVM

Q9. It returns a vector of dual weights (a_i's) and a vector of margin errors ξ_i's. Dual SM-SVM has a refined upper bound on the a_i's as $0 \leq a_i \leq C$.

Q10. In the dual SM-SVM, we have a constraint $0 \leq a_i \leq C$. Support vectors with their multipliers at the upper bound C are called bounded support vectors, and those for which $0 < a_i < C$ are called unbounded SVs.

Q11. It is computational convenience. The classical primal SVM is solved as a quadratic programming problem with as many variables as the dimension of training data. The dual SVM has as many constraints as the training size. Thus it is much faster. Classifier of the dual-SVM is a linear combination of dot products of support vectors and the new data vector x. $f(x) = \theta \left(\sum_{i=1}^{|SV|} \alpha_i y_i (x_i.x) + b \right)$ where $\theta()$ is the sign function.

Q12. We are given $K(x, y) = (1 + x.y)^2$. Let (x_1, x_2) and (y_1, y_2) be two arbitrary data points. Then $(1 + x.y)^2 = (1 + x_1 y_1 + x_2 y_2)^2$. Next proceed as in Q3 to prove that it is indeed an applicable mapping.

Q13. Let (x_1, x_2) and (y_1, y_2) be two arbitrary data points. Then $(x.y)^2 = (x_1 y_1 + x_2 y_2)^2 = (x_1^2 y_1^2 + x_2^2 y_2^2 + 2 x_1 x_2 y_1 y_2) = [x_1^2, \sqrt{2} x_1 x_2, x_2^2][y_1^2, \sqrt{2} y_1 y_2, y_2^2]$.

Q14. $(x_r + x_s)/2 = (1,4,0,6)/2$, from which $\hat{b} = .5 * [4 * 1 + 3 * 4 + 7 * 0 - 1 * 6] = 5$

Q16. With 3 classes:– An SVM to predict if the stock price at the end of a trading day will be (up, steady, down), blood pressure risk of a patient=(low, medium, high). With 4 classes:– predicting the brought-in condition of an emergency patient as (stable, severe, critical, deadly)

Q17. Asymmetric classification arises when the classes are not of equal importance. The weighted SVM is the best choice.

Q18. Separate out all those pairwise linearly separable classes into one group (say G1). Train separate binary SVMs on each of them. Train a different NL SVM (or SM SVM) on the rest of the group. Each new data instance to be classified is first put through the classifiers corresponding to G1. If it cannot be classified successfully, pass it on to the other SVMs.

Q19. (i) $b_2 - b_1/||w|| = 2/\sqrt{16 + 9} = 2/5$, (ii) $2/\sqrt{1.44 + 2.56} = 1$

Q20. See text.

Q21. (i) $\binom{5}{2}$ OVO and 5 OVA, (ii) $\binom{n}{2}$ OVO and n OVA.

Chapter 14

Q1. (a) F, (b) F, (c) T, (d) T (e) F, (f) T, (g) T, (h) T (i) F, (j) T, (k) T, (l) F (m) T

Q2. SVD is the recommended choice in IR because it decomposes the available information into document-space and term-space in the transformed domain where each axis is weighted according to the relative importance of terms as evidenced by the diagonal matrix of SVD. Two data vectors that are very similar in input space are thus mapped to two vectors in close proximity in the feature space by the SVD. This property is not well maintained by other factorisation techniques.

Q3. Automatic document indexing, topic segmentation, uni-language and cross-language IR, inter-language translations, and clustering documents using similarity measures. Yes

Q4. Better performance (due to reduced dimensionality), re-uses factored matrices, generalisation capability, execution speed, cross-language retrievals, reduced synonymy and polysemy.

Q5. Theoretically Yes, when query contains just one keyword at position j and j^{th} row of $U_k S_k^{-1}$ contain all zeroes except at one position (say at i). In this case the query will get mapped to align exactly with i^{th} axis.

Q6. A user query is first vectorised using the same weighting used in forming the TD matrix (if there is just one keyword in the query, a boolean weighting will work). It is then projected onto the latent semantic space as described in section §14.4. This gives us a pseudo-document. All matching vectors in the feature space that have a reasonable similarity (as judged by a similarity metric like the cosine similarity) are then filtered out and returned to the user.

Q7. Words or phrases in natural languages with the same meaning are called synonyms. Examples of synonym words in English are (student, pupil), (plane, aircraft, jet), (petrol, gasoline, benzene), (drug, medicine, medication). Words that have multiple meanings are called polysems. As examples in English, 'mouse' is a pointing device in computing and a rodent elsewhere, 'real' can mean the opposite of 'complex' (in Maths) or opposite of 'virtual' in real-life. (a) noun-verb polysem (b) pronoun-verb (c) noun-verb polysem

Q8. (i) When the TD matrix is large and sparse, (ii) Speed of execution is important, (iii) memory resources are limited (especially for multiple simultaneous user sessions), Dimensionality reduction is achieved when the eigen values tail-off rapidly and becomes negligible after a few values. LSI achieves reduced dimension by discarding all singular values of TD matrix below a threshold, and discarding corresponding rows and columns from U and V matrices. If k is the cutoff limit, the savings are n(n-k)+(m*m-k*k)+n(n-k) = 2n(n-k)+$(m^2 - k^2)$.

Q9. LSI facilitates relevance feedback searching and hierarchical searching. The factored matrices returned by SVD can be saved and reused in query sessions, as long as the document corpora is static. Results returned to the user are automatically segmented using user selections. This information is used in subsequent queries to filter out narrowly focused content.

Q10. (i) When a query contain just one keyword or phrase, (ii) When eponyms are searched for, (iii) when just two keywords are present and their expected co-occurrence probabilities are equal.

Q11. We find the cosine similarity between $\hat{q}$ and each of the vectors in V^* as CS={.23/ .55812, .08/.18708, .655/1.1101 }={.4121, .42762, .59004}. As the cosine similarity is maximum for the third document, we retrieve it as the most matching one.

Q13. Suppose two or more singular values are exactly identical. There are two cases to consider– (i) the identical values are large (and above the cutoff threshold) (ii) the identical values are small (and below the cutoff threshold). In the first case we can drop all duplicates except just one of them. As an example, if three singular values are 15, we keep just one 15 and drop the other two. The reason for this is as follows. When there are multiple duplicates, the S matrix will have identical values along its diagonal. This implies that the inverse of S (which is also diagonal) will have identical values. In addition, identical eigen values result in identical eigen vectors. Hence the V matrix will have as many identical columns as there are identical eigen values. As U = W V T, this matrix also will have exactly identical columns. In addition, the mapped pseudo-query will have exactly identical components. The query execution will then involve summing identical components and searching in duplicated dimensions in the feature space. The solution is to drop all except one of the identical singular values. In the second case we cannot speedup the query execution. Rearranging rows/cols does not affect the solution.

Q14. Cosine similarity of two parallel vectors is 1, as the angle between them is zero and $\cos(0)$ =1. Cosine similarity of two perpendicular vectors is 0 as the angle is 90 and $\cos(90)$ =0.

Chapter 15

Q1. (a) T, (b) T, (c) T, (d) F, (e) F, (f) T, (g) F, (h) F (i) T

Q2. i) Stemming (also called lemmatising) reduces a word to its root (or stem) in the language with an intention to improve the scope for similarity matching and dimensionality reduction (ii) mining citation data for patterns and trends (iii) Those words that best represent or capture the key concepts, ideas or information in a document are called the *facets* of a document

Q3. Linguistic pre-processing used for parts of speech identification, word disambiguation, acronym expansion etc to make subsequent analysis easier.

Q4. Duplicate documents will have exactly identical columns. Documents with overlaps will almost identical columns if boolean weighting is used. For other types of weighting, we may have to find correlations between column vectors to see the extend of overlap. Two almost identical columns of the TD matrix with large dimensionality indicate that the corresponding documents are almost similar. The correlation between the corresponding columns of documents with overlapping content will be very high (say between .9 and 1.0).

Q5. Text summarisation and visualisation, text categorisation, text clustering, trend detection in text, colocation, text-based recommendation systems, document similarity matching, etc. XML documents stripped off tags can be used.

Q6. DF, TF-IDF, edit distance, CNC, TS. X-means algorithm.

Q7. Hypertext mining looks for patterns in text, hyperlinks, text markups, etc. It is the mining of complex web content created using a markup language (HTML, XML, etc). An implicit assumption in text mining is that the data are either in pure text form, or has been cleansed from other forms. Most common techniques for text mining include classification, link analysis, and clustering, [Q8.] see text

Q9. Precision (P) and recall (R) are the popular measures of query performance. Recall measures the completeness of a search. Higher recall values indicate lesser missing documents. It is defined as the following ratio:– Recall = number of relevant documents retrieved/entire relevant documents on the web. Recall and Precision does not depend upon each other.

Q10. Precision = number of relevant documents retrieved/ total number of documents retrieved = 250/330 =.757576. Recall = number of relevant documents retrieved/entire relevant documents on the web = 250/770 = 0.324675.

Q11. F=2a/(2a+b+c). We are given that a=(b+c)/2, or equivalently 2a = b+c. Substitution in the above gives F = (b+c)/(b+c+ b+c) = 1/2 = 2a/(2a+2a). Range of G is [0,1). From the data in Q15, we get the value of G as $250/\sqrt{(250 + 80) * (250 + 520)}$ = .49594975

Q12. F=2PR/(P+R). Dividing numerator and denominator by PR gives F = 2/(1/P+1/R), which is the HM of precision and recall. $(F_{\beta_1}/F_{\beta_2}) - 1 = (\beta R + P) - (\beta P + R)/(\beta P + R) = (\beta - 1)(R - P)$.

Q13. Metadata can be stored internally in an application, externally in separate files (as text file, either in flat form or in structured form (eg: XML format)) or in metadata catalogs/metadata marts. One xml document can store the metadata of multiple files.

Metadata mining extracts unknown patterns and trends from metadata files. They are most often utilised for ontology extraction, summarisation, classification, and clustering. There are many categorisations of metadata (eg: Content dependent and independent metadata). They are used for web content mining, especially so with distributed collaborative data mining.

Q14. Since the average time spent by a user in page 7 is 18, total time spent by 20 users is 18*20 = 360 seconds or 6 minutes approximately. Similarly, total time spent by 20 users in page 10 is 19*20 = 380 seconds or 6 minutes and 20 seconds approx. The average time for an arbitrary user to reach page P10 from home page P0 cannot be found from the information given.

Q15. All of them except D) and G)

Q16. unstructured:– textified songs, dictations, talks, medical or legal transcriptions; semi-structured:– textified lectures, presentations, data extracted from spreadsheets or application programs, structured:– markup data in HTML, XML, WML, VRML and other formats.

Q17. In text data each unit (keyword) has a syntactic meaning. Stream of binary digits 0 and 1 can also be considered as text data, but that too carry syntactic meaning as there could be dependency between adjacent bits.

Q18. C, [Q19.] A) and C), [Q20.], [Q21.] see text

Q23. when data are taken from text-warehouses or from specially indexed text corpora.

Author Index

Subject Index